Helmut Schenkel-Brunner

Human Blood Groups

Chemical and Biochemical Basis
of Antigen Specificity

Second,
completely revised edition

SpringerWienNewYork

Dr. Helmut Schenkel-Brunner
Institut für Medizinische Biochemie
Universität Wien, Vienna, Austria

Editorial Consultant: Alexandra Salvini-Plawen
Camera-ready copies provided by the author
Printed by Druckerei Theiss GmbH, A-9400 Wolfsberg
Graphic design: Ecke Bonk
Printed on acid-free and chlorine-free bleached paper
SPIN 10764127

With 173 Figures

ISBN 3-211-83471-0 Springer-Verlag Wien New York
ISBN 3-211-82705-6 1st ed. Springer-Verlag Wien New York

Foreword

As President and organiser of the 26th Congress of the International Society of Blood Transfusion 'ISBT 2000', I am pleased to introduce this work on the biochemistry of human blood groups, which will serve as a valuable complement to blood group serology work. For this new edition the author has singlehandedly covered virtually all blood group research published before the current year. It is my firm belief that all participants in the ISBT conference will find this book a highly useful addition to their scientific libraries.

Vienna, April 2000

W.R. Mayr
Congress President

Preface

This revised and updated edition of the monograph on the chemistry and biochemistry of human blood groups is being published to honour the 100th anniversary of Dr. Karl Landsteiner's discovery of human blood groups here in Vienna.

Five years have passed since publication of the first edition, and in this time research on the biochemistry of human blood groups has reached staggering proportions. Thanks to extensive molecular biological applications, research groups have identified the bearer molecules of even more blood groups, and are now able to clarify the molecular basis of a vast number of blood group specificities.

Human blood group research has become one of the most fruitful of all biochemical research fields. While this development is immensely gratifying to a researcher like myself, who has worked in human blood groups for the past 34 years, coping with the close to overwhelming number of publications has proved a highly challenging task.

In order to keep the number of pages of this edition from reaching a condition close to critical mass, my criteria in selecting the literature for this edition have had to be far more stringent than those applied for the first edition. Thus it has not been possible to offer a more thorough discussion of the serology involved; for this, the interested reader is encouraged to refer to Geoff Daniels' book "Human Blood Groups" (Blackwell Science, Ltd.). I have only briefly touched on the physiological aspects of blood groups (e.g. their function in cell adhesion and regulation of tissue formation in embryonal development; their role in defending tissues from viruses, microorganisms, and parasites; as well as the thousands of investigations on blood groups in malignantly transformed tissue). I hope, nevertheless, that readers will find the references I have selected useful in establishing a good grasp of this field.

Like the first edition, the second is a single-author opus. The author is aware that mistakes cannot have been completely avoided, and will gratefully acknowledge all suggestions and constructive criticism.

My thanks go to DiaMed Diagnostica for its financial support of the publishing of, and to Sandra Salvini-Plawen for the English language editing of this second edition.

Vienna, May 2000   Helmut Schenkel-Brunner

Preface to the First Edition

Although a few books covering primarily serological aspects of human blood groups are available, it became clear to me in the course of my research that no compendium of the non-serological aspects of human blood group systems exists. This book has been written to facilitate access to the vast number of publications scattered throughout the literature in both chemical and medical journals on the chemistry, biochemistry, and molecular biology of blood groups. It is designed as a concise survey for use by blood bankers and researchers in biochemistry, blood group serology, immunohaemotology, forensic medicine, population genetics, and anthropology; the text is supplemented by numerous illustrations and tables.

This volume encompasses the entire field of blood group serology and provides a comprehensive survey of present knowledge in the field. The serological aspects have been kept to a minimum. I have emphasised the chemical, biochemical and molecular genetic basis of blood group specificity and given full consideration to molecular biology investigations, in particular to those on the structure of blood group genes and the structural basis of alleles and rare blood group variants. The book covers the latest developments in research and discusses literature up to the beginning of 1995.

The generosity of the publisher has made it possible for me to include a fairly detailed reference list, which nevertheless – in view of the vast number of publications on the various aspects of blood group science – had to be restricted to an absolute minimum. Since this book is conceived both as a textbook for newcomers in the field and as a reference book for experts, I have done my best to facilitate studies on each topic: all groups working in a given field were mentioned, and reference made in each case to at least the first relevant paper and the most recent publication.

This work is the culmination of years of study: I wrote my doctoral thesis on the biosynthesis of blood group A determinant after completing my studies in chemistry, physics, and zoology at the University of Vienna. During my further investigations into the biosynthesis of blood group ABH and Lewis substances, the structure of glycoconjugates, and the distribution and function of animal lectins, I have been able to follow developments from the very first models of the ABH determinants advanced by Watkins and Kabat in 1966 to the most recent investigations on the molecular basis of blood group allelism and the elucidation of the structure of blood group genes. These investigations offered insight into genetic mechanisms and cast light on interrelations which classical serological methods could not explain. New methods made it possible to locate rare blood groups on previously unknown membrane constituents, thus providing a means of investigating their function and in some cases finding the cause of diseases. No wonder this field has fascinated me ever since my student days!

Let me stress here that this entire monograph is a single-author synthesis of the current state of blood group chemistry and biochemistry. The task was not confined to evaluating the extensive literature and compiling the results in a comprehensible and logical form: it also comprised typing, proof-reading, desktop publishing, and preparing the camera-ready manuscript to keep printing costs down. I am aware that misinterpretations and mistakes may have crept into the manuscript, although every effort has been made to avoid them. All relevant criticism, suggestions and corrections will be gratefully acknowledged and taken into consideration in a second edition.

My special thanks go to Sandra Salvini-Plawen, who is by training a translator specialising in scientific texts. She not only transformed my English original into correct English but also in the course of our many fruitful discussions helped to make the text more accessible to non-specialists.

Contents

Contents

Contents

xiii

Contents

Contents

1 Introduction

Blood Groups – A Brief Discourse

At the turn of this century Karl Landsteiner discovered the blood group **ABO** system of human erythrocytes. The importance of blood group antigens in clinical transfusion practice and in haemolytic disease of the newborn was soon recognised, and thorough investigations led to the discovery of a vast number of red cell antigenic characters. More than 600 erythrocyte antigens have been defined thus far(see [4]); while some of these antigens are common red cell characteristics in a given population, others may be extremely rare and found only in one family. Significant differences in statistical frequency have been observed between the different races.

In addition to their clinical relevance, blood group characters are extensively used as serological markers in genetical and anthropological investigations [8]. Further, the regulatory function of blood group antigens in embryonic development and cell differentiation as well as in malignant cell transformation has repeatedly been discussed.

The blood group antigens show great diversity of distribution. They may be confined to erythrocytes or found on other blood and tissue cells, some are carried by water-soluble substances in tissue fluids and secretions. Some are found only in humans; others also occur in some non-human primates, still others occur widespread throughout the animal kingdom and are even found in plants and microorganisms.

Genetically related erythrocyte antigens are combined into blood group systems. To qualify as a blood group system, the encoding genes must be inherited independently of all other genes and must also have a unique chromosomal assignment. Thus, the antigens of a blood group system are controlled either by a single polymorphic gene or by contiguous homologous genes usually derived from a common ancestor.

The assignment of antigens to a blood group system is mainly based on classical family studies and population statistics demonstrating either an antithetical relationship between a pair of antigens (e.g. **A** and **B** or **M** and **N**) or linkage disequilibrium between the encoding genes (e.g. **MN** and **Ss**). In rare cases certain low-frequency antigens (e.g. para-**Kell** and para-**Lutheran** antigens) have been placed within systems after the defining antibody failed to react with appropriate 'null type' red cells.

1

Chemical and biochemical investigations

The blood group specificity of any given erythrocyte is determined by the chemical structure of the antigen determinants located on proteins and glycoconjugates (i.e. glycolipids and glycoproteins) of the erythrocyte membrane.

First attempts to establish the chemical basis of erythrocyte blood group antigenicity failed due to the inability of classical analytical methods to work with highly hydrophobic membrane constituents which occur only in infinitesimal quantities.

Thanks to the fact that most carbohydrate-based blood group antigens are not confined to erythrocytes but are also found in large quantities in water-soluble material, the chemical structures of blood group **ABO** and **Lewis** antigens were the first to be determined in the early 1960s. Subsequent biosynthesis experiments studied the specificity of glycosyltransferases; the correlation between expression of these enzymes and blood group phenotype provided the first insight into the genetic regulation of blood group specificity.

The genetic polymorphism of membrane proteins was much more difficult to establish. The first polymorphic erythrocyte membrane protein to be studied by classical protein sequencing was glycophorin A, the protein carrying the blood group **MN** antigens. In contrast to glycophorin A, which is highly hydrophilic, most protein-based antigens are located on highly hydrophobic membrane proteins which occur only in very small quantities. The development of sensitive and sophisticated immunological methods enabled investigations thanks to which it became possible to assign blood group specificities to certain proteins, and to isolate the respective membrane constituents. Later, modern molecular biological methods used in analysis of cDNAs greatly facilitated the determination of peptide sequences and subsequently the investigation of the amino acid polymorphism characteristic for specific blood group antigens.

The application of modern molecular biological methods also provides precise information on gene structure. Thus far many different mechanisms responsible for blood group polymorphism have been discovered. In most cases point mutations have been detected which lead to amino acid substitutions, formation or loss of splicing or initiation sites, or insertion of a stop codon. Further variability is caused by gene duplication, hybridisation of closely linked homologous genes, loss or duplication of an exon, or insertion of DNA segments (see e.g. **MNS** system).

Recent attempts to define new antigen specificities based solely on structural data and not on reactions with respective antisera (e.g. **Dr**[b] and **Rg3**) are frowned upon by classical serologists.

2

Methods of blood group determination

Blood groups in the classical sense are polymorphic serological characteristics of the erythrocyte membrane defined by their reaction with a specific antiserum. The reagents which have been used to specify red cell antigens are either

(1) **'Naturally occurring antibodies'**
  (a) Alloantibodies (also termed isoantibodies) which occur spontaneously within the **ABO** system and in rare phenotypes of other blood group systems;
  (b) monoclonal antibodies produced in certain pathological conditions (e.g. the cold agglutinins in autoimmune diseases);
(2) **immune antibodies** induced by pregnancies, blood transfusions or active immunisation of humans and animals;
(3) **lectins** of plant and animal origin. Due to their easy accessibility and their high epitope specificity these reagents have been widely used in structural investigations. In laboratory use they are being replaced by
(4) **hybridoma antibodies**[1] which can be produced in great quantities and show a clear-cut specificity comparable to that of the highly specific alloantibodies.

Modern molecular biological methods have recently been introduced into blood group research: the polymerase chain reaction (= PCR) to determine blood group genotypes (e.g. [9]), and DNA restriction fragment length polymorphism to establish the chromosomal location of blood group genes [10].

Blood group terminology

It was initially left to the discretion of the investigators to designate the blood group specificities and phenotypes – the antigen characters were first named by a single letter code (e.g. **A, B, M, N, P**). Later, abbreviated names of the blood or serum donors were used (e.g. **Fy** for 'Duffy' or **Jk** for 'J. Kidd'). Alternative specificities were designated either by superscript letters (a, b, or c) or subscript numbers according to their frequency or the order of their discovery (see [4]). The classical nomenclature is thus far from being systematic.

---

[1] Only 'naturally occurring' monoclonal antibodies in autoimmunehaemolytic diseases (e.g. anti-I's and anti-i's) will be so called in this work. Monoclonal antibodies produced by the hybridoma technique will be termed 'hybridoma antibodies' throughout this book.

Table 1.1: Designation for blood group systems and antigen specificities

Name	System symbol	System number	Antigen number within system														
			001	002	003	004	005	006	007	008	009	010	011	012	013	014	015
ABO [a]	ABO	001	A	B	A,B	A1											
MNS [a]	MNS	002	M	N	S	s	U	He	Mi^a	M^c	Vw	Mur	M^g	V^r	M^e	Mt^a	St^a
P	P1	003	P1														
Rh [b]	RH	004	D	C	E	c	e	f	Ce	C^w	C^x	V	E^w	G	Rh^A	Rh^B	Rh^c
Lutheran [c]	LU	005	Lu^a	Lu^b	Lu^{ab}	Lu4	Lu5	Lu6	Lu7	Lu8	Mull	...	Lu11	Much	Hughes	Lu14	...
Kell [d]	KEL	006	K	k	Kp^a	Kp^b	Ku	Js^a	Js^b	...	...	Ul^a	Cote	Bøc	K13	San	...
Lewis	LE	007	Le^a	Le^b	Le^{ab}												
Duffy	FY	008	Fy^a	Fy^b	Fy^3	Fy^4	Fy^5	Fy6									
Kidd	JK	009	Jk^a	Jk^b	Jk3												
Diego [e]	DI	010	Di^a	Di^b	Wr^a	Wr^b	Wd^a	Rb^a	WARR	ELO	Wu	Bp^a	Mo^a	Hg^a	Vg^a	Sw^a	BOW
Yt	YT	011	Yt^a	Yt^b													
Xg	XG	012	Xg^a														
Scianna	SC	013	Sc1	Sc2	Sc3												
Dombrock	DO	014	Do^a	Do^b	Gy^a	Hy	Jo^a										
Colton	CO	015	Co^a	Co^b	Co^{ab}												
Landsteiner-Wiener	LW	016	...	...	...	...	LW^a	LW^{ab}	LW^b								
Chido/Rodgers	CH/RG	017	Ch1	Ch2	Ch3	Ch4	Ch5	Ch6	WH				Rg1	Rg2			
Hh	H	018	H														
Kx	XK	019	Kx														
Gerbich	GE	020	...	Ge2	Ge3	Ge4	Wb	Ls^a	An^a	Dh^a							
Cromer	CROM	021	Cr^a	Tc^a	Tc^b	Tc^c	Dr^a	Es^a	IFC	WES^a	WES^b	UMC					
Knops	KN	022	Kn^a	Kn^b	McC^a	Sl^a	Yk^a										
Indian	IN	023	In^a	In^b													

[a] Antigens MNS16 to MNS40: see *Table 9.2.* [b] Antigens RH16 to RH52: see *Table 13.1.* [c] Antigens LU16 to LU20: see *Table 19.1.* [d] Antigens KEL16 to KEL26: see *Table 17.1.* [e] Antigens DI16 to DI21: see *Table 20.1.*
... indicates a number which is now obsolete; the antigen has either been transferred to another System or to a Collection, or has altogether been removed because of inadequate documentation. These numbers may not be reused.
From Daniels et al. [1,2].

Table 1.2: Collections of antigens – specificities with serological, biochemical, or genetic connections.

Name	System symbol	System number	Antigen number within system		
			001	002	003
Cost	COST	205	Cs^a	Cs^b	
Ii	I	207	I	i	
Er	ER	208	Er^a	Er^b	
	GLOB	209	P	P^k	LKE
		210	Le^c	Le^d	

From Daniels et al. [1,2].

Following a recommendation by the *International Society of Blood Transfusion* a uniform blood group terminology system has been proposed by the 'ISBT Working Party on Automation and Data Processing' [2,3,5-7]. The Working Party did not intend to do away with the current terminology but rather to establish a parallel standard terminology which would be in accordance with the genetic basis of blood groups and suitable for on-line use.

Blood group antigens have been classified into three categories, viz. Systems, Collections, and Series:

(1) **Systems** include antigen specificities which are encoded by alleles of a single gene locus or of contiguous, largely homologous genes. A total of 23 blood group systems containing 210 antigens have been established to date *(Table 1.1)*.

(2) **Collections** comprise antigens which are clearly related on a serological, genetical, or biochemical basis but do not yet satisfy the criteria mandatory for blood group systems as defined above *(Table 1.2)*.

(3) **Series** have been defined as two 'holding files' for antigens which can not yet be assigned to any of the known Systems or Collections: the 700-series comprises antigens of low incidence (<1% in Europids), and the 901-series antigens of high incidence (>99% in Europids) *(Table 1.3)*.

Table 1.3: Antigens not assigned to Systems or Collections: the 700 series (low-incidence antigens) and the 901 series (high-incidence antigens).

Number	Name	Symbol	Number	Name	Symbol
700 002	Batty	By	700 039	Milne	
700 003	Christiansen	Chr^a	700 040	Rasmussen	RASM
700 005	Biles	Bi	700 043	Oldeide	Ol^a
700 006	Box	Bx^a	700 044		JFV
700 015	Radin	Rd	700 045	Katagiri	Kg
700 017	Torkildsen	To^a	700 049		HJK
700 047	Jones	JONES	700 050		HOFM
700 018	Peters	Pt^a	700 052		SARA
700 019	Reid	Re^a	700 053		LOCR
700 021	Jensen	Je^a	700 054		REIT
700 028	Livesey	Li^a			
901 001		Vel	901 009	Anton	AnWj
901 002	Langereis	Lan	901 011	Raph	MER2
901 003	August	At^a	901 012	Sid	Sd^a
901 005		Jr^a	901 013	Duclos	
901 006		Ok^a	901 014		PEL
901 007		JMH	901 015		ABTI
901 008		Emm			

From Daniels et al. [1,2].

According to the terminology recommended by the ISBT Working Party each authentic blood group specificity is characterised by a six-digit identification number, in which the first three numbers represent the blood group system or the antigen series, and the last three numbers indicate the specificity. The **A** antigen of the **ABO** system is thus termed **001 001**, or in the abbreviated form **1.1**.

As an alternative to the above, a combined alphabetical / numerical system has been proposed which would be primarily for textual and verbal use. In this case the blood group character is designated by the symbol of the System, Collection or Series

(1 to 4 capital letters of the Latin alphabet) immediately followed by the antigen number. The **Rh** specificities, for example, number **RH1** to **RH48**.

In cases where a specificity has been assigned to another category, or the specifying antibody or the antigen-positive red cells are no longer available, or the specificity has been removed altogether because of inadequate documentation, the original numbers have become obsolete and may not be reused.

Blood group **phenotypes** are designated by the group symbol followed by a colon and the specificity numbers of the defining antigens, separated by commas; absence of a specificity is indicated by a preceding minus sign. The phenotype **DCce** (= **R₁r**) is thus termed **RH:1,2,-3,4,5**.

Alleles or **haplotypes** are italicised or underlined. The group symbol and the list of the specificity numbers are separated by an asterisk or a space. The **CDe** (= **R'**) haplotype, for example, becomes **RH*1,2,5** (or **RH 1,2,5**), and **cde** (= r) becomes **RH*4,5** (or **RH 4,5**). In **genotype** designations the alleles are separated by a stroke, e.g. **KEL*2,4/2,4**; zero is used for an apparently silent allele, thus the genotype **CᵃᵉCo** in traditional terms would be termed **CO*1/0**.

Since the ISBT terminology has only recently been established, the majority of blood group literature still uses the classical terminology. Therefore, in order to avoid confusion and to facilitate literature study, the familiar terms will be kept throughout this monograph. The ISBT terms will of course be indicated in each chapter.

References

1. DANIELS, G. L., ANSTEE, D. J., CARTRON, J. P., DAHR, W., HENRY, S., ISSITT, P. D., JORGENSEN, J., JUDD, W. J., KORNSTAD, L., LEVENE, C., LOMAS-FRANCIS, C., LUBENKO, A., MALLORY, D., MOULDS, J. M., MOULDS, J. J., OKUBO, Y., OVERBEEKE, M., REID, M. E., ROUGER, P., SEIDL, S., SISTONEN, P., WENDEL, S. & ZELINSKI, T. (1996): Terminology for red cell surface antigens - Makuhari report. *Vox Sang.* 71, 246-248.
2. DANIELS, G. L., ANSTEE, D. J., CARTRON, J. P., DAHR, W., ISSITT, P. D., JORGENSEN, J., KORNSTAD, L., LEVENE, C., LOMAS-FRANCIS, C., LUBENKO, A., MALLORY, D., MOULDS, J. J., OKUBO, Y., OVERBEEKE, M., REID, M. E., ROUGER, P., SEIDL, S., SISTONEN, P., WENDEL, S., WOODFIELD, G. & ZELINSKI, T. (1995): Blood group terminology 1995 - ISBT working party on terminology for red cell surface antigens. *Vox Sang.* 69, 265-279.
3. DANIELS, G. L., MOULDS, J. J., ANSTEE, D. J., BIRD, G. W. G., BRODHEIM, E., CARTRON, J. P., DAHR, W., ENGELFRIET, C. P., ISSITT, P. D., JØRGENSEN, J., KORNSTAD, L., LEWIS, M., LEVENE, C., LUBENKO, A., MALLORY, D., MOREL, P., NORDHAGEN, R., OKUBO, Y., REID, M., ROUGER, P., SALMON, C., SEIDL, S., SISTONEN, P., WENDEL, S., WOODFIELD, G. & ZELINSKI, T. (1993): ISBT Working Party on Terminology for Red Cell Surface Antigens - Sao Paulo report. *Vox Sang.* 65, 77-80.
4. ISSITT, P. D. (1985): *Applied Blood Group Serology.* Montgomery Scientific Publications,Miami, FL.
5. LEWIS, M., ALLEN, F. H., ANSTEE, D. J., BIRD, G. W. G., BRODHEIM, E., CONTRERAS, M., CROOKSTON, M.,

DAHR, W., ENGELFRIET, C. P., GILES, C. M., ISSITT, P. D., JØRGENSEN, J., KRONSTAD, L., LEIKOLA, J., LUBENKO, A., MARSH, W. L., MOORE, B. P. L., MOREL, P., MOULDS, J. J., NEVANLINNA, H., NORDHAGEN, R., ROSENFIELD, R. E., SABO, B. & SALMON (1985): ISBT workshop party on terminology for red cell surface antigens: Munich Report. *Vox Sang. 49*, 171-175.

6. LEWIS, M., ANSTEE, D. J., BIRD, G. W. G., BRODHEIM, E., CARTRON, J. P., CONTRERAS, M., DAHR, W., DANIELS, G. L., ENGELFRIET, C. P., ISSITT, P. D., JØRGENSEN, J., KORNSTAD, L., LUBENKO, A., McCREARY, J., MOREL, P., MOULDS, J. J., NEVANLINNA, H., NORDHAGEN, R., OKUBO, Y., PEHTA, J., ROUGER, P., RUBINSTEIN, P., SALMON, C., SEIDL, S., SISTONEN, P., TIPPETT, P., WALKER, R. H., G., W., YOUNG, S. & ZELINSKI, T. (1991): ISBT Working Party on Terminology for Red Cell Surface Antigens: Los Angeles report. *Vox Sang. 61*, 158-160.

7. LEWIS, M., ANSTEE, D. J., BRID, G. W. G., BRODHEIM, E., CARTRON, J. P., CONTRERAS, M., CROOKSTON, M. C., DAHR, W., DANIELS, G. L., ENGELFRIET, C. P., GILES, C. M., ISSITT, P. D., JØRGENSEN, J., KORNSTAD, L., LUBENKO, A., MARSH, W. L., McCREARY, J., MOORE, B. P. L., MOREL, P., MOULDS, J. J., NEVANLINNA, H., NORDHAGEN, R., OKUBO, Y., ROSENFIELD, R. E., ROUGER, P., RUBINSTEIN, P., SALMON, C., SEIDL, S., SISTONEN, P., TIPPETT, P., WALKER, R. H., WOODFIELD, G. & YOUNG, S. (1990): Blood group terminology 1990. *Vox Sang. 58*, 152-169.

8. MOURANT, A. E. (1983): *Blood Relations - Blood Groups and Anthropology*. University Press, London, Oxford.

9. UGOZZOLI, L. & WALLACE, R. B. (1992): Application of an allele-specific polymerase chain reaction to the direct determination of ABO blood group genotypes. *Genomics 12*, 670-674.

10. ZELINSKI, T. (1991): The use of DNA restriction fragment length polymorphisms in conjunction with blood group serology. *Transfusion 31*, 762-770.

2 Glycoconjugates

The antigenic properties of erythrocytes and tissue cells, and the serological characteristics of body secretions are based almost exclusively on specific protein or oligosaccharide structures. But whereas proteins are usually thoroughly discussed in chemistry and biochemistry courses, carbohydrates and glycoconjugates are only briefly (and it seems reluctantly) mentioned. A brief survey of the structures and properties of this substance class is therefore appropriate.

Glycoconjugates are complex molecules consisting of either carbohydrate and lipid (→ glycolipids) or carbohydrate and protein (→ glycoproteins). The carbohydrate moiety of the blood group specific glycoconjugates is composed of D-galactose, D-glucose, D-mannose, N-acetyl-D-galactosamine, N-acetyl-D-glucosamine, L-fucose, and N-acetyl-D-neuraminic acid (= sialic acid) (see *Fig. 2.1*). Other monosaccharides – such as xylose, occur only in plant material or in glycoproteins not discussed here.

2.1 Glycolipids

Within the context of blood group antigens only those glycosphingolipids in which the oligosaccharide chain is borne by a ceramide residue are of interest (see *Fig. 2.2*). Ceramides are N-acyl derivatives of sphingosine (hence the name 'glyco-sphingo-lipids'), an amino alcohol containing a long unsaturated carbohydrate chain; derivatives of N-acyl-sphinganine (dihydro-sphingosine) and N-acyl-phytosphingosine (4-hydroxy-sphinganine) are also found in smaller quantities. The fatty acids linked to the amino group of the sphingosine residue are either saturated or unsaturated molecules containing 16 to 26 carbon atoms; the α-C-atom is frequently hydroxylated. The oligosaccharide unit is linked covalently to the hydroxyl group on carbon-1 of the sphingosine residue.

The carbohydrate moieties of glycosphingolipids are extremely variable. Seven basic types of oligosaccharide units – based on the first three or four monosaccharide residues – have been defined: the **globo**, **isoglobo**, **muco**, **lacto**,

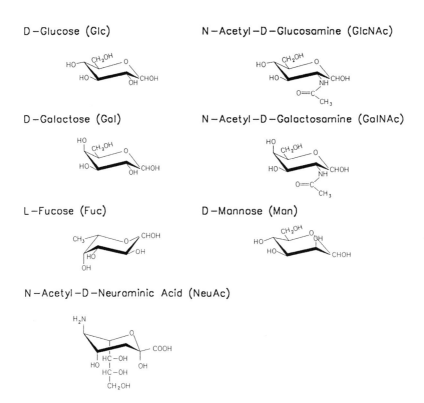

Figure 2.1: Structures of monosaccharides occurring in glycoconjugates.

neolacto, ganglio, and **gala** chains ([3], see *Fig. 2.3*)[1]. These basic chains can be extended by additional N-acetyllactosamine units to form fairly complex oligo-saccharide structures which are often branched. These extended chains may ultimately contain up to 60 monosaccharide units (see **'polyglycosyl-ceramides'** in *Sects. 5.3.2 and 7.3.1*). The terminal galactose residue usually carries $\alpha 2 \rightarrow 3$ linked N-acetyl-neuraminic acid or, less frequently, $\alpha 1 \rightarrow 2$ fucose, or $\alpha 1 \rightarrow 3$ galactose. The chemical structures of common erythrocyte membrane glycosphingolipids are presented in *Fig. 4.5*.

---

[1] The isoglobo, muco, and gala series of oligosaccharide chains do not carry any known blood group character.

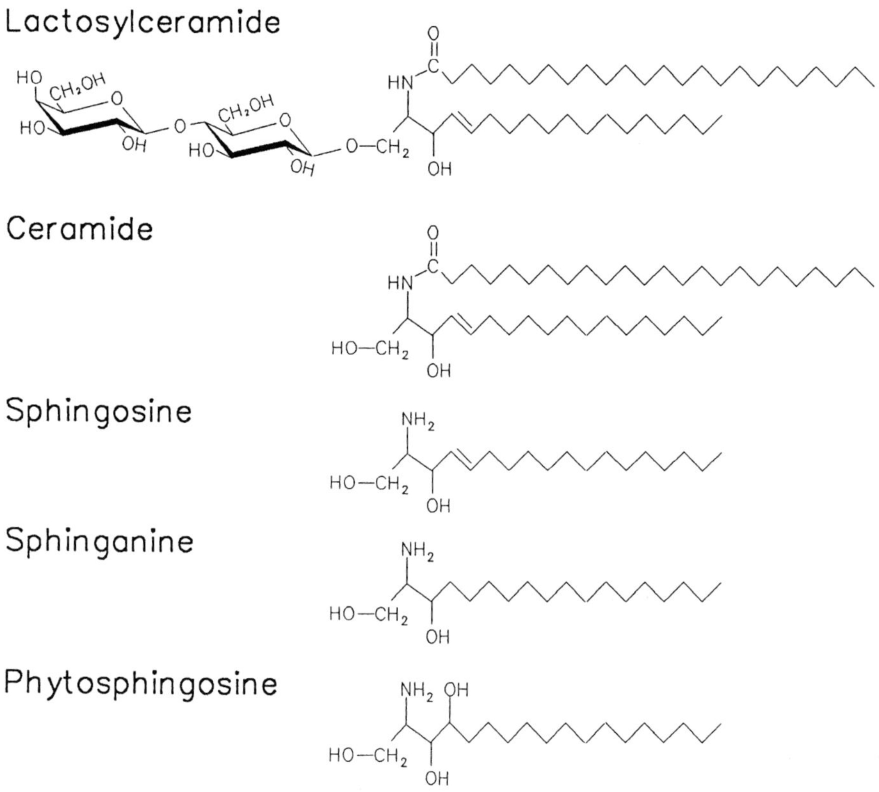

Lactosylceramide

Ceramide

Sphingosine

Sphinganine

Phytosphingosine

Figure 2.2: Chemical structures of a glycosphingolipid, of ceramide, and various sphingosines.

The fact that expression of distinct glycolipids is strictly regulated during cellular differentiation and malignant transformation is proof of the important roles of glycolipids in normal cellular biology [1]. Moreover, glycolipids are also the initial attachment sites for several pathogenic organisms. It is known that bacteria, viruses, and bacterial toxins recognise host glycolipids as their initial attachment sites [2,6,7].

Globo series (Gb)

GalNAc $\xrightarrow{\beta1\rightarrow3}$ Gal $\xrightarrow{\alpha1\rightarrow4}$ Gal $\xrightarrow{\beta1\rightarrow4}$ Glc $\xrightarrow{\beta1\rightarrow1}$ Ceramide

Isoglobo series (iGb)

GalNAc $\xrightarrow{\beta1\rightarrow3}$ Gal $\xrightarrow{\alpha1\rightarrow3}$ Gal $\xrightarrow{\beta1\rightarrow4}$ Glc $\xrightarrow{\beta1\rightarrow1}$ Ceramide

Muco series (Mc)

Gal $\xrightarrow{\beta1\rightarrow3}$ Gal $\xrightarrow{\beta1\rightarrow4}$ Gal $\xrightarrow{\beta1\rightarrow4}$ Glc $\xrightarrow{\beta1\rightarrow1}$ Ceramide

Lacto series (Lc)

Gal $\xrightarrow{\beta1\rightarrow3}$ GlcNAc $\xrightarrow{\beta1\rightarrow3}$ Gal $\xrightarrow{\beta1\rightarrow4}$ Glc $\xrightarrow{\beta1\rightarrow1}$ Ceramide

Neolacto series (nLc)

Gal $\xrightarrow{\beta1\rightarrow4}$ GlcNAc $\xrightarrow{\beta1\rightarrow3}$ Gal $\xrightarrow{\beta1\rightarrow4}$ Glc $\xrightarrow{\beta1\rightarrow1}$ Ceramide

Ganglio series (Gg)

Gal $\xrightarrow{\beta1\rightarrow4}$ GlcNAc $\xrightarrow{\beta1\rightarrow3}$ Gal $\xrightarrow{\beta1\rightarrow4}$ Glc $\xrightarrow{\beta1\rightarrow1}$ Ceramide

Gala series (Ga)

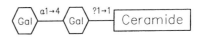

Gal $\xrightarrow{\alpha1\rightarrow4}$ Gal $\xrightarrow{?1\rightarrow1}$ Ceramide

Figure 2.3: Classes of glycosphingolipids
Semisystematic nomenclature as proposed by the 'IUPAC-IUB Commission on Biochemical Nomenclature' [3].

2.2 Glycoproteins

In glycoproteins the carbohydrate units are carried by a polypeptide backbone. The molecular masses of glycoproteins range from approximately 16,000 Dalton up to several million Dalton, and the carbohydrate content of these units varies from 1% to more than 85%.

The oligosaccharide chains may be attached to the protein backbone either by N-glycosidic or by O-glycosidic linkages. In all cases the overall structure of an oligosaccharide chain is basically determined by the chain type to which it belongs. The exact monosaccharide sequence, however, shows a certain degree of variation – which is a consequence of the mechanism of the biosynthesis of the oligosaccharide chains (see *Sect. 3.3*, **'microheterogeneity'**).

N-glycosidically linked oligosaccharide units are found primarily in glycoproteins of serum and cell membranes.

The N-linked chains are bound to the amido group of asparagine residues via β-N-acetylglucosamine. The carbohydrate binding site on the peptide chain shows the characteristic amino acid sequence **Asn-X-Thr/Ser** (**X** can be any amino acid with the probable exception of proline and aspartic acid). All asparagine-linked carbohydrate chains share the common core structure, **Man$_3$GlcNAc$_2$**. On the basis of their peripheral sections three main categories can be distinguished as shown in *Fig. 2.4*, [4]:

- the **high-mannose type** oligosaccharides, containing two to six additional mannose residues linked to the pentasaccharide core,
- the **complex-type** oligosaccharide units, characterized by two to four, and in rare cases five, branches of N-acetyllactosamine bound to the mannose residues of the pentasaccharide core. These disaccharide units are often elongated by additional N-acetyllactosamine residues and may carry short N-acetyl-lactosamine side branches (the so-called 'erythroglycans', see *Sect. 5.3.2*). The complex-type chains further contain an α1→6 fucose attached to the innermost N-acetylglucosamine residue. The terminal galactosyl groups usually carry α2→3 linked N-acetylneuraminic acid residues, or, less frequently, α1→2 fucose, α1→3 galactose, or poly-N-acetylneuraminic acid chains (**NeuAcα2→8)$_n$NeuAcα2→3** (n = 2–3). The subterminal N-acetylglucosamine is often substituted with α1→3 fucose[2].

---

[2] In keratan-sulfate of the cornea the N-acetylglucosamine and to a lesser extent the Carbon-6 of the galactose residues as well are sulfated. These types of glycoproteins carry no blood group character whatsoever.

High Mannose—type

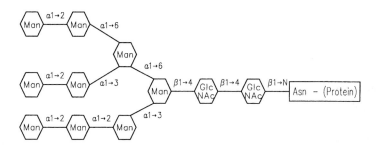

Complex—type

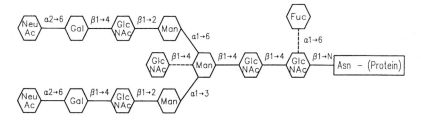

Hybrid—type

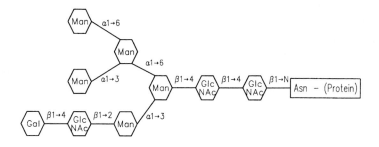

Figure 2.4: Structures of N-glycosidically linked oligosaccharide chains occurring in glycoproteins.

– the **hybrid-type** carbohydrate units have features of both the high-mannose and the complex-type oligosaccharide chains. In most cases they contain 'bisecting' $\beta1\rightarrow4$ N-acetyllactosamine linked to the β-mannose residue.

O-glycosidically linked carbohydrate chains occur mainly in secretions of exocrine glands and in mucins produced by mucus cells and pathological cysts (= 'mucin-type' linkage) and are found less frequently in membrane glycoproteins (see for example 'glycophorins').

The O-linked carbohydrate chains are attached to the hydroxyl groups of serine and threonine residues of a protein backbone via α-N-acetylgalactosamine[3]. The size of the oligosaccharide units extends from single N-acetylgalactosaminyl residues or a **NeuAcα2→6GalNAc** disaccharide as found in ovine submaxillary mucin, to the highly complex chains containing galactose, N-acetylglucosamine, and N-acetyl-neuraminic acid of human water-soluble blood group substances (see e.g. *Fig. 5.13)* [5]. In contrast to N-linked oligosaccharide units, O-linked chains lack mannose.

Three regions of the oligosaccharide chains can be distinguished according to Hounsell and Feizi's proposal (see [5]) – the core close to the peptide, the backbone, and the peripheral parts. The general features of these chain structures are presented in *Fig. 2.5* (see also the structure of 'water-soluble blood group substances', *Sect. 5.3.4*).

The mucin substances, which are the main carriers of the O-linked chains, are either membrane-bound or are secreted into the lumina of the gastrointestinal, respiratory, and reproductive tracts.

Mucins are heavily glycosylated molecules, and carbohydrate may account for up to 85% of the molecular mass in some cases. The protein backbones are characterised by a high content of serine and threonine residues which carry the oligosaccharide chains.

The membrane-bound mucins contain a hydrophobic membrane-spanning domain. They protect the mucosa against pathogens and can also function as cell adhesion molecules.

The secreted molecules are considerably larger, the protein core in some cases containing more than 5000 amino acid residues. A heavily glycosylated central region is flanked by less glycosylated cysteine-rich domains. End-to-end assembly via disulphide bonds forms linear oligomers or polymers of 500–30,000 kDa.

---

[3] Oligosaccharide chains linked via fucose to serine/threonine (epidermal growth factor and some serum glycoproteins), via galactose or glucose to hydroxyproline (collagens), or via xylose to serine (proteoglycans) carry no known blood group determinant.

Cores

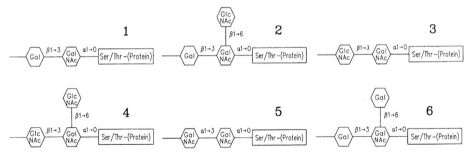

Backbones

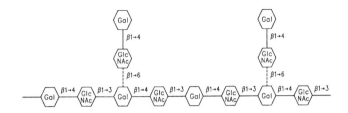

Terminal structures (with possible substituents)

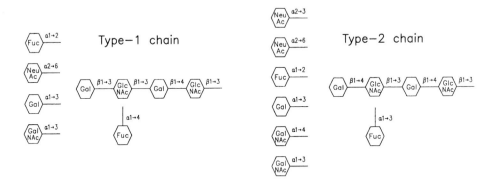

Figure 2.5: Basic types of O-glycosidically linked oligosaccharide chains occurring in glycoproteins.
According to Hounsell and Feizi (see [5]).

The high density of oligosaccharide side chains at the carrier protein forces the polypeptide backbone into an extended, linear conformation which is responsible for both the high viscosity of mucin solutions and the viscoelastic properties of mucus gels. These hydrophilic gels readily stick to all kinds of surfaces and function as physical barriers protecting the underlying tissue against harmful influences of the environment (e.g low pH, proteases etc.) as well as from attack of microorganisms. Furthermore, the lubricating properties of mucous layers are important for protecting tissues against mechnical damage. In order to meet the specific needs evolving from the local conditions, each epithelial tissue contains distinct mucin species.

References

1. FREDMAN, P. (1993): Glycosphingolipid tumor antigens. *Adv. Lipid Res. 25*, 213-234.
2. HANSSON, G. C., KARLSSON, K. A., LARSON, G., STRÖMBERG, N. & THURIN, J. (1985): Carbohydrate-specific adhesion of bacteria to thin-layer chromatograms: a rationalized approach to the study of host cell glycolipid receptors. *Anal. Biochem. 146*, 158-163.
3. IUPAC – IUB COMMISSION ON BIOCHEMICAL NOMENCLATURE (CBN) (1977): The nomenclature of lipids. Recommendations 1976. *Eur. J. Biochem. 79*, 11-21.
4. KORNFELD, R. & KORNFELD, S. (1985): Assembly of asparagine-linked oligosaccharides. *Annu. Rev. Biochem. 54*, 631-664.
5. ROUSSEL, P., LAMBLIN, G., LHERMITTE, M., HOUDRET, N., LAFITTE, J. J., PERINI, J. M., KLEIN, A. & SCHARFMAN, A. (1988): The complexity of mucins. *Biochimie 70*, 1471-1482.
6. STRÖMBERG, N., DEAL, C., NYBERG, G., NORMARK, S., SO, M. & KARLSSON, K. A. (1988): Identification of carbohydrate structures that are possible receptors for Neisseria gonorrhoeae. *Proc. Natl. Acad. Sci. USA 85*, 4902-4906.
7. STRÖMBERG, N., RYD, M., LINDBERG, A. A. & KARLSSON, K. A. (1988): Studies on the binding of bacteria to glycolipids. Two species of Propionibacterium apparently recognize separate epitopes on lactose of lactosylceramide. *FEBS Lett. 232*, 193-198.

3 Biosynthesis of Glycoconjugates

The basic aspects of molecular genetics and protein biosynthesis are covered by all introductory courses on biochemistry and molecular biology and will not be discussed in this book. Little attention is paid, however, to the biosynthesis of glyco-conjugates. Since these substances are of pivotal importance for blood group serology, the formation of glycoconjugates will be discussed below in detail.

In contrast to proteins, which are assembled according to a 'template' mechanism, the carbohydrate units of glycoproteins and glycolipids are assembled stepwise by the coordinated action of highly specific glycosyltransferases. These enzymes attach monosaccharide residues either to protein or lipid acceptors or to growing cabohydrate chains, and each single glycosyl transfer forms the acceptor substrate for the next transfer step in the sequence. The sugar nucleotides UDP-galactose, UDP-N-acetylglucosamine, UDP-N-acetylgalactosamine, GDP-mannose, GDP-fucose, and CMP-N-acetylneuraminic acid act as glycosyl donors. The proper function of the glycosyltransferases depends on the presence of metal ions – fucosyltransferases need Mg^{++}, whereas galactosyl-, N-acetylglucosaminyl-, N-acetyl-neuraminyl-, and N-acetylgalactosaminyltransferases need Mn^{++} which in some cases may be replaced by other divalent cations, such as Cd^{++}, Co^{++}, Zn^{++} or Ca^{++}.

The scheme below depicts the formation of an N-acetyllactosamine unit:

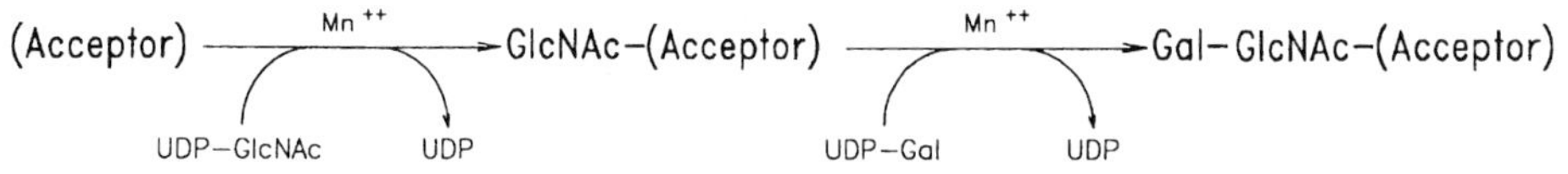

The glycosylation pattern of a glycoconjugate is defined by: (1) the set of glycosyltransferases provided by transcriptional and translational control mechanisms, and (2) by the organisation of the enzymes within the membranes of the endoplasmic reticulum and Golgi apparatus where the biosynthesis of oligosaccharide units takes place [12,16].

As consequences of this mechanism the sequence of an oligosaccharide chain is not defined as precisely as that of a peptide chain: a large variety of glycolipids with

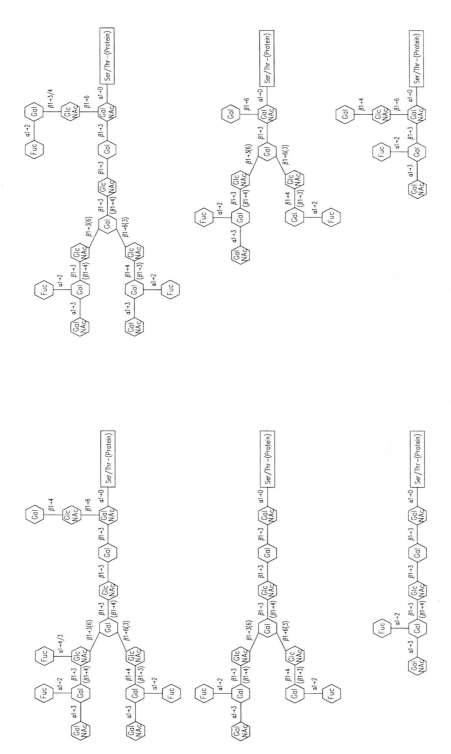

Figure 3.1: Neutral oligosaccharides isolated from blood group A active human gastric mucin, an example of microheterogeneity.
According to Slomiany et al. [22].

differing carbohydrate moieties is synthesised by one set of transferases. The fact that in glycoproteins, carbohydrate chains of various length, differing degrees of branching and different position of branching points are found even in highly 'homogeneous' preparations is often overlooked. This so-called **micro-heterogeneity** or **peripheral heterogeneity** is restricted to the characteristic overall structure determined by the specificities of the transferases involved (see *Fig. 3.1)*.

In general, one gene encodes one transferase, ansd so a series of different genes controls the biosynthesis of a complete carbohydrate chain. Though most of the transferase genes are independent of each other, all glycosyltransferases thus far investigated showed a characteristic topology: peptide chains consisting of a short N-terminal cytoplasmic segment, a hydrophobic trans-membrane domain, and a large C-terminal portion in which the catalytic domain is localised. Many of the peptide sequences – as well as the point mutations which change enzyme specificity – have already been described (see **ABH**-transferases, *Sect. 5.4.5)*.

3.1 Biosynthesis of Glycoproteins

3.1.1 Biosynthesis of N-linked Carbohydrate Chains

The carbohydrate chains of glycoproteins bound via N-acetylglucosamine to the amido nitrogen of asparagine residues of the protein backbone are not assembled as a whole on the carrier protein itself; they are synthesised in two separate stages (see *Fig. 3.2)*[1]:

In the first stage a complex branched oligosaccharide unit is assembled in the membrane of the rough endoplasmic reticulum by sequential transfer of N-acetylglucosamine, mannose, and glucose residues onto dolichol pyrophosphate as an intermediate carrier molecule. The oligosaccharide thereby formed has the average composition **$Glc_3Man_9GlcNAc_2$** (*Fig. 3.3)*.

In the second stage this oligosaccharide unit is moved from the lipid molecule onto the protein. As discussed in *Chap. 2*, the characteristic sequences for the glycosyl acceptor region within the polypeptide chain are **Asn-X-Thr** and **Asn-X-Ser**, whereby **X** can be any amino acid, with the probable exceptions of proline and aspartic acid. The transfer of the carbohydrate unit is facilitated by an increased nucleophilicity of the amido electron pair induced by the formation of an H-bridge between the amido group of the aspartic acid and the hydroxyl group of serine or threonine (*Fig. 3.4*, [2]).

---

[1] Detailed information on the exact mechanism of each single step can be found in Kornfeld and Kornfeld's review article [13].

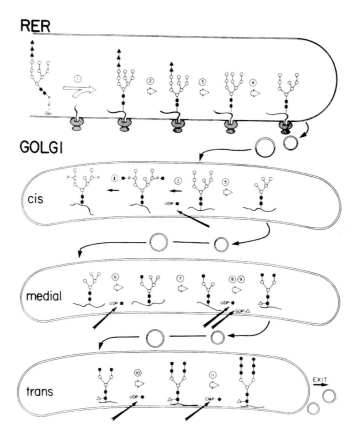

Figure 3.2: Assembly of asparagine-linked carbohydrate chains.
Schematic pathway of oligosaccharide processing within the rough endoplasmic reticulum and Golgi apparatus.

■ : N-acetylglucosamine, ○ : mannose, ▲ : glucose, △ : fucose, ● : galactose,

◆ : N-acetylneuraminic acid.

Reproduced from Kornfeld and Kornfeld [13], by permission of the Annual Reviews, Inc., Palo Alto, CA (©1985).

The enzymatically catalysed transfer of the carbohydrate unit takes place in the ergastoplasma at the very moment when the nascent protein chain penetrates the membrane of the rough endoplasmic reticulum. It must be assumed that optimum accessibility of the aspartic acid is provided only during chain growth before the protein adopts its specific secondary structure.

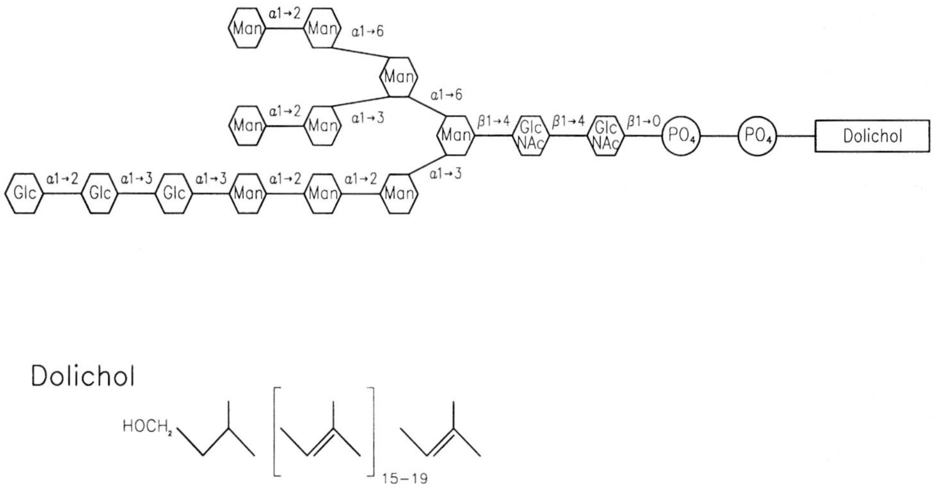

Figure 3.3: Dolichol and the precursor oligosaccharide as bound to dolichol pyrophosphate.

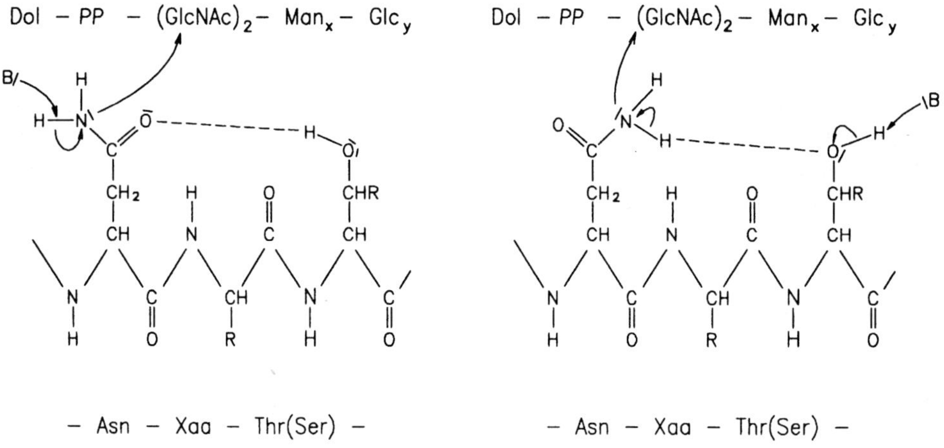

Figure 3.4: Transfer of the lipid-bound precursor oligosaccharide to protein.
Proposed catalytic mechanisms involving a hydrogen bond between the side chains of asparagine and the hydroxy amino acid in the 'acceptor sequence' Asn-Xaa-Thr(Ser).
Reproduced from Bause and Legler [2] by permission of the Biochemical Society and Portland Press, London.

This entire oligosaccharide unit is subsequently processed to its final structure in the course of the movement of the glycoprotein through the cell compartments. While the protein is still connected to the ribosome, the three glucosyl residues and at least one mannosyl residue are split off by specific membrane-bound glycosidases in the rough endoplasmic reticulum. Subsequently the glycoprotein is transferred to the cisternae of the Golgi apparatus.

In the case of **type I** oligosaccharide structures ('high-mannose' type), these chains are not subjected to any further changes and are transferred via the medial-Golgi cisternae from the cis-Golgi into the trans-Golgi cisternae, and from there into the trans-Golgi reticulum. A specific peptide sequence at the N-terminus probably signals the incorporation of the glycoprotein into the membrane or its being packed into vesicles pending its release into the cytoplasm or into the surrounding medium as required (see e.g. [7]).

In **type II** oligosaccharide chains ('complex type') the mannose-rich basic unit is processed further to its final structure during transport of the glycoprotein through the Golgi apparatus (*Fig. 3.2*): in the medial cisternae the peripheral five mannose residues are split off and subsequently one or two N-acetylglucosamine units are attached to the residual terminal mannose residues. Here, too, the $\alpha1{\to}6$ fucose residue is linked to the first N-acetylglucosamine. The trans-Golgi cisternae contain $\beta1,3$- and $\beta1,4$-galactosyl-, $\beta1,3$-N-acetylglucosaminyl-, and $\alpha1,2$- and $\alpha1,3$-fucosyltransferases, all of which synthesise the peripheral segments of the chains [13]. The $\alpha2,3/6$-N-acetylneuraminyltransferases which attach the chain-terminating sialic acid residues are found in the trans-Golgi reticulum [21,23] (see also [20]).

The factors which control the synthesis of a distinct chain type have not yet been investigated. It must be assumed, however, that the accessibility of the carbohydrate chain to the action of the mannosidases determines the pathway according to which a high-mannose type chain or a complex-type chain is formed (cf. [8,11]).

3.1.2 Biosynthesis of O-linked Carbohydrate Chains

Mucin-type carbohydrate chains are bound via α-N-acetylgalactosamine to the hydroxyl groups of Ser and Thr residues of the carrier protein (see *Chap. 2*).

The initial step in O-glycosylation is the transfer of N-acetylgalactosamine onto serine/threonine, which is catalysed by highly specific N-acetylgalactosaminyltransferases located in the cis-Golgi cisternae (see e.g. [18,19]). The sequential assembly of the oligosaccharide chain takes place later in the trans-Golgi cisternae and in the trans-Golgi reticulum and is presumably mediated by the same enzymes which are involved in the synthesis of N-linked chains.

In contrast to the specific attachment sites of N-linked carbohydrate structures, there is no consensus sequence for the addition of O-linked N-acetylgalactosamine residues to proteins. However, a significantly increased frequency of proline (and to a lesser extent of serine, threonine, and alanine as well) has been observed in the vicinity of O-glycosylation sites [25,27]. Furthermore, clusters of 2, 3, or 4 adjacent serine and threonine residues are a characteristic feature of highly O-glycosylated regions of glycoproteins.

3.2 Biosynthesis of Glycosphingolipids

The biosynthesis of glycosphingolipids is not as well investigated as that of the glycoproteins. Nevertheless, it is generally agreed that the carbohydrate moiety of the glycosphingolipids is assembled by an analogous sequential mechanism (cf. [5,15]).

Basu and co-workers have isolated and characterised a series of glycolipid glycosyltransferases [1]. On the basis of their investigations two groups of enzymes have been distinguished – the first group is able only to transfer the monosaccharide onto glycolipids; the second is able to use both glycolipids and glycoproteins (and oligosaccharides) as acceptor substrates. It is not yet known if the transferases of these two groups are different enzymes encoded by different genes or if they arise from a common gene by alternative splicing: the possibility that in some cases the acceptor substrate specificity of a transferase is changed by specific co-factors cannot be excluded.

3.3 Regulation of Chain Structure

Three factors determine the sequence and configuration of a carbohydrate chain: (a) the substrate specificity of the enzymes, (b) competition among two or more glycosyltransferases for a common acceptor substrate [3,16], and (c) the cellular expression of specific glycosyltransferases. Other factors, such as availability of substrate, transfer of chain-terminating monosaccharide residues (mainly N-acetylneuraminic acid or fucose, see below), and processing of sugar chains [13] also contribute to the diversity of carbohydrate chains.

(a) Substrate Specificity of the Glycosyltransferases

Glycosyltransferases are highly specific for the monosaccharide transferred, its anomeric configuration, and its position on the glycosyl acceptor; only to a lesser degree is the transfer activity influenced by the chemical structure of the acceptor molecule.

The substrate specificity of the glycosyltransferases will be discussed below by presenting selected examples:

The polypeptide N-acetylgalactosaminyltransferases attach N-acetylgalactos-amine residues onto serine and threonine of a protein carrier. Though the characteristics of all enzymes of this specificity thus far described are fairly similar, the single transferase reacted only with a few distinct acceptor proteins (cf. [6,24]). The specificity is obviously determined by the primary and secondary structure of the polypeptide chain, i.e. by accessibility and local conformation of the acceptor site.

Enzymes transferring monosaccharide residues onto growing oligosaccharide chains show greater variations of acceptor substrate specificity: some transferases recognise a monosaccharide residue of only a defined anomeric configuration, others need just a partial structure of the acceptor site, whereas some transferases require larger segments of the acceptor molecule (i.e. the subterminal sugar or the amino acid as well) for a successful glycosyl transfer.

In most cases sterical conditions of the acceptor molecule are of essential importance:

− the **H** gene-encoded α1,2-fucosyltransferase (*Sect. 5.4.2*) is able to transfer fucose residues onto a series of acceptor substrates, i.e. mainly glyco-conjugates with terminal galactose, but also free galactose and inositol as well. However, the transferase is fairly sensitive to the conformation of the acceptor substrate − the enzyme shows a significantly different transfer activity towards type-1 and type-2 chain precursors, whereas substrates in which the subterminal monosaccharide residue is substituted by fucose cannot be used as fucosyl acceptors at all (see *Sect. 6.4.2*);
− the glycosyl transfer of the blood group **A** gene-encoded α1,3-N-acetylgalactos-aminyltransferase and the blood group **B** gene-encoded α1,3-galactosyl-transferase (*Sect. 5.4.1*) is restricted to acceptors carrying a terminal **H** determinant (i.e. the disaccharide **Fucα1→2Gal**). However, due to the close similarity between N-acetylgalactosamine and galactose (the sugars differ only in the substituent of carbon-2) the enzymes show a significant cross-reactivity towards the donor substrate − the **A**-transferase is able to transfer small amounts of galactose residues, and the **B**-transferase transfers small amounts of N-acetylgalactosamine residues onto the blood group **H** precursor. This unspecificity may result in the unexpected production of 'aberrant' blood group antigens under altered physiological conditions such as those found in cancer cells;

– the β1,4-galactosyltransferase attaches galactose residues not only onto various acceptors with terminal N-acetylglucosamine, but also onto free N-acetylglucosamine and to a lesser degree onto glucose. In the presence of α-lactalbumin, however, the affinity towards glucose is highly increased, and in mammary glands this transferase is mainly involved in the production of lactose ($\rightarrow$ 'lactose synthase'). Until now the β1,4-galactosyltransferase is the only known example of an enzyme the specificity of which is modified by a co-factor [4].

(b) Regulation of the Glycosylation Pattern by Enzyme Competition

The glycosyltransferase system of a cell organelle frequently contains two or more enzymes which compete for the same acceptor substrate. In these cases the glycosylation pattern of a cell is determined by the relative enzyme activities of the competing transferases, i.e. by the order according to which the enzymes act upon an acceptor substrate. The joint influence of transferases may range from enhancing or decreasing transfer activity of the other enzyme (e.g. the assembly of a branched poly-N-acetyllactosamine chain presented in *Sect. 7.5*) to a total block of a glycosylation step (as represented by the chain-terminating function of N-acetyl-neuraminic acid and fucose [17,26]). A model for the coordinated action of several glycosyltransferases synthesising an O-glycosidically linked carbohydrate chain is presented in *Fig. 3.5* [3].

(c) Expression of Specific Glycosyltransferases

Changes in composition and structure of cellular glycoconjugates during embryonic development and cell maturation (see *Sects. 6.5.3* and *7.4*) are based on the regulated expression of specific glycosyltransferases. An example: the appearance of the 'stage-specific embryonic antigens', **Le**[x] and sialoyl-**Le**[x], both of which are essential in cell interaction and cell sorting, depends on the presence of the myeloid type α1,3-fucosyltransferase. This transferase is expressed in all tissues of five- to ten-week-old embryos. In older embryos the enzyme is gradually replaced by plasma- or **Lewis**-type α1,3-fucosyltransferases, and ultimately attains the enzyme pattern of the adult stage (see *Sect. 6.5.5*).

In the cells of colonic mucosa of human adults the selected expression of glycosyltransferases is reflected by the characteristic distribution of glycosphingolipids in different cell types. Studies by Holgersson et al. [10] showed that epithelial cells contain mainly neutral glycosphingolipids, i.e. monoglycosylceramide and the **Le**[a]

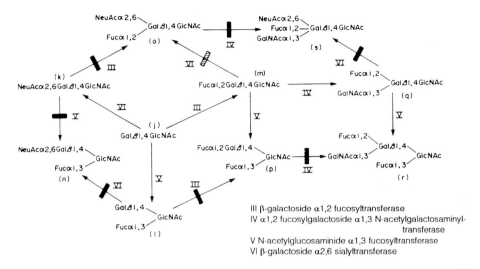

Figure 3.5: Proposed biosynthestic pathways for oligosaccharides formed by coordinated action of glycosyltransferases.
Solid bars designate a reaction which cannot proceed, and *hatched bars* denote a reaction which proceeds very slowly.
Reproduced from Beyer et al. [3] by permission of the authors.

active ceramide pentasaccharide, as wll as small quantities of neutral glycolipids of the lacto- and neolacto-series with **ABH-**, **Le^b-**, and **Le^y** activities.

Non-epithelial cells, on the other hand, contain almost exclusively acidic glyco-sphingolipids of the ganglio series.

And since malignant transformation is usually accompanied by disorders in the biosynthetic pathway of glycoconjugates, cellular glycosylation patterns also show some marked alterations associated with oncogenesis. A complete spectrum of alterations has been observed: in some tumours the activity of certain glycosyl-transferases is considerably increased, in others it is greatly decreased; in some tumours glycosyltransferases are expressed which are active only in an early stage of cellular development. Though these phenomena are relatively unspecific, some glycosyltransferases, such as galactosyl- and fucosyltransferases, as well as the 'aberrant' oligosaccharide structures formed by the changed glycosyltransferase pattern have repeatedly been used as tumour markers for cancer diagnosis [9,14].

References

1. BASU, S., BASU, M., DAS, K. K., DAUSSIN, F., SCHAEPER, R. J., BANERJEE, P., KHAN, F. A. & SUZUKI, I. (1988): Solubilized glycosyltransferases and biosynthesis in vitro of glycolipids. *Biochimie 70*, 1551-1563.
2. BAUSE, E. & LEGLER, G. (1981): The role of the hydroxy amino acid in the triplet sequence Asn-Xaa-Thr(Ser) for the N-glycosylation step during glycoprotein biosynthesis. *Biochem. J. 195*, 639-644.
3. BEYER, T. A., REARICK, J. I., PAULSON, J. C., PRIEELS, J. P., SADLER, J. E. & HILL, R. L. (1979): Biosynthesis of mammalian glycoproteins. Glycosylation pathways in the synthesis of the nonreducing terminal sequences. *J. Biol. Chem. 254*, 12531-12541.
4. BREW, K. (1970): Lactose synthetase: evolutionary origins, structure, and control. In: *Essays in Biochemistry* (P. N. Campbell & Dickens, eds.). Academic Press, London, New York, Vol. 6, pp. 93-118.
5. BURGER, K. N. J., VAN DER BIJL, P. & VAN MEER, G. (1996): Topology of sphingolipid galactosyl-transferases in ER and Golgi: transbilayer movement of monohexosyl sphingolipids is required for higher glycosphingolipid biosynthesis. *J. Cell Biol. 133*, 15-28.
6. CLAUSEN, H. & BENNETT, E. P. (1996): A family of UDP-GalNAc: polypeptide N-acetyl-galactos-aminyl-transferases control the initiation of mucin-type O-linked glycosylation. *Glycobiology 6*, 635-646.
7. COLLEY, K. J., LEE, E. U., ADLER, B., BROWNE, J. K. & PAULSON, J. C. (1989): Conversion of Golgi apparatus sialyltransferase to a secretory protein by replacement of the NH_2-terminal signal anchor with a signal peptide. *J. Biol. Chem. 264*, 17619-17622.
8. FAYE, L., STURM, A., BOLLINI, R., VITALE, A. & CHRISPEELS, M. J. (1986): The position of the oligo-saccharide side-chains of phytohemagglutinin and their accessibility to glycosidases determines their subsequent processing in the Golgi. *Eur. J. Biochem. 158*, 655-661.
9. HAKOMORI, S. I. (1989): Aberrant glycosylation in tumors and tumor-associated carbohydrate antigens. *Adv. Cancer. Res. 52*, 257-331.
10. HOLGERSSON, J., STROMBERG, N. & BREIMER, M. E. (1988): Glycolipids of human large intestine: difference in glycolipid expression related to anatomical localization, epithelial/non-epithelial tissue and the ABO, Le, and Se phenotypes of the donors. *Biochimie 70*, 1565-1574.
11. HSIEH, P., ROSNER, M. R. & ROBBINS, P. W. (1983): Selective cleavage by endo-β-N-acetyl-glucosaminidase H at individual glycosylation sites of Sindbis virion envelope glycoproteins. *J. Biol. Chem. 258*, 2555-2561.
12. IVATT, R. J. (1981): Regulation of glycoprotein biosynthesis by formation of specific glycosyl-transferase complexes. *Proc. Natl. Acad. Sci. USA 78*, 4021-4025.
13. KORNFELD, R. & KORNFELD, S. (1985): Assembly of asparagine-linked oligosaccharides. *Annu. Rev. Biochem. 54*, 631-664.
14. LA MONT, J. T. & ISSELBACHER, K. J. (1975): Alterations in glycosyltransferase activity in human colon cancer. *J. Natl. Cancer Inst. 54*, 53-56.
15. MACCIONI, H. J. F., DANIOTTI, J. L. & MARTINA, J. A. (1999): Organization of ganglioside synthesis in the Golgi apparatus. *Biochim. Biophys. Acta 1437*, 101-118.
16. PAULSON, J. C. & COLLEY, K. J. (1989): Glycosyltransferases. Structure, localization, and control of cell type-specific glycosylation. *J. Biol. Chem. 264*, 17615-17618.
17. PILLER, F. & CARTRON, J. P. (1983): UDP-GlcNAc:Galβ1-4Glc(NAc)β1-3N-acetylglucosaminyltrans-ferase. Identification and characterization in human serum. *J. Biol. Chem. 258*, 12293-12299.
18. PILLER, V., PILLER, F., KLIER, F. G. & FUKUDA, M. (1989): O-glycosylation of leucosialin in K562 cells. Evidence for initiation and elongation in early Golgi compartments. *Eur. J. Biochem. 183*, 123-135.
19. ROTH, J. (1984): Cytochemical localization of terminal N-acetyl-D-galactosamine residues in cellular compartments of intestinal goblet cells: implications for the topology of O-glycosylation. *J. Cell Biol. 98*, 399-406.

20. ROTH, J., GREENWELL, P. & WATKINS, W. M. (1988): Immunolocalization of blood group A gene specified α1,3 N-acetylgalactosaminyltransferase and blood group A substance in the trans-tubular network of the Golgi apparatus and mucus of intestinal goblet cells. *Eur. J. Cell Biol. 46*, 105-112.

21. ROTH, J., TAATJES, D. J., LUCOCQ, J. M., WEINSTEIN, J. & PAULSON, J. C. (1985): Demonstration of an extensive trans-tubular network continuous within the Golgi apparatus stack that may function in glycosylation. *Cell 43*, 287-295.

22. SLOMIANY, A., ZDEBSKA, E. & SLOMIANY, B. L. (1984): Structures of the neutral oligosaccharides isolated from A-active human gastric mucin. *J. Biol. Chem. 259*, 14743-14749.

23. TAATJES, D. J., ROTH, J., WEINSTEIN, J. & PAULSON, J. C. (1988): Post-Golgi apparatus localization and regional expression of rat intestinal sialyltransferase detected by immunoelectron microscopy with polypeptide epitope-purified antibody. *J. Biol. Chem. 263*, 6302-6309.

24. WANDALL, H. H., HASSAN, H., MIRGORODSKAYA, E., KRISTENSEN, A. K., ROEPSTORFF, P., BENNETT, E. P., NIELSEN, P. A., HOLLINGSWORTH M. A., BURCHELL, J., TAYLOR-PAPADIMITRIOU, J. & CLAUSEN, H. (1997): Substrate specificities of three members of the human UDP-N-acetyl-α-D-galactosamine: polypeptide N-acetylgalactosaminyltransferase family, GalNAc-T1, -T2, and -T3. *J. Biol. Chem. 272*, 23503-23514.

25. WILSON, I. B. H., GAVEL, Y. & VON HEIJNE, G. (1991): Amino acid distributions around O-linked glycosylation sites. *Biochem. J. 275*, 529-534.

26. YATES, A. D. & WATKINS, W. M. (1983): Enzymes involved in the biosynthesis of glycoconjugates. A UDP-2-acetamido-2-deoxy-D-glucose:β-D-galactopyranosyl-(1→4)-saccharide (1→3)-2-acetamido-2-deoxy-β-D-glucopyranosyltransferase in human serum. *Carbohydr. Res. 120*, 251-268.

27. YOSHIDA, A., SUZUKI, M., IKENAGA, H. & TAKEUCHI, M. (1997): Discovery of the shortest sequence motif for high level mucin-type O-glycosylation. *J. Biol. Chem. 272*, 16884-16888.

4 Erythrocyte Membrane

The majority of the blood group characters treated in this monograph are defined as serological properties of the erythrocyte membrane. The expression of these characters is highly influenced by the structure and prevailing constitution of the membrane. It is therefore appropriate to present a concise survey on membrane composition and structure here.

Plasma membranes in general do not just represent the boundary between a cell and its surroundings; they also play an essential role in regulating various cell functions. In addition to their function as a permeability barrier with a regulated active transport, the membranes also act as a 'switchboard', converting extracellular information into intracellularly active messages. In order to maintain all these functions, various enzymes and transport proteins, different receptors for external stimuli such as hormones and growth factors, as well as functional sites regulating the social behaviour of the cells, are incorporated in the membrane matrix. Furthermore, binding sites for viruses, toxins, pharmaca, and other foreign agents are found on the membrane surface. All these organisational units and 'membrane organelles' are kept within the cell membrane in a strictly defined topological arrangement providing optimum function. And further, proteins and glycoconjugate components of the cell membrane act as antigens representing binding sites for various antibodies and lectins.

Erythrocytes of mammals do not contain a nucleus and represent cells with a restricted metabolism. Despite its highly reduced functions, the erythrocyte membrane in its essential elements resembles a normal biological membrane (see review articles: [11,19,34,51]).

4.1 Constituents of the Erythrocyte Membrane

Human erythrocyte membranes consist of 43.6% lipid, 49.2% protein, and 7.2% carbohydrate [42].

The main membrane constituent classes are discussed below.

Table 4.1: Lipid content of the human erythrocyte membrane.*

Lipid constituent	mg x 10^{10} per cell[a]	% of total lipid
Total lipid	**4.94**	**100**
Cholesterol	1.25	25.3
Cholesteryl esters ⎫ Free fatty acids ⎬ Glycerides ⎭	0.17	3.5
Glycosphingolipids	0.53	10.5
Phospholipids	2.99	60.5
Sphingomyelin	0.77	15.6
Phosphatidylcholine (= lecithin)	0.86	17.4
Phosphatidylethanolamine	0.85	17.2
Phosphatidylserine	0.40	8.1
Lysolecithin	0.03	0.6
Phosphatidic acids	0.02	0.4
Phosphatidylinositol	0.04	0.8
Polyglycerolphospholipids	0.02	0.4

* Mean values of the data published by van Deenen and de Gier [58].
[a] The haemoglobin-free ghost has an average mass of 1.5×10^{-9} mg [60].

4.1.1 Membrane Lipids

The lipid portion of the erythrocyte membrane is composed of approximately 60% phospholipids, 25% cholesterol and cholesterol esters, 11% glycosphingolipids, and 4% glycerides and free fatty acids (*Table 4.1*).

The **phospholipids** are derived either from glycerol ($\rightarrow$ phosphoglycerides) or from sphingosine ($\rightarrow$ sphingomyelin).

The **phosphoglycerides** (= phosphatidylglycerols) are glycerol derivatives in which the hydroxyl groups at Carbon 1 and Carbon 2 are esterified to fatty acids, and the hydroxyl group at Carbon 3 to phosphoric acid ($\rightarrow$ phosphatidic acid). The most

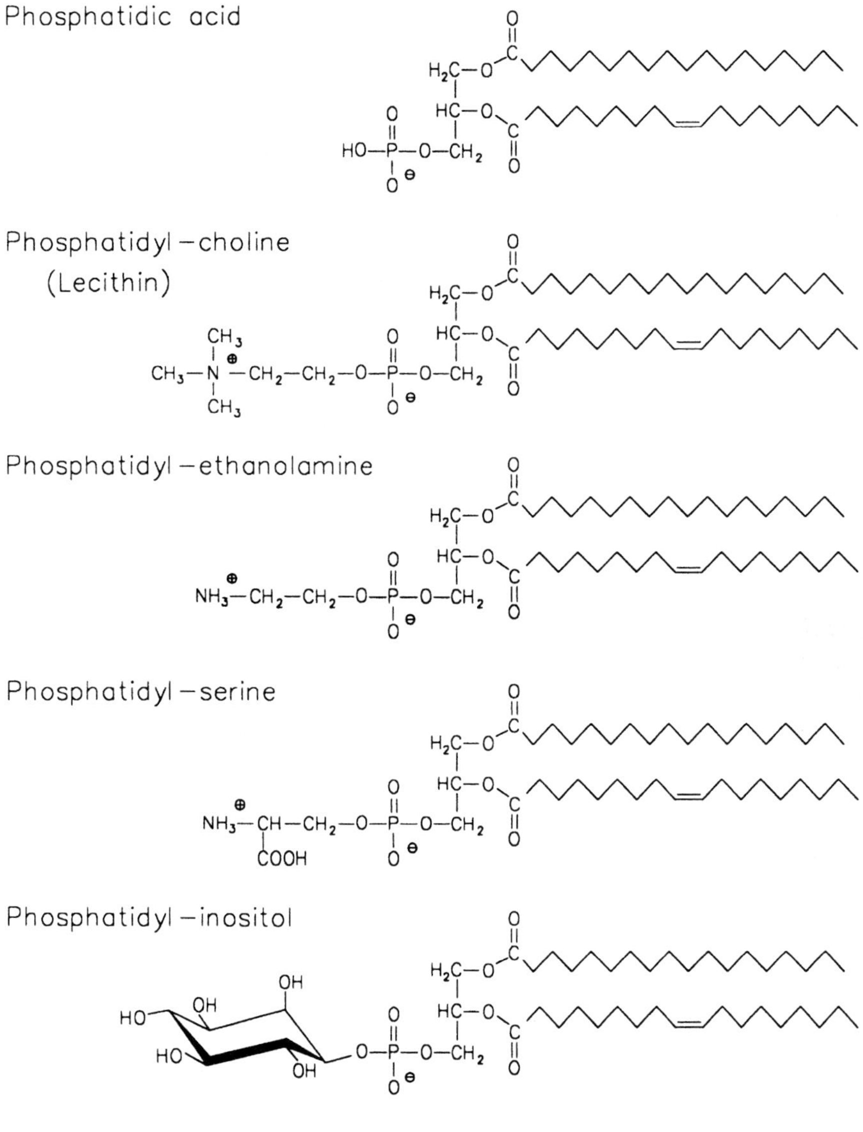

Phosphatidic acid

Phosphatidyl −choline
    (Lecithin)

Phosphatidyl −ethanolamine

Phosphatidyl −serine

Phosphatidyl −inositol

Figure 4.1: Chemical structures of phosphatidylglycerols of the human erythrocyte membrane.

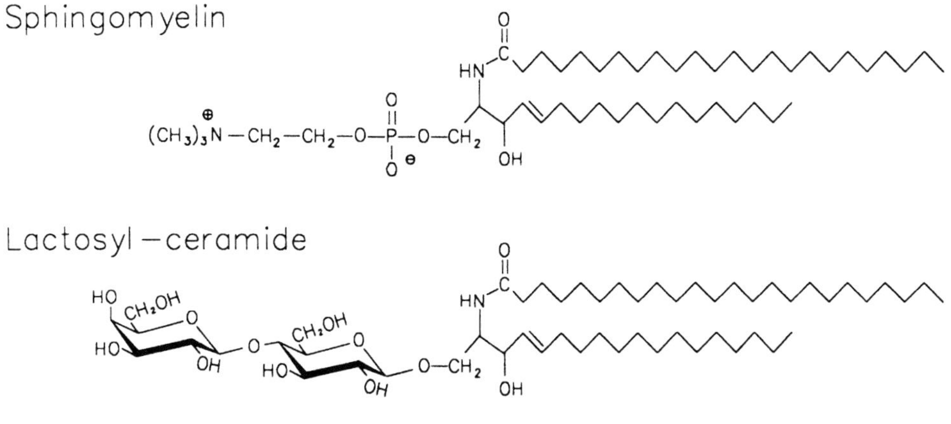

Sphingomyelin

Lactosyl−ceramide

Figure 4.2: Chemical structures of sphingomyelin and a glycosphingolipid.

common fatty acids are saturated or mono-unsaturated molecules with 16 or 18 carbon atoms; the unsaturated acids are normally attached to carbon 2 of the glycerol. The phosphate residue of phosphatidic acid is esterified to the hydroxyl group of an alcohol, such as choline (→ phosphatidyl-choline = lecithin), serine (→ phosphatidyl-serine), ethanolamine (→ phosphatidyl-ethanolamine) or myo-inositol (→ phosphatidyl-inositol). The more complex polyglycerol-phospholipids are found only in very low concentrations. *Fig. 4.1* shows the chemical structures of phosphoglycerides occurring in erythrocyte membranes.

The **sphingomyelins** are derivatives of sphingosine, an amino alcohol with a long, unsaturated hydrocarbon chain (*Fig. 2.3*). The amino group of the sphingosine residue is connected to a fatty acid by an amide bond (→ ceramide), and the primary hydroxyl group is esterified to phosphorylcholine (*Fig. 4.2*).
Ceramide is also the basic constituent of **glyco-sphingolipids** (see *Sect. 2.1*). In this substance class the primary hydoxyl group of the sphingosine is linked to an oligosaccharide unit (*Fig. 4.2*).

Cholesterol represents another important constituent of the erythrocyte membrane. The hydroxyl group at carbon 3 can be esterified to a fatty acid (→ cholesterol ester).

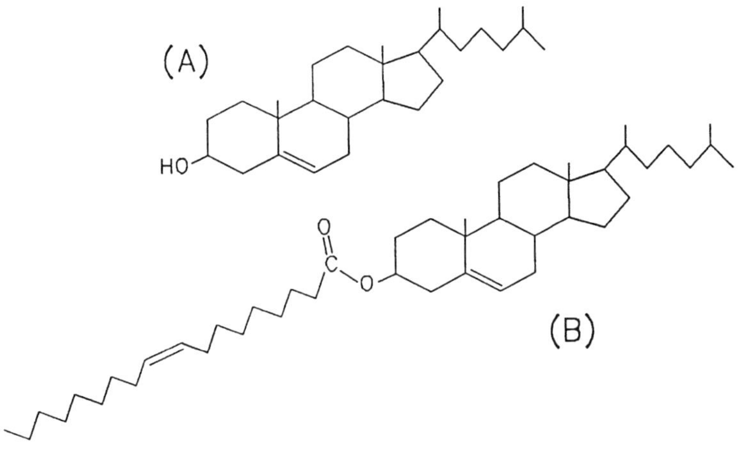

Figure 4.3: Cholesterol (A) and a cholesterol ester (B).

The chemical structures of cholesterol and a cholesterol ester are presented in *Fig. 4.3.*

4.1.2 Membrane Proteins

Erythrocyte membranes, when solubilised in sodium dodecylsulfate (SDS) and subjected to polyacrylamide-gel electrophoresis (PAGE), separate into a series of polypeptide bands with apparent molecular masses between 15 and 250 kDa [18,51]. Proteins are visualised by staining with Coomassie Blue or Amidoschwarz 6B (*Fig. 4.4.A*), whereas glycoproteins are characterised by their strong reaction with periodic acid Schiff reagent (= PAS reagent) (*Fig. 4.4.B*).

Three types of membrane proteins can be distinguished:

(A) The **extrinsic proteins** (or peripheral proteins) are located on the cytoplasmic surface of the membrane. They are only loosely bound to the membrane and can be easily removed in low ionic strength medium, in the presence of chelating agents, or under alkaline conditions. This group of membrane proteins is represented by the proteins of bands 1, 2, 2.1, 2.2, 2.3, 4.1, 4.2, 5, 6, and 7 [51];

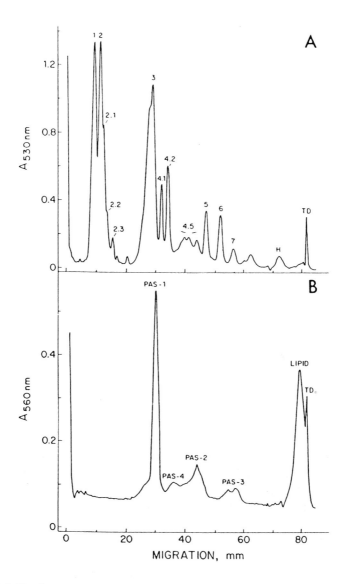

Figure 4.4: Erythrocyte membrane proteins.
Separation by SDS polyacrylamide-gel electrophoresis and densitometric scannning of the gels:
(A) stained for proteins with Coomassie Blue, **(B)** stained for carbohydrate with PAS reagent.
H = haemoglobin, **TD** = 'tracking dye'. For nomenclature of bands compare *Tables 4.2* and *4.4*.
Reproduced from Steck [51] by permission of the author and of the Rockefeller University Press, New York.

(B) The **intrinsic proteins** (or integral proteins) are embedded in the membrane matrix and span the bilayer. They can be solubilised only by detergents, or by treatment of the membranes with 'chaotropic agents' (such as iodide and rhodanide), or organic solvents. These proteins are essential membrane constituents and are represented by the proteins of bands 3 and 4.5, and the sialoglycoproteins of the PAS bands 1 to 4 [51].

All intrinsic membrane proteins are characterised by at least two hydrophilic domains mainly composed of polar amino acids, and by at least one hydrophobic domain predominantly composed of apolar amino acids. The hydrophilic sections of the proteins are in contact with the surrounding water phases − the extracellular domain is generally glycosylated [52], whereas the cytoplasmic domain is usually phosphorylated [29]. The hydrophobic domains of the membrane proteins are integrated in the lipid bilayer and interact with the hydrocarbon chains of the lipids and/or the hydrophobic regions of other proteins. The non-covalent bonds thus formed are fairly strong and therefore upon isolation intrinsic membrane proteins in most cases will remain associated with lipids; when completely freed of lipids these proteins easily aggregate and become insoluble in water.

(C) Another class of membrane proteins is attached to the outer surface of the cell membrane through a glycosyl-phosphatidylinositol unit (= **'GPI-anchored membrane proteins'**) [31,43]. This glycolipid anchor is composed of phosphatidyl-inositol, one non-acetylated glucosamine and three mannose residues, and phosphatidylethanolamine; the anchor unit is connected to the C-terminal amino acid of the protein via an amide bond (*Fig. 4.5*). This complex molecule is attached to the cell membrane by the fatty acid residues of the phosphatidylinositol moiety embedded in the outer leaflet of the bilayer.

Most of these proteins can be removed from the erythrocyte membrane by treatment with phosphatidylinositol-specific phospholipase C of bacterial and protozoan origin; these enzymes cleave the inositol from the diacylglycerol unit [20]. However, some phosphatidylinositol-linked proteins of human erythrocytes, such as acetylcholinesterase or complement regulatory proteins, show varying degrees of resistance to exogenous phospholipase C. This resistance has been attributed to an additional hydrophobic anchor provided by palmitoylation of the inositol ring [41].

The wide variety of phosphatidylinositol-anchored membrane proteins of human blood cells includes complement control proteins (decay accelerating factor, 'membrane inhibitor of reactive lysis' (CD59), and 'C8 binding protein'), immune control proteins (LFA-3 and CD16, the major Fc receptor), and various enzymes (e.g. acetylcholinesterase, alkaline phosphatase, and 5'-nucleotidase) (see [39,43]). Further, it has been shown that antigens of at least five different blood group systems, viz. **Cromer** (*Chap. 21*), **Dombrock** (*Chap. 22*), **Cartwright** (*Chap. 23*), and **John Milton Hagen (JMH)** (*Chap. 30*), are located on GPI-linked membrane proteins [56].

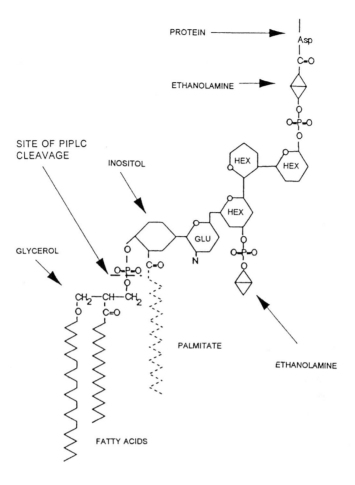

Figure 4.5: Structure of the glycosylphosphatidylinositol anchor of human erythrocyte membrane proteins.
Reproduced from Rosse [43] by permission of the author and of W.B. Saunders Co., Philadelphia, PA, USA.

The phosphatidylinositol-linked proteins may also appear as soluble molecules or as proteins inserted into the lipid bilayer by a transmembrane domain. Various investigations have suggested that these forms of the molecule arise from alternative splicing of the mRNA transcripts.

The abnormal erythrocytes of patients suffering from paroxysmal nocturnal haemoglobinuria (PNH) are deficient in all phosphatidylinositol-anchored proteins

[43,45]. Red cells of this rare acquired haemolytic anaemia are abnormally sensitive towards complement-mediated lysis, which is probably due to the reduced amount of complement-regulating proteins. However, not all erythrocytes are equally affected by this disorder – blood from PNH patients normally contains variable proportions of cell populations that differ in their susceptibility towards haemolytic action of complement. Three types of erythrocytes have been defined based on their complement sensitivity: normal or near-normal cells (= PNH I cells), moderately sensitive cells with 3- to 5-fold increased sensitivity to lysis (= PNH II cells), and abnormally sensitive cells with 25- to 30-fold increased sensitivity (= PNH III cells) [44]. These cells show different degrees of deficiency in phosphatidylinositol-linked proteins: for example, PNH III erythrocytes lack the decay accelerating factor completely, whereas PNH II cells contain a reduced amount of this protein, and PNH I cells have close to normal values [61].

Lack of phosphatidylinositol-linked proteins in the PNH syndrome is not confined to erythrocytes; it is also found in leucocytes and platelets [35]. This fact shows that paroxysmal nocturnal haemoglobinuria is the result of a somatic mutation in a pluripotent haematopoietic stem cell [43].

The genetic lesion causing this haematological disorder resides in most if not in all cases in the gene encoding the α1,6-N-acetylglucosaminyltransferase, the enzyme which catalyses an essential step in the biosynthesis of the glycosyl-phosphoinositol unit [4,21,53]. This gene is probably identical to the *PIG-A* ('phosphatidylinositoglycan-class A') gene [8] which has been mapped to the X chromosome at position p22.1 [54]. The finding of four distinct somatic mutations of the *PIG-A* gene in a single patient suggests that paroxysmal nocturnal haemoglobinuria in some cases can be a polyclonal rather that a monoclonar disease [17].

The Proteins of the Erythrocyte Membrane

Many erythrocyte membrane proteins are fairly well studied and will be presented below (a survey of their properties is shown in *Table 4.2*, and a survey of their functions within the membrane in *Table 4.3*):

Spectrin (or Tectin A) is the major constituent of the membrane skeleton; it accounts for approximately 25% of total red cell membrane proteins or ~35% of the membrane skeleton mass.

The spectrin fraction consists of two fibrous proteins which can be resolved on SDS-polyacrylamide gels into two closely adjacent bands, **band-1** (α-chain) and **band-2** (β-chain); the respective molecular masses are 240 and 220 kDa.

Table 4.2: Properties and associations of erythrocyte membrane protein fractions.

Protein fraction	Apparent molecular mass	Copies per cell ($\times 10^{-5}$)	Status of association in intact membrane
Band-1 (spectrin)	240,000	2.2	Dimers of band-1 and band-2 form tetramers
Band-2 (spectrin)	220,000		and oligomers
Band-2.1 (ankyrin)	210,000	1.1	Monomer
Band-2.2 (ankyrin)	183,000		
Band-2.3 (ankyrin)	65,000		
Band-3	95,000	12	Dimer ($\leftrightarrow$ Tetramer)
Band-4.1 a	80,000	2.3	Dimer(?)
Band-4.1 b	78,000		
Band-4.2	72,000	2.3	Tetramer
Band-4.5/1	59,000	1.3	
Band-4.5/2	–	1.4	
Band-4.5/3	–	0.7	
Band-4.5/4	–	1.0	
Band-4.5/5	–	1.3	
Band-4.5/6	52,000	1.3	
Band-4.9	48,000	1	Trimer(?)
Band-5 (actin)	43,000	5.1	Oligomer (10–15)
Band-6	36,000	4.1	Tetramer
Band-7	29,000	5	
Glycophorin A	37,000	9	Dimer
Glycophorin B	24,000	3	
Glycophorin C	35,000	1	
Glycophorin D	27,000	0.2	

According to Haest [19] by permission of the author and of Elsevier Science Publishers BV, Amsterdam.

Table 4.3: Function of erythrocyte membrane protein fractions.

Protein fraction	Function	In intact membrane associated with
Band-1 (spectrin)	Membrane skeleton	Band-3 and 4.1
Band-2 (spectrin)		Ankyrin
Band-2.1 (ankyrin) Band-2.2 (ankyrin) Band-2.3 (ankyrin)	Connect membrane skeleton with intrinsic membrane domain	Band-2 and 3
Band-3	Transport system for inorganic anions; skeleton attachment site	Band-1, ankyrin, band-4.2, band-6, and glycophorin A
Band-4.1 a Band-4.1 b	Connect membrane skeleton with intrinsic membrane domain; mediate spectrin-actin interaction	Spectrin, glycophorin C
Band-4.2	unknown	Band-3
Band-4.5/1 Band-4.5/2 Band-4.5/3 Band-4.5/4 Band-4.5/5 Band-4.5/6	Possible transfer system for monosaccharides, L-lactate and nucleosides	
Band-4.9	Membrane skeleton	Band-2
Band-5 (actin)	Membrane skeleton	Spectrin, band-4.1
Band-6	Glyceraldehyde-3-P-dehydrogenase	Spectrin – actin, Band-3
Band-7		
Glycophorin A		Band-3
Glycophorin B		
Glycophorin C	Skeleton attachment site	Band-4.1

According to Haest ([19], modified) by permission of the author and of Elsevier Science Publishers BV, Amsterdam.

Ankyrin (band-2.1 protein) is a peripheral membrane protein with a molecular mass of 210 kDa. Because it binds to both spectrin and band-3 protein it is largely responsible for maintaining the close association of the membrane skeletal proteins with the lipid bilayer.

Band-2.2 and **band-2.3** proteins are probably ankyrin analogues.

Band-3 protein is the predominant integral polypeptide of the erythrocyte membrane and is responsible for the anion transport (mainly of chloride and bicarbonate). In polyacrylamidegel electrophoresis the molecule migrates as a diffuse band within the region of 88–98 kDa; this molecular mass dispersion is probably caused by heterogeneity in glycosylation [57].

The current structural model predicts a membrane protein with two cytoplasmic domains and a membrane domain consisting of 12 to 14 transmembrane helices (see *Fig. 20.2)* [33,40,55]. The N-terminal cytoplasmic domain of the band 3 protein links the membrane to the underlying spectrin-based membrane skeleton and binds glycolytic enzymes and haemoglobin [32]. In the **Diego** blood group system (see *Chap. 20)* the antigens are located on the extracellular part of the molecule, and the N-glycan attached to the fourth extracellular loop carries blood group **ABH** and **Ii** determinants, as well as receptors for lectins, such as concanavalin A and *Ricinus communis* agglutinin.

Band 3 in the membrane exists as a mixed population of dimers and teramers [13]. Association of band 3 with the cytoskeleton is critical for the maintenance of red cell morphology [28,38]. Mutations of the band-3 protein are connected with Southeast Asian ovalocytosis [24,47], hereditary spherocytosis [23,25-27,46], spherocytic hemolysis [22], and congenital acanthocytosis [12].

Further, the 'senescent cell antigen', is generated by modification (oxidation?) of the band-3 protein [7,30]. This antigen appears on the surface of senescent and damaged cells marking them for removal by initiating specific IgG autoantibody binding. There is also some evidence that band-3 protein acts as a receptor during invasion of human erythrocytes by *Plasmodium falciparum* [36].

Band-4.1 protein is a globular protein with a molecular mass in the range of 78 to 82 kDa; in gel electrophoresis using discontinuous buffer systems it migrates as a closely spaced doublet (4.1.a and 4.1.b). This membrane protein is associated with spectrin, actin, and the cytoplasmic domains of glycophorin A and C and thus represents an additional binding site of the membrane skeleton to the membrane matrix. It probably plays a pivotal role in maintaining membrane elasticity and erythrocyte shape. Lack of this protein is one major cause of elliptocytosis.

Band-4.2 protein is found in the erythrocyte membrane as a homo-tetramer closely associated with band-3 protein. Its function is not yet known; deficiency or lack of this protein, however, is connected with several types of inherited haemolytic anaemias [15].

Band-4.5 represents a group of some 6 proteins with molecular masses between 52 and 59 kDa. One protein of this group is responsible for the transport of monosaccharides and nucleosides through the membrane, other peptides are probably proteolytic degradation products of band-4.1 protein [14]. The glucose transporter (GLUT1) contains 12 transmembrane segments and a central cytosolic loop. *In situ* it occurs as a mixture of dimers and tetramers [5].

The band-4.5 proteins are not associated with the membrane skeleton and have no known function in constructing membrane architecture.

Band-4.9 protein is a phosphoprotein with a molecular mass of 48 kDa, which interacts with the actin of the membrane skeleton and regulates its degree of polymerisation.

Actin (band-5 protein) oligomers of 10–15 units are essential components of the membrane skeleton.

Band-6 protein is a glyceraldehyde-3-phosphate dehydrogenase with a pivotal function in erythrocyte metabolism.

The function of **band-7** protein is thus far unknown. It is absent in hereditary stomatocytosis (see [15]).

The **sialoglycoproteins** represent a group of membrane constituents, which due to their high content of neuraminic acid are characterised by a strong reaction with PAS reagent. The most important sialoglycoproteins to be mentioned here are glycophorin A, glycophorin B, and glycophorin C; further PAS-positive peptides are present only in trace amounts (see *Table 4.4)*. These membrane constituents are integral membrane proteins with molecular masses in the region of 24–37 kDa. At the amino termini the proteins are heavily glycosylated and provide 30–45% of total erythrocyte carbohydrate; almost 90% of the neuraminic acid present in the erythrocyte membrane is bound to these membrane glycoproteins, making them responsible for most of the negative charge of erythrocytes.

Various serological characters of the erythrocyte membrane are located on glycophorin molecules: glycophorin A and glycophorin B carry blood group **MNS** determinants, the T antigens, receptors for influenza viruses, malaria parasites, *Coli*

Table 4.4: The sialoglycoproteins of human erythrocyte membranes.

PAS band	Glycophorin
A	
PAS-1	Glycophorin A Dimer
F	
PAS-4 or B	Glycophorin A/B Dimer
C	Glycophorin B Dimer
PAS-2	Glycophorin A Monomer
PAS-2' or D	Glycophorin C
L	
K_1 and K_2	
E	Glycophorin D
PAS-3	Glycophorin B Monomer
J	

See also *Table 9.3*.

bacteria, as well as binding sites for various lectins (e.g. as WGA and PHA). Glycophorin C is responsible for the blood group **Gerbich** phenotype. Erythrocytes lacking glycophorin A and/or glycophorin B (see the blood group **MNS** variants **En(a–)**, **S–s–**, and **M^k** in *Sect. 9.6)* show no functional disorders. Glycophorin C, however, is an essential factor for maintaining membrane architecture; absence of this sialoglyco-protein is another cause of elliptocytosis [3].

Glycophorin A and glycophorin B will be discussed in *Sect. 9.3*, glycophorin C in *Chap. 17*.

4.1.3 Carbohydrate Moiety of Glycoconjugates

The outer surface of the membrane is heavily glycosylated. Oligosaccharide chains of highly variable structure are linked mainly to proteins, and only about 7% are carried by sphingolipids.

The carbohydrate moiety of human erythrocyte membranes consists of 55.5% neutral sugars (D-glucose, D-galactose, D-mannose, and L-fucose), 27.8% hexosamine (N-acetyl-D-glucosamine and N-acetyl-D-galactosamine), and 16.7% N-acetyl-D-neuraminic acid [42].

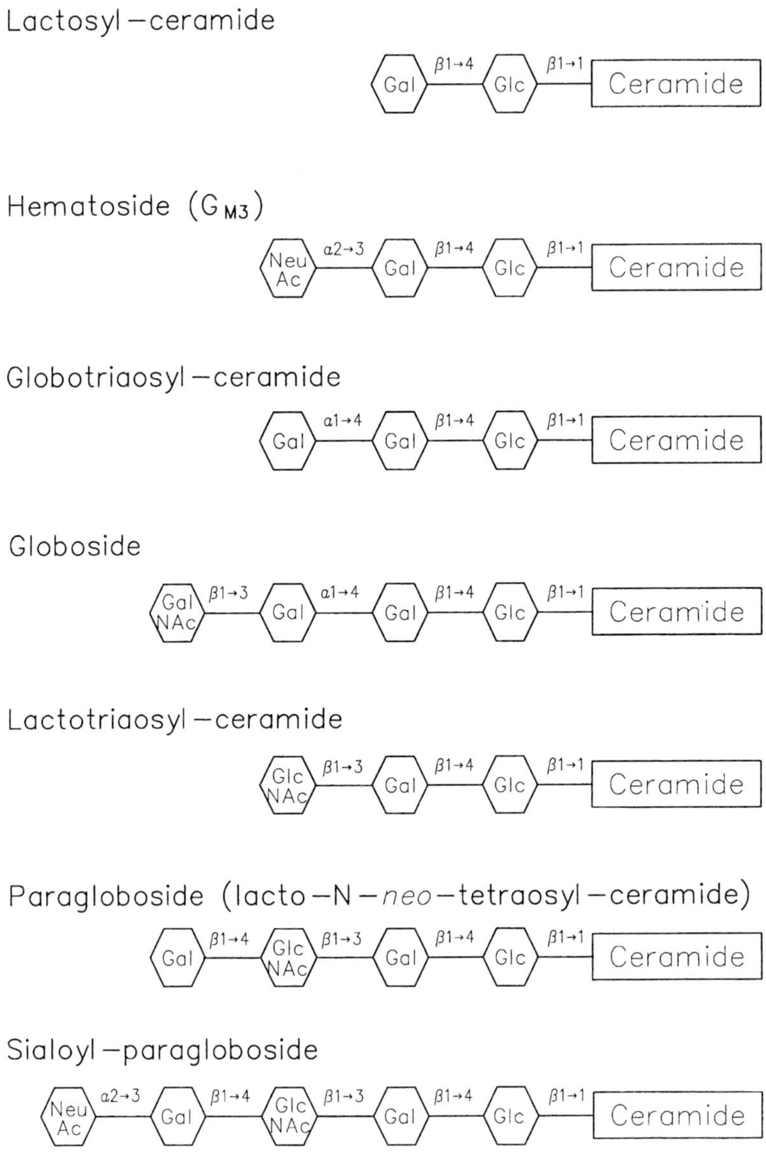
Lactosyl-ceramide
Gal β1→4 Glc β1→1 Ceramide
Hematoside (G M3)
Neu Ac α2→3 Gal β1→4 Glc β1→1 Ceramide
Globotriaosyl-ceramide
Gal α1→4 Gal β1→4 Glc β1→1 Ceramide
Globoside
Gal NAc β1→3 Gal α1→4 Gal β1→4 Glc β1→1 Ceramide
Lactotriaosyl-ceramide
Glc NAc β1→3 Gal β1→4 Glc β1→1 Ceramide
Paragloboside (lacto-N-neo-tetraosyl-ceramide)
Gal β1→4 Glc NAc β1→3 Gal β1→4 Glc β1→1 Ceramide
Sialoyl-paragloboside
Neu Ac α2→3 Gal β1→4 Glc NAc β1→3 Gal β1→4 Glc β1→1 Ceramide

Figure 4.6: Chemical structures of glycosphingolipids of the human erythrocyte membrane.

(a) The Oligosaccharide Chains of Glycosphingolipids

The membranes of human erythrocytes contain glycosphingolipids of the globo-, ganglio-, and neolacto-series carrying short oligosaccharide chains generally composed of 5–10 monosaccharide units; the glycolipids of the neolacto-series, however, may contain fairly complex and heavily branched oligosaccharide chains composed of up to 60 monosaccharide units (so-called polyglycosyl-ceramides or megalo-glycolipids, see *Sect. 5.3.2).*

The chemical structures of the main erythrocyte membrane glycosphingolipids are presented in *Fig. 4.6.* Examples of blood group active glycosphingolipids are found in *Figs. 5.5–8* (**ABH** substances), *Fig. 8.1* (**P** substances), and *Figs. 7.1* and *7.3* (**I** and i substances).

(b) The Oligosaccharide Chains of Glycoproteins

The erythrocyte membrane glycoproteins contain both N-linked and O-linked oligosaccharide side chains (see *Sect. 2.2).* Examples of carbohydrate structures occurring in membrane glycoproteins are presented in *Fig. 5.10* (band-3 protein) and *Fig. 9.3* (glycophorin A).

4.2 Molecular Architecture of the Erythrocyte Membrane

According to the model proposed by Singer and Nicolson [49] the matrix of a biological membrane is formed by lipid material arranged in a bilayer about 5 nm thick. The two single layers are oriented opposite each other, the hydrophobic hydrocarbon chains being in the inner side of the membrane and the hydrophilic components (polar and charged groups) at the outside of the membrane where they interact with the surrounding water molecules. The hydrocarbon chains of the lipids within the membrane bilayer are oriented parallel to one another.

Incorporation of cholesterol, as well as the asymmetric arrangement of the lipids (see below) prevent the formation of pseudo-crystalline areas or non-lamellar structures, such as H_{II} ('inverted hexagonal') phases. In the semi-liquid state thus attained the membrane constituents are able to move laterally within the membrane plane, whereas movement vertical to the surface is impaired.

The phospholipids are not evenly distributed between the two lipid layers [37]: phosphatidylserine, phosphatidylethanolamine, and phosphatidylinositol are mainly

Figure 4.7: Freeze-fractured erythrocyte membrane.
Top: extracellular surface of the membrane, bottom: fracture face of the protoplasmic leaflet of the membrane with intramembrane particles (x 156,000).
Reproduced from Weinstein et al. [59] by permission of John Wiley & Sons, Inc., New York (©1978).

found in the inner layer, whereas lecithin (phosphatidylcholine) and sphingomyelin are located mainly in the outer (extracellular) layer. Cholesterol, due to its rapid trans-bilayer movement, is probably evenly partitioned within this lipid matrix. There is evidence that this asymmetric distribution of phospholipids within the bilayer is maintained by specific interaction of the lipids with membrane proteins − to be discussed below.

A number of intrinsic membrane proteins aggregated to larger protein complexes is embedded in this lipid bilayer. The basic structure of such a protein complex is a tetramer of band-3 protein associated with a glycophorin A dimer (*Fig. 4.8.a*). The cytoplasmic portion of the band-3 protein is bound to band-4.2 protein, haemoglobin, and various enzymes of the glycolytic metabolism (e.g. aldolase and glyceraldehyde-3-phosphate dehydrogenase). These aggregates can be visualised by the freeze-fracture electron microscopy method. The average diameter of these so-called **intramembrane particles** is approximately 7 nm ([59], *Fig.4.7*).

The hydrophobic domains of the proteins are located within the membrane matrix and are closely associated with lipids: phosphatidylserine and phosphatidylinositol, for example, interact strongly with glycophorin A, and cholesterol with band-3 protein. It is generally assumed that the formation of such protein–lipid complexes is essential both in providing a suitable conformation of the membrane proteins and in maintaining a defined localisation within the membrane, which may be crucial for its functioning.

On its cytoplasmic side the cell membrane is supported by a **membrane skeleton** which represents a highly interconnected network formed of spectrin, actin, band-4.1 protein, and band-4.9 protein [6,14,19]:

The basic constituent of the membrane skeleton is an $\alpha\beta$ dimer of spectrin. This represents a highly flexible filamentous structure of ~100 nm in length. Spectrin tetramers arising from head-to-head association of two dimers form a dense, two-dimensional network due to the cooperative action of actin-oligomers, band-4.1, and band-4.9 proteins (*Fig. 4.8.b–d*).

This network is tightly anchored to the membrane matrix via non-covalent (i.e. reversible) bonds of moderate affinity. Such bonds occur between band-4.1 protein and glycophorin A and C, and between spectrin and band-3 protein via ankyrin (band-2.1 protein). In addition, hydrophobic and electrostatic interaction is observed between the components of the membrane skeleton and various lipids of the membrane matrix.

These protein interactions are subjected to various control mechanisms: polyphosphoinositols increase the affinity of band-4.1 protein towards glycophorin [2], whereas 2,3-diphosphoglycerol impairs the binding between spectrin and actin [48]. Regulation of the organisation of the skeleton proteins by ATP and Ca^{++} ions is also the subject of some discussion [14].

One of the main functions of the membrane skeleton is the maintenance of the characteristic shape of the erythrocyte even after extreme deformation during passage through blood capillaries [16]. Moreover, the underlying skeletal network greatly restricts lateral diffusion of the membrane constituents. Factors at work here function partly by reversible association with proteins and lipids, partly sterically by limiting the mobility of cytoplasmic domains of intrinsic proteins. The membrane skeleton thus not only stabilises the lipid double-layer, but also maintains an arrangement of the membrane organelles suitable for optimum function of various physiological processes of the cell.

The serological expression of various surface antigens is worth mention here as iot is controlled by the composition of the lipid moiety. Studies by Borochov et al. [9,10] have revealed that the accessibility of the membrane proteins is influenced by the microviscosity of the lipid matrix: a high cholesterol-to-phospholipid ratio, which increases membrane viscosity, leads to an increase of the protein exposure to the aqueous surrounding. In contrast, cholesterol depletion, which lowers microviscosity, decreases the accessibility of several membrane proteins (see also **Rh** antigens, *Chap. 13)*.

The attachment of the membrane skeleton to the membrane matrix can be released in weakly alkaline buffer of low ionic strength. Membranes isolated under these conditions are very sensitive to fragmentation by shear stress. This phenomenon is used to obtain stable sealed membrane vesicles. The direction of vesiculation, i.e. the orientation of the membrane layers, depends on the medium used [50]: in the presence of divalent cations (normally Mg^{++}) the original orientation is maintained ('rightside-out' (= RO) vesicles), whereas in the absence of divalent cations the membrane curvature is inverted ('inside-out' (= IO) vesicles).

4.3 Architecture of the Membranes of Nucleated Cells

The membranes of nucleated tissue cells are also based on a lipid bilayer matrix in which various membrane proteins are embedded. As expected, the membrane proteins of nucleated cells show great variability, reflecting the higher biochemical complexity of these membranes.

Although in tissue cells a series of proteins have been found which are chemically, morphologically, and immunologically very similar to the above-mentioned constituents of the erythrocyte membrane framework (see [6,14]), no explicit membrane skeleton has been detected. Nucleated cells instead contain a **cytoskeleton** which controls cell shape and cell movement, as well as the exact localisation and overall organisation of intracellular organelles, nuclei and other cytoplasmic inclusions.

The cytoskeleton is composed of three main species of fibrillar structures – microfilaments, microtubules, and intermediate filaments [1,11]:

(a) Microfilaments are found in all nucleated cells as thin filamentous structures with a diameter of approximately 6 nm. They are formed of aggregated actin. The configuration and the physiological behaviour of the filaments is controlled by the attachment of various, so-called actin-binding proteins, of which more than a dozen have already been described. Two forms of microfilaments can be distinguished morphologically – 'lattice microfilaments', which represent a loose network of short fibres often associated with the cytoplasmic side of the plasma membrane, and 'stress microfilaments', which are located as bundles beneath the plasma membrane (in the case of fibroblasts most of them are arranged parallel to the long axis of the cell). The microfilaments are primarily responsible for maintaining cell shape and locomotion.

(a)

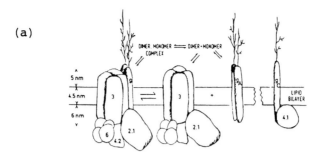

(b)

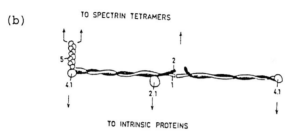

(c)

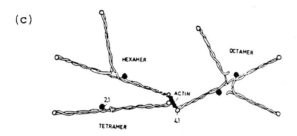

(d)

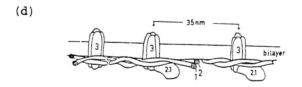

Figure 4.8: Structure of the erythrocyte membrane
(a) Formation of 'intramembrane particles' by association of intrinsic membrane proteins (band-3 protein and glycophorin A) with extrinsic membrane proteins (ankyrin, band-4.1, band-4.2, and band-6 proteins),
(b) formation of the basic unit of the membrane skeleton by association of spectrin dimers with band-4.1 protein, actin (band-5 protein), and ankyrin (band-2.1 protein), and
(c) subsequent assembly of the skeletal network by formation of spectrin oligomers and by actin–spectrin association,
(d) binding of spectrin tetramers to band-3 protein via ankyrin (band-2.1 protein).
Reproduced from Haest [19] by permission of the author and of Elsevier Science Publishers BV, Amsterdam.

49

(b) Microtubules are large cylindrical structures of aggregated α- and β-tubulin; the tubules have a diameter of 25 nm and a lumen of about 14 nm. Various proteins attached to the microtubules control organisation and activity of these structures. In most nucleated mammalian cells the microtubules form a network originating at the centrosome and loosely associated with the cell membrane. The main function of the microtubules presumably lies in cell organisation and intracellular transport of cell constituents.

(c) Intermediate filaments have an average diameter of 8–10 nm and are found in most mammalian cells, often in close proximity to microtubules. They form a network throughout the entire cytoplasm. In some cases they are bundled, in others they occur as isolated filaments. Based on their chemical composition five basic types of intermediary filaments can be distinguished: vimentin fibres in mesenchymatous cells, keratin fibres in epithelial cells, desmin fibres in muscle cells, neurofilaments in neurons, and gliafilaments in glia cells. These filaments have probably a support function in cells subjected to mechanical stress.

Since the membrane skeleton is absent in tissue cells or in some cases only weakly expressed, the membrane proteins of nucleated cells are fairly mobile and can move independently of each other in the membrane plane. In cells treated with specific antibodies or lectins the membrane proteins form aggregates ('patching') which subsequently move to a cell pole ('capping') and are finally taken up by the cell by pinocytosis.

References

1. ALBERTS, B., BRAY, D., LEWIS, J., RAFF, M., ROBERTS, K. & WATSON, J. D. (1989): *Molekularbiologie der Zelle*. VCH Verlagsgesellschaft mbH, Weinheim, Germany.
2. ANDERSON, R. A. & MARCHESI, V. T. (1985): Regulation of the association of membrane skeletal protein 4.1 with glycophorin by a polyphosphoinositide. *Nature 318*, 295-298.
3. ANSTEE, D. J., PARSONS, S. F., RIDGWELL, K., TANNER, M. J. A., MERRY, A. H., THOMSON, E. E., JUDSON, P. A., JOHNSON, P., BATES, S. & FRASER, I. D. (1984): Two individuals with elliptocytic red cells apparently lack three minor erythrocyte membrane sialoglycoproteins. *Biochem. J. 218*, 615-619.
4. ARMSTRONG, C., SCHUBERT, J., UEDA, E., KNEZ, J. J., GELPERIN, D., HIROSE, S., SILBER, R., HOLLAN, S., SCHMIDT, R. E. & MEDOF, M. E. (1992): Affected paroxysmal nocturnal hemoglobinuria T lymphocytes harbor a common defect in assembly of N-acetyl-D-glucosamine inositol phospholipid corresponding to that in class A Thy-1⁻ murine lymphoma mutants. *J. Biol. Chem. 267*, 25347-25351.
5. BELL, G. I., BURANT, C. F., TAKEDA, J. & GOULD, G. W. (1993): Structure and function of mammalian facilitative sugar transporters. *J Biol Chem 268*, 19161-19164.
6. BENNETT, V. (1985): The membrane skeleton of human erythrocytes and its implications for more complex cells. *Annu. Rev. Biochem. 54*, 273-304.

7. BEPPU, M., MIZUKAMI, A., ANDO, K. & KIKUGAWA, K. (1992): Antigenic determinants of senescent antigen of human erythrocytes are located in sialylated carbohydrate chains of Band 3 glycoprotein. *J. Biol. Chem. 267*, 14691-14696.
8. BESSLER, M., MASON, P. J., HILLMEN, P., MIYATA, T., YAMADA, N., TAKEDA, J., LUZZATTO, L. & KINOSHITA, T. (1994): Paroxysmal nocturnal haemoglobinuria (PNH) is caused by somatic mutations in the PIG-A gene. *EMBO J. 13*, 110-117.
9. BOROCHOV, H., ABBOTT, R. E., SCHACHTER, D. & SHINITZKY, M. (1979): Modulation of erythrocyte membrane proteins by membrane cholesterol and lipid fluidity. *Biochemistry 18*, 251-255.
10. BOROCHOV, H. & SHINITZKY, M. (1976): Vertical displacement of membrane proteins mediated by changes in microviscosity. *Proc. Natl. Acad. Sci. USA 73*, 4526-4530.
11. BRETSCHER, M. S. & RAFF, M. C. (1975): Mammalian plasma membranes. *Nature 258*, 43-49.
12. BRUCE, L. J., KAY, M. M. B., LAWRENCE, C. & TANNER, M. J. A. (1993): Band-3-HT, a human red-cell variant associated with acanthocytosis and increased anion transport, carries the mutation Pro[868] → Leu in the membrane domain of band-3. *Biochem. J. 293*, 317-320.
13. CASEY, J. R. & REITHMEIER, R. A. F. (1991): Analysis of the oligomeric state of band 3, the anion transport protein of the human erythrocyte membrane, by size exclusion high performance liquid chromatography. *J. Biol. Chem. 266*, 15726-15737.
14. COHEN, C. M. (1983): The molecular organization of the red cell membrane skeleton. *Sem. Hematol. 20*, 141-158.
15. COHEN, C. M. & GASCARD, P. (1992): Regulation and post-translational modification of erythrocyte membrane and membrane-skeletal proteins. *Sem. Hematol. 29*, 244-292.
16. ELGSAETER, A., STOKKE, B. T., MIKKELSEN, A. & BRANTON, D. (1986): The molecular basis of erythrocyte shape. *Science 234*, 1217-1223.
17. ENDO, M., WARE, R. E., VREEKE, T. M., SINGH, S. P., HOWARD, T. A., TOMITA, A., HOLGUIN, M. H. & PARKER, C. J. (1996): Molecular basis of the heterogeneity of expression of glycosyl phosphatidylinositol anchored proteins in paroxysmal nocturnal hemoglobinuria. *Blood 87*, 2546-2557.
18. FAIRBANKS, G., STECK, T. L. & WALLACH, D. F. H. (1971): Electrophoretic analysis of the major polypeptides of the human erythrocyte membrane. *Biochemistry 10*, 2606-2617.
19. HAEST, C. W. M. (1982): Interactions between membrane skeleton proteins and the intrinsic domain of the erythrocyte membrane. *Biochim. Biophys. Acta 694*, 331-352.
20. HERELD, D., KRAKOW, J. L., BANGS, J. D., HART, G. W. & ENGLUND, P. T. (1986): A phospholipase C from Trypanosoma brucei which selectively cleaves the glycolipid on the variant surface glycoprotein. *J. Biol. Chem. 261*, 13813-13818.
21. HILLMEN, P., BESSLER, M., MASON, P. J., WATKINS, W. M. & LUZZATTO, L. (1993): Specific defect in N-acetylglucosamine incorporation in the biosynthesis of the glycosylphosphatidylinositol anchor in cloned cell lines from patients with paroxysmal nocturnal hemoglobinuria. *Proc. Natl. Acad. Sci. USA 90*, 5272-5276.
22. IWASE, S., IDEGUCHI, H., TAKAO, M., HORIGUCHI-YAMADA, J., IWASAKI, M., TAKAHARA, S., SEKIKAWA, T., MOCHIZUKI, S. & YAMADA, H. (1998): Band 3 Tokyo: Thr[837]→Ala[837] substitution in erythrocyte band 3 protein associated with spherocytic hemolysis. *Acta Haematol. 100*, 200-203.
23. JAROLIM, P., MURRAY, J. L., RUBIN, H. L., TAYLOR, W. M., PRCHAL, J. T., BALLAS, S. K., SNYDER, L. M., CHROBAK, L., MELROSE, W. D., BRABEC, V. & PALEK, J. (1996): Characterization of 13 novel band 3 gene defects in hereditary spherocytosis with band 3 deficiency. *Blood 88*, 4366-4374.
24. JAROLIM, P., PALEK, J., AMATO, D., HASSAN, K., SAPAK, R., NURSE, G. T., RUBIN, H. L., ZHAI, S., SAHR, K. E. & LIU, S. C. (1991): Deletion in erythrocyte band 3 gene in malaria-resistant Southeast Asian ovalocytosis. *Proc. Natl. Acad. Sci. USA 88*, 11022-11026.
25. JAROLIM, P., PALEK, J., RUBIN, H. L., PRCHAL, J. T., KORSGREN, C. & COHEN, C. M. (1992): Band 3 Tuscaloosa: Pro[327] →Arg[327] substitution in the cytoplasmic domain of erythrocyte band 3 protein associated with spherocytic hemolytic anemia and partial deficiency of protein 4.2. *Blood 80*, 523-529.
26. JAROLIM, P., RUBIN, H. L., BRABEC, V., CHROBAK, L., ZOLOTAREV, A. S., ALPER, S. L., BRUGNARA, C.,

WICHTERLE, H. & PALEK, J. (1995): Mutations of conserved arginines in the membrane domain of erythroid band 3 lead to a decrease in membrane-associated band 3 and to the phenotype of hereditary spherocytosis. *Blood 85*, 634-640.

27. JAROLIM, P., RUBIN, H. L., LIU, S., CHO, M. R., BRABEC, V., DERICK, L. H., YI, S. J., SAAD, S. T., ALPER, S. & BRUGNARA, C. (1994): Duplication of 10 nucleotides in the erythroid band 3 (AE1) gene in a kindred with hereditary spherocytosis and band 3 protein deficiency (band 3 PRAGUE). *J. Clin. Invest. 93*, 121-130.

28. JAY, D. G. (1976): Glycosylation site of band 3: the human erythrocyte anion-exchange protein. *Biochemistry 25*, 554-556.

29. JOHNSON, R. M., McGOWAN, G. M., MORSE, P. D. & DZANDU, J. K. (1982): Proteolytic analysis of the topological arrangement of red cell phosphoproteins. *Biochemistry 21*, 3599-3604.

30. KAY, M. M. (1993): Generation of senescent cell antigen on old cells initiates IgG binding to a neoantigen. *Cell. Mol. Biol. 39*, 131-153.

31. KINOSHITA, T., INOUE, N. & TAKEDA, J. (1995): Defective glycosyl phosphatidylinositol anchor synthesis and paroxysmal nocturnal hemoglobinuria. *Adv. Immunol. 60*, 57-103.

32. LOW, P. S. (1986): Structure and function of the cytoplasmic domain of band 3: center of erythrocyte membrane-peripheral protein interactions. *Biochim. Biophys. Acta 860*, 145-167.

33. LUX, S. E., JOHN, K. M., KOPITO, R. R. & LODISH, H. F. (1989): Cloning and characterization of band 3, the human erythrocyte anion-exchange protein (AE1). *Proc. Natl. Acad. Sci. USA 86*, 9089-9093.

34. MARCHESI, V. T., FURTHMAYR, H. & TOMITA, M. (1976): The red cell membrane. *Annu. Rev. Biochem. 45*, 667-685.

35. NAVENOT, J. M., BERNARD, D., HAROUSSEAU, J. L., MULLER, J. Y. & BLANCHARD, D. (1996): Expression of glycosyl-phosphatidylinositol-linked glycoproteins in blood cells from paroxysmal nocturnal haemoglobinuria patients. A flow cytometry study using CD55, CD58 and CD59 monoclonal antibodies. *Leuk. Lymphoma 21*, 143-151.

36. OKOYE, V. C. N. & BENNETT, V. N. (1985): Plasmodium falciparum malaria: band 3 as a possible receptor during inasion of human erythrocytes. *Science 227*, 169-171.

37. OP DEN KAMP, J. A. (1979): Lipid asymmetry in membranes. *Annu. Rev. Biochem. 48*, 47-71.

38. PETERS, L. L., SHIVDASANI, R. A., LIU, S. C., HANSPAL, M., JOHN, K. M., GONZALEZ, J. M., BRUGNARA, C., GWYNN, B., MOHANDAS, N., ALPER, S. L., ORKIN, S. H. & LUX, S. E. (1996): Anion exchanger 1 (band 3) is required to prevent erythrocyte membrane surface loss but not to form the membrane skeleton. *Cell 86*, 917-927.

39. PRATT, J. C. & GAULTON, G. N. (1993): Multifunctional roles of glycosyl-phosphatidylinositol lipids. *DNA Cell Biol. 12*, 861-869.

40. REITHMEIER, R. A. F. (1993): The erythrocyte anion transporter (band 3). *Curr. Opin. Struct. Biol. 3*, 515-523.

41. ROBERTS, W. L., MYHER, J. J., KUKSIS, A., LOW, M. G. & ROSENBERRY, T. L. (1988): Lipid analysis of the glycoinositol phospholipid membrane anchor of human erythrocyte acetylcholinesterase. Palmitoylation of inositol results in resistance to phosphatidylinositol-specific phospholipase C. *J. Biol. Chem. 263*, 18766-18775.

42. ROSENBERG, S. A. & GUIDOTTI, G. J. (1968): The protein of human erythrocyte membranes. I. Preparation, solubilization and partial characterization. *J. Biol. Chem. 243*, 1985-1992.

43. ROSSE, W. F. (1990): Phosphatidylinositol-linked proteins and paroxysmal nocturnal hemoglobinuria. *Blood 75*, 1595-1601.

44. ROSSE, W. F. & DACIE, J. V. (1966): Immune lysis of normal human and paroxysmal nocturnal hemoglobinuria (PNH) red blood cells. I. The sensitivity of PNH red cells to lysis by complement and specific antibody. *J. Clin. Invest. 45*, 736-748.

45. ROTOLI, B., BESSLER, M., ALFINITO, F. & DEL VECCHIO, L. (1993): Membrane proteins in paroxysmal nocturnal haemoglobinuria. *Blood Rev. 7*, 75-86.

46. RYBICKY, A. C., QIU, J. J., MUSTO, S., ROSEN, N. L., NAGEL, R. L. & SCHWARTZ, R. S. (1993): Human erythrocyte protein 4.2 deficiency associated with hemolytic anemia and a homozygous [40]glutamic

acid → lysine substitution in the cytoplasmic domain of band 3 (band 3[montefiore]). *Blood 81*, 2155-2165.

47. SCHOFIELD, A. E., TANNER, M. J., PINDER, J. C., CLOUGH, B., BAYLEY, P. M., NASH, G. B., DLUZEWSKI, A. R., REARDON, D. M., COX, T. M. & WILSON, R. J. (1992): Basis of unique red cell membrane properties in hereditary ovalocytosis. *J. Mol. Biol. 223*, 949-958.

48. SHEETZ, M. P. (1983): Membrane skeletal dynamics: role in modulation of red cell deformability, mobility of transmembrane proteins and shape. *Sem. Hematol. 20*, 175-188.

49. SINGER, S. J. & NICOLSON, G. L. (1972): The fluid mosaic model of the structure of cell membranes. *Science 175*, 720-731.

50. STECK, T. J. & KANT, J. A. (1974): Preparation of impermeable ghosts and inside-out vesicles from human erythrocyte membranes. In: *Methods in Enzymology* (S. Fleisher & Packer, eds.). New York, NY. *Vol. 31A*, pp. 172-180.

51. STECK, T. L. (1974): The organization of proteins in the human red blood cell membrane. A review. *J. Cell Biol. 62*, 1-19.

52. STECK, T. L. & DAWSON, G. (1974): Topographical distribution of complex carbohydrates in the erythrocyte membrane. *J. Biol. Chem. 249*, 2135-2142.

53. TAKAHASHI, M., TAKEDA, J., HIROSE, S., HYMAN, R., INOUE, N., MIYATA, T., UEDA, E., KITANI, T., MEDOF, E. & KINOSHITA, T. (1993): Deficient biosynthesis of N-acetylglucosaminyl-phosphatidylinositol, the first intermediate of glycosyl phosphatidylinositol anchor biosynthesis, in cell lines established from patients with paroxysmal nocturnal hemoglobinuria. *J. Exp. Med. 177*, 517-521.

54. TAKEDA, J., MIYATA, T., KAWAGOE, K., IIDA, Y., ENDO, Y., FUJITA, T. & TAKAHASHI, M. (1993): Deficiency of the GPI anchor caused by somatic mutation of the PIG-A gene in paroxysmal nocturnal hemoglobinuria. *Cell 73*, 703-711.

55. TANNER, M. J. A., MARTIN, P. G. & HIGH, S. (1988): The complete amino acid sequence of the human erythrocyte membrane anion-transport protein deduced from the cDNA sequence. *Biochem. J. 256*, 703-712.

56. TELEN, M. J., ROSSE, W. F., PARKER, C. J., MOULDS, M. K. & MOULDS, J. J. (1990): Evidence that several high-frequency human blood group antigens reside on phosphatidylinositol-linked erythrocyte membrane proteins. *Blood 75*, 1404-1407.

57. TSUJI, T., IRIMURA, T. & OSAWA, T. (1981): Heterogeneity in the carbohydrate moiety of band 3 glycoprotein of human erythrocyte membranes. *Carbohydr. Res. 92*, 328-332.

58. VAN DEENEN, L. L. M. & DE GIER, J. (1974): Lipids of the red cell membrane. In: *The Red Blood Cell* (D. M. Surgenour, ed.). Academic Press Inc., New York, NY, pp. 147-211.

59. WEINSTEIN, R. S., KHODADAD, J. K. & STECK, T. L. (1978): Fine structure of the band 3 protein in human red cell membranes: freeze-fracture studies. *J. Supramol. Struct. 8*, 325-335.

60. WINTROBE, M. M. (1967): *Clinical Hematology*. Philadelphia, PA, USA.

61. YOMTOVIAN, R., PRINCE, G. M. & MEDOF, M. E. (1993): The molecular basis for paroxysmal nocturnal hemoglobinuria. *Transfus. 33*, 852-873.

5 ABO(H) System

The **ABO** system was discovered in 1900 by Karl Landsteiner. Based on his assumption of two erythrocyte antigens, **A** and **B**, Landsteiner found it possible to classify human individuals into four blood groups depending whether one, both, or neither of these antigenic characters were present on the cells [272,273]. He also recognised that the sera of individuals lacking one or both antigens usually contain anti-**A** and/or anti-**B** isoagglutinins (see *Table 5.1*).

In the course of further investigations many subgroups and phenotypes have been detected which are defined both by quantitative differences in antigen content as well as by a characteristic tissue distribution of the antigens.

The blood group **H** system comprises only one character, **H** (ISBT Nr. 018 001) and is genetically independent of the **ABO** system. The antigen, however, represents the precursor substance in the biosynthetic pathway leading to **A** and **B** determinant structures. Based on this close chemical relationship the **H** system is now included in the discussion of the **ABO** system (→ '**ABO(H)** System').

The antigenic determinants of the **ABO(H)** system are not confined to erythrocytes; they are also found on tissue cells and they occur in various tissue fluids and secretions where they are bound to water-soluble substances.

Table 5.1: Definition of ABO blood groups

Blood group	ISBT Nr.[1]	Frequency in Europids [*]	Antigens on erythrocytes	Antibodies in serum
A	ABO1	41.7 %	A	anti-B
B	ABO2	8.5 %	B	anti-A
AB	ABO3	3.0 %	A and B	–
O		46.7 %	–	anti-A and anti-B

[1] In numerical terms 001 001 to 001 003.
[*] According to Race and Sanger [394].

Furthermore, substances with serological specificities identical or closely related to human **ABH** substances are found widely distributed in animals. Phylogenetic investigations have revealed that the **ABH** antigens are primarily characteristics of tissue cells (**ABH** antigens are therefore often referred to as 'histo-blood group antigens' [68]). Erythrocytes are the most recent cells in the evolutionary tree to acquire these antigens, since only humans and anthropoid apes express **ABH** antigens on their red cells [364].

ABH or **ABH**-like determinants have also been detected in plants and microorganisms; these determinants, however, are not under discussion here.

5.1 Genetics

The characters defined in the **ABO(H)** system are subject to strict genetic control and thus are inherited as simple Mendelian dominant characters [394,395].

Occurrence and distribution of **ABH** specificities are controlled by three independent but closely interrelated gene systems, *ABO* [20,99], *Hh* [338,525], and *Sese* [440,441]:

The **ABO Gene**

The three alleles *A*, *B*, and *O* of this gene locus account for the four main blood groups, **A**, **B**, **AB**, and **O**. The alleles *A* and *B* control the formation of **A** and **B** specificity on erythrocytes and in secretions, whereas the inactive *O* allele does not give rise to any blood group character. The relationship between *ABO* genotype and observed erythrocyte phenotype is presented in *Table 5.2*. The majority of **A** and **B** subgroups defined to date is controlled by variants of *A* and *B* genes.

The *ABO* locus has been localised on chromosome 9, position 9q34.1–q34.2 [17,322].

The **Hh Gene**

The blood group **H** character is found on erythrocytes and in secretions of blood group **O** individuals and has thus been considered the product of the *O* gene. **H** determinants, however, are also present, though not in quantity, in **A**, **B**, and **AB** subjects. Since individuals of genotypes *AA*, *BB*, and *AB* lack the *O* gene (see *Table 5.2*), Watkins and Morgan [525] concluded that this specificity must be the product of an independent gene, which they called *H*.

H substance represents the precursor which, under the influence of genes *A* and *B*, is converted into blood group **A** or **B** substance, respectively. Since the inactive *O* allele does not alter **H** substance, the presence of unchanged **H** determinants is characteristic for **O** individuals. There is furthermore a reciprocal relationship between

Table 5.2: Correlation of ABO genotype to erythrocyte phenotype.

Genotype	Phenotype
AA and AO	A
BB and BO	B
AB	AB
OO	O

H activity and **A** and **B**: **H** activity of erythrocytes decreases in strength in the order O, **A₂**, **B** and **A₁** [394].

An inactive allele of the *H* gene (= *h*) has been defined which, when present on both chromosomes, results in the complete lack of **H** activity on erythrocytes and in secretions producing the rare **Bombay** and **Parabombay phenotypes**. The detection of further blood group **H** deficient phenotypes led to the assumption that additional variants of the *H* gene exist.

The *H* locus (= *FUT1*) has been localised on the long arm of chromosome 19 at position q13.3 [11,404,412]

The **Secretor Gene (Se/se)**

ABH antigens are not confined to erythrocytes but are also found in various secretions and tissue fluids. Similarly, the ability to secrete **ABH** active substances is under genetic control and inherited by the allelic genes, *Se* and *se*, which are genetically independent of *ABO* and *Hh*. Individuals containing the allele *Se* in single or double dose, the so-called **ABH-secretors** (less accurately designated as 'secretors'), are able to secrete **ABH** active material according to their genotype. **ABH-nonsecretors** ('nonsecretors') are homozygous for the recessive allele *se* and, although they carry **ABH** antigens on erythrocytes, are not able to secrete **ABH** specific substances.

It should be noted that the **secretor**-deficient allele found in 20–25% of Mongolids is a '**weak-secretor**' allele which controls the secretion of minute amounts of **ABH** substances [195].

The *SE* locus (= *FUT2*) is found on chromosome 19q13.3; it is closely linked to the *H* locus [11,404,412]

Possible combinations of these genes and the resulting **ABH** phenotypes are presented in *Table 5.3*.

Molecular biological studies on the structure of these genes are discussed in *Sects. 5.4.5.* and *5.4.6.*

Table 5.3: Occurrence of ABH antigens on erythrocytes and in secretions of individuals with different genotypes

Secretor status	Frequency[*]	Genotype	Phenotype	
			Antigens on the erythrocytes	Specificity of secretions
Secretors	77 %	A B H Se O	A B H	A(H) B(H) H
Non-secretors	23 %	A B H sese O	A B H	– – –

[*] In Europid populations, according to Race and Sanger [394].

5.2 Antisera and Lectins

Origin and specificity of **ABH** specific antisera will be discussed here only briefly. Details and literature references are found in the monographs edited by Race and Sanger [394] and Issitt [214].

ABH specific hybridoma antibodies have been used to an increasing degree for blood group work: routine blood grouping, isolation of blood group substances, characterisation of blood group specific structures, and in immunohistochemical investigations to localise the respective determinants on cell surfaces and in tissues. Examples of **ABH** specific hybridoma antibodies are presented in *Table 5.4*.

Further information on the occurrence and specificity of **ABH** specific lectins may be found in various review articles (see e.g. [32,151,306]).

5.2.1 Anti-A and Anti-B Reagents

Iso-Agglutinins (Allo-Antibodies):
The **ABO** system represents one of the few blood group systems in which antibodies towards the different antigenic characters are regularly present in serum (see *Table 5.1*). These 'naturally' occurring antibodies are normally of the IgM type and are characterised by a high serological specificity. As mentioned above, either

Table 5.4: Examples of ABH specific hybridoma antibodies[*]

Antibody	Type	Blood group specificity	Binding specificity	References
AH21	IgM	Anti-A	chain type-1	[1]
CLH6	IgM	Anti-A	chain type-1	[125]
HH4	IgG3	Anti-A	chain type-2	[74]
AH16	IgG3	Anti-A	chain type-1 + 2	[1]
TH-1	IgG2a	Anti-A	chain type-3	[75]
KB-26.5	IgG3	Anti-A	chain type-3 + 4	[284]
TS-1	IgG1	Anti-A	chain type-3 + 4	[284]
3E-7	IgM	Anti-B	chain type-1 + 2	[46,290]
1A-12	IgM	Anti-B	chain type-1 + 2	[46,290]
$E_1$15-2		Anti-B	chain type-2	[183]
101	IgG	Anti-H	chain type-1	[119]
102	IgM	Anti-H	chain type-2	[119]
BE2	IgM	Anti-H	chain type-2	[563]
TRA-1-85		Anti-H	chain type-2	[109]
G10		Anti-H	chain type-2	[235]
H11	IgM	Anti-H	chain type-2	[243]
MBr1		Anti-H	chain type-3 + 4	[45,72]

[*] Further **ABH** specific antibodies are described by Chen et al. [62], Furukawa et al. [130], Gooi et al. [156], Lowe et al. [301], Lubenko and Ivanyi [302], McGowan et al. [323], Messeter et al. [325], Nakajima et al. [348], Nemec et al. [351], Oriol et al. [370], and Voak et al. [511]; many are commercially available. Hybridoma antibodies binding to the difucosyl derivatives of **A** and **B** (anti-**ALe^b**, anti-**BLe^b**, anti-**AY**, and anti-**BY**) are discussed in *Sect. 6.2.*

they are directed towards an **ABH** antigen absent in the individual, or they are not reactive with the subject's own erythrocytes or tissues under normal conditions.

The anti-**A** and anti-**B** reactivity of secretions, such as milk, saliva, tear fluid, cervical secretion, and ascites fluid is probably caused by IgA antibodies (studies on such antibodies are discussed in Race and Sanger [394], p. 45).

Immune Antibodies:

Immune antibodies with anti-**A** and anti-**B** specificity are produced mainly after transfusions with **ABO** incompatible blood or in the course of pregnancies with **ABO** incompatible fetuses. Active immunisation of volunteers with erythrocytes or water-soluble blood group substances of suitable specificity also yields highly specific antisera. The immune antibodies generated in this manner are primarily of the IgG class.

The anti-**A** and anti-**B** specificities of IgG antibodies frequently found in **O** individuals cannot be separated by serum absorption. It is therefore assumed that these so-called anti-**(A+B)** are cross-reacting antibodies which bind to a structure common to blood group **A** and **B** determinants [90,442]. As will be discussed in *Sect. 5.3.3*, at least some of these antibodies react with the type-1 chain **A** and **B** substances found in plasma.

The presence of **I** or **i** antigens is essential for optimum reactivity of most **ABH** specific antibodies occurring in autoimmune sera. Examples of such antibodies having anti-**AI**, anti-**A₁I**, anti-**BI**, anti-**(A+B)I**, and anti-**HI** specificities are presented in *Sect. 7.1*.

Specific anti-**A** and anti-**B** antibodies have been prepared by immunising rabbits with erythrocytes or blood group substances of suitable specificity (see e.g. [157]).

Hybridoma Antibodies:

A variety of monoclonal antibodies with anti-**A**, anti-**B**, and anti-**(A+B)** specificities have been produced using the hybridoma technique. Examples of such hybridoma antibodies as described in literature are shown in *Table 5.4*.

Lectins:

In the seeds of leguminous plants a great number of anti-**A** specific lectins has been described. The lectins most frequently used are from *Dolichos biflorus* [30,107,174,383], *Vicia cracca* [233,405], and *Phaseolus lunatus limensis* (lima beans) [36,174,455]. Suitably diluted, these lectins also represent excellent anti-**A₁** reagents (see below). More recently anti-**A** specific lectins have been detected in various fungus species [132].

Pure anti-**B** specific lectins have thus far been found only in fungi, such as *Polyporus* [= *Fomes*] *fomentarius* [308] or *Marasmius oreades* [104].

Some plant lectins agglutinate both **A** and **B** erythrocytes. Among those 'bifunctional' lectins the 'lectin I' of *Bandeiraea* (= *Griffonia*) *simplicifolia* [307] is best characterised [153,343,383,537]. It is a tetrameric molecule built of two different subunits − one of them reacting primarily with blood group **A** structures (subunit A) and the other binding to blood group **B** determinants (subunit B). As expected, the five possible isolectins formed by these subunits (A_4, A_3B, A_2B_2, AB_3, and B_4) show specificities ranging from specific anti-**A** to anti-**(A+B)** and specific anti-**B**; the different isolectin molecules can be separated by electrophoresis and affinity chromatography. Further examples of bi-functional lectins are the anti-**(A+B)** from *Sophora japonica* [266], the anti-**(B+H)** of various *Evonymus* species [98,295,373,374,381,443,553], and the anti-**(A+N)** from *Moluccella laevis* [31a] (see also *Sect. 9.2)*. With few exceptions these specificities cannot be separated from each other.

Great amounts of anti-**A** specific lectins have been detected in the albumen glands of helicid snails [378]. Among those lectins the anti-**A**$_{HP}$ from *Helix pomatia* [173,174,383,388] is most thoroughly characterised and has been widely used as an anti-**A** reagent. It should be emphasised, however, that the high blood group **A** specificity of this lectin is confined to its reaction with erythrocytes. In soluble systems anti-**A**$_{HP}$ binds to a series of blood group inactive glycoproteins (e.g. [173]) and can be purified by affinity chromatography on Sephadex®, a dextran cross-linked with epichlorohydrin [213].

Lectins with anti-**B** specificity have been detected in the female gonads of fish species, i.e. roach (*Rutilus rutilus*) [444], herring (*Clupea harengus*) [506], and various salmonids [219,264,387,497]. However, the distinct reactivity of these lectins with blood group **P** erythrocytes (see *Sect. 8.1)* cannot be separated from anti-**B** activity [3].

5.2.2 Anti-A₁ Reagents

Isoagglutinins with anti-**A**$_1$ specificity are fairly frequently found in the sera of **A**$_2$**B** individuals [394]. Anti-**A**$_1$ sera can also be prepared by absorption of normal anti-**A** with **A**$_2$ erythrocytes or water-soluble blood group substances.

More recently a hybridoma antibody with anti-**A**$_1$ specificity has been described ('*AbS12*', [130]).

An excellent anti-**A**$_1$ reagent is found in the lectin of *Dolichos biflorus* [31].

5.2.3 Anti-H Reagents

Iso-Agglutinins and Other 'Naturally' Occurring Antibodies

Highly specific anti-**H** isoagglutinins are occasionally found in the sera of **A**$_1$ or **A**$_1$**B** donors. In most cases, however, they are only weakly active and represent mixtures of anti-**H** and anti-**HI** antibodies [152,509] (see also *Sect. 7.1)*. The fairly strong anti-**H** agglutinins present in the sera of '**Bombay**' or **para-Bombay** individuals (see *Sect. 5.5.6)* are generally accompanied by anti-**A** and anti-**B**, neither of which can readily be separated from the anti-**H** specific antibodies.

Anti-**H** specific antibodies occur also in non-immune sera of various mammals (dog, cat, pig, goat, cattle, rabbit, etc.) and in chickens [152,394]. Anti-**H** have also been prepared by immunisation of rabbits, goats, and chickens with **O** erythrocytes or blood group **H** substance [25,152]. The sera of eels (*Anguilla anguilla* and *A. japonica* [321,472]) show strong anti-**HI** specificity [64]. More recent investigations revealed that eel agglutinin reacts preferentially with fucosylated type 1 chain antigens [10].

Hybridoma Antibodies

Examples of anti-**H** specific hybridoma antibodies are presented in *Table 5.4.*

Lectins

A series of anti-**H** agglutinins has been detected in plants. The 'lectin I' from the common gorse, *Ulex europaeus*, is especially well characterised [59,199,320,380] and most widely used both as anti-**H** and as a reagent to distinguish between **A**$_2$ and **A**$_1$ erythrocytes [504]. *Ulex* anti-**H** binds exclusively to type-2 chain **H** determinants (see *Sect. 5.3)* [10]. Anti-**H** specific lectins also occur in the seeds of other leguminous plant species, such as *Lotus tetragonolobus* [379,405], *Cytisus sessilifolius* [257,321,384, 405], and *Laburnum alpinum* [258,405]. Anti-**H** specific lectins have also been found in various fungus species [132].

Strong anti-**H** specific lectins have been detected in the female gonads of perches (e.g. *Perca fluviatilis*) [389].

5.3 Blood Group ABH Substances

Reports on the first efforts to isolate blood group active substances from erythrocytes were published soon after the discovery of the **ABO** system. Researchers had treated red cell membranes with organic solvents; this yielded **A** and **B** active material, the so-called '**alcohol-soluble blood group substances**'. The serological activity of these extracts was extremely low, however, and the results of different groups conflicting and not readily reproducible, so the experiments were soon abandoned. A review of these investigations can be found in Kabat's 'Blood Group Substances' [226].

Considerable impetus to the investigations on the chemical structures of the **ABH** determinants came from the detection of blood group **ABH** active material in tissue fluids and secretions of humans, and in the gastric mucosal linings of certain animals ('**water-soluble blood group substances**'). The groups of Morgan (Lister Institute

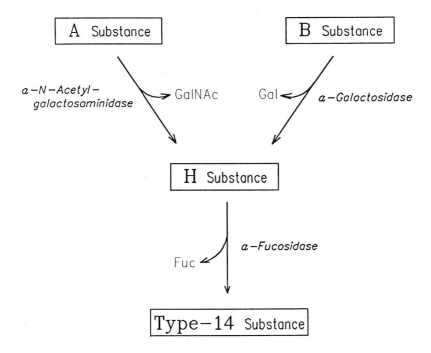

Figure 5.1: Changes in blood group specificity as induced in water-soluble blood group substances by treatment with glycosidases.
According to Watkins [521].

of Preventive Medicine, London) and Kabat (Columbia University, New York) began to study the structures of the **ABH** determinants on soluble material in the late 1940s. Their work was based on the assumption − later proved correct − that, due to the identical immunological reaction, the essential characteristics of the **ABH** epitopes of water-soluble material must be closely similar, if not identical, to those of the erythrocyte blood group substances.

5.3.1 Structures of the ABH Determinants

First investigations demonstrated that the secreted blood group substances are glycoproteins in which the **ABH** determinant structures are located in the carbohydrate moiety. Haemagglutination inhibition and immunoprecipitation inhibition experiments indicated that N-acetylgalactosamine, galactose, and fucose are essentially associated with **A**, **B**, and **H** specificities respectively. Degradation of blood group substances with specific glycosidases showed that removal of the immunodominant sugar residue is accompanied by a change in serological specificity (*Fig. 5.1)*: when **A** or **B** substances are treated with α-N-acetylgalactosaminidase or α-galactosidase, respectively, **H** activity is revealed[1]. Further, treatment of H substance with α-fucosidase developed or enhanced cross-reactivity with anti-type 14 pneumococcal serum ('**type 14 substance**', see below). Structural studies on oligosaccharides obtained by mild acid degradation of the blood group substances or by hydrolysis in alkaline borohydride provided additional information on the inner core regions of the carbohydrate chain not directly involved in blood group specificity. For further information on these investigations Watkins's detailed review should be consulted [521].

More recently, increasingly sophisticated analytical methods have permitted research on the blood group substances of erythrocytes and tissue cells.

Based on these investigations six main types of **ABH** determinants have been defined. They are classified according to the terminal disaccharide structure of the precursor chain [365]:

Type 1:   **Galβ1→3GlcNAcβ1→X-**
Type 2:   **Galβ1→4GlcNAcβ1→X-**
Type 3:   **Galβ1→3GalNAcα1→X-**
Type 4:   **Galβ1→3GalNAcβ1→X-**
Type 5:   **Galβ1→3Galβ1→X-**
Type 6:   **Galβ1→4Glcβ1→X-**

These precursor structures do not show any **ABH** specificity. The type-2 chain terminus, however, is responsible for the cross reactivity of partially degraded blood group substances with anti-type 14 pneumococcal sera ('**type-14 substance**')[2] [227,294]. The **H** determinant is characterised by an α-fucose attached to carbon-2 of the terminal β-galactose unit. As is evident in *Fig. 5.2,* the **A** and **B** specific structures are represented by α1→3 N-acetylgalactosamine or α1→3 galactose residue linked to the subterminal β-galactose unit of the **H** determinant, respectively. The difucosylated

---

[1] Treatment with α-galactosidase from green coffee beans have already been used for a large-scale conversion of blood group **B** erythrocytes into **O** cells [574].

[2] In addition, the type-2 chain is an integral part of blood group **I** and **i** specific structures (see *Sect. 7.2)*, whereas the type-1 chain also represents the basic structure of blood group **Lewis** substances (*Sect. 6.3)*.

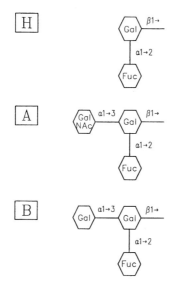

Figure 5.2: Structures of blood group ABH determinants.

A and **B** determinants of type-1 and type-2, i.e. **ALe**[b] and **BLe**[b], or **AY** and **BY**, respectively, will be discussed in *Chap. 6*.

These blood group **ABH** specific structures are carried by glycolipids and glycoproteins and may occur as subunits of free oligosaccharides of milk and urine.

These structures show that the immunodeterminant groups of the **ABH** antigens are characterised by only one single monosaccharide residue. Antibodies, with the exception of the 'immunodominant' sugar, usually recognise two or three additional monosaccharide units, whereas lectins bind to a smaller epitope comprising only the immunodominant monosaccharide residue or a part thereof (see Kabat [228]). The carbohydrate chain core structures, the carrier molecule and, in the case of membrane-bound blood group substances, the cell membranes do not change blood group specificity, but may influence serological reactivity by controlling the steric arrangement of the determinant groups.

Chemical Synthesis of ABH Determinant Oligosaccharides

Large quantities of **ABH** specific oligosaccharides have been prepared by chemical synthesis (c.f. [200,216,293]). These oligosaccharides, bound to water-

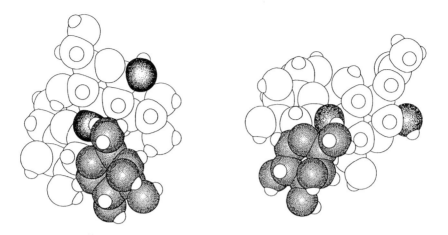

Figure 5.3: Three-dimensional models of H type-1 (left) and H type-2 (right) determinants.
Molecular models based on the torsion angles published by Rao et al. [400].
Black: glycosidic oxygens, shaded: fucose residue.

insoluble matrices, are widely used for isolation and purification of antibodies (e.g. the Synsorb® products offered by Chembiomed, Edmonton, Alberta, Canada).

Three-Dimensional Structures of the ABH Determinant Oligosaccharides

Based on NMR spectroscopy studies on **ABH** specific oligosaccharides and conformational energy calculations, models for the three-dimensional structures of the **ABH** determinants have been established [47,212,292,354,400,411]. The minimum energy conformations of **H type-1** and **H type-2** epitopes are presented in *Fig. 5.3*, those of **A** specific glycosphingolipids of types 1–4 in *Fig. 5.4*.

Membrane glycolipids in particular show drastic differences in their orientation of the oligosaccharide chains: in type-1 substances the carbohydrate chain extends almost perpendicularly to the membrane plane, whereas in type 2, 3, and 4 substances the terminal part of the oligosaccharide chains is more or less parallel to the membrane. The $\alpha1{\rightarrow}2$ fucosyl residue in type 3 and 4 glycolipids is directed towards the environment, whereas in type 2 glycolipids it faces the membrane. Since the specific orientations of the oligosaccharide chains are largely maintained (in particular in the case of membrane glycolipids [354]), the differences in the

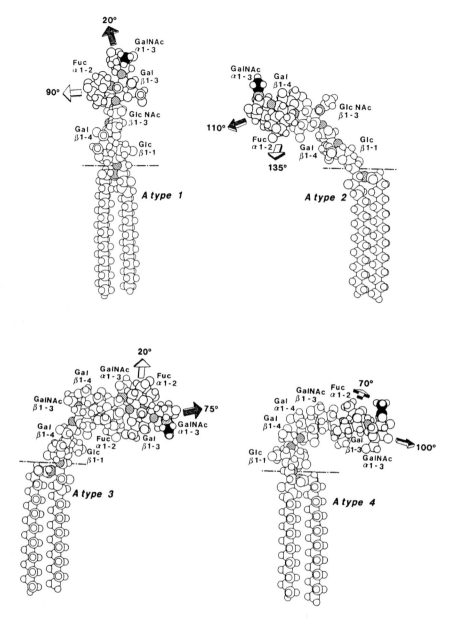

Figure 5.4: Three-dimensional models of A type-1, A type-2, A type-3 and A type-4 specific glycosphingolipids.

Reproduced from Nyholm et al. [354] by permission of John Wiley & Sons Ltd., Chichester, UK.

presentation of the **ABH** determinants might explain the chain-type specificity observed on various hybridoma antibodies (*Table 5.4*). Furthermore, the different substrate specificities of A^1 and A^2 gene-dependent N-acetylgalactosaminyltransferases (see *Sect. 5.5.1*) are obviously based on the distinct accessibility of the **H** epitopes.

5.3.2 Blood Group ABH Substances of Erythrocytes

Since the first **ABH** substances isolated from human erythrocytes were glycolipids, it was assumed at the time that blood group **ABH** activity of red cells was determined exclusively by this class of membrane constituents. Polyglycosyl-ceramides in particular were considered the main carriers of serological activity [86]. When several groups obtained evidence that glycoproteins also contribute to blood group **ABH** activity of red cells [484,532] the results were dismissed as 'artefacts'. Finally it did become possible to demonstrate beyond any doubt the presence of **ABH** determinants in glycoproteins of the erythrocyte membrane [114,126]. Among the blood group **ABH** glycolipids of erythrocytes, substances with type-2 chains are predominant [165]; type-3 and type-4 chain glycolipids are present only in insignificant amounts, and type-1 chain glycolipids are not endogenous erythrocyte substances but are absorbed from the plasma (see below). In **ABH** active glycoproteins of the erythrocyte membrane only type-2 determinants have been detected thus far [125,485].

In the following section the different classes of blood group substances occurring in the erythrocyte membrane will be discussed in more detail.

(A) Glycolipids

(a) Type-2 Chain (Lacto-Series) Glycosphingolipids

The type-2 chain glycosphingolipids of the erythrocyte membrane vary significantly in structure: from compounds with very simple, short oligosaccharide chains to molecules with complex and highly branched carbohydrate structures.

ABH active glycolipids with short carbohydrate chains are present only in small quantities in the erythrocyte membranes; the highest recoveries range from 1.8 to 2.6 mg of pure substance per litre of packed erythrocytes [177]. Nevertheless, it was from this very kind of substance that amounts sufficient for structural studies were first isolated (see Hakomori & Strycharz [171]). This class of **ABH** substances has been characterised as lacto-series glycosphingolipids containing 5 to 15 monosaccharide

Neolactotetraosylceramide (Paragloboside)

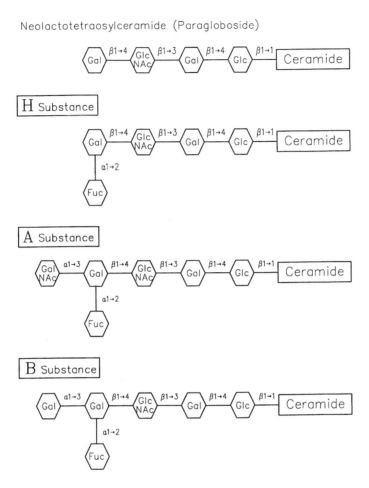

Figure 5.5: Glycosphingolipids of human erythrocyte membranes.
Structures of the most basic blood group ABH active substances.

units per molecule. They all have in common neolactotetraosylceramide (= paragloboside) as their precursor substance (*Fig. 5.5*). This basic molecule can be extended by additional N-acetyl-lactosamine units forming more complex oligosaccharide chains.

Structures of four **A** substances [74,127,170], five **B** substances [175,179,262], and three **H** substances [175,262,475,518] have been established to date. The

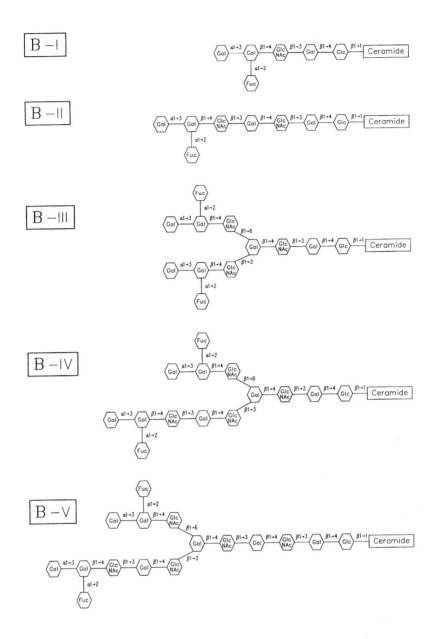

Figure 5.6: Structures of blood group B active glycosphingolipids of human erythrocyte membranes (substances B-I to B-V).
According to Hanfland [175,179].

glycolipids differ only in the carbohydrate basic chain, i.e. in the length of the oligosaccharide unit and in the degree of branching; variations in the lipid moiety (different fatty acids and sphingosines) occur independently of the respective molecule type.

In *Fig. 5.6* the structures of the **B** active glycosphingolipids with short carbohydrate chains thus far characterised are shown (the **A** and **H** specific glycolipids have analogous structures and are not included in this figure for that reason).

Since most preparations of **A** substances with branched carbohydrate chains are strongly reactive with anti-**H** reagents, it must be assumed that normally not all the **H** determinants are converted to **A** substance. Molecules with terminal neuraminic acid on one chain ending were shown to react with anti-**Fl** sera ([519], see *Chap. 12*). **ABH** active glycosphingolipids with branched oligosaccharide chains reveal distinct blood group **I** activity (see *Chap. 7)* [518].

In the course of their investigations [140,141] Gardas and Koscielak found blood group **ABH** (and **I**) active glycosphingolipids with highly complex oligosaccharide chains. Because of their high carbohydrate content these substances are water-soluble but are readily absorbed *in vitro* by erythrocytes. These molecules cannot, however, be considered a separate class of glycosphingolipids since lacto-series glyco-sphingolipids of the erythrocyte membrane contain a spectrum of molecules ranging from glycosphingolipids with short carbohydrate chains to glycolipids with complex oligosaccharide units. Nevertheless, the designation 'polyglycosyl-ceramides' (or 'megaloglycolipids') proposed by Gardas and Koscielak should be retained for practical reasons.

Analogous to the short-chain lacto-series glycosphingolipids, the polyglycosyl-ceramides are composed of fucose, galactose, N-acetylglucosamine, glucose and ceramide. The general formula for these substances is:

$$\text{Fuc}_{(3-4)} \cdot \text{Gal}_n \cdot \text{GlcNAc}_{(n-2)} \cdot \text{Glc}_1 \cdot \text{Ceramide}_1 \qquad (n = 10-27)$$

A active substances contain two or three N-acetylgalactosamine residues per molecule [260].

The oligosaccharide chains of the polyglycosyl-ceramides are highly complex structures comprising up to 60 monosaccharide units; the average molecule contains some 30–35 glycosyl residues. The carbohydrate backbone chain consists of repeating units of N-acetyllactosamine units ($\rightarrow$3**Gal** β1$\rightarrow$4**GlcNAc** β1$\rightarrow$); short N-acetyl-lactosamine side chains may be attached to carbon 6 of the galactose residues (with an average of one branch at every second galactose residue). Three to four **ABH**

Figure 5.7: Polyglycosyl-ceramides of human erythrocyte membranes.
Model of the backbone structure according to Koscielak et al. [260].

determinants, and about two unsubstituted β-galactosyl residues on an average molecule with 30 to 35 monosaccharide units have been deduced from the estimated number of terminal groups [260].

A schematic model of a polyglycosyl-ceramide molecule is presented in *Fig. 5.7*.

(b) Type-3 Chain Glycosphingolipids

The simplest structure of a type-3 chain glycosphingolipid [72,75] is an **A** type-2 ceramide hexasaccharide extended by a terminal **A** or **H** determinant group (see *Fig. 5.8*). Further glycolipids with long-chain carbohydrates which react with a hybridoma antibody towards **A type-3** determinants (*TH-1*) have been detected in extracts of human erythrocyte membranes [75]; these substances probably differ in the length of the N-acetyllactosamine backbone (cf. the lacto-series glycosphingolipids, above). Moreover, the isolation of incomplete chains, i.e. galactosyl-**A** and sialoylgalactosyl-**A**, has been reported [76]. The extended **A** structure has been found so far only in glycolipids [75].

Type-3 chain glycolipids have been found only in group **A** erythrocytes − **A type-3** substance is characteristic for A_1 erythrocytes and is present in only trace amounts in A_2 cells. The precursor, **H type-3**, occurs in greater quantities in A_2 erythrocytes than in A_1 erythrocytes; it is absent, however, in **O** or **B** cells [75]. These findings led to the assumption that **A type-3** may be identical with the hypothesised 'A_1 substance' (see *Sect. 5.5.1*).

(c) Type-4 Chain Glycosphingolipids

In **type-4** chain glycolipids [77,230] **A** or **H** determinants are attached to the terminal β-N-acetylgalactosamine residue of globoside (*Fig. 5.8*). '**Globo-A**' is found

71

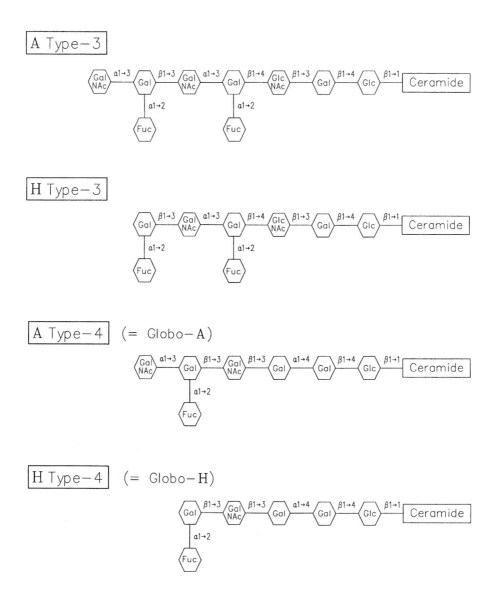

Figure 5.8: Blood group A and H active glycosphingolipids of human erythrocyte membranes: structures of type-3 and type-4 substances.

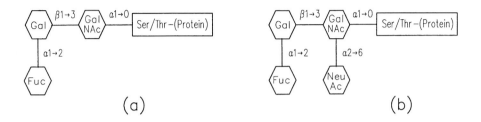

Figure 5.9: Glycoproteins of human erythrocyte membranes.
O-glycosidically linked carbohydrate side chains carrying blood group H determinant $\alpha 1 \rightarrow 2$ fucose:
(a) H active chain, **(b)** sialylated H chain.
According to Takasaki et al. [485].

only in blood group A_1 erythrocytes, **'globo-H'** in O or A_2 cells; **'globo-B'** substance has not yet been found.

Although type-4 **A** and **H** substances represent only very minor components of erythrocyte membranes, they are present in large quantities in human kidneys and occur in certain tumour tissues (see below).

(B) Glycoproteins

(a) O-Glycosidically Linked Carbohydrate Chains

Blood group **H** active glycopeptides with O-linked carbohydrate chains have been prepared from membrane glycoproteins which Takasaki et al. [484,485] have isolated from human **O** erythrocytes. The **H** specific oligosaccharide unit was identified as the trisaccharide **Fuc$\alpha 1 \rightarrow 2$Gal$\beta 1 \rightarrow 3$GalNAc–** *(Fig. 5.9.a)*. In the majority of these **H** chains the N-acetylgalactosamine carries an $\alpha 2 \rightarrow 6$ neuraminic acid residue *(Fig. 5.9.b)* – this structure, however, does not seem to have any serological activity whatsoever. A carbohydrate unit of more than 4,000 Dalton, has not yet been further investigated.

These O-linked chains are presumably carried by glycophorins of the red cell membrane (see *Sect. 9.3*). Assuming a quantity of $\sim 10^6$ glycophorin molecules per erythrocyte *(Sect. 9.3)* and an estimated quantity of about 170,000 **ABH** sites present on O-linked chains (see below), it appears that only about one fifth of all glycophorin molecules contain an **H** active chain.

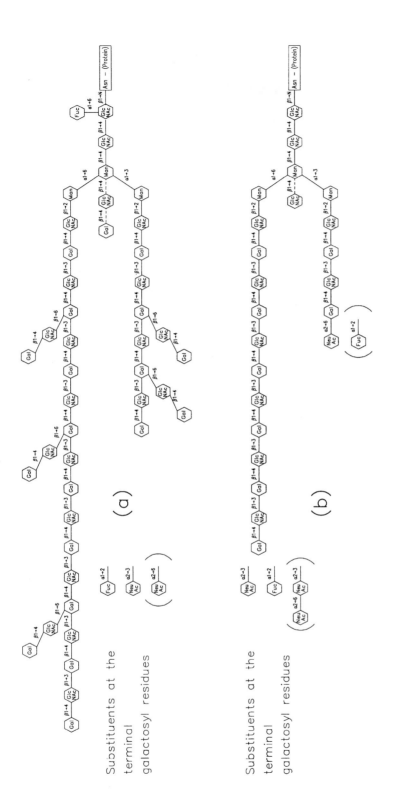

Figure 5.10: Glycoproteins of human erythrocyte membranes.

Structures of the N-glycosidically linked lactosaminoglycan chain of the band 3 protein: **(a)** adult erythrocytes [125], **(b)** foetal erythrocytes [124].

(b) N-Glycosidically Linked Carbohydrate Chains

ABH determinants have also been found on highly complex carbohydrate chains N-linked to intrinsic erythrocyte membrane constituents [114,218,267]. These substances[3] are pronase-sensitive *in situ* and have been identified as band-3 and band-4.5 proteins [126,232].

Investigations by Fukuda et al. [125] revealed that the carbohydrate units of band-3 protein are lactosaminoglycan-type oligosaccharide chains (see *Sect. 2.2)*: the core structure is composed of one chitobiose unit and three mannose residues, and carries two poly-N-acetyllactosamine chains: a long one of 10–12 lactosamine units and a shorter one of ~5–6[4]). Both chains are branched − the longer chain has three to five N-acetyllactosamine side chains attached to carbon-6 of galactose, the short chain one or two. About half of the terminal galactose residues carry α1→2 fucose, α2→3 N-acetylneuraminic acid, or α2→6 N-acetylneuraminic acid residues (see *Fig. 5.10.a)*.

According to the studies performed by Viitala et al. [508] on oligosaccharides isolated from **AB** erythrocytes, **A** or **B** determinants are located on separate carbohydrate chains. This finding may have great potential to advance further investigations on the biosynthesis of complex carbohydrate structures.

Number of Blood Group ABH Determinants per Erythrocyte

Several techniques have been applied to determine the number of **ABH** antigenic sites per erythrocyte:

(a) Quantitative immunoabsorption [103].
(b) Use of radioactively labelled UDP-N-acetylgalactosamine to estimate the amount of free **H** determinants by *in vitro* conversion into **A** determinants (see *Sect. 5.4.1)* [431].
(c) Calculation of the quantity of immunodominant α-galactose in **B** erythrocytes [185].

The data obtained by these methods are listed in *Table 5.5*. Though the values differ somewhat, the average number of **ABH** sites per erythrocyte should be in the range of 1.5–2.0 million.

---

[3] The glycopeptides purified from pronase digests of erythrocyte ghosts have been termed 'erythroglycans' [218].

[4] In addition, core structures with four mannose residues and three poly-N-acetyllactosamine chains have been found.

Table 5.5: Number of ABH determinants per erythrocyte

Blood group of erythrocytes	A sites	B sites	H sites
A₁	$0.81–1.17 \times 10^6$ [a]	–	$0.07–0.17 \times 10^6$ [c][e]
B	–	$0.61–0.83 \times 10^6$ [a] 2.2×10^6 [b]	$0.40–0.47 \times 10^6$ [c][e]
O	–		$1.59–1.74 \times 10^6$ [c] $1.7–1.9 \times 10^6$ [d]

[a] Quantitative immunoabsorption, from Economidou et al. [103].
[b] Assay of the immunodominant α-galactose, from Harpaz et al. [185].
[c] Conversion of **H** sites into **A** determinant structures using the **A** transferase from human serum, from Schenkel-Brunner [431]. Using the **A** transferase from hog gastric mucosa [437,446] a value of 3×10^6 **ABH** sites was obtained (since the hog enzyme is not defined as exactly as the human transferase these results are probably due to non-specific transfer of N-acetylgalactosamine residues).
[d] Lectin absorption, from Matsumoto et al. [321].
[e] Schenkel-Brunner (1981), unpublished results.

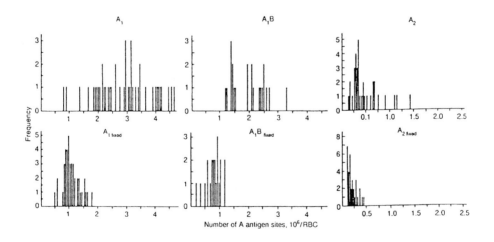

Figure 5.11: Antigen densities on human erythrocytes.

Histograms of the number of **A** antigen sites on erythrocyte samples of three different **ABO** phenotypes (native erythrocytes and cells fixed with glutaraldehyde).
Reproduced from Berneman et al. [18] by permission of S. Karger AG, Basel.

The antigen densities obtained by the approaches mentioned above represent only 'average' values. Normal human blood contains erythrocyte populations with highly differing numbers of **ABH** antigenic sites [18,407,409,464] (see *Fig. 5.11)*. This variability is presumably due to differences in the content of the blood group **ABH** substances [210] and reflects a physiological heterogeneity within the erythroid precursor cells of the bone marrow [407]. Extreme variations in the density of antigen sites among different populations of erythrocytes of an individual are also found in those **ABH** variants characterised by mixed-field agglutination, e.g. A_3, B_3, A_x, and A_{end} (see *Sect. 5.5)*.

The Distribution of ABH Determinants among the Different Classes of Blood Group Substances

The number of blood group **ABH** determinants localised in the different classes of blood group active glycoproteins and glycolipids has been established by two methods:

(a) Labelling of the **H** determinants of native **O** erythrocytes by *in vitro* conversion into **A** determinants by incubation with highly purified A_1-transferase and radioactively labelled UDP-N-acetylgalactosamine. After subsequent fractionation of the membrane constituents the distribution of radioactivity can be determined [431,535].

(b) Separation of the erythrocyte membrane constituents by SDS-poly-acrylamide-gel electrophoresis and subsequent testing of the protein bands for **ABH** activity using radioactively labelled lectins [112].

Though the values obtained by the two methods mentioned above differ significantly, the approximate distribution of **ABH** determinants among the different membrane constituents can, nevertheless, be estimated with some accuracy (*Table 5.6)*

The glycolipids with short carbohydrate chains which originally were considered 'the' blood group **ABH** substances of human erythrocytes constitute only about 5% of the total number of **ABH** sites, whereas the polyglycosyl-ceramides constitute ~20%. The major carriers of blood group **ABH** determinants are glycoproteins, about 60% of which are located on highly branched N-linked oligosaccharide chains of erythroglycans (bands 3 and 4.5) and 10–15% on O-linked chains probably glycophorins).

These data, of course, do not provide any information regarding the contribution of the different substances to the serological activity of an intact erythrocyte.

Table 5.6: Distribution of erythrocyte ABH determinants among erythrocyte membrane constituents.

Membrane constitutent	% of the total blood group determinants
Glycolipids with short carbohydrate chains	4–5
Polyglycosyl-ceramides	15–20
Glycoproteins with O-glycosidically linked carbohydrate chains [a]	15–25
Glycoproteins with N-glycosidically linked carbohydrate chains [b]	60–70

[a] **A** activity is nearly equally distributed between bands PAS-1 and PAS-2 [112].
[b] **A** activity is distributed between band 3 and 4.5 in a ratio of about 1 : 1.5 [112].
Compiled from the results of different groups: Finne [112], Schenkel-Brunner [431] and Wilczynska et al. [535].

5.3.3 Blood Group ABH Substances of Human Plasma

Small quantities of **ABH** active glycosphingolipids have been detected in human plasma. Like the blood group **Lewis** antigens (see *Sect. 6.3*), they are not derived from the red cell membrane but are passively acquired by the erythrocytes in a reversible process [406,496]. It is generally assumed that this class of glyco-sphingolipids makes only a minor contribution to the **ABH** activity of erythrocytes. Absorbed plasma **ABH** substances, however, provide part of the **ABH** activity of lymphocytes [367,369,396] and thrombocytes [81,101,237].

It is worth noting that cells which absorbed **A** and **B** glycolipids from the plasma are agglutinated by the anti-**(A+B)** antibodies present in most group **O** individuals, but rarely react with anti-**A** or anti-**B** from **B** or **A** donors [406,496].

The **ABH** substances from plasma are type-1 chain glycosphingolipids which, like the blood group **Lewis** substances discussed in *Sect. 6.3*, are derived from lactotetraosylceramide as the basic precursor glycolipid:

Jowall et al. [221] detected an **A** active ceramide hexasaccharide with type-1 chains in the plasma of **A₁Le(a–b–)** secretors, a substance which is probably identical to the **A type-1** glycolipid isolated from erythrocyte membranes by Clausen et al. [73].

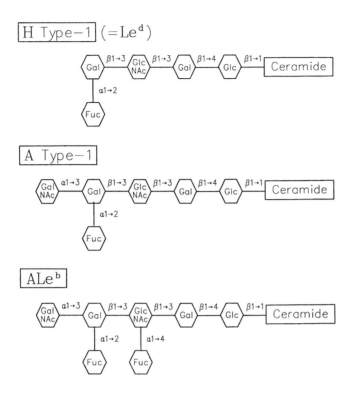

Figure 5.12: Structures of glycosphingolipids of human plasma.

The **Le^d** active glycosphingolipid described by the groups of Hanfland [178] and Lemieux [291] is actually an **H type-1** substance. These structures and the serum glycolipid with the 'hybrid determinant' **ALe^b** are shown in *Fig. 5.12*. The **Le^d** (= **H** type 1) and **ALe^b** substances will be discussed in *Sect. 6.3*.

The quantity of **A** and **B** active glycosphingolipids in plasma is affected by the secretor status of the individual (see *Sect. 5.4.2*): in **ABH**-secretors significantly more **A** and **B** activity is found than in **ABH**-nonsecretors [496]. Furthermore, the serological specificity of the plasma substances is influenced by the blood group *Lewis* gene: in **Lewis**-positive individuals the **A** and **H** active glycolipids are converted primarily into **ALe^b** and **Le^b** substances (see *Sect. 6.3*).

79

The exact cellular origin of the **A** and **B** active plasma glycosphingolipids is not yet known. Type-1 chain glycosphingolipids are not integral compounds of the erythrocyte membranes, and expression of these substances is controlled by the *secretor* gene: these findings suggest that these glycosphingolipids are derived from tissues of entodermal origin.

5.3.4 Secreted Blood Group Substances (Mucins)

As mentioned at the beginning of this chapter, **ABH** active substances are not confined to erythrocytes. They are found throughout the body in mucus-secreting tissues of the gastrointestinal, respiratory, and genital tract and their secretions, and are known as 'water-soluble blood group substances'. Large quantities of **ABH** active mucins have been isolated from the pathological fluid of pseudomucinous ovarian cysts (*Kystoma pseudomucinosum*) [521].

Mucin substances are oligomers of ~300 kDa glycoprotein subunits linked together end-to-end by disulfide bonds [87]. The complexes apparently weigh up to several million daltons, depending on origin and purification method used [521]; even highly purified preparations show a considerable degree of polydispersity. The substances contain 80–90% carbohydrate.

Sedimentation analyses and electron microscopic investigations showed that mucin glycoproteins are semirigid thread-like molecules with a diameter of 3 nm and a length of 50 to more than 500 nm [149,458].

(a) The Carbohydrate Moiety

The carbohydrate moiety of water-soluble blood group substances is composed of five monosaccharides – galactose, N-acetylglucosamine, N-acetylgalactosamine, fucose, and N-acetylneuraminic acid. As expected, **A** substances contain an increased amount of N-acetylgalactosamine. N-acetylneuraminic acid is not part of any **ABH** epitope.

The oligosaccharide units of mucins are joined by O-glycosidic linkages to serine and threonine residues of the protein backbone via an α-N-acetylgalactosamine residue [521]. The overall structures of the carbohydrate chains closely resemble those of **ABH** active glycolipids. In contrast to the **ABH** substances of the erythrocytes, the mucins contain both type-1 and type-2 chains [521]. In some tissues (primarily in salivary glands [416]) most of the subterminal N-acetylglucosamine residues are fucosylated, and thus **Le**b and **Le**y determinants are found instead of **H** type-1 and **H** type-2, respectively (see *Chap. 6*). Type-3 chains analogous to those found in

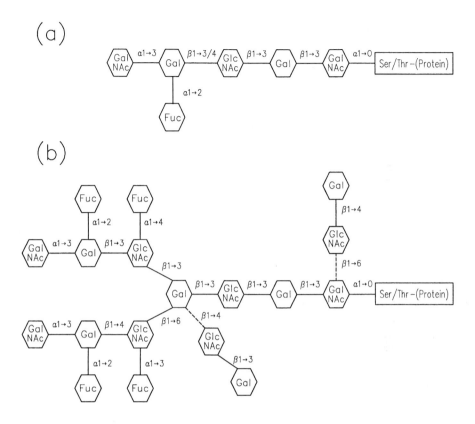

Figure 5.13: Carbohydrate chains of blood group A active glycoproteins (from ovarian cysts).
(a) As proposed by Watkins [520], (b) as proposed by Kabat [297].

erythrocyte glycolipids have not been detected in glycoproteins [71]. It has also not yet been established if the structure **GalNAcα1→3Galβ1→3GalNAcα1→OProtein** found in ovarian cyst mucin [94] is immunologically related to the type-3 chain of glycolipids [287].

The first models of the oligosaccharide chains found in secreted blood group substances were established by Watkins [520] (see *Fig. 5.13.a)*. The proposed monosaccharide sequences represent the simplest chain structures able to carry the serologically active segments at the non-reducing ends. However, the core regions of the chains not directly involved in blood group specificity are better described by the

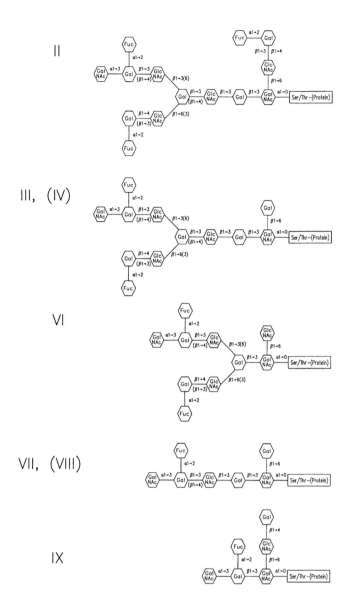

Figure 5.14: Blood group A active carbohydrate structures as found in water-soluble blood group substances.
Examples for oligosaccharide side chains isolated from the mucin of human gastric mucosa (numbering corresponds to that of the original paper).
According to Slomiany et al. [460].

Table 5.7: Composition of water-soluble blood group substances from ovarian cysts.

Compound	Quantity [a]
Aspartic Acid	3.7
Threonine	27.5
Serine	17.5
Glutamic Acid	4.3
Proline	13.6
Glycine	5.8
Alanine	9.7
Cysteine	0.4
Valine	4.5
Isoleucine	2.2
Leucine	2.3
Tyrosine	0.3
Phenylalanine	1.0
Lysine	1.7
Histidine	2.4
Arginine	2.8
Reducing Sugar	**53.0**
Hexosamine	25.8 [b]
Fucose	18.4
N-Acetylneuraminic acid	2.5

[a] Average values from Watkins [521] (µmol of amino acid per 100 µmol of total amino acids, or mg sugar per 100 mg of total carbohydrate) by permission of Elsevier Science Publishers, Amsterdam.
[b] The value for **A** substances is ~32.

more complex models presented by Lloyd & Kabat [297] (see *Fig. 5.13.b)*. As already discussed in *Chap. 3* the mechanism of biosynthesis does not result in the production of exactly defined chain sequences. The variety of oligosaccharide chain structures found in ovarian cyst substances and gastric mucin (e.g. [95,460]) clearly illustrates the high degree of microheterogeneity (see *Fig. 5.14)*.

(b) Protein Moiety

The peptide moiety of the secreted blood group glycoproteins shows an unusual amino acid composition: threonine and serine are the predominant constituents –

together with proline and alanine they make up about two thirds of the amino acids present [93,390] *(Table 5.7)*. Aromatic and sulphur-containing amino acids occur only in insignificant amounts – in most preparations they are removed by proteolytic digestion in the course of the purification procedure [87,93].

First investigations on ovarian cyst glycoproteins performed by Goodwin & Watkins [155] revealed that there are different regions in the protein backbone: some domains are rich in aspartyl and glutamyl residues and poor in carbohydrate side chains, while heavily glycosylated segments rich in seryl and threonyl residues are also found. The amino acid sequence of peptides isolated from the glycosylated region of the glycoprotein showed a close packing of serine and threonine residues (see *Fig. 5.15*).

The findings by Donald [93] of a 1 : 1 (mol) ratio of galactosamine to hydroxy amino acids in glycopeptides from **B** and **H** active glycoproteins (in which N-acetylgalactos-amine occupies only the linkage position between the oligosaccharide and the carrier protein) showed that in the carbohydrate-rich domains of the glycoprotein nearly all serine and threonine residues carry oligosaccharide chains.

```
H₂N - Thr - Thr - Ser
H₂N - Thr - Ser - Thr - Ser
H₂N - Thr - Pro - Thr - Ser
H₂N - Pro - Thr - Thr - Ser
H₂N - Thr - Pro - Thr - Ser - Ser
H₂N - Pro - Thr - Thr - Thr - Pro - Ser
H₂N - Ala - Pro - Thr - Thr - Ser - Gly - Ser
```

Sequences deduced from mixtures of peptides differing at only one position:

```
H₂N - Pro - Thr - Ala - Ser - Thr
H₂N - Thr - Thr - Pro - Ser
H₂N - Pro - Thr - Thr - Thr - Pro - Ser - Thr
H₂N - Thr - Thr - Ala - Ser - Thr
H₂N - Ser - Pro - Thr - Ser - Thr - Ser
H₂N - Ser - Ala - Thr - Ser
```

Figure 5.15: Some peptide sequences in the glycosylated regions of the protein backbone of ovarian cyst blood group substances.
According to Goodwin and Watkins [155].

Lacto−N−biose I (LNB−I)

Lactose

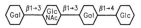

Lacto−N−tetraose (LNT)

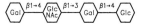

Lacto−N−neotetraose (neolactotetraose)

Lacto−N−neotetraose (neolactotetraose)

2'−Fucosyllactose

Lacto−N−fucopentaose I (LNF−I)

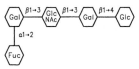

Lacto−N−difucohexaose IV (LND−IV)

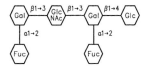

Figure 5.16: Oligosaccharides as found in human milk.
Examples of blood group H specific molecules and important precursors.
According to Kobata [244].

5.3.5 Blood Group ABH Active Oligosaccharides from Milk and Urine

A series of **ABH** active oligosaccharides have been isolated from human milk [244] and urine [303]. The molecules are of differing structure and complexity and, with few exceptions, are derivatives of the disaccharide lactose (**Gal** $\beta 1 \rightarrow$**4Glc**). The majority of the **ABH** determinants are based on a type-1 core structure: type-2 chains are only rarely found [555]. Examples of oligosaccharides isolated from milk are presented in *Fig. 5.16*.

5.3.6 Other Sources of Blood Group ABH Active Material

Human Tissues and Secretions

Many glandular tissues contain great quantities of **ABH** active glycolipids. Pancreas [43], small intestine [33,42], and meconium [234] are good sources for such substances. An increased content of blood group **B** substance was reported in the pancreas of a blood group **B** patient suffering from Fabry's disease, a defect in α-galactosidase [528].

Type-1 as well as type-2 chain substances are present in the pancreas [38]. The glycolipids of the small intestinal mucosa contain exclusively type-1 chains [38], whereas the glycoproteins carry predominantly type-2 chains ([113], see *Fig. 5.17*). Globo-**A** is the prevalent substance in the kidney of **A** individuals where it makes up about half of the glycolipid-based **A** antigens [41,44]. A more recent investigation [448] suggests that the renal globo-**A** glycolipid acts as a receptor for the attachment of uropathogenic *Escherichia coli* strains. In blood group **B** donors, only traces of globo-**B** have been detected [203].

In addition to the oligosaccharides mentioned above, **ABH** active material of higher molecular mass (about 150–200 kDa) has been detected in urine [115,240]. The substances, assumed to be glycoproteins, have not yet been characterised.

Animal Tissues

Blood group **ABH** reactive material is also found in many animal species:

A and **H** active glycosphingolipids have been isolated from the intestine of dogs [465,466], and rats [35,39,40,181], as well as from hog gastric mucosa (e.g. [459,461-463,505]). In some cases the substances contain **B**-like determinants lacking the **H** specific $\alpha 1 \rightarrow 2$ fucose, such as the **B** active glycolipids of rabbit erythrocytes

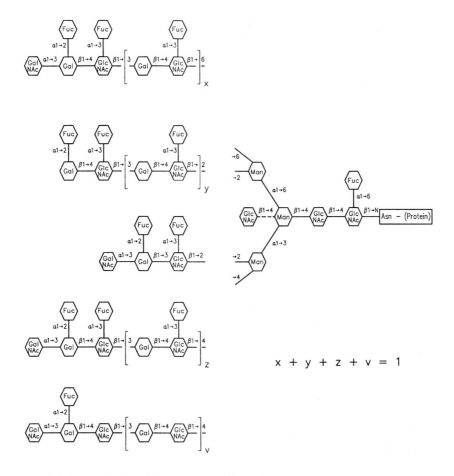

Figure 5.17: Glycoproteins of human small intestine.
Comprehensive structures of blood group A active carbohydrate units as proposed by Finne et al. [113].

[176,180]. In other cases chain types not yet detected in human material have been found, such as ganglio- and isoglobo-series glycosphingolipids, or a glycolipid with a structure derived from lactosylceramide (e.g. [182,184,204]).

Good sources for the isolation of water-soluble blood group substances are the gastric mucosa of pigs (**A** and/or **H**), horses (**A** and/or **B**), and cattle (**A**, **B**, and **H**) [226]; the **A/H** substances found in hog gastric mucin in particular have repeatedly been used for investigations on structure and biosynthesis of **ABH** determinants.

Strong **ABH** activity has also been detected in the egg jelly of amphibian spawn, e.g. toads (*Bufo bufo, B. viridis*, and *B. calamita* [434,538]), water frogs (*Rana* *esculenta* and *R. ridibunda* [Kothbauer, 1978 #1503]), brown frog (*Rana latastei*) [265], and treefrog (*Hyla arborea* [435]).

Plants and Microorganisms

The occurrence of **ABH** active material has repeatedly been reported in plants [470,552,554] and microorganisms [344,471,473]. Only in a few cases. however, have the structures of these **ABH** and **ABH**-like antigens been established.

5.3.7 Distribution of Blood Group ABH Active Material in Human Tissues

Immunohistological investigations[5] have revealed that in tissues of ecto- and mesodermal origin the expression of **ABH** antigens is independent of the secretor status, whereas in tissues of entodermal origin the synthesis of **ABH** antigens is controlled by the *secretor* gene [363,368]. **ABH** characters, however, are not equally expressed on all cell types: practically all epidermis cells and epithelial cells of **ectodermal** origin contain **ABH** antigens. In the nervous system − which is derived from the ectoderm − only the primary sensory neurons show **ABH** activity, whereas sensory cells derived from the central nervous system lack **ABH**. On **mesodermal** cells, also phylogenetically derived from the ectoderm, **ABH** antigens are expressed on the endothelial cells of the blood vessels, on erythrocytes, and on other blood cells. **ABH** activity is absent in connective tissue, muscle, and bones. The epithelial cells of the gastrointestinal, respiratory, and urogenital tract are of **entodermal** origin, and the expression of the **ABH** antigens depends on the secretor status of the individual. In stomach, small intestine, and proximal parts of the colon of **ABH**-secretors, **ABH** substances are localised in the cytoplasm of surface epithelium and mucous glands; beginning in the descending colon the **ABH** activity gradually decreases towards the rectum, where it can no longer be detected [483,536]. In kidneys, the cells of the urinary epithelium expressing **ABH** antigens are governed by the *secretor* gene, whereas glomeruli, proximal and convoluted tubules, all of which are derived from the mesoderm, are independent of secretor status.

---

[5] Note that the results of immunohistological and immunohistochemical investigations depend strongly on the antibodies used; thus it is virtually impossible to compare the data obtained by different groups.

It has also been found that ecto- and mesodermally derived tissues produce type-2 chain antigens virtually exclusively, whereas entodermally derived tissues produce mainly type-1 chain substances [83,331-333].

5.3.8 Occurrence of ABH Antigens in the Course of Embryonic Development

The **ABH** antigens show distinct spatial and temporal expression patterns throughout the mammalian developmental process. They can be detected in the earliest stages of development. The distribution in embryonic tissue, however, differs from that in adult tissue. In most tissues the **ABH** activity initially present disappears in the course of development and finally, in the mature organism, becomes restricted to specific cell types. However, under certain pathological conditions (as in some cases of malignant transformation, see below), it may re-appear, yielding aberrant epitope patterns. These changes often mirror the glycoconjugate arrays displayed during the embryonic development of the transformed tissue.

(a) On Erythrocytes

Blood group **ABH** activity can be detected on the erythrocytes of embryos in the fifth week after fertilisation [469,482]; the cells, however, show much less **ABH** activity

Table 5.8: Number of ABH determinants on foetal erythrocytes.

Blood group of erythrocytes	A sites[a]	H sites[b]
A₁ (cord)	0.25–0.37 x 10^6	0.07 x 10^6
A₁ (adult)	0.81–1.17 x 10^6	0.07–0.17 x 10^6 [c]
H (cord)	–	0.29–0.35 x 10^6
H (adult)	–	1.59–1.74 x 10^6

[a] Quantitative immunoabsorption, from Economidou et al. [103].
[b] Conversion of the **H** sites into **A** determinants using the **A₁**-transferase from human serum, from Schenkel-Brunner: [431] cells from adults, [430] foetal cells.
[c] Schenkel-Brunner (1981), unpublished results.

than the erythrocytes of adults [79,410]. This weak serological activity is due to a decreased **ABH** site density on embryonal red cells — 250,000 to 370,000 blood group **A** sites on cord cells from A_1 individuals compared to 1 to 1.5 x 10^6 on the cells of A_1 adults *(Table 5.8)*. The serological activity increases soon after birth and reaches its full expression in the third year of life [162].

The reduced **ABH** site density of cord blood erythrocytes is mainly due to the fact that the 'branching enzyme' is not expressed in erythrocyte precursors of human fetuses (see also *Chap. 7)*. Thus, glycosphingolipids [128,263,280,517] and erythroglycans [124] of cord cells contain predominantly short and unbranched (i.e. monovalent) carbohydrate chains (see *Fig. 5.10.b)*. It has also been noted that glycolipids of embryonal erythrocytes are significantly higher sialylated than those of red cells from adults [128], resulting in lower production of neutral **ABH** reactive glycosphingolipids.

(b) In Tissue Cells

Histological investigations on human embryos and fetuses by Szulman [481,482] showed that **ABH** antigens are represented in the smallest embryos studied (four weeks after fertilisation). The antigens have been found on endothelial cell membranes throughout the cardiovascular system, as well as on all epithelial cells with the exception of those of the central nervous system, adrenal glands, and liver. The antigens on the endothelium and the stratified epithelia of integument, oesophagus, and the lower urinary tract are permanent and remain for life. On all other cells the **ABH** antigens gradually disappear in the course of morphologic differentiation and functional maturation of the organ concerned; the final status of the distribution of cell wall **ABH** antigens is reached approximately during the 12th week of gestation.

The mucus-secreting apparatus begins to function in the ninth week after fertilisation. In **ABH** secretors the **ABH** antigens first appear in the salivary glands and in the stomach, followed in fixed sequence by the rest of the gastrointestinal tract, the respiratory system, and the pancreas. In the colon the **ABH** antigens gradually decrease in the course of further development and cannot be detected in adult individuals.

It should be mentioned that the **ABH** antigens found in amniotic fluid [4] and the **ABH** active amnion cells [123] can only originate from the embryo.

5.3.9 ABH Antigens in Malignantly Transformed Human Tissues

In most cases of malignant transformation the expression of **ABH** antigens is changed: in many tumour tissues **ABH** activity is lost during functional and morphological de-differentiation of the cells, whereas in tissues in which **ABH** antigens are not expressed in adults but are present in fetuses (e.g. thyroid gland, liver, and distal colon), **ABH** activity may be re-expressed in malignantly transformed cells. In tumours of blood group **O** and **B** individuals, incompatible **A**-like antigens occasionally appear (see *Sects. 5.4.1* and *5.4.4*). Further, globo-**H**, which cannot be detected in normal tissue, has been found in teratocarcinoma [231] and breast carcinoma cells [45] of blood group **O** individuals, and ganglio-**H** (fucosyl-G_{M1}) has been traced in human small cell lung carcinoma tissue [353]. For more details on the change of **ABH** characteristics in the course of tumour formation see the review articles by Hakomori [166,168,169].

5.4 Biosynthesis of ABH Antigens and Its Genetic Regulation

The structures described in the preceding chapter indicate that the **ABH** determinants are characterised by a single monosaccharide residue. Removal of this so-called immunodominant sugar with the aid of specific glycosidases results, as stated above, in the loss of the respective blood group character and in the appearance of a new specificity (see *Fig. 5.1*). Based on these results, Watkins and Morgan [525], and Ceppellini [60] hypothesised that the biosynthesis of the **ABH** determinants proceeds in reverse order, i.e. by sequential assembly of the respective monosaccharide residues through the action of specific glycosyltransferases. According to this scheme, a type-2 precursor chain built under the control of a series of common transferases is converted into the **H** determinant structure by the *H* gene-encoded α1,2-fucosyltransferase. The **H** epitope may then be converted into **A** or **B** determinants by the action of the *A* or *B* gene products, α1,3-N-acetyl-galactosaminyl- and α1,3-galactosyltransferase, respectively (see *Fig. 5.18*). This proves that the blood group genes are not responsible for the synthesis of the entire blood group substance; rather, they are responsible only for the transfer of the respective immunodominant monosaccharide to a suitable acceptor substance.

As discussed in *Chap. 3*, this mechanism is a general scheme for the biosynthesis of glycoconjugates and oligosaccharides − and which in the case of the biosynthesis of **ABH** antigens has been verified in virtually all its details.

The substrate specificity of the glycosyltransferases is concentrated on only a small section of the acceptor molecule. It is generally assumed that in most cases one single gene product acts on different classes of blood group active substances (i.e. glycoproteins, glycolipids, and oligosaccharides with various types of carbohydrate chains).

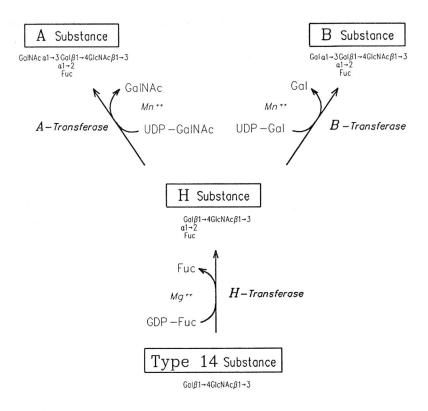

Figure 5.18: Biosynthetic pathway leading to the formation of ABH determinants.

The monosaccharide sequence in the chains is determined by the substrate specificity of the glycosyltransferases involved. Moreover, the synthesis of an oligosaccharide chain is interrupted when one step in the biosynthesis sequence is blocked; if more transferases compete for one substrate, other chain types may occur in concentrations greater than normal. For example, in **ABH**-nonsecretors and in 'Bombay' individuals the formation of **H** specific structures is blocked and no **A** or **B** substance can be produced; the unchanged precursor chains are instead sialylated by a sialyltransferase.

The glycosyltransferases participating in the biosynthesis of blood group antigens are found virtually everywhere in the human organism, and glandular tissues (gastric mucosa and salivary glands in particular) have proved excellent enzyme sources. The enzymes are also found in soluble form in serum, milk, and ovarian cyst fluid.

Glycosyltransferases with analogous specificities have also been detected in animals: **A**-transferase in canine trachea [9], **B**-transferase in submaxillary glands and gastric mucosa of baboons [393,575], in rabbit gastric mucosa [21,575], and in the sera of the Japanese tortoise *Clemmys japonica* [319] and of bull frogs (*Rana catesbeiana*) [318], **A**- and **H**-transferase in submaxillary glands and gastric mucosa of pigs [24,161,447,502], **H**-transferase in bovine spleen [14], and in the small intestine of the rat [16,314].

A- [446,447,500] and H-transferases [23,24] isolated from gastric mucosa and submaxillary glands of pigs have been used for studies on the biosynthesis of **ABH** antigens. It is to be expected, however, that substrate specificity and kinetic properties of the transferases of animal origin differ from the respective characteristics of human enzymes. For this reason, only the human *ABO*, *Hh*, and *Sese* gene products will be presented in the following section. Further characteristics of these transferases, especially those of the enzyme variants in **ABH** subgroups, are discussed in *Sect. 5.5*.

5.4.1 Products of the ABO Gene System

The product of the **A gene** is an α1,3-N-acetylgalactosaminyltransferase, which forms blood group **A** specific structures by transferring N-acetylgalactosamine residues onto **H** determinants:

$$
\begin{array}{l}
\text{Gal } \beta 1\rightarrow 3/4 \text{ GlcNAc } \beta 1\rightarrow 3 \text{ R} \\
\quad {\scriptstyle \alpha 1\rightarrow 2} \\
\text{Fuc}
\end{array}
\xrightarrow[\substack{\text{UDP-GalNAc}\quad\text{UDP}}]{\text{Mn}^{++}}
\begin{array}{l}
\text{GalNAc } \alpha 1\rightarrow 3 \text{ Gal } \beta 1\rightarrow 3/4 \text{ GlcNAc } \beta 1\rightarrow 3 \text{ R} \\
\quad\quad\quad\quad\quad {\scriptstyle \alpha 1\rightarrow 2} \\
\quad\quad\quad\quad\quad \text{Fuc}
\end{array}
$$

The product of the **B gene** is an α1,3-galactosyltransferase, which forms **B** specific structures by transferring α-galactose onto **H** determinants:

$$
\begin{array}{l}
\text{Gal } \beta 1\rightarrow 3/4 \text{ GlcNAc } \beta 1\rightarrow 3 \text{ R} \\
\quad {\scriptstyle \alpha 1\rightarrow 2} \\
\text{Fuc}
\end{array}
\xrightarrow[\substack{\text{UDP-Gal}\quad\text{UDP}}]{\text{Mn}^{++}}
\begin{array}{l}
\text{Gal } \alpha 1\rightarrow 3 \text{ Gal } \beta 1\rightarrow 3/4 \text{ GlcNAc } \beta 1\rightarrow 3 \text{ R} \\
\quad\quad\quad {\scriptstyle \alpha 1\rightarrow 2} \\
\quad\quad\quad \text{Fuc}
\end{array}
$$

The silent allele *O* carries no genetic information for a functional glycosyltransferase.

The qualitative and quantitative assay of the **A**- and **B**-transferases is based on their capacity to transfer radioactively labelled N-acetylgalactosamine or galactose from the respective sugar nucleotide (UDP-GalNAc or UDP-Gal) onto suitable acceptor substrates, such as 2'-fucosyllactose [393] or **H** active glycoproteins from gastric mucosa or ovarian cysts (e.g. [160,500]). Furthermore, the **H** substances of the erythrocyte membrane can be converted *in situ* into **A** or **B** substances by *in vitro* treatment of **O** cells with **A**- or **B**-transferase in the presence of sugar nucleotides [437,438]. The appearance of the new blood group specificity can then be determined by the direct haemagglutination test, and the quantity of determinants synthesised can be assayed semi-quantitatively by serum titration [439].

The **A** and **B** gene-dependent glycosyltransferases have been found in bone marrow [488], erythrocyte membranes [239,261], serum [239,345,346,426,427], submaxillary glands [188,385,393], gastric mucosa [438,501], gut mucosa [350], lung tissue [78], ovarian cysts (linings and fluids) [187], saliva [242,253], milk [246,247,375], and urine [254,492]. The occurrence of these enzymes is dependent on the genotype of the individual – the α1,3-N-acetylgalactosaminyltransferase is expressed only in persons with an *A* gene, while the α1,3-galactosyltransferase is expressed only in persons with a *B* gene; the enzymes are absent in blood group **O** subjects. The presence of both enzymes in the whole organism is independent of the **ABH**-secretor status.

A- and **B**-enzymes have been purified from human milk [245,375], serum [51,345,346,487,530], gut mucosa [350], and lung tissue [78]. The simplest and most efficient method for the isolation of serum **A**-transferase is based on its property to bind to Agarose-gel (Sepharose 4B®) – the adsorbed enzyme is readily eluted by UDP [530]. The **B**-transferase does not bind to Agarose-gel [346].

The transferases are of equal molecular weight – the value of 35–40 kDa determined by classical methods [78,350,487] has recently been verified by sequencing cDNAs encoding these enzymes (see *Sect. 5.4.5*). In serum the transferases are found as dimers [345,346,531], while in **AB** individuals **A**- and **B**-transferases may form hetero-dimers [347].

Both **A**- and **B**-transferases are highly specific towards the blood group **H** structure as the sugar acceptor – the enzymes react exclusively with those glycoconjugates and oligosaccharides carrying the terminal group **Fuc**α1$\rightarrow$2**Gal**β1$\rightarrow$X. According to published data they show approximately equal affinities towards type-1 and type-2 **H** structures. Type-1 and type-2 precursor chains lacking terminal α1$\rightarrow$2 fucose are not active as substrates. Of less significance here is the monosaccharide unit following the β-galactose residue. **Le**^{Y} or **Le**^{b} structures, however, in which the

N-acetylglucosamine residue is substituted with $\alpha1\rightarrow3$ and $\alpha1\rightarrow4$ fucose respectively (see *Chap. 6*), do not act as acceptors, probably due to steric hindrance or change in molecular configuration.

These enzymes are also highly specific for the sugar nucleotide. It has been noticed, however, that **B**-transferase when incubated at elevated pH-values is able to transfer **A** specific α-N-acetylgalactosamine onto **H** structures [160,556]; similarly, **A**-transferase transfers traceable quantities of **B** determinant α-galactose to low-molecular-weight acceptors [350,487,557]. It is a fact that A_1 erythrocytes contain low quantities of **B** antigen [154,512]; cells from donors with extremely high **B**-transferase activity (B_{sup}, see *Sect. 5.5.2*) react weakly but nevertheless distinctly with anti-**A** reagents [15]. These findings clearly show that such side reactions may also occur *in vivo*. Studies on enzyme kinetics of the **A**-transferase yielded similar K_m values for UDP-N-acetylgalactosamine and UDP-galactose, but showed a decreased v_{max} value for the galactose transfer [557]; the **B**-transferase showed similar transfer velocity for N-acetylgalactosamine and galactose, but a greatly increased (>25 x) K_m value for UDP-N-acetylgalactosamine [160] (see *Table 5.9*). Affinities of the transferases towards the alternative co-substrate are also confirmed by the finding that **A**-transferase can be inhibited by UDP-galactose [392] and **B**-transferase by UDP-N-acetylgalactosamine [51].

Table 5.9: A^1 and *B* **gene-specified transferases, capacity to transfer galactose and N-acetylgalactosamine, and the kinetic constants with UDP-Gal and UDP-GalNAc as donor substrates.**

		A_1-Transferase	B-Transferase
Transfer[a] of	^{14}C-GalNAc	12,944	2,585
	^{14}C-Gal	1,145	40,584
K_m (µM)	UDP-GalNAc	20.4	285
	UDP-Gal	19.2	11
v_{max} (pmol/h/ml)	UDP-GalNAc	870	195
	UDP-Gal	3.2	220

[a] Amount of labelled monosaccharide (cpm) incorporated into lacto-N-fucopentaose I; different incubation mixtures were used for A_1 and **B** transferases.
From the investigations of the group of Watkins [160,557].

It is obvious that these side reactions do not play a significant role in the biosynthesis of **A** and **B** antigens in normal tissue. Under the changed physiological conditions in transformed cells, however, the use of a different donor substrate may be favoured, so that **A**- and **B**-transferases would then produce detectable amounts of the alternative determinant [160]. These overlapping substrate specificities of **A**- and **B**-transferases may thus provide one explanation for the aberrant expression of **A** antigens repeatedly reported in cancer tissue from blood group **B** and **O** patients (see [167]).

Several groups have succeeded in producing polyclonal rabbit antibodies [80,489,490,569] and hybridoma antibodies [529] towards the active sites of **A**- and **B**-transferase. The antibodies specifically inhibited the activity of the respective enzyme, and most of the antibodies were cross-reactive and inactivated both transferases. The formation of antibodies towards **A**- and **B**-transferases has recently been observed in patients who have received **ABO** incompatible liver or bone marrow transplants [12,13,255,317,330,414].

In 1979 Yoshida et al. [569] produced an antibody towards **A**-transferase which neutralised the activity of **A**- and **B**-transferases. They also reported that the same antibody combined with material from the plasma of blood group **O** individuals. This 'O-cross-reacting material' (= O-CRM), was believed to be the enzymatically inactive product of the *O* gene; it has also been found in the serum of *AO* and *BO* subjects [561].

However, other groups were not able to confirm these findings. Studies using a rabbit polyclonal antiserum [80] and hybridoma antibodies [529] did not support the presence of 'cross-reactive' material in **O** individuals. Nevertheless, recent molecular biological studies indicate that some *O* alleles may produce a protein immunologically related to **A**- or **B**-transferases (see *Sect. 5.4.5*).

Greenwell et al. [159] have detected low but significant **A**-transferase activity in pooled plasma and secretory tissues of blood group **O** individuals. This phenomenon may also be due to those *O* alleles which code for immunologically related but enzymatically slightly active forms of the **A/B** transferases (see *Sect. 5.4.4*).

5.4.2 Products of the Gene Systems *Hh* and *Sese*

The products of the genes *H* and *Se* are α1,2-fucosyltransferases which form **H** specific structures by attaching α-fucose residues to the C-2 position of terminal β-galactose:

$$\text{Gal } \beta 1\rightarrow 3/4 \text{ GlcNAc } \beta 1\rightarrow 3 \text{ R} \xrightarrow{\quad Mg^{++} \quad} \text{Gal } \beta 1\rightarrow 3/4 \text{ GlcNAc } \beta 1\rightarrow 3 \text{ R}$$

GDP-Fuc   Fuc

$\alpha 1\rightarrow 2$

Fuc

The alleles **h** and **se** either are amorphous or code for enzymatically inactive proteins (see also *Sect. 5.5.6*, 'H-deficient variants').

The activity of $\alpha 1,2$-fucosyltransferases is calculated by measuring the rate of transfer of radioactive fucose from GDP-fucose to substances with terminal, non-reducing β-D-galactose. Generally, oligosaccharides such as lactose, N-acetyl-lactosamine, lacto-N-biose I and lacto-N-tetraose have been used as acceptor substrates [65,558]; phenyl-β-galactoside proved a very specific substrate as it reacts only with $\alpha 1,2$-fucosyltransferases [66]. In some cases defucosylated blood group **H** substance [161] or neuraminidase-treated orosomucoid (α_1-acid glycoprotein) [342] were used as fucosyl acceptors.

Enzymes with $\alpha 1,2$-fucosyltransferase activity have been found in tracheal epithelium [63], bone marrow [376], erythrocyte stroma [339], thrombocytes [56,456], and lymphocytes [56,158], submaxillary glands [65], gastric mucosal linings [65], serum [342,433], milk [386,454, and saliva [Yazawa, 1980 #2682]. $\alpha 1,2$-fucosyl-transferase has been highly purified from human serum [271,425,559] and gastric mucosa [316][6].

In contrast to $\alpha 1,3$-fucosyltransferases, $\alpha 1,2$-fucosyltransferases are selectively inhibited by the sulfhydryl reagent N-ethylmaleinimide [67,316].

As discussed in *Sect. 5.1* the **secretor** gene system **Se/se** controls the ability to secrete water-soluble **ABH** substances. In this system the active allele, **Se**, causes the secretion of **ABH** active material; individuals with the inactive allele **se** in double dose do not secrete **ABH** active substances. The **secretor** gene also regulates the occurrence of **ABH** active glycolipids in plasma and of **ABH** active oligosaccharides in milk and urine. It has no influence, however, on the occurrence of **ABH** determinants in haematopoietic tissue.

First investigations of the distribution of $\alpha 1,2$-fucosyltransferase in the human organism demonstrated that the presence of the enzyme in serum [342,433], erythrocyte stroma [56], and bone marrow [376] is independent of secretor status (only

---

[6] An $\alpha 1,2$-fucosyltransferase which reacts preferentially with **H type-1** precursors has also been found in the slime mold *Dictyostelium discoideum* [499,527].

individuals with the rare **H** deficient variants lack the transferase, see *Sect. 5.5.6).* However, in mucus-secreting tissue (e.g. submaxillary glands [65]), saliva [558], and in milk [454], α1,2-fucosyltransferase is expressed only in **ABH**-secretors and is absent in nonsecretors.

These findings are in full accordance with the results of serological and genetical investigations, according to which the inability of **ABH**-nonsecretors to secrete **A** and **B** substances is based on the absence of **H** specific acceptor substances rather than on the lack of the *A /B* gene products.

In the classical model put forward by Watkins and Morgan [523,525] the α1,2-fucosyltransferase is encoded by one structural gene, *H*, the expression of which in secretory tissues is controlled by the gene system, *Se/se*. Later a second regulatory gene, *Z/z*, was postulated which, art least in theory, was able to control the expression of the *H* gene in haematopoietic tissue [468] (see *Sect. 5.5.6).*

Type 1

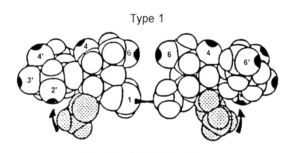

βDGal(1'→3)βDGlcNAc-CH₂

Type 2

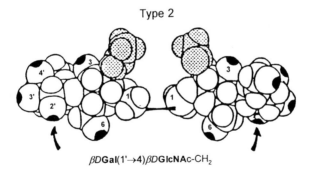

βDGal(1'→4)βDGlcNAc-CH₂

Figure 5.19: Accessibility of the terminal β-galactose in type-1 and type-2 precursor chains to the 2'-hydroxyl group (note arrows).

Reproduced from Le Pendu et al. [288] by permission of The University of Chicago Press, USA.

According to a more recent model advanced by Oriol [366], however, the *secretor* gene is not a regulatory but a structural gene, which codes for a second α1,2-fucosyltransferase different from the **H**-enzyme. This assumption has been based mainly on the stereochemical investigations on blood group **ABH** determinants reported by Lemieux and coworkers [290]. These studies disclosed distinct differences in the three-dimensional structures of type-1 and type-2 precursor chains − in the case of type-2 chains the hydroxyl group at carbon-2 of the terminal β-galactose unit is easily accessible, whereas in type-1 chains the access to the C-2 hydroxyl group is restricted by the acetyl group of the subterminal N-acetylglucosamine (see *Fig. 5.19)*. These basic stereochemical differences suggested the existence of two separate α1,2-fucosyltransferases synthesising type-1 and type-2 **H** substances. The theory has also been supported by genetical investigations on **Bombay** phenotype individuals (*Sect. 5.5.6)*.

On the basis of the immuno-histochemical data referred to in *Sect. 5.3.7*, it has been proposed that the enzyme encoded by the *secretor* gene is expressed in tissues of entodermal origin, whereas the transferase encoded by the *H* gene is restricted to ectodermally and mesodermally derived cells. Furthermore, it has been postulated that the **secretor**-transferase transfers α1$\rightarrow$2 fucosyl residues mainly onto type-1 chain substrates, while the **H**-transferase reacts preferentially with type-2 chain precursors. The model put forward by Oriol, moreover, suggested a close coupling of the genes *H* and *Se*.

First biochemical evidence for the existence of two different fucosyltransferases has been obtained by Le Pendu et al. [289]. The authors detected a small amount of α1,2-fucosyltransferase in the serum of **Parabombay** (**H**$_z$) individuals (see *Sect. 5.5.6)*; this transferase was not identical with the enzyme present in **H** nonsecretors [283]. Further results supporting the existence of two α1,2-fucosyltransferases were reported by Kumazaki et al. [270] who compared the α1,2-fucosyltransferases from serum and milk, as well as by Betteridge and Watkins [22] in their study on the enzymes from submaxillary glands of **ABH**-secretors and nonsecretors. More recently Sarnesto et al. [424,425] have been able to separate two different α1,2-fucosyltransferases from normal human serum.

In all cases the transferases showed significant differences in kinetic properties, in their affinity towards donor and acceptor substrates, as well as in their transfer activity, which was in accordance with the 'requirements' of Oriol's theory: the **secretor**-transferase showed a significantly lower affinity for GDP-fucose, phenyl-β-D-galactoside, and type-2 oligosaccharide acceptors than the **H**-transferase, but preferentially utilised type-1 and type-3 chain acceptors [283]. The **H**-enzyme, on the other hand, showed a distinct preference for type-2 chain acceptors (see *Table 5.10)*. Further, the enzymes also differed in several physicochemical properties (cf.[236]), the **secretor**-transferase being more sensitive to heat than the **H**-transferase [283].

Table 5.10: Apparent K$_m$ and relative v$_{max}$ values for α1,2-fucosyltransferases obtained from submaxillary glands of ABH-secretors and ABH-nonsecretors [*]

| Acceptor substrate used | α1,2-fucosyltransferases from | | | |
| | secretors | | nonsecretors | |
	K$_m$	v$_{max}$	K$_m$	v$_{max}$
Lacto-N-biose I (type-1)	1.0	100	3.7	120
Lacto-N-tetraose (type-1)	0.7	130	8.9	120
N-acetyllactosamine (type-2)	1.1	40	0.7	890
Lacto-N-neotetraose (type-2)	1.1	20	1.3	250
Phenyl β-galactoside	2.1	100	1.6	100

[*] Mean values of K$_m$ (mM) for three separate submaxillary gland preparations, v$_{max}$ relative to that obtained for phenyl β-D-galactoside.
From Betteridge et al. [22] by permission of Chapman and Hall, Ltd., London.

5.4.3 Cellular Origin of the Plasma Glycosyltransferases

Investigations on subjects with haematopoietic chimaerism, and patients with bone marrow grafts or pathological modifications of the **ABH** antigens showed that erythropoietic tissue contributes only 20–30% of the total **A**- and **B**-transferases in serum, whereas the majority of the transferases is of hitherto unknown origin [427,539]. In contrast, the **H**-transferase is derived mainly from haematopoietic tissues (platelets? [457]); the small amount of *secretor*-transferase (about 5–10% of the α1,2-fucosyltransferase activity) detected in plasma probably originates from entodermally derived tissue [12,289,425].

5.4.4 Expression of Glycosyltransferases in Malignantly Transformed Tissues

As discussed in *Sect. 5.3.9* malignant transformation is associated either with loss or re-expression of cell-bound **ABH** antigens.

Investigations on the distribution of the **A/B** gene-dependent glycosyltransferases by Mandel et al. [310] showed that the loss of **A** and **B** antigens in malignant bladder and oral epithelia was accompanied by loss of these enzymes, whereas in colon carcinomata no change in **A/B**-transferase expression was noticed.

Lower expression of serum α1,2-fucosyltransferase has been reported in patients with acute leukemia [269]. Here it has been shown, however, that the changes in serum enzyme level result from changes in platelet numbers and not from alterations in the cellular expression of the transferase in leukemia [457].

In those colon adenocarcinomas in which **ABH** antigens are accumulated or *de novo* expressed, a dramatic increase in the expression of **H**-transferase has been reported [372,479]. The authors noticed that expression of the enzyme is highly correlated with tumour progression.

Antigens have repeatedly been detected in tumour specimens of incompatible **A** (see [167-169]). Clausen et al. [69] reported that in 10–15 % of **O** and **B** tumours containing 'real' **A** antigens (mono- and difucosylated type-1 **A** glycolipids), incompatible **A**-transferase is expressed. The possible genetic background of this phenomenon will be discussed in *Sect. 5.4.5.*

5.4.5 Molecular Biological Investigations on the Glycosyltransferases Encoded by A, B, and O Genes

The coding region of the human *ABO* gene consists of seven exons spanning over 18 kb of the genomic DNA [17,544]. The exons range in size from 28 to 688 bp, with most of the coding sequence (~77%) located in exons 6 and 7.

Nucleotide sequence analysis of cDNA encoding the **A₁-transferase** revealed a coding region of 1062 base pairs. The derived protein is a polypeptide of 354 amino acids with a calculated molecular mass of 41 kDa [78,548]. Hydrophobicity plot analysis of the transferase showed a protein of type II membrane topology, i.e. a polypeptide with a short N-terminal segment, a transmembraneous hydrophobic region, a short proteolytically sensitive stem region and a large C-terminal domain in which the catalytically active region is located. The soluble form of the enzyme starts at Ala^{54} and thus does not contain a transmembrane domain [549]. The protein is glycosylated [78] containing one N-glycosylation site at position 113. The peptide sequence of the most frequently occurring *A¹* allele (*ABO*A101*) is presented in *(Fig. 5.20)*.

⇓

```
  1 MAEVLRTLAG KPKCHALRPM ILFLIMLVLV LFGYGVLSPR SLMPGSLERG FCMAVREPDH  60

 51 LQRVSLPRMV YPQPKVLTPC RKDVLVVTPW LAPIVWEGTF NIDILNEQFR LQNTTIGLTV 120

121 FAIKKYVAFL KLFLETAEKH FMVGHRVHYY VFTDQPAAVP RVTLGTGRQL SVLEVRAYKR 180

181 WQDVSMRRME MISDFCERRF LSEVDYLVCV DVDMEFRDHV GVEILTPLFG TLHPGFYGSS 240

241 REAFTYERRP QSQAYIPKDE GDFYYLGGFF GGSVQEVQRL TRACHQAMMV DQANGIEAVW 300

301 HDESHLNKYL LRHKPTKVLS PEYLWDQQLL GQPAVLRKLR FTAVPKNHQA VRNP
```

Figure 5.20: Amino acid sequence of A-transferase as derived from cDNA sequencing.
The hydrophobic segment representing the putative transmembrane domain is underlined; **N** : potential
N-glycosylation site, ⇓ : N-terminus of soluble form of the A-transferase.
After Yamamoto et al. [547]. The peptide sequence of the **A** transferase is deposited in the EMBL/ Gen Bank data
library (accession number J05175).

The allele encoding the **B-transferase** differs from the allele encoding the
A_1-transferase at seven nucleotides, viz. three silent mutations ($A^{297} \rightarrow G$, $C^{657} \rightarrow T$, and
$G^{930} \rightarrow A$), and four structural mutations ($C^{526} \rightarrow G$, $G^{703} \rightarrow A$, $C^{796} \rightarrow A$, and $G^{803} \rightarrow C$)
which cause amino acid changes between **A**- and **B**-transferases, i.e. Arg$^{176} \rightarrow$ Gly,
Gly$^{235} \rightarrow$ Ser, Leu$^{266} \rightarrow$ Met, and Gly$^{268} \rightarrow$ Ala, respectively [547] *(Fig. 5.21)*.

In subsequent papers Yamamoto et al. [541,542] studied the influence of the four
amino acid substitutions on the sugar-nucleotide donor specificities of **A**- and
B-transferases (see *Table 5.11*)[7]. According to gene reconstruction experiments and
studies of enzyme expression in DNA transfected HeLa cells, the third and the fourth
amino acid substitutions (Leu/Met at position 266 and Gly/Ala at position 268) are
crucial in determining sugar-nucleotide specificity: transferase constructs in which the
last two positions were *AA* or *BB* expressed only **A**- or **B**-transferase, respectively,
while *BA* constructs expressed both **A**- and **B**-transferase. The second amino acid

---

[7] Extensive investigations on the kinetic parameters of *AB* constructs have been published by Seto
et al. [449,450].

Table 5.11: Expression of blood group A and B antigens on HeLa cells transfected with DNA of A/B transferase cDNA-chimera constructs.

Plasmid DNA[a]	Amino acid sequence				Specificity of transfected cells
	176	235	266	268	
pAAAA	Arg	Gly	Leu	Gly	A
pAAAB	Arg	Gly	Leu	Ala	A
pAABA	Arg	Gly	Met	Gly	AB
pAABB	Arg	Gly	Met	Ala	B
pABAA	Arg	Ser	Leu	Gly	A
pABAB	Arg	Ser	Leu	Ala	A(B)
pABBA	Arg	Ser	Met	Gly	AB
pABBB	Arg	Ser	Met	Ala	B
pBAAA	Gly	Gly	Leu	Gly	A
pBAAB	Gly	Gly	Leu	Ala	A
pBABA	Gly	Gly	Met	Gly	AB
pBABB	Gly	Gly	Met	Ala	B
pBBAA	Gly	Ser	Leu	Gly	A
pBBAB	Gly	Ser	Leu	Ala	A(B)
pBBBA	Gly	Ser	Met	Gly	AB
pBBBB	Gly	Ser	Met	Ala	B

[a] Name of the construct indicating the A or B specific codon located in the four variable positions. According to Yamamoto et al. [541].

substitution (Gly/Ser at position 235) was found much less influential: in the case of *AB* constructs the amino acid at the second position influenced the specificity of the expressed transferase − when it was *A*, only **A** activity was observed, whereas when it was *B*, both **A** and weak **B** activity were present. The first substitution (Arg/Gly at position 176) may be pivotal for kinetic activity [449,450].

Computer analyses have shown that the last two substitutions in the **B**-transferase decrease the flexibility of the protein in this region, which, considering the size difference between galactose and N-acetylgalactosamine, may be the crucial factor in determining sugar specificity of the enzyme [541].

103

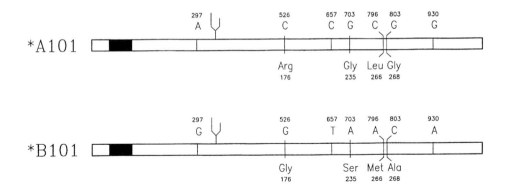

Figure 5.21: Amino acid sequences of A_1 and B alleles.
■: transmembrane domain, ⊃-: N-glycan.
Provisional nomenclature of the genes according to Ogasawara et al. [355].
According to Yamamoto et al. [547].

More recently, three further A_1 alleles have been detected *(Fig. 5.21)*:

- one allele very frequent in Mongolids (~ 83% of the A_1 individuals) had a $C^{467} \rightarrow T$ substitution causing a proline → leucine exchange at position 156 (provisional notation **ABO*A102)** [129,355]. Enzyme tests showed that this variant shows an activity reduced by about 10% compared to the most frequent A^1 allele *(ABO*A101)* [356]. This mutation has also been found in some other **A** variants (see e.g. A_2 and *cis-AB* in *Sects. 5.5.1* and *5.5.4*, respectively).
- one had the same sequence as the **B** gene at nucleotide 526 (C → G, $Arg^{176} \rightarrow Gly$) but was identical to the **A** gene at positions 703 and 796 [477];
- a hybrid allele comprising exon 6 of the **B** allele *(ABO*B101)* and exon 7 of the O^1 allele *(ABO*O101)* [480]. Since the O^1 allele has the consensus sequence of the A^1 allele (see below), the hybrid gene encodes a transferase with A_1 specificity.

An entire series of different **O alleles** have thus far been detected. In the cases investigated the cDNAs were found almost identical to A^1 allelic cDNA, but contained point mutations which result in expression of enzymatically inactive proteins *(Fig. 5.22)*:

(A) The most common mutant, provisionally termed O^1, is identified by a critical single-base (= G) deletion at nucleotide 261 in exon 6, that is to say in the coding

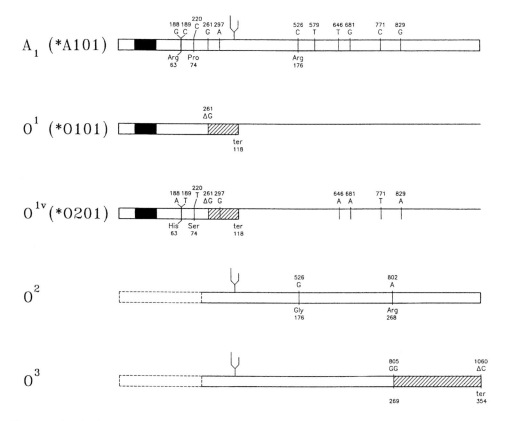

Figure 5.22: Amino acid sequences of the most frequent *O* alleles.
⬚ = sequence not determined, ■ = transmembrane domain,
▨ = additional domain at the C-terminus due to frameshift,
⊃– = N-glycan.
Provisional nomenclature as proposed by Olsson et al. [358] and Ogasawara et al. [355] (in brackets).
According to Yamamoto et al. [545,547] and Olsson et al. [358,359]

region near the N-terminus of the protein. This deletion causes a shift in the reading frame and creates a premature stop codon at nucleotides 352 – 354. When transcribed, this allele encoded an entirely different 117 amino acid protein without enzymatic activity [547].

Subsequent investigations revealed variations of this mutant which, in comparison with the *A¹* allele are characterised by further point mutations both in the translated region as well as in the untranslated region after the premature stop codon [355,359,547] (refer to *O¹ᵛ*; here there are nine additional mutations compared to the *O¹* sequence [359]).

(B) A less common *O* allele[8], provisionally termed O^2, differs from the A^1 consensus allele in three point mutations – a silent mutation (A → G) at nucleotide 297 and two structural mutations at nucleotide 526 (C → G) and 802 (G → A); the latter mutations cause an arginine → glycine substitution at codon 176 and a glycine → arginine substitution at codon 268, respectively [164,545]. The base exchange at nucleotide 297 and the glycine residue at position 176 are identical to that found in the B-transferase, whereas it was shown that the Gly^{268} → Arg substitution eliminated the enzymatic activity of the protein [542,545].

(C) An *O* allele (provisionally termed O^3) found in a family of Swedish origin [358] shows the C → T substitution at nucleotide 467 and the cytosine deletion at position 1060, both of which are characteristic for the A_2 allele. Here too the guanosine insertion at position 798 – 804 which is characteristic for the A_{el} allele can be recognised.

(D) More recently a series of *O* alleles had been found which may be formed by crossing-over or (possibly) by gene conversion events between known alleles in the **ABO** system [143,361,480], e.g. O^{1v}-**B**, **B**-O^{1v}, O^1-A^2, O^1-O^{1v}, and O^1-O^{1v}-O^1. It can be assumed that the frequency of such hybrid alleles is actually much higher than initially thought since many recombinants are identical to known alleles. The junctions of all these hybrid alleles are located within intron-6, the intron-7 / exon-7 boundaries, or exon-7.

It is almost certain that some *O* alleles are transcribed, and the encoded proteins will most likely cross-react with antibodies towards **A**- or **B**-transferases. The detection of 'O-cross-reacting material' (O-CRM, see *Sect. 5.4.1*) hitherto unexplained and unconfirmed,may well be based on the existence of such *O* alleles.

The investigations on the peptide sequences of **A**- and **B**-transferase variants are discussed below (A_2, A_3, and A_x in *Sect. 5.5.1*, B_3 in *Sect. 5.5.2*, and **cis-AB** in *Sect. 5.5.4)*.

Incompatible A antigens in cancer

Incompatible **A** antigens have been repeatedly observed in human tumours of blood group **O** and **B** individuals [166] (see also *Sect. 5.4.4)*. More recent

---

[8] This *O* allele constitutes approximately 4% of *O* alleles in Europids and Negrids [118,164]; it has not been found in individuals of Mongolid ancestry [118,129,355].

investigations using hybridoma antibodies towards blood group determinants or **A**-transferase, as well as enzymatic assays, haveclearly identified these structures as true **A** determinants [69,85,186,326].

These newly generated **A** antigens may also be responsible for the approximately 10% lower incidence of gastric and ovarial adenocarcinoma in blood group **O** and **B** individuals than in type **A** individuals (cf. [69]): the increased risk may be due to the lack of anti-**A** as a means to destroy cancer tissue developing incompatible **A**.

Thus far the *ABO* gene found present in incompatible **A** cancer cells has not yet been the focus of major studies. Nevertheless, several mechanisms can be postulated which might restore the reading frame of the *O* and *B* alleles. Since the *B* allele and most of the *O* alleles are highly homologous to the *A¹* allele, active **A**-transferases could be produced by fairly simple processes:

(a) Additional point mutations (i.e. nucleotide exchange, nucleotide deletions or insertions):

The *O¹* allele is characterised by a single nucleotide deletion in exon 6 near the N-terminal of the coding sequence. A base insertion in close proximity of the deletion would restore the codon frame, in some cases rendering an active **A**-transferase.

The *O²* allele contains an amino acid substitution at position 268 inactivating the resulting enzyme. A single missense mutation (A → G at nucl 802) would change the amino acid residue at position 268 from Arg to Gly which is characteristic for **A**-transferase.

Likewise, missense mutation(s) of the *B* allele at amino acids 266 and/or 268 to those of the *A* allele may lead to incompatible **A** expression.

(b) Alternative splicing:

Unusual cDNA's which reflects different splicing patterns have been isolated from various organs [17] as well as from several tumour cell lines [544] – some contain intron sequences probably caused by premature splicing, and in other cases exons are eliminated. The results indicate the presence of cryptic promotor(s) and splicing donor and acceptor site(s) which may be used under certain circumstances. It is not yet known, however, to what extent these alternative transcripts are translated and if the encoded proteins are enzymatically active.

For example, one possibilty to correct the reading frame of the *O¹* allele would be to eliminate exon 6, which contains the premature stop codon. However, a transcript lacking only exon 6 will be inactive because a new stop codon is formed when joining coding exons 5 and 7. Nevertheless, alternative splicing leading to a transcript without exons 5 and 6 would encode a truncated but possibly enzymatically active transferase [544].

(c) Translational frame shifting involves the synthesis of a single protein from two separate reading frames on a RNA template [82].

(d) Unequal crossing-over or gene conversion events may also produce active **A**-transferases. Examples for hybrid **ABO** alleles have been presented above (see in the section **O** alleles above).

Nevertheless, the most likely candidate for the expression of incompatible **A** in cancer tissue is the O^2 allele. Since it is higly homologous to the A^1 allele, the encoded protein may be able to exhibit weak **A**-transferase activity, which may be enhanced by the changed metabolic conditions in cancer cells, and/or by co-factors present in tumour cells, but absent in normal cells, or by **allelic enhancement** (see *Sect, 5.5*.3).

ABO Pseudogenes:

The **B**-like structure **Galα1→3Galβl→4GlcNAc–** is widely expressed in mammalian species but absent in Catarrhines (Old World monkeys and apes) and humans [138]. Correspondingly, 'naturally occurring' antibodies towards this epitope are regularly found in human sera, constituting as much as 1% of the circulating IgG molecules [134,136].

cDNA containing the coding sequence for an α1,3-galactosyltransferase has been recently isolated from mouse [224,278], cattle [223], and pig [476] sources, and from a New World monkey [191]; the predicted protein sequences share up to 80% identity.

Southern blot analysis on human genomic DNA revealed DNA sequences homologous to that of the animal α1,3-galactosyltransferase [222,223,279,543]. One of these genomic DNA fragments ('*HGT-10*') has been assigned to chromosome 9q33–q34, which means it is closely linked to the human **ABO** blood group gene, while fragment '*HGT-2*' maps to chromosome 12q14–q15 [452].

The sequencing of the putative coding regions proved that these genes are non-functional pseudogenes containing frameshift mutations and nonsense codons precluding the synthesis of functional enzymes [191,222,279,543]. A significant similarity (up to 50%) in nucleotide sequence to the human blood group **ABO** gene suggests that these pseudogenes and the **ABO** gene have derived from a common ancestral gene through gene duplication followed by divergence (see also [139]).

The following pivotal functions of human antibodies towards the **Galα1→3Gal** determinant have already been discussed:

- Exposure of cryptic **Galα1→3Gal** (or sterically similar) determinants has been connected with the identification and subsequent removal of senescent human red blood cells and pathologic erythrocytes by anti-**Gal** antibodies [133,135];
- this determinant may also play a role in various autoimmune disorders, such as autoimmune thyroiditis or rheumatoid arthritis [106];
- **Galα1→3Gal** expression by animal organs is assumed to be a major barrier for xenotransplantation to human recipients: This is due to the fact that natural anti-(**Galα1→3Gal**) antibodies cause hyperacute rejection of the grafted tissue [371,423];
- mammalian C-type C retroviruses are inactivated by these antibodies [486]. The antibody may thus inhibit transmission of oncoviruses from other mammals to humans;
- the antibodies may also confer immune protection by complement lysis of certain bacteria [137,172];
- last but not least, reference must be made to a study which showed that human antibodies towards the **Galα1→3Gal** determinant inhibit the *in vitro* growth of the malaria parasite *Plasmodium falciparum*, probably by binding to α-galactosyl epitopes on merozoite surface molecules [399]. The presence of this antibody may thus provide partial protection against malaria infection. Elevated titers of anti-Gal have in fact been found in persons living in highly malaria-endemic regions and in patients with acute *P. falciparum* malaria [401].

Evolution of the *ABO* gene

As discussed above, **ABH** antigens are not restricted to humans; they are present universally in various organisms such as bacteria, plants, and animals (see *Sect. 5.3.6)*. When a human **A**-transferase cDNA probe was used to search for homologous sequence(s) in other organisms, distinct hybridization was observed only with DNA from the mammalian species already tested; all other organisms showed no such hybridisation [256]. Nevertheless, considering the universal presence of **ABH** substance in nature it must be assumed that animals outside the mammal phylum have analogous genes with only minor homology.

In primates the nucleotide sequences of the genes homologous to the human *ABO* gene are highly conserved [256,315]. In particular, in the deduced peptide sequences the amino acid residues corresponding to the codons 266 and 268 of the human **A**- and **B**- transferases conformed with the **ABO** blood group in all cases, that is to say, Leu/Gly as characteristic for **A** and Met/Ala for **B** transferase. The *O* alleles, however, are species-specific and result from independent silencing mutations [238].

As mentioned above, **ABH** substances are present both on erythrocytes and in secretions only in humans and anthropoid apes (chimpanzee, gorilla, orangutan, and gibbon). In all other mammalian species these substances are found only in secretions.

Blood Grouping by Molecular Biological Methods

Various PCR techniques have been developed to identify **ABO** phenotypes and genotypes (refer to [129,143,220,356,357,477,503,516]. These techniques may complement conventional serological methods, as they are extremely useful in linkage analysis, paternity testing, and individualisation in forensic work. However, the results of these analyses are reliable only when nucleotide substitutions are clearly related to specific phenotypes. Unexpected variations in the genes (e.g. rare blood group variants, mutations specific for ethnic groups) might give false results. Thus, for accurate determination of **ABO** phenotypes the analysis of several regions of the genes may be necessary.

5.4.6 Molecular Biological Investigations on the Glycosyltransferases Encoded by the Hh and Secretor Genes

H-Transferase

A gene-transfer scheme was used to isolate a human genomic DNA segment which determines the expression of an α1,2-fucosyltransferase [105,398].

Based on the open reading frame detected in the cDNA corresponding to this genome sequence, a protein of 365 amino acids with a calculated molecular mass of 41,249 Da has been predicted [277]. The primary structure of the derived polypeptide is shown in *Fig. 5.23*. Hydropathy analysis also indicated in this case a chain topology of a type II transmembrane glycoprotein characteristic for membrane-bound glycosyl-transferases, i.e. an N-terminal cytoplasmic (8 residues), a hydrophobic trans-membrane domain (17 residues) flanked by basic amino acids, and a C-terminal domain (340 amino acids) with two potential N-glycosylation sites (residues 65 and 327).

When expressed in α1,2-fucosyltransferase-deficient COS-1 cells, this cDNA directed the synthesis of cell surface **H**-determinants and of an α1,2-fucosyltransferase with properties similar to those found for the human **H**-transferase. The transfection studies further showed that the transferase is membrane-bound and that only trace amounts of the enzyme are released from the COS-1 cells.

```
  1  MWLRSHRQLC LAFLLVCVLS VIFFLHIHQD SFPHGLGLSI LCPDRRLVTP PVAIFCLPGT    60

 61  AMGPNASSSC PQHPASLSGT WTVYPNGRFG NQMGQYATLL ALAQLNGRRA FILPAMHAAL   120

121  APVFRITLPV LAPEVDSRTP WRELQLHDWM SEEYADLRDP FLKLSGFPCS WTFFHHLREQ   180

181  IRREFTLHDH LREEAQSVLG QLRLGRTGDR PRTFVGVHVR RGDYLQVMPQ RWKGVVGDSA   240

241  YLRQAMDWFR ARHEAPVFVV TSNGMEWCKE NIDTSQGDVT FAGDGQEATP WKDFALLTQC   300

301  NHTIMTIGTF GFWAAYLAGG DTVYLANFTL PDSEFLKIFK PEAAFLPEWV GINADLSPLW   360

361  TLAKP
```

Figure 5.23: Polypeptide sequence of the *H* (= *FUT1)* gene encoded α1,2-fucosyl-transferase as derived from cDNA sequencing.
N : potential N-glycosylation sites, the putative transmembrane domain is underlined..
According to Larsen et al. [277].
The peptide sequence of the transferase is deposited in the EMBL/Gen Bank data library (accession number M35531).

FUT1-specific mRNA has been found in bone marrow cells and erythroleukaemic cells [249], but the gene product was virtually absent from normal digestive mucosa [479]. However, a dramatic increase of the *FUT1* gene product has been detected in adenocarcinomas of the colon and several gastric, colonic, and ovarian cancer cell lines [249,479], reflecting the *de novo* synthesis and accumulation of type 1 and type 2 fucosylated lactoseries structures in carcinomas (see also *Sects. 5.3.9* and *5.4.4)*.

The *H* (= *FUT1*) gene consists of four exons; the protein coding region is located in exon 4 [249]. Further, two distinct transcription-initiation sites have been detected which, by alternative splicing, generate several forms of the *FUT1* transcript [249]. The authors assume that the different forms of the *FUT1* mRNA may be associated with time- and tissue-specific expression of the gene.

Investigations on the *H* gene variants of H-deficient mutants are discussed in *Sect. 5.5.6*.

Secretor-Transferase

Based on the findings that the **H**- and **Secretor**-transferases exhibit similar catalytic properties, it seemed probable that these two enzymes maintain sufficient sequence homologies to allow isolation of the secretor locus by a cross-hybridisation approach.

When chromosome 19 cosmid libraries were screened using the human H-transferase (*FUTI*) cDNA, two distinct sequences were detected which cross-hybridize with *FUTI* [412]. The sequences which have been provisionally termed *Secl* and *Sec2* (for 'Secretor candidate 1 and 2') are in close vicinity to the *H* locus − they are separated by 12 kb, and are 65.5 kb and 35 kb apart, respectively, from the *FUTI* gene. Both DNA segments share substantial DNA sequence similarity to the human H-transferase gene [236]:

The cDNA sequence of *Secl* showed an open reading frame which is, however, disrupted by translational frameshift and nonsense mutations. The finding that expression of this segment in α1,2-fucosyltransferase deficient COS-7 cells generated no detectable α1,2-fucosyltransferase activity, as well as the fact that no transcripts corresponding to the *Secl* gene could be detected in human tissues suggest that the *Secl* segment represents a pseudogene. The same authors discuss the possibility that a *Secl* allele 'corrected' by mutation or RNA editing, may explain the observation that a cultured human carcinoma cell line expresses an αl,2-fucosyltransferase catalytically distinct from the **H**- and **Se**-transferases [34]. Chandrasekaran et al. [61], however, suggest that this α1,2-fucosyltransferase activity is associated with the **Lewis** gene-associated α1,3/4-fucosyltransferase (see *Sect. 6.4.2*).

DNA sequencing of segment *Sec2* yielded an open reading frame containing two closely-spaced, in-frame methionine codons, both of which represent appropriate candidates for an initiation codon. Therefore, alternative splicing events might lead to the synthesis of two different polypeptides differing by 11 amino acids in the N-terminus but having identical catalytic domains. The two putative isoforms contain 332 and 343 amino acids respectively, and share 68% sequence identity with the C-terminal 292 residues of the human **H**-transferase. The 14-residue hydrophobic segment close to the N-terminus is characteristic for a type-2 transmembrane topology typical for mammalian glycosyltransferases. The predicted polypeptides contain three potential N-glycosylation sites at positions 177, 271, and 297 *(Fig. 5.24)*.

Extracts of α1,2-fucosyltransferase-deficient COS-7 cells transfected with *Sec2* DNA segments showed considerable α1,2-fucosyltransferase activity. The catalytic properties (i.e. K_m-values for acceptor substrates and GDP-fucose, and pH-activity profile) of the *Sec2*-encoded α1,2-fucosyltransferase were virtually identical to those reported for the α1,2-fucosyltransferase encoded by the human **Secretor** locus [236].

The high α1,2-fucosyltransferase activity found in the growth medium showed that the **Secretor**-transferase is readily released from the cells; the **H**-transferase, in contrast, remains largely cell-associated when expressed in COS-7 cells (see above) [277].

```
                                              M  LVVQMPFSFP

      ▾
  1 MAHFILFVFT VSTIFHVQQR LAKIQAMWEL PVQIPVLAST SKALGPSQLR GMWTINAIGR  60

 61 LGNQMGEYAT LYALAKMNGR PAFIPAQMHS TLAPIFRITL PVLHSATASR IPWQNYHLND 120

121 WMEEEYRHIP GEYVRFTGYP CSWTFYHHLR QEILQEFTLH DHVREEAQKF LRGLQVNGSR 180

181 PGTFVGVHVR RGDYVHVMPK VWKGVVADRR YLQQALDWFR ARYSSLIFVV TSNGMAWCRE 240

241 NIDTSHGDVV FAGDGIEGSP AKDFALLTQC NHTIMTIGTF GIWAAYLTGG DTIYLANYTL 300

301 PDSPFLKIFK PEAAFLPEWT GIAADLSPLL KH
```

Figure 5.24: Polypeptide sequence of the *secretor* (= *FUT2*) gene encoded α1,2-fucosyltransferase as derived from cDNA sequencing.
The amino acids are numbered from the methionine residue initiating the short isoform of the protein (▾). **N** : potential N-glycosylation sites; the hydrophobic segment representing the putative transmembrane domain is underlined.
According to Kelly et al. [236]. The peptide sequence of the transferase is deposited in the EMBL/Gen Bank data library (accession number U17894).

The *Se* (= *FUT2*) gene consists of 2 exons – exon 1 is located about 7 kb upstream of exon 2 which contains the total protein coding region [251].

Northern blot analyses demonstrated mRNA transcripts in small intestine, colon, and lung, but not in liver or kidney [412] and further in ovarian, gastric, and colonic cancer cell lines, but not in erythroleukaemic cell lines [251]. These results are consistent with previous observations that the **Secretor**-transferase is found only in human tissues of entodermal origin (see *Sect. 5.3.7*).

Approximately 20% of randomly-selected individuals presumably of Europid ancestry were found homozygous for a $G^{428} \rightarrow A$ nonsense mutation which changes the Trp^{143} codon[9] to a stop codon, thus truncating the enzymatically active domain on the C-terminal segment of the fucosyltransferase[10] [236] (provisional notation *'se1'*) *(Fig. 5.25)*.

---

[9] Numbered from the methionine residue initiating the short isoform of the protein.

[10] A second nucleotide exchange found in this allele ($G^{740} \rightarrow A$) would lead to a $Gly^{247} \rightarrow Ser$ exchange in the nontranslated section after the premature stop codon (i.e. at position 247). This mutation was shown to be functionally neutral when inserted into a 'wild type' cDNA [236].

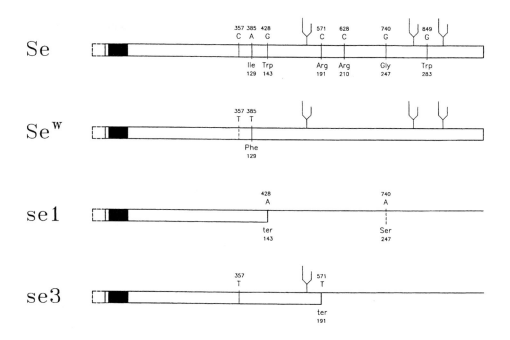

Figure 5.25: Amino acid sequences of the most frequent alleles of the *Secretor* gene.
■ = transmembrane domain, ⟳ = N-glycan.
Provisional nomenclature as proposed by Koda et al. [250].
According to Henry et al. [193,194] and Kelly et al. [236].

When cDNA of this mutant was tested by transfection for its ability to encode a functional enzyme, no α-2-fucosyltransferase activity was detected in fucosyltransferase-deficient COS-7 cells. The investigations further showed that each of six unrelated non-secretor individuals tested were homozygous for this null allele, whereas secretor-positive persons maintained at least one functional wild-type allele. The fact that the frequency of this null allele closely corresponded to the frequency of the non-secretor phenotype in most human populations provided further evidence for the **Sec2** gene to encode the **Secretor**-transferase.

Population studies on the **Secretor** gene revealed that the $G^{428} \rightarrow A$ nonsense mutation represents the common null allele in Europids and probably also in Negrids, but is absent in individuals of Mongolid ancestry. Another null allele has been identified in the **FUT2** gene of Mongolid non-secretors. It is characterised by a $C^{571} \rightarrow T$ mutation which changes the arginine codon at position 191 to a stop codon [194,250,570]. This allele (provisional notation '*se3*') seems to be the common null allele in most Mongolid populations *(Fig. 5.25)*.

Further investigations have detected a series of defect mutations which are, however, confined to smaller ethnic groups:

- in the Taiwanese indigenous Paiwan population a $G^{849} \rightarrow A$ nonsense mutation has been found which changes the tryptophane codon at position 283 to a stop codon [570];
- in the Ami tribe of Taiwan an allele with a three-nucleotide deletion (GTGGT to GT) in the region of nucleotides 685-689 has been detected which leads to a loss of Val^{230} [571]:
- in Japanese non-secretors a *se* allele has been described characterised by a $C \rightarrow T$ exchange at nucleotide 628 thus altering the arginine codon at position 210 to a stop codon [250] (provisional notation '*se4*'):
- in a *se* allele found in the Xhosa population of Africa a single base (C) deletion at nucleotide 778 leads to a frame shift producing a stop codon at codon 275 [296];
- another *se* allele with a gene frequency of 0.057 in Japanese is a fusion gene consisting of the 5'-region of the *Sec1* pseudogene and the 3'-region of the functional *Se* gene (provisional notation '*se5*') [250]. The hybrid allele was obviously generated by homologous but unequal crossover between bases 253 and 313 of the pseudogene and between bases 211 and 271 of the *Se* gene[11]. Transfection experiments showed that this allele encodes a weak $\alpha1,2$-fucosyltransferase exhibiting ~20% of the activity of the *Se* gene product; the transfected cells, however, did not express H antigen on their cell membranes [250]. The fusion gene must therefore be considered a non-secretor allele;
- in H-deficient nonsecretor individuals (see **Bombay** phenotype, *Sect. 5.5.6)* of Indian origin complete deletion of exon 2 of the *FUT2* gene has been detected. This allele has probably been formed by an unequal crossing-over event in which the 5' break point of the gene deletion was located within the 7 kb intron of *FUT2* and the 3' break point within the 35 kb sequence between *FUT2* and *FUT1* [248].

Further various silent mutations in the *Secretor* gene which have no effect on the protein sequence have been described – a $C^{357} \rightarrow T$ substitution frequently found in all ethnic groups, $C^{480} \rightarrow T$ and $A^{375} \rightarrow G$ substitutions in African Xhosas [296], and two mutations within the 3' untranslated region ($G^{1009} \rightarrow A$ and $C^{1011} \rightarrow T$) detected in a high frequency in Mongolid donors [250,268,570,572].

---

[11] The nucleotide sequence of the fusion gene has been deposited in the GenBank database (accession number D82933).

A 'weak' *Secretor* allele (*Se^w*) has been postulated to account for the **Le(a+b+)** blood group phenotype which is extremely rare (or probably absent) in Europids but is observed in relatively high frequency (up to 25%) in individuals of Mongolid ancestry [195,493] (see *Chap. 6*). A weak **Secretor**-transferase would leave unsubstituted a larger proportion of the precursor chains. These could then be transformed into **Le^a** by the **Lewis**-transferase. Further evidence for the assumption that the **Se^w** phenotype is caused by an 'inefficient' **Secretor**-transferase has been deduced from the finding that **Le(a+b+)** is associated with poor expression of salivary **ABH** substances ('partial secretor phenotype' [196]).

When the secretor gene of Indonesian, Taiwanese, and Japanese **Le(a+b+)** donors was investigated, an $A^{385} \rightarrow T$ point mutation was found in the coding sequence which results in an Ile to Phe substitution at position 129 [192,193,229,250,268,572] (*Fig. 5.25*). This *Se^w* allele was present in double dose in **Le(a+b+)** individuals, but not in **Le(a-b+)** individuals. An *Se^w* allele frequency of 0.436 among Japanese donors has been determined [250].

Transfection experiments performed using fucosyltransferase-deficient COS cells showed that the $Ile^{129} \rightarrow Phe$ substitution does not cause complete inactivation of the enzyme – a reduced amount of **H** type 1 and 2 epitopes had been detected at the surface of transfected cells [193,250], and the cell lysates retained about < 5% of the transferase activity found in the lysates of wild-type transfected cells [268]. Further investigations showed that the acceptor substrate pattern and the K_m values are very similar to those of the wild-type transferase, whereas the v_{max} value of the *Se^w* mutant decreases to about one fifth of the wild-type mutant [193].

Three mutations, viz. $A^{40} \rightarrow G$ ($Ile^{14} \rightarrow Val$), $C^{379} \rightarrow T$ ($Arg^{127} \rightarrow Cys$), $G^{481} \rightarrow A$ ($Asp^{161} \rightarrow Asn$), have been found in Xhosas; they reduce transferase activity by 20 – 50% compared to the control [296]. The probands, however, type as normal **Se**-positive individuals.

These studies show that the *Se* gene exhibits a high degree of heterogeneity, and distribution and frequency of the different *Se* alleles vary markedly among different ethnic populations. Thus, this high polymorphism of the *FUT2* gene may become a useful tool in understanding the evolution of human populations.

Evolution

It is interesting to note that homologous $\alpha1,2$-fucosyltransferases have also been detected in mammals; this may be of some significance at evolutionary studies.

With regards to rats: Piau et al. [382] isolated two distinct cDNA clones from a rat colon genomic cDNA library. The clones hybridised with a human H-transferase cDNA probe. Both sequences showed considerable sequence similarity to the human H-transferase cDNA.

With regards to rabbits: Hitoshi et al. [201,202] isolated three DNA clones encoding α1,2-fucosyltransferases (*RFT-I*, *RFT-II*, and *RFT-III*) from a rabbit genomic cDNA library. The deduced amino acid sequences of *RFT-I* (373 residues), *RFT-II* (354 residues), and *RFT-III* (347 residues) showed high (60–80%) similarity with the primary structures of human α1,2-fucosyltransferases. DNA sequence analyses suggested a homology of *RFT-I* to the human *H* gene, *RFT-II* to the *Sec1* pseudogene, and *RFT-III* to the human *Secretor* gene. Transfection experiments using COS-7 cells confirmed that *RFT-I* encoded a transferase with an activity closely resmbling that of human H-transferase, whereas *RFT-II* and *RFT-III* encoded enzymes resembling human **Secretor**-transferase. The distribution of the enzymes was also similar to that of the homologous human enzymes – the *RFT-I* gene was expressed in tissues of mesodermal origin (in contrast to the human H-transferase, however, it was limited to the brain of adult animals). The *RFT-II* and *RFT-III* genes were expressed in tissues of entodermal origin (salivary and lactating mammary glands, or gastrointestinal tract, respectively).

With regards to pig and mouse: more recently, genes highly homologous to human *FUT1*, *FUT2*, and the pseudogene (all located in a gene cluster) have been detected in a porcine [324,495] and a mouse genomic library [92].

In lower vertebrates (i.e. reptilia and amphibia) **ABH** antigens are found exclusiveley in entodermal tissue and only in mammals are they found in the endothelium as well. It can thus be assumed that the *H* gene is derived from the *Se* gene by gene duplication at the appearance of mammals in the course of evolutionary development [282].

5.4.7 Biosynthesis of Type-14 Substance

In the course of studies on the biosynthesis of blood group substances performed at the Institute of Biochemistry at the University of Vienna it has been observed that the lactose synthase of human milk is able to transfer galactosyl residues onto water-soluble blood group substances [429]. Subsequently it could be shown that this β1,4-galactosyltransferase is involved in the biosynthesis of the **type-14 structure** (see *Sect. 5.3.1)*: blood group **H** substance which had been defucosylated and β-galactosidase-treated showed no reaction with anti-**type-14** Pneumococcal sera;

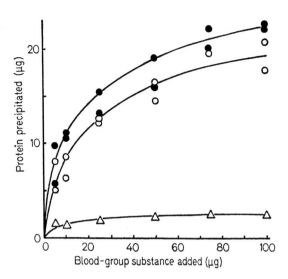

Figure 5.26: In vitro synthesis of 'type 14 substance' by β1,4-galactosyltransferase isolated from human milk.
○ : 'type-14 substance'; Δ : 'type-14 substance' treated with β-galactosidase (= 'type-14 precursor'); ● : 'type-14 precursor' treated with β1,4-galactosyltransferase.
Quantitative immunoprecipitation using horse anti-type-14 Pneumococcal serum ('horse 615').
Reproduced from Schenkel-Brunner et al. [439].

after incubation with lactose synthase in the presence of UDP-galactose, it became possible to restore the **type-14** activity[12] (see *Fig. 5.26*).

It has been discussed in connection with the biosynthesis scheme for blood group **ABH** substances that the oligosaccharide core is assembled by the concerted action of a series of common genes. The investigation described above made it possible to identify one of these gene products as lactose synthase which, in the presence of α-lactalbumin, is responsible for the biosynthesis of lactose in mammary glands.

5.5 ABH Subgroups and Variants

Over the past decades, a number of subgroups and variants of blood groups **A**, **B**, and **H** have been defined which are based on serological properties of erythrocytes

---

[12] The structure formed with this reaction also showed strong blood group I activity (I Ma) (see *Sect. 7.5*).

and secreted blood group substances clearly differing from those of the 'normal' **A**, **B**, and **H** phenotypes. In these subgroups the **A**, **B**, or **H** activity of the erythrocytes is either significantly reduced or cannot even be detected; nevertheless, the red cells, without exception, are able to absorb the respective antibodies. In addition, a decrease in blood group **ABH** activity is usually observed in the secretions of *Se* individuals. Occasionally, homologous isoagglutinins are found in serum (e.g. anti-**A** in **A** variants, anti-**B** in **B** variants, anti-**H** in **H** variants); the antibodies, however, do not react with the individual's own erythrocytes.

With the exception of subgroup **A**$_2$ which is fairly common in Europids (±22% of all blood group **A** individuals [394]), the **ABH** subgroups are a very rare phenomenon indeed. The average frequency in Europids is about 1 in 50,000; in other races, however, some variants may occur more frequently.

The majority of the **ABH** subgroups are passed on by rare alleles on the *ABO* and *H* locus. Molecular biological investigations revealed that most of the gene variants are generated from common alleles by single nucleotide substitutions, deletions, or insertions; others have obviously originated from recombination or gene conversion-like events.

Such gene variants encode aberrant, i.e. less efficient, glycosyltransferases or almost totally inactive protein mutants. Further important factors influencing the phenotypical expression of **ABH** phenotypes are gene interaction, i.e. control by regulator genes (such as *Yy* or *Zz*), inhibition or enhancement of **A** or **B** activity in **AB** heterozygotes, or suppression of a determinant, caused by blocking of one step in its sequential biosynthesis – result of a mutation in another gene locus. Finally, the expression of a blood group phenotype can be influenced by changes in membrane architecture not controlled by the blood group genes.

Phenotypical variations within most of the subgroups suggest a genetic heterogeneity, and it must be assumed, therefore, that different *ABH* alleles may generate a similar **ABH** variant phenotype.

Number of Blood Group Determinants on Erythrocytes of Different ABH Subgroups

The erythrocytes of the different **A** subgroups vary greatly in density of blood group **A** determinants (see *Table 5.12*): assays by quantitative immunoabsorption method have yielded values ranging from 1–1.5 million on **A**$_1$ erythrocytes to fewer than 2,000 on **A**$_m$ and **A**$_{el}$ cells.

However, the antigen site densities presented in *Table 5.12* must be taken as average values and not as indicators of the antigen content of a single erythrocyte. As mentioned in *Sect. 5.2.2*, normal blood contains erythrocyte populations with different

Table 5.12: Number of blood group A determinants on erythrocytes from adults belonging to different A subgroups.

Blood group		Number of A determinants	
A_1		810,000 –	1,170,000
A_2		160,000 –	440,000
A_3	average value	40,600 –	118,000
	agglutinated cells	98,700 –	187,000
	non-agglutinated cells	21,300 –	38,000
A_x	average value	7,500 –	10,500
	agglutinated cells	10,800 –	22,500
	non-agglutinated cells	7,300 –	8,800
A_{end}	average value	2,100 –	2,700
	agglutinated cells	69,000 –	137,000
	non-agglutinated cells	0	
A_m		100 –	1,900
A_{el}		100 –	1,400
A_y		100 –	1,900

The values were solely obtained by serum absorption: A_1 and A_2 by Economidou et al. [103], those of the 'weak A' variants by Cartron et al. [53].

antigen densities. Mixed field agglutination is observed in cases where only few cells show high antigen content and the majority of cells a low content (see A_3 and A_{end}). When **A** antigens are expressed on only few erythrocytes (e.g. less than 5% in the cases of A_m and A_{el} variants [190]) the cells are no longer agglutinated by anti-**A**, but are still able to bind the antibody.

The number of **B** and **H** antigens on erythrocytes of **B** or **H** variant individuals have not yet been determined.

5.5.1 A Subgroups

Subgroups A_1 and A_2

A_1 and A_2 are inherited by respective alleles at the *ABO* locus [100,121,494]. The subgroups are serologically well- characterised (see *Table 5.13*): A_1 erythrocytes

Table 5.13: Serological characterisation of A subgroups

Sub-group	Reaction of erythrocytes[a,b] with				Antibodies in serum[c]		Blood group substance in Saliva of ABH-secretors[e] (Ratio A:H)[e]
	anti-A	anti-(A+B)	anti-A$_1$[d]	anti-H	anti-A$_1$	anti-A	
A$_1$	++++	++++	+++	−	−[f]	−	A + H (3.1±1.2)
A$_2$	+++	+++	−	+++	+/−[g]	−	A + H (2.1±0.5)
A$_{Int}$	++++	++++	++	+++			A + H
A$_3$	++[M]	++[M]	−	++++	+/−	−	A + H (1.2±0.5)
A$_x$	−/(+)	+/++	−	++++	+	−/(+)	A$_x$[h] + H (0.5)
A$_m$	(+)/−	(+)	−	++++	−	−	A + H (3.2±1.0)
A$_{end}$	(+)[M][i]	(+)[M]	−	++++	+/−	−	H
A$_{finn}$	(+)[M]	(+)[M]	−	++++			H
A$_{bantu}$	+/++[M]	+/++[M]	−	++++			H
A$_{el}$	−	−	−	++++	++	+/−	H
A$_y$	−	−	−	++++	−	−	(A)+ H (0.5 − 1.0)
A$_{lae}$	−	−	+++	++++			H

[M] Mixed-field agglutination (few small agglutinates in a pool of free red cells),
[a] all erythrocytes bind anti-A, no reaction with anti-B,
[b] degree of agglutination ranging from ++++ (strong) to + (weak) and (+) (very weak); − (no agglutination),
[c] + antibody present, (+) reduced amount of antibody, − antibody absent; all sera contain anti-B,
[d] anti-A$_1$ lectin of *Dolichos biflorus*,
[e] after Cartron [53],
[f] in the case of A$_1$B frequently anti-H,
[g] in the case of A$_2$B frequently anti-A$_1$,
[h] A substance detected by inhibition of the donor's own red cell agglutination by anti-A sera,
[i] slow agglutination (2−3 min).

From Salmon & Cartron, CRC Handbook Series in Clinical Laboratory Science (D. Seligson, ed.), Table 12, p. 80. By permission of CRC Press, Inc. USA react with anti-**A** and anti-**A₁** reagents, **A₂** erythrocytes are agglutinated only by anti-**A** and show distinct **H** activity. Most sera of **A₂** individuals and almost all sera of **A₂B** individuals contain anti-**A₁**.

A₂ erythrocytes further differ from **A₁** cells in density of **A** determinants, viz. some 300,000 on **A₂** cells as compared to more than 1 million determinants on **A₁** erythrocytes (*Table 5.12*).

Serological differences between secreted blood group substances from **A₁** and **A₂** secretors are less pronounced [241,337].

A group of serologists [328] have assumed that a **qualitative** difference between **A₁** and **A₂** exists, based primarily on the presence of anti-**A₁** antibodies which specifically agglutinate **A₁** cells (see *Sect. 5.2.2*). According to this view, **A₁** and **A₂** cells are characterised by specific antigenic patterns. The reduced number of **A** determinants and the presence of large quantities of **H** determinants on **A₂** erythrocytes may be explained by different specificities of the *A*-gene-dependent N-acetylgalactos-aminyltransferases, the **A₁**-enzyme being able to transform all types of **H** determinants into **A**, while the **A₂**-enzyme is able to transform only a few classes of **H** substances.

A lively debate between proponents of the above theory and the proponents of a seemingly contradictory second theory originally advanced by Watkins has been in progress for the past decades: based on the findings that anti-**A₁** agglutinins are absorbed in significant quantities by **A₂** cells (see also [121,298]) the latterit have proposed that there is no basic difference in specificity between anti-**A** and anti-**A₁** reagents, and that **A₁** and **A₂** subgroups are characterised solely on a **quantitative** basis (e.g. [94,298,309,507,524]). According to this hypothesis, the **A** determinants in both subgroups are equally distributed among all types of blood group **ABH** substances. Any reduced transfer activity of the **A₂**-enzyme, however, results in only partial conversion of **H** to **A** in the time 'available' for the formation of membrane-bound **ABH** substances.

It is this author's conviction that each of these two theories explains a different aspect of the **A₁/A₂** problem:

To date, studies on the **A**-transferases of **A₁** and **A₂** individuals show that both enzymes are N-acetylgalactosaminyltransferases, which transfer $\alpha 1 \rightarrow 3$ N-acetyl-galactosaminyl residues specifically onto terminal **H** determinants. The enzymes, however, differ significantly in their kinetic properties, such as reaction velocity, cation requirement, K_m values for acceptor substrates, pH optimum, and isoelectric point ([427,428,498], see *Table 5.14*).

Table 5.14: Some characteristics of A₁- and A₂-transferases.

			A₁-transferase	A₂-transferase
pH Optimum[a] (enzyme from serum)			5.5–6.5	7–8
Isoelectric point[b] (enzyme from serum)			9–10 [c]	6–7
(enzyme from cyst fluid)			9.5–10	9.5–10
K_m (mM)	2'-FL[d]	pH 5.8	1.15	
(enzyme from serum)		pH 7.2	0.1	3.6
	LNF-1[d]	pH 5.8	3.6	
		pH 7.2	0.7	5.6

[a] Isoelectric point after Topping and Watkins [498]
[b] pH optima and K_M values after Schachter et al. [428]
[c] In a more recent investigation a small but distinct peak has been found at pH 5.5–6.5 [523]
[d] **2'-FL** = 2'-Fucosyllactose, **LNF-1** = Lacto-N-fucopentaose I

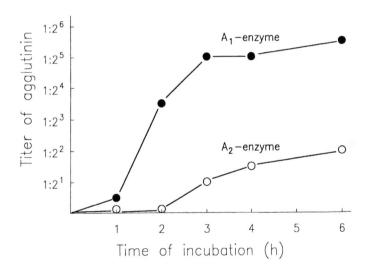

Figure 5.27: Enzymatic transfer of blood group A specificity onto O erythrocytes by the α1,3-N-acetylgalactosaminyltransferase isolated from A₁ and A₂ individuals.
The solutions of **A₁**-transferase and **A₂**-transferase have been adjusted to show approximately equal transfer activity towards water-soluble **H** substance.
Redrawn from Schenkel-Brunner and Tuppy [439].

An example: the α1,3-N-acetylgalactosaminyltransferase activity in sera from A_1 individuals is 5–10 times higher than that in A_2 subjects [427]. Incorporation experiments on erythrocytes conducted in examination of the above theories showed that although both A_1- and A_2-transferases isolated from gastric mucosa or serum are able to convert **O** erythrocytes into **A** cells [439], the incorporation of N-acetyl-galactosamine and thus the formation of **A** determinants catalysed by the A_2-transferase was much slower than the sugar transfer catalysed by the A_1-transferase (*Fig. 5.27*). As was expected, **O** and A_2 cells were easily converted to A_1 erythrocytes by the action of A_1-transferase. However, after prolonged incubation the A_2-transferase was also capable of transforming **O** and A_2 erythrocytes into A_1 specific cells (*Fig. 5.28*) [432,439].

The *in vitro* production of A_1 specific erythrocytes from **O** and A_2 cells by A_2-transferase action clearly indicates that quantitative aspects influence the serological differentiation between A_1 and A_2. According to a hypothesis advanced by Mäkelä et al. [309] the anti-A_1 antibodies are of the IgM type showing low affinity towards **A** epitopes. Due to the high receptor density of A_1 erythrocytes, the ten valence IgM molecules are able to form multiple bonds and are capable of agglutinating these cells in spite of their weak affinity. The low antigen density on A_2 erythrocytes, on the other hand, allows only one bond per cell – not enough for agglutination. The agglutinability of IgG antibodies and those IgM antibodies showing higher affinity for **A** sites must therefore be influenced only to a lesser degree by the receptor density. Indeed, evidence for the validity of this hypothesis has been provided by investigations on IgM hybridoma antibody *AbS12* and *Dolichos biflorus* lectin [131]). These studies revealed that these anti-A_1 agglutinins react only with erythrocytes showing a sufficiently high density of **A** determinants.

On the other hand, results of kinetic investigations using various types of **H** acceptor substrates have clearly indicated qualitative differences between A_1- and A_2-transferases. First, the transfer activities towards acceptor substrates of increasing complexity (oligosaccharides – glycoproteins – erythrocyte substances) show that the A_2-transferase depends much more on the accessibility of the acceptor **H** epitopes than does the A_1-enzyme (see [439,523]. Second, the studies by Clausen et al. [70] revealed great variations in affinity towards type-3 and type-4 chain acceptor substrates: in contrast to A_1-transferase, the A_2-transferase showed only very low transfer activity towards **H** active glycolipids with type-3 and type-4 chains; the conversion by A_2-transferase of type-2 chain **H** to type-2 chain **A** was less restricted. This difference in specificity explains previous findings according to which type-3 and type-4 **A** substances are confined to membranes of A_1 erythrocytes [75,77], whereas A_2 cells contain only type-3 and type-4 **H** substances [70,72].

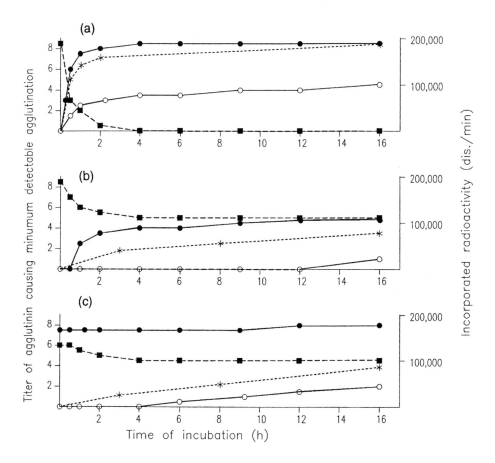

Figure 5.28: Enzymatic transfer of blood group A and A₁ specificity onto O and A₂ erythrocytes by the α1,3-N-acetylgalactosaminyltransferases of A₁ and A₂ individuals.
(a) O erythrocytes treated with **A₁** transferase, (b) O erythrocytes treated with **A₂** transferase, (c) **A₂** erythrocytes treated with **A₂** transferase.
●—● reaction with anti-**A**;   ○—○ reaction with anti-**A₁**;   ■—■ reaction with anti-**H**;
— incorporated radioactivity.

Redrawn from Schenkel-Brunner [432].

The **A type-3** glycolipid may indeed be the long sought for 'A₁ substance'; The influence of **A type-4** substances, although likewise restricted to A₁-cells, is probably insignificant, as the substances are found in erythrocytes only in trace amounts. In fact, it has been shown that hybridoma antibodies towards **A** type-3 epitopes (*KB-26.5,*

TS-1 [287], and *TH-1* [75]) agglutinate A_1 erythrocytes specifically. However, the affinity of naturally occurring anti-A_1 antibodies towards **A** type-3 glycosphingolipids must be tested as ultimate proof of this assumption.

The differences between the secreted blood group glycoproteins of A_1 and A_2 individuals are less well-studied. It can be assumed, however, that they are also based on both quantitative (number of **A** determinants) and qualitative differences (type-1 vs. type-2 chains or branched vs. unbranched carbohydrate chains) – refer to [241,337].

Molecular Biological Investigations

The first molecular genetic investigations on cDNA (coding exons 6 and 7) from A_2 individuals revealed an *A^2* allele characterised by a single base deletion at one of three consecutive cytosine residues at nucleotides 1059–1061, that is to say in the coding sequence near the C-terminus of the A-transferase (see *Fig. 5.29* and *ABO*A105* [13] in *Fig. 5.30)* [356,549]. The frame-shift caused by this nucleotide deletion results in an altered *A* gene encoding a protein with an additional domain of 21 amino acids at the C-terminus. A proline → leucine substitution at position 156 caused by a C → T exchange at nucleotide 467 has been found in all A_2 specimens investigated. It has also been found in one *A^1* allele from a human colon adenocarcinoma cell line [548] and in some other **A** subgroups (e.g. **cis-AB**) [360]. This mutation does not in any way affect enzyme activity or sugar-nucleotide donor specificity, as previous transfection experiments have shown [541].

In order to prove that this altered gene encodes an enzyme with characteristics corresponding to those of the A_2-transferase, the same authors [549] have introduced this nucleotide deletion into an A_1-transferase cDNA expression construct. DNA transfection into HeLa cells showed that the construct containing the single nucleotide deletion did in fact express A-transferase; the activity of the enzyme, however, was reduced to approximately 2–3% of the value found in cells transfected with A_1-specific cDNA. The restricted acceptor and donor substrate specificity (see *Tables 5.14* and *5.15)* and weaker A-transferase activity of the A_2-enzyme is probably due to steric hindrance by the additional C-terminal domain.

More recently, three further alleles have been detected in A_2 individuals of Japanese origin *(ABO*A106, ABO*A107*, and *ABO*R101)* [356]. *ABO*A106* and *ABO*A107* are characterised by single substitutions at nucleotide 1054 (C → T and

---

[13] Preliminary notation of the genes according to Ogasawara et al. [355].

Nucleotide sequence:

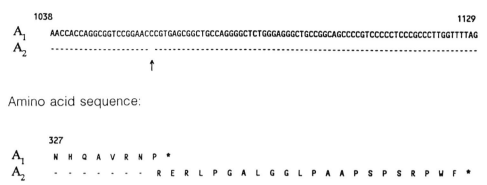

Amino acid sequence:

```
       327
A₁    N  H  Q  A  V  R  N  P *
A₂    -  -  -  -  -  -  -  -   R  E  R  L  P  G  A  L  G  G  L  P  A  A  P  S  P  S  R  P  W  F *
```

Figure 5.29: Nucleotide sequence and deduced amino acid sequences of A^1 and A^2 alleles in the N-terminal segment.
Nucleotide numbers begin with initiation code of membrane-bound A_1-transferase.
– : identical nucleotide or amino acid, ⋆ : termination codon, ↑ : nucleotide deletion.
From Yamamoto et al. [549] by permission of the authors and of Academic Press, Inc., San Diego, CA.

C → G, respectively) resulting in an amino acid substitution at position 352 (Arg → Trp and Arg → Gly, respectively). It should be noted that the Arg[352] → Trp exchange has also been found in a B_3 allele (see *Sect. 5.5.2*). The third fairly rare allele, *ABO*R101*, has six nucleotide substitutions compared to the common A_1 allele *(ABO*A101)*, i.e. three silent mutations (A → G at nucleotide 297, C → T at nucleotide 657, and C → T at nucleotide 771) and three missense mutations (C → G at nucleotide 526, G → A at nucleotide 703, and G → A at nucleotide 829). The latter result in three amino acid substitutions (Arg → Gly at position 176, Gly → Ser at position 235, and Val → Met at position 277, respectively). The nucleotide sequence of *ABO*R101* was found to be identical to that of *ABO*B101* upstream of nucleotide 703 and to that of *ABO*O201* downstream of nucleotide 771. Based on these findings, the allele *ABO*R101* may have originated by recombination between *ABO*B101* and *ABO*O201* in the region of nucleotides 703–771.

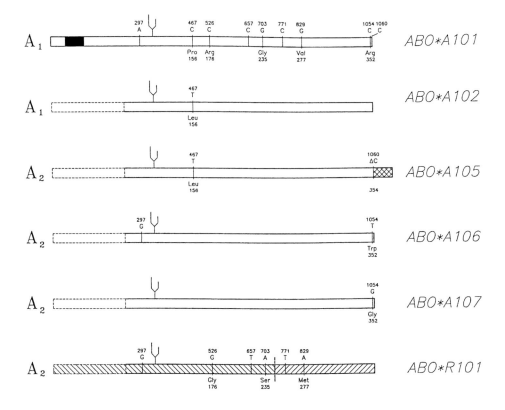

Figure 5.30: Amino acid sequences of A_2 alleles.
⌐⌐ = sequence not determined, ▇ = transmembrane domain,
▨ = additional domain at the C-terminus due to frameshift,
⊃– = N-glycan, ▨ = sequence of *ABO*B101*, ▨ = sequence of *ABO*O201*,
Provisional nomenclature as proposed by Ogasawara et al. [356].

Subgroup A $_{int}$

A$_{int}$ ('**A** intermediate') [274] is a fairly heterogeneous subgroup which occurs frequently in Negrids, where nearly 30% of the blood group **A** individuals have this subgroup – compared to only 3% in other races [163]. **A$_{int}$** erythrocytes are agglutinated both by anti-**A$_1$** and anti-**H**. In most cases their reaction with anti-**H** is similar to that of **A$_2$** cells, whereas the agglutination of **A$_{int}$** erythrocytes by anti-**A$_1$** reagents is normally considerably weaker than that of **A$_1$** cells.

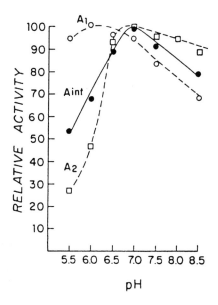

Figure 5.31: Effect of pH on the α1,3-N-acetylgalactosaminyltransferases from A $_{int}$, A$_1$, and A$_2$ individuals.
Reproduced from Yoshida et al. [564] by permission of The University of Chicago Press, USA.

Table 5.15: Kinetic properties of the A-transferase present in the serum of A $_{int}$ individuals.
(The corresponding values of **A**$_1$ and **A**$_2$ enzymes are given in comparison).

Blood group	K_m (μM)	
	UDP-GalNAc	2'-fucosyllactose
A$_{int}$	10 ± 0.7	7,000 ± 300
A$_1$	9 ± 1.2	200 ± 30
A$_2$	66 ± 5	2,100 ± 400

From Yoshida et al. [564] by permission of The University of Chicago Press, USA.

Characterisation of the **A**-transferase in the serum of A_{int} individuals by Yoshida et al. [564] revealed that the A_{int} subgroup is inherited by a variant of the **A** gene. The α-N-acetylgalactosaminyltransferase encoded by this allele differs clearly from A_1- and A_2-transferases in pH optimum *(Fig. 5.31)* and kinetic properties: the A_{int}-transferase has a strong affinity towards UDP-N-acetylgalactosamine and a very low affinity towards 2'-fucosyllactose; in comparison, A_1-transferase has a strong affinity and the A_2-transferase a low affinity towards both substrates *(Table 5.15)*.

The reduced number of **A** sites found in A_{int} cells reflects the low affinity of the A_{int}-transferase towards the **H** specific acceptor substances. Based on the fact that A_{int} erythrocytes are agglutinated by anti-A_1 reagents, closer investigation of the specificity of the A_{int}-transferase towards **H** type-3 substance (see also A_1/A_2) may provide additional information concerning the subgroup A_1 and A_2 differentiation.

Preliminary molecular biological studies on exons 6 and 7 of the *ABO* gene revealed different genetical backgrounds of this subgroup [360]: the A_{int} individuals thus far investigated were genotyped either as A^1O^1, A^1A^2, or A^2A^2. The decreased A_1 activity of A_{int} erythrocytes might thus be based on mutations of the *ABO* gene which lower the transferase activity. Decreased activity might also be caused by the action of other factors which prevent a normal *A* gene from expressing the carbohydrate antigens normally found in A_1 individuals.

'Weak A' phenotypes

The so-called '**weak A**' phenotypes are characterised by a highly reduced expression of the **A** antigen on erythrocytes and in secretions. The red cells of these variants are only weakly agglutinated by anti-**A** or anti-**(A+B)** or show no visible reaction; nevertheless, the cells are able to absorb anti-**A** in all cases. Saliva of **ABH** secretors exhibits strong **H** activity but only weak **A** activity, and in some subgroups none at all. In some cases the serum contains anti-**A** or, more frequently, anti-A_1.

As mentioned above, the various **A** variants are not uniform and homogeneous phenotypes. For practical reasons, the **A** variants are combined into groups with similar characteristics. A survey of the serological features of the 'weak **A**' variants is presented in *Table 5.13*. The change in serological activity occasionally observed in **AB** heterozygotes where 'weak' **A** variants are combined with 'normal' **B** is discussed in *Sect. 5.5.3*.

Subgroup A_3 erythrocytes [122,402] are characterised by mixed field agglutination – only small aggregates are observed when treated with anti-**A** or anti-**(A+B)**, whereas the majority of the erythrocytes is not agglutinated. Saliva of

ABH-secretors of this phenotype usually shows distinct **A** activity, and serum occasionally contains anti-**A**$_1$.

The number of **A** sites per erythrocyte is highly variable, and the values range from 20,000–40,000 for non-agglutinated cells to 100,000–190,000 for agglutinated cells [53] (see *Table 5.12)*. The inherited ability of **A**$_3$ subjects to produce erythrocyte populations with extremely variable **A** site densities follows the Mendelian laws.

Studies on the transferase activity in serum of different **A**$_3$ individuals have revealed that the **A**$_3$ subgroup is not homogeneous [52,349,522]. Based on the enzyme characteristics three different types of **A**$_3$ have been distinguished:

(a) In one group of donors a serum transferase with fairly high activity (i.e. ~50% of the value obtained for **A**$_1$ serum) and a pH optimum of 6.0 was found. In contrast to the enzymes from groups (b) and (c) the transferase was able to convert **O** erythrocytes into **A** cells; it also could be found in stroma of anti-**A** agglutinable erythrocytes; the enzyme could not be detected in non-agglutinable cells. The transferase present in this group of **A**$_3$ individuals is obviously a modified **A**$_1$-transferase.

(b) In a second group a weakly active transferase entailing 1–16% of the activity found in **A**$_1$ individuals was detected. The enzyme with a pH optimum of 7.0 is probably a modified **A**$_2$-transferase.

(c) No enzyme activity was detected in some **A**$_3$ subjects. This group presumably contains a modified **A**$_2$-transferase with very low activity.

Yamamoto et al. [551] have carried out preliminary investigations on cDNA from four **A**$_3$ individuals, focusing on the nucleotide sequence of the last two coding exons of the **A**-transferase. The deduced peptide sequences reflect the heterogeneity of the **A**$_3$ variant:

– in two cases a single-base substitution (G → A at nucleotide 871) was observed which led to an Asp → Asn exchange at position 291 [551]. Preliminary transfection experiments showed that this amino acid substitution diminished the transferase activity [540] *(Fig. 5.32)*;
– in all other cases investigated the amino acid sequence of the **A**$_3$-transferase in the two major exons was identical to that of the **A**$_1$-transferase [360,551].

The regulatory mechanism which produces erythrocytes with greatly differing **A** site densities is still unknown. Nevertheless, the finding that the **A**-transferase is virtually absent in **A**$_3$ erythrocyte membranes suggests that the proper function of the enzyme within the glycosyltransferase complex synthesising cell-bound blood group **A** substances is impaired by changes in the membrane anchor domain (see also **B**$_3$

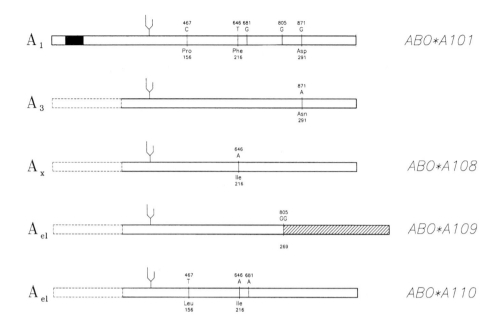

Figure 5.32: Amino acid sequences of some *A* variant alleles.
☐ = sequence not determined,　▬ = transmembrane domain,
▨ = additional domain at the C-terminus due to frameshift,
⊃‒ = N-glycan.
Provisional nomenclature as proposed by Ogasawara et al. [356].

variant, *Sect. 5.5.3)*. The crucial variations may therefore be expected in the transmembrane region, the amino acid sequence of which has not yet been investigated. Further, mutations in the promoter region or the occurrence of different spliceoforms of the **ABO** gene may also play a role.

A_x erythrocytes [116,513,560] are agglutinated only very weakly by anti-**A** sera from **B** individuals, yet they show a distinct reaction with anti-**(A+B)** from **O** individuals. More recently a hybridoma antibody that agglutinates A_x cells has been described [335]. The number of **A** sites per cell ranges between 1,400 and 10,000 as has been determined [53]. Although saliva of secretors exhibits extremely low **A** activity, it nevertheless distinctly inhibits the agglutination of the donor's own red cells by anti-**A** reagents. The donor's sera usually contain anti-A_1 – only in rare cases is anti-**A** also found.

Under prevailing conditions for transferase tests only traces of **A** enzyme have been detected in serum of A_x individuals [52,522]. Concentration of the enzyme from large volumes of serum on Sepharose 4B®, however, has demonstrated the presence of very low levels of **A**-transferase in the serum of A_x donors (0.004–0.08% of the level found in group A_2 control sera) [159]. In some rare cases significant **A**-transferase activity has been detected in $A_x B$ heterozygotes (see *Sect. 5.5.3*).

Analysis of one cDNA sample encoding exons 6 and 7 of the A_x-transferase revealed a sequence identical to that encoding A_1-transferase with the exception of a T → A exchange at nucleotide 646 which causes a Phe → Ile substitution at position 216 [356,546] (***ABO*A108***, see *Fig. 5.32*). The finding that other cases of A_x contained either A_1 or A_2 sequences in the last two coding exons lacking the T^{646} → A mutant reflects a considerable heterogeneity of this subgroup [360].

A_m erythrocytes [206,207,418,534] are not agglutinated by anti-**A** from **B** individuals or anti-**(A+B)** from **O** individuals. However, the cells are still able to absorb anti-**A**. An average number of 1,200 **A** sites per A_m erythrocyte has been determined [53]. Scanning immune electron microscopy revealed that less than 5% of A_m erythrocytes expressed **A** antigens (some of which showed extremely strong labelling), whereas the majority of the cells did not react with anti-**A** serum [190]. Saliva of A_m secretors shows strong **A** activity, comparable to that found in A_1 and A_2 individuals. Serum of A_m subjects contains no anti-**A**, but in some cases anti-A_1 has been observed.

In the serum of A_m individuals the **A**-transferase is easily detected [52,55]; it yields 30–50% of the values found in A_1 control sera. The pH optimum of 6.0 found in most cases suggests a variation of the A_1-transferase ($\rightarrow A_m^{A1}$); in one family, however, a modified A_2-transferase was found with a pH optimum of 7.0 ($\rightarrow A_m^{A2}$). In contrast to A_1-phenotype, only very low activity of the A_m^{A1}-transferase is found in erythrocyte stroma of A_m individuals [54]. A_m erythrocytes can be converted into normal **A** cells either by *in vitro* treatment with A_1-transferase, or even by the donor's own **A**-transferase [55].

The serological and biochemical investigations reveal that the A_m phenotype is complex and may be based on at least three genetic models:

– alleles at the *ABO* locus, which are characterised by the inability of the respective transferases to synthesise membrane-borne **A** antigens (see [207]). This inability may be caused either by a change in the domain responsible for the binding of the transferase to the enzyme complex which synthesises the membrane-bound blood group substances, or by a cis-effect in an operator or promotor gene at the *ABO* locus (see [58]);

- a gene independent of the **ABO** locus which controls the synthesis of **A** antigen on the red cell [526]; the **Y** gene enables the normal synthesis of **A** on the erythrocytes, and the recessive **y** allele, in the homozygous state prevents that synthesis but allows the normal production of **A** substance in secretions (see [97]),
- a dominant suppressor gene inherited independently of the **ABO** locus which prevents the expression of **A** antigen on the cells [413]

The fact that A_m erythrocytes can be converted into **A** reactive cells even by the donor's own **A**-transferase is worth noting as it in marked contrast to the original hypothesis that the A_m variant is caused by a change in the erythrocyte membrane controlled by a gene independent of the **ABO** locus (**Yy**). Thus far no molecular biological investigations on the A_m variant have been performed.

A_y erythrocytes [84,88,526] are not agglutinated by anti-**A** sera and can fix only a very small amount of anti-**A**. A_y cells contain only an average amount of 1,200 **A** sites [53]. Serum lacks anti-**A**, but in some cases anti-A_1 is found, and saliva shows weak but distinct **A** activity.

Serum of some A_y individuals contains an extremely weak α1,3-N-acetylgalactosaminyltransferase which can be detected only by very sensitive methods, i.e. enrichment on Agarose gel or prolonged incubation [54]; in other cases no enzyme activity could be detected [259].

Since the A_y phenotype is inherited as a recessive trait, Weiner et al. [526] suggested the existence of a modificator gene (Y^A/y^A) closely linked to the **ABO** locus. This gene may control the expression of the **A**-transferase, but not that of the **B**-transferase. According to this hypothesis, A_y individuals should be homozygous for the silent allele y^A. More recent genetic studies performed by Koscielak et al. [259], however, favour an allele of the **A** locus as the cause of A_y.

A_{end} erythrocytes [145,334,478] react like abnormally weak A_3 cells: they are only weakly agglutinated by anti-**A** and anti-**(A+B)** showing mixed-field agglutination. The agglutinated cells (± 5 %) contain 70,000–140,000 **A** determinants; on non-agglutinable cells no blood group **A** sites can be detected [57]. The agglutination of A_{end} cells by anti-**A** is not enhanced by protease treatment. Saliva of **ABH**-secretors lacks **A** activity. Serum of A_{end} subjects occasionally contains anti-**A** and/or anti-A_1, both of which react only at low temperatures [215].

No **A** transferase activity has been detected in A_{end} sera [52].

Preliminary molecular biological studies on exons 6 and 7 of the **ABO** gene of three unrelated individuals suggested that A_{end} is a variant of the A^2 gene [360].

A_{finn} erythrocytes [329,352] react similarly to A_{end} cells. In contrast to the latter the agglutinability of A_{finn} cells by anti-**A** is enhanced by protease treatment.

To date no enzyme tests have been performed using this variant.

A_{bantu} [37] is found fairly frequently among Bantus (4%), Hottentotts, and Bushmen (8%). This variant is phenotypically expressed as weak A_3 showing mixed-field agglutination. In saliva only **H** but no **A** activity can be detected. Serum of A_{bantu} subjects contains anti-A_1.

No transferase tests have been carried out on this group to date.

A_{el} erythrocytes [403,467,478] are not agglutinated by anti-**A** or anti-**(A+B)**; they are, nevertheless, able to fix anti-**A**. Serum frequently contains anti-A_1, and in rare cases anti-**A**. In saliva of **ABH**-secretors only **H** activity is detectable. An average value of 700 **A** sites has been determined on A_{el} erythrocytes [53]. Scanning immune electron microscopy revealed that only less than 5% of the erythrocytes expressed **A** antigens (see A_m also variant) [190].

No **A** transferase activity has been detected in A_{el} sera [52,522].

Molecular biological studies on exons 6 and 7 of the **ABO** gene revealed the complexity of the A_{el} variant *(Fig. 5.32)*:

- the allele of six A^{el} probands was characterised by a single nucleotide insertion, the sequence of seven guanosine residues at nucleotides 798-804 found in the A_1 consensus sequence being increased to eight [356,362] (*ABO*109*). This insertion causes a shift in the reading frame changing the peptide sequence immediately after the glycine residue at position 268 (pivotal for nucleotide sugar binding to the transferase). It also further extends the translated protein by 37 amino acids as compared to the consensus A^1 gene product. This additional domain alters the enzyme structure and, analogous to the A_2-enzyme, should reduce the transferase activity considerably;
- one allele (*ABO*A110*) differed from the A^1 allele at three positions: (1) a C → T exchange at nucleotide 467 leading to a Phe → Leu substitution at position 156, (2) a T → A exchange at nucleotide 646 leading to a Phe → Ile substitution at position 216, and (3) a synonymous mutation at nucleotide 681 (G → A) [356]. According to the authors this gene may have originated by a gene conversion event between *ABO*A102* and *ABO*A201*;
- in two other cases the alleles could not be distinguished from *ABO*A102* [356].

The A_{lae} character (*l*ectin, *a*bsorption *e*lution of anti-**A**) described by Schuh et al. [445] is probably not an **A** variant, but a weak form of the **Cad** blood group (see *Chap. 11*).

5.5.2 Variations in Blood Group B

Variations in 'Normal' B

Serological investigations have disclosed a relatively high degree of variability in the agglutination of blood group **B** erythrocytes. Based on the reactivity of the cells with anti-**B** sera under standard conditions, Gibbs et al. [150] described three subgroups within normal group **B** persons; one subgroup with very high **B** content (also described as **B*** or **B** superactive [327,565,566]) has been found only in non-Europids.

Investigations by Badet et al. [6,7] on the activities of the *B*-gene-encoded α1,3-galactosyltransferase in the serum of individuals with normal **B** phenotype provide further evidence for the existence of three groups of 'normal' **B** *(Fig. 5.33)*: two distinct groups representing 84% and 16% of the population have been distinguished in Europids, whereas the third group with greatly increased transferase activity has been detected only in Negrid donors. The differences in transferase activity are not related to the secretor status and do not reflect a gene-dose effect; however, a significant correlation between enzyme activity and red cell agglutinability has been noted, demonstrating a certain degree of polymorphism of the common **B** locus.

Erythrocytes of the **B(A)** phenotype type as normal **B** but show distinct reaction with an anti-**A** hybridoma antibody (*BioClone MH04*, Ortho Diagnostics, Raritan, NJ; published by Treacy & Stroup, cited in [546]). The **A** activity can be removed by treatment of the cells with α-N-acetylgalactosaminidase [154]. Serum of **B(A)** subjects contains anti-**A**.

Enzyme studies showed a distinct **A**-transferase activity in the serum of two **B(A)** donors; the enzyme, like normal **A**-enzyme, was adsorbed onto Sepharose 4B [281].

Analysis of cDNA encoding the last two coding exons of the *ABO* gene of a **B(A)** individual revealed a nucleotide sequence identical to that of the 'normal' **B**-allele. There are, however, two exceptions characteristic for the *A^1* allele − a cytosine at nucleotide 657 and a guanosine at nucleotide 703, the latter forming a glycine codon at position 235 [546] − see *Fig. 5.34*. This amino acid substitution is located at the second of the four sites which discriminate human **A**- and **B**-transferases; as shown above, the amino acid at this position has minute but nevertheless significant influence on the recognition and/or binding of the donor nucleotide sugar donor substrate (see *Sect. 5.4.5*).

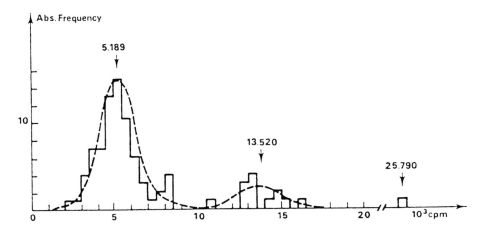

Figure 5.33: α1,3-galactosyltransferase (B-transferase) activity in the sera of blood group B donors.
Based on enzymatic activity (towards 2'-fucosyllactose at pH 6.5) three groups can be distinguished –
B$_I$, B$_{II}$, and B$_{III}$.
Reproduced from Badet et al. [7] by permission of Masson et Cie., Paris.

B Variants

The **B** variants are much rarer than the **A** variants. They are characterised by weak reaction of the erythrocytes with anti-**B** or anti-**(A+B)** sera, as well as by the ability of the cells to absorb anti-**B**. Saliva of **ABH**-secretors exhibits strong **H** and generally distinct **B** activity. In some variants the serum may contain anti-**B** antibodies, which, however, do not react with the donor's own erythrocytes under normal conditions.

In general, the **B** subgroups are not as well defined as the analogous **A** subgroups, and their classification is often controversial. The characteristics of the **B** variants thus far described can be seen in *Table 5.16*. The change in the expression of **B** activity occasionally found in **AB** heterozygous subjects where 'weak' **B** alleles are combined with normal **A** will be discussed in *Sect. 5.5.3*.

The subgroup **B$_3$** [417,533] is characterised by mixed-field agglutination of the erythrocytes. No anti-**B** is present in serum; saliva shows strong **H** and distinct **B** activity.

Tab. 5.16: Serological characterisation of B subgroups.

Sub-group	Reaction of erythrocytes[a,b] with			Antibodies in serum[c]			Blood group substance in saliva of ABH-secretors
	anti-B	anti-(A+B)	anti-H	anti-A₁	anti-A	anti-B	
B	++++	++++	+	+	+	−	B + H
B₃	+++[M]	+++[M]	+++	+	+	−	(B) + H
Bₓ	(+)	(+)	+++	+	+	(+)[d]	(B)[e] + H
Bₘ	−	−	+++	+	+	−	B + (H)
Bₑₗ	−	−	+++	+	+	+/−	H

[M] Mixed-field agglutination
[a] Degree of agglutination ranging from ++++ (strong) to + (weak) and (+) (very weak); − (no agglutination),
[b] The red cells of all subgroups bind anti-**B**,
[c] **+** Antibody present, **(+)** reduced amounts of antibody, **−** antibody absent
[d] Occasionally anti-**B** antibodies are present which do not react with the donor's own red cells
[e] **B** substance detected by inhibition of the donor's own red cell agglutination by anti-**B** sera.
From Salmon & Cartron, CRC Handbook Series in Clinical Laboratory Science (D. Seligson, ed.). Table 24, p. 89) by permission of CRC Press, Inc., Boca Raton, USA

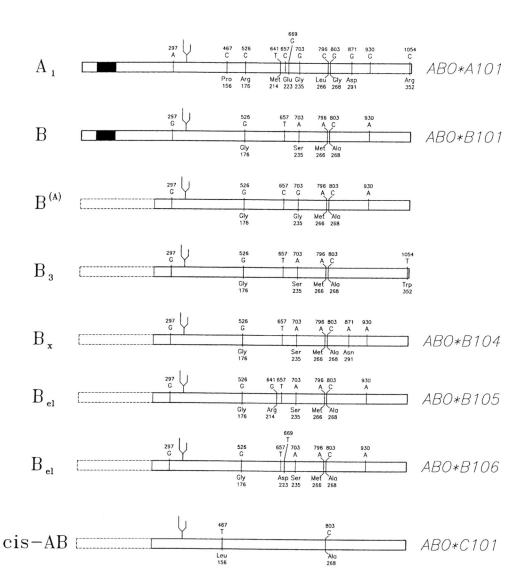

Figure 5.34: Amino acid sequences of some *B* variant alleles.

[....] = sequence not determined, ■ = transmembrane domain,
⊃— = N-glycan.
Provisional nomenclature as proposed by Ogasawara et al. [356].

The **B**-transfera$e has been clearly detected in serum of **B₃** individuals, but cannot be found in the membranes of their erythrocytes [5].

First investigations on the last two coding exons of the **ABO** gene from three **B₃** individuals indicated a heterogeneity in this **B** variant: one **B₃** sample showed a single-base substitution (C → T at nucleotide 1054) leading to an Arg → Trp exchange at position 352 of the **B**-transferase, whereas in two other cases in this region no differences to the 'normal' **B**-enzyme could be found [551] *(Fig. 5.34)*. As discussed in *Sect. 5.5.1* (subgroup **A₃**), the exact nature of the enzyme defect leading to extreme differences in antigenic site density among the erythrocytes of a single individual is thus far unknown. However, the crucial mutations will probably be found in the membrane anchor domain, which awaits investigation.

B_x erythrocytes [417] are weakly agglutinated by anti-**B** sera from **A** individuals, but react markedly with anti-**(A+B)** from **O** individuals (see **A_x** variant). The water-soluble **B** substances found in the saliva of **B_x** secretors inhibit the agglutination of the donor's own red cells by anti-**B**. Serum of **B_x** subjects contains only weak anti-**B**.

No **B** transferase activity has been detected in serum and erythrocyte membranes of **B_x** individuals [5].

Thus far the **B** gene of only one **B_x** propositus has been examined (exons 6 and 7) [356]. The allele (**ABO*B104**) had a single substitution (G → A) at nucleotide 871 causing an aspatic acid → asparagine exchange at position 291. The same mutation was previously found in the **A** gene of **A₃** individuals [551].

B_el erythrocytes [417] are not agglutinated by anti-**B** sera, but the cells are nevertheless able to absorb anti-**B**. Only 2–3% of the erythrocytes in a proposita who typed as **B_el** (or **B_m**) expressed **B** antigen [189]. Saliva of **ABH**-secretors shows strong **H** but no **B** activity, and serum occasionally contains weak anti-**B**.

In serum and red cell membranes of **B_el** individuals no **B** transferase activity has been detected [5].

Preliminary investigations on the **B** gene (exons 6 and 7) of two individuals attributed to the **B_el** variant [356] revealed a heterogeneity of this subgroup: one allele differed from the wild type **B** allele by a T → G exchange at nucleotide 641 resulting in a methionine → arginine substitution at position 214 (**ABO*B105**), and the other by a G → T exchange at nucleotide 669 resulting in a glutamic acid → aspartic acid substitution at position 223 (**ABO*B106**).

B_m erythrocytes [211,312] are not agglutinated by anti-**B** sera from **A** individuals, and only a few samples of anti-**(A+B)** sera show a weak reaction; in this case the cells are also able to absorb anti-**B**. Saliva of **ABH**-secretors shows strong **H** and **B** activity; serum of B_m individuals lacks anti-**B**.

The α1,3-galactosyltransferase has been detected in serum [253,261,491], red cell membranes [261], saliva [253,491], and urine [254,492] of B_m individuals. In all cases the activity was distinctly reduced.

B_m and A_1B_m erythrocytes contain great amounts of free **H** sites, which can be converted into **B** sites by *in vitro* treatment with **B**-transferase derived from normal **B** individuals. By using the **B**-enzyme isolated from B_m individuals, α1→3 galactosyl residues can be transferred onto B_m erythrocytes by conversion into cells reacting with anti-**B** sera [252,261,491,522].

The possible causes for the appearance of the B_m variant are a subject of discussion of the subgroup A_m above.

5.5.3 Expression of A and B Antigens in AB Heterozygotes

In *AB* heterozygotes mutual interaction of the genes *A* and *B* is frequently observed. The serological activity of the blood group characters can be depressed ('allele depression') or enhanced ('allele enhancement'). The effect of this gene interaction is especially well expressed when 'weak' **A** or **B** variants are combined with 'normal' **B** or **A**, respectively.

(a) Allele Depression

In *AB* heterozygous individuals the **A** and/or **B** activities of erythrocytes are usually lower than those found in **A** or **B** subjects. Thus A_1B erythrocytes show far fewer **A** and **B** sites than A_1 or **B** cells *(Table 5.17)*. When 'strong' **A** or **B** is combined with 'weak' **B** or **A**, the allele depression may change the phenotypical expression of the weak antigen. For example, the genotype A_2B is sometimes expressed as A_3B [2], and A_xB and A_1B_3 erythrocytes often show extremely reduced **A** and **B** activity, respectively [419].

At the enzyme level the phenomenon of allele depression is based on the competition of **A**- and **B**-transferases for the **H** determinant structure as their common precursor substrate. A 'strong' allele – one which produces a highly active transferase – may restrain the expression of a 'weak' allele, which codes for a less active enzyme. The expression of the A^1 gene is influenced only in rare cases: e.g. in A_1B individuals who produce only small amounts of **H** precursor [510] or A_1B^* subjects [120,563,565] may type as A_2B.

Table 5.17: Number of ABH determinants per erythrocyte in AB heterozygous individuals.
(The values found for cells from **A** and **B** homozygous subjects are given in comparison).

Blood group of erythrocytes	A sites	B sites
A₁B	420,000– 850,000	310,000–560,000
A₂B	120,000	
A₁	810,000–117,0000	–
A₂	240,000– 290,000	–
B	–	610,000–830,000

Values obtained by quantitative immunoabsorption, according to Economidou et al. [103].

(b) Allele Enhancement (Allele Complementation)

In rare cases the expression of a weak *A* or *B* gene is enhanced when combined with a 'strong' *B* or *A* gene in an **AB** heterozygote; the increase in serological activity is observed both on erythrocytes and in soluble blood group substances of secretions. No combination of weak *A* or *B* genes with the *O* allele yields this effect.

A series of family studies (e.g. *Duv.* [422], *G.* [49,117], *Lap.* [96]) showed that in some *AₓB* heterozygote cases, an **Aₓ** allele may be expressed as **A₂** (see *Fig. 5.35*). This increase in serological activity reflects an increase in antigen density from an average of 11,200 **A** sites on the erythrocytes of an **Aₓ** individual to 96,000 in an **AₓB** heterozygote of the same family (*Lap.*) [419]. The allele enhancement phenomenon is also found in water-soluble substances of saliva (see *Table 5.18*).

Similarly, it has been observed that combining the *Bₓ* gene with an *A¹* gene may enhance its expression and produce **A₁B** cells with nearly normal **B** activity [209]. This elevated **B** activity results in a four-fold increase in the number of agglutinated cells. In this case, too, the saliva of an **A₂Bₓ** heterozygote shows increased **B** activity as compared to 'normal' **Bₓ** subjects *(Fig. 5.36)*. A similar phenomenon was observed in a **B_{el}** family [110].

First studies on the **A**- and **B**-transferases in the serum of individuals showing allele enhancement revealed that the activity of the 'weak' enzyme is distinctly increased in **AB** heterozygotes: as discussed in the preceding chapter, the **A**-enzyme

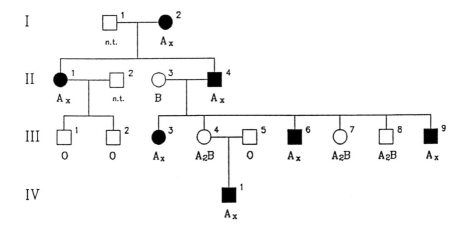

Figure 5.35: Allele enhancement – an example of A$_x$ phenotype transmitted by an A$_2$B parent.
Tree of the **Lap** family.
Redrawn from Ducos et al. [96] by permission of S. Karger AG, Basel.

Table 5.18: A case of allele enhancement (the Lap family): number of A sites and activity of the α1,3-N-acetylgalactosaminyltransferase (= A-transferase).

Subject[a]	Number of A sites	Blood group activity in saliva	Activity of A-transferase[b]
III-6 (A$_x$)	11,200	H	27
III-7 (A$_2$B)	96,000	A + H	2,400
III-8 (A$_2$B)			2,700
III-9 (A$_x$)			6
A$_x$ control	4,800		<100
A$_2$B control	120,000		72,000

[a] See *Fig. 5.35*
[b] Enzymatic activity at pH 7.0 towards 2'-fucosyllactose as acceptor
From Salmon and Cartron [419] by permission of CRC Press, Inc., Boca Raton, FL, USA.

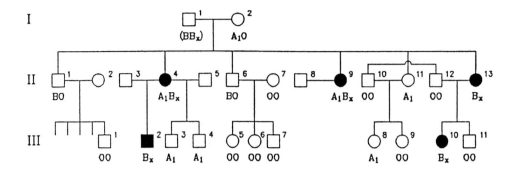

Figure 5.36: Allele enhancement – a case of complementation of the B$_x$ gene expression due to A^1 gene.
Tree of the **Slez** family.

Redrawn from Hrubisko et al. [209] by permission of the authors and of Marcel Dekker, Inc., New York.

Table 5.19: A case of allele enhancement (the Slez family): blood group B specificity and activity of the α1,3-galactosyltransferase (= B-transferase).

Subject[a]	Number of agglutinated erythrocytes	Blood group B activity in saliva	Activity of B-transferase[b]
II-13 (B$_x$)	17 %	−	81
III-2 (B$_x$)	14 %	−	273
III-10 (B$_x$)	20 %	−	208
II-4 (A$_1$B$_x$)	68 %	+	1,979
II-9 (A$_1$B$_x$)	69 %	+	2,091
II-11 (A$_1$B)		++	43,356
B(I) control		+++	25,590
B(II) control		+++	64,365

[a] See *Fig. 5.36*
[b] Acceptor not indicated
From Hrubisko et al. [209] by permission of the authors and of Marcel Dekker, Inc., New York.

cannot be detected in A_xO subjects. However, A-transferase activity was easily detectable in an A_xB individual of the above-mentioned family *Lap.* showing A_2B phenotype [419]. An analogous result was obtained with the B transferase in the B_x family *Slez.* [209] (see *Table 5.19*).

Based on the assumption that oligosaccharide chains are synthesised by transferase complexes (see *Chap. 3.3)* it has been suggested that the defect of a 'weak' transferase may be partially compensated by 'normal' A- or B-enzymes present in the complex. Further investigations are called for before an exact explanation of the phenomenon of genetical complementation can be presented.

5.5.4 Cis AB

Cis AB individuals [304,451] transmit **A** and **B** characters as one genetical unit, so that *AB* x *O* matings can produce both **AB** and **O** offspring. The assumption must therefore be valid that a single chromosome controls the formation of **A** and **B** determinants.

Cis AB erythrocytes are characterised by weak or very weak reactivity with anti-**A** and anti-**B** sera – the **A** activity corresponds to that of A_2 cells, and the expression of the **B** character is far less pronounced than in A_2B erythrocytes (mixed field agglutination is frequently observed). **Cis AB** erythrocytes are strongly agglutinated by anti-**H** reagents.

The values estimated for the density of **A** sites on **cis AB** erythrocytes vary between 170,000 and 580,000 [300].

In most **cis AB** sera weak anti-**B** is found, which, however, is not reactive with the donor's own erythrocytes. Saliva shows strong **H** activity and distinct **A** activity; **B** activity is very weak and can be detected only by agglutination inhibition tests using the donor's own erythrocytes.

The serological characteristics of **cis AB** erythrocytes from unrelated donors are variable; within a given family the expression of **A** and **B** activity was found to be generally homogeneous [8,300].

Cis AB phenotype may develop according to two different mechanisms – either a rare allele at the *ABO* locus, or an unequal crossing-over event:

Cis AB type 1 originated from a structural mutation of the *A* or *B* allele, giving rise to a 'bifunctional' enzyme variant capable of transferring both α-N-acetyl-galactosamine and α-galactose to **H** specific structures, synthesising **A** and **B** determinants.

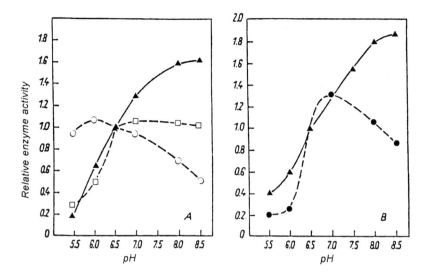

Figure 5.37: Cis AB type-1 (caused by a variant glycosyltransferase).
pH profile of **A**- and **B**-transferase activities in the plasma of **cis AB**, **A₁**, **A₂** und **B** subjects.

O : A₁-transferase, □ : A₂-transferase, ● : B-transferase, ▲ : cisAB-transferase.

Reproduced from Yoshida [562] by permission of Akademie Verlag, Berlin.

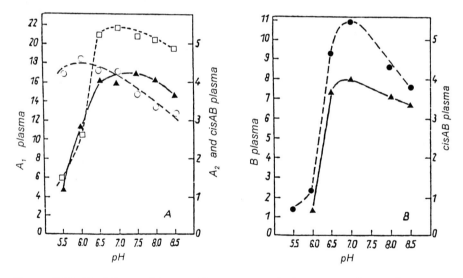

Figure 5.38: Cis AB type-2 (caused by an unequal chromosomal crossing-over event).
pH profile of **A**- and **B**-transferase activities in the plasma of **cis AB**, **A₁**, **A₂**, and **B** subjects.

O : A₁-transferase, □ : A₂-transferase, ● : B-transferase, ▲ : cisAB-transferase.

Values are given as % monosaccharide transferred.

Reproduced from Yoshida [562] by permission of Akademie Verlag, Berlin.

Because N-acetylgalactosamine and galactose differ only in the substituent of Carbon-2, a small structural change in the active site of the transferase might be sufficient to change specificity, particularly since the transferases show a certain affinity towards the alternative cosubstrate (see *Sect. 5.4.1*).

Yoshida et al.'s [568] investigations on the serum transferases of various **cis AB** individuals by have revealed a **cis AB type 1** phenotype in two families: plasma of these individuals contained both N-acetylgalactosaminyl- and galactosyltransferase activities. The respective activities in unfractionated plasma were <10% that of the average A_1 or **B** plasma. The pH activity profiles of both transferases differed distinctly from those of A_1, A_2, and **B** enzymes, exhibiting higher activities at alkaline pH *(Fig. 5.37)*. This finding, as well as the fact that both activities were completely adsorbed on Agarose gel (see *Sect. 5.4.1*), clearly indicates that **A** and **B** enzyme activities are properties of a single transferase molecule.

Further cases of **cis AB** studied by other groups have revealed similar results [8,377,415], and all these cases can thus be assumed to belong to **cis AB type 1**.

Cis AB type 2 arose from unequal crossing-over within the *ABO* locus, resulting in a chromosome which contains both *A* and *B* genes side by side in *cis* position.

Thus far, type 2 **cis AB** has been detected in only one family [567]. Investigations on the serum transferases showed in this case, as predicted, two clearly separable enzymes with pH profiles and kinetic data identical to those of A_2 and **B** transferase, respectively *(Fig. 5.38)*.

Investigations on the coding region in the last two coding exons of *ABO* gene from **cis AB type 1** individuals showed that the amino acid sequence of the **cis AB**-transferase differed from the A_1-enzyme in two positions: the Pro $\rightarrow$ Leu substitution at position 156 which is characteristic A_2-transferase, and the Gly $\rightarrow$ Ala substitution at position 268 which is characteristic for **B**-transferase (*ABO*C101*) [550] *(Fig. 5.34)*. As discussed above, the amino substitutions at positions 266 and 268 are crucial for determining sugar nucleotide specificity of the transferase (see *Sect. 5.4.4*); in fact, DNA transfection experiments showed that the construct pAAAB which possesses the first three amino acid substitutions of A_1-transferase and the last substitution of **B**-transferase expressed small amounts of **B** antigen in HeLa cells ([541], see *Table 5.11*).

5.5.5 Acquired B

Acquired B [50] (or pseudo-**B** [313]) is not an inherited antigen. It represents a transient somatic change in the serological specificity of erythrocytes in the course of bacterial infections.

The acquisition of a **B**-like character is most frequently associated with carcinoma of the colon or rectum, or various infections of the intestinal tract. It often disappears after recovery of the patient (*Fig. 5.39*, see also [142]); there have been a few reports of **acquired B** antigen in apparently healthy individuals [275,311].

The appearance of **acquired B** is occasionally accompanied by **T** and/or **Tk** transformation (see e.g. [48,147,225]).

The **acquired B** transformation is confined to blood group **A** erythrocytes. It occurs almost exclusively in **A₁** individuals; in only a few cases has it been reported in A_2 subjects [394]. A_1 **acquired B** erythrocytes are agglutinated by most anti-**B** sera obtained from A_2 and **O** individuals. They show only weak or no reaction with anti-**B** sera from A_1 subjects. A human serum antibody, a rabbit immune antibody, and a set of hybridoma antibodies which react specifically with '**acquired B**' cells have also been described [198,217].

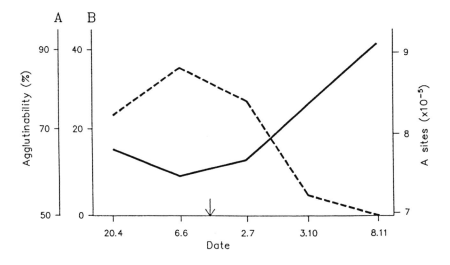

Figure 5.39: Reciprocal correlation between A and B reactivities of A₁ acquired B erythrocytes with time.
↓ : Right hemicolectomy.
Reproduced from Gerbal et al. [146] by permission of S. Karger AG, Basel.

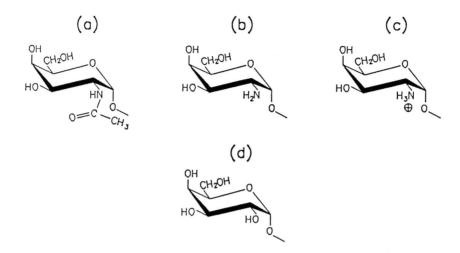

Figure 5.40: Structures of (a) N-acetylgalactosamine, (b) galactosamine, (c) galactosamine-ammonium ion, and (d) galactose.

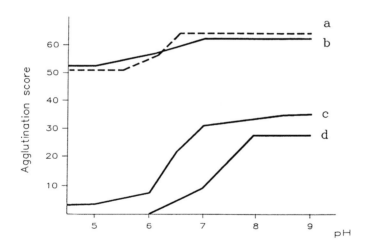

Figure 5.41: Acquired B erythrocytes – variation of agglutination with pH.
(a) agglutination of A_1 **acquired B** cells by anti-**A**,
(b) agglutination of 'normal' **B** cells by anti-**B**,
(c) and **(d)** agglutination of two A_1 **acquired B** cell samples by anti-**B**.
Reproduced from Gerbal et al. [146] by permission of S. Karger AG, Basel.

149

The transformation to 'acquired B' does not affect all cells to an equal extent: in a single blood sample, erythrocytes with strong 'B' activity and no longer agglutinated by anti-A sera can be separated from cells with only weak 'B' reactivity. Furthermore, a reciprocal correlation between the reaction of the erythrocytes with anti-B and the A activity of the cells is observed [146], i.e. 'B' reactivity develops as A activity declines (Fig. 5.39).

The serum of **acquired B** individuals contains anti-B which reacts with all normal B erythrocytes but fails to agglutinate **acquired B** cells. In saliva of **ABH**-secretors only A and H activity is found.

Biochemical investigations on **acquired B** transformed cells revealed that the development of the B-like character on A_1 cells is caused by bacterial deacetylases released into the circulatory system. The **acquired B** transformation can also be induced in vitro by incubation of A_1 cells with patient's serum [474], bacterial culture filtrates, and purified deacetylases from certain strains of Escherichia coli, Clostridium tertium (K12) [147], or Bacillus cereus [408]. Furthermore, it has been observed that A_1 cells transfused into persons with **acquired B** soon develop 'B' activity [311].

The bacterial deacetylases convert the group **A** determinant sugar, N-acetylgalactosamine, into galactosamine. Since the sterical configuration of this reaction product is similar to that of the group **B** determinant sugar galactose (Fig. 5.40), some anti-B antibodies may bind to the altered antigen. Rahuel et al. [397] confirmed this by showing the trisaccharide **GalNα1$\rightarrow$3[Fucα1$\rightarrow$2]Gal** and galactosamine as distinct inhibitors of **acquired B** cell agglutination by anti-**acquired B** sera (see also [217]).

After in vitro acetylation of **acquired B** erythrocyte the 'B' activity is abolished and normal A_1 activity restored [148]. Moreover, the agglutination of **acquired B** cells disappears at pH values below 6 when the amino group of galactosamine is converted to an NH_3^+-group (Fig. 5.41) [146].

The rarity of the **acquired B** transformation phenomenon is probably due to the fact that most type **A** individuals have antibodies reacting with the **acquired B** antigen, thus limiting the lifetime of the modified erythrocytes in the circulatory system. It must also be assumed that only few of the E. coli strains involved in bacterial infections of the digestive tract contain a suitable deacetylase.

5.5.6 H Deficient Variants

In 1952 the first **H** deficient variant, the 'Bombay' phenotype, was described by Bhende et al. [29]. This blood group phenotype was defined by a total lack of **ABH** activity on erythrocytes and in secretions. In the years since, a series of further examples of **H** deficient variants have been discovered. Similar to findings for for **A** and **B** variants, **H** deficiency of erythrocytes is fairly heterogeneous: it can vary from complete lack of **H**, to phenotypes with low but clearly detectable amounts of **H**. Recent molecular biological investigations form the basis of two main classes of **H** deficient phenotypes *(Table 5.20)* [515] – the **Bombay phenotype** which includes all H-deficient **ABH**-nonsecretors (genotype *h/h se/se*), and the **Parabombay phenotype** which includes all H-deficient **ABH**-secretors (genotype *h/h Se*)[14]; a further, extremely rare case of H deficiency (H$_m$) is obviously not influenced by the *H* gene.

The **Bombay phenotype** (O$_h$) [29] is extremely rare in Europids (1 case in 312,081) [514]. In the Marathi population around Bombay, however, it occurs at a frequency of about one in 7,600 [27].

Originally, **Bombay** individuals were characterised by the complete absence of **ABH** activity both on erythrocytes and in secretions. Their red blood cells are not agglutinated by anti-**A**, anti-**B**, or anti-**H** reagents under the conditions normally used in blood group serology [26]. Later however, the serum absorption/elution method [276] or testing by papain treated erythrocytes [89] made possible the detection of traces of **A** and/or **B** substances ('cryptic **ABH** substances') on the erythrocytes of many **Bombay** individuals carrying **A** and/or **B** genes. The designations O$_h$^{A}, O$_h$^{B}, and O$_h$^{AB} indicated the presence of an **A** and/or **B** gene. In very rare cases human anti-**H** antibodies were fixed ('**atypical O$_h$**' erythrocytes). Saliva and plasma of O$_h$ individuals are totally lacking in **ABH** substances. Serum contains anti-**A**, anti-**B**, as well as anti-**H** antibodies reacting with both type-1 and type-2 chain substances [286]. O$_h$ cells, probably due to the absence of **ABH** specific terminal sugars, often exhibit a stronger reaction with anti-**I** and anti-**i** sera (see *Chap. 7*) than do normal **O** erythrocytes [91,102,299,336].

Investigations on the *H* gene-specified α1,2-fucosyltransferase in serum showed that this enzyme – found in all normal sera and erythrocyte stroma (see *Sect. 5.4.2)* – cannot be detected in **Bombay** individuals [339,342,433]]. Depending on the *ABO* genotype The **A** and **B** transferases, however, are present in normal activity [391]. Since O$_h$ individuals represent **ABH**-nonsecretors (genotype *h/h se/se*), they are not able to synthesise **H** determinants. Consequently, even in the presence of **A** and/or **B** gene products **A** or **B** substances cannot be formed on erythrocytes and in secretions.

---

[14] In older literature the **Parabombay** phenotype is defined by a low but distinct **A** and **B** activity on the erythrocytes regardless of the presence or absence of **ABH** antigens in secretions (termed A$_h$ and B$_h$, see e.g. [420]).

Table 5.20: Genotypes of H-deficient phenotypes.

Genotype	Trivial name	ABH substances in	
		Erythroid cells	Secretions
H/H, Se/–	ABH-Secretor	+	+
H/H, se/se	ABH-nonsecretor	+	–
h/h, Se/–	Parabombay	–	+
h/h, se/se	Bombay	–	–

The occasional finding of 'cryptic' **ABH** antigens and the ability of some O_h cells to fix anti-**H** mentioned above, however, led to the assumption that *H* gene mutants produce weakly active transferases (to be discussed below).

The absence of H-transferase in **Bombay** individuals suggested that O_h erythrocytes contain great amounts of the **H** precursor substance(s) (i.e. unsubstituted type-2 chains). However, it has been shown that enzymatic transfer of $\alpha1\rightarrow2$ fucose onto O_h erythrocytes and conversion into **H**, and subsequently into **A** or **B** active cells, occurs only after neuraminidase treatment of the cells [23,436]. This finding showed that in the case of O_h erythrocytes, the precursors of the **ABH** substances which also represent substrates for neuraminyltransferase(s) are completely sialylated.

A further class of H-deficient nonsecretors is represented by the so-called **Reunion phenotype** described by Le Pendu and co-workers [144,285]. Erythrocytes of this variant show weak but distinct **ABH** activity; the serum contains weak anti-**H** antibodies directed mainly towards **H** type-1 structures [286]. All probands thus far tested were **ABH**-nonsecretors.

Serum of **Reunion**-type individuals contains a weakly active $\alpha1,2$-fucosyltransferase [284].

Table 5.21: Mutations in the FUT1 transferase found in H-deficient individuals

Nucleotide exchange	amino acid position	References
$C^{35} \rightarrow T$	$Ala^{12} \rightarrow Val$	[573]
$C^{349} \rightarrow T$	$His^{117} \rightarrow Tyr$	[111]
$G^{442} \rightarrow T$	$Asp^{148} \rightarrow Tyr$	[229]
$T^{460} \rightarrow C$	$Tyr^{154} \rightarrow His$	[229,515,573]
$A^{461} \rightarrow G$	$Tyr^{154} \rightarrow Cys$	[514]
$T^{491} \rightarrow A$	$Leu^{164} \rightarrow His$	[235]
$G^{513} \rightarrow C$	$Trp^{171} \rightarrow Cys$	[514]
$C^{658} \rightarrow T$	$Arg^{220} \rightarrow Cys$	[573]
$G^{695} \rightarrow A$	$Trp^{232} \rightarrow ter$	[229]
$T^{721} \rightarrow C$	$Tyr^{241} \rightarrow His$	[229]
$T^{725} \rightarrow G$	$Leu^{242} \rightarrow Arg$	[11,248]
$T^{776} \rightarrow A$	$Val^{259} \rightarrow Glu$	[514]
$C^{826} \rightarrow T$	$Gln^{276} \rightarrow ter$	[235]
$C^{944} \rightarrow T$	$Ala^{315} \rightarrow Val$	[514]
$C^{948} \rightarrow G$	$Tyr^{316} \rightarrow ter$	[235]
$A^{980} \rightarrow C$	$Asn^{327} \rightarrow Thr$	[573]
$G^{1042} \rightarrow A$	$Glu^{348} \rightarrow Lys$	[229,515]
$G^{1047} \rightarrow C$	$Trp^{349} \rightarrow Cys$	[514]
AG deletion at nucleotides 547-552		[573]
TT deletion at nucleotides 880-882		[573]
CT deletion at nucleotides 969-970		[514]
G deletion at nucleotide 990		[229]

Parabombay individuals[15] (genotype *h/h Se*) do not express **ABH** antigens on erythrocyte membranes because of the lack in the H-transferase. They do, however, show normal **ABH** activity in saliva. As in the case of **Bombay** erythrocytes, a weak **ABH** activity, caused by slightly active *H* alleles, is occasionally detected on the

[15] This phenotype has originally been described as H_z [420], $O_m^{\,h}$ [468], O_{Hm} [208], or AH_z ($A_m^{\,h}$) [28,468], BH_z ($B^m_{\,h}$) [108] or ABH_z.

erythrocytes. Some **H** activity on the red cells may also be attributed to absorption of **H** active glycolipids from the plasma (see [19]). Sera of **Parabombay** subjects usually contain weak anti-**H** (or anti-**HI**) (see also [305]).

Enzyme investigations revealed that both serum and erythrocyte membranes of **Parabombay** individuals lack the *H* gene-dependent α1,2-fucosyltransferase [339,433], whereas **A**- and **B**-transferases are not impaired.

Molecular Biological Investigations

Molecular biological investigations on the *H* gene of H-deficient individuals showed a great number of mutations, most of them producing silent alleles which when transcribed encode inactive fucosyltransferases. Some alleles, however, were shown to encode weakly active enzymes which are obviously responsible for the weak **H** expression in atypical **Bombay** and **Parabombay** phenotypes.

The mutations in the highly polymorphic *H* gene of H-deficient individuals thus far described are shown in *Table 5.21*. It can be seen that most alleles are characterised by nucleotide substitutions which cause amino acid exchanges or create premature stop codons. Some alleles show base deletions within the coding region, thus changing the reading frame after the deleted nucleotide and encoding truncated proteins.

In most cases in which transfection experiments have been performed the variant genes were shown to encode inactive forms of the **H**-transferase.

Some mutations, however, did not cause complete inactivation of the encoded enzymes. The [241]His allele in particular showed distinct **H**-transferase activity when transfected into COS-1 cells [229]; weak transferase activity was also detected with the Tyr[148] and His[154] alleles, as well as with the mutant characterised by a G deletion at nucleotide 990 [229,515,573]. In these cases the reduced enzyme activity correlated with a weak **ABH** activity on the donor's erythrocytes and in the donor's saliva. These alleles may thus be consided weak variants of the *H* gene as they synthesise only a small amount of **H** determinants.

In the *H* gene of the partially H-deficient individuals of the **Reunion phenotype** a $C^{349} \rightarrow$ T mutation inducing a His $\rightarrow$ Tyr change at position 117 has been found [111]. Transfection experiments revealed a transferase with substrate specificity and K_m values similar to those of the wild type enzyme but with a v_{max} reduced by $\pm$ 90%.

In one **H**-deficient proband a silent allele of the *H* gene was detected in which a coding region identical to that of the wild type allele and no alteration in the splicing sites of the intron were found [514]. The defect in expressing the *H* gene may therefore reside in the promoter or in other non-coding DNA sequences of the gene.

H_m

The H_m phenotype [205,421] (also described as O_{Hm}, or in the presence of an **A** or **B** gene as A_{Hm} or B_{Hm}) is characterised by weak expression or lack of erythrocyte **ABH** activity; the red cells are able to fix only minute amounts of human anti-**H** sera (the anti-**H** lectin from *Ulex europaeus* does not bind). Secretions and plasma, on the other hand, show normal **ABH** characteristics. In the sera of H_m individuals no anti-**H** or anti-**HI** antibodies are detected, but anti-**A** and/or anti-**B** are present depending on the genotype. Thus on a serological basis, phenotype H_m closely resembles the H_z variant. In contrast to H_z, however, H_m is inherited as a dominant trait.

First biochemical investigations showed an α1,2-fucosyltransferase present in serum and erythrocyte membranes of H_m individuals [340]; this enzyme, however, has not yet been characterised.

The occurrence of the H_m phenotype is explained by the existence of a dominant inhibitor gene which might affect the synthesis of the **H** precursor substance. This is still the subject of discussion. Investigations continue, but no significant results have been published to date.

A further **H**-deficient variant (*D.H.*) with normal levels of **A**-, **B**-, and **H**-transferase in serum and erythrocyte membranes has been described by Herron et al. [197]. The proposita showed weak expression of **A** and **B** and lacked **H** activity on red cells and in saliva. A similar case has been reported by Shechter et al. [453].

In these cases a block (or an inhibitory mechanism) may exist; it could lie further back in the biochemical pathway and affect the synthesis of the donor substrate, GDP-Fucose. This explanation is further supported by the extremely weak **Lewis** activity of saliva and red cells despite the presence of normal levels of the *Le* gene-specified α1,4-fucosyltransferase in the proposita. The synthesis of the oligosaccharide precursors necessary for **H** antigen formation is obviously not impaired, since both P_1 and **I** antigens, which are based on type-2 oligosaccharide chains, have been detected on the patient's erythrocytes.

References

1. ABE, K., LEVERY, S. B. & HAKOMORI, S. I. (1984): The antibody specific to type 1 chain blood group A determinant. *J. Immunol. 132*, 1951-1954.

2. ALTER, A. A. & ROSENFIELD, R. E. (1964): The nature of some subtypes of A. *Blood 23*, 605-620.
3. ANSTEE, D. J., HOLT, P. D. J. & PARDOE, G. I. (1973): Agglutinins from fish ova defining blood groups B and P. *Vox Sang. 25*, 347-360.
4. ARCILLA, M. B. & STURGEON, P. (1972): Lewis and ABH substances in amniotic fluid obtained by amniocentesis. *Pediatr. Res. 6*, 853-858.
5. BADET, J., HUET, M., MULET, C., LOPEZ, M., ROPARS, C. & SALMON, C. (1980): B-gene specified 3-α-D galactosyltransferase activity in human B blood group variants. *FEBS Lett. 122*, 25-28.
6. BADET, J., ROPARS, C., CARTRON, J. P., DOINEL, C. & SALMON, C. (1976): Groups of α-D-galactosyltransferase activity in sera of individuals with normal B phenotype. II. Relationship between transferase activity and red cell agglutinability. *Vox Sang 30*, 105-113.
7. BADET, J., ROPARS, C., CARTRON, J. P. & SALMON, C. (1974): Groups of α-galactosyl transferase activity in sera of individuals with normal B phenotype. *Biomedicine 21*, 230-232.
8. BADET, J., ROPARS, C. & SALMON, C. (1978): α-N-acetyl-D-galactosaminyl- and α-D-galactosyltransferase activities in sera of cis AB blood group individuals. *J. Immunogenet. 5*, 221-231.
9. BAKER, A. P., GRIGGS, L. J., MUNRO, J. R. & FINKELSTEIN, J. A. (1973): Blood-group A active glycoproteins of respiratory mucus and their synthesis by an N-acetylgalactosaminyltransferase. *J. Biol. Chem. 248*, 880-883.
10. BALDUS, S. E., THIELE, J., PARK, Y. O., HANISCH, F. G., BARA, J. & FISCHER, R. (1996): Characterization of the binding specificity of Anguilla anguilla agglutinin (AAA) in comparison to Ulex europaeus agglutinin I (UEA-I). *Glycoconj. J. 13*, 585-590.
11. BALL, S. P., TONGUE, N., GIBAUD, A., LE PENDU, J., MOLLICONE, R., GÉRARD, G. & ORIOL, R. (1991): The human chromosome 19 linkage group FUT1 (H), FUT2 (SE), LE, LU, PEPD, C3, APOC2, D19S7, and D19S9. *Ann. Hum. Genet. 55*, 225-233.
12. BARBOLLA, L., MOJENA, M. & BOSCA, L. (1988): Presence of antibody to A- and B-transferases in minor incompatible bone marrow transplants. *Brit. J. Haematol. 70*, 471-476.
13. BARBOLLA, L., MOJENA, M., CIENFUEGOS, J. A. & ESCARTIN, P. (1988): Presence of an inhibitor of glycosyltransferase activity in a patient following an ABO incompatible liver transplant. *Brit. J. Haematol. 69*, 93-96.
14. BASU, S., BASU, M. & CHIEN, J. L. (1975): Enzymatic synthesis of a blood group H-related glycosphingolipid by an α-fucosyltransferase from bovine spleen. *J. Biol. Chem. 250*, 2956-2962.
15. BECK, M. L., YATES, A. D., HARDMAN, J. T. & KOWALSKI, M. A. (1987): Consequences of overlapping substrate specificity of glycosyltransferases. (Abstract). *Transfusion 27*, 535.
16. BELLA, A. & KIM, Y. S. (1971): Biosynthesis of intestinal glycoprotein: a study of an α(1→2) fucosyltransferase in rat small intestinal mucosa. *Arch. Biochem. Biophys. 147*, 753-761.
17. BENNETT, E. P., STEFFENSEN, R., CLAUSEN, H., WEGHUIS, D. O. & VAN KESSEL, A. G. (1995): Genomic cloning of the human histo-blood group ABO locus. *Biochem. Biophys. Res. Commun. 206*, 318-325.
18. BERNEMAN, Z. N., VAN BOCKSTAELE, W. M., VAN ZAELEN, C., COLE-DERGENT, J., MUYLLE, L. & PEETERMANS, M. E. (1991): Flow-cytometric analysis of erythrocytic blood group A antigen density profile. *Vox Sang. 61*, 265-274.
19. BERNOCO, M., DANILOVS, J., TERASAKI, P. I., CARTRON, J. P., MOLLICONE, R., LE PENDU, J. & ORIOL, R. (1985): Detection of combined ABH and Lewis glycosphingolipids in sera of H-deficient donors. *Vox Sang. 49*, 58-66.
20. BERNSTEIN, F. (1924): Ergebnisse einer biostatistischen zusammenfassenden Betrachtung über die erblichen Blutstrukturen des Menschen. *Klin. Wochenschr. 3*, 1495-1497.
21. BETTERIDGE, A. & WATKINS, W. M. (1983): Two α-3-D-galactosyltransferases in rabbit stomach mucosa with different acceptor substrate specificities. *Eur. J. Biochem. 132*, 29-35.
22. BETTERIDGE, A. & WATKINS, W. M. (1985): Variant forms of α-2-L-fucosyltransferase in human submaxillary glands from blood group ABH "secretor" and "non-secretor" individuals. *Glycoconj. J. 2*, 61-78.
23. BEYER, T. A. & HILL, R. L. (1980): Enzymatic properties of the β-galactoside α1→2 fucosyltransferase

from porcine submaxillary gland. *J. Biol. Chem. 255*, 5373-5379.

24. BEYER, T. A., SADLER, J. E. & HILL, R. L. (1980): Purification to homogeneity of the H blood group β-galactoside α1→2 fucosyltransferase from porcine submaxillary gland. *J. Biol. Chem. 255*, 5364-5372.

25. BHATIA, H. M. (1964): Serological specificity of anti-H blood group antibodies. *Ind. J. Med. Res. 52*, 5-14.

26. BHATIA, H. M. (1977): Serologic reactions of ABO and O$_h$ (Bombay) phenotypes due to variations in H antigens. *Human Blood Groups - Proceedings of the 5th International Convocation on Immunologty, Buffalo 1976*. S. Karger, Basel, pp. 296-305.

27. BHATIA, H. M. & SATHE, M. S. (1974): Incidence of 'Bombay' (O$_h$) phenotype and weaker variants of A and B antigen in Bombay (India). *Vox Sang. 27*, 524-532.

28. BHATIA, H. M. & SOLOMON, J. M. (1967): Further observations on A^h_m and O^h_m phenotypes. *Vox Sang. 12*, 457-460.

29. BHENDE, Y. M., DESHPANDE, C. K., BHATIA, H. M., SANGER, R., RACE, R. R., MORGAN, W. T. J. & WATKINS, W. M. (1952): A "new" blood-group character related to the ABO system. *Lancet i*, 903-904.

30. BIRD, G. W. G. (1951): Specific agglutinating activity for human red blood corpuscles in extracts of Dolichos biflorus. *Curr. Sci. 20*, 298-299.

31. BIRD, G. W. G. (1952): Relationship of the blood sub-groups A$_1$, A$_2$and A$_1$B,A$_2$B to haemagglutinins present in the seeds of Dolichos biflorus. *Nature 170*, 674.

31a. BIRD, G. W. G. & WINGHAM, J. (1970): Agglutinins for antigens of two different human blood group systems in the seeds of Moluccella laevis. *Vox Sang. 18*, 235-239.

32. BIRD, G. W. G. (1977): Lectins. In: *CRC Handbook Series in Clinical Laboratory Science*. Section D, Blood Banking. (D. Seligson, T. J. Greenwalt, and E. A. Steane, eds.). CRC Press Inc., Cleveland, Ohio, Vol. 1, pp. 459-473.

33. BJÖRK, S., BREIMER, M. E., HANSSON, G. C., KARLSSON, K. A. & LEFFLER, H. (1987): Structures of blood group glycosphingolipids of human small intestine. A relation between the expression of fucolipids of epithelial cells and the ABO, Le, and Se phenotype of the donor. *J. Biol. Chem. 262*, 6758-6765.

34. BLASZCZYK-THURIN, M., SARNESTO, A., THURIN, J., HIDSGAUL, O. & KOPROWSKI, H. (1988): Biosynthetic pathways for the Le^b and Y glycolipids in the gastric carcinoma cell line KATO III as analyzed by a novel assay. *Biochem. Biophys. Res. Commun. 151*, 100-108.

35. BOUHOURS, D., HANSSON, G. C. & BOUHOURS, J. F. (1995): Structure and genetic polymorphism of blood group A-active glycosphingolipids of the rat large intestine. *Biochim. Biophys. Acta 1255*, 131-140.

36. BOYD, W. C. & REGUERA, R. (1949): Hemagglutinating substances for human cells in various plants. *J. Immunol. 62*, 333-339.

37. BRAIN, P. (1966): Subgroups of A in the South African Bantu. *Vox Sang. 11*, 686-698.

38. BREIMER, M. E. (1984): Tissue specificity of glycosphingolipids as expressed in pancreas and small intestine of blood group A and B human individuals. *Arch. Biochem. Biophys. 228*, 71-85.

39. BREIMER, M. E., FALK, K. E., HANSSON, G. C. & KARLSSON, K. A. (1982): Structural identification of two ten-sugar branched chain glycosphingolipids of blood group H type present in epithelial cells of rat small intestine. *J. Biol. Chem. 257*, 50-59.

40. BREIMER, M. E., HANSSON, G. C., KARLSSON, K. A. & LEFFLER, H. (1982): Isolation and partial characterization of blood group A and H active glycosphingolipids of rat small intestine. *J. Biol. Chem. 257*, 906-912.

41. BREIMER, M. E. & JOVALL, P. A. (1985): Structural characterization of a blood group A heptaglycosylceramide with globoseries structure: the major glycolipid based blood group A antigen of human kidney. *FEBS Lett. 179*, 165-172.

42. BREIMER, M. E., KARLSSON, K. A., LARSON, G. & MCKIBBIN, J. M. (1983): Chemical characterization of a blood group H type pentaglycosylceramide of human small intestine. *Chem. Phys. Lipids 33*, 135-144.

43. BREIMER, M. E., KARLSSON, K. A. & SAMUELSSON, B. E. (1981): Characterization of a blood group H type pentaglycosylceramide of human pancreas based on a type 1 carbohydrate chain. *J. Biol. Chem. 256*, 3810-3816.

44. BREIMER, M. E. & SAMUELSSON, B. E. (1986): The specific distribution of glycolipid-based blood group A antigens in human kidney related to A_1/A_2, Lewis, and secretor status of single individuals. *Transplantation 42*, 88-91.

45. BREMER, E. G., LEVERY, S. B., SONNINO, S., GHIDONI, R., CANEVARI, F., KANNAGI, R. & HAKOMORI, S. I. (1984): Characterization of a glycosphingolipid antigen defined by the monoclonal antibody MBr1 expressed in normal and neoplastic epithelial cells of human mammary gland. *J. Biol. Chem. 259*, 14773-14777.

46. BUNDLE, D. R., GIDNEY, M. A. J., KASSAM, N. & RAHMAN, A. F. R. (1982): Hybridomas specific for carbohydrates; synthetic human blood group antigens for the production, selection, and characterization of monoclonal typing reagents. *J. Immunol. 129*, 678-682.

47. BUSH, C. A., YAN, Z. Y. & RAO, B. N. N. (1986): Conformational energy calculations and proton nuclear Overhauser enhancements reveal a unique conformation for blood group A oligosaccharides. *J. Amer. Chem. Soc. 108*, 6168-6173.

48. BYRNE, U., BROWN, A., ROPARS, C. & MOORE, B. P. L. (1979): Acquired B antigen, Tk activation and A_1 destroying enzyme activity in a patient with septicaemia. *Vox Sang. 36*, 208-212.

49. CAHAN, A., JACK, J. A., SCUDDER, J., SARGENT, M., SANGER, R. & RACE, R. R. (1957): A family in which A_x is transmitted through a person of the blood group A_2B. *Vox Sang. 2*, 8-15.

50. CAMERON, C., GRAHAM, F., DUNSFORD, I., SICKLES, G., MACPHERSON, C. R., CAHAN, A., SANGER, R. & RACE, R. R. (1959): Acquisition of a B-like antigen by red blood cells. *Brit. Med. J. ii*, 29-32.

51. CARNE, L. R. & WATKINS, W. M. (1977): Human blood-group B specified α-3-galactosyltransferase: purification of the enzyme in serum by biospecific adsorption onto blood-group O erythrocyte membranes. *Biochem. Biophys. Res. Commun. 77*, 700-707.

52. CARTRON, J. P. (1976): Etude des propriétés α-N-acetylgalactosaminyltransférasiques des sérums de sujets A et "A faible". *Rev. Fr. Transfus. 19*, 67-88.

53. CARTRON, J. P. (1976): Etude quantitative et thermodynamique des phénotypes érythrocytaires "A faible". *Rev. Fr. Transfus. 19*, 35-54.

54. CARTRON, J. P., BADET, J., MULET, C. & SALMON, C. (1978): Study of the α-N-acetylgalactosaminyltransferases in sera and red cell membranes of human A subgroups. *J. Immunogenet. 5*, 107-116.

55. CARTRON, J. P., GERBAL, A., BADET, J., ROPARS, C. & SALMON, C. (1975): Assay of α-N-acetylgalactosaminyltransferases in human sera. Further evidence for several types of A_m individuals. *Vox Sang. 28*, 347-365.

56. CARTRON, J. P., MULET, C., BAUVOIS, B., RAHUEL, C. & SALMON, C. (1980): ABH and Lewis glycosyltransferases in human red cells, lymphocytes, and platelets. *Blood Transfus. Immunohaematol. 23*, 271-282.

57. CARTRON, J. P., REYES, F., GOURDIN, M. F., GARRETTA, M. & SALMON, C. (1977): Antigen site distribution among 'weak A' red cell populations. A study of A_3, A_x, and A_{end} variants. *Immunology 32*, 233-244.

58. CARTRON, J. P., ROPARS, C., CALKOVSKA, Z. & SALMON, C. (1976): Detection of A_1A_2 and $A_2A_m^{A1}$ heterozygotes among human A blood group phenotyes. *J. Immunogenet. 3*, 155-161.

59. CAZAL, P. & LALAURIE, M. (1952): Recherches sur quelques phyto-agglutinines specifiques des groupes sanguins ABO. *Acta Haematol. 8*, 73-80.

60. CEPPELLINI, R. (1959): Physiological genetics of human blood factors. In: *Ciba Foundation Symposium on Biochemistry of Human Genetics* (G. E. W. Wolstenholme, and C. M. O'Connor, eds.). Churchill, London, pp. 242-261.

61. CHANDRASEKARAN, E. V., JAIN, R. K., RHODES, J. M., SRNKA, C. A., LARSEN, R. D. & MATTA, K. L. (1995): Expression of blood group Lewis b determinant from Lewis a: association of this novel α(1,2)-L-fucosylating activity with the Lewis type α(1,3/4)-L-fucosyltransferase. *Biochemistry 34*, 4748-4756.

62. CHEN, H. T. & KABAT, E. A. (1985): Immunochemical studies on blood groups. The combining site specificities of mouse monoclonal hybridoma anti-A and anti-B. *J. Biol. Chem. 260*, 13208-13217.

63. CHENG, P. W. (1986): Mucin biosynthesis: enzymic properties of human-tracheal epithelial GDP-L-fucose:β-D-galactoside α-(1→2)-L-fucosyltransferase. *Carbohydr. Res. 149*, 253-261.

64. CHESSIN, L. N. & McGINNISS, M. (1968): Further evidence for the serologic association of the O(H) and I blood groups. *Vox Sang. 14*, 194-201.

65. CHESTER, M. A. & WATKINS, W. M. (1969): α-L-Fucosyltransferases in human submaxillary gland and stomach tissues associated with the H, Le^a and Le^b blood group characters and ABH secretor status. *Biochem. Biophys. Res. Commun. 34*, 835-842.

66. CHESTER, M. A., YATES, A. D. & WATKINS, W. M. (1976): Phenyl β-D-galactopyranoside as an acceptor substrate for the blood-group H gene-associated guanosine diphosphate-L-fucose: β-D-galactosyl α-2-L-fucosyltransferase. *Eur. J. Biochem. 69*, 583-592.

67. CHOU, T. H., MURPHY, C. & KESSEL, D. (1977): Selective inhibition of a plasma fucosyltransferase by N-ethylmaleinimide. *Biochem. Biophys. Res. Commun. 74*, 1001-1006.

68. CLAUSEN, H. & HAKOMORI, S. I. (1989): ABH and related histo-blood group antigens; immunochemical differences in carrier isotypes and their distribution. *Vox Sang. 56*, 1-20.

69. CLAUSEN, H., HAKOMORI, S. I., GRAEM, N. & DABELSTEEN, E. (1986): Incompatible A antigen expressed in tumors of blood group O individuals: Immunochemical, immunohistologic, and enzymatic characterization. *J. Immunol. 136*, 326-330.

70. CLAUSEN, H., HOLMES, E. & HAKOMORI, S. I. (1986): Novel blood group H glycolipid antigens exclusively expressed in blood group A and AB erythrocytes (type 3 chain H). II. Differential conversion of different H substrates by A$_1$ and A$_2$ enzymes, and type 3 chain H expression in relation to secretor status. *J. Biol. Chem. 261*, 1388-1392.

71. CLAUSEN, H., LEVERY, S. B., DABELSTEEN, E. & HAKOMORI, S. I. (1987): Blood group ABH antigens. A new sereis of blood group A-associated structures (genetic regulation and tissue distribution). *Transplant. Proc. 19*, 4408-4412.

72. CLAUSEN, H., LEVERY, S. B., KANNAGI, R. & HAKOMORI, S. I. (1986): Novel blood group H glycolipid antigens exclusively expressed in blood group A and AB erythrocytes (type 3 chain H). I. Isolation and chemical characterization. *J. Biol. Chem. 261*, 1380-1387.

73. CLAUSEN, H., LEVERY, S. B., McKIBBIN, J. M. & HAKOMORI, S. I. (1985): Blood group A determinants with mono- and difucosyl type 1 chain in human erythrocyte membranes. *Biochemistry 24*, 3578-3586.

74. CLAUSEN, H., LEVERY, S. B., NUDELMAN, E., BALDWIN, M. & HAKOMORI, S. I. (1986): Further characterization of type 2 and type 3 chain blood group A glycosphingolipids from human erythrocyte membranes. *Biochemistry 25*, 7075-7085.

75. CLAUSEN, H., LEVERY, S. B., NUDELMAN, E., TSUCHIYA, S. & HAKOMORI, S. I. (1985): Repetitive A epitope (type 3 chain A) defined by blood group A$_1$-specific monoclonal antibody TH-1: chemical basis of qualitative A$_1$ and A$_2$ distinction. *Proc. Natl. Acad. Sci. USA 82*, 1199-1203.

76. CLAUSEN, H., LEVERY, S. B., NUDELMAN, E. D., STROUD, M., SALYAN, M. E. K. & HAKOMORI, S. I. (1987): Isolation and characterization of novel glycolipids with blood group A-related structures: galactosyl-A and sialosylgalactosyl A. *J. Biol. Chem. 262*, 14228-14234.

77. CLAUSEN, H., WATANABE, K., KANNAGI, R., LEVERY, S. B., NUDELMAN, E., ARAO-TOMONO, Y. & HAKOMORI, S. I. (1984): Blood group A glycolipid (A^x) with globo-series structure which is specific for blood group A$_1$ erythrocytes: one of the chemical bases for A$_1$ and A$_2$ distinction. *Biochem. Biophys. Res. Commun. 124*, 523-529.

78. CLAUSEN, H., WHITE, T., TAKIO, K., TITANI, K., STROUD, M., HOLMES, E., KARKOV, J., THIM, L. & HAKOMORI, S. I. (1990): Isolation to homogeneity and partial characterization of a histo-blood group A defined Fucα1→2Galα1→3-N-acetylgalactosaminyltransferase from human lung tissue. *J. Biol. Chem. 265*, 1139-1145.

79. CONSTANDOULAKIS, M. & KAY, H. E. M. (1962): A and B antigens of the human foetal erythrocyte. *Brit. J. Haematol. 8*, 57-63.

80. COOK, G. A., GREENWELL, P. A. & WATKINS, W. M. (1982): A rabbit antibody to the

blood-group-A-gene-specified α-3-N-acetylgalactosaminyltransferase. *Biochem. Soc. Trans. 10*, 446-447.

81. COOMBS, R. R. A. & BEDFORD, D. (1955): The A and B antigens on human platelets demonstrated by means of mixed erythrocyte-platelet agglutination. *Vox Sang. 5*, 111-115.

82. CRAIGEN, W. J. & CASKEY, C. T. (1987): Translational frameshifting: where will it stop? *Cell 50*, 1-2.

83. DABELSTEEN, E., GRAEM, N., CLAUSEN, H. & HAKOMORI, S. I. (1988): Structural variations of blood group A antigens in human normal colon and carcinomas. *Cancer Res. 48*, 181-187.

84. DARNBOROUGH, J., VOAK, D. & PEPPER, R. M. (1973): Observations on a new example of the A_m phenotype which demonstrates reduced A secretion. *Vox Sang. 24*, 216-227.

85. DAVID, L., LEITAO, D., SOBRINHO-SIMOES, M., BENNETT, E. P., WHITE, T., MANDEL, U., DABELSTEEN, E. & CLAUSEN, H. (1993): Biosynthetic basis of incompatible histo-blood group A antigen expression: anti-A transferase antibodies reactive with gastric cancer tissue of type O individuals. *Cancer Res. 53*, 5494-5500.

86. DEJTER-JUSZYNSKI, M., HARPAZ, N., FLOWERS, H. M. & SHARON, N. (1978): Blood-group ABH specific macroglycolipids of human erythrocytes: isolation in high yield from a crude membrane glycoprotein fraction. *Eur. J. Biochem. 83*, 363-373.

87. DEKKER, J., AELMANS, P. H. & STROUS, G. J. (1991): The oligomeric structure of rat and human gastric mucins. *Biochem. J. 277*, 423-427.

88. DODD, B. E. & GILBEY, B. E. (1957): An unusual variant of group A. *Vox Sang. 2*, 390-398.

89. DODD, B. E. & LINCOLN, P. J. (1978): Serological studies of the H activity of O_h red cells with various anti-H reagents. *Vox Sang. 35*, 168-173.

90. DODD, B. E., LINCOLN, P. J. & BOORMAN, K. E. (1967): The cross-reacting antibodies of group O sera: immunological studies and possible explanation of the observed facts. *Immunology 12*, 39-52.

91. DOINEL, C. (1976): Antigenicité I des hématies Bombay. *Rev. Fr. Transfus. 19*, 185-191.

92. DOMINO, S. E., HIRAIWA, N. & LOWE, J. B. (1997): Molecular cloning, chromosomal assignment, and tissue-specific expression of a murine $\alpha(1,2)$fucosyltransferase expressed in thymic and epididymal epithelial cells. *Biochem. J. 327*, 105-115.

93. DONALD, A. S. R. (1973): The products of pronase digestion of purified blood-group specific glycoproteins. *Biochim. Biophys. Acta 317*, 420-436.

94. DONALD, A. S. R. (1981): A-active trisaccharides isolated from A_1 and A_2 blood group specific glycoproteins. *Eur. J. Biochem. 120*, 243-249.

95. DUA, V. K., RAO, B. N. N., WU, S. S., DUBE, V. E. & BUSH, C. A. (1986): Characterization of the oligosaccharide alditols from ovarian cyst mucin glycoproteins of blood group A using high pressure liquid chromatography (HPLC) and high field ^1H NMR spectroscopy. *J. Biol. Chem. 261*, 1599-1608.

96. DUCOS, J., MARTY, Y. & RUFFIE, J. (1975): A case of A_x phenotype transmitted by an A_2B parent. *Vox Sang. 29*, 390-393.

97. DUCOS, J., MARTY, Y. & RUFFIE, J. (1975): A family with one child of phenotype A_m providing further evidence for the existence of the modifyer genes Yy. *Vox Sang. 28*, 456-459.

98. DUK, M., MITRA, D., LISOWSKA, E., KABAT, E. A., SHARON, N. & LIS, H. (1992): Immunochemical studies on the combining site of the A + N blood type specific Moluccella laevis lectin. *Carbohydr. Res. 236*, 245-258.

99. DUNGERN, E. & HIRSZFELD, L. (1910): Über Vererbung gruppenspezifischer Strukturen des Blutes. *Z. Immun.-Forsch. 6*, 284-292.

100. DUNGERN, E. & HIRSZFELD, L. (1911): Über gruppenspezifische Strukturen des Blutes. III. *Z. Immun.-Forsch. 8*, 526-562.

101. DUNSTAN, R. A., SIMPSON, M. B., KNOWLES, R. W. & ROSSE, W. F. (1985): The origin of ABH antigens on human platelets. *Blood 65*, 615-619.

102. DZIERZKOWA-BORODEJ, W., MEINHARD, W., NESTOROWICZ, S. & PIROG, J. (1972): Successful elution of anti-A and certain anti-H reagents from two "Bombay" (O^A_h) blood samples and investigation of isoagglutinins in their sera. *Arch. Immunol. Ther. Exp. 20*, 841-849.

103. ECONOMIDOU, J., HUGHES-JONES, N. C. & GARDNER, B. (1967): Quantitative measurements

concerning A and B antigen sites. *Vox Sang. 12*, 321-328.

104. ELO, J., ESTOLA, E. & MALMSTROM, N. (1951): On phytagglutinins present in mushrooms. *Ann. Med. Exp. Biol. Fenn. 29*, 297-308.

105. ERNST, L. K., RAJAN, V. P., LARSEN, R. D., RUFF, M. M. & LOWE, J. B. (1989): Stable expression of blood group H determinants and GDP-L-fucose:β-D-galactoside 2-α-L-fucosyltransferase in mouse cells after transfection with human DNA. *J. Biol. Chem. 264*, 3436-3447.

106. ETIENNE-DECERF, J., MALAISE, M., MAHIEU, P. & WINAND, R. (1987): Elevated anti-α-galactosyl antibody titres. A marker of progression in autoimmune thyroid disorders and in endocrine ophthalmopathy? *Acta Endocrinol. 115*, 67-74.

107. ETZLER, M. E. & KABAT, E. A. (1970): Purification and characterization of a lectin (plant hemagglutinin) with blood group A specificity from Dolichos biflorus. *Biochemistry 9*, 869-877.

108. FAWCETT, K. J., ECKSTEIN, E. G., INNELLA, F. & YOKOYAMA, M. (1970): Four examples of B^h_m blood in one family. *Vox Sang. 19*, 457-467.

109. FEIZI, T. (1985): Demonstration by monoclonal antibodies that carbohydrate structures of glycoproteins and glycolipids are onco-developmental antigens. *Nature 314*, 53-57.

110. FENG, C. S., COOK, J. L., BEATTIE, K. M., KAO, Y. S., WALLACE, M. E. & DE JONGH, D. S. (1984): Variant of type B blood in an El Salvador family. Expression of a variant B gene enhanced by the presence of an A_2 gene. *Transfusion 24*, 264-266.

111. FERNANDEZ-MATEOS, P., CAILLEAU, A., HENRY, S., COSTACHE, M., ELMGREN, A., SVENSSON, L., LARSON, G., SAMUELSSON, B. E., ORIOL, R. & MOLLICONE, R. (1998): Point mutations and deletion responsible for the Bombay H_{null} and the Reunion H_{weak} blood groups. *Vox Sang. 75*, 37-46.

112. FINNE, J. (1980): Identification of the blood group ABH-active glycoprotein components of human erythrocyte membrane. *Eur. J. Biochem. 104*, 181-189.

113. FINNE, J., BREIMER, M. E., HANSSON, G. C., KARLSSON, K. A., LEFFLER, H., VLIEGENTHART, J. F. G. & VAN HALBEEK, H. (1989): Novel polyfucosylated N-linked glycopeptides with blood group A, H, X, and Y determinants from human small intestinal epithelial cells. *J. Biol. Chem. 264*, 5720-5735.

114. FINNE, J., KRUSIUS, T., RAUVALA, H., KEKOMÄKI, R. & MYLLYLÄ, G. (1978): Alkali-stable blood group A- and B-active poly(glycosyl)-peptides from human erythrocyte membrane. *FEBS Lett. 89*, 111-115.

115. FIORI, A., PANARI, G., ROSSI, G. & DE MERCURIO, D. (1978): Polymorphism of A, B, and H substances in human urine. *J. Chromat. 145*, 41-62.

116. FISCHER, W. & HAHN, F. (1935): Über auffallende Schwäche der gruppenspezifischen Reaktionsfähigkeit bei einem Erwachsenen. *Z. Immun.-Forsch. 84*, 177-188.

117. FISHER, N. & CAHAN, A. (1962): An addition to the family in which A_x is transmitted through a person of the blood group A_2B. *Vox Sang. 7*, 484.

118. FRANCO, R. F., SIMOES, B. P. & ZAGO, M. A. (1995): Relative frequencies of the two O alleles of the histo-blood ABH system in different racial groups. *Vox Sang. 69*, 50-52.

119. FREDMAN, P., RICHERT, N. D., MAGNANI, J. L., WILLINGHAM, M. C., PASTAN, I. & GINSBURG, V. (1983): A monoclonal antibody that precipitates the glycoprotein receptor for epidermal growth factor is directed against the human blood group H type 1 antigen. *J. Biol. Chem. 258*, 11206-11210.

120. FREDRICK, J., HUNTER, J., GREENWELL, P., WINTER, K. & GOTTSCHALL, J. L. (1985): The A^1B genotype expressed as A_2B on the red cells of individuals with strong B gene-specific transferases. Results from two paternity cases. *Transfusion 25*, 30-33.

121. FRIEDENREICH, V. (1931): Über die Serologie der Untergruppen A_1 und A_2. *Z. Immun.-Forsch. 71*, 283-313.

122. FRIEDENREICH, V. (1936): Eine bisher unbekannte Blutgruppeneigenschaft (A_3). *Z. Immun.-Forsch. 89*, 409-422.

123. FRIEDHOFF, F. & KUHNS, W. J. (1968): Detection and characterization of blood group antigens on untransformed human amnion cells. *Transfusion 8*, 244-249.

124. FUKUDA, M., DELL, A. & FUKUDA, M. N. (1984): Structure of fetal lactosaminoglycan. The carbohydrate moiety of band 3 isolated from human umbilical cord erythrocytes. *J. Biol. Chem. 259*, 4782-4791.

125. Fukuda, M., Dell, A., Oates, J. E. & Fukuda, M. N. (1984): Structure of branched lactosaminoglycan, the carbohydrate moiety of band 3 isolated from adult human erythrocytes. *J. Biol. Chem. 259*, 8260-8273.
126. Fukuda, M. N., Fukuda, M. & Hakomori, S. I. (1979): Cell surface modification by endo-β-galactosidase. Change of blood group activities and release of oligosaccharides from glycoproteins and glycosphingolipids of human erythrocytes. *J. Biol. Chem. 254*, 5458-5465.
127. Fukuda, M. N. & Hakomori, S. I. (1982): Structures of branched blood group A-active glycosphingolipids in human erythrocytes and polymorphism of A- and H-glycolipids in A_1 and A_2 subgroups. *J. Biol. Chem. 257*, 446-455.
128. Fukuda, M. N. & Levery, S. B. (1983): Glycolipids of fetal, newborn, and adult erythrocytes. Glycolipid pattern and structural study of H_3-glycolipid from newborn erythrocytes. *Biochemistry 22*, 5034-5040.
129. Fukumori, Y., Ohnoki, S., Shibata, H. & Nishimukai, H. (1996): Suballeles of the ABO blood group system in a Japanese population. *Hum. Hered. 46*, 221-223.
130. Furukawa, K., Clausen, H., Hakomori, S. I., Sakamoto, J., Look, K., Lundblad, A., Mattes, M. J. & Lloyd, K. O. (1985): Analysis of the specificity of five murine anti-blood group A monoclonal antibodies, including one that identifies type 3 and type 4 A determinants. *Biochemistry 24*, 7820-7826.
131. Furukawa, K., Mattes, M. J. & Lloyd, K. O. (1985): A_1 and A_2 erythrocytes can be distinguished by reagents that do not detect structural differences between the two cell types. *J. Immunol. 135*, 4090-4094.
132. Furukawa, K., Ying, R., Nakajima, T. & Matsuki, T. (1995): Hemagglutinins in fungus extracts and their blood group specificity. *Exp. Clin. Immunogenet. 12*, 223-231.
133. Galili, U. (1988): The natural anti-Gal antibody, the B-like antigen, and human red cell aging. *Blood Cells 14*, 205-228.
134. Galili, U., Clark, M. R., Shohet, S. B., Buehler, J. & Macher, B. A. (1987): Evolutionary relationship between the natural anti-Gal antibody and the Galα1→3Gal epitope in primates. *Proc. Natl. Acad. Sci. USA 84*, 1369-1373.
135. Galili, U., Flechner, I., Knyszynski, A., Danon, D. & Rachmilewitz, E. A. (1986): The natural anti-α-galactosyl IgG on human normal senescent red blood cells. *Brit. J. Haematol. 62*, 317-324.
136. Galili, U., Macher, B. A., Buehler, J. & Shohet, S. B. (1985): Human natural anti-alpha-galactosyl IgG. II. The specific recognition of alpha (1→3)-linked galactose residues. *J. Exp. Med. 162*, 573-582.
137. Galili, U., Mandrell, R. E., Hamadeh, R. M., Shohet, S. B. & Griffiss, J. M. (1988): Interaction between human natural anti-α-galactosyl immunoglobulin G and bacteria of the human flora. *Infect. Immun. 56*, 1730-1737.
138. Galili, U., Shohet, S. B., Kobrin, E., Stults, C. L. M. & Macher, B. A. (1988): Man, apes, and Old World monkeys differ from other mammals in the expression of α-galactosyl epitopes on nucleated cells. *J. Biol. Chem. 263*, 17755-17762.
139. Galili, U. & Swanson, K. (1991): Gene sequences suggest inactivation of α-1,3-galactosyltransferase in catarrhines after the divergence of apes from monkeys. *Proc. Natl. Acad. Sci. USA 88*, 7401-7404.
140. Gardas, A. & Koscielak, J. (1973): New form of A-, B-, and H-blood-group active substances extracted from erythrocyte membranes. *Eur. J. Biochem. 32*, 178-187.
141. Gardas, A. & Koscielak, J. (1974): Megaloglycolipids - unusually complex glycosphingolipids of human erythrocyte membrane with A, B, H, and I blood group specificity. *FEBS Lett. 42*, 101-104.
142. Garratty, G., Willbanks, E. & Petz, L. D. (1971): An acquired B antigen associated with Proteus vulgaris infection. *Vox Sang. 21*, 45-56.
143. Gassner, C., Schmarda, A., Nussbaumer, W. & Schonitzer, D. (1996): ABO glycosyltransferase genotyping by polymerase chain reaction using sequence-specific primers. *Blood 88*, 1852-1856.
144. Gerard, G., Vitrac, D., Le Pendu, J., Muller, A. & Oriol, R. (1982): H-deficient blood groups (Bombay) of Reunion Island. *Amer. J. Hum. Genet. 34*, 937-947.

145. GERBAL, A., LIBERGE, G., CARTRON, J. P. & SALMON, C. (1970): Les phénotypes A_end: étude immunologique et génétique. *Rev. Fr. Transfus. 13*, 243-250.

146. GERBAL, A., MASLET, C. & SALMON, C. (1975): Immunological aspects of the acquired B antigen. *Vox Sang. 28*, 398-403.

147. GERBAL, A. & ROPARS, C. (1976): L'antigène B acquis. *Rev. Fr. Transfus. 19*, 127-133.

148. GERBAL, A., ROPARS, C., GERBAL, R., CARTRON, J. P., MASLET, C. & SALMON, C. (1976): Acquired B antigen disappearance by in vitro acetylation associated with A_1 activity restoration. *Vox Sang. 31*, 64-66.

149. GIBBONS, R. A., CREETH, J. M. & DENBOROUGH, M. A. (1970): Biophysical characteristics of the blood group substances. In: *Blood and Tissue Antigens* (D. Aminoff, ed.). Academic Press, New York - London, pp. 307-324.

150. GIBBS, M. B., AKEROYD, J. H. & ZAPF, J. J. (1961): Quantitative subgroups of the antigen B in man and their occurrence in three racial groups. *Nature 192*, 1196-1197.

151. GOLD, E. R. & BALDING, P. (1975): *Receptor-Specific Proteins. Plant and Animal Lectins.* Excerpta Medica, Amsterdam.

152. GOLD, E. R. & BHATIA, H. M. (1964): Observations on the H antigen. *Vox Sang. 9*, 625-628.

153. GOLDSTEIN, I. J., BLAKE, D. A., EBISU, S., WILLIAMS, T. J. & MURPHY, L. A. (1981): Carbohydrate binding studies on the Bandeiraea simplicifolia I isolectins. *J. Biol. Chem. 256*, 3890-3893.

154. GOLDSTEIN, J., LENNY, L., DAVIES, D. & VOAK, D. (1989): Further evidence for the presence of A antigen on group B erythrocytes through the use of specific exoglycosidases. *Vox Sang. 57*, 142-146.

155. GOODWIN, S. D. & WATKINS, W. M. (1974): The peptide moiety of blood group specific glycoproteins. Some amino-acid sequences in the regions carrying the carbohydrate chains. *Eur. J. Biochem. 47*, 371-382.

156. GOOI, H. C., HOUNSELL, E. F., PICARD, J. K., LOWE, A. D., VOAK, D., LENNOX, E. S. & FEIZI, T. (1985): Differing reactions of monoclonal anti-A antibodies with oligosaccharides related to blood group A. *J. Biol. Chem. 260*, 13218-13224.

157. GREENBURY, C. L., MOORE, D. H. & NUNN, L. A. C. (1963): Reaction of 7S and 19S components of immune rabbit antisera with human group A and AB red cells. *Immunology 6*, 421-433.

158. GREENWELL, P., BALL, M. G. & WATKINS, W. M. (1983): Fucosyltransferase activity in human lymphocytes and granulocytes. Blood group H-gene-specified α-2-L-fucosyltransferase is a discriminatory marker of peripheral blood lymphocytes. *FEBS Lett. 164*, 314-317.

159. GREENWELL, P. & WATKINS, W. M. (1987): Demonstration of fucosyl α1-2 galactoside α1-3-N-acetylgalactosaminyltransferase activity in the serum and tissues of blood group O individuals. *Proceedings of the 9th International Symposium on Glycoconjugates, Lille*, Abstract E-34.

160. GREENWELL, P., YATES, A. D. & WATKINS, W. M. (1986): UDP-N-acetyl-D-galactosamine as a donor substrate for the glycosyltransferase encoded by the B gene at the human blood group ABO locus. *Carbohydr. Res. 149*, 149-170.

161. GROLLMAN, A. P. & MARCUS, D. M. (1966): Enzymatic incorporation of fucose into blood group H substance. *Biochem. Biophys. Res. Commun. 25*, 542-548.

162. GRUNDBACHER, F. J. (1964): Changes in the human A antigen of erythrocytes with the individual's age. *Nature 204*, 192-194.

163. GRUNDBACHER, F. J. & SUMMERLIN, D. C. (1971): Inherited differences in blood group A subtypes in Caucasians and Negroes. *Hum. Hered. 21*, 88-96.

164. GRUNNET, N., STEFFENSEN, R., BENNETT, E. P. & CLAUSEN, H. (1994): Evaluation of histo-blood group ABO genotyping in a Danish population: Frequency of a novel O allele defined as O-2. *Vox Sang. 67*, 210-215.

165. HAKOMORI, S. I. (1981): Blood group ABH and Ii antigens of human erythrocytes: chemistry, polymorphism, and their developmental change. *Sem. Hematol. 18*, 39-44.

166. HAKOMORI, S. I. (1984): Tumor associated carbohydrate antigens. *Annu. Rev. Immunol. 2*, 103-126.

168. HAKOMORI, S. I. (1985): Aberrant glycosylation in cancer cell membranes as focused on glycolipids:

overview and perspective. *Cancer Res. 45*, 2405-2414.

169. HAKOMORI, S. I. (1989): Aberrant glycosylation in tumors and tumor-associated carbohydrate antigens. *Adv. Cancer. Res. 52*, 257-331.

170. HAKOMORI, S. I., STELLNER, K. & WATANABE, K. (1972): Four antigenic variants of blood group A glycolipid: examples of highly complex, branched chain glycolipid of animal cell membrane. *Biochem. Biophys. Res. Commun. 49*, 1061-1068.

171. HAKOMORI, S. I. & STRYCHARZ, G. D. (1968): Investigations on cellular blood-group substances. I. Isolation and chemical composition of blood-group ABH and Le[b] isoantigens of sphingoglycolipid nature. *Biochemistry 7*, 1279-1286.

172. HAMADEH, R. M., JARVIS, G. A., GALILI, U., MANDRELL, R. E., ZHOU, P. & GRIFFISS, J. M. (1992): Human natural anti-Gal IgG regulates alternative complement pathway activation on bacterial surfaces. *J. Clin. Invest. 89*, 1223-1235.

173. HAMMARSTRÖM, S. & KABAT, E. A. (1969): Purification and characterization of a blood-group A reactive hemagglutinin from the snail Helix pomatia and a study of its combining site. *Biochemistry 8*, 2696-2705.

174. HAMMARSTRÖM, S., MURPHY, L. A., GOLDSTEIN, I. J. & ETZLER, M. E. (1977): Carbohydrate binding specificity of four N-acetyl-D-galactosamine- "specific" lectins: Helix pomatia A hemagglutinin, soy bean agglutinin, lima bean lectin, and Dolichos biflorus lectin. *Biochemistry 16*, 2750-2755.

175. HANFLAND, P. (1975): Characterization of B and H blood group active glycosphingolipids from human B erythrocyte membranes. *Chem. Phys. Lipids 15*, 105-124.

176. HANFLAND, P., EGGE, H., DABROWSKI, U., KUHN, S., ROELCKE, D. & DABROWSKI, J. (1981): Isolation and characterization of an I-active ceramide decasaccharide from rabbit erythrocyte membranes. *Biochemistry 20*, 5310-5319.

177. HANFLAND, P. & EGLI, H. (1975): Quantitative isolation and purification of blood group-active glycosphingolipids from human B erythrocytes. *Vox Sang. 28*, 438-452.

178. HANFLAND, P. & GRAHAM, H. A. (1981): Immunochemistry of the Lewis-blood-group system: partial characterization of Le[a]-, Le[b]-, and H-type 1 (Le[dH])-blood-group active glycosphingolipids from human plasma. *Arch. Biochem. Biophys. 210*, 383-395.

179. HANFLAND, P., KORDOWICZ, M., NIERMANN, H., EGGE, H., DABROWSKI, U., PETER-KATALINIC, J. & DABROWSKI, J. (1984): Purification and structures of branched blood-group-B-active glycosphingolipids from human erythrocyte membranes. *Eur. J. Biochem. 145*, 531-542.

180. HANFLAND, P., KORDOWICZ, M., PETER-KATALINIC, J., EGGE, H., DABROWSKI, J. & DABROWSKI, U. (1988): Structure elucidation of blood group B-like and I-active ceramide eicosa- and pentacosasaccharides from rabbit erythrocyte membranes by combined gas chromatography - mass spectromerty, electron-impact and fast-atom-bombardment mass spectrometry, and two-dimensional correlated, relayed-coherence transfer, and nuclear Overhauser effect 500-MHz [1]H-N.M.R. spectroscopy. *Carbohydr. Res. 178*, 1-21.

181. HANSSON, G. C. (1983): The structure of two blood group A-active glycosphingolipids with 12 sugars and a branched chain present in the epithelial cells of rat small intestine. *J. Biol. Chem. 258*, 9612-9615.

182. HANSSON, G. C., BOUHOURS, J. F. & ANGSTRÖM, J. (1987): Characterization of neutral blood group B-active glycosphingolipids of rat gastric mucosa. A novel type of blood group active glycosphingolipid based on isogloboside. *J. Biol. Chem. 262*, 13135-13141.

183. HANSSON, G. C., KARLSSON, K. A., LARSON, G., MCKIBBIN, J. M., BLASZCZYK, M., HERLYN, M., STEPLEWSKI, Z. & KOPROWSKI, H. (1983): Mouse monoclonal antibodies against human cancer cell lines with specificities for blood group and related antigens. *J. Biol. Chem. 258*, 4091-4097.

184. HANSSON, G. C., KARLSSON, K. A. & THURIN, J. (1980): Glycolipids of rat large intestine. Characterization of a novel blood group-B active tetraglycosylceramide absent from small intestine. *Biochim. Biophys. Acta 620*, 270-280.

185. HARPAZ, N., FLOWERS, H. M. & SHARON, N. (1975): Studies on B-antigenic sites of human erythrocytes by use of coffee bean α-galactosidase. *Arch. Biochem. Biophys. 170*, 676-683.

186. HATTORI, H., UEMURA, K. I., ISHIHARA, H. & OGATA, H. (1992): Glycolipid of human pancreatic cancer;

the appearance of neolacto-series (type 2 chain) glycolipid and the presence of incompatible blood group antigen in tumor tissues. *Biochim. Biophys. Acta 1125*, 21-27.

187. HEARN, V. M., RACE, C. & WATKINS, W. M. (1972): α-N-acetylgalactosaminyl- and α-galactosyltransferases in human ovarian cyst epithelial linings and fluids. *Biochem. Biophys. Res. Commun. 46*, 948-956.

188. HEARN, V. M., SMITH, Z. G. & WATKINS, W. M. (1968): An α-N-acetyl-D-galactosaminyltransferase associated with the human blood group A character. *Biochem. J. 109*, 315-317.

189. HEIER, H. E., KORNSTAD, L., NAMORK, E., ØSTGARD, P. & SANDIN, R. (1992): Expression of B and H antigens on red cells from a group B$_{weak}$ individual studied by serologic and scanning electron microscopic techniques. *Immunohematology 8*, 94-99.

190. HEIER, H. E., NAMORK, E., CALKOVSKA, Z., SANDIN, R. & KORNSTAD, L. (1994): Expression of A antigens on erythrocytes of weak blood group A subgroups. *Vox Sang. 66*, 231-236.

191. HENION, T. R., MACHER, B. A., ANARAKI, F. & GALILI, U. (1994): Defining the minimal size of catalytically active primate α1,3 galactosyltransferase: structure-function studies on the recombinant truncated enzyme. *Glycobiology 4*, 193-201.

192. HENRY, S., MOLLICONE, R., FERNANDEZ, P., SAMUELSSON, B., ORIOL, R. & LARSON, G. (1996): Homozygous expression of a missense mutation at nucleotide 385 in the FUT2 gene associates with the Le(a+b+) partial- secretor phenotype in an Indonesian family. *Biochem. Biophys. Res. Commun. 219*, 675-678.

193. HENRY, S., MOLLICONE, R., FERNANDEZ, P., SAMUELSSON, B., ORIOL, R. & LARSON, G. (1996): Molecular basis for erythrocyte Le(a+b+) and salivary ABH partial-secretor phenotypes: expression of a FUT2 secretor allele with an A→T mutation at nucleotide 385 correlates with reduced α(1,2)fucosyltransferase activity. *Glycoconj. J. 13*, 985-993.

194. HENRY, S., MOLLICONE, R., LOWE, J. B., SAMUELSSON, B. & LARSON, G. (1996): A second nonsecretor allele of the blood group α(1,2)fucosyltransferase gene (FUT2). *Vox Sang. 70*, 21-25.

195. HENRY, S., ORIOL, R. & SAMUELSSON, B. (1995): Lewis histo-blood group system and associated secretory phenotypes. *Vox Sang. 69*, 166-182.

196. HENRY, S. M., BENNY, A. G. & WOODFILED, D. G. (1990): Investigation of Lewis phenotypes on Polynesians: evidence of a weak secretor phenotype. *Vox Sang. 58*, 61-66.

197. HERRON, R., GREENWELL, P., WETSWOOD, M. C., RACE, C., SMITH, D. S. & WATKINS, W. M. (1980): An H-deficient blood with normal H transferase levels. *Vox Sang. 39*, 186-194.

198. HERRON, R., YOUNG, D., CLARK, M., SMITH, D. S., GILES, C. M., POOLE, J. & LIEW, Y. W. (1982): A specific antibody for cells with acquired B antigen. *Transfusion 22*, 525-527.

199. HINDSGAUL, O., KHARE, D. P., BACH, M. & LEMIEUX, R. U. (1985): Molecular recognition. III. The binding of the H-type 2 human blood group determinant by the lectin I of Ulex europaeus. *Can. J. Chem. 63*, 2653-2658.

200. HINDSGAUL, O., NORBERG, T., LE PENDU, J. & LEMIEUX, R. U. (1982): Synthesis of type 2 human blood group antigenic determinants. The H, X, and Y haptens and variations of the H type 2 determinants as probes for the combining site of the lextin I of Ulex europaeus. *Carbohydr. Res. 109*, 109-142.

201. HITOSHI, S., KUSUNOKI, S., KANAZAWA, I. & TSUJI, S. (1995): Molecular cloning and expression of two types of rabbit β-galactoside α1,2-fucosyltransferase. *J. Biol. Chem. 270*, 8844-8850.

202. HITOSHI, S., KUSUNOKI, S., KANAZAWA, I. & TSUJI, S. (1996): Molecular cloning and expression of a third type of rabbit GDP-L-fucose:β-D-galactoside 2-α-L- fucosyltransferase. *J. Biol. Chem. 271*, 16975-16981.

203. HOLGERSSON, J., BACKER, A. E., BREIMER, M. E., GUSTAVSSON, M. L., JOVALL, P. A., KARLSSON, H., PIMLOTT, W. & SAMUELSSON, B. E. (1992): The blood group-B type-4 heptaglycosylceramide is a minor blood group-B structure in human B-kidneys in contrast to the corresponding A-type-4 compound in A-kidneys - structural and in vitro biosynthetic studies. *Biochim. Biophys. Acta 1180*, 33-43.

204. HOLGERSSON, J., JOVALL, P. A., SAMUELSSON, B. E. & BREIMER, M. E. (1990): Structural characterization of non-acid glycosphingolipids in kidneys of single blood group-O and group-A

pigs. *J. Biochem. 108*, 766-777.

205. HRUBISKO, M. (1976): Deficient H types. *Rev. Fr. Transfus. 19*, 157-174.
206. HRUBISKO, M., CALKOVSKÁ, Z., MERGANCOVÁ, O. & GALLOVÁ, K. (1966): Beobachtungen über Varianten des Blutgruppensystems ABO. I. Studie der Variante A_m. *Blut 13*, 137-142.
207. HRUBISKO, M., CALKOVSKÁ, Z., MERGANCOVÁ, O. & GALLOVÁ, K. (1966): Beobachtungen über Varianten des Blutgruppensystems ABO. II. Beitrag zur Erblichkeit der A_m- Variante. *Blut 13*, 232-239.
208. HRUBISKO, M. & MERGANCOVA, O. (1966): Beobachtungen über Varianten des Blutgruppensystems ABO. III. Die neuen Variationen O_{Hm} und A_{Hm}. Ein Beitrag zur Frage der Biosynthese der Blutgruppen-Antigene. *Blut 13*, 278-285.
209. HRUBISKO, M., MERGANCOVÁ, O., PRODANOV, P., HAMMEROVÁ, T. & RACKOVÁ, M. (1980): Interalleleic competition and complementation in the ABO blood group system. *Immunol. Commun. 9*, 139-153.
210. HUMMEL, K. & KLEMENT, G. (1970): Untersuchungen über die Heterogenität des Bestandes an A- und H-Rezeptoren individueller A_1 und A_2 Erythrocyten. *Z. Immun.-Forsch. 140*, 221-258.
211. IKEMOTO, S. & FURUHATA, T. (1971): Serology and genetics of a new blood type B_m. *Nat. New Bi. 231*, 184-185.
212. IMBERTY, A., MIKROS, E., KOCA, J., MOLLICONE, R., ORIOL, R. & PEREZ, S. (1995): Computer simulation of histo-blood group oligosaccharides: energy maps of all constituting disaccharides and potential energy surfaces of 14 ABH and Lewis carbohydrate antigens. *Glycoconj. J. 12*, 331-349.
213. ISHIYAMA, I. & UHLENBRUCK, G. (1972): Some problems concerning the absorption mechanism of anti-A-agglutinin from Helix pomatia onto Sephadex G-200. *Z. Immun.-Forsch. 143*, 147-155.
214. ISSITT, P. D. (1985): The ABO blood group system. In: *Applied Blood Group Serology*. Montgomery Scientific Publications, Miami, Florida, USA, pp. 132-168.
216. JACQUINET, J. C. & SINAY, P. (1977): Synthesis of blood-group substances. 6. Synthesis of O-α-L-fucopyranosyl-(1-2)-O-β-D-galactopyranosyl-(1-4)-O-(α-L-fucopyranosyl-(1-3))-2-acet-amido-2-deoxy-α-D-glucopyranoside, the postulated Lewis d antigenic determinant. *J. Org. Chem. 42*, 720-724.
217. JANVIER, D., VEAUX, S., REVIRON, M., GUIGNIER, F. & BENBUNAN, M. (1990): Serological characterization of murine monoclonal antibodies directed against acquired B red cells. *Vox Sang. 59*, 92-95.
218. JÄRNEFELDT, J., RUSH, J., LI, Y. T. & LAINE, R. A. (1978): Erythroglycan, a high molecular weight glycopeptide with the repeating structure [galactosyl(1-4)-2-deoxy-2-acetamidoglucosyl(1-3)] comprising more than one-third of the protein-bound carbohydrate of human erythrocyte stroma. *J. Biol. Chem. 253*, 8006-8009.
219. JAROSCH, K., SCHNITZLER, S., PROKOP, O. & UHLENBRUCK, G. (1967): Anti-B (Anti-B_{sal}) in Forelleneiern. *Z. Ärztl. Fortbildung 61*, 758-759.
220. JOHNSON, P. H. & HOPKINSON, D. A. (1992): Detection of ABO blood group polymorphism by denaturing gradient gel electrophoresis. *Hum. Mol. Genet. 1*, 341-344.
221. JOVALL, P. A., LINDSTRÖM, K., PASCHER, I., PIMLOTT, W. & SAMUELSSON, B. E. (1987): Identification of a blood group A active hexaglycosylceramide with a type 1 carbohydrate chain in plasma of an A_1 Le(a-b-) secretor. *Arch. Biochem. Biophys. 257*, 409-415.
222. JOZIASSE, D. H., SHAPER, J. H., JABS, E. W. & SHAPER, N. L. (1991): Characterzation of an α1→3-galactosyltransferase homologue on human chromosome 12 that is organized as a processed pseudogene. *J. Biol. Chem. 266*, 6991-6998.
223. JOZIASSE, D. H., SHAPER, J. H., VAN DEN EIJNDEN, D. H., VAN TUNEN, A. J. & SHAPER, N. L. (1989): Bovine α1→3-galactosyltransferase: isolation and characterizatiohn of a cDNA clone. Identification of homologous sequences in human genomic DNA. *J. Biol. Chem. 264*, 14290-14297.
224. JOZIASSE, D. H., SHAPER, N. L., KIM, D. Y., VAN DEN EIJNDEN, D. H. & SHAPER, J. H. (1992): Murine α1,3-galactosyltransferase. A single gene locus specifies four isoforms of the enzyme by alternative splicing. *J. Biol. Chem. 267*, 5534-5541.
225. JUDD, W. J., McGUIRE-MALLORY, N., ANDERSON, K. M., HEATH, E. J., SWANSON, J., GRAY, J. M. & OBERMAN, H. A. (1979): Concomitant T- and Tk activation associated with acquired B antigens.

Transfusion 19, 293-298.
226. KABAT, E. A. (1956): *Blood Group Substances. Their Chemistry and Immunochemistry*. Academic Press, New York.
227. KABAT, E. A. (1962): Immunochemical studies on blood groups XXIX. Action of various oligosaccharides from human milk in inhibiting the cross reactions of type-XIV antipneumococcal sera with partially hydrolyzed blood-group substances (P1 fractions). *Arch. Biochem. Biophys. 1*, 181-186.
228. KABAT, E. A. (1971): *Einführung in die Immunchemie und Immunologie*. Springer-Verlag, Berlin-Heidelberg-New York.
229. KANEKO, M., NISHIHARA, S., SHINYA, N., KUDO, T., IWASAKI, H., SENO, T., OKUBO, Y. & NARIMATSU, H. (1997): Wide variety of point mutations in the H gene of Bombay and para-Bombay individuals that inactivate H enzyme. *Blood 90*, 839-849.
230. KANNAGI, R., LEVERY, S. B. & HAKOMORI, S. I. (1984): Blood group H antigen with globo-series structure. Isolation and characterization from human blood group O erythrocytes. *FEBS Lett. 175*, 397-401.
231. KANNAGI, R., LEVERY, S. B., ISHIGAMI, F., HAKOMORI, S. I., SHEVINSKY, L. H., KNOWLES, B. B. & SOLTER, D. (1983): New globoseries glycosphingolipids in human teratocarcinoma reactive with the monoclonal antibody directed to a developmentally regulated antigen, stage-specific embryonic antigen 3. *J. Biol. Chem. 258*, 8934-8942.
232. KARHI, K. K. & GAHMBERG, C. G. (1980): Identification of blood group A-active glycoproteins in the human erythrocyte membrane. *Biochim. Biophys. Acta 622*, 344-354.
233. KARHI, K. K. & GAHMBERG, C. G. (1980): Isolation and characterization of the blood group A-specific lectin from Vicia cracca. *Biochim. Biophys. Acta 622*, 337-343.
234. KARLSSON, K. A. & LARSON, G. (1981): Molecular characterization of cell surface antigens of fetal tissue. Detailed analysis of glycosphingolipids of meconium of a human O Le(a-b+) secretor. *J. Biol. Chem. 256*, 3512-3524.
235. KELLY, R. J., ERNST, L. K., LARSEN, R. D., BRYANT, J. G., ROBINSON, J. S. & LOWE, J. B. (1994): Molecular basis for H blood group deficiency in Bombay (O_h) and para-Bombay individuals. *Proc. Natl. Acad. Sci. USA 91*, 5843-5847.
236. KELLY, R. J., ROUQUIER, S., GIORGI, D., LENNON, G. G. & LOWE, J. B. (1995): Sequence and expression of a candidate for the human Secretor blood group α(1,2)fucosyltransferase gene (FUT2). Homozygosity for an enzyme-inactivating nonsense mutation commonly correlates with the non-secretor phenotype. *J. Biol. Chem. 270*, 4640-4649.
237. KELTON, J. G., HAMID, C., AKER, S. & BLAJCHMAN, A. (1982): The amount of blood group A substance on platelets is proportional to the amount in the plasma. *Blood 59*, 980-985.
238. KERMARREC, N., ROUBINET, F., APOIL, P. A. & BLANCHER, A. (1999): Comparison of allele O sequences of the human and non-human primate ABO system. *Immunogenetics 49*, 517-526.
239. KIM, Y. S., PERDOMO, J., BELLA, A. & NORDBERG, J. (1971): N-acetyl-D-galactosaminyltransferase in human serum and erythrocyte membranes. *Proc. Natl. Acad. Sci. USA 68*, 1753-1756.
240. KING, J. S., FILEDEN, M. L., GOODMAN, H. O. & BOYCE, W. H. (1961): Total nondialyzable solids in human urine. X. Isolation and characterization of non ultrafiltrable material with blood-group substance activity. *Arch. Biochem. Biophys. 95*, 310-315.
241. KISAILUS, E. C. & KABAT, E. A. (1978): Immunochemical studies on blood-groups. LXVI. Competitive binding assays of A_1 and A_2 blood-group substances with insolubilized anti-A serum and insolubilized A agglutinin from Dolichos biflorus. *J. Exp. Med. 147*, 830-843.
242. KISHI, K., TAKIZAWA, H. & ISEKI, S. (1977): Isoelectric analysis of B gene-associated α-galactosyltransferases in human serum and saliva. *Proc. J p. Acad. 53*, 172-177.
243. KNOWLES, R. W., BAI, Y., DANIELS, G. L. & WATKINS, W. M. (1982): Monoclonal anti-type 2 H: an antibody detecting a precursor of the A and B blood group antigens. *J. Immunogenet. 9*, 69-76.
244. KOBATA, A. (1977): Milk glycoproteins and oligosaccharides. In: *The Glycoconjugates*. (M. I. Horowitz, and W. Pigman, eds.). Academic Press, New York, Vol. 1, pp. 423-440.
245. KOBATA, A. & GINSBURG, V. (1970): Uridine diphosphate-N-acetyl-D-galactosamine:D-galactose

α-3-N-acetyl-D-galactosaminyltransferase, a product of the gene that determines blood type A in man. *J. Biol. Chem. 245*, 1484-1490.

246. KOBATA, A., GROLLMAN, E. F. & GINSBURG, V. (1968): An enzymic basis for blood type A in humans. *Arch. Biochem. Biophys. 124*, 609-612.

247. KOBATA, A., GROLLMAN, E. F. & GINSBURG, V. (1968): An enzymatic basis for blood type B in humans. *Biochem. Biophys. Res. Commun. 32*, 272-277.

248. KODA, Y., SOEJIMA, M., JOHNSON, P. H., SMART, E. & KIMURA, H. (1997): Missense mutation of FUT1 and deletion of FUT2 are responsible for Indian Bombay phenotype of ABO blood group system. *Biochem. Biophys. Res. Commun. 238*, 21-25.

249. KODA, Y., SOEJIMA, M. & KIMURA, H. (1997): Structure and expression of H-type GDP-L-fucose:β-D-galactoside 2-α-L-fucosyltransferase gene (FUT1) - Two transcription start sites and alternative splicing generate several forms of FUT1 mRNA. *J. Biol. Chem. 272*, 7501-7505.

250. KODA, Y., SOEJIMA, M., LIU, Y. H. & KIMURA, H. (1996): Molecular basis for secretor type α(1,2)-fucosyltransferase gene deficiency in a Japanese population: a fusion gene generated by unequal crossover responsible for the enzyme deficiency. *Amer. J. Hum. Genet. 59*, 343-350.

251. KODA, Y., SOEJIMA, M., WANG, B. J. & KIMURA, H. (1997): Structure and expression of the gene encoding secretor-type galactoside 2-α-L-fucosyltransferase (FUT2). *Eur. J. Biochem. 246*, 750-755.

252. KOGURE, T. (1975): The action of group B_m or cisAB sera on group O red cells in the presence of UDP-D-galactose. *Vox Sang. 29*, 51-58.

253. KOGURE, T. & FURUKAWA, K. (1976): Enzymatic conversion of human group O red cells into group B active cells by α-D-galactosyltransferase of sera and salivas from group B and its variant types. *J. Immunogenet. 3*, 147-154.

254. KOGURE, T. & FURUKAWA, K. (1980): Detection and activity of blood group B gene-associated α-galactosyltransferase in human urine. *J. Immunogenet. 7*, 375-380.

255. KOMINATO, Y., FUJIKURA, T., TAKIZAWA, T., HAYASHI, K., MORI, T., MATSUE, K., YASUE, S. & MATSUDA, T. (1990): Antibody to blood group glycosyltransferases in a patient transplanted with an ABO incompatible bone marrow. *Exp. Clin. Immunogenet. 7*, 85-90.

256. KOMINATO, Y., MCNEILL, P. D., YAMAMOTO, M., RUSSELL, M., HAKOMORI, S. & YAMAMOTO, F. (1992): Animal histo blood group ABO genes. *Biochem. Biophys. Res. Commun. 189*, 154-164.

257. KONAMI, Y., YAMAMOTO, K. & OSAWA, T. (1991): Purification and characterization of two types of Cytisus sessilifolius anti-H(O) lectins by affinity chromatography. *Biol. Chem. Hoppe-Seyler 372*, 103-111.

258. KONAMI, Y., YAMAMOTO, K., TSUJI, T., MATSUMOTO, I. & OSAWA, T. (1983): Purification and characterization of two types of Laburnum alpinum anti-H(O) hemagglutinin by affinity chromatography. *Hoppe-Seyler's Z. Physiol. Chem. 364*, 397-405.

259. KOSCIELAK, J., LENKIEWICZ, B., ZIELENSKI, J. & SEYFRIED, H. (1986): Weak A phenotypes possibly caused by mutation. *Vox Sang. 50*, 187-190.

260. KOSCIELAK, J., MILLER-PODRAZA, H., KRAUZE, R. & PIASEK, A. (1976): Isolation and characterization of poly(glycosyl)ceramides (megaloglycolipids) with A, H, and I blood group activities. *Eur. J. Biochem. 71*, 9-18.

261. KOSCIELAK, J., PACUSKA, T. & DZIERZKOWA-BORODEJ, W. (1976): Activity of B-gene-specified galactosyltransferase in individuals with B_m phenotypes. *Vox Sang. 30*, 58-67.

262. KOSCIELAK, J., PIASEK, A., GORNIAK, H., GARDAS, A. & GREGOR, A. (1973): Structures of fucose-containing glycolipids with H and B blood-group activity and of sialic acid and glucosamine-containing glycolipid of human erythrocyte membrane. *Eur. J. Biochem. 37*, 214-225.

263. KOSCIELAK, J., ZDEBSKA, E., WILCZYNSKA, Z., MILLER-PODRAZA, H. & DZIERZKOWA-BORODEJ, W. (1979): Immunochemistry of Ii-active glycosphingolipids of erythrocytes. *Eur. J. Biochem. 96*, 331-337.

264. KOTHBAUER, H. & SCHENKEL-BRUNNER, H. (1975): Hemagglutinins in fish eggs: comparative studies on different Salmonidae species. *Comp. Biochem. Physiol. 50A*, 27-29.

265. KOTHBAUER, H. & SCHENKEL-BRUNNER, H. (1978): Immunchemische Untersuchungen an

Braunfrosch-Laich: Vergleichende Untersuchungen an Rana temporaria, Rana arvalis, Rana dalmatina und Rana latastei. *Z. Zool. Syst. Evolut.-Forsch. 16*, 144-148.

266. KRÜPE, M. & BRAUN, C. (1952): Über ein pflanzliches Hämagglutinin gegen menschliche B-Blutzellen. *Naturwiss. 39*, 284-285.

267. KRUSIUS, T., FINNE, J. & RAUVALA, H. (1978): The poly(glycosyl) chains of glycoproteins. Characterization of a novel type of glycoprotein saccharides from human erythrocyte membrane. *Eur. J. Biochem. 92*, 298-300.

268. KUDO, T., IWASAKI, H., NISHIHARA, S., SHINYA, N., ANDO, T., NARIMATSU, I. & NARIMATSU, H. (1996): Molecular genetic analysis of the human Lewis histo-blood group system. 2. Secretor gene inactivation by a novel single missense mutation A385T in Japanese nonsecretor individuals. *J. Biol. Chem. 271*, 9830-9837.

269. KUHNS, W. J., OLIVER, R. T., WATKINS, W. M. & GREENWELL, P. (1980): Leukemia-induced alterations of serum glycosyltransferase enzymes. *Cancer Res. 40*, 268-275.

270. KUMAZAKI, T. & YOSHIDA, A. (1984): Biochemical evidence that secretor gene, Se, is a structural gene encoding a specific fucosyltransferase. *Proc. Natl. Acad. Sci. USA 81*, 4193-4197.

271. KYPRIANOU, P., BETTERIDGE, A., DONALD, A. S. R. & WATKINS, W. M. (1990): Purification of the blood group H gene associated α-2-L-fucosyltransferase from human plasma. *Glycoconj. J. 7*, 573-588.

272. LANDSTEINER, K. (1900): Zur Kenntnis der antifermentativen, lytischen und agglutinierenden Wirkungen des Blutserums und der Lymphe. *Zentralbl. Bakteriol. 27*, 357-362.

273. LANDSTEINER, K. (1901): Über Agglutinationserscheinungen normalen menschlichen Blutes. *Wien. Klin. Wochenschr. 14*, 1132-1134.

274. LANDSTEINER, K. & LEVINE, P. (1930): Differentiation of a type of human blood by means of normal animal serum. *J. Immunol.*, 87-94.

275. LANSET, S. & ROPARTZ, C. (1971): A second example of acquired B-like antigen in a healthy person. *Vox Sang. 20*, 82-84.

276. LANSET, S., ROPARTZ, C., ROUSSEAU, P. Y., GUERBET, Y. & SALMON, C. (1966): Une famille comportant les phénotypes Bombay: O^{AB}_h et O^B_h. *Transfusion (Paris) 9*, 255-258.

277. LARSEN, R. D., ERNST, L. K., NAIR, R. P. & LOWE, J. B. (1990): Molecular cloning, sequence, and expression of a human GDP-L-fucose : β-D-galactoside 2-α-L-fucosyltransferase cDNA that can form the H blood group antigen. *Proc. Natl. Acad. Sci. USA 87*, 6674-6678.

278. LARSEN, R. D., RAJAN, V. P., RUFF, M. M., KUKOWSKA-LATALLO, J., CUMMINGS, R. D. & LOWE, J. B. (1989): Isolation of a cDNA encoding a murine UDP-galactose: β-D-galactosyl-1,4-N-acetyl-D-glucosaminide α-1,3-galactosyltransferase: expression cloning by gene transfer. *Proc. Natl. Acad. Sci. USA 86*, 8227-8231.

279. LARSEN, R. D., RIVERA-MARRERO, C. A., ERNST, L. K., CUMMINGS, R. D. & LOWE, J. B. (1990): Frameshift and nonsense mutations in a human genomic sequence homologous to a murine UDP-Gal:β-D-Gal(1,4)-D-GlcNAc α(1,3)-galactosyltransferase cDNA. *J. Biol. Chem. 265*, 7055-7061.

280. LARSON, G. & SAMUELSSON, B. E. (1980): Blood-group type glycosphingolipids of human cord blood erythrocytes. *J. Biochem. 88*, 647-657.

281. LAU, P., SERERAT, S., BEATTY, J., OILSCHLAGER, R. & KINI, J. (1990): Group A variants defined with a monoclonal anti-A reagent. *Transfusion 30*, 142-145.

282. LE PENDU, J. (1989): A hypothesis on the dual significance of ABH, Lewis and related antigens. *J. Immunogenet. 16*, 53-61.

283. LE PENDU, J., CARTRON, J. P., LEMIEUX, R. U. & ORIOL, R. (1985): The presence of at least two different H-blood-group-related β-D-Gal α-2-L-fucosyltransferases in human serum and the genetics of blood group H substances. *Amer. J. Hum. Genet. 37*, 749-760.

284. LE PENDU, J., CLAMAGIRAND-MULET, C., CARTRON, J. P., GERARD, G., VITRAC, D. & ORIOL, R. (1983): H-deficient blood groups of Reunion island. III. α-L-Fucosyltransferase activity in sera of homozygous and heterozygous individuals. *Amer. J. Hum. Genet. 35*, 497-507.

285. LE PENDU, J., GERARD, G., VITRAC, D., JUSZCZAK, G., LIBERGE, G., ROUGER, P., SALMON, C., LAMBERT, F., DALIX, A. M. & ORIOL, R. (1983): H-deficient blood groups of Reunion island. II. Differences

between Indians (Bombay phenotype) and whites (Reunion phenotype). *Amer. J. Hum. Genet. 35*, 484-496.

286. LE PENDU, J., LAMBERT, F., GERARD, G., VITRAC, D., MOLLICONE, R. & ORIOL, R. (1986): On the specificity of human anti-H antibodies. *Vox Sang. 50*, 223-226.

287. LE PENDU, J., LAMBERT, F., SAMUELSSON, B., BREIMER, M. E., SEITZ, R. C., URDANIZ, M. P., SUESA, N., RATCLIFFE, M., FRANCOIS, A., POSCHMANN, A., VINAS, J. & ORIOL, R. (1986): Monoclonal antibodies specific for type 3 and type 4 chain-based blood group determinants: relationship to the A1 and A2 subgroups. *Glycoconj. J. 3*, 255-271.

288. LE PENDU, J., LEMIEUX, R. U., LAMBERT, F., DALIX, A. M. & ORIOL, R. (1982): Distribution of H type 1 and H type 2 antigenic determinants in human sera and saliva. *Amer. J. Hum. Genet. 34*, 402-415.

289. LE PENDU, J., ORIOL, R., JUSZCZAK, G., LIBERGE, G., ROUGER, P., SALMON, C. & CARTRON, J. P. (1983): α-2-L-fucosyltransferase activity in sera of individuals with H-deficient red cells and normal H antigen in secretions. *Vox Sang. 44*, 360-365.

290. LEMIEUX, R. U. (1978): Human blood groups and carbohydrate chemistry. *Chem. Soc. Rev. 7*, 423-452.

291. LEMIEUX, R. U., BAKER, D. A., WEINSTEIN, W. M. & SWITZER, C. M. (1981): Artificial antigens. Antibody preparations for the localization of Lewis determinants in tissues. *Biochemistry 20*, 199-205.

292. LEMIEUX, R. U., BOCK, K., DELBAERE, L. T. J., KOTO, S. & RAO, V. S. (1980): The conformations of oligosaccharides related to the ABH and Lewis human blood group determinants. *Can. J. Chem. 58*, 631-653.

293. LEMIEUX, R. U. & DRIGUEZ, H. (1975): The chemical synthesis of 2-O-(α-L-fucopyranosyl)-3-O-(α-D-galactopyranosyl)-D-galactose. The terminal structure of the blood-group B antigenic determinant. *J. Amer. Chem. Soc. 97*, 4069-4075.

294. LINDBERG, B., LONNGREN, J. & POWELL, D. A. (1977): Structural studies on the specific type-14 pneumococcal polysaccharide. *Carbohydr. Res. 58*, 177-186.

295. LIS, H., LATTER, H., ADAR, R. & SHARON, N. (1988): Isolation of two blood type A and N specific isolectins from Moluccella laevis seeds. *FEBS Lett. 233*, 191-195.

296. LIU, Y., KODA, Y., SOEJIMA, M., PANG, H., SCHLAPHOFF, T., DU TOIT, E. & KIMURA, H. (1998): Extensive polymorphism of the FUT2 gene in an African (Xhosa) population of South Africa. *Hum. Genet. 103*, 204-210.

297. LLOYD, K. O. & KABAT, E. A. (1968): Immunochemical studies on blood groups. XLI. Proposed structures for the carbohydrate portions of blood group A, B, H, Lewis[a], and Lewis[b] substances. *Proc. Natl. Acad. Sci. USA 61*, 1470-1477.

298. LOPEZ, M., BENALI, J., CARTRON, J. P. & SALMON, C. (1980): Some notes on the specificity of anti-A$_1$ reagents. *Vox Sang. 39*, 271-276.

299. LOPEZ, M., GERBAL, A. & SALMON, C. (1972): Excés d'antigène I dans les érythrocytes de phénotypes O$_h$, A$_h$ et B$_h$. *Rev. Fr. Transfus. 15*, 187-193.

300. LOPEZ, M., LIBERGE, G., GERBAL, A., BROCTEUR, J. & SALMON, C. (1976): Cis AB blood groups. Immunologic, thermodynamic, and quantitative studies of ABH antigen. *Biomedicine 24*, 265-271.

301. LOWE, A. D., LENNOX, E. S. & VOAK, D. (1984): A new monoclonal anti-A. Culture supernatants with a performance of hyperimmune human reagents. *Vox Sang. 46*, 29-35.

302. LUBENKO, A. & IVANYI, J. (1986): Epitope specificity of blood-group-A-reactive murine monoclonal antibodies. *Vox Sang. 51*, 136-142.

303. LUNDBLAD, A. (1977): Urinary glycoproteins, glycopeptides, and oligosaccharides. In: *The Glycoconjugates* (M. I. Horowitz, and W. Pigman, eds.). Academic Press, New York, Vol. 1, pp. 441-458.

304. MADSEN, G. & HEISTÖ, H. (1968): A Korean family showing inheritance of A and B on the same chromosome. *Vox Sang. 14*, 211-217.

305. MAK, K. H., LUBENKO, A., GREENWELL, P., VOAK, D., YAN, K. F. & POOLE, J. (1996): Serologic characteristics of H-deficient phenotypes among Chinese in Hong Kong. *Transfusion 36*, 994-999.

306. MÄKELÄ, O. (1957): Studies in hemagglutinins of Leguminosae seeds. *Ann. Med. Exp. Biol. Fenn.* *35*, 1-133.
307. MÄKELÄ, O. & MÄKELÄ, P. (1956): Some new blood group specific phytagglutinins. *Ann. Med. Exp. Biol. Fenn.* 34, 402-404.
308. MÄKELÄ, O., MÄKELÄ, P. & KRÜPE, M. (1959): Zur Spezifität der Anti-B Phythämagglutinine. *Z. Immun.-Forsch. 117*, 220-229.
309. MÄKELÄ, O., RUOSLATHI, E. & EHNHOLM, C. (1969): Subtypes of human ABO blood-groups and subtype-specific antibodies. *J. Immunol. 102*, 763-771.
310. MANDEL, U., LANGKILDE, N. C., ØRNTOFT, T. F., THERKILDSEN, M. H., KARKOV, J., REIBEL, J., WHITE, T., CLAUSEN, H. & DABELSTEEN, E. (1992): Expression of histo-blood-group-A/B-gene-defined glycosyltransferases in normal and malignant epithelia: correlation with A/B-carbohydrate expression. *Int. J. Cancer 52*, 7-12.
311. MARSH, W. C. (1960): The pseudo-B antigen. A study of its development. *Vox Sang. 5*, 387-397.
312. MARSH, W. L., FERRARI, M., NICHOLS, M. E., FERNANDEZ, G. & COOPER, K. (1973): B_m^H: a weak B antigen variant. *Vox Sang. 25*, 341-346.
313. MARSH, W. L., JENKINS, W. J. & WALTHER, W. W. (1959): Pseudo B: an acquired group antigen. *Brit. Med. J. ii*, 63-66.
314. MARTIN, A., BIOL, M. C., ARRAMBIDE, E., RICHARD, M. & LOUISOT, P. (1981): Paramètres cinétiques d'une fucosyl-transférase intestinale soluble purifiée. *Biochimie 63*, 241-245.
315. MARTINKO, J. M., VINCEK, V., KLEIN, D. & KLEIN, J. (1993): Primate ABO glycosyltransferases - evidence for trans-species evolution. *Immunogenetics 37*, 274-278.
316. MASUTANI, H. & KIMURA, H. (1995): Purification and characterization of secretory-type GDP-L-fucose:β-D-galactoside 2-α-L-fucosyltransferase from human gastric mucosa. *J. Biochem. 118*, 541-545.
317. MATSUE, K., YASUE, S., MATSUDA, T., IWABUCHI, K., OHTSUKA, M., UEDA, M., KONDO, K., SHIOBARA, S., MORI, T., KOIZUMI, S. I., YAMAGAMI, M. & HARADA, M. (1989): Plasma glycosyltransferase activity after ABO-incompatible bone marrow transplantation and development of an inhibitor for glycosyltransferase activity. *Exp. Hematol. 17*, 827-831.
318. MATSUKURA, Y. (1976): α-Galactosyltransferase activity in the serum of frogs (Rana catesbeiana). *Z. Immun.-Forsch. 152*, 260-265.
319. MATSUKURA, Y. (1976): On the blood group B gene-specified α-galactosyltransferase in the serum of the Japanese tortoise (Clemmys japonica). *Immunology 31*, 571-575.
320. MATSUMOTO, I. & OSAWA, T. (1969): Purification of and anti-H(O) phytohaemagglutinin of Ulex europaeus. *Biochim. Biophys. Acta 194*, 180-189.
321. MATSUMOTO, I. & OSAWA, T. (1974): Specific purification of eel serum and Cytisus sessilifolius anti-H hemagglutinins by affinity chromatography and their binding to human erythrocytes. *Biochemistry 13*, 582-588.
322. MCALPINE, P. J., SHOWS, T. B., BOUCHEIX, C., STRANC, L. C., BERENT, T. G., PAKSTIS, A. J. & DOUTE, R. C. (1989): Report of the nomenclature committee and the catalog of mapped genes. *Cytogenet. Cell Genet. 51*, 13-66.
323. MCGOWAN, A., TOD, A., CHIRNSIDE, A., GREEN, C., MCCOLL, K., MOORE, S., YAP, P. L., MCCLELLAND, D. B. L., MCCANN, M. C., MICKLEM, L. R. & JAMES, K. (1989): Stability of murine monoclonal anti-A, anti-B, and anti-A,B ABO grouping reagents and a multi-centre evaluation of their performance in routine use. *Vox Sang. 56*, 122-130.
324. MEIJERINK, E., FRIES, R., VÖGELI, P., MASABANDA, J., WIGGER, G., STRICKER, C., NEUENSCHWANDER, S., BERTSCHINGER, H. U. & STRANZINGER, G. (1997): Two α(1,2) fucosyltransferase genes on porcine Chromosome 6q11 are closely linked to the blood group inhibitor (S) and Escherichia coli F18 receptor (ECF18R) loci. *Mamm. Genome 8*, 736-741.
325. MESSETER, L., BRODIN, T., CHESTER, M. A., LOW, B. & LUNDBLAD, A. (1984): Mouse monoclonal antibodies with anti-A, anti-B, and anti-A,B specificities; some superior to human polyclonal ABO reagents. *Vox Sang. 46*, 185-194.
326. METOKI, R., KAKUDO, K., TSUJI, Y., TENG, N., CLAUSEN, H. & HAKOMORI, S. I. (1989): Deletion of

histo-blood group A and B antigens and expression of incompatible A antigen in ovarian cancer. *J. Natl. Cancer Inst. 81*, 1151-1157.

327. MILNER, L. V. & CALITZ, F. (1968): Quantitative studies of the erythrocytic B antigen in South African Caucasian, Bantu, and Asiatic blood donors. *Transfusion 8*, 277-282.

328. MOHN, J. F., CUNNINGHAM, R. K. & BATES, J. E. (1977): Qualitative distinctions between subgroups A_1 and A_2. In: *Human Blood Groups - Proceedings of the 5th International Convocation on Immunologty, Buffalo 1976.* (J.F. Mohn, R.W. Plunkett, R.K. Cunningham, and R. R. Lambert, eds.). S. Karger, Basel, pp. 361-325.

329. MOHN, J. F., CUNNINGHAM, R. K., PIRKOLA, A., FURUHJELM, U. & NEVANLINNA, H. R. (1973): An inherited blood group A variant in the Finnish population. I. Basic characteristics. *Vox Sang. 25*, 193-211.

330. MOJENA, M. & BOSCÁ, L. (1989): Identification of an anti-A and anti-B blood group glycosyltransferase antibody after incompatible bone marrow transplant. *Blood 74*, 1134-1138.

331. MOLLICONE, R., BARA, J., LE PENDU, J. & ORIOL, R. (1985): Immunohistologic pattern of type 1 (Le^a, Le^b) and type 2 (X, Y, H) blood-group related antigens in the human pyloric and duodenal mucosae. *Lab. Invest. 53*, 219-227.

332. MOLLICONE, R., CAILLARD, T., LE PENDU, J., FRANCOIS, A., SANSONETTI, N., VILLARROYA, H. & ORIOL, R. (1988): Expression of ABH and X (Le^x) antigens on platelets and lymphocytes. *Blood 71*, 1113-1119.

333. MOLLICONE, R., LE PENDU, J., BARA, J. & ORIOL, R. (1986): Heterogeneity of the ABH antigenic determinants expressed in human pyloric and duodenal mucosae. *Glycoconj. J. 3*, 187-202.

334. MOORE, B. P. L., NEWSTEAD, P. H. & MARSON, A. (1961): A weak inherited group A phenotype. *Vox Sang. 6*, 624-626.

335. MOORE, S., CHIRNSIDE, A., MICKLEM, L. R., MCCLELLAND, D. B. L. & JAMES, K. (1984): A mouse monoclonal antibody with anti-A,(B) specificity which agglutinates A_x cells. *Vox Sang. 47*, 427-434.

336. MOORES, P. P., ISSITT, P. D., PAVONE, B. G. & MCKEEVER, B. G. (1975): Some observations on "Bombay" bloods, with comments on evidence for the existence of two different O_h phenotypes. *Transfusion 15*, 237-243.

337. MORENO, C., LUNDBLAD, A. & KABAT, E. A. (1971): Immunochemical studies on blood-groups. LI. A comparative study of the reaction of A_1 and A_2 blood-group glycoproteins with human anti-A. *J. Exp. Med. 134*, 439-457.

338. MORGAN, W. T. J. & WATKINS, W. M. (1948): The detection of a product of the blood group O gene and the relationship of the so-called O substance to the agglutinogens A and AB. *Brit. J. Exp. Pathol. 29*, 159-173.

339. MULET, C., CARTRON, J. P., BADET, J. & SALMON, C. (1977): Activity of 2-α-L-fucosyltransferase in human sera and red cell membranes. A study of common ABH blood donors, rare 'Bombay' and 'Parabombay' individuals. *FEBS Lett. 84*, 74-78.

340. MULET, C., CARTRON, J. P., LOPEZ, M. & SALMON, C. (1978): ABH glycosyltransferase levels in sera and red cell membranes from H_z and H_m varaint bloods. *FEBS Lett. 90*, 233-238.

341. MULET, C., CARTRON, J. P., SCHENKEL-BRUNNER, H., DUCHET, H., SINAY, P. & SALMON, C. (1979): Probable biosynthetic pathway for the synthesis of the B antigen from B_h variants. *Vox Sang. 37*, 272-280.

342. MUNRO, J. R. & SCHACHTER, H. (1973): The presence of two GDP-L-fucose:glycoprotein fucosyltransferases in human serum. *Arch. Biochem. Biophys. 156*, 534-542.

343. MURPHY, L. A. & GOLDSTEIN, I. J. (1977): Five α-D-galactopyranosyl-binding isolectins from Bandeiraea simplicifolia seeds. *J. Biol. Chem. 252*, 4739-4742.

344. MUSCHEL, L. H. & OSAWA, E. (1959): Human blood group substance B in Escherichia coli O 86. *Proc. Soc. Exp. Biol. 101*, 614-617.

345. NAGAI, M., DAVÉ, V., KAPLAN, B. E. & YOSHIDA, A. (1978): Human blood group glycosyltransferases. I. Purification of N-acetylgalactosaminyltransferase. *J. Biol. Chem. 253*, 377-379.

346. NAGAI, M., DAVÉ, V., MUENSCH, H. & YOSHIDA, A. (1978): Human blood group glycosyltransferase.

II. Purification of galactosyltransferase. *J. Biol. Chem. 253*, 380-381.

347. NAGAI, M. & YOSHIDA, A. (1978): Possible existence of hybrid glycosyltransferase in heterozygous blood group AB subjects. *Vox Sang. 35*, 378-381.

348. NAKAJIMA, T., YAZAWA, S., MIYAZAKI, S. & FURUKAWA, K. (1993): Immunochemical characterization of anti-H monoclonal antibodies obtained from a mouse immunized with human saliva. *J. Immunol. Methods 159*, 261-267.

349. NAKAMURA, I., TAKIZAWA, H. & NISHINO, K. (1989): A_3 phenotype with A^1 gene-specified enzyme character in serum. *Exp. Clin. Immunogenet. 6*, 143-149.

350. NAVARATNAM, N., FINDLAY, J. B. C., KEEN, J. N. & WATKINS, W. M. (1990): Purification, properties and partial amino acid sequence of the blood-group-A-gene-associated α-3-N-acetylgalactosaminyltransferase from human gut mucosal tissue. *Biochem. J. 271*, 93-98.

351. NEMEC, M., DRIMALOVA, D., HOREJSI, V., VANAK, J., BARTEK, J. & VIKLICKY, V. (1987): Murine monoclonal antibodies to human A erythrocytes: differential reactivity with N-acetyl-D-galactosamine. *Vox Sang. 52*, 125-128.

352. NEVANLINNA, H. R. & PIRKOLA, A. (1973): An inherited blood group A variant in the Finnish population. II. Population studies. *Vox Sang. 24*, 404-416.

353. NILSSON, O., MANSSON, J. E., BREZICKA, T., HOLMGREN, J., LINDHOLM, L., SORENSON, S., YNGVASON, F. & SVENNERHOLM, L. (1984): Fucosyl G_{M1}, a ganglioside associated with small cell lung carcinomas. *Glycoconj. J. 1*, 43-49.

354. NYHOLM, P. G., SAMUELSSON, B. E., BREIMER, M. & PASCHER, I. (1989): Conformational analysis of blood group A-active glycosphingolipids using HSEA-calculations. The possible significance of the core oligosaccharide chain for the presentation and recognition of the A-determinant. *J. Mol. Recognit. 2*, 103-113.

355. OGASAWARA, K., BANNAI, M., SAITOU, N., YABE, R., NAKATA, K., TAKENAKA, M., FUJISAWA, K., UCHIKAWA, M., ISHIKAWA, Y., JUJI, T. & TOKUNAGA, K. (1996): Extensive polymorphism of ABO blood group gene: three major lineages of the alleles for the common ABO phenotypes. *Hum. Genet. 97*, 777-783.

356. OGASAWARA, K., YABE, R., UCHIKAWA, M., SAITOU, N., BANNAI, M., NAKATA, K., TAKENAKA, M., FUJISAWA, K., ISHIKAWA, Y., JUJI, T. & TOKUNAGA, K. (1996): Molecular genetic analysis of variant phenotypes of the ABO blood group system. *Blood 88*, 2732-2737.

357. OLSSON, M. L. & CHESTER, M. A. (1995): A rapid and simple ABO genotype screening method using a novel B/O² versus A/O² discriminating nucleotide substitution at the ABO hocus. *Vox Sang. 69*, 242-247.

358. OLSSON, M. L. & CHESTER, M. A. (1996): Evidence for a new type of O allele at the ABO locus, due to a combination of the A² nucleotide deletion and the A^{el} nucleotide insertion. *Vox Sang. 71*, 113-117.

359. OLSSON, M. L. & CHESTER, M. A. (1996): Frequent occurrence of a variant O^1 gene at the blood group ABO locus. *Vox Sang. 70*, 26-30.

360. OLSSON, M. L. & CHESTER, M. A. (1996): Polymorphisms at the ABO locus in subgroup A individuals. *Transfusion 36*, 309-313.

361. OLSSON, M. L., GUERREIRO, J. F., ZAGO, M. A. & CHESTER, M. A. (1997): Molecular analysis of the O alleles at the blood group ABO locus in populations of different ethnic origin reveals novel crossing-over events and point mutations. *Biochem. Biophys. Res. Commun. 234*, 779-782.

362. OLSSON, M. L., THURESSON, B. & CHESTER, M. A. (1995): An A^{el} allele-specific nucleotide insertion at the blood group ABO locus and its detection using a sequence-specific polymerase chain reaction. *Biochem. Biophys. Res. Commun. 216*, 642-647.

363. ORIOL, R. (1987): ABH and related tissue antigens. *Biochem. Soc. Trans. 15*, 596-599.

364. ORIOL, R. (1987): Tissular expression of ABH and Lewis antigens in humans and animals: expected value of different animal models in the study of ABO-incompatible organ transplants. *Transpl. Proc. 19*, 4416-4420.

365. ORIOL, R. (1990): Genetic control of the fucosylation of ABH precursor chains. Evidence for new epistatic interactions in different cells and tissues. *J. Immunogenet. 17*, 235-245.

366. ORIOL, R., DANILOVS, J. & HAWKINS, B. R. (1981): A new genetic model proposing that the Se gene is a structural gene closely linked to the H gene. *Amer. J. Hum. Genet. 33*, 421-431.

367. ORIOL, R., DANILOVS, J., LEMIEUX, R. U., TERASAKI, P. I. & BERNOCO, D. (1980): Lymphocytotoxic definition of combined ABH and Lewis antigens and their transfer from sera to lymphocytes. *Hum. Immunol. 1*, 195-205.

368. ORIOL, R., LE PENDU, J. & MOLLICONE, R. (1986): Genetics of ABO, H, Lewis, X, and related antigens. *Vox Sang. 51*, 161-171.

369. ORIOL, R., LE PENDU, J. & SPARKES, R. (1981): Insights into the expression of ABH and Lewis antigens through human bone marrow transplantation. *Amer. J. Hum. Genet. 33*, 551-560.

370. ORIOL, R., SAMUELSSON, B. E. & MESSETER, L. (1990): ABO antibodies. Serological behaviour and immuno-chemical characterization. *J. Immunogenet. 17*, 279-299.

371. ORIOL, R., YE, Y., KOREN, E. & COOPER, D. K. (1993): Carbohydrate antigens of pig tissues reacting with human natural antibodies as potential targets for hyperacute vascular rejection in pig-to-man organ xenotransplantation. *Transplantation 56*, 1433-1442.

372. ØRNTOFT, T. F., GREENWELL, P., CLAUSEN, H. & WATKINS, W. M. (1991): Regulation of the oncodevelopmental expression of type 1 chain ABH and Lewis(b) blood group antigens in human colon by α-2-L-fucosylation. *Gut 32*, 287-293.

373. OTTENSOOSER, F., SATO, M., SATO, R. & KURATA, H. (1972): Lectin specificity induced by a fungus. *Vox Sang. 22*, 354-358.

374. OTTENSOOSER, F., SATO, R. & SATO, M. (1968): A new anti-B lectin. *Transfusion 8*, 44-46.

375. PACUSKA, T. & KOSCIELAK, J. (1972): The biosynthesis of blood-group-B character on human O-erythrocytes by a soluble α-galactosyltransferase from milk. *Eur. J. Biochem. 31*, 574-577.

376. PACUSKA, T. & KOSCIELAK, J. (1974): α1→2 Fucosyltransferase of human bone marrow. *FEBS Lett. 41*, 348-351.

377. PACUSKA, T., KOSCIELAK, J., SEYFRIED, H. & WALEWSKA, I. (1975): Biochemical, serological, and family studies in individuals with cis AB phenotypes. *Vox Sang. 29*, 292-300.

378. PEMBERTON, R. T. (1974): Anti-A and anti-B of gastropod origin. *Ann. N.Y. Acad. Sci. 234*, 95-121.

379. PEREIRA, M. E. A. & KABAT, E. A. (1974): Specificity of purified hemagglutinin (lectin) from Lotus tetragonolobus. *Biochemistry 13*, 3184-3192.

380. PEREIRA, M. E. A., KISAILUS, E. C., GRUEZO, F. & KABAT, E. A. (1978): Immunochemical studies on the combining site of the blood group H-specific lectin 1 from Ulex europaeus seeds. *Arch. Biochem. Biophys. 185*, 108-115.

381. PETRYNIAK, J. & GOLDSTEIN, I. J. (1986): Immunochemical studies on the interaction between synthetic glycoconjugates and α-L-fucosyl binding lectins. *Biochemistry 25*, 2829-2838.

382. PIAU, J. P., LABARRIERE, N., DABOUIS, G. & DENIS, M. G. (1994): Evidence for two distinct α(1,2)-fucosyltransferase genes differentially expressed throughout the rat colon. *Biochem. J. 300*, 623-626.

383. PILLER, V., PILLER, F. & CARTRON, J. P. (1990): Comparison of the carbohydrate-binding specificities of seven N-acetyl-D-galactosamine-recognizing lectins. *Eur. J. Biochem. 191*, 461-466.

384. PLATO, C. C. & GERSHOWITZ, H. (1961): Specific differences in the inhibition titers of the anti-H lectins from Cytisus sessilifolius and Ulex europaeus. *Vox Sang. 6*, 336-347.

385. PORETZ, R. D. & WATKINS, W. M. (1972): Galactosyltransferases in human submaxillary glands and stomach mucosa associated with the biosynthesis of blood group B specific glycoproteins. *Eur. J. Biochem. 25*, 455-462.

386. PRIEELS, J. P., BEYERS, T. & HILL, R. L. (1977): Human milk fucosyltransferases. *Biochem. Soc. Trans. 5*, 838-839.

387. PROKOP, O., SCHLESINGER, D. & GESERICK, G. (1967): Thermostabiles B-Agglutinin aus Konserven von Lachskaviar. *Z. Immun.-Forsch. 132*, 491-494.

388. PROKOP, O., SCHLESINGER, D. & RACKWITZ, A. (1965): Über eine thermostabile 'antibody-like substance' (Anti-A$_{hel}$) bei Helix pomatia und deren Herkunft. *Z. Immun.-Forsch. 129*, 402-412.

389. PROKOP, O., SCHNITZLER, S. & UHLENBRUCK, G. (1967): Über einen kräftigen H-Antikörper bei zwei Vertretern der Percidae, aufgefunden im Rogen der Tiere. *Acta Biol. Med. Germ. 18*, K7.

390. PUSZTAI, A. & MORGAN, W. T. J. (1963): Studies in immunochemistry. 22. The amino acid composition of the human blood group A, B, H, and Le^a specific substances. *Biochem. J. 88*, 546-555.

391. RACE, C. & WATKINS, W. M. (1972): The enzymic products of the human A and B blood group genes in the serum of 'Bombay' O$_h$ donors. *FEBS Lett. 27*, 125-130.

392. RACE, C. & WATKINS, W. M. (1974): Inhibition of the blood group A$_1$ and A$_2$ gene-specified N-acetyl-α-D-galactosaminyltransferases by uridine diphosphate D-galactose. *Carbohydr. Res. 37*, 239-244.

393. RACE, C., ZIDERMAN, D. & WATKINS, W. M. (1968): An α-D-galactosyltransferase associated with the blood-group B character. *Biochem. J. 107*, 733-735.

394. RACE, R. R. & SANGER, R. (1975): The ABO blood groups. In: *Blood Groups in Man*. Blackwell Scientific Publications, Oxford, pp. 8-91.

395. RACE, R. R. & SANGER, R. (1975): Secretors and Non-secretors. In: *Blood Groups in Man*. Blackwell Scientific Publications, Oxford, pp. 311-322.

396. RACHKEWICH, R. A., CROOKSTON, M. C., TILLEY, C. A. & WHERRETT, J. R. (1978): Evidence that blood group A antigen on lymphocytes is derived from the plasma. *J. Immunogenet. 5*, 25-29.

397. RAHUEL, C., LUBINEAU, A., DAVID, S., SALMON, C. & CARTRON, J. P. (1983): Acquired B antigen: further studies using synthetic oligosaccharides. *Blood Transfus. Immunohaematol. 26*, 347-358.

398. RAJAN, V. P., LARSEN, R. D., AJMERA, S., ERNST, L. K. & LOWE, J. B. (1989): A cloned human DNA restriction fragment determines expression of a GDP-L-fucose:β-D-galactoside 2-α-L-fucosyltransferase in transfected cells. Evidence for isolation and transfer of the human H blood group locus. *J. Biol. Chem. 264*, 11158-11167.

399. RAMASAMY, R. & RAJAKARUNA, R. (1997): Association of malaria with inactivation of α1,3-galactosyl transferase in catarrhines. *Biochim. Biophys. Acta 1360*, 241-246.

400. RAO, B. N. N., DUA, V. K. & BUSH, C. A. (1985): Conformations of blood group H-active oligosaccharides of ovarian cyst mucins. *Biopolymers 24*, 2207-2229.

401. RAVINDRAN, B., SATAPATHY, A. K. & DAS, M. K. (1988): Naturally-occurring anti-α-galactosyl antibodies in human Plasmodium falciparum infections - a possible role for autoantibodies in malaria. *Immunol. Lett.* 137-141.

402. REED, T. E. (1964): The frequency and nature of blood group A$_3$. *Transfusion 4*, 457-460.

403. REED, T. E. & MOORE, B. P. L. (1964): A new variant of blood group A. *Vox Sang. 9*, 363-366.

404. REGUIGNE-ARNOULD, I., COUILLIN, P., MOLLICONE, R., FAURÉ, S., FLETCHER, A., KELLY, R. J., LOWE, J. B. & ORIOL, R. (1995): Relative positions of two clusters of human α-L- fucosyltransferases in 19q (FUT1-FUT2) and 19p (FUT6-FUT3- FUT5) within the microsatellite genetic map of chromosome 19. *Cytogenet. Cell Genet. 71*, 158-162.

405. RENKONEN, K. O. (1948): Studies on hemagglutinins present in seeds of some representatives of the family of Leguminosae. *Ann. Med. Exp. Biol. Fenn. 26*, 66-72.

406. RENTON, P. H. & HANCOCK, J. A. (1962): Uptake of A and B antigens by transfused group O erythrocytes. *Vox Sang. 7*, 33-38.

407. REYES, F., GOURDIN, M. F., LEJONC, J. L., CARTRON, J. P., BRETON-GORIUS, J. & DREYFUS, B. (1976): The heterogeneity of erythrocyte antigen distribution in human normal phenotypes: an immuno-electron microscopy study. *Brit. J. Haematol. 34*, 613-621.

408. RIESS, H., ECKSTEIN, R., BINSACK, T., RUCKDESCHEL, G. & MEMPEL, W. (1988): Acquired B antigen associated with infection by Bacillus cereus: in vivo and in vitro transformation of A$_1$ red cells. *Blut 56*, 237-238.

409. ROCHANT, H., TONTHAT, H., HENRI, A., TITEUX, M. & DREYFUS, B. (1976): Abnormal distribution of erythrocytes A$_1$ antigen in preleukemia as demonstrated by an immunofluorescence technique. *Blood Cells 2*, 237-255.

410. ROMANO, E. L., MOLLISON, P. L. & LINARES, J. (1978): Number of B sites generated on group O red cells from adults and newborn infants. *Vox Sang. 34*, 14-17.

411. ROSEVEAR, P. R., NUNEZ, H. A. & BARKER, R. (1982): Synthesis and solution conformation of the type 2 blood group oligosaccharide αLFuc(1→2)βDGal(1→4)βDGlcNAc. *Biochemistry 21*,

1421-1431.

412. ROUQUIER, S., LOWE, J. B., KELLY, R. J., FERTITTA, A. L., LENNON, G. C. & GIORGI, D. (1995): Molecular cloning of a human genomic region containing the H blood group $\alpha(1,2)$fucosyltransferase gene and two H locus- related DNA restriction fragments - Isolation of a candidate for the human Secretor blood group locus. *J. Biol. Chem. 270*, 4632-4639.

413. RUBINSTEIN, P., ALLEN, F. H. & ROSENFIELD, R. E. (1973): A dominant suppressor of A and B. *Vox Sang. 25*, 377-381.

414. RYDBERG, L. & SAMUELSSON, B. E. (1991): Presence of glycosyltransferase inhibitors in the sera of patients with long-term surviving ABO incompatible (A_2 to O) kidney grafts. *Transfus. Med. 1*, 177-182.

415. SABO, B. H., BUSH, M., GERMAN, J., CARNE, L. R., YATES, A. D. & WATKINS, W. M. (1978): The cis AB phenotype in three generations of one family: serological, enzymatic, and cytogenetic studies. *J. Immunogenet. 5*, 87-106.

416. SAKAMOTO, J., YIN, B. W. T. & LLOYD, K. O. (1984): Analysis of the expression of H, Lewis, X, Y, and precursor blood group determinants in saliva and red cells using a panel of mouse monoclonal antibodies. *Mol. Immunol. 21*, 1093-1098.

417. SALMON, C. (1976): Les phénotypes B faibles B_3, B_x, B_{el}: classification pratique proposée. *Rev. Fr. Transfus. Immuno-Hématol. 19*, 89-104.

418. SALMON, C., BORIN, P. & ANDRÉ, R. (1958): Le groupe sanguin A_m dans deux générations d'une même famille. *Rev. Hématol. 13*, 529-532.

419. SALMON, C. & CARTRON, J. P. (1977): Interactions in AB heterozygotes. In: *CRC Handbook Series in Clinical Laboratory Science*. Section D, Blood Banking. (D. Seligson, T. J. Greenwalt, and E. A. Steane, eds.). CRC Press Inc., Cleveland, Ohio, Vol. 1, pp. 131-138.

420. SALMON, C., CARTRON, J. P., ROUGER, P., LIBERGE, G., JUSZCZAK, G., MULET, C. & LOPEZ, M. (1980): H deficient phenotypes: a proposed practical classification Bombay A_h, H_z, H_m. *Blood Transfus. Immunohaematol. 23*, 233-248.

421. SALMON, C., JUSZCZAK, G., LIBERGE, G., LOPEZ, M., CARTRON, J. P. & KLING, C. (1978): Une famille où un phénotype 'Hm' est transmis a travers trois générations. *Rev. Fr. Transfus. Immuno-Hématol. 21*, 21-27.

422. SALMON, C., SALMON, D. & REVIRON, J. (1965): Etude immunologique et génétique de la variabilité du phénotype A_x. *Nouv. Rev. Fr. Hématol. 5*, 275-280.

423. SANDRIN, M. S. H. A., DABKOWSKI, P. L. & MCKENZIE, I. F. (1993): Anti-pig IgM antibodies in human serum react predominantly with Gal(α1-3)Gal epitopes. *Proc. Natl. Acad. Sci. USA 90*, 11391-11395.

424. SARNESTO, A., KÖHLIN, T., HINDSGAUL, O., THURIN, J. & BLASZCZYK-THURIN, M. (1992): Purification of the secretor-type β-galactoside $\alpha1\rightarrow2$- fucosyltransferase from human serum. *J. Biol. Chem. 267*, 2737-2744.

425. SARNESTO, A., KÖHLIN, T., THURIN, J. & BLASZCZYK-THURIN, M. (1990): Purification of H gene-encoded β-galactoside $\alpha1\rightarrow2$ fucosyltransferase from human serum. *J. Biol. Chem. 265*, 15067-15075.

426. SAWICKA, T. (1973): Isolation and characterization of N-acetylgalactosaminyltransferase from human serum of blood group A_1. *Bull. Acad. Sci. Sér. Sci. Biol. Cl. II 21*, 491-498.

427. SCHACHTER, H., MICHAELS, M. A., CROOKSTON, M. C., TILLEY, C. A. & CROOKSTON, J. (1971): A quantitative difference in the activity of blood group A-specific N-acetylgalactosaminyltransferase in serum from A_1 and A_2 human subjects. *Biochem. Biophys. Res. Commun. 45*, 1011-1018.

428. SCHACHTER, H., MICHAELS, M. A., TILLEY, C. A., CROOKSTON, M. C. & CROOKSTON, J. H. (1973): Qualitative differences in the N-acetyl-D-galactosaminyltransferases produced by human A_1 and A_2 genes. *Proc. Natl. Acad. Sci. USA 70*, 220-224.

429. SCHENKEL-BRUNNER, H. (1973): Incorporation of galactose into blood-group (ABH) precursor substance by lactose synthetase from human milk. *Eur. J. Biochem. 33*, 30-35.

430. SCHENKEL-BRUNNER, H. (1980): Blood group ABH antigens on human cord red cells. Number of H antigenic sites and their distribution among different classes of membrane constituents. *Vox Sang. 38*, 310-314.

431. Schenkel-Brunner, H. (1980): Blood-group-ABH antigens of human erythrocytes. Quantitative studies on the distribution of H antigenic sites among different classes of membrane components. *Eur. J. Biochem. 104*, 529-534.

432. Schenkel-Brunner, H. (1982): Studies on blood-groups A_1 and A_2. Further evidence for the predominant influence of quantitative differences in the number of A antigenic sites present on A_1 and A_2 erythrocytes. *Eur. J. Biochem. 122*, 511-514.

433. Schenkel-Brunner, H., Chester, M. A. & Watkins, W. M. (1972): α-L-Fucosyltransferases in human serum from donors of different ABO, secretor, and Lewis blood group phenotypes. *Eur. J. Biochem. 30*, 269-277.

434. Schenkel-Brunner, H. & Kothbauer, H. (1976): Immunochemical investigations on toad (Bufo) eggs: comparative studies on three species (B. bufo, B. viridis, B. calamita). *J. Immunogenet. 3*, 395-399.

435. Schenkel-Brunner, H. & Kothbauer, H. (1978): Immunchemische Untersuchungen an Laubfrosch-Laich: Zur Unterscheidung von Hyla arborea und Hyla meridionalis. *Zool. Anz. 201*, 289-292.

436. Schenkel-Brunner, H., Prohaska, R. & Tuppy, H. (1975): Action of glycosyl transferases upon "Bombay" (O_h) erythrocytes. Conversion to cells showing blood-group H and A specificities. *Eur. J. Biochem. 56*, 591-594.

437. Schenkel-Brunner, H. & Tuppy, H. (1969): Enzymatic conversion of human O into A erythrocytes and of A into AB erythrocytes. *Nature 223*, 1272-1273.

438. Schenkel-Brunner, H. & Tuppy, H. (1970): Enzymes from human gastric mucosa conferring blood-group A and B upon erythrocytes. *Eur. J. Biochem. 17*, 218-222.

439. Schenkel-Brunner, H. & Tuppy, H. (1973): Enzymatic conversion of human blood-group-O erythrocytes into A_2 and A_1 cells by α-N-acetyl-D-galactosaminyl transferases of blood-group-A individuals. *Eur. J. Biochem. 34*, 125-128.

440. Schiff, F. & Sasaki, H. (1932): Der Ausscheidungstypus, ein auf serologischem Wege nachweisbares mendelndes Merkmal. *Klin. Wochenschr. 11*, 1426-1429.

441. Schiff, F. & Sasaki, H. (1932): Über die Vererbung des serologischen Ausscheidungstypus. *Z. Immun.-Forsch. 77*, 129-139.

442. Schiffman, G. & Howe, C. (1965): The specificity of blood group A-B cross-reacting antibody. *J. Immunol. 94*, 197-204.

443. Schmidt, G. (1954): Die Hämagglutination, im besonderen menschlicher B-Blutzellen, durch Extrakte aus Samen von Evonymus vulgaris (Pfaffenhütchen). *Z. Immun.-Forsch. 111*, 432-439.

444. Schnitzler, S., Müller, G. & Prokop, O. (1967): Ein "neuer" Antikörper, Anti-P_{rut}, aufgefunden im Rogen von Rutilus rutilus. *Z. Immun.-Forsch. 134*, 45-53.

445. Schuh, V., Vyas, G. N. & Fudenberg, H. H. (1972): Study of a French family with a new variant of blood group A: A_{lae}. *Amer. J. Hum. Genet. 24*, 11-16.

446. Schwyzer, M. & Hill, R. L. (1977): Porcine A blood group specific N-acetylgalactosaminyltransferase. II. Enzymatic properties. *J. Biol. Chem. 252*, 2346-2355.

447. Schwyzer, M. & Hill, R. L. (1977): Porcine A blood group-specific N-acetylgalactosaminyltransferase. I. Purification from porcine submaxillary glands. *J. Biol. Chem. 252*, 2238-2345.

448. Senior, D., Baker, N., Cedergren, B., Falk, P., Larson, G., Lindstedt, R. & Eden, S. C. (1988): Globo-A - a new receptor specificity for attaching Eschericha coli. *FEBS Lett. 237*, 123-127.

449. Seto, N. O. L., Compston, C. A., Evans, S. V., Bundle, D. R., Narang, S. A. & Palcic, M. M. (1999): Donor substrate specificity of recombinant human blood group A, B and hybrid A/B glycosyltransferases expressed in Escherichia coli. *Eur. J. Biochem. 259*, 770-775.

450. Seto, N. O. L., Palcic, M. M., Compston, C. A., Li, H., Bundle, D. R. & Narang, S. A. (1997): Sequential interchange offour amino acids from blood group B to blood group A glycosyltransferase boosts catalytic activity and progressively modifies substrate recognition in human recombinant enzymes. *J. Biol. Chem. 272*, 14133-14138.

451. Seyfried, H., Walewska, I. & Werblinska, B. (1964): Unusual inheritance of ABO group in a family

with weak B antigens. *Vox Sang. 9*, 268-277.

452. SHAPER, N. L., LIN, S. P., JOZIASSE, D. H., KIM, D. Y. & YANG-FENG, T. L. (1992): Assignment of two human α-1,3-galactosyltransferase gene sequences (GGTA1 and GGTA1P) to chromosomes 9q33-q34 and 12q14-q15. *Genomics 12*, 613-615.

453. SHECHTER, Y., ETZIONI, A., LEVENE, C. & GREENWELL, P. (1995): A Bombay individual lacking H and Le antigens but expressing normal levels of α-2- and α-4-fucosyltransferases. *Transfusion 35*, 773-776.

454. SHEN, L., GROLLMAN, E. F. & GINSBURG, V. (1968): An enzymatic basis for secretor status and blood group substance specificity in humans. *Proc. Natl. Acad. Sci. USA 59*, 224-230.

455. SIKDER, S. K., KABAT, E. A., ROBERTS, D. D. & GOLDSTEIN, I. J. (1986): Immunochemical studies on the combining site of the blood group A-specific lima bean lectin. *Carbohydr. Res. 151*, 247-260.

456. SKACEL, P. O. & WATKINS, W. M. (1987): Fucosyltransferase expression in human platelets and leucocytes. *Glycoconj. J. 4*, 267-272.

457. SKACEL, P. O. & WATKINS, W. M. (1988): Significance of altered α-2-L-fucosyltransferase levels in serum of leukemic patients. *Cancer Res. 48*, 3998-4001.

458. SLAYTER, H. S., COOPER, A. G. & BROWN, M. C. (1974): Electron microscopy and physical parameters of human blood group i, A, B, and H antigens. *Biochemistry 13*, 3365-3371.

459. SLOMIANY, A., SLOMIANY, B. L. & HOROWITZ, M. I. (1974): Structural study of the blood group A active glycolipids of hog gastric mucosa. *J. Biol. Chem. 249*, 1225-1230.

460. SLOMIANY, A., ZDEBSKA, E. & SLOMIANY, B. L. (1984): Structures of the neutral oligosaccharides isolated from A-active human gastric mucin. *J. Biol. Chem. 259*, 14743-14749.

461. SLOMIANY, B. L. & SLOMIANY, A. (1977): Branched blood-group A fucolipids of hog gastric mucosa. *Biochim. Biophys. Acta 486*, 531-540.

462. SLOMIANY, B. L. & SLOMIANY, A. (1978): Blood-group-(A+H) complex fucolipids of hog gastric mucosa. *Eur. J. Biochem. 90*, 39-49.

463. SLOMIANY, B. L., SLOMIANY, A. & MURTY, V. L. N. (1979): Partial characterization of the highly complex fucolipids from gastric mucosa. *Biochem. Biophys. Res. Commun. 88*, 1092-1097.

464. SMALLEY, C. E. & TUCKER, E. M. (1983): Blood group A antigen site distribution and immunoglobulin binding in relation to red cell age. *Brit. J. Haematol. 54*, 209-219.

465. SMITH, E. L., MCKIBBIN, J. M., KARLSSON, K. A., PASCHER, I. & SAMUELSSON, B. E. (1975): Characterization by mass spectrometry of blood group A active glycolipids from human and dog small intestine. *Biochemistry 14*, 2120-2124.

466. SMITH, E. L., MCKIBBIN, J. M., KARLSSON, K. A., PASCHER, I., SAMUELSSON, B. E. & LI, S. C. (1975): Characterization of dog small intestinal fucolipids with human blood group H activity. *Biochemistry 14*, 3370-3376.

467. SOLOMON, J. M. & STURGEON, P. (1964): Quantitative studies of the phenotype A_{el}. *Vox Sang. 9*, 476-486.

468. SOLOMON, J. M., WAGGONER, R. & LEYSHON, W. C. (1965): A quantitative immunogenetic study of gene suppression involving A_1 and H antigens of the erythrocyte without affecting secreted blood group substances. The ABH phenotypes $A_m{}^h$ and $O_m{}^h$. *Blood 25*, 470-485.

469. SPEISER, P. (1959): Über die bisher jüngste menschliche Frucht (27mm/22g) an der bereits die Erbmerkmale A, M, N, s, Fy(a+), C, c, D, E, e, Jk(a+?) im Blut festgestellt werden konnten. *Wien. Klin. Wochenschr. 71*, 549-551.

470. SPRINGER, G. F. (1956): Inhibition of blood-group agglutinins by substances occurring in plants. *J. Immunol. 76*, 399-407.

471. SPRINGER, G. F. (1966): Relation of microbes to blood-group active substances. *Angew. Chem. (Int. Edn.) 5*, 909-920.

472. SPRINGER, G. F. & DESAI, P. R. (1971): Monosaccharides as specific precipitinogens of eel anti-human blood-group H(O) antibody. *Biochemistry 10*, 3749-3761.

473. SPRINGER, G. F., WILLIAMSON, P. & BRANDES, W. C. (1961): Blood group activity of gram-negative bacteria. *J. Exp. Med. 113*, 1077-1093.

474. STAYBOLDT, C., REARDEN, A. & LANE, T. A. (1987): B antigen acquired by normal A_1 red cells

exposed to a patient's serum. *Transfusion 27*, 41-44.

475. STELLNER, K., WATANABE, K. & HAKOMORI, S. I. (1973): Isolation and characterization of glycosphingolipids with blood group H specificity from membranes of human erythrocytes. *Biochemistry 12*, 656-661.

476. STRAHAN, K. M., GU, F., PREECE, A. F., GUSTAVSSON, I., ANDERSSON, L. & GUSTAFSSON, K. (1995): cDna sequence and chromosome localization of pig $\alpha1,3$ galactosyltransferase. *Immunogenetics 41*, 101-105.

477. STRONCEK, D. F., KONZ, R., CLAY, M. E., HOUCHINS, J. P. & McCULLOUGH, J. (1995): Determination of ABO glycosyltransferase genotypes by use of polymerase chain reaction and restriction enzymes. *Transfusion 35*, 231-240.

478. STURGEON, P., MOORE, B. P. L. & WEINER, W. (1964): Notations for two weak A variants: A_{end} and A_{el}. *Vox Sang. 9*, 214-215.

479. SUN, J., THURIN, J., COOPER, H. S., WANG, P., MACKIEWICZ, M., STEPLEWISKI, Z. & BLASZCZYK-THURIN, M. (1995): Elevated expression of H type GDP-L-fucose:β-D-galactoside α-2-L-fucosyltransferase is associated with human colon adenocarcinoma progression. *Proc. Natl. Acad. Sci. USA 92*, 5724-5728.

480. SUZUKI, K., IWATA, M., TSUJI, H., TAKAGI, T., TAMURA, A., ISHIMOTO, G., ITO, S., MATSUI, K. & MIYAZAKI, T. (1997): A de novo recombination in the ABO blood group gene and evidence for the occurrence of recombination products. *Hum. Genet. 99*, 454-461.

481. SZULMAN, A. E. (1964): The histological distribution of the blood group substances in man as disclosed by immunofluorescence. III. The A, B, and H antigens in embryos and fetuses from 18 mm length. *J. Exp. Med. 119*, 503-515.

482. SZULMAN, A. E. (1965): The ABH antigens in human tissues and secretions during embryonal development. *J. Histochem. Cytochem. 13*, 752-754.

483. SZULMAN, A. E. (1966): Chemistry, distribution, and function of blood group substances. *Annu. Rev. Med. 17*, 307-322.

484. TAKASAKI, S. & KOBATA, A. (1976): Chemical characterization and distribution of ABO blood-group active glycoproteins in human erythrocyte membranes. *J. Biol. Chem. 251*, 3610-3615.

485. TAKASAKI, S., YAMASHITA, K. & KOBATA, A. (1978): The sugar chain structures of ABO blood group active glycoproteins obtained from human erythrocyte membrane. *J. Biol. Chem. 253*, 6086-6091.

486. TAKEUCHI, Y., PORTER, C. D., STRAHAN, K. M., PREECE, A. F., GUSTAFSSON, K., COSSET, F. L., WEISS, R. A. & COLLINS, M. K. (1996): Sensitization of cells and retroviruses to human serum by $\alpha1,3$-galactosyltransferase. *Nature 379*, 85-88.

487. TAKEYA, A., HOSOMI, O. & ISHIURA, M. (1990): Complete purification and characterization of α-3-N-acetylgalactosaminyltransferase encoded by the human blood group A gene. *J. Biochem. 107*, 360-368.

488. TAKIZAWA, H. & ISEKI, S. (1974): Biosynthesis of A and B blood-group substances of human erythrocytes and saliva. *Jp. J. Hum. Genet. 19*, 147-156.

489. TAKIZAWA, H. & ISEKI, S. (1982): Cross-reacting antibodies to human blood group A and B glycosyltransferases. *Proc. Jp. Acad. 58*, 65-68.

490. TAKIZAWA, H. & ISEKI, S. (1983): Immunological specificity of blood group A glycosyltransferase demonstrated on the oligosaccharide-acceptor. *Proc. Jp. Acad. 59B*, 247-250.

491. TAKIZAWA, H., KISHI, K. & ISEKI, S. (1978): Biochemical and serological studies on alpha-galactosyltransferases in the sera and salivas from blood group B_m and A_1B_m individuals. *Proc. Jp. Acad. Sci. 54*, 402-407.

492. TAKIZAWA, H., KISHI, K. & ISEKI, S. (1979): On the B gene-associated α-galactosyltransferase in human urine. *Proc. Jp. Acad. 55*, 362-367.

493. TANEGASHIMA, A., NISHI, K., FUKUNAGA, T., RAND, S. & BRINKMANN, B. (1996): Ethnic differences in the expression of blood group antigens in the salivary gland secretory cells from German and Japanese non-secretor individuals. *Glycoconj. J. 13*, 537-545.

494. THOMSEN, O., FRIEDENREICH, V. & WORSAAE, E. (1930): Über die Möglichkeit der Existenz zweier neuer Blutgruppen; auch ein Beitrag zur Beleuchtung sogenannter Untergruppen. *Acta Pathol.*

Microbiol. Scand. 7, 157-190.

495. THURIN, J. & BLASZCZYK-THURIN, M. (1995): Porcine submaxillary gland GDP-L-fucose:β-D-galactoside α-2-L-fucosyltransferase is likely a counterpart of the human Secretor gene-encoded blood group transferase. *J. Biol. Chem. 270*, 26577-26580.

496. TILLEY, C. A., CROOKSTON, M. C., BROWN, B. L. & WHERRETT, J. R. (1975): A and B and A_1Le^b substances in glycosphingolipid fractions of human serum. *Vox Sang. 28*, 25-33.

497. TODD, G. M. (1971): Blood group antibodies in Salmonidae roe. *Vox Sang. 21*, 451-454.

498. TOPPING, M. D. & WATKINS, W. M. (1975): Isoelectric points of the human blood group A^1, A^2, and B gene-associated glycosyltransferases in ovarian cyst fluids and serum. *Biochem. Biophys. Res. Commun. 64*, 89-96.

499. TRINCHERA, M. & BOZZARO, S. (1996): Dictyostelium cytosolic fucosyltransferase synthesizes H type 1 trisaccharide in vitro. *FEBS Lett. 395*, 68-72.

500. TUPPY, H. & SCHENKEL-BRUNNER, H. (1969): Formation of blood-group A substance from H substance by an α-N-acetylgalactosaminyl transferase. *Eur. J. Biochem. 10*, 152-157.

501. TUPPY, H. & SCHENKEL-BRUNNER, H. (1969): Occurrence and assay of alpha-N-acetylgalactosaminyl transferase in the gastric mucosa of humans belonging to blood-group A. *Vox Sang. 17*, 139-142.

502. TUPPY, H. & STAUDENBAUER, W. L. (1966): Microsomal incorporation of N-acetyl-D-galactosamine into blood group substance. *Nature 210*, 316-317.

503. UGOZZOLI, L. & WALLACE, R. B. (1992): Application of an allele-specific polymerase chain reaction to the direct determination of ABO blood group genotypes. *Genomics 12*, 670-674.

504. VAN ARSDEL, P. P. (1958): The usefulness of the plant-lectin, Ulex europaeus, in a large-scale blood group study. *Vox Sang. 3*, 448-455.

505. VAN HALBEEK, H., DORLAND, L., VLIEGENTHART, J. F. G., KOCHETKOV, N. K., ARBATSKY, N. P. & DEREVITSKAYA, V. A. (1982): Characterization of the primary structure and the microheterogeneity of the carbohydrate chains of porcine blood-group H substance by 500 mHz ¹H-NMR spectroscopy. *Eur. J. Biochem. 127*, 21-29.

506. VETTER, O. (1969): Beobachtungen an einem 'A-like' Blutgruppenprinzip in Clupea harengus L. *Acta Biol. Med. Germ. 22*, 427-429.

507. VIITALA, J., FINNE, J. & KRUSIUS, T. (1982): Blood group A and H determinants in polyglycosyl peptides of A_1 and A_2 erythrocytes. *Eur. J. Biochem. 126*, 401-406.

508. VIITALA, J., KARHI, K. K., GAHMBERG, C. G., FINNE, J., JÄRNEFELDT, J., MYLLYLÄ, G. & KRUSIUS, T. (1981): Blood-group A and B determinants are located in different polyglycosyl peptides isolated from human erythrocytes of blood group AB. *Eur. J. Biochem. 113*, 259-265.

509. VOAK, D., LODGE, T. W., HOPKINS, J. & BOWLEY, C. C. (1968): A study of the antibodies of the H'O'I-B complex with special reference to their occurrence in notation. *Vox Sang. 15*, 353-366.

510. VOAK, D., LODGE, T. W., STAPLETON, R. R., FOGG, H. & ROBERTS, H. E. (1970): The incidence of H deficient A_2 and A_2B bloods and family studies on the AH/ABH status of an A_{int} and some new variant blood types ($A_{int\,H}^{A1}$, $A_{2H}w^{A1}$, $A_2B_Hw^{A1B}$, and $A_2B_Hw^{A2B}$). *Vox Sang. 19*, 73-84.

511. VOAK, D., SACKS, S., ALDERSON, T., TAKEY, F., LENNOX, E., JARVIS, J., MILSTEIN, C. & DARNBOROUGH, J. (1980): Monoclonal anti-A from a hybrid-myeloma: evaluation as a blood grouping reagent. *Vox Sang. 39*, 134-140.

512. VOAK, D., SONNEBORN, H. & YATES, A. (1992): The A_1 (B)-phenomenon:a monoclonal anti-B (BS-85) demonstrates low levels of B-determinants on A_1 red cells. *Transfus. Med. 2*, 119-127.

513. VOS, G. H. (1964): Five examples of red cells with the A_x subgroup of blood group A. *Vox Sang. 9*, 160-167.

514. WAGNER, F. F. & FLEGEL, W. A. (1997): Polymorphism of the h allele and the population frequency of sporadic nonfunctional alleles. *Transfusion 37*, 284-290.

515. WANG, B. J., KODA, Y., SOEJIMA, M. & KIMURA, H. (1997): Two missense mutations of H type α(1,2)fucosyltransferase gene (FUT1) responsible for para-bombay phenotype. *Vox Sang. 72*, 31-35.

516. WATANABE, G., UMETSU, K., YUASA, I. & SUZUKI, T. (1997): Amplified product length polymorphism (APLP): a novel strategy for genotyping the ABO blood group. *Hum. Genet. 99*, 34-37.

517. WATANABE, K. & HAKOMORI, S. I. (1976): Status of blood group carbohydrate chains in ontogenesis and in oncogenesis. *J. Exp. Med. 144*, 644-653.
518. WATANABE, K., LAINE, R. A. & HAKOMORI, S. I. (1975): On neutral fucoglycolipids having long, branched carbohydrate chains: H-active and I-active glycosphingolipids of human erythrocyte membranes. *Biochemistry 14*, 2725-2733.
519. WATANABE, K., POWELL, M. & HAKOMORI, S. I. (1978): Isolation and characterization of a novel fucoganglioside of human erythrocyte membranes. *J. Biol. Chem. 253*, 8962-8967.
520. WATKINS, W. M. (1966): Blood group substances. *Science 152*, 172-181.
521. WATKINS, W. M. (1972): Blood group specific substances. In: *Glycoproteins: Their Composition, Structure and Function* (A. Gottschalk, ed.) Elsevier, Amsterdam, pp. 830-891.
522. WATKINS, W. M. (1978): Blood group gene specified glycosyltransferases in rare ABO groups and in leukemia. *Rev. Fr. Transfus. 21*, 201-228.
523. WATKINS, W. M. (1980): Biochemistry and genetics of the ABO, Lewis and P blood group systems. In: *Advances in Human Genetics.* Vol. 10, pp. 1-136.
524. WATKINS, W. M. & MORGAN, W. T. J. (1957): The A and H character of the blood group substances secreted by persons belonging to group A_2. *Acta Genet. Statist. Med. 6*, 521-526.
525. WATKINS, W. M. & MORGAN, W. T. J. (1959): Possible genetical pathways for the biosynthesis of blood group mucopolysaccharides. *Vox Sang. 4*, 97-119.
526. WEINER, W., LEWIS, H. B. M., MOORES, P., SANGER, R. & RACE, R. R. (1957): A gene, y, modifying the ABO group antigen A. *Vox Sang. 2*, 25-37.
527. WEST, C. M., SCOTT-WARD, T., TENG-UMNUAY, P., VAN DER WEL, H., KOZAROV, E. & HUYNH, A. (1996): Purification and characterization of an α 1,2-L-fucosyltransferase, which modifies the cytosolic protein FP21, from the cytosol of Dictyostelium. *J. Biol. Chem. 271*, 12024-12035.
528. WHERRETT, J. R. & HAKOMORI, S. I. (1973): Characterization of a blood group B glycolipid, accumulating in the pancreas of a patient with Fabry's disease. *J. Biol. Chem. 248*, 3046-3051.
529. WHITE, T., MANDEL, U., ØRNTOFT, T. F., DABELSTEEN, E., KARKOV, J., KUBEJA, M., HAKOMORI, S. & CLAUSEN, H. (1990): Murine monoclonal antibodies directed to the human histo-blood group A transferase (UDP-GalNAc:Fucα1→2Gal α1→3-N-acetylgalactosaminyltransferase) and the presence therein on N-linked histo-blood group A determinant. *Biochemistry 29*, 2740-2747.
530. WHITEHEAD, J. S., BELLA, A. & KIM, Y. S. (1974): An N-acetylgalactosaminyltransferase from human blood group A plasma. I. Purification and agarose binding properties. *J. Biol. Chem. 249*, 3442-3447.
531. WHITEHEAD, J. S., BELLA, A. & KIM, Y. S. (1974): An N-acetylgalactosaminyltransferase from human blood group A plasma. II. Kinetic and physicochemical properties. *J. Biol. Chem. 249*, 3448-3452.
532. WHITTEMORE, N. B., TRABOLD, N. C., REED, C. F. & WEED, R. I. (1969): Solubilized glycoprotein from human erythrocyte membranes possessing blood group A, B, and H activity. *Vox Sang. 17*, 289-299.
533. WIENER, A. S. & CIOFFI, A. F. (1972): A group B analogue of subgroup A_3. *Amer. J. Clin. Pathol. 58*, 693-697.
534. WIENER, A. S. & GORDON, E. B. (1956): A hitherto undescribed human blood group. *Brit. J. Haematol. 2*, 305-309.
535. WILCZYNSKA, Z., MILLER-PODRAZA, H. & KOSCIELAK, J. (1980): The contribution of different glycoconjugates to the total ABH blood-group activity of human erythrocytes. *FEBS Lett. 112*, 277-279.
536. WILEY, E. L., MURPHY, P., MENDELSOHN, G. & EGGLESTON, J. C. (1981): Distribution of blood group substances in normal human colon. *Amer. J. Clin. Pathol. 76*, 806-809.
537. WOOD, C., KABAT, E. A., MURPHY, L. A. & GOLDSTEIN, I. J. (1979): Immunochemical studies of the combining sites of the two isolectins, A_4 and B_4, isolated from Bandeiraea simplicifolia. *Arch. Biochem. Biophys. 198*, 1-11.
538. WRANN, M., SCHENKEL-BRUNNER, H. & KOTHBAUER, H. (1978): Blood-group A and H specific structures in toad (Bufo) spawn. Comparative studies on three species (Bufo bufo, Bufo viridis, Bufo calamita). *Z. Immun.-Forsch. 154*, 471-474

539. WROBEL, D. M., MC, D. I., RACE, C. & WATKINS, W. M. (1974): "True" genotype of chimeric twins revealed by blood-group gene products in plasma. *Vox Sang. 27*, 395-402.

540. YAMAMOTO, F. (1995): Molecular genetics of the ABO histo-blood group system. *Vox Sang. 69*, 1-7.

541. YAMAMOTO, F. & HAKOMORI, S. I. (1990): Sugar-nucleotide donor specificity of histo-blood-group-A and group-B transferases is based on amino acid substitutions. *J. Biol. Chem. 265*, 19257-19262.

542. YAMAMOTO, F. & MCNEILL, P. D. (1996): Amino acid residue at codon 268 determines both activity and nucleotide-sugar donor substrate specificity of human histo- blood group A and B transferases.In vitro mutagenesis study. *J. Biol. Chem. 271*, 10515-10520.

543. YAMAMOTO, F., MCNEILL, P. D. & HAKOMORI, S. (1991): Identification in human genomic DNA of the sequence homologous but not identical to either the histo-blood group ABH genes or $\alpha1{\rightarrow}3$ galactosyltransferase pseudogene. *Biochem. Biophys. Res. Commun. 175*, 986-994.

544. YAMAMOTO, F., MCNEILL, P. D. & HAKOMORI, S. (1995): Genomic organization of human histo-blood group ABO genes. *Glycobiology 5*, 51-58.

545. YAMAMOTO, F., MCNEILL, P. D., YAMAMOTO, M., HAKOMORI, S., BROMILOW, I. M. & DUGUID, J. K. M. (1993): Molecular genetic analysis of the ABO blood group system .4. Another type of O allele. *Vox Sang. 64*, 175-178.

546. YAMAMOTO, F., MCNEILL, P. D., YAMAMOTO, M., HAKOMORI, S. & HARRIS, T. (1993): Molecular genetic analysis of the ABO blood group system. 3. A_X and B(A) Alleles. *Vox Sang. 64*, 171-174.

547. YAMAMOTO, F. I., CLAUSEN, H., WHITE, T., MARKEN, J. & HAKOMORI, S. I. (1990): Molecular genetic basis of the histo-blood group ABO system. *Nature 345*, 229-233.

548. YAMAMOTO, F. I., MARKEN, J., TSUJI, T., WHITE, T., CLAUSEN, H. & HAKOMORI, S. I. (1990): Cloning and characterization of DNA complementary to human UDP-GalNAc: Fuc$\alpha1{\rightarrow}2$Gal $\alpha1{\rightarrow}3$GalNAc transferase (histo-blood group A transferase) mRNA. *J. Biol. Chem. 265*, 1146-1151.

549. YAMAMOTO, F. I., MCNEILL, P. D. & HAKOMORI, S. I. (1992): Human histo-blood group A_2 transferase coded by A_2 allele, one of the A subtypes, is characterized by a single base deletion in the coding sequence, which results in an additional domain at the carboxyl terminal. *Biochem. Biophys. Res. Commun. 187*, 366-374.

550. YAMAMOTO, F. I., MCNEILL, P. D., KOMINATO, Y., YAMAMOTO, M., HAKOMORI, S. I., ISHIMOTO, S., NISHIDA, S., SHIMA, M. & FUJIMURA, Y. (1993): Molecular genetic analysis of the ABO blood group system. 2. cis-AB alleles. *Vox Sang. 64*, 120-123.

551. YAMAMOTO, F. I., MCNEILL, P. D., YAMAMOTO, M., HAKOMORI, S. I., HARRIS, T., JUDD, W. J. & DAVENPORT, R. D. (1993): Molecular genetic analysis of the ABO blood group system. 1. Weak subgroups: A_3 and B_3 alleles. *Vox Sang. 64*, 116-119.

552. YAMAMOTO, S. (1977): The occurrence of materials cross-reacting with anti-A and anti-B agglutinins in fruit or seed extracts of higher plants. *J. Immunogenet. 4*, 325-330.

553. YAMAMOTO, S. & SAKAI, I. (1981): Composition and immunochemical properties of glycoproteins with anti-B agglutinin activity isolated from Euonymus sieboldiana seeds. *J. Immunogenet. 8*, 271-279.

554. YAMAMOTO, S., SAKAI, I. & ISEKI, S. (1981): Purification, composition, and immunochemical properties of arabinogalactan-protein H active glycopeptides from Euonymus sieboldiana seeds. *Immunol. Commun. 10*, 215-236.

555. YAMASHITA, K., TACHIBANA, Y., TAKASAKI, S. & KOBATA, A. (1976): ABO blood group determinants with branched cores. *Nature 262*, 702-703.

556. YATES, A. D., FEENEY, J., DONALD, A. S. R. & WATKINS, W. M. (1984): Characterisation of a blood-group A-active tetrasaccharide synthesised by a blood-group B gene-specified glycosyltransferase. *Carbohydr. Res. 130*, 251-260.

557. YATES, A. D. & WATKINS, W. M. (1982): The biosynthesis of blood group B determinants by the blood group A gene-specified α-3-N-acetyl-D-galactosaminyltransferase. *Biochem. Biophys. Res. Commun. 109*, 958-965.

558. YAZAWA, S. & FURUKAWA, K. (1980): α-L-Fucosyltransferases related to biosynthesis of blood group substances in human saliva. *J. Immunogenet. 7*, 137-148.

559. YAZAWA, S. & FURUKAWA, K. (1983): Immunochemical properties of human plasma α1→2fucosyltransferase specified by blood group H-gene. *J. Immunogenet. 10*, 349-360.

560. YOKOYAMA, M. & PLOCINIK, B. (1965): Serologic and immunochemical characterization of A$_x$ blood. *Vox Sang. 10*, 149-160.

561. YOSHIDA, A. (1980): Identification of genotypes of blood group A and B. *Blood 55*, 119-123.

562. YOSHIDA, A. (1981): Genetic mechanism of blood group (ABO) expression. *Acta Biol. Med. Germ. 40*, 927-941.

563. YOSHIDA, A. (1983): The existence of atypical blood group galactosyltransferase which causes an expression of A$_2$ character in A^{1}B red blood cells. *Amer. J. Hum. Genet. 35*, 1117-1125.

564. YOSHIDA, A., DAVÉ, V., BRANCH, D. R., YAMAGUCHI, H. & OKUBO, Y. (1982): An enzyme basis for blood type A intermediate status. *Amer. J. Hum. Genet. 34*, 919-924.

565. YOSHIDA, A., DAVÉ, V. & HAMILTON, H. B. (1988): Imbalance of blood group A subtypes and the existence of superactive B* gene in Japanese in Hiroshima and Nagasaki. *Amer. J. Hum. Genet. 43*, 422-428.

566. YOSHIDA, A., DAVÉ, V. & PRCHAL, J. (1985): Uncertainty in identification of blood group A subtypes by agglutination test. *Hum. Hered. 35*, 1-6.

567. YOSHIDA, A., YAMAGUCHI, H. & OKUBO, N. (1980): Genetic mechanism of cis-AB inheritance. I. A case associated with unequal chromosomal crossing over. *Amer. J. Hum. Genet. 32*, 332-338.

568. YOSHIDA, A., YAMAGUCHI, H. & OKUBO, Y. (1980): Genetic mechanism of cis-AB inheritance. II. Cases associated with structural mutation of blood group glycosyltransferase. *Amer. J. Hum. Genet. 32*, 645-650.

569. YOSHIDA, A., YAMAGUCHI, Y. F. & DAVÉ, V. (1979): Immunologic homology of human blood group glycosyltransferases and genetic background of blood group (ABO) determination. *Blood 54*, 344-350.

570. YU, L. C., BROADBERRY, R. E., YANG, Y. H., CHEN, Y. H. & LIN, M. (1996): Heterogeneity of the human Secretor α(1,2)fucosyltransferase gene among Lewis(a+b-) non-secretors. *Biochem. Biophys. Res. Commun. 222*, 390-394.

571. YU, L. C., LEE, H. L., CHU, C. C., BROADBERRY, R. E. & LIN, M. (1999): A newly identified nonsecretor allele of the human histo-blood group α(1,2)fucosyltransferase gene (FUT2). *Vox Sang. 76*, 115-119.

572. YU, L. C., YANG, Y. H., BROADBERRY, R. E., CHEN, Y. H., CHAN, Y. S. & LIN, M. (1995): Correlation of a missense mutation in the human Secretor α1,2-fucosyltransferase gene with the Lewis(a+b+) phenotype: a potential molecular basis for the weak Secretor allele (Se^w). *Biochem. J. 312*, 329-332.

573. YU, L. C., YANG, Y. H., BROADBERRY, R. E., CHEN, Y. H. & LIN, M. (1997): Heterogeneity of the human H blood group α(1,2)fucosyltransferase gene among para-Bombay individuals. *Vox Sang. 72*, 36-40.

574. ZHU, A., LENG, L., MONAHAN, C., ZHANG, Z. F., HURST, R., LENNY, L. & GOLDSTEIN, J. (1996): Characterization of recombinant α-galactosidase for use in seroconversion from blood group B to O of human erythrocytes. *Arch. Biochem. Biophys. 327*, 324-329.

575. ZIDERMAN, D., GOMPERTZ, S., SMITH, Z. G. & WATKINS, W. M. (1967): Glycosyl transferases in mammalian gastric mucosal linings. *Biochem. Biophys. Res. Commun. 29*, 56-61.

6 Lewis System and Antigens Le^x and Le^y

The identification of the **Le**^{a} character is generally attributed to Mourant in 1946 [264], although Ueyama and Furuhata had described it as the 'T-antigen' in 1939 (see Race and Sanger [302]). Since that time further characters of the blood group **Lewis** system have been defined – **Le**^{b} [5], **Le**^{c} [125], **Le**^{d} [299], and the embryonal **Lewis** antigen, **Le**^{ab} (originally termed **Le**^{x}) [7][1].

The blood group **Lewis** antigens are found on the erythrocytes of humans and a few species of non-human primates [263,369]. Similar to the antigens of the **ABO(H)** system (*Chap. 5*), the **Lewis** characters are not confined to red blood cells but are found wide-spread in various bodily secretions and fluids.

Most structures of **Lewis** reactive substances detected in plants [383] have not yet been investigated. The isolation from mung bean seedlings of an α1,4-fucosyl-transferase, which is able to transfer fucose onto type-1 chain acceptors forming **Le**^{a} structures [69,347], provides strong evidence for the presence of **Lewis** determinants in plants. More recently, the **Le**^{a} carbohydrate motif has been detected in a glycoprotein from *Vaccinium myrtillus L.* [251]. In other plant studies, however, the **Lewis** activity is possibly due to an unspecific reaction of anti-**Lewis** antibodies with cross-reacting material.

Three **Lewis** phenotypes are present in Europids: **Le(a+b–)**, **Le(a–b+)**, and **Le(a–b–)** [302] (*Table 6.1*). The **Le(a+b+)** phenotype is extremely rare in Europids [265] but is found fairly commonly in some non-Europids (e.g. Taiwanese [44,226], Japanese [353], Polynesians [152], and Australian Aborigines [33,368]); in these cases, 25% of the population may be of this phenotype.

The antigens **Le**^{x} and **Le**^{y} are serological characters widely distributed in human and animal organisms [131]. Though first described as **X** and **Y**, the terminology **Le**^{x} and **Le**^{y} has been generally accepted. As a consequence, the **Lewis** character of

---

[1] ISBT terminology (see *Chap. 1*): **Le**^{a} = LE1 (007 001), **Le**^{b} = LE2 (007 002), **Le**^{c} = 210 001, **Le**^{d} = 210 002.

184

Table 6.1: Blood group Lewis phenotype of erythrocytes and secretions in individuals of different Lewis and secretor genotype.

Genotype		Lewis specificity of erythrocytes	Lewis specificity in secretions	Frequency in Europids[1]
Le	Se	Le(a–b+c–d–)	Le^b, (Le^a, Le^{ab})	72 %
Le	sese	Le(a+b–c–d–)	Le^a, Le^{ab}	22 %
lele	Se	Le(a–b–c–d+)	Le^d	5 %[2]
lele	sese	Le(a–b–c+d–)	Le^c	1 %[2]

[1] According to Issitt [168].

[2] The **Le(a-b-)** phenotype is uncommon among Europids but up to 5 times more common among Negrids [261]. Further, the **Lewis**-negative phenotype refers strictly to the serological reactions of erythrocytes, since small amounts of **Le^a** and **Le^b** antigens may be found in **Le(a-b-)** individuals (see *Sect. 6.3.1*).

embryonal erythrocytes, originally described as **Le^x** [7], had to be changed to **Le^{ab}**. Because **Le^x** and **Le^y** are limited to tissues and secretions and are not expressed on erythrocytes, they are not considered blood group properties in the strict sense. Nevertheless, due to their close chemical and functional relationship with the **Lewis** antigens, these antigens will be discussed here.

Evidence is steadily increasing that the **Le^a** and **Le^x** antigens and their sialoyl derivatives in particular are involved in cell adhesion events and play an important role in cell recognition during various physiological and pathological processes. They are, for example important regulators in mammalian embryonic development and are essential factors in recruiting circulating leucocytes to sites of inflammation; the accumulation of **Le^a**, **Le^x**, and **Le^y** antigens in cancer tissues also suggests an essential role of these antigens in tumour metastasis [89].

6.1 Genetics

The blood group **Lewis** system is genetically independent of all other blood group systems [58,123]. However, it was discovered that the expression of the different

Lewis phenotypes was closely connected with the **ABH**-secretor status of the individual [57,122][2]. The formation of the five **Lewis** determinants mentioned above is controlled by the gene systems *Le /le* and *Se /se*. In *Table 6.1* the correlation between genotype and **Lewis** phenotype is presented: Le^a character is found only on the erythrocytes of **Lewis**-positive **ABH**-nonsecretors, Le^b only on the cells of **Lewis**-positive **ABH**-secretors. Red bloodcells of **Lewis**-negative **ABH**-secretors show Le^d specificity, whereas those of nonsecretors show Le^c and the character Le^{ab} is present on all erythrocytes from **Lewis**-positive adults and newborns. The **Lewis** characteristics of the body secretions, with the exception of significant Le^a activity in saliva of Le^b individuals, are identical to those of the erythrocytes.

As is the case in Le^b, the synthesis of Le^y from the Le^x determinant is controlled by the *secretor* gene.

The **Lewis** (= *FUT3*) gene is located on the short arm of chromosome 19 at position 13.3 [22,207,248], the *secretor* gene on chromosome 19, position q13.3 (see *Sect. 5.1*).

At least five genes account for the formation of Le^x antigenic structures: (1) the **Lewis** (= *FUT3*) gene; (2) *FUT5*, which encodes an unspecified type of an α1,3-fucosyltransferase, and (3) *FUT6*, which controls the synthesis of the 'plasma-type' Le^x. These three genes form a cluster of homologous genes on chromosome 19 at position p13.3 in the order *telomere – FUT6 – FUT3 – FUT5 – centromere* [305]. (4) *FUT4*, the gene responsible for the synthesis of the 'myeloid-type' Le^x, is located on chromosome 11 at q21 [304,362], and (5) *FUT7*, which encodes the leukocyte α1,3-fucosyltransferase and has been assigned to chromosome 9 at q34.4 [306].

6.2 Antisera and Lectins

Serologically and clinically significant human anti-**Lewis** antibodies are mostly of the IgM type [168]. Those IgG antibodies which are frequently found [157,342] are generally of low affinity and thus of little practical importance. In several cases good antisera have been produced by immunisation of animals with **Lewis**-active material.

The reactivity of most anti-**Lewis** sera towards erythrocytes is fairly weak; agglutination, however, can be enhanced by ficin-treatment of the cells.

---

[2] For further information on the **ABH**-secretor status and the *secretor* gene, see *Sects. 5.1* and *5.4.2*.

6.2.1 Descriptive Listing of Antisera with Their Specificities

Anti-Le^a

Anti-Le^a is a fairly common antibody in the sera of Le(a–b–) non-O ABH-secretors (i.e. blood groups A$_1$, B, or A$_1$B, see [168,302]); in rare cases [68,180] antibodies of this specificity have been found in Le(a–b+) individuals. Anti-Le^a is often accompanied by anti-Le^b.

High-quality anti-Le^a reagents have also been obtained through immunisation of rabbits, goats, or chickens with Le^a-active material such as the saliva of ABH-nonsecretors [189], Le^a substance from ovarian cysts [20], erythrocytes coated with Le^a substance [224], or Le^a-active oligosaccharides coupled to carrier proteins [219,220,243,245]).

Anti-Le^a sera agglutinate erythrocytes of *Le /sese* individuals independent of their ABO group. They are inhibited by saliva of Le(a+b–) and Le(a–b+) individuals.

Many anti-Le^a sera contain lymphocytotoxic antibodies [79].

Hybridoma antibodies towards the sialoyl-Le^a determinant have also been described to date [107,195,222].

Anti-Le^b

Anti-Le^b antibodies occur less frequently than anti-Le^a. They are found mainly in Lewis-negative ABH-nonsecretors, and frequently appear together with anti-Le^a [168,302]. Anti-Le^b is also produced in goats after immunisation with Le^b-active material [219,243,245,391].

Two types of anti-Le^b sera have been defined according to their inhibition pattern with saliva from ABH-secretors and nonsecretors and their ability to agglutinate O and A$_1$ erythrocytes [38,57]:

– anti-Le^{bL} sera are inhibited by saliva of Lewis-positive secretors and agglutinate all Le(a–b+) erythrocytes irrespective of their ABO group,
– anti-Le^{bH} sera agglutinate preferentially Le(a–b+) erythrocytes showing strong H activity, i.e. O and A$_2$ cells (A$_1$ or B erythrocytes react only weakly or not at all). The sera are inhibited by saliva of ABH-secretors regardless of their Lewis phenotype. Anti-Le^{bH} sera are found more frequently than anti-Le^{bL}.

It must be stressed, however, that this classification is still arbitrary, since the anti-Le^b sera show a range of specificities extending from anti-H (= anti-HLe^d, see below) via anti-HLe^b and anti-Le^{bH} to anti-Le^b (= anti-Le^{bL}) (see also [204]).

GS-IV, a lectin isolated from the seeds of *Bandeiraea simplicifolia* (= *Griffonia simplicifolia*) binds preferentially to Le^b determinants, but also shows significant affinity towards the Le^y Antigen (see *Sect. 6.5*) [182,325,343].

Anti-Le^c

Only a few examples of anti-**Le**^{c} antibodies have been detected in human serum [125,230]. Anti-**Le**^{c} is produced, however, in goats and rabbits after immunisation with saliva of **Lewis**-negative nonsecretors [116,167,298].

Antibodies of this specificity agglutinate **Le(a–b–)** erythrocytes of **ABH**-nonsecretors irrespective of **ABO** group. They are inhibited by nonsecretor saliva and also bind, though weakly, to saliva substances of **Le(a+b–)** **ABH**-nonsecretors and **Le(a–b–)** secretors. However, the results of various investigations suggest that 'anti-**Le**^{c}' designates a family of antibodies which react with several antigens present in significant amounts on **Le**^{c} erythrocytes.

Anti-Le^d

Anti-**Le**^{d} specific antibodies have not yet been detected in humans but have been prepared repeatedly by immunising goats either with erythrocytes or saliva of **Lewis**-negative **ABH**-secretors [116,298,299], or with **Le**^{d}-active oligosaccharides attached to suitable carrier molecules [219].

Anti-**Le**^{d} sera agglutinate **Le(a–b–)** erythrocytes of **ABH**-secretors and are inhibited by their saliva. They react, however, preferentially with blood group **O** and **A**$_2$ cells (hence the preferred designations anti-**Le**^{dH} or anti-**HLe**^{d}).

Anti-Le^{ab}

Anti-**Le**^{ab} (originally described as anti-**Le**^{x}) is frequently found in human sera [3,7,179]. The antibody occurs mainly in **Le(a–b–)** secretors of blood groups **A**$_1$, **B**, or **A**$_1$**B**, and is often accompanied by anti-**Le**^{a} and/or anti-**Le**^{b}. In contrast to the sera described above, anti-**Le**^{ab} sera agglutinate not only 90% of foetal erythrocyte samples from umbilical cord blood, but also all **Le(a+b–)** and **Le(a–b+)** erythrocytes of adults. They are distinctly inhibited by **Le**^{a} saliva, whereas **Le**^{b} saliva inhibits them only weakly. These antibodies often show a weak reactivity towards secretions of **Le(a–b–)** nonsecretors [6,7]. Since the reactivities towards **Le(a+b–)** and **Le(a–b+)** cells, as well as towards foetal erythrocytes, cannot be separated [7,13,179], it is assumed that anti-**Le**^{ab} is a distinct type of antibody [320].

Anti-A$_1$Le^b

Thus far only very few examples of Anti-**A**$_1$**Le**^{b} antibodies have been found in human sera [124,322]. These antibodies react with erythrocytes and secretions of **A**$_1$ **Le(a–b+)** or **A**$_1$**B** **Le(a–b+)** individuals, but show no reaction with cells or saliva of **Le(b+)** persons of blood groups **O**, **A**$_2$, or **B**.

Several lymphocytotoxic sera with the specificities anti-**A**$_1$**Le**^{b} and anti-**BLe**^{b} have also been described (e.g. [173,247,275,286,294]).

Anti-ALe[dH]

The first antiserum with this specificity ('Magard Serum') was obtained from an **A₂ Le(a–b+)** individual [2]. The serum agglutinated mainly **A₁ Le(a–b–)** erythrocytes of secretors and was inhibited by their saliva; cells obtained from **A₂ Le(a–b–)** secretors reacted only weakly. This antibody is probably directed towards a type-1 chain **A** determinant (see *Sect. 5.3.1*).

Nielsen et al. [275] present a description of a lymphocytotoxic antibody with anti-**ALe**[d] specificity.

Anti-ILe[bH]

Only one example of an anti-**ILe**[bH] antibody has been detected in an **A₁ Le(a–b–)** individual [360]. The serum reacted with **O** and **A₂ Le(a–b+)** erythrocytes of blood group **I** persons and was inhibited by their saliva.

However, the fact that, in contrast to all other blood group **Lewis** substances (see below), **ILe**[b] is not absorbed from the serum suggests that this character is not associated with the blood group **Lewis** system.

Anti-Lewis Hybridoma Antibodies

A series of anti-**Lewis** specific hybridoma antibodies has been described – anti-**Le**[a] [29,98,111,387], anti-**Le**[b] [29,45,98,111,252], anti-**A₁Le**[b] [64,113], anti-**Le**[d] (= anti-**H** type-1) [99], and anti-**A₁Le**[d] (= anti-**A** type-1) [169].

A detailed characterisation of twelve anti-**Le**[a] and anti-**Le**[b] hybridoma antibodies is presented by Good et al. [111].

Anti-Le[x] and Anti-Le[y] Reagents

All anti-**Le**[x] (cf. [46,77,145,163,341]) and anti-**Le**[y] reagents (cf. [1,28,48,185, 213,229,349]) thus far described are hybridoma antibodies. Most of them have been prepared by immunising mice with various tumour cells. Reports of anti-**Le**[x] antibodies have been published which react preferentially with mono-, di-, or trifucosyl structures (see [103]), or with **Le**[x] antigen located on branched oigosaccharide chains [113]. More recently, an anti-**Le**[y] hybridoma antibody (*KH1*) has been described which binds specifically to a trifucosyl-**Le**[y] determinant [181]. Furthermore, it is now possible to prepare monoclonal antibodies which bind to the determinants **ALe**[y] [64,113] and to sialylated **Le**[x] structures [77,105,106].

The lectin of *Griffonia simplicifolia* – originally described as an anti-**Le**[b] reagent – shows a significant affinity for the **Le**[y] antigen [325,343].

189

6.3 Blood Group Lewis Substances

6.3.1 Occurrence of Blood Group Lewis Substances

Sneath and Sneath [339,340] observed that **Le(a+b–)** erythrocytes became **Le(a+b+)** specific when transfused into an **Le(a–b+)** recipient. The same erythrocytes, however, gradually lost their **Lewis** activity when transfused to an **Le(a–b–)** subject. It has been possible to reproduce this effect *in vitro* by incubation of erythrocytes with sera of different **Lewis** specificities [239,339] (see *Table 6.2)*. Later it was shown that the specificities **ALe**[b] [363], **Le**[ab] [13], **Le**[c] and **Le**[d] [156] are also taken up from serum.

Analogous 'natural experiments' have been reported in chimaeric twins [70,274]. As a consequence of placental vascular anastomoses and interchange of primordial blood cells between the embryos, haematopoietic chimaeras have two distinct populations of blood cells. In all cases investigated the grafted cells preserved their genetically defined serological properties, but adopted the **Lewis** character of the host.

These studies revealed beyond all doubt that the blood group **Lewis** substances of human erythrocytes are not derived from the erythrocytes themselves or from their precursor cells, but are instead absorbed from the plasma.

In repeating the above-mentioned absorption experiments with lipoprotein and glycolipid fractions isolated from human serum, Marcus and Cass [244] were able to show that the blood group **Lewis** substances of the erythrocytes are glycolipids.

It has further been demonstrated that these glycolipids also bind to other blood cells and are thus responsible for the **Lewis** activity of lymphocytes and thrombocytes (e.g. [4,79,80])[(3)].

Quantitative investigations on **Le**[b] glycolipid in human serum showed an average concentration of 0.9 µg per ml and a distribution between serum and erythrocytes in a ratio of about 2 : 1 [308].

The ability of these plasma glycolipids to bind spontaneously to erythrocyte membranes is frequently used as a simple and sensitive test for blood group **Lewis**-active plasma glycolipids ('quantitative passive haemagglutination'). Hanfland's studies [137] revealed that this method detects quantities of less than 0.01 µg of these substances. In fact, in a few cases an amount of 0.0001 µg of a **Le**[b]-active glycolipid was sufficient to convert 9×10^7 **O Le(a–b–)** erythrocytes into **Le**[b]-positive cells (this quantity corresponds to approximately 400 molecules absorbed per cell).

It is evident that the **Lewis**-active glycosphingolipids found in human serum must originate from tissue of entodermal origin in which the synthesis of **ABH**-active glyco-

---

[(3)] See also *Sect. 5.3.3* for a discussion of blood group **A** and **B** specific plasma glyco-sphingolipids.

Table 6.2: *In vitro* uptake from human plasma of blood group Lewis substances by erythrocytes.

Erythrocyte blood group	Plasma donor blood group	Reaction of erythrocytes with	
		anti–Le^a	anti–Le^b
	Le(a+b−)	+	−
Le(a+b−)	Le(a−b+)	+	+
	Le(a−b−)	(+)	−
	Le(a+b−)	+	+
Le(a−b+)	Le(a−b+)	−	+
	Le(a−b−)	−	(+)
	Le(a+b−)	+	−
Le(a−b−)	Le(a−b+)	−	+
	Le(a−b−)	−	−

Le(a+b−), Le(a−b+), and Le(a−b−) erythrocytes were incubated for 24 h with plasma from Le(a+b−), Le(a−b+), and Le(a−b−) donors.
According to Sneath and Sneath [339].

conjugates is controlled by the *secretor* gene (see *Sect. 6.4.2*). The proposal that the kidneys may be the site of synthesis can be rejected on the basis of findings of Oriol and co-workers (cited in [374]), according to which transplanted kidneys influence only the **Lewis** specificity of the urine substances, but not that of the plasma lipids. Further, due to its high content of **Lewis**-active glycolipids, intestinal mucosa has repeatedly been discussed as the probable source of these substances (see [287]). Nevertheless, the exact origin of the **Lewis** glycosphingolipids of human serum has yet to be determined.

Substances carrying **Lewis** determinants are not confined to erythrocytes and blood plasma but are also found in various tissues and body fluids. **Lewis**-specific glycosphingolipids have been isolated from gastric and intestinal mucosa and from pancreas. Saliva, gastric juice, amniotic fluid, and ovarian cyst fluid are good sources of blood group **Lewis**-active glycoproteins; urine and milk contain **Lewis**-active oligosaccharides, and meconium is rich in serologically active glycoproteins and glycolipids. Great quantities of **Le**^{a} and, in some cases, of **Le**^{b} glycosphingolipids have been isolated from various adenocarcinomas of the gastrointestinal tract as well (see *Sect. 6.3.4*).

Occurrence and distribution of **Lewis** antigens in the human organism is largely age-dependent.

Erythrocytes of **Lewis**-positive newborns react only with anti-**Le**^{ab} antibodies [179], though in some cases weak **Le**^{a} activity is detectable [71]. Clearly recognisable **Le**^{a} develops about 6–10 days after birth. In **Le(a–b+)** individuals **Le**^{b} activity appears in the second month, whereas **Le**^{a} activity decreases concurrently, in most cases disappearing during the second year of life. The **Le(a+b+)** transition state, however, may last up to 7 or even 14 years of age [39,71,179]. As mentioned above, the **Le(a+b+)** phenotype is only rarely found in adult Europids, but does occur rather frequently in non-Europids.

Le^a, **Le**^{b}, and **Le**^{ab} characters are fully expressed at birth in secretions and body fluids [212]. Szulman and Marcus [354] were able to detect **Lewis** antigens together with **ABH** antigens in glandular epithelia (salivary gland, gastric mucosa, intestine) of embryos of 35–40 mm length. **Lewis**-active material is also found in amniotic fluid [11,13]; the serologic specificity corresponded to the **Lewis** phenotype of the fetus in all cases.

It has been repeatedly observed that in **Lewis**-positive women the **Le**^{a} or **Le**^{b} activity of erythrocytes is significantly reduced during gestation – in some cases the **Lewis** activity is lost altogether and the red cells may type as **Le(a–b–)**, whereas the **Lewis** activity in saliva remains unaffected [135,359].

According to investigations on pregnant women by Hammar et al. [135] the concentration of plasma glycosphingolipids changes only slightly, but the ratio of lipoprotein to red cell mass increases more than fourfold during pregnancy. Under these conditions a higher percentage of the **Lewis** glycosphingolipids may be associated with plasma lipoprotein and only a lesser percentage adsorbed to erythrocytes, thus resulting in weak or even negative reactions of the red cells with anti-**Lewis** antibodies.

A significant reduction of **Le**^{a} or **Le**^{b} activity of erythrocytes has been observed in **Lewis**-positive individuals suffering from various cancers [211,386], as well as in some cases of chronic hepatitis, liver cirrhosis, alcoholic pancreatitis, and hydatid cyst disease [240,350]. The tumour-associated changes in **Lewis** antigen expression may be due to the formation of irregular anti-**Lewis** antibodies induced by the increased synthesis of **Lewis**-specific substances in transformed tissue [297]. In the majority of cases the change from **Lewis**-positive to **Lewis**-negative phenotype has been shown to occur in *Le/le* heterozygous individuals, who, due to a gene-dosage effect, produce less **Lewis**-determinant structures [289].

The term 'Lewis-negative' refers strictly to the erythrocyte phenotype. So-called 'non-genuine **Lewis**-negative' or 'weak **Lewis**' (= **Le**^{w}) individuals express small amounts of **Lewis**-active substances in tissues and secretions (e.g. serum [150,151], saliva glands and saliva [12,242], as well as epithelial cells of small intestine, colon, and bladder [27,148,288]). The amounts of **Le**^{a} and **Le**^{b} glycosphingolipids detected in **Lewis**-negative individuals were only 1–5% of the normal **Lewis** glycolipids present in **Lewis**-positive individuals [27,150]. These investigations clearly show that the enzymes encoded by some inactivating mutant alleles are not totally inactive but are still able to produce small amounts of antigen.

It has been shown that binding of bacteria – for example such as *Staphylococcus aureus* [310,311] – to epithelial cells is correlated with the amount of **Le**^{a} antigen present on the cells. Since **Le**^{a} is predominantly expressed on the tissues of infants during the first months of life, the authors suggest a relevance of colonisation by toxogenic bacteria containing **Le**^{a}-specific adhesins to *sudden infant death syndrome*.

6.3.2 Chemical Structures of the Blood Group Lewis Determinants

The structures of **Le**^{a} and **Le**^{b} determinants present in **Lewis**-active glycoproteins of ovarian cysts and gastric mucosa, or carried by oligosaccharides of milk and urine, have been established by Morgan and Watkins (Lister Institute of Preventive Medicine, London) and Kabat (Columbia University, New York). A detailed survey on these investigations is found in the review article by Watkins [373]. More recent investigations on **Le**^{a}- and **Le**^{b}-active material from serum and different tissues are referred to below. The structures of the antigens **Le**^{c} and **Le**^{d} have been investigated by Hanfland and co-workers. A series of studies have likewise been performed on the chemical nature of **Le**^{ab}; the exact structure of this antigen, however, is still unknown.

As shown in *Fig. 6.1*, the **Lewis** antigens are derived from a common **precursor structure**, a type-1 oligosaccharide chain[4]. The terminal disaccharide unit, **Galβ1→3GlcNAc** carries the carbohydrate residues characteristic of the different **Lewis** properties:

---

[4] See *Sect. 5.3.1.*

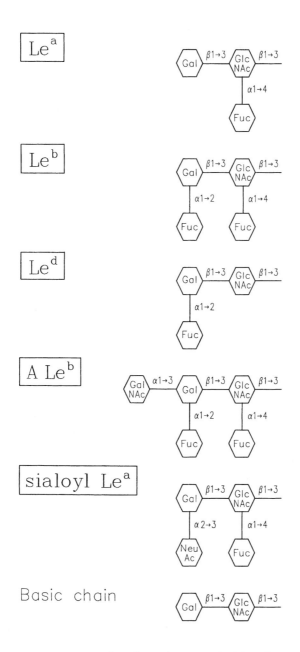

Figure 6.1: Structures of Leᵃ, Leᵇ, Leᵈ, ALeᵇ, and sialoyl-Leᵃ determinant groups and of the type-1 basic chain as their common precursor.

The **Le^a** determinant is characterised by an α-fucose attached to carbon 4 of the subterminal N-acetylglucosamine residue.

The **Le^b** specificity is determined by a precursor chain substituted with an **Le^a**-specific $\alpha1\rightarrow4$ fucose and a blood group **H** determinant $\alpha1\rightarrow2$ fucose. This antigen thus represents a hybrid epitope composed of **Lewis** and **H** determinant structures.

The **Le^d** antigen is characterised by α-fucose at carbon 2 of the terminal galactose residue [83,117,138]. **Le^d** is therefore identical with a type-1 blood group **H** determinant (see *Sect. 5.3.1*).

The **ALe^b** structure is represented by an **Le^b** determinant elongated by a blood group **A** specific $\alpha1\rightarrow3$ N-acetylgalactosamine residue. In erythrocytes and in small intestine this terminal structure is carried by glycolipids [63,227]. Oligosaccharides containing this structure have been detected in the urine of **A$_1$Le^b** individuals [232]. A glycolipid with the **BLe^b** determinant has been isolated from human small intestine [37], and a corresponding oligosaccharide was found in urine of **BLe^b** individuals [232].

Since **Le^c** specificity is found mainly on erythrocytes and in saliva of **Lewis**-negative nonsecretors it has repeatedly been assumed that anti-**Le^c** reagents bind to the unchanged precursor chain [116,215]. Inhibition tests and quantitative passive haemagglutination with lactotetraosylceramide (see *Fig. 6.6)*, however, clearly show that at least one sample of anti-**Le^c** (serum *Arm.*) does not bind to this structure [139]. A significant inhibition of this serum by 3-fucosyllactose (**Gal$\beta1\rightarrow4$GlcNAc3$\leftarrow1\alpha$Fuc**) [125] and lacto-N-fucopentaitol III (**Gal$\beta1\rightarrow4$[Fuc$\alpha1\rightarrow3$] GlcNAc$\beta1\rightarrow3$Gal$\beta1\rightarrow4$Sorbitol**) [109] suggested that **Le^c** specificity is determined by a 3-fucosyllactosaminyl residue, i.e. by the **Le^x** determinant based on a type 2 chain structure (see *Sect. 6.5)*. More recently Hanfland et al. [141] reported the isolation of two **Le^c**-active glycosphingolipids from the plasma of **O Le(a–b–)** nonsecretors. The investigations on these substances revealed that the immunodominant structure of the **Le^c** determinant is in fact represented by an **Le^x**-specific 3-fucosyl-N-acetyl-lactosamine, which is, however, attached to a type-1 chain based oligosaccharide sequence (*Fig. 6.2)*.

The exact structure of the **Le^{ab}** antigen has not yet been established. First serological investigations [13] have revealed that the **Le^{ab}** determinant must be a product of the *Le* gene. Serum absorption tests showed that at least some anti-**Le^{ab}** sera bind not only to immunoadsorbents carrying **Le^a** and **Le^b** epitopes, but also to those containing the partial structure **Fuc$\alpha1\rightarrow4$GlcNAc-**, which is common to both antigens [17,320]. This result led to the assumption that the **Le^{ab}** character is probably

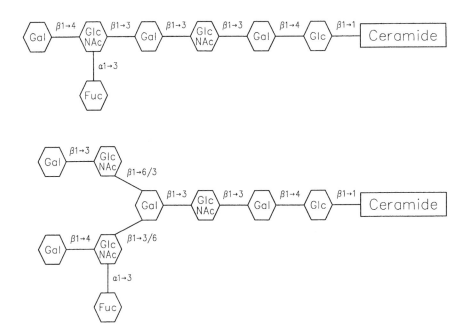

Figure 6.2: Structures of Le^c active glycosphingolipids.
After Hanfland et al. [141].

not determined by a typical chemical structure, but is based on the ability of the anti-**Le**^{ab} antibodies to react with the actual blood group **Lewis** determinant, **Fuc**$\alpha1{\rightarrow}$**4GlcNAc**.

Nevertheless, it has yet to be established why anti-**Le**^{ab} sera, in contrast to anti-**Le**^{a} (and anti-**Le**^{b}) sera, readily agglutinate foetal erythrocytes of **Lewis**-positive individuals. Since anti-**Le**^{ab} antibodies bind to a smaller epitope than normal anti-**Le**^{a} antibodies, they are probably less influenced by the sterical conditions on the erythrocyte surface and thus able to agglutinate foetal erythrocytes containing only traces of **Le**^{a} substance [320]. However, there is a possibility that in an early stage of embryonic development a distinct **Le**^{ab} (or **Le**^{a}?) substance is formed which shows only weak affinity towards the usual anti-**Le**^{a} sera but readily binds to anti-**Le**^{ab} sera. In this connection investigations of the anti-**Le**^{ab} properties of two monoclonal antibodies which show the above-mentioned specificity would be of interest [387].

It has been shown that several hybridoma antibodies directed towards tumour tissues (e.g. '19-9'and '52a' [154,203]) bind specifically to sialylated **Le**^{a} structure *(Fig. 6.1)*. Various amounts of this antigen have been detected on glycoproteins and glyco-

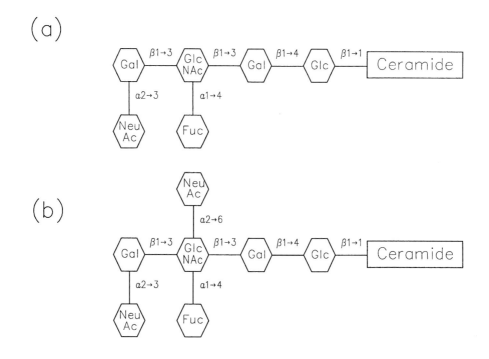

Figure 6.3: Structures of sialylated Le^a glycolipids from cancer tissue.
(a) 'Gastrointestinal Cancer Antigen' (**GICA** or **Ca 19-9**) [236] and (b) disialoyl-Le^a substance [282].

lipids of tissue cells and secretions of **Lewis**-positive individuals [19,144,196]. Oligo-saccharides carrying the sialoyl-**Le**^{a} determinant have been isolated from human milk [191,193,194].

Great quantities of sialoyl-**Le**^{a} antigen have been detected in tumours of the gastrointestinal tract (stomach, colon, and pancreas) of **Lewis**-positive individuals [18,192,236]. Furthermore, the antigen can be detected in serum of tumour patients, so that sialoyl-**Le**^{a} is widely discussed as a 'tumour marker' (**CA 19-9** or **GICA** = 'gastrointestinal cancer antigen') [16,154,202,236,237][5]. Increased amounts of

---

[5] Due to a gene dosage effect, the serum cut-off level of the antigen, above which level cancer may be suspected, is lower in *Le/le* heterozygous individuals [289].

sialoyl-**Le**^{a} antigen have been found in serum and saliva of cystic fibrosis patients as well [377,378].

The structure of a glycolipid carrying the sialoyl-**Le**^{a} epitope isolated from human adenocarcinoma cells [236] is presented in *Fig. 6.3.a*. A further sialylated **Le**^{a} antigen – disialoyl-**Le**^{a}, which reacts with the hybridoma antibody *FH7* – has been described in adenocarcinomas of human colon by Nudelman et al. [282] (*Fig. 6.3.b*).

6.3.3 Sterical Configuration of the Blood Group Lewis Determinants

Based on theoretical considerations and on data from NMR spectroscopy, three-dimensional conformations of the **Le**^{a}, **Le**^{b}, and **Le**^{d} epitopes have been proposed by several investigators [24,51,218]. According to these molecular modelling studies the **Lewis** determinants adopt relatively rigid structures in solution: in the case of the **Le**^{b} determinant the two α-fucosyl residues approach each other so closely that the two methyl groups almost touch each other over the β-side of the galactose unit; the individual orientations of the fucose groups are also maintained in **Le**^{a} and **Le**^{d} epitopes which contain only one fucosyl residue each. The topography of the **Le**^{y} determinant (see below, *Sect. 6.5*) is very similar to that of **Le**^{b} (*Fig. 6.4*).

The heterogeneity and cross-reactivity observed among different **Lewis** antibodies has been discussed in publications by the groups led by Lemieux [219] and Oriol [97,111] (see *Fig. 6.5*). Since anti-**Le**^{a} antibodies obviously bind to an epitope comprising $\alpha1{\rightarrow}4$ fucose and $\beta1{\rightarrow}3$ galactose, they exhibit strong cross-reactivity with type-1 precursor structures, and only a few samples bind to **Le**^{b} antigen. The cross-reactivity of anti-**Le**^{bL} sera with **Le**^{a} antigens and of anti-**Le**^{bH} sera with **Le**^{d} (H type-1) antigens undoubtedly reflects the nature of the immunodominant group at the binding sites which are $\alpha1{\rightarrow}4$ or $\alpha1{\rightarrow}2$ fucose, respectively. The highly similar conformation of **Le**^{b} and **Le**^{y} antigens easily explains the strong cross-reaction of most anti-**Le**^{b} reagents with the **Le**^{y} specific structure. In particular, the investigations by Good et al. [111] provide further evidence for the theory that the anti-**Lewis** sera represent a series of antibodies with specificities gradually changing from **Le**^{a} via **Le**^{bL} to **Le**^{bH}.

6.3.4 Structures of the Blood Group Lewis Substances

(a) Lewis-active Glycolipids

Blood group **Lewis**-specific glycosphingolipids are not confined to blood plasma – they have been detected in intestinal mucosa [30,250,336] and meconium

198

Le^a

Le^d(H)

Le^b

Y

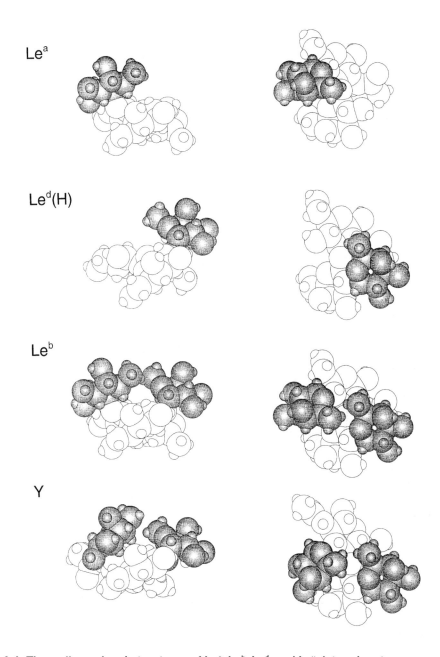

Figure 6.4: Three-dimensional structures of Le^a, Le^b, Le^d, and Le^y determinant groups.
Left: side view, right: top view.
Molecular models as based on the torsion angles published by Lemieux [218,219]..

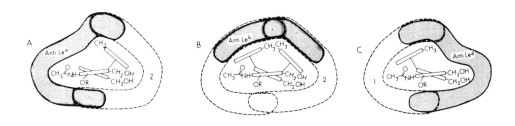

Figure 6.5: Schematic structures for the Le^a (A), Le^b (B), and Le^d (C) determinant groups.

Assumed binding regions of the respective antibodies to the epitopes.
Reproduced from Lemieux [219], with permission of the American Chemical Society (©1981).

as well [8,188]; moreover, they are frequently found in tumour tissues, especially in adenocarcinomas of the gastrointestinal tract [30,129,329].

Thus far a series of glycosphingolipids with various **Lewis**-specificities has been isolated and characterised (e.g. **Le^a** [72,83,90,138,140,336]), **Le^b** (e.g. [72,83,90,129, 137,138]), **ALe^b** [63,227], and **Le^d** [72,83,138]). In all these cases the serologically determinant carbohydrate residues are carried by lacto-N-tetraosyl-ceramide. Molecules with longer oligosaccharide chains containing one or two α1→4 fucosyl residues have been isolated from human cancer tissue ([30,129,186,351,352], see *Fig. 6.6*. The glycolipids show a distinct variability in the lipid moiety which is a characteristic of the origin of these glycolipids (e.g. the **Lewis** substances of human serum can be separated in up to three or four fractions depending on their ceramide component [137,138,140]).

(b) Lewis-active Glycoproteins

The main sources of **Lewis** active glycoproteins are the 'water-soluble blood group substances' occurring in saliva, gastric juice, and ovarian cyst fluid. The characteristics of these substances have been discussed in detail in *Sect. 5.3.4*: they are glycoproteins containing various amounts of type-1 chains, which may carry **Lewis** determinants (see Watkins [373]). *Figure 6.7* presents examples of **Le^a** and **Le^b** active oligosaccharide chains as found in ovarian cyst glycoproteins [358] and gastric mucosal mucin [334] (refer also to the chain model proposed by Kabat, *Fig. 5.13.b*).

Considerable quantities of **Lewis** active 12-kDa glycoproteins have been detected in urine [327,328].

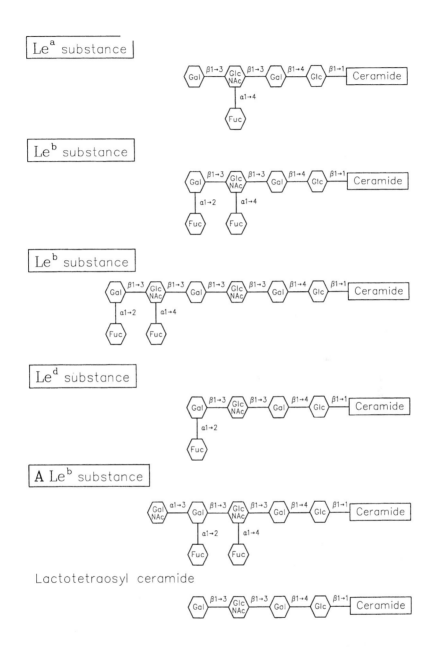

Figure 6.6: Structures of blood group Lewis-specific glycosphingolipids of human serum and cancer tissues, and of lactotetraosylceramide as their common precursor.

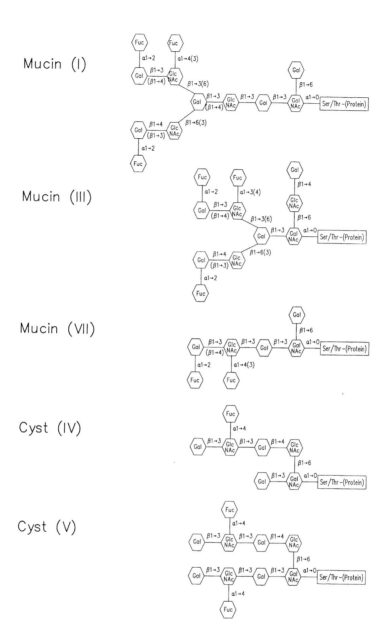

Figure 6.7: Structures of Le^a and Le^b active oligosaccharide units from human gastric mucosal mucin and ovarian cyst fluid glycoprotein.

According to Slomiany et al. [334] (structures I,III,and VII) and Tanaka et al. [358] (structures IV and V).

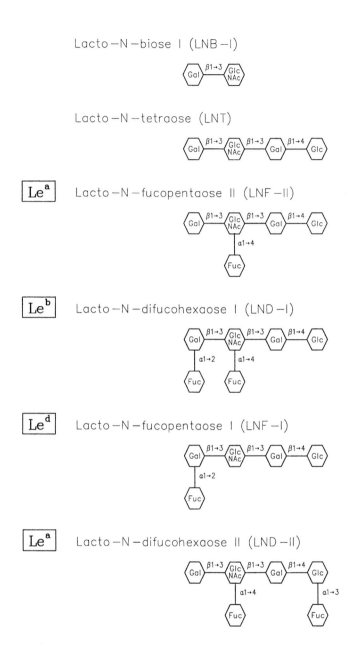

Figure 6.8: Human milk oligosaccharides with blood group Lewis-specific structures.
According to Kobata [199].

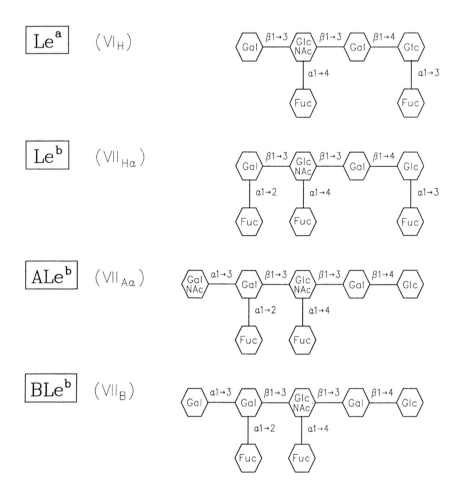

Figure 6.9: Human urine oligosaccharides with blood group Lewis-specific structures.
According to Lundblad [232].

Leᵃ specific glycoprotein material has also been isolated from meconium by Coté and Valet [67].

There is evidence that **Le**ᵃ and **Le**ᵇ determinants linked to glycoproteins occur in human adenocarcinoma cells [30].

(c) Lewis-active Oligosaccharides

Both milk and urine contain oligosaccharides which carry blood group **Lewis** determinants [49,73,82,121,199,232]. Examples of such oligosaccharides are presented in *Figs. 6.8* and *6.9*.

6.3.5 Chemical Synthesis of Blood Group Lewis Determinants

Oligosaccharides with **Le^a**, **Le^b**, and **Le^d** determinant structures have also been prepared by chemical synthesis (e.g. [35,221,303]). More recently Sato et al. [318] have described the total synthesis of **Le^b** active lacto-N-difucohexaosyl-l-ceramide.

These substances exhibit the expected serological activity and, when fixed onto suitable water-insoluble carriers, are able to induce the formation of the respective antibodies in animals [219,220]. When coupled to inert carriers (e.g. Synsorb®; Chembiomed, Edmonton, Alberta, Canada) these oligosaccharides can be used successfully for the purification of antibodies.

6.4 Biosynthesis of Blood Group Lewis Determinants

6.4.1 Enzymes

The **Le^a**, **Le^b**, and **Le^d** determinants are formed from type-1 chain precursor substances by the action of two transferases, an α1,4-fucosyltransferase and an α1,2-fucosyltransferase.

(a) The product of the *Le* gene is a fucosyltransferase (= *FucTF-III*), which links fucose residues to carbon 4 of the subterminal N-acetylglucosamine. This enzyme is involved in the synthesis of **Le^a** and **Le^b** determinants and catalyses the following reaction:

$$\text{Gal } \beta1{\to}3 \text{ GlcNAc } \beta1{\to}3 \text{ R} \xrightarrow[\substack{\text{GDP-Fuc} \qquad \text{GDP}}]{\text{Mg}^{++}} \text{Gal } \beta1{\to}3 \underset{\substack{\alpha1{\to}4 \\ \text{Fuc}}}{\text{GlcNAc}} \beta1{\to}3 \text{ R}$$

Enzymes with the above specificity have been found in glandular tissues (submaxillary gland and gastric mucosa [60,301]), in kidney and gall bladder [285], saliva [178,384], milk [88,120,172,300], and in human colonic adenocarcinoma (Colo 205) cells [158]; however, it has not been possible to detect this transferase in serum [268,319], erythrocyte stroma, thrombocytes, granulocytes, or lymphocytes [56,119].

The enzyme isolated from human milk is a 98 kDa protein, which represents a dimer of two 44 kDa subunits [88].

Investigations on the substrate specificity revealed that the transferase can use not only unsubstituted type-1 structures, but also **H** type-1 and $\alpha2\rightarrow3$ sialylated type-1 structures as fucosyl acceptors [175]. Though highly purified enzyme preparations acted only on type-1 chains [65,175], *FucTF-III*, when expressed in BHK-21 cells, was able to fucosylate small amounts of selected mono- and difucosyl type-2 structures [114] forming mainly sialoyl-**Le**^{x} determinants (see *Sect. 6.5*).

The properties of the $\alpha1,4$-fucosyltransferase isolated from the culture medium of a human epidermoid carcinoma cell line [176] closely resembled those of the enzyme isolated from human milk.

(b) The second enzyme involved in the biosynthesis of **Le**^{b} and **Le**^{d} determinants is the product of the *secretor* gene ('**secretor**-transferase' = *FucTF-II*, see *Sect. 5.1*). This enzyme is an $\alpha1,2$-fucosyltransferase which transfers α-fucose to carbon 2 of the terminal galactose residue of a type-1 chain:

$$\text{Gal } \beta1\rightarrow3 \text{ GlcNAc } \beta1\rightarrow3 \text{ R} \xrightarrow[\text{GDP-Fuc} \quad \text{Fuc}]{\text{Mg}^{++}} \begin{array}{l} \text{Gal } \beta1\rightarrow3 \text{ GlcNAc } \beta1\rightarrow3 \text{ R} \\ \quad \alpha1\rightarrow2 \\ \quad \text{Fuc} \end{array}$$

The **secretor**-transferase has been found in glandular epithelia [60], saliva [384], and in milk [323] of **ABH**-secretors; the enzyme is absent in nonsecretors. Serum [268,319], bone marrow [291], and erythrocyte membranes [56] lack the *secretor*-transferase but contain the $\alpha1,2$-fucosyltransferase encoded by the *H* gene. This enzyme transfers fucosyl residues mainly onto type-2 chains, whereas the **secretor** transferase can use both type-1 and type-2 chains as acceptor substrates, slightly favouring type-1 chains. The properties of these two α1,2-fucosyltransferases have been discussed in detail in *Sect. 5.4.2*.

The occurrence of the α1,2- and α1,4-fucosyltransferases is strictly gene-dependent, and moreover, the enzymes show a characteristic distribution within the organism:

– the **secretor** gene-encoded α1,2-fucosyltransferase is found exclusively in **ABH**-secretors; **ABH**-nonsecretors lack the enzyme. Furthermore, the enzyme is expressed only in tissues and organs of entodermal origin and is absent in ecto- and mesodermal tissues (see *Sect. 5.3.7)*;

– the **Lewis** gene-encoded α1,4-fucosyltransferase, as expected, is found only in **Lewis**-positive subjects and is absent in **Le(a–b–)** persons. Further, this enzyme occurs only in tissues in which the **secretor**-transferase is expressed; in systems which are not controlled by the **secretor** gene the **Lewis**-transferase is absent.

The restriction of the α1,4-fucosyltransferase to systems in which the **secretor** gene is expressed is in accordance to the view that the **Lewis** substances derive from entodermal tissues (see *Sect. 6.3.1)*. Moreover, the gene-dependency of the two fucosyltransferases further illustrates the close correlation between **Lewis phenotype** and **secretor status** discussed above (see *Sects. 6.1* and *6.4.2)*.

Molecular Biological Investigations on the *Lewis* gene-encoded α1,3/4-fucosyltransferase

The cDNA encoding the **Lewis**-transferase has been isolated by Kukowska-Latallo et al. from a human cell line [207]. The primary structure of the enzyme deduced from the nucleotide sequence revealed a 361 amino acid transmembrane protein with a domain structure typical for mammalian glycosyltransferases – the enzyme molecule is composed of a short N-terminal cytoplasmic segment (15 amino acids), a single transmembrane region (19 amino acids), and a fairly long C-terminal portion (327 amino acids) in which the catalytically active domain is located *(Fig. 6.10)*; a stem region of approximately 29 amino acids is located between the transmembrane domain and the actual catalytic domain [380]. The protein shows two potential N-glycosylation sites. Transfection experiments using COS cells revealed that the cloned cDNA directed the *de novo* synthesis of a fucosyltransferase acting on both type-1 and type-2 acceptors and the *de novo* expression of both **Le^a** and **Le^x** antigens on the cell surface.

```
  1 MDPLGAAKPQ WPWRRCLAAL LFQLLVAVCF FSYLRVSRDD ATGSPRAPSG SSRQDTTPTR  60

 61 PTLLILLWTW PFHIPVALSR CSEMVPGTAD CHITADRKVY PQADTVIVHH WDIMSNPKSR 120

121 LPPSPRPQGQ RWIWFNLEPP PNCQHLEALD RYFNLTMSYR SDSDIFTPYG WLEPWSGQPA 180

181 HPPLNLSAKT ELVAWAVSNW KPDSARVRYY QSLQAHLKVD VYGRSHKPLP KGTMMETLSR 240

241 YKFYLAFENS LHPDYITEKL WRNALEAWAV PVVLGPSRSN YERFLPPDAF IHVDDFQSPK 300

301 DLARYLQELD KDHARYLSYF RWRETLRPRS FSWALDFCKA CWKLQQESRY QTVRSIAAWF 360

361 T
```

Figure 6.10: Amino acid sequence of the α1,3/4-fucosyltransferase encoded by the Lewis gene (= FUT3) as derived from cDNA sequencing.
Underlined: hydrophobic segment representing the putative transmembrane domain; *N*: potential N-glycosylation sites.
After Kukowska-Latallo et al. [207]. The sequence data are deposited in the EMBL/GenBank data libraries under the accession number X53578.

In the *Le* gene of **Lewis**-deficient individuals a series of different point mutations have thus far been detected (*Fig. 6.11*):

- $T^{59} \rightarrow G$ causing a Leu $\rightarrow$ Arg substitution at position 20 [85,201,279,289],
- $T^{202} \rightarrow C$ causing a Trp $\rightarrow$ Arg substitution at position 68 [85,289],
- $C^{314} \rightarrow T$ causing a Thr $\rightarrow$ Met substitution at position 105 [85,87,289],
- $C^{445} \rightarrow A$ causing a Leu $\rightarrow$ Met substitution at position 146 [289],
- $G^{484} \rightarrow A$ causing an Asp $\rightarrow$ Asn substitution at position 162 [293],
- $G^{508} \rightarrow A$ causing a Gly $\rightarrow$ Ser substitution at position 170 [201,278,279,289],
- $G^{667} \rightarrow A$ causing a Gly $\rightarrow$ Arg substitution at position 223 [293],
- $G^{808} \rightarrow A$ causing a Val $\rightarrow$ Met substitution at position 270 [293],
- $A^{1007} \rightarrow C$ causing an Asp $\rightarrow$ Ala substitution at position 336 [279],
- $T^{1067} \rightarrow A$ causing a Ile $\rightarrow$ Lys substitution at position 356 [85,260,278,289].

The occurrence of these point mutations varies between different races, despite the similarity of incidence (~10%) of **Lewis**-negative individuals [85,201,260,278, 279,289]: the $Gly^{170} \rightarrow Ser$ substitution is most common in Japanese, but fairly rare in Europids, whereas the $Trp^{68} \rightarrow Arg$, $Thr^{105} \rightarrow Met$, and $Leu^{146} \rightarrow Met$ mutations have thus far been found only in Europids; the $Leu^{20} \rightarrow Arg$, $Gly^{170} \rightarrow Ser$, and $Ile^{356} \rightarrow Lys$ exchanges occur in both races. In Indonesia they frequently occur as single mutations, whereas in Japan and Europe the mutations at positions 20 and 170, at 68 and 105, and at 20

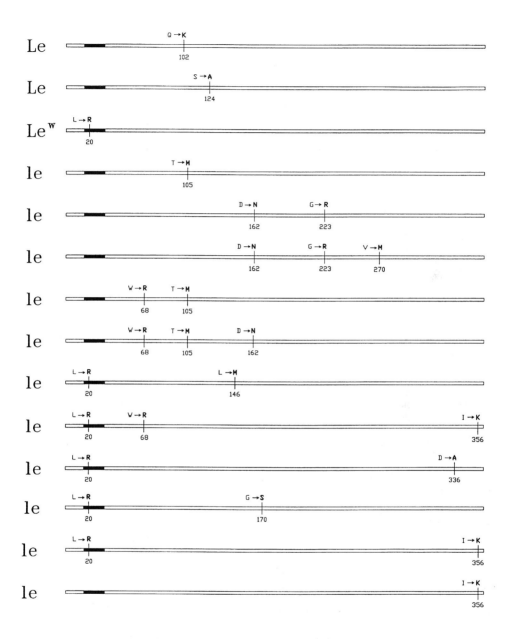

Figure 6.11: Point mutations in the *Le* gene of Lewis-deficient individuals.
■ : transmembrane domain.

and 356 were found co-located to the same allele in most cases. The $Asp^{162} \rightarrow Asn$, $Gly^{223} \rightarrow Arg$, and $Val^{270} \rightarrow Met$ substitutions have been detected in the African Xhosa population [293].

Transfection experiments performed with chimaeric *FUT3* constructs containing these mutations have shown that the $Gly^{170} \rightarrow Ser$, $Leu^{146} \rightarrow Met$, $Gly^{223} \rightarrow Arg$, $Val^{270} \rightarrow Met$ and $Asp^{336} \rightarrow Ala$ substitutions resulted in defective **Lewis**-transferases [201,260,278,279,292]. The construct containing the $Asp^{162} \rightarrow Asn$ reduced the transferase activity to about 20% as compared to the wild-type enzyme [292].

The $Leu^{20} \rightarrow Arg$ substitution is located in the transmembrane segment of the enzyme. As expected, this mutation does not affect the catalytic characteristics of the enzyme. Transfection experiments, however, revealed a significantly reduced level of the mutant enzyme in the transfected COS cells ([260], also [201,277-279]). These investigations suggest that this mutation may lead to an inappropriate anchoring of the transferase in the Golgi membrane without altering its catalytic activity. The finding by the same authors of this mutation in double dose in **Le(a–b–)** Indonesians, who expressed low amounts of **Lewis** antigen in saliva indicated that the $Leu \rightarrow Arg$ mutation at position 20 may account for the so-called 'weak **Lewis**' (= **Le^w**) phenotype [12,288].

Analogous experiments have shown that the $Trp^{68} \rightarrow Arg$ mutation decreased the enzyme activity (v_{max}) to less than 1% of that of the 'wild type' *FUT3* allele, but did not affect the enzyme's affinity for either the acceptor or nucleotide sugar donor substrates. Though the $Thr^{105} \rightarrow Met$ mutation itself did not change transferase activity, the construct containing the $Trp^{68} \rightarrow Arg$ and $Thr^{105} \rightarrow Met$ mutations in combination totally lacked enzyme activity - the gene, however, was actively transcribed [86].

Similarly, the $Ile^{356} \rightarrow Lys$ mutation reduced the **Lewis**-enzyme activity to less than 10% of that of the 'wild type' transferase. Its combination with the $Leu^{20} \rightarrow Arg$ mutation, however, led to the production of an enzymatically inactive protein [260,278].

These data show that, in addition to the $Leu^{20} \rightarrow Arg$ substitution, the occurrence of partially inactivating genetic mutations may also be responsible for the frequently observed expression of **Lewis** antigens in different tissues and body fluids of **Lewis**-negative individuals, and the presence of occasional low $\alpha 1,3/4$-fucosyltransferase activity in their tissues [288].

Two other mutations, viz. $C^{304} \rightarrow A$ causing a $Gln \rightarrow Lys$ substitution at position 102 and $T^{370} \rightarrow G$ causing a $Ser \rightarrow Ala$ substitution at position 124 have been detected in South African Europids [293]. The mutant genes, however, encoded functional enzymes (*Fig. 6.11*).

The characteristics of the **secretor**-transferase (= *FucTF-II*) responsible for the synthesis of **Le^b** determinents are presented in *Sect. 5.4.6*.

6.4.2 Biosynthetic Pathways

Degradation experiments on **Le^b** substances using specific fucosidases (see [373]) originally led to the assumption that the **Le^b** determinant was synthesised via **Le^a**. Investigations on the glycosyltransferases involved [323,374,384] did reveal, however, that the **secretor**-transferase does not use **Le^a** substance as a substrate, whereas the **Lewis**-transferase readily transfers $\alpha1{\rightarrow}4$ fucose residues onto the type-1 chain precursor and the **H** type-1 determinant. Furthermore, it was found that the **A**-transferase was not able to attach $\alpha1{\rightarrow}3$ N-acetylgalactosamine residues to the **Le^b** structure [200]. **Le^b** and **ALe^b** substances are thus formed via **Le^d** (**H** type-1) and **ALe^d**, respectively. The biosynthetic pathways for the synthesis of blood group **Lewis** specific structures are shown in *Fig. 6.12*.

This biosynthesis scheme further shows that the quantitative ratio between the different antigens is influenced by the competition of the enzymes for their common precursor substance:

- both the **secretor**- and the **Lewis**-transferase attach fucose residues to type-1 chain precursors. Since erythrocytes of **Lewis**-positive secretors contain mainly **Le^b** active glycosphingolipids and only minute amounts of **Le^a** substance, it must be assumed that the **secretor**-transferase is generally more active than the **Lewis** transferase and is thus more efficient in competing for the common precursor substrate;
- the rare **Le(a+b+)** individuals contain *Se* gene variants, (= *Se^w*), which code for only weakly active $\alpha1,2$-fucosyltransferases (see *Sect. 5.4.6*). In this case the **Lewis**-transferase competes more effectively for the type-1 chain precursor and thus favours the formation of **Le^a**, which does not undergo any further conversion. The weak **ABH** activity found in saliva of **Le(a+b+)** individuals [149,153,353] corroborates this assumption;
- moreover, **Lewis**-positive secretors of blood group **A$_1$** contain mainly **ALe^b** substances [227], whereas in the corresponding **A$_2$** individuals only **Le^b** substances are detectable. The highly active **A$_1$**-transferase, which is obviously more efficient than the **Lewis**-transferase, forms preferentially **ALe^d**. This is converted secondarily into **ALe^b** by the action of the **Lewis**-enzyme. Due to the lower activity of the **A$_2$**-transferase in **A$_2$** individuals, the **Le^d** epitopes are preferentially converted into **Le^b**, which cannot be transformed into **ALe^b**.

These considerations are valid mainly for the biosynthesis of **Lewis** active plasma glycosphingolipids. In secretions the competition between **A**- and **B**-transferases on the one hand and **Lewis**-transferase on the other is generally not so clearly expressed [214,216].

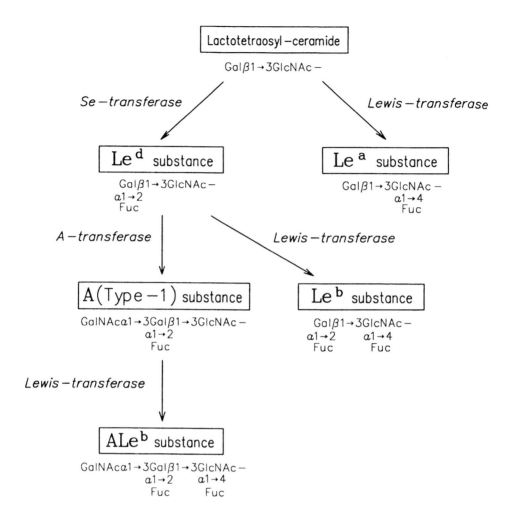

Figure 6.12: Biosynthetic pathways for the formation of blood group Lewis-specific structures.

Further, in carcinoma tissue and carcinoma cell lines the existence of an alternate biosynthetic path leading from the precursor via **Le^a** to **Le^b** has been reported [31,171]. More recent investigations suggested that this $\alpha1,2$-fucosyltransferase activity which converts the blood group **Le^a** determinant into **Le^b** may be associated with the **Lewis** gene encoded $\alpha1,3/4$-fucosyltransferase [59,271].

The sialoyl-**Le**^{a} structure is synthesised by transfer of neuraminic acid residues to a type-1 precursor by an $\alpha2,3$-N-acetylneuraminyltransferase, followed by addition of fucose by the **Lewis**-transferase. The transfer of neuraminyl residues to the **Le**^{a} determinant has not been observed [146].

6.5 Antigens Le^x and Le^y

6.5.1 Chemical Structure of Le^x and Le^y Antigens

Both **Le**^{x} and **Le**^{y} determinants are based on a type-2 chain core structure: the **Le**^{x} character is determined by an $\alpha1 \rightarrow 3$ fucose bound to the subterminal N-acetylglucosamine residues. Like **Le**^{b}, the **Le**^{y} character is a hybrid antigen in which the type-2 precursor chain carries an **Le**^{x} determinant $\alpha1 \rightarrow 3$ fucose and an **H** determinant $\alpha1 \rightarrow 2$ fucose *(Fig. 6.13)*. This means that **Le**^{x} and **Le**^{y} are type-2 isomers of **Le**^{a} and **Le**^{b} determinants, respectively [133]. Moreover, parallel to the corresponding structures in the **Lewis** system, sialylated **Le**^{x} [106] and the difucosylated determinants **ALe**^{y} [94,332,335] and **BLe**^{y} [9,32] have been detected.

6.5.2 Le^x and Le^y Substances

Large amounts of **glycolipids** carrying **Le**^{x} and **Le**^{y} determinants have been isolated from tumour tissues. These tissues are primarily adenomas or adenocarcinomas of the gastrointestinal tract or mammary glands which have proved good sources of **Le**^{x} [132,134,357] and **Le**^{y} [32,223,283] reactive material. **Le**^{x} specific glycolipids have also been found in human granulocytes[6] [102,270]. Dog small intestine [250,337] and hog gastric mucosa [333] also serve as non-human sources of **Le**^{x} and **Le**^{y} active glycolipids.

The structures presented in *Fig. 6.14* show that the **Le**^{x} and **Le**^{y} specific glycolipids are a group of polylactosaminyl-glycosphingolipids which differ in their number of N-acetyllactosamine residues. They generally contain and in rare cases exceed one to three $\alpha1 \rightarrow 3$ fucosyl residues.

Le^x and **Le**^{y} determinants are also carried by **glycoproteins**. They have been found in water-soluble substances of saliva [313], gastric mucosal mucins [334] *(Fig. 6.7)*, and ovarian cyst fluid [228] *(Fig. 5.13.b)*. They are also found attached to cell-

---

[6] The **Le**^{x} specific glycolipids isolated from 'erythrocyte membranes' by Kannagi et al. [187] presumably originate from contaminating granulocytes.

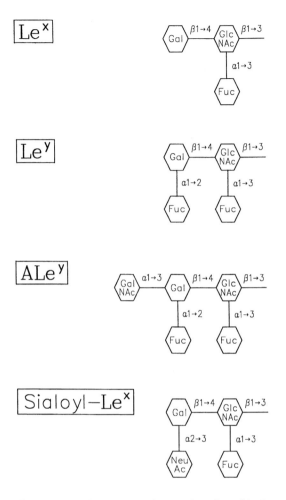

Figure 6.13: Structures of the determinants Le^x, Le^y, ALe^y, and sialoyl-Le^x.

bound polylactosaminoglycans of the granulocyte membrane [101,344] (see *Fig. 6.15)*, of small intestinal epithelial cells [94], and of various carcinoma cells [307].

Oligosaccharides carrying **Le^x** and **Le^y** specific structures have been detected in milk [121,199] and urine [232] (see *Fig. 6.16)*. Further examples of **Le^x** and **Le^y** specific oligosaccharides have been published by Bruntz et al. [49] and Egge et al. [82].

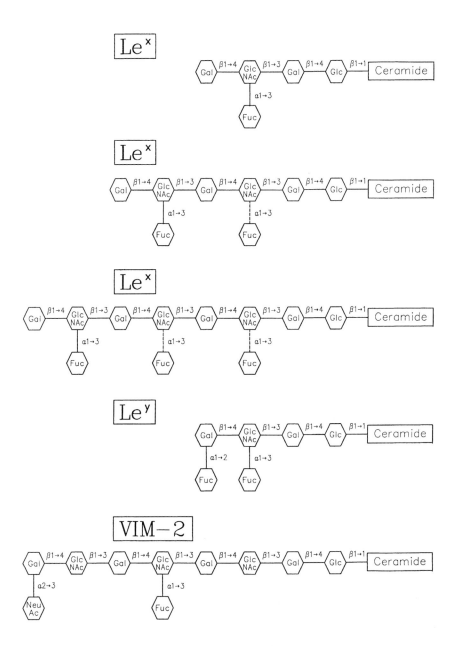

Figure 6.14: Le^x and Le^y specific glycosphingolipids found in human tissues.
According to Abe et al. [1], Fukuda et al. [102], Hakomori et al. [134], and Kannagi et al. [187].

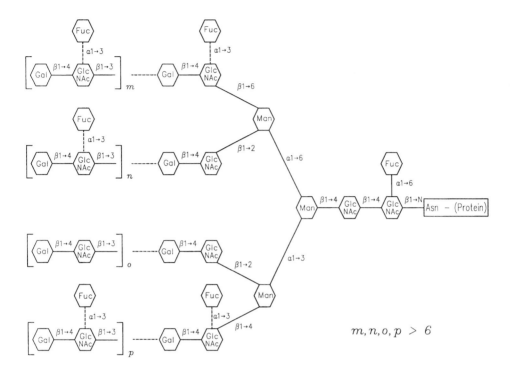

Figure 6.15: Carbohydrate chains of polylactosaminoglycans from the membrane of human granulocytes.
According to Spooncer et al. [344].

The **sialoyl-Le**ˣ determinant is an antigen characteristic of granulocytes [101,106,270] and the epithelium of the proximal kidney tubuli [106]. Mucin-bound sialoyl-Leˣ has also been detected in amniotic fluid [142]. The same antigen has been found in elevated concentrations in pathologically changed tissues of cystic fibrosis patients [210] and in malignantly transformed tissues [105,106,357]. Human milk contains oligosaccharides carrying the sialoyl-Leˣ determinant [372].

A related structure recognised by the hybridoma antibody **VIM-2** (*Fig. 6.14,* [234]) has been detected on the surface of myeloid cells.

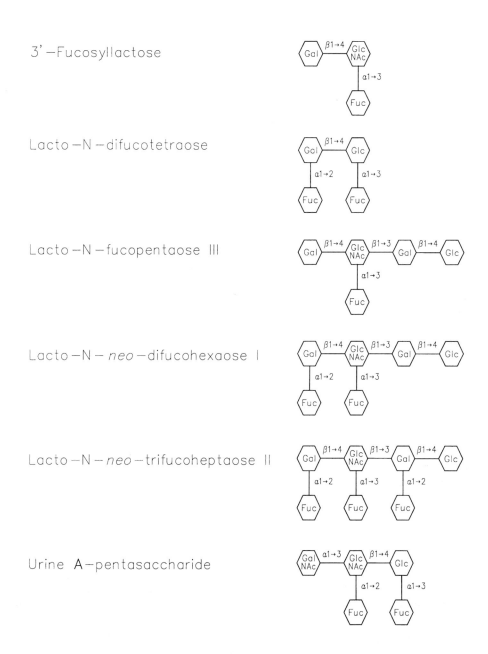

Figure 6.16: Oligosaccharides from human milk and urine containing α→3 linked fucose.
According to Kobata [199] and Lundblad [232].

6.5.3 Chemical Synthesis of Le^x and Le^y Determinants

Le^x [155,170,276,317] and **Le^y** specific [155] oligosaccharides have been prepared by chemical synthesis. And more recently the total syntheses of **Le^x**-active lacto-N-fucopentaosyl-III-ceramide and sialoyl-**Le^x** have been described [21,183].

6.5.4 Biosynthesis of Le^x and Le^y Determinants

The **Le^x determinants** are formed by an α1,3-fucosyltransferase which transfers fucose to the subterminal N-acetylglucosamine of a type-2 precursor chain:

$$
\text{Gal } \beta1{\to}4 \text{ GlcNAc } \beta1{\to}3 \text{ R} \xrightarrow{\text{Mg}^{++}} \text{Gal } \beta1{\to}4 \text{ GlcNAc } \beta1{\to}3 \text{ R}
$$

GDP–Fuc   Fuc

$\alpha1{\to}3$

Fuc

α1,3-Fucosyltransferases have been detected in a variety of human tissues and body fluids: gastric mucosa and submaxillary glands [60], leukocytes [331], lymphocytes and granulocytes [119], milk [88,177,300], serum [268,315,319], saliva [178,384], and amniotic fluid [143], as well as in various human tumour cell lines [158,160,269]. The presence of α1,3-fucosyltransferases in animals, plants and bacteria has also been reported (see [346]).

Investigations at molecular biological level have revealed the occurrence of at least six different α1,3-fucosyltransferases in humans. Each enzyme shows a typical tissue-specific expression pattern within the organism and is characterised by a well-defined capacity to transfer α-fucosyl residues onto distinct oligosaccharide acceptors:

- *FucTF-III*, the blood group **Lewis**-type α1,3/4-fucosyltransferase (discussed above, *Sect. 6.4.1)*, found in milk, gall bladder, gastric mucosa, and kidney;
- *FucTF-IV*, the myeloid-type α1,3-fucosyltransferase, expressed in myeloid cell lines (leukocytes, granulocytes, monocytes, and lymphoblasts), and brain [256];
- *FucTF-V*, an unspecified type of α1,3-fucosyltransferase; to date little is known about its tissue expression pattern [375];
- *FucTF-VI*, the plasma-type α1,3-fucosyltransferase occurring in blood plasma, renal proximal convoluted tubules, and hepatocytes [174,257,259];

- *FucTF-VII*, the leukocyte-type transferase: its expression is limited to the leukocyte cell line [62,272,316]; and
- *FucTF-IX*, which is expressed in the glandular compartments of the stomach, in spleen and peripheral blood leukocytes [184].

FucTF-III, encoded by the *Le* gene, is an α1,4-fucosyltransferase fucosylating the respective type-1 acceptor substrates to generate **Le^a**, **Le^b**, and sialoyl-**Le^a** determinants (see *Sect. 6.4*). When expressed in BHK-1 cells, however, the enzyme is also able to synthesise small amounts of sialyol-**Le^x** structure [114]. Nevertheless, it seems unlikely that *FucTF-III* is able to synthesise significant amounts of type-2 chain derivatives *in vivo*.

FucTF-IV transfers fucose onto unsubstituted type-2 chains and **H** type-2 determinants and thus forms **Le^x** and **Le^y** structures [95,114]. It is only weakly reactive with sialylated type-2 structures and generates only insignificant quantities of sialoyl-**Le^x** determinant [231]; α1,3-fucosylation of polylactosaminoglycan chains occurs also at internal GlcNAc residues [136].

FucTF-V and *FucTF-VI* share distinct substrate specificities in that they are both able to transfer α1,3-fucose residues onto unsubstituted, as well as α1,2-fucosylated, and α2,3-sialylated type-2 oligosaccharide acceptors, thereby creating **Le^x**, **Le^y**, and sialoyl-**Le^x** determinants, respectively [76,376]. *FucTF-VI* is responsible for the α1,3-fucosylation of serum glycoproteins (e.g. α_1-acid glycoprotein) in the liver, the organ which is the major source of α1,3-fucosyltransferase activity in plasma [41].

FucTF-VII preferentially fucosylates type-2 acceptors with terminal α2,3-linked sialic acid [43,114,326]), by synthesising not only the sialoyl-**Le^x**, but also the sialoyl-dimeric-**Le^x** determinant and their 6-SO$_4$ derivatives as well [197,241,338,370]. Subsequent studies have established that both FucTF-IV and FucTF-VII are required for the biosynthesis of multifucosylated sialylated polylactosamines [43].

FucTF-IX is able to synthesise **Le^x** and **Le^y** epitopes, but cannot form sialoyl-**Le^x** [184].

The **Le^y** determinants are synthesised by an α1,2-fucosyltransferase which transfers fucose onto terminal galactose residues of type-2 chains:

$$Gal\ \beta1{\to}4\ GlcNAc\ \beta1{\to}3\ R \xrightarrow{Mg^{++}} Gal\ \beta1{\to}4\ GlcNAc\ \beta1{\to}3\ R$$

GDP–Fuc   Fuc α1→2 Fuc

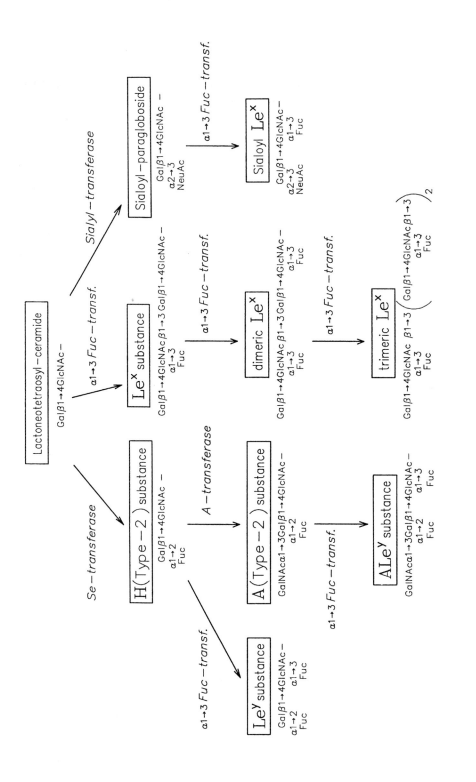

Figure 6.17: Biosynthetic pathways for the formation of X, Y, and sialoyl-X specific structures.

In every case the expression of Le^y antigen depends on the secretor status of the individual [47,313]: therefore, this fucosyl transfer must be controlled by the **secretor**-transferase (see *Chaps. 5.4.2 and 6.4*) and not, as expected, by the *H* gene-dependent enzyme (cf. [365]).

Experiments on the biosynthesis of the Le^y antigen showed that the Le^x structure does not serve as an acceptor substrate for the $\alpha1,2$-fucosyltransferase [31]. Thus, it has been assumed that the Le^y antigen is synthesised exclusively via the **H type-2** intermediate *(Fig. 6.16)*.

The biosynthesis of the sialoyl-Le^x determinant proceeds via sialylation of the type-2 precursor structure, followed by $\alpha1{\rightarrow}3$ fucosylation at the subterminal N-acetylglucosamine residue [143,161,255] *(Fig. 6.17)*.

Molecular Biological Investigations on the $\alpha1,3$-Fucosyltransferases

The $\alpha1,3$-fucosyltransferases *FucTF-III*, *FucTF-V*, and *FucTF-VI* are encoded by three highly homologous genes (*FUT3*, *FUT5*, and *FUT6*, respectively) which form a cluster on the short arm of chromosome 19 at p13.3 [248] as already mentioned in *Sect. 6.1*. Sequence analyses revealed that these enzymes differ in their amino acid sequence by 10–15% only [375,376] *(Fig. 6.18)*.

The view that the *FUT3-FUT5-FUT6* gene cluster is derived from a common ancestor has recently been confirmed by the discovery of a single gene in the bovine genome in a position corresponding to that in the human gene cluster ([290], see also [66]). This bovine gene shares about 68% of the amino acid sequence with human **F**UT3, *FUT5*, and *FUT6* gene products. The fucosyltransferase encoded by this same gene shows an acceptor specificity similar to that of the *FUT6* product in that it synthesises Le^x and sialoyl-Le^x but not type-1 structures; the tissue distribution, however, is most similar to that of the human *FUT3* product.

FucTF-VI, the plasma-type $\alpha1,3$-fucosyltransferase [205,376], is a glycoprotein with an apparent molecular mass of 45 kDa [315].

Most humans contain the *FucTF-VI* in plasma. In rare cases[7] this enzyme is not expressed at all. In *FucTF-VI* deficient individuals two different mutations have been detected which inactivate the enzyme: (1) a $G^{739} \rightarrow A$ exchange leading to a Glu $\rightarrow$ Lys substitution at position 247, and (2) a $C^{945} \rightarrow A$ exchange changing the Tyr codon at position 315 to a premature stop codon, which truncates the C-terminus by 45 amino acids. A $C^{370} \rightarrow T$ exchange causing a Pro $\rightarrow$ Ser substitution at position 124 frequently

---

[7] Although nearly 10% of an Indonesian population on Java do not express this enzyme.

FucTF-III

```
  1 MDPLGAAKPQ WPWRRCLAAL LFQLLVAVCF FSYLRVSRDD ATGSPRAPSG SSRQDTTPTR  60
 61 PTLLILLWTW PFHIPVALSR CSEMVPGTAD CHITADRKVY PQADTVIVHH WDIMSNPKSR 120
121 LPPSPRPQGQ RWIWFNLEPP PNCQHLEALD RYFNLTMSYR SDSDIFTPYG WLEPWSGQPA 180
181 HPPLNLSAKT ELVAWAVSNW KPDSARVRYY QSLQAHLKVD VYGRSHKPLP KGTMMETLSR 240
241 YKFYLAFENS LHPDYITEKL WRNALEAWAV PVVLGPSRSN YERFLPPDAF IHVDDFQSPK 300
301 DLARYLQELD KDHARYLSYF RWRETLRPRS FSWALDFCKA CWKLQQESRY QTVRSIAAWF 360
361 T
```

FucTF-IV

```
  1 MGAPWGSPTA AAGGRRGWRR GRGLPWTVCV LAAAGLTCTA LITYACWGQL PPLPWASPTP  60
 61 SRPVGVLLWW EPFGGRDSAP RPPPDCPLRF NISGCRLLTD RASYGEAQAV LFHHRDLVKG 120
121 PPDWPPPWGI QAHTAEEVDL RVLDYEEAAA AAEALATSSP RPPGQRWVVM NFESPSHSPG 180
181 LRSLASNLFN WTLSYRADSD VFVPYGYLYP RSHPGDPPSG LAPPLSRKQG LVAWVVSHWD 240
241 ERQARVRYYH QLSQHVTVDV FGRGGPGQPV PEIGLLHTVA RYKFYLAFEN SQHLDYITEK 300
301 LWRNALLAGA VPVVLGPDRA NYERFVPRGA FIHVDDFPSA SSLASYLLFL DRNPAVYRRY 360
361 FHWRRSYAVH ITSFWDEPWC RVCQAVQRAG DRPKSIRNLA SWFER
```

FucTF-V

```
  1 MDPLGPAKPQ WLWRRCLAGL LFQLLVAVCF FSYLRVSRDD ATGSPRPGLM AVEPVTGAPN  60
 61 GSRCQDSMAT PAHPTLLILL WTWPFNTPVA LPRCSEMVPG AADCNITADS SVYPQADAVI 120
121 VHHWDIMYNP SANLPPPTRP QGQRWIWFSM ESPSNCRHLE ALDGYFNLTM SYRSDSDIFT 180
181 PYGWLEPWSG QPAHPPLNLS AKTELVAWAV SNWKPDSARV RYYQSLQAHL KVDVYGRSHK 240
241 PLPKGTMMET LSRYKFYLAF ENSLHPDYIT EKLWRNALEA WAVPVVLGPS RSNYERFLPP 300
301 DAFIHVDDFQ SPKDLARYLQ ELDKDHARYL SYFRWRETLR PRSFSWALAF CKACWKLQQE 360
361 SRYQTVRSIA AWFT
```

FucTF-VI

```
  1 MDPLGPAKPQ WSWRCCLTTL LFQLLMAVCF FSYLRVSQDD PTVYPNGSRF PDSTGTPAHS  60
 61 IPLILLWTWP FNKPIALPRC SEMVPGTADC NITADRKVYP QADAVIVHHR EVMYNPSAQL 120
121 PRSPRRQGQR WIWFSMESPS HCWQLKAMDG YFNLTMSYRS DSDIFTPYGW LEPWSGQPAH 180
181 PPLNLSAKTE LVAWAVSNWG PNSARVRYYQ SLQAHLKVDV YGRSHKPLPQ GTMMETLSRY 240
241 KFYLAFENSL HPDYITEKLW RNALEAWAVP VVLGPSRSNY ERFLPPDAFI HVDDFQSPKD 300
301 LARYLQELDK DHARYLSYFR WRETLRPRSF SWALAFCKAC WKLQEESRYQ TRGIAAWFT
```

FucTF-VII

```
  1 MNNAGHGPTR RLRGLGVLAG VALLAALWLL WLLGSAPRGT PAPQPTITIL VWHWPFTDQP  60
 61 PELPSDTCTR YGIARCHLSA NRSLLASADA VVFHHRELQT RRSHLPLAQR PRGQPWVWAS 120
121 MESPSHTHGL SHLRGIFNWV LSYRRDSDIF VPYGRLEPHW GPSPPLPAKS RVAAWVVSNF 180
181 QERQLRARLY RQLAPHLRVD VFGRANGRPL CASCLVPTVA QYRFYLSFEN SQHRDYITEK 240
241 FWRNALVAGT VPVVLGPPRA TYEAFVPADA FVHVDDFGSA RELAAFLTGM NESRYQRFFA 300
301 WRDRLRVRLF TDWRERFCAI CDRYPHLPRS QVYEDLEGWF QA
```

Figure 6.18: Peptide sequences of human α1,3-fucosyltransferases.
Underlined: putative transmembrane domains, *N* : potential N-glycosylation sites.
The sequences are deposited in the EMBL/GenBank data library (accession numbers X53578, M58597, M81485, M98825, and U08112, respectively).

found in the same probands did not affect the substrate specificity of the enzyme but resulted in an enzyme with two- to three-times higher activity than the wild-type enzyme [259]. Almost all **Lewis**-positive individuals lacking the plasma-type transferase have been typed as **Le(a–b–)**; this provides additional evidence for a close genetic relationship between the *Le* locus and the gene encoding the plasma-type fucosyltransferase [52]. Thus far no clinical abnormalities have been detected in the *FucTF-VI* deficient individuals.

FucTF-IV, the myeloid-type α1,3-fucosyltransferase [95], is encoded by the *FUT4* gene at chromosome 11q21. The gene codes for a protein of 405 amino acids and 2 N-glycosylation sites. Despite a different chromosomal localisation, the amino acid sequence deduced from cDNA analysis [110,208,231] revealed substantial similarity to corresponding positions of the **Lewis**- and plasma-type fucosyltransferases, (about 60% homology with the transferases encoded by the chromosome-19 cluster.

FucTF-VII, the α1,3-fucosyltransferase encoded by the *FUT7* gene at chromosome 9q34.4, is a type II transmembrane protein of 341 amino acids [272,316]. It is identical to 47% with the *FUT4* gene product and to about 42% with the transferases encoded by the chromosome-19 cluster.

Littel is known about the structural features of the fucosyltransferases that dictate their different acceptor substrate specificities. Nevertheless, all fucosyltransferases seem to have a similar substrate-binding pocket with some of the amino acid residues conserved within these pockets [40,76]. Domain swapping experiments localised the catalytic domain of the transferases encoded by the chromosome 19 cluster in a 'hypervariable' peptide segment of some 50 amino acids just outside the membrane-spanning domain (i.e. between residues 103 and 153 of *FucTF-III*, and residues 116 and 166 of *FucTF-VI*) [217]. Subsequent investigations have shown that there are two amino acids, i.e. His[73] and Ile[74] of *FucTF-III*, responsible for type 1 acceptor substrate specificity [273]. A Trp[111] $\rightarrow$ Arg substitution changed the specificity of *FucTF-III* from H-type 1 to H-type 2 acceptors [81].

Further, the use of N-ethylmaleinimide has been pivotal in the characterisation and classification of fucosyltransferase isotypes [235]. *FucTF-III*, *FucTF-V*, and *FucTF-VI* are inactivated by N-ethylmaleinimide, a reagent which forms a covalent linkage with the sulfhydryl groups of cystine residues. Sequence alignments of the fucosyltransferases indicated a conserved cysteine residue in *FucTF-III* (Cys[143]), in *FucTF-V* (Cys[156]), and in *FucTF-VI* (Cys[142]). The fact that the presence of GDP and GDP-fucose protected the enzymes from inactivation suggests that this cysteine residue is one of the key amino acids involved in the binding of the nucleotide donor, GDP-fucose, to the transferase molecule and/or transferring the fucosyl residue to the carbohydrate acceptor [42, 162]. *FucTF-IV* and *FucTF-VII*, in which the cystein residue is replaced by serine and threonine, respectively, are insensitive to this reagent [42].

Similar experiments suggested the requirement of basic amino acids such as histidine and lysine as essential for α1,3-fucosyltransferase activity [42,162,324].

6.5.5 Physiological Role of Le^a and Le^x and Their Derivatives.

Cell-surface expression of **Le**^{x} and sialoyl-**Le**^{x} is strictly controlled during embryonic development, cell differentiation, and malignant transformation [93,100,126]. For this reason **Le**^{x} is considered a differentiation antigen, i.e. one which has an important signal function in cell interaction and cell sorting during mammalian embryonic development (hence the designation **SSEA-1** = 'Stage-Specific Embryonic Antigen-1' [112,341]).

Le^x is highly expressed in the morula stage of mouse embryos, yet decreases after compaction, whichmay mean that this antigen plays a pivotal role during pre-implantation embryogenesis [341]. Strong evidence that the compaction process is induced by carbohydrate-carbohydrate interaction between the **Le**^{x} determinants themselves has been presented [84]. These investigations are the basis of discussions whether cell adhesion in general may be initiated by **Le**^{x}/**Le**^{x} interaction and subsequently reinforced by stronger adhesion mechanisms [108,128].

Immunohistochemical studies on gastrointestinal and urogenital epithelia of human embryos in various stages of development revealed that the **Le**^{x} antigen appeared after day 40 from fertilisation, reached maximum expression between days 50 and 70 and gradually regressed upon further differentiation and development [104]. In adults the **Le**^{x} antigen is restricted to neutrophiles, brain, the epithelium of proximal kidney tubuli [54,96,104], and a few cell types in gastrointestinal tissue [312].

These adaptations reflect changes of fucosyltransferase activities during these dynamic processes [206,235]:

Investigations [54,256] have shown that ubiquitous expression of myeloid-type fucosyltransferase is characteristic for early stage (5- to 10-week-old) embryos. In the course of further development the other α1,3-fucosyltransferases (mainly **Lewis**- and plasma-type enzymes) are progressively expressed, until the tissues finally reach the corresponding enzyme patterns of the adult stage. In adults, the **Lewis**- and plasma-type α1,3-fucosyltransferases are present at high levels in organs containing a large proportion of epithelial cells. *FUT5* expression was lower and restricted to fewer cell types [53], and *FUT4* gene remains expressed only in myeloid cells and brain [257].

Another attempt to correlate the cell surface expression of **Le**^{x} and sialoyl-**Le**^{x} structures revealed that the decrease of **Le**^{x} and increase of sialoyl-**Le**^{x} in the course of dimethyl sulfoxide-induced differentiation of the human leukemic cell line, *HL-60* [330], is accompanied by a change from *FUT4* expression to *FUT7* expression [62].

Sialoyl and sulfatyl derivatives of **Le**^{x} and **Le**^{a} are important ligands for a family of adhesion molecules, the selectins, which have key roles in homing of lymphocytes into secondary lymphoid organs and in acute inflammatory responses, as a series of investigations revealed [26,36,190,367]. It is well known that lymphocyte infiltration is also a 'point of no return' in acute rejections of solid organ transplants.

Adhesion and extravasation of lymphocytes from the blood stream is a multistep process initiated by rolling and tethering of lymphocytes on the endothelium mainly by the binding of selectins to their carbohydrate ligands. Stable arrest and trans-endothelial migration of the lymphocytes is subsequently effected by integrins of the leucocyte cell surface binding to the 'intracellular adhesion molecules' (ICAM-1 and ICAM-2) expressed on the surface of the endothelial cells [25,225,233,382].

Three types of selectins have thus far been characterised − **(1)** E-selectin expressed on cytokine-activated vascular endothelial cells, **(2)** P-selectin on activated endothelial cells and thrombocytes, and **(3)** L-selectin on leukocytes[8] [106,249,284, 309,356].

The specificities of the different selectins are similar but not identical. For example, E-selectin binds preferentially to sialoyl-**Le**^{x} and sialoyl-**Le**^{a}, whereas P- and L-selectins show a broader specificity, and L-selectin binds mainly to sulfated sialoyl-**Le**^{x} [166,364]. Other studies suggest that glycoconjugates with multivalent (dimeric and trimeric) sialoyl-**Le**^{x} structures are required for biologically relevant recognition [309]. An essential role in selectin specificity is also played by the protein on which the carbohydrate is expressed [281,314,348,390].

The investigations conducted so far suggest that oligosaccharide diversity is not random, rather, it exhibits precise tissue-specific and developmental expression patterns. Distinct cell lines obviously differ in their capacity to construct different fucosylated molecules. It can further be assumed that expression of the different ligands is also regulated according to physiological status.

Le^a, **Le**^{x}, and their sialoyl derivatives, as well as **Le**^{b} and **Le**^{y}, accumulate in various tumour tissues, as noted at the beginning of this section. This is presumably a result of de-differentiation of the malignantly transformed cells [47,106,159,312]. There is valid evidence that adhesion of tumour cells to endothelial cells, mediated by

---

[8] E-selectin has also been known as ELAM-1 ('endothelial cell-leucocyte adhesion molecule-1'), and P-selectin as PADGEM ('platelet activation-dependent granule external membrane protein').

the interaction between the sialylated Le^a and Le^x antigens and E-selectin, represents an important factor in haematogenous metastasis of tumour cells [128,238,356,388]. Discussion continues regarding to what extent Le^x, sialoyl-Le^x, and sialoyl-Le^a (= CA-19.9, see *Sect. 6.3.2)* together with other so-called 'oncofoetal' (or 'onco-developmental') antigens [112,127]) such as **CEA**, **SSEA-3**, and **SSEA-4**, can be used as tumour markers (see also [92,130]).

Investigations on different tumours revealed an elevated expression of the $\alpha 1,3$- and $\alpha 1,3/4$-fucosyltransferase genes [23,385]. Surveys of human leukemia and epithelial carcinoma cell lines suggest that *FUT3* and *FUT6* transcripts are expressed at high levels in transformed epithelia [316,381] (see also [164,371]). The myeloid-type fucosyltransferase is also well-expressed in many tumour cell lines [256]: all lymphoblastoid cell lines and half of the tumour epithelial cell lines tested expressed the myeloid-like pattern of enzyme found in normal embryonic tissues. *FUT5* transcripts were totally missing in epithelilal cancer cell lines, but have been found in melanoma cell lines [209]; more recently *FUT5* transcripts were also found in human liver, testicle, and colon tissues [53].

Le^x, Le^y, and, to a lesser extent, Le^a, Le^b and/or H type-1 antigens have been detected in different strains of the human pathogen, *Helicobacter pylori* [14,15,246,262]. This gram-negative bacterium causes chronic gastritis [379] and gastric and duodenal ulcers [115], and has been linked to gastric lymphoma [296] and gastric adenocarcinoma [280,295].

Pathogenesis of *Helicobacter*-induced diseases is obviously a complex process. In the case of Le^x-producing strains, the Le^x antigen generated by the bacterium mimicks host cell antigens – this should protect the pathogen against the host's defense mechanism [10]. In addition, the bacterial Le^x antigens have been shown to induce anti-Le^x antibodies, which cross-react with human gastric mucosa and may provoke *H. pyloris*-associated gastritis [10]. It is also possible, however, that anti-Le^x antibodies bind to the bacteria, thereby inducing their ingestion by polymorphonuclear leukocytes and possibly limiting the growth of *H. pylori* on gastric mucosa [61]. Some strains of *Helicobacter* contained neutrophil-activating proteins which bind to sialylated type-2 oligosaccharide chains [253,254,361]. These proteins are likely to be involved in the early stages of the disease processes, which are characterised by accumulation of neutrophils in the superficial gastric mucosa. An adhesin which reacts preferentially with Le^b active glycoconjugates to mediate the attachment of *Helicobacter pylori* to human gastric mucosal epithelium, thus promoting *H. pylori* colonisation of gastric tissue has been found by several groups [34,91,165].

A comparable molecular mimicry mechanism in the human blood fluke *Schistosoma mansoni* has been reported; this mechanism synthesises glycoconjugates carrying the **Le**^{x} antigen [75,198,345]. Here, too, the fucosylated oligosaccharides may act as anti-inflammatory agents by inhibiting leukocyte-endothelial cell interactions at the site of parasite attachment [50,266,267,355], and thus being one of the several mechanisms used by the schistosome to circumvent host defense [74]. Anti-**Le**^{x} antibodies produced in infected humans and animals can induce partial resistance towards re-infection [55,78], but by inducing complement-dependent cytolysis of granulocytes may also be responsible for the moderate neutropenia observed during schistosome infection [366].

References

1. ABE, K., MCKIBBIN, J. M. & HAKOMORI, S. I. (1983): The monoclonal antibody directed to difucosylated type 2 chain (Fucα1→2Galβ1→4[Fucα1→3]GlcNAc; Y determinant). *J. Biol. Chem. 258*, 11793-11797.
2. ANDERSEN, J. (1958): Modifying influence of the secretor gene on the development of the ABH substance. A contribution to the conception of the Lewis group system. *Vox Sang. 3*, 251-261.
3. ANDERSEN, J. (1960): Serological studies on two Lewis sera. *Acta Pathol. Microbiol. Scand. 48*, 374-384.
4. ANDO, B. & IBAYASHI, H. (1986): Identification of anti-Le^a by platelet complement fixation. *Vox Sang. 50*, 169-173.
5. ANDRESEN, P. H. (1948): The blood group system L. A new blood group L$_2$. A case of epistasy within the blood groups. *Acta Pathol. Microbiol. Scand.* 25, 728-731.
6. ANDRESEN, P. H. (1972): Demonstration of Le^x substance in the saliva of ABH non-secretors. *Vox Sang. 23*, 262-269.
7. ANDRESEN, P. H. & JORDAL, K. (1949): An incomplete agglutinin related to the L-(Lewis) system. *Acta Pathol. Microbiol. Scand.* 26, 636-638.
8. ÅNGSTRÖM, J., FALK, K. E., KARLSSON, K. A. & LARSON, G. (1982): Chemical fingerprinting of glycosphingolipids in meconium of a human blood group O Le(a-b+) secretor. *Biochim. Biophys. Acta 710*, 428-436.
9. ÅNGSTRÖM, J., FALK, P., HANSSON, G. J., HOLGERSSON, J., KARLSSON, H., KARLSSON, K. A., STROMBERG, N. & THURIN, J. (1987): The mono- and difucosyl blood group B glycosphingolipids of rat large intestine differ in type of core saccharide. *Biochim. Biophys. Acta 926*, 79-86.
10. APPELMELK, B. J., SIMOONS-SMIT, I., NEGRINI, R., MORAN, A. P., ASPINALL, G. O., FORTE, J. G., DE VRIES, T., QUAN, H., VERBOOM, T., MAASKANT, J. J., GHIARA, P., KUIPERS, E. J., BLOEMENA, E., TADEMA, T. M., TOWNSEND, R. R., TYAGARAJAN, K., CROTHERS, J. M., MONTEIRO, M. A., SAVIO, A. & DE GRAAFF, J. (1996): Potential role of molecular mimicry between Helicobacter pylori lipopolysaccharide and host Lewis blood group antigens in autoimmunity. *Infect. Immun. 64*, 2031-2040.
11. ARCILLA, M. B. & STURGEON, P. (1972): Lewis and ABH substances in amniotic fluid obtained by amniocentesis. *Pediatr. Res. 6*, 853-858.
12. ARCILLA, M. B. & STURGEON, P. (1973): Studies on the secretion of blood group substances. II. Observations on the red cell phenotype Le(a-b-x-). *Vox Sang. 25*, 72-87.
13. ARCILLA, M. B. & STURGEON, P. (1974): Le^x, the spurned antigen of the Lewis blood group system. *Vox Sang. 26*, 425-438.
14. ASPINALL, G. O. & MONTEIRO, M. A. (1996): Lipopolysaccharides of Helicobacter pylori strains P466 and MO19: structures of the O antigen and core oligosaccharide regions. *Biochemistry 35*, 2498-2504.
15. ASPINALL, G. O., MONTEIRO, M.A., PANG, H., WALSH, E.J. & MORAN, A.P. (1996): Lipopolysaccharide

of the Helicobacter pylori type strain NCTC 11637 (ATCC 43504): structure of the O antigen chain and core oligosaccharide regions. *Biochemistry 35*, 2489-2497.

16. ATKINSON, B. F., ERNST, C. S., HERLYN, M., STEPLEWSKI, Z., SEARS, H. F. & KOPROWSKI, H. (1982): Gastrointestinal cancer-associated antigen in immunoperoxidase assay. *Cancer Res. 42*, 4820-4823.

17. AUBUCHON, J. P., DAVEY, R. J., ANDERSON, H. J., PATEL, S., STAVELY, L. M., SCHOEPPNER, S. L. & ZAFIROPULOS, D. (1986): Specificity and clinical significance of anti-Le^x. *Transfusion 26*, 302-303.

18. BAECKSTRÖM, D., HANSSON, G. C., NILSSON, O., JOHANSSON, C., GENDLER, S. J. & LINDHOLM, L. (1991): Purification and characterization of a membrane-bound and a secreted mucin-type glycoprotein carrying the carcinoma- associated sialyl-Le^a epitope on distinct core proteins. *J. Biol. Chem. 266*, 21537-21547.

19. BAECKSTRÖM, D., KARLSSON, N. & HANSSON, G. C. (1994): Purification and characterization of sialyl-Le^a-carrying mucins of human bile; evidence for the presence of MUC1 and MUC3 apoproteins. *J. Biol. Chem. 269*, 14430-14437.

20. BAER, H., NAYLOR, I., GIBBEL, N. & ROSENFIELD, R. E. (1959): The production of precipitating antibody in chickens to a substance present in the fluids of non-secretors of blood groups A, B, and O. *J. Immunol. 82*, 183-189.

21. BALL, G. E., O'NEILL, R. A., SCHULTZ, J. E., LOWE, J. B., WESTON, B. W., NAGY, J. O., BROWN, E. G., HOBBS, C. J. & BEDNARSKI, M. D. (1992): Synthesis and structural analysis using 2-D NMR of sialyl Lewis-X (SLe^x) and Lewis-X (Le^x) oligosaccharides - ligands related to E-selectin (ELAM-1) binding. *J. Amer. Chem. Soc. 114*, 5449-5451.

22. BALL, S. P., TONGUE, N., GIBAUD, A., LE PENDU, J., MOLLICONE, R., GÉRARD, G. & ORIOL, R. (1991): The human chromosome 19 linkage group FUT1 (H), FUT2 (SE), LE, LU, PEPD, C3, APOC2, D19S7, and D19S9. *Ann. Hum. Genet. 55*, 225-233.

23. BAUER, C. H., REUTTER, W. G., ERHART, K. P., KOTTGEN, E. & GEROK, W. (1978): Decrease of human serum fucosyltransferase as an indicator of successful tumor therapy. *Science 201*, 1232-1233.

24. BECHTEL, B., WAND, A. J., WROBLEWSKI, K., KOPROWSKI, H. & THURIN, J. (1990): Conformational analysis of the tumor-associated carbohydrate antigen 19-9 and its Le^a blood group antigen component as related to the specificity of monoclonal antibody CO19-9. *J. Biol. Chem. 265*, 2028-2037.

25. BERLIN, C., BARGATZE, R. F., CAMPBELL, J. J., VON ANDRIAN, U. H., SZABO, M. C., HASSLEN, S. R., NELSON, R. D., BERG, E. L., ERLANDSEN, S. L. & BUTCHER, E. C. (1995): α4 integrins mediate lymphocyte attachment and rolling under physiologic flow. *Cell 80*, 413-422.

26. BIRD, M. I., FOSTER, M. R., PRIEST, R. & MALHOTRA, R. (1997): Selectins: physiological and pathophysiological roles. *Biochem. Soc. Trans. 25*, 1199-1206.

27. BJÖRK, S., BREIMER, M. E., HANSSON, G. C., KARLSSON, K. A. & LEFFLER, H. (1987): Structures of blood group glycosphingolipids of human small intestine. A relation between the expression of fucolipids of epithelial cells and the ABO, Le, and Se phenotype of the donor. *J. Biol. Chem. 262*, 6758-6765.

28. BLAINEAU, C., LE PENDU, J., ARNAUD, D., CONNAN, F. & AVNER, P. (1983): The glycosidic antigen recognized by a novel monoclonal antibody, 75.12, is developmentally regulated on mouse embryonal carcinoma cells. *EMBO J. 2*, 2217-2222.

29. BLASZCZYK, M., HANSSON, G. C., KARLSSON, K. A., LARSON, G., STROMBERG, N., THURIN, J., HERLYN, M., STEPLEWSKI, Z. & KOPROWSKI, H. (1984): Lewis blood group antigens defined by monoclonal anti-colon carcinoma antibodies. *Arch. Biochem. Biophys. 233*, 161-168.

30. BLASZCZYK, M., PAK, K. Y., HERLYN, M., SEARS, H. F. & STEPLEWSKI, Z. (1985): Characterization of Lewis antigens in normal colon and gastrointestinal adenocarcinomas. *Proc. Natl. Acad. Sci. USA 82*, 3552-3556.

31. BLASZCZYK-THURIN, M., SARNESTO, A., THURIN, J., HIDSGAUL, O. & KOPROWSKI, H. (1988): Biosynthetic pathways for the Le^b and Y glycolipids in the gastric carcinoma cell line KATO III as analyzed by a novel assay. *Biochem. Biophys. Res. Commun. 151*, 100-108.

32. BLASZCZYK-THURIN, M., THURIN, J., HINDSGAUL, O., KARLSSON, K. A., STEPLEWSKI, Z. & KOPROWSKI, H. (1987): Y and blood group B type 2 glycolipid antigens accumulate in a human gastric carcinoma cell line as detected by monoclonal antibody. Isolation and characterization by mass spectrometry and NMR spectroscopy. *J. Biol. Chem. 262*, 372-379.

33. BOETTCHER, B. & KENNY, R. (1971): A quantitative study of Le^a, A, and H antigens in salivas of Australian Caucasians and Aborigines. *Hum. Hered. 21*, 334-345.

34. BORÉN, T., FALK, P., ROTH, K. A., LARSON, G. & NORMARK, S. (1993): Attachment of Helicobacter pylori to human gastric epithelium mediated by blood group antigens. *Science 262*, 1892-1895.

35. BOVIN, N. V. & KHORLIN, A. Y. (1985): Convenient synthon for preparing Le and ABH blood group determinants. *Bioorg. Khim. 11*, 826-829.

36. BRANDLEY, B. K., SWIEDLER, S. J. & ROBBINS, P. W. (1990): Carbohydrate ligands of the LEC cell adhesion molecules. *Cell 63*, 861-863.

37. BREIMER, M. E., KARLSSON, K. A. & SAMUELSSON, B. E. (1982): Characterization of a human intestinal difucosyl heptaglycosylceramide with a blood group B determinant and a type 1 carbohydrate chain. *J. Biol. Chem. 257*, 1079-1085.

38. BRENDEMOEN, O. J. (1950): Further studies of agglutination and inhibition in the Le^a-Le^b system. *J. Lab. Clin. Med. 36*, 335-341.

39. BRENDEMOEN, O. J. (1961): Development of the Lewis blood group in the newborn. *Acta Pathol. Microbiol. Scand. 52*, 55-58.

40. BRETON, C., ORIOL, R. & IMBERTY, A. (1998): Conserved structural features in eukaryotic and prokaryotic fucosyltransferases. *Glycobiology 8*, 87-94.

41. BRINKMAN-VAN DER LINDEN, E. C. M., MOLLICONE, R., ORIOL, R., LARSON, G., VAN DEN EIJNDEN, D. H. & VAN DIJK, W. (1996): A missense mutation in the FUT6 gene results in total absence of α3-fucosylation of human α_1-acid glycoprotein. *J. Biol. Chem. 271*, 14492-14495.

42. BRITTEN, C. J. & BIRD, M. I. (1997): Chemical modification of an α3-fucosyltransferase; definition of amino acid residues essential for enzyme activity. *Biochim. Biophys. Acta 1334*, 57-64.

43. BRITTEN, C. J., VAN DEN EIJNDEN, D. H., MCDOWELL, W., KELLY, V. A., WITHAM, S. J., EDBROOKE, M. R., BIRD, M. I., DE VRIES, T. & SMITHERS, N. (1998): Acceptor specificity of the human leukocyte α3 fucosyltransferase: role of FucT-VII in the generation of selectin ligands. *Glycobiology 8*, 321-327.

44. BROADBERRY, E. R. & LIN-CHU, M. (1991): The Lewis blood group system among Chinese in Taiwan. *Hum. Hered. 41*, 290-294.

45. BROCKHAUS, M., MAGNANI, J. L., BLASZCZYK, M., STEPLEWSKI, Z., KOPROWSKI, H., KARLSSON, K. A., LARSON, G. & GINSBURG, V. (1981): Monoclonal antibodies directed against the human Le^b blood group antigen. *J. Biol. Chem. 256*, 13223-13225.

46. BROCKHAUS, M., MAGNANI, J. L., HERLYN, M., BLASZCZYK, M., STEPLEWSKI, Z., KOPROWSKI, H. & GINSBURG, V. (1982): Monoclonal antibodies directed against the sugar sequence of lacto-N-fucopentaose III are obtained from mice immunized with human tumors. *Arch. Biochem. Biophys. 217*, 647-651.

47. BROWN, A., ELLIS, I. O., EMBLETON, M. J., BALDWIN, R. W., TURNER, D. R. & HARDCASTLE, J. D. (1984): Immunohistochemical localization of Y hapten and the structurally related H type-2 blood-group antigen on large-bowel tumours and normal adult tissues. *Int. J. Cancer 33*, 727-736.

48. BROWN, A., FEIZI, T., GOOI, H. C., EMBLETON, M. J., PICARD, J. K. & BALDWIN, R. W. (1983): A monoclonal antibody against human colonic adenoma recognizes difucosylated type-2 blood-group chains. *Biosci. Rep. 3*, 163-170.

49. BRUNTZ, R., DABROWSKI, U., DABROWSKI, J., EBERSOLD, A., PETER-KATALINIC, J. & EGGE, H. (1988): Fucose-containing oligosaccharides from human milk from a donor of blood group O Le^a nonsecretor. *Biol. Chem. Hoppe-Seyler 369*, 257-273.

50. BUERKE, M., WEYRICH, A. S., ZHENG, Z., GAETA, F. C., FORREST, M. J. & LEFER, A. M. (1994): Sialyl Lewis^x-containing oligosaccharide attenuates myocardial reperfusion injury in cats. *J. Clin. Invest. 93*, 1140-1148.

51. CAGAS, P. & BUSH, C. A. (1990): Determination of the conformation of Lewis blood group

oligosaccharides by simuloation of two-dimensional nuclear Overhauser data. *Biopolymers 30*, 1123-1138.

52. CAILLARD, T., LE PENDU, J., VENTURA, M., MADA, M., RAULT, G., MANNONI, P. & ORIOL, R. (1988): Failure of expression of α-3-L-fucosyltransferase in human serum is coincident wit the absence of the X (or Le^x)antigen in the kidney but not on leucocytes. *Exp. Clin. Immunogenet. 5*, 15-23.

53. CAMERON, H. S., SZCZEPANIAK, D. & WESTON, B. W. (1995): Expression of human chromosome 19p α(1,3)-fucosyltransferase genes in normal tissues. Alternative splicing, polyadenylation, and isoforms. *J. Biol. Chem. 270*, 20112-20122.

54. CANDELIER, J. J., MOLLICONE, R., MENNESSON, B., BERGEMER, A. M., HENRY, S., COULLIN, P. & ORIOL, R. (1993): α-3-Fucosyltransferases and their glycoconjugate antigen products in the developing human kidney. *Lab. Invest. 69*, 449-459.

55. CAPRON, A., DESSAINT, J. P., CAPRON, M., OUMA, J. H. & BUTTERWORTH, A. E. (1987): Immunity to schistosomes: progress toward vaccine. *Science 238*, 1065-1072.

56. CARTRON, J. P., MULET, C., BAUVOIS, B., RAHUEL, C. & SALMON, C. (1980): ABH and Lewis glycosyltransferases in human red cells, lymphocytes, and platelets. *Blood Transfus. Immunohaematol. 23*, 271-282.

57. CEPPELLINI, R., DUNN, L. C. & INNELLA, F. (1959): Immunogenetica II. Annalisi genetica formale de caratteri Lewis con particolare riguardo alla natura epistatica della specificita serologica Le^b. *Fol. Hered. Pathol. 8*, 261-296.

58. CEPPELLINI, R. & SINISCALCO, M. (1955): Una nuova ipotesi genetica per il sistema Lewis secretore e suoi riflessi nei riguardi di alcune evidence di linkage con altri loci. *Rev. Ist. Sieroterap. Ital. 30*, 431-445.

59. CHANDRASEKARAN, E. V., JAIN, R. K., RHODES, J. M., SRNKA, C. A., LARSEN, R. D. & MATTA, K. L. (1995): Expression of blood group Lewis b determinant from Lewis a: association of this novel α(1,2)-L-fucosylating activity with the Lewis type α(1,3/4)-L-fucosyltransferase. *Biochemistry 34*, 4748-4756.

60. CHESTER, M. A. & WATKINS, W. M. (1969): α-L-Fucosyltransferases in human submaxillary gland and stomach tissues associated with the H, Le^a and Le^b blood group characters and ABH secretor status. *Biochem. Biophys. Res. Commun. 34*, 835-842.

61. CHMIELA, M., WADSTROM, T., FOLKESSON, H., MALECKA, I. P., CZKWIANIANC, E., RECHCINSKI, T. & RUDNICKA, W. (1998): Anti-Lewis X antibody and Lewis X anti-Lewis X immune complexes in Helicobacter pylori infection. *Immunol. Lett. 61*, 119-125.

62. CLARKE, J. L. & WATKINS, W. M. (1996): α1,3-L-Fucosyltransferase expression in developing human myeloid cells Antigenic, enzymatic, and mRNA analyses. *J. Biol. Chem. 271*, 10317-10328.

63. CLAUSEN, H., LEVERY, S. B., MCKIBBIN, J. M. & HAKOMORI, S. I. (1985): Blood group A determinants with mono- and difucosyl type 1 chain in human erythrocyte membranes. *Biochemistry 24*, 3578-3586.

64. CLAUSEN, H., MCKIBBIN, J. M. & HAKOMORI, S. I. (1985): Monoclonal antibodies defining blood group A variants with difucosyl type 1 chain (ALe^b) and difucosyl type 2 chain (ALe^y). *Biochemistry 24*, 6190-6194.

65. COSTA, J., COSTA, M. T., GRABENHORST, E., NIMTZ, M. & CONRADT, H. S. (1998): Construction of recombinant BHK cell lines expressing wild-type and mutants of human alpha 1,3/4-fucosyltransferase. *Carbohydr. Polym. 37*, 287-290.

66. COSTACHE, M., APOIL, P. A., CAILLEAU, A., ELMGREN, A., LARSON, G., HENRY, S., BLANCHER, A., IORDACHESCU, D., ORIOL, R. & MOLLICONE, R. (1997): Evolution of fucosyltransferase genes in vertebrates. *J. Biol. Chem. 272*, 29721-29728.

67. CÔTÉ, R. H. & VALET, J. P. (1976): Isolation, composition, and reactivity of the neutral glycoproteins from human meconiums with specificities of the ABO and Lewis systems. *Biochem. J. 153*, 63-73.

68. COWLES, J. W., SPITALNIK, S. L. & BLUMBERG, N. (1986): Detection of anti-Le^a in Le(a-b+) individuals by kinetic ELISA. *Vox Sang. 50*, 164-168.

69. CRAWLEY, S. C., HINDSGAUL, O., RATCLIFFE, R. M., LAMONTAGNE, L. R. & PALCIC, M. M. (1989): A plant fucosyltransferase with human Lewis blood-group specificity. *Carbohydr. Res. 193*, 249-256.

70. CROOKSTON, M. C., TILLEY, C. A. & CROOKSTON, J. H. (1970): Human blood chimaera with seeming breakdown of immune tolerance. *Lancet ii*, 1110-1112.

71. CUTBUSH, M., GIBLETT, E. R. & MOLLISON, P. L. (1956): Demonstration of the phenotype Le(a+b+) in infants and adults. *Brit. J. Haematol. 2*, 210-213.

72. DABROWSKI, J., HANFLAND, P., EGGE, H. & DABROWSKI, U. (1981): Immunochemistry of the Lewis-blood-group system: proton nuclear magnetic resonance study of plasmatic Lewis-blood-group-active glycosphingolipids and related substances. *Arch. Biochem. Biophys. 210*, 405-411.

73. DAKOUR, J., LUNDBLAD, A. & ZOPF, D. (1988): Detection and isolation of oligosaccharides with Le[a] and Le[b] blood group activities by affinity chromatography using monoclonal antibodies. *Arch. Biochem. Biophys. 264*, 203-213.

74. DAMIAN, R. T. (1987): The exploitation of host immune responses by parasites. *J. Parasitol. 73*, 3-13.

75. DE BOSE-BOYD, R., NYAME, A. K. & CUMMINGS, R. D. (1996): Schistosoma mansoni: characterization of an α1-3 fucosyltransferase in adult parasites. *Exp. Parasitol. 82*, 1-10.

76. DE VRIES, T., SRNKA, C. A., PALCIC, M. M., SWIEDLER, S. J., VAN DEN EIJNDEN, D. H. & MACHER, B. A. (1995): Acceptor specificity of different length constructs of human recombinant α1,3/4-fucosyltransferases. Replacement of the stem region and the transmembrane domain of fucosyltransferase V by protein A results in an enzyme with GDP-fucose hydrolyzing activity. *J. Biol. Chem. 270*, 8712-8722.

77. DINH, Q., WENG, N. P., KISO, M., ISHIDA, H., HASEGAWA, A. & MARCUS, D. M. (1996): High affinity antibodies against Le[x] and sialyl Le[x] from a phage display library. *J. Immunol. 157*, 732-738.

78. DISSOUS, C., GRZYCH, J. M. & CAPRON, A. (1986): Schistosoma mansoni shares a protective oligosaccharide epitope with freshwater and marine snails. *Nature 323*, 443-445.

79. DORF, M. E., EGURO, S. Y., CABRERA, G., YUNIS, E. J., SWANSON, J. & AMOS, D. B. (1972): Detection of cytotoxic non-HL-A antisera. I. Relationship to anti-Le[a]. *Vox Sang. 22*, 447-456.

80. DUNSTAN, R. A., SIMPSON, M. B. & ROSSE, W. F. (1985): Le[a] blood group antigen on human platelets. *Amer. J. Clin. Pathol. 83*, 90-94.

81. DUPUY, F., PETIT, J. M., MOLLICONE, R., ORIOL, R., JULIEN, R. & MAFTAH, A. (1999): A single amino acid in the hypervariable stem domain of vertebrate α1,3/1,4-fucosyltransferases determines the type 1 / type 2 transfer. Characterization of acceptor substrate specificity of the Lewis enzyme by site-directed mutagenesis. *J. Biol. Chem. 274*, 12257-12262.

82. EGGE, H., DELL, A. & VON NICOLAI, H. (1983): Fucose containing oligosaccharides from human milk. I. Separation and identification of new constituents. *Arch. Biochem. Biophys. 224*, 235-253.

83. EGGE, H. & HANFLAND, P. (1981): Immunochemistry of the Lewis-blood-group-system. Mass spectrometric analysis of permethylated Le[a]-, Le[b]-, and H-type 1 (Le[dH]) blood-group active and related glycosphingolipids from human plasma. *Arch. Biochem. Biophys. 210*, 396-404.

84. EGGENS, I., FENDERSON, B., TOYOKUNI, T., DEAN, B., STROUD, M. & HAKOMORI, S. I. (1989): Specific interaction between Le[x] and Le[x] determinants. A possible basis for cell recognition in preimplanataion embryos and in embryonal carcinoma cells. *J. Biol. Chem. 264*, 9476-9484.

85. ELMGREN, A., BÖRJESON, C., SVENSSON, L., RYDBERG, L. & LARSON, G. (1996): DNA sequencing and screening for point mutations in the human Lewis (FUT3) gene enables molecular genotyping of the human Lewis blood group system. *Vox Sang. 70*, 97-103.

86. ELMGREN, A., MOLLICONE, R., COSTACHE, M., BÖRJESON, C., ORIOL, R., HARRINGTON, J. & LARSON, G. (1997): Significance of individual point mutations, T202C and C314T, in the human Lewis (FUT3) gene for expression of Lewis antigens by the human α(1,3/1,4)-fucosyltransferase, Fuc-TIII. *J. Biol. Chem. 272*, 21994-21998.

87. ELMGREN, A., RYDBERG, L. & LARSON, G. (1993): Genotypic heterogeneity among Lewis-negative individuals. *Biochem. Biophys. Res. Commun. 196*, 515-520.

88. EPPENBERGER-CASTORI, S., LOTSCHER, H. & FINNE, J. (1989): Purification of the N-acetylglucosaminide α(1-3/4)fucosyltransferase of human milk. *Glycoconj. J. 6*, 101-114.

89. ERNST, C., ATKINSON, B., WYSOCKA, M., BLASZCZYK, M., HERLYN, M., SEARS, H., STEPLEWSKI, Z. &

KOPROWSKI, H. (1984): Monoclonal antibody localization of Lewis antigens in fixed tissue. *Lab. Invest. 50*, 394-400.

90. FALK, K. E., KARLSSON, K. A. & SAMUELSSON, B. E. (1979): Proton nuclear magnetic resonance analysis of anomeric structure of glycosphingolipids. Lewis-active and Lewis-like substances. *Arch. Biochem. Biophys. 192*, 191-202.

91. FALK, P. G., BRY, L., HOLGERSSON, J. & GORDON, J. I. (1995): Expression of a human α-1,3/4-fucosyltransferase in the pit cell lineage of FVB/N mouse stomach results in production of Le[b]-containing glycoconjugates: a potential transgenic mouse model for studying Helicobacter pylori infection. *Proc. Natl. Acad. Sci. USA 92*, 1515-1519.

92. FEIZI, T. (1985): Demonstration by monoclonal antibodies that carbohydrate structures of glycoproteins and glycolipids are onco-developmental antigens. *Nature 314*, 53-57.

93. FEIZI, T. & CHILDS, R. A. (1985): Carbohydrate structures of glycoproteins and glycolipids as differentiation antigens, tumour-associated antigens and components of receptor systems. *Trends Biochem. Sci. 10*, 24-29.

94. FINNE, J., BREIMER, M. E., HANSSON, G. C., KARLSSON, K. A., LEFFLER, H., VLIEGENTHART, J. F. G. & VAN HALBEEK, H. (1989): Novel polyfucosylated N-linked glycopeptides with blood group A, H, X, and Y determinants from human small intestinal epithelial cells. *J. Biol. Chem. 264*, 5720-5735.

95. FOSTER, C. S., GILLIES, D. R. B. & GLICK, M. C. (1991): Purification and characterization of GDP-L-Fuc-N-acetyl-β- D-glucosaminide α1→3 fucosyltransferase from human neuroblastoma cells. Unusual substrate specificities of the tumor enzyme. *J. Biol. Chem. 266*, 3526-3531.

96. FOX, N., DAMJANOV, I., KNOWLES, B. B. & SOLTER, D. (1983): Immunohistochemical localization of the mouse stage-specific embryonic antigen 1 in human tissues and tumors. *Cancer Res. 43*, 669-678.

97. FRANÇOIS, A., SANSONETTI, N., MOLLICONE, R., LE PENDU, J., GALTON, J., JAULMES, B. & ORIOL, R. (1986): Heterogeneity of Lewis antibodies. A comparison of the reaction of human and animal reagents with synthetic oligosaccharides. *Vox Sang. 50*, 227-234.

98. FRASER, R. H., ALLAN, E. K., MURPHY, M. T., INGLIS, G. & MITCHELL, R. (1990): Monoclonal anti-Le[a] and anti-Le[b]: serological and immunochemical characterization. *2nd Int. Workshop on Human Red Blood Cell Antigens*, Lund, Sweden, Abstract p. 73.

99. FREDMAN, P., RICHERT, N. D., MAGNANI, J. L., WILLINGHAM, M. C., PASTAN, I. & GINSBURG, V. (1983): A monoclonal antibody that precipitates the glycoprotein receptor for epidermal growth factor is directed against the human blood group H type 1 antigen. *J. Biol. Chem. 258*, 11206-11210.

100. FUKUDA, M. (1985): Cell surface glycoconjugates as onco-differentiation markers in hematopoietic cells. *Biochim. Biophys. Acta 780*, 119-150.

101. FUKUDA, M., SPOONCER, E., OATES, J. E., DELL & KLOCK, J. C. (1984): Structure of sialylated fucosyl lactosaminoglycan isolated from human granulocytes. *J. Biol. Chem. 259*, 10925-10935.

102. FUKUDA, M. N., DELL, A., OATES, J. E., WU, P., KLOCK, J. C. & FUKUDA, M. (1985): Structures of glycosphingolipids isolated from human granulocytes. The presence of a series of linear poly-N-acetyllactosaminylceramide and its significance in glycolipids of whole blood cells. *J. Biol. Chem. 260*, 1067-1082.

103. FUKUSHI, Y., HAKOMORI, S. I., NUDELMAN, E. & COCHRAN, N. (1984): Novel fucolipids accumulating in human adenocarcinoma. II. Selective isolation of hybridoma antibodies that differentially recognize mono-, di-, and trifucosylated type 2 chain. *J. Biol. Chem. 259*, 4681-4685.

104. FUKUSHI, Y., HAKOMORI, S. I. & SHEPARD, T. (1984): Localization and alteration of mono-, di-, and trifucosyl α1→3 type 2 chain structures during human embryogenesis and in human cancer. *J. Exp. Med. 160*, 506-520.

105. FUKUSHI, Y., NUDELMAN, E., LEVERY, S. B., HAKOMORI, S. I. & RAUVALA, H. (1984): Novel fucolipids accumulating in human adenocarcinoma. III. A hybridoma antibody (FH6) defining a human cancer-associated difucoganglioside (VI[3]-NeuAcV[3]III[3]Fuc$_2$nLc$_6$). *J. Biol. Chem. 259*, 10511-10517.

106. FUKUSHIMA, K., HIROTA, M., TERASAKI, P. A., WAKISAKA, A., TOGASHI, H., CHIA, D., SUYAMA, N., FUKUSHI, Y., NUDELMAN, E. M. & HAKOMORI, S. I. (1984): Characterization of sialosylated Lewis[x] as a new tumor-associated antigen. *Cancer Res. 44*, 5279-5285.

107. FUKUTA, S., MAGNANI, J. L., GAUR, P. K. & GINSBURG, V. (1987): Monoclonal antibody CC3C195, which detects cancer-associated antigens in serum, binds to the human Le[a] blood group antigen and to its sialylated derivative. *Arch. Biochem. Biophys. 255*, 214-216.

108. GEYER, A., GEGE, C. & SCHMIDT, R. R. (1999): Carbohydrate-carbohydrate recognition between Lewis[x] glycoconjugates. *Angew. Chem. Int. Ed. 38*, 1466-1468.

109. GINSBURG, V., MCGINISS, M. H. & ZOPF, D. A. (1980): Biochemical basis for some bloood groups. In: *Immunobiology of the Erythrocyte* (S. G. Sandler, J. Nusbacher, and M. S. Schanfield, eds.). Alan R. Liss Inc., New York, pp. 45-53.

110. GOELTZ, S. E., HESSION, C., GOFF, D., GRIFFITHS, B., TIZARD, R., NEWMAN, B., CHI-ROSSO, G. & LOBB, R. (1990): ELFT: a gene that directs the expression of an ELAM-1 ligand. *Cell 63*, 1349-1356.

111. GOOD, A. H., YAU, O., LAMONTAGNE, L. R. & ORIOL, R. (1992): Serological and chemical specificities of twelve monoclonal anti- Le[a] and anti-Le[b] antibodies. *Vox Sang. 62*, 180-189.

112. GOOI, H. C., FEIZI, T., KAPADIA, A., KNOWLES, B. B., SOLTER, D. & EVANS, M. J. (1981): Stage-specific embryonic antigen involves $\alpha 1 \rightarrow 3$ fucosylated type 2 blood group chains. *Nature 292*, 156-158.

113. GOOI, H. C., PICARD, J. K., HOUNSELL, E. F., GREGORIOU, M., REES, A. R. & FEIZI, T. (1985): Monoclonal antibody (EGR/G49) reactive with the epidermal growth factor receptor of A431 cells recognizes the blood group ALe[b] and ALe[y] structures. *Mol.. Immunol. 22*, 689-694.

114. GRABENHORST, E., NIMTZ, M., COSTA, J. & CONRADT, H. S. (1998): In vivo specificity of human $\alpha 1,3/4$-fucosyltransferases III-VII in the biosynthesis of Lewis[x] and sialyl Lewis[x] motifs on complex-type N-glycans.Coexpression studies from BHK-21 cells together with human β-trace protein. *J. Biol. Chem. 273*, 30985-30994.

115. GRAHAM, D. Y. (1991): Helicobacter pylori: its epidemiology and its role in duodenal ulcer disease. *J. Gastroenterol. Hepatol. 6*, 105-113.

116. GRAHAM, H. A., HIRSCH, H. F. & DAVIES, D. M. (1977): Genetics and immunochemical relationships between soluble and cell-bound antigens of the Lewis system. In: *Human Blood Groups*. (J. F. Mohn, R. W. Plunkett, R. K. Cunningham, and R. M. Lambert, eds.). S. Karger, Basel, pp. 257-267.

117. GRAHAM, H. A., SINA, P., HIRSCH, H. F. & JACQUINET, J. C. (1978): Inhibition of anti-Le[dH] and Ulex anti-H lectin with oligosaccharides. *XVth Congress of the International Society of Blood Transfusion*, Paris, Abstract p. 542.

118. GREEN, P. J., TAMATANI, T., WATANABE, T., MIYASAKA, M., HASEGAWA, A., KISO, M., YUEN, C. T., STOLL, M. S. & FEIZI, T. (1992): High affinity binding of the leucocyte adhesion molecule L-selectin to 3'-sulphated-Le[a] and 3'-sulphated-Le[x] oligosaccharides and the predominance of sulphate in this interaction demonstrated by binding studies with a series of lipid-linked oligosaccharides. *Biochem. Biophys. Res. Commun. 188*, 244-251.

119. GREENWELL, P., BALL, M. G. & WATKINS, W. M. (1983): Fucosyltransferase activity in human lymphocytes and granulocytes. Blood group H-gene-specified α-2-L-fucosyltransferase is a discriminatory marker of peripheral blood lymphocytes. *FEBS Lett. 164*, 314-317.

120. GROLLMAN, E. F., KOBATA, A. & GINSBURG, V. (1969): An enzymatic basis for Lewis blood types in man. *J. Clin. Invest. 48*, 1489-1494.

121. GRÖNBERG, G., LIPNIUNAS, P., LUNDGREN, T., LINDH, F. & NILSSON, B. (1992): Structural analysis of five new monosialylated oligosaccharides from human milk. *Arch. Biochem. Biophys. 296*, 597-610.

122. GRUBB, R. (1948): Correlation between Lewis blood group and secretor character in man. *Nature 162*, 933.

123. GRUBB, R. (1951): Observations on the human group system Lewis. *Acta Pathol. Microbiol. Scand. 28*, 61-81.

124. GUNDOLF, F. (1973): Anti-A$_1$Le[b] in serum of a person of a blood group A$_{1h}$. *Vox Sang. 25*, 411-419.

125. GUNSON, H. H. & LATHAM, V. (1972): An agglutinin in human serum reacting with cells from Le(a-b-) non-secretor individuals. *Vox Sang. 22*, 344-353.

126. Hakomori, S. I. (1986): Tumor-associated glycolipid antigens, their metabolism and organization. *Chem. Phys. Lipids 42*, 209-233.

127. Hakomori, S. I. (1989): Aberrant glycosylation in tumors and tumor-associated carbohydrate antigens. *Adv. Cancer. Res. 52*, 257-331.

128. Hakomori, S. I. (1992): Le[x] and related structures as adhesion molecules. *Histochem. J. 24*, 771-776.

129. Hakomori, S. I. & Andrews, H. D. (1970): Sphingolipids with Le[b] activity and the copresence of Le[a] and Le[b] glycolipids in human tumor tissue. *Biochim. Biophys. Acta 202*, 225-228.

130. Hakomori, S. I. & Kannagi, R. (1983): Glycosphingolipids as tumor-associated and differentiation markers. *J. Natl. Cancer Inst. 71*, 231-251.

131. Hakomori, S. I. & Kobata, A. (1974): Blood group antigens. In: *The Antigens* (M. S. Sela, ed.). Academic Press, New York, Vol. 2, pp. 79-140.

132. Hakomori, S. I., Nudelman, E., Kannagi, R. & Levery, S. B. (1982): The common structure in fucosyllactosaminolipids accumulating in human adenocarcinomas, and its possible absence in normal tissue. *Biochem. Biophys. Res. Commun. 109*, 36-44.

133. Hakomori, S. I., Nudelman, E., Levery, S., Solter, D. & Knowles, B. B. (1981): The hapten structure of a developmentally regulated glycolipid antigen (SSEA-1) isolated from human erythrocytes and adenocarcinoma: a preliminary note. *Biochem. Biophys. Res. Commun. 100*, 1578-1586.

134. Hakomori, S. I., Nudelman, E., Levery, S. B. & Kannagi, R. (1984): Novel fucolipids accumulating in human adenocarcinoma. I. Glycolipids with di- or trifucosylated type 2 chain. *J. Biol. Chem. 259*, 4672-4680.

135. Hammar, L., Månsson, S., Rohr, T., Chester, M. A., Ginsburg, V., Lundblad, A. & Zopf, D. (1981): Lewis phenotype of erythrocytes and Le[b]-active glycolipid in serum of pregnant women. *Vox Sang. 40*, 27-33.

136. Handa, K., Withers, D. A. & Hakomori, S. (1998): The $\alpha1{\rightarrow}3$ fucosylation at the penultimate GlcNAc catalyzed by fucosyltransferase VII is blocked by internally fucosylated residue in sialosyl long-chain poly-LacNAc: enzymatic basis for expression of physiological E-selectin epitope. *Biochem. Biophys. Res. Commun. 243*, 199-204.

137. Hanfland, P. (1978): Isolation and purification of Lewis blood-group active glycosphingolipids from the plasma of human O Le[b] individuals. *Eur. J. Biochem. 87*, 161-170.

138. Hanfland, P. & Graham, H. A. (1981): Immunochemistry of the Lewis-blood-group system: partial characterization of Le[a]-, Le[b]-, and H-type 1 (Le[dH])-blood-group active glycosphingolipids from human plasma. *Arch. Biochem. Biophys. 210*, 383-395.

139. Hanfland, P., Graham, H. A., Crawford, R. J. & Schenkel-Brunner, H. (1982): Immunochemistry of the Lewis blood group system. Investigations on the Le[c] antigen. *FEBS Lett. 142*, 77-80.

140. Hanfland, P., Kladetzky, R. G. & Egli, G. (1978): Isolation and purification of Le[a] blood-group active and related glycosphingolipids from human plasma of blood group A Le[a] individuals. *Chem. Phys. Lipids 22*, 141-151.

141. Hanfland, P., Kordowicz, M., Peter-Katalinic, J., Pfannschmidt, G., Crawford, R. J., Graham, H. A. & Egge, H. (1986): Immunochemistry of the Lewis blood-group system: isolation and structures of Lewis-c active and related glycosphingolipids from the plasma of blood-group O Le(a-b-) nonsecretors. *Arch. Biochem. Biophys. 246*, 655-672.

142. Hanisch, F. G., Egge, H., Peter-Katalinic, J. & Uhlenbruck, G. (1986): Primary structure of a major sialyl-saccharide alditol from human amniotic mucins expressing the tumor-associated sialyl-X antigenic determinant. *FEBS Lett. 200*, 42-46.

143. Hanisch, F. G., Mitsakos, A., Schroten, H. & Uhlenbruck, G. (1988): Biosynthesis of cancer-associated sialyl-X antigen by a $(1{\rightarrow}3)$-α-L-fucosyltransferase of human amniotic fluid. *Carbohydr. Res. 178*, 23-28.

144. Hanisch, F. G., Uhlenbruck, G. & Dienst, C. (1984): Structure of tumor-associated carbohydrate antigen Ca 19-9 on human seminal-plasma glycoproteins from healthy donors. *Eur. J. Biochem.*

144, 467-473.

145. HANSSON, G. C., KARLSSON, K. A., LARSON, G., McKIBBIN, J. M., BLASZCZYK, M., HERLYN, M., STEPLEWSKI, Z. & KOPROWSKI, H. (1983): Mouse monoclonal antibodies against human cancer cell lines with specificities for blood group and related antigens. *J. Biol. Chem. 258*, 4091-4097.

146. HANSSON, G. C. & ZOPF, D. (1985): Biosynthesis of the cancer-associated sialyl-Le[a] antigen. *J. Biol. Chem. 260*, 9388-9392.

147. HEMMERICH, S., BERTOZZI, C. R., LEFFLER, H. & ROSEN, S. D. (1994): Identification of the sulfated monosaccharides of GlyCAM-1, an endothelial-derived ligand for L-selectin. *Biochemistry 33*, 4820-4829.

148. HENRY, S., JOVALL, P. Å., GHARDASHKHANI, S., ELMGREN, A., MARTINSSON, T., LARSON, G. & SAMUELSSON, B. (1997): Structural and immunochemical identification of Le[a], Le[b], H type 1, and related glycolipids in small intestinal mucosa of a group O Le(a-b-) nonsecretor. *Glycoconj. J. 14*, 209-223.

149. HENRY, S. M., BENNY, A. G. & WOODFILED, D. G. (1990): Investigation of Lewis phenotypes on Polynesians: evidence of a weak secretor phenotype. *Vox Sang. 58*, 61-66.

150. HENRY, S. M., JOVALL, P. A., GHARDASHKHANI, S., GUSTAVSSON, M. L. & SAMUELSSON, B. E. (1995): Structural and immunochemical identification of Le[b] glycolipids in the plasma of a group O Le(a-b-) secretor. *Glycocon. J. 12*, 309-317.

151. HENRY, S. M., ORIOL, R. & SAMUELSSON, B. E. (1994): Detection and characterization of Lewis antigens in plasma of Lewis-negative individuals. Evidence of chain extension as a result of reduced fucosyltransferase competition. *Vox Sang. 67*, 387-396.

152. HENRY, S. M., SIMPSON, L. A. & WOODFIELD, D. G. (1988): The Le(a+b+) phenotype in Polynesians. *Hum. Hered. 38*, 111-116.

153. HENRY, S. M., WOODFIELD, D. G., SAMUELSSON, B. E. & ORIOL, R. (1993): Plasma and red-cell glycolipid patterns of Le(a+b+) and Le(a+b-) Polynesians as further evidence of the weak secretor gene Se[w]. *Vox Sang. 65*, 62-69.

154. HERLYN, M., SEARS, STEPLEWSKI, Z. & KOPROWSKI, H. (1982): Monoclonal antibody detection of a circulating tumor-associated antigen. I. Presence of antigen in sera of patients with colorectal, gastric, and pancreatic carcinoma. *J. Clin. Immunol. 2*, 135-140.

155. HINDSGAUL, O., NORBERG, T., LE PENDU, J. & LEMIEUX, R. U. (1982): Synthesis of type 2 human blood group antigenic determinants. The H, X, and Y haptens and variations of the H type 2 determinants as probes for the combining site of the lextin I of Ulex europaeus. *Carbohydr. Res. 109*, 109-142.

156. HIRSCH, H. F. & GRAHAM, H. A. (1980): Adsorption of Le[c] and Le[d] from plasma onto red blood cells. *Transfusion 20*, 474-475.

157. HOLBURN, A. M. (1974): IgG Anti-Le[a]. *Brit. J. Haematol. 27*, 489-500.

158. HOLMES, E. H. & LEVERY, S. B. (1989): Biosynthesis fo fucose containing lacto-series glycolipids in human colonic adenocyrcinoma Colo 205 cells. *Arch. Biochem. Biophys. 274*, 633-647.

159. HOLMES, E. H., OSTRANDER, G. K., CLAUSEN, H. & GRAEM, N. (1987): Oncofetal expression of Le[x] carbohydrate antigens in human colonic adenocarcinomas. Regulation through type 2 core chain synthesis rather than fucosylation. *J. Biol. Chem. 262*, 11331-11338.

160. HOLMES, E. H., OSTRANDER, G. K. & HAKOMORI, S. I. (1985): Enzymatic basis for the accumulation of glycolipids with X and dimeric X determinants in human lung cancer cells (NCI-H69). *J. Biol. Chem. 260*, 7619-7627.

161. HOLMES, E. H., OSTRANDER, G. K. & HAKOMORI, S. I. (1986): Biosynthesis of the sialyl-Le[x] determinant carried by type 2 chain glycosphingolipids ($IV^3NeuAcIII^3FucnLc_4$, $VI^3NeuAcV^3FucnLc_6$, and $VI^3NeuAcIII^3V^3Fuc_2nLc_6$) in human lung carcinoma PC9 cells. *J. Biol. Chem. 261*, 3737-3743.

162. HOLMES, E. H., XU, Z. H., SHERWOOD, A. L. & MACHER, B. A. (1995): Structure-function analysis of human $\alpha1\rightarrow3$ fucosyltransferases. A GDP-fucose-protected, N-ethylmaleimide-sensitive site in FucT-III and FucT-V corresponds to Ser^{178} in FucT-IV. *J. Biol. Chem. 270*, 8145-8151.

163. HUANG, L. C., BROCKHAUS, M., MAGNANI, J. L., CUTTITTA, F., ROSEN, S., MINNA, J. D. & GINSBURG, V. (1983): Many monoclonal antibodies with an apparent specificity for certain lung cancers are

235

directed against a sugar sequence found in lacto-N-fucopentaose III. *Arch. Biochem. Biophys.* *220*, 318-320.

164. IKEHARA, Y., NISHIHARA, S., KUDO, T., HIRAGA, T., MOROZUMI, K., HATTORI, T. & NARIMATSU, H. (1998): The aberrant expression of Lewis a antigen in intestinal metaplastic cells of gastric mucosa is caused by augmentation of Lewis enzyme expression. *Glycoconj. J. 15*, 799-807.

165. ILVER, D., ARNQVIST, A., ÖGREN, J., FRICK, I. M., KERSULYTE, D., INCECIK, E. T., BERG, D. E., COVACCI, A., ENGSTRAND, L. & BORÉN, T. (1998): Helicobacter pylori adhesin binding fucosylated histo-blood group antigens revealed by retagging. *Science 279*, 373-377.

166. IMAI, Y., LASKY, L. A. & ROSEN, S. D. (1993): Sulphation requirement for GlyCAM-1, an endothelial ligand for L-selectin. *Nature 361*, 555-557.

167. ISEKI, S., MASAKI, S. & SHIBASAKI, K. (1957): Studies on Lewis blood group system. I. Le^c blood group factor. *Proc. Imp. Acad. Japan 33*, 492-497.

168. ISSITT, P. D. (1985): The Lewis system. In: *Applied Blood Group Serology*. Montgomery Scientific Publications, Miami, Florida, USA, pp. 169-191.

169. IWAKI, Y., KASAI, M., TERASAKI, P. I., BERNOKO, D., PARK, M. S., CICCIARELLI, J., HEINTZ, R., SAXTON, R. E., BURK, M. W. & MORTON, D. L. (1982): Monoclonal antibody against A_1Lewis^d antigen produced by the hybridoma immunized with a pulmonary carcinoma. *Cancer Res. 42*, 409-411.

170. JACQUINET, J. C. & SINAY, P. (1977): Synthesis of blood-group substances. 6. Synthesis of O-α-L-fucopyranosyl-(1-2)-O-β-D-galactopyranosyl-(1-4)-O-(α-L-fucopyranosyl-(1-3))-2-acet-amido-2-deoxy-α-D-glucopyranoside, the postulated Lewis d antigenic determinant. *J. Org. Chem. 42*, 720-724.

171. JAIN, R. K., PAWAR, S. M., CHANDRASEKARAN, E. V., PISKORZ, C. F. & MATTA, K. L. (1993): Synthesis of Gal-β-1→3(Fuc-α-1→4)GlcNAc-β-OR as potential acceptors for a new member of the α-1,2-L-fucosyltransferase family. *Bioorg. Med. Chem. Lett. 3*, 1333-1338.

172. JARKOWSKY, Z., MARCUS, D. M. & GROLLMAN, E. F. (1970): Fucosyltransferases found in human milk. Product of the Lewis blood group gene. *Biochemistry 9*, 1123-1128.

173. JEANNET, M., SCHAPIRA, M. & MAGNIN, C. (1974): Mise en évidence d'anticorps lymphocytotoxiques dirigés contre les antigènes A et B et contre des antigènes d'histocompatibilité non HL-A. *Schweiz. Med. Wochenschr. 104*, 152.

174. JOHNSON, P. H., DONALD, A. S. R., CLARKE, J. L. & WATKINS, W. M. (1995): Purification, properties and possible gene assignment of an α1,3-fucosyltransferase expressed in human liver. *Glycoconj. J. 12*, 879-893.

175. JOHNSON, P. H., DONALD, A. S. R., FEENEY, J. & WATKINS, W. M. (1992): Reassessment of the acceptor specificity and general properties of the Lewis blood-group gene associated α-3/4-fucosyltransferase purified from human milk. *Glycoconj. J. 9*, 251-264.

176. JOHNSON, P. H., DONALD, A. S. R. & WATKINS, W. M. (1993): Purification and properties of the α-3/4-L-fucosyltransferase released into the culture medium during the growth of the human A431 epidermoid carcinoma cell line. *Glycoconj. J. 10*, 152-164.

177. JOHNSON, P. H., WATKINS, W. M. & DONALD, A. S. R. (1987). Further purification of the Le gene associated alpha-L-fucosyltransferase from human milk. *Proceedings of the 9th International Symposium on Glycoconjugates*, Lille, E 107.

178. JOHNSON, P. H., YATES, A. D. & WATKINS, W. M. (1981): Human salivary fucosyltransferases: evidence for two distinct α-3-L-fucosyltransferase activities one of which is associated with the Lewis blood group Le gene. *Biochem. Biophys. Res. Commun. 100*, 1611-1618.

179. JORDAL, K. (1956): The Lewis blood groups in children. *Acta Pathol. Microbiol. Scand. 39*, 399-406.

180. JUDD, W. J., STEINER, E. A., FRIEDMAN, B. A. & OBERMAN, H. A. (1978): Anti-Le^a as an autoantibody in the serum of a Le(a-b+) individual. *Transfusion 18*, 436-440.

181. KAIZU, T., LEVERY, S. B., NUDELMAN, E., STENKAMP, R. E. & HAKOMORI, S. I. (1986): Novel fucolipids of human adenocarcinoma: monoclonal antibody specific for trifucosyl Le^y ($III^3FucV^3FucVI^2FucnLc_6$) and a possible three-dimensional epitope structure. *J. Biol. Chem. 261*,

11254-11258.

182. KALADAS, P. M., KABAT, E. A., SHIBATA, S. & GOLDSTEIN, I. J. (1983): Immunochemical studies on the binding specificity of the blood group Le^b specific lectin Griffonia simplicifolia IV. *Arch. Biochem. Biophys. 223*, 309-318.

183. KAMEYAMA, A., ISHIDA, H., KISO, M. & HASEGAWA, A. (1991): Synthetic studies on sialoglycoconjugates. 22. Total synthesis of tumor-associated ganglioside, sialyl Lewis X. *J. Carbohydr. Chem. 10*, 549-560.

184. KANEKO, M., KUDO, T., IWASAKI, H., IKEHARA, Y., NISHIHARA, S., NAKAGAWA, S., SASAKI, K., SHIINA, T., INOKO, H., SAITOU, N. & NARIMATSU, H. (1999): α1,3-fucoslytransferase IX (Fuc-TIX) is very highly conserved between human and mouse; molecular cloning, characterization and tissue distribution of human Fuc-TIX. *FEBS Lett. 452*, 237-242.

185. KANEKO, T., IBA, Y., ZENITA, K., SHIGETA, K., NAKANO, K., ITOH, W., KUROSAWA, Y., KANNAGI, R. & YASUKAWA, K. (1993): Preparation of mouse-human chimeric antibody to an embryonic carbohydrate antigen, Lewis Y. *J. Biochem. 113*, 114-117.

186. KANNAGI, R., LEVERY, S. B. & HAKOMORI, S. I. (1985): Le^a-active heptaglycosylceramide, a hybrid of type 1 and type 2 chain, and the pattern of glycolipids with Le^a, Le^b, X (Le^x), and Y(Le^y) determinants in human blood cell membranes (ghosts). Evidence that type 2 chain can elongate repetively but type 1 chain cannot. *J. Biol. Chem. 260*, 6410-6415.

187. KANNAGI, R., NUDELMAN, E., LEVERY, S. B. & HAKOMORI, S. I. (1982): A series of human erythrocyte glycosphingolipids reacting to the monoclonal antibody directed to a developmentally regulated antigen, SSEA-1. *J. Biol. Chem. 257*, 14865-14874.

188. KARLSSON, K. A. & LARSON, G. (1981): Molecular characterization of cell surface antigens of fetal tissue. Detailed analysis of glycosphingolipids of meconium of a human O Le(a-b+) secretor. *J. Biol. Chem. 256*, 3512-3524.

189. KERDE, C., BRUNK, R., FÜNFHAUSEN, G. & PROKOP, O. (1960): Über die Herstellung von Anti-Lewis-Seren an Capra hircus. *Z. Immun.-Forsch. 119*, 462-468.

190. KERR, M. A. & STOCKS, S. C. (1992): The role of CD15-(Le^x)-related carbohydrates in neutrophil adhesion. *Histochem. J. 24*, 811-826.

191. KITAGAWA, H., NAKADA, H., FUKUI, S., FUNAKOSHI, I., KAWASAKI, T., YAMASHINA, I., TATE, S. & INAGAKI, F. (1991): Novel oligosaccharides with the sialyl-Le^a structure in human milk. *Biochemistry 30*, 2869-2876.

192. KITAGAWA, H., NAKADA, H., FUKUI, S., KAWASAKI, T. & YAMASHINA, I. (1991): Characterization of mucin-type oligosaccharides with the sialyl-Le^a structure from human colorectal adenocarcinoma cells. *Biochem. Biophys. Res. Commun. 178*, 1429-1436.

193. KITAGAWA, H., NAKADA, H., KUROSAKA, A., HIRAIWA, N., NUMATA, Y., FUKUI, S., FUNAKOSHI, I., KAWASAKI, T., YAMASHINA, I., SHIMADA, I. & INAGAKI, F. (1989): Three novel oligosaccharides with the sialyl-Le^a structure in human milk: isolation by immunoaffinity chromatography. *Biochemistry 28*, 8891-8897.

194. KITAGAWA, H., NAKADA, H., NUMATA, Y., KUROSAKA, A., FUKUI, S., FUNAKOSHI, I., KAWASAKI, T., SHIMADA, I., INAGAKI, F. & YAMASHINA, I. (1990): Occurrence of tetra- and pentasaccharides with the sialyl-Le^a structure in human milk. *J. Biol. Chem. 265*, 4859-4862.

195. KITAGAWA, H., NAKADA, H., NUMATA, Y., KUROSAKA, A., FUKUI, S., FUNAKOSHI, I., KAWASAKI, T. & YAMASHINA, I. (1988): A monoclonal antibody that recognizes sialyl-Le^a oligosaccharide, but is distinct from NS 19-9 as to epitope recognition. *J. Biochem. 104*, 817-821.

196. KLUG, T. L., LEDONNE, N. C., GREBER, T. F. & ZURAWSKI, V. R. (1988): Purification and composition of a novel gastrointestinal tumor-associated glycoprotein expressing sialylated lacto-N-fucopentaose-II (CA 19-9). *Cancer Res. 48*, 1505-1511.

197. KNIBBS, R. N., CRAIG, R. A., MALY, P., SMITH, P. L., WOLBER, F. M., FAULKNER, N. E., LOWE, J. B. & STOOLMAN, L. M. (1998): α(1,3)-fucosyltransferase VII-dependent synthesis of P- and E-selectin ligands on cultured T lymphoblasts. *J. Immunol. 161*, 6305-6315.

198. KO, A. I., DRÄGER, U. C. & HARN, D. A. (1990): A Schistosoma mansoni epitope recognized by a protective monoclonal antibody is identical to the stage-specific embryonic antigen 1. *Proc. Natl.*

Acad. Sci. USA 87, 4159-4163.

199. Kobata, A. (1977): Milk glycoproteins and oligosaccharides. In: *The Glycoconjugates* (M. I. Horowitz and W. Pigman, eds.). Vol. 1, pp. 423-440.

200. Kobata, A. & Ginsburg, V. (1970): Uridine diphosphate-N-acetyl-D-galactosamine:D-galactose α-3-N-acetyl-D-galactosaminyltransferase, a product of the gene that determines blood type A in man. *J. Biol. Chem. 245*, 1484-1490.

201. Koda, Y., Kimura, H. & Mekada, E. (1993): Analysis of Lewis fucosyltransferase genes from the human gastric mucosa of Lewis-positive and Lewis-negative individuals. *Blood 82*, 2915-2919.

202. Koprowski, H., Herlyn, M., Steplewski, Z. & Sears, H. (1981): Specific antigen in serum of patients with colon carcinoma. *Science 212*, 53-55.

203. Koprowski, H., Steplewski, Z., Mitchell, K., Herlyn, M., Herlyn, D. & Fuhrer, J. P. (1979): Colorectal carcinoma antigens detected by hybidoma antibodies. *Somat. Cell Genet 5*, 957-972.

204. Kornstad, I. (1969): Anti-Le[b] in the serum of Le(a+b-) and Le(a-b-) persons: absorption studies with erythrocytes of different ABO and Lewis phenotypes. *Vox Sang. 16*, 124-129.

205. Koszdin, K. L. & Bowen, B. R. (1992): The cloning and expression of a human α-1,3 fucosyltransferase capable of forming the E-selectin ligand. *Biochem. Biophys. Res. Commun. 187*, 152-157.

206. Kuijpers, T. W. (1993): Terminal glycosyltransferase activity - a selective role in cell adhesion. *Blood 81*, 873-882.

207. Kukowska-Latallo, J. F., Larsen, R. D., Nair, R. P. & Lowe, J. B. (1990): A cloned human cDNA determines expression of a mouse stage-specific embryonic antigen and the Lewis blood group α(1,3/1,4) fucosyltransferase. *Genes Dev. 4*, 1288-1303.

208. Kumar, R., Potvin, B., Muller, W. A. & Stanley, P. (1991): Cloning of a human α(1,3)-fucosyl-transferase gene that encodes ELFT but does not confer ELAM-1 recognition on Chinese hamster ovary cell transfectants. *J. Biol. Chem. 266*, 21777-21783.

209. Kunzendorf, U., Krugerkrasagakes, S., Netter, M., Hock, H., Walz, G. & Diamantstein, T. (1994): A sialyl-Le(X)-negative melanoma cell line binds to E selectin but not to P selectin. *Cancer Res. 54*, 1109-1112.

210. Lamblin, G., Biersman, A., Klein, A., Roussel, P., van Halbeek, H. & Vliegenthart, J. F. G. (1984): Primary structure determination of five sialylated oligosaccharides derived from bronchial mucus glycoproteins of patients suffering from cystic fibrosis. The occurrence of the NeuAcα(2→3)Galβ(1→4)[Fucα(1→3)]GlcNAcβ(1→*) structural element revealed by 500-MHz ^{1}H NMR spectroscopy. *J. Biol. Chem. 259*, 9051-9058.

211. Langkilde, N. C., Wolf, H., Meldgard, P. & Orntoft, T. F. (1991): Frequency and mechanism of Lewis antigen expression in human urinary bladder and colon carcinoma patients. *Brit. J. Cancer 63*, 583-586.

212. Lawler, S. D. & Marshall, R. (1961): Lewis and secretor characters in infancy. *Vox Sang. 6*, 541-554.

213. Le Pendu, J., Fredman, P., Richter, N. D., Magnani, J. L., Willingham, M. C., Pastan, I., Oriol, R. & Ginsburg, V. (1985): Monoclonal antibody 101 that precipitates the glycoprotein receptor for epidermal growth factor is directed against the Y antigen, not the H type 1 antigen. *Carbohydr. Res. 141*, 347-349.

214. Le Pendu, J., Lemieux, R. U., Dalix, A. M., Lambert, F. & Oriol, R. (1983): Competition between ABO and Le gene specified enzymes. I. A Lewis related difference in the amount of A antigen in saliva of A$_1$ and A$_2$ secretors. *Vox Sang. 45*, 349-358.

215. Le Pendu, J., Lemieux, R. U. & Oriol, R. (1982): Purification of anti-Le[c] antibodies with specificity for βDGal(1→3)βDGlcNAcO- using a synthetic immunoadsorbent. *Vox Sang 43*, 188-195.

216. Le Pendu, J., Oriol, R., Lambert, F., Dalix, A. M. & Lemieux, R. U. (1983): Competition between ABO and Le gene specified enzymes. II. Quantitative analysis of A and B antigens in saliva of ABH nonsecretors. *Vox Sang. 45*, 421-425.

217. Legault, D. J., Kelly, R. J., Natsuka, Y. & Lowe, J. B. (1995): Human α(1,3/1,4)-fucosyl-transferases discriminate between different oligosaccharide acceptor substrates through a

discrete peptide fragment. *J. Biol. Chem. 270*, 20987-20996.

218. LEMIEUX, R. U. (1978): Human blood groups and carbohydrate chemistry. *Chem. Soc. Rev. 7*, 423-452.

219. LEMIEUX, R. U., BAKER, D. A., WEINSTEIN, W. M. & SWITZER, C. M. (1981): Artificial antigens. Antibody preparations for the localization of Lewis determinants in tissues. *Biochemistry 20*, 199-205.

220. LEMIEUX, R. U., BUNDLE, D. R. & BAKER, D. A. (1975): The properties of a "synthetic" antigen related to the human blood-group Lewis a. *J. Amer. Chem. Soc. 97*, 4076-4083.

221. LEMIEUX, R. U. & DRIGUEZ, H. J. (1975): The chemical synthesis of 2-acetamido-2-deoxy-4-O-(α-L-fucopyranosyl)-3-O-(β-D-galactopyranosyl)-D-glucose. The Lewis a blood group antigenic determinant. *J. Amer. Chem. Soc.* 97, 4063-4068.

222. LEMIEUX, R. U., HINDSGAUL, O., BIRD, P., NARASIMHAN, S. & YOUNG, W. W. (1988): The binding of the Lewis-a human blood group determinant by two hybridoma monoclonal anti-Le^a antibodies. *Carbohydr. Res. 178*, 293-305.

223. LEVERY, S. B., NUDELMAN, E. D., ANDERSEN, N. H. & HAKOMORI, S. I. (1986): ^{1}H-N.M.R. analysis of glycolipids possessing mono- and multi-meric X and Y haptens: characterization of two novel extended Y structures from human adenocarcinoma. *Carbohydr. Res. 151*, 311-328.

224. LEVINE, P. & CELANO, M. (1960): The antigenicity of Lewis (Le^a) substance in saliva coated on to tanned red cells. *Vox Sang. 5*, 53-61.

225. LEY, K., BULLARD, D. C., ARBONES, M. L., BOSSE, R., VESTWEBER, D., TEDDER, T. F. & BEAUDET, A. L. (1995): Sequential contribution of L- and P-selectin to leukocyte rolling in vivo. *J. Exp. Med. 181*, 669-675.

226. LIN, M. & SHIEH, S. H. (1994): Postnatal development of red cell Le^a and Le^b antigens in Chinese infants. *Vox Sang. 66*, 137-140.

227. LINDSTRÖM, K., BREIMER, M. E., JOVALL, P. A., LANNE, B., PIMLOTT, W. & SAMUELSSON, B. E. (1992): Non-acid glycosphingolipid expression in plasma of an A$_1$ Le(a- b+) secretor human individual: identification of an ALe^b heptaglycosylceramide as major blood group component. *J. Biochem. 111*, 337-345.

228. LLOYD, K. O. & KABAT, E. A. (1968): Immunochemical studies on blood groups. XLI. Proposed structures for the carbohydrate portions of blood group A, B, H, Lewis^a, and Lewis^b substances. *Proc. Natl. Acad. Sci. USA 61*, 1470-1477.

229. LLOYD, K. O., LARSON, G., STRÖMBERG, N., THURIN, J. & KARLSSON, K. A. (1983): Mouse monoclonal antobody F-3 recognizes the difucosyl type-2 blood group structure. *Immunogenetics 17*, 537-541.

230. LODGE, T. W., ANDERSEN, J. & GOLD, E. R. (1965): Observations on antibodies reacting with adult and cold Le(a-b-) cells, with O$_h$ Le(a-b-) cells and a soluble antigen present in certain salivas. *Vox Sang. 10*, 73-81.

231. LOWE, J. B., KUKOWSKA-LATALLO, J. F., NAIR, R. P., LARSEN, R. D., MARKS, R. M., MACHER, B. A., KELLY, R. J. & ERNST, L. K. (1991): Molecular cloning of a human fucosyltransferase gene that determines expression of the Lewis x and VIM-2 epitopes but not ELAM-1 dependent cell adhesion. *J. Biol. Chem. 266*, 17467-17477.

232. LUNDBLAD, A. (1977): Urinary glycoproteins, glycopeptides, and oligosaccharides. In: *The Glycoconjugates* (M. I. Horowitz and W. Pigman, eds.). Vol. 1, pp. 441-458.

233. LUSCINSKAS, F. W., DING, H. & LICHTMAN, A. H. (1995): P-selectin and vascular cell adhesion molecule 1 mediate rolling and arrest, respectively, of CD4+ T lymphocytes on tumor necrosis factor alpha-activated vascular endothelium under flow. *J. Exp. Med. 181*, 1179-1186.

234. MACHER, B. A., BUEHLER, J., SCUDDER, P., KNAPP, W. & FEIZI, T. (1988): A novel carbohydrate, differentiation antigen on fucogangliosides of human myeloid cells recognized by monoclonal antibody VIM-2. *J. Biol. Chem. 263*, 10186-10191.

235. MACHER, B. A., HOLMES, E. H., SWIEDLER, S. J., STULTS, C. L. & SRNKA, C. A. (1991): Human α 1-3 fucosyltransferases. *Glycobiology 1*, 577-584.

236. MAGNANI, J. L., NILSSON, B., BROCKHAUS, M., ZOPF, D., STEPLEWSKI, Z., KOPROWSKI, H. &

GINSBURG, V. (1982): A monoclonal antibody-defined antigen associated with gastrointestinal cancer is a ganglioside containing sialylated lacto-N-fucopentaose II. *J. Biol. Chem. 257*, 14365-14369.

237. MAGNANI, J. L., STEPLEWSKI, Z., KOPROWSKI, H. & GINSBURG, V. (1983): The gastrointestinal and pancreatic cancer-associated antigen detected by monoclonal antibody 19-9 in the sera of patients is a mucin. *Cancer Res. 43*, 5489-5492.

238. MAJURI, M. L., NIEMELÄ, R., TIISALA, S., RENKONEN, O. & RENKONEN, R. (1995): Expression and function of α2,3-sialyl- and α1,3/1,4-fucosyltransferases in colon adenocarcinoma cell lines: role in synthesis of E-selectin counter-receptors. *Int. J. Cancer63*, 551-559.

239. MÄKELÄ, O. & MÄKELÄ, P. (1956): Le^b antigen. Studies on its occurrence in red cells, plasma, and saliva. *Ann. Med. Exp. Biol. Fenn. 34*, 157-162.

240. MAKNI, S., DALIX, A. M., CAILLARD, T., COMPAGNON, B., LE PENDU, J., AYED, K. & ORIOL, R. (1987): Discordance between red cell and saliva Lewis phenotypes in patients with hydatid cysts. *Exp. Clin. Immunogenet. 4*, 136-143.

241. MALÝ, P., THALL, A. D., PETRYNIAK, B., ROGERS, G. E., SMITH, P. L., MARKS, R. M., KELLY, R. J., GERSTEN, K. M., CHENG, G. Y., SAUNDERS, T. L., CAMPER, S. A., CAMPHAUSEN, R. T., SULLIVAN, F. X., ISOGAI, Y., HINDSGAUL, O., VON ANDRIAN, U. H. & LOWE, J. B. (1996): The α(1,3)fucosyltransferase Fuc-TVII controls leukocyte trafficking through an essential role in L-, E-, and P- selectin ligand biosynthesis. *Cell 86*, 643-653.

242. MANDEL, U., ORNTOFT, T. F., HOLMES, E. H., SORENSEN, H., CLAUSEN, H., HAKOMORI, S. I. & DABELSTEEN, E. (1991): Lewis blood group antigens in salivary glands and stratified epithelium - lack of regulation of Lewis antigen expression in ductal and buccal mucosal lining epithelia. *Vox Sang. 61*, 205-214.

243. MARCUS, D. M., BASTANI, A., ROSENFIELD, R. E. & GROLLMAN, A. P. (1967): Studies on blood group substances. II. Hemagglutinating properties of caprine antisera to human Le^a and Le^b blood group substances. *Transfusion 7*, 277-280.

244. MARCUS, D. M. & CASS, L. E. (1969): Glycosphingolipids with Lewis blood group activity: uptake by human erythrocytes. *Science 164*, 553-555.

245. MARCUS, D. M. & GROLLMAN, A. P. (1966): Studies of blood group substances. I. Caprine precipitating antisera to human Le^a and Le^b blood group substances. *J. Immunol. 97*, 867-875.

246. MARTIN, S. L., EDBROOKE, M. R., HODGMAN, T. C., VAN DEN EIJNDEN, D. H. & BIRD, M. I. (1997): Lewis X biosynthesis in Helicobacter pylori. Molecular cloning of an α(1,3)-fucosyltransferase gene. *J. Biol. Chem. 272*, 21349-21356.

247. MAYR, W. R. & MAYR, D. (1974): A lymphocytotoxic antibody associated with ABO blood group and ABH secretor status. *J. Immunogenet. 1*, 43-48.

248. MCCURLEY, R. S., RECINOS, A., OLSEN, A. S., GINGRICH, J. C., SZCZEPANIAK, D., CAMERON, H. S., KRAUSS, R. & WESTON, B. W. (1995): Physical maps of human α(1,3)fucosyltransferase genes FUT3-FUT6 on chromosomes 19p13.3 and 11q21. *Genomics 26*, 142-146.

249. MCEVER, R. P., MOOORE, K. L. & CUMMINGS, R. D. (1995): Leukocyte trafficking mediated by selectin-carbohydrate interactions. *J. Biol. Chem. 270*, 11025-11028.

250. MCKIBBIN, J. M., SPENCER, W. A., SMITH, E. L., MANSSON, J. E., KARLSSON, K. A., SAMUELSSON, B. E., LI, Y. T. & LI, S. C. (1982): Lewis blood group fucolipids and their isomers from human and canine intestine. *J. Biol. Chem. 257*, 755-760.

251. MELO, N. S., NIMTZ, M., CONRADT, H. S., FEVEREIRO, P. S. & COSTA, J. (1997): Identification of the human Lewis^a carbohydrate motif in a secretory peroxidase from a plant cell suspension culture (Vaccinium myrtillus L.). *FEBS Lett. 415*, 186-191.

252. MESSETER, L., BRODIN, T., CHESTER, M. A., KARLSSON, K. A., ZOPF, D. & LUNDBLAD, A. (1984): Immunochemical characterization of a monoclonal anti-Le^b blood grouping reagent. *Vox Sang. 46*, 66-74.

253. MILLER-PODRAZA, H., ABUL MILH, M., TENEBERG, S. & KARLSSON, K. A. (1997): Binding of Helicobacter pylori to sialic acid-containing glycolipids of various origins separated on thin-layer chromatograms. *Infect. Immun. 65*, 2480-2482.

254. MILLER-PODRAZA, H., MILH, M. A., BERGSTRÖM, J. & KARLSSON, K. A. (1996): Recognition of glycoconjugates by Helicobacter pylori: an apparently high-affinity binding of human polyglycosylceramides, a second sialic acid-based specificity. *Glycoconj. J. 13*, 453-460.

255. MITSAKOS, A., HANISCH, F. G. & UHLENBRUCK, G. (1988): Biosynthesis of the cancer-associated sialyl-Le^x determinant in human amniotic fluid. *Biol. Chem. Hoppe-Seyler 369*, 661-665.

256. MOLLICONE, R., CANDELIER, J. J., MENNESSON, B., COUILLIN, P., VENOT, A. P. & ORIOL, R. (1992): Five specificity patterns of (1→3)-α-L-fucosyltransferase activity defined by use of synthetic oligosaccharide acceptors. Differential expression of the enzymes during human embryonic development and in adult tissues. *Carbohydr. Res. 228*, 265-276.

257. MOLLICONE, R., GIBAUD, A., FRANCOIS, A., RATCLIFFE, M. & ORIOL, R. (1990): Acceptor specificity and tissue distribution of three human α-3-fucosyltransferases. *Eur. J. Biochem. 191*, 169-176.

259. MOLLICONE, R., REGUIGNE, I., FLETCHER, A., AZIZ, A., RUSTAM, M., WESTON, B. W., KELLY, R. J., LOWE, J. B. & ORIOL, R. (1994): Molecular basis for plasma α(1,3)-fucosyltransferase gene deficiency (FUT6). *J. Biol. Chem. 269*, 12662-12671.

260. MOLLICONE, R., REGUIGNE, I., KELLY, R. J., FLETCHER, A., WATT, J., CHATFIELD, S., AZIZ, A., CAMERON, H. S., WESTON, B. W., LOWE, J. B. & ORIOL, R. (1994): Molecular basis for Lewis α(1,3/1,4)-fucosyltransferase gene deficiency (FUT3) found in Lewis-negative Indonesian pedigrees. *J. Biol. Chem. 269*, 20987-20994.

261. MOLLISON, P. L., ENGELFRIET, P. & CONTRERAS, M. (1992): *Blood Transfusion in Clinical Medicine.* 9th edn., Blackwell, Oxford, UK.

262. MONTEIRO, M. A., CHAN, K. H. N., RASKO, D. A., TAYLOR, D. E., ZHENG, P. Y., APPELMELK, B. J., WIRTH, H. P., YANG, M. Q., BLASER, M. J., HYNES, S. O., MORAN, A. P. & PERRY, M. B. (1998): Simultaneous expression of type 1 and type 2 Lewis blood group antigens by Helicobacter pylori lipopolysaccharides. *J. Biol. Chem. 273*, 11533-11543.

263. MOOR-JANKOWSKI, J., WIENER, A. S. & ROGERS, C. M. (1964): Human blood group factors in non-human primates. *Nature 202*, 663-665.

264. MOURANT, A. E. (1946): A 'new' human blood group antigen of frequent occurrence. *Nature 158*, 237-238.

265. MOURANT, A. E., KOPEC, A. C. & DOMANIEWSKA-SOBCZAK, K. (1976): *The distribution of the human blood groups and other polymorphisms.* 2nd edn. Oxford University Press, London.

266. MULLIGAN, M. S., LOWE, J. B., LARSEN, R. D., PAULSON, J., ZHENG, Z. L., DEFREES, S., MAEMURA, K., FUKUDA, M. & WARD, P. A. (1993): Protective effects of sialylated oligosaccharides in immune complex-induced acute lung injury. *J. Exp. Med. 178*, 623-631.

267. MULLIGAN, M. S., PAULSON, J. C., DE FREES, S., ZHENG, Z. L., LOWE, J. B. & WARD, P. A. (1993): Protective effects of oligosaccharides in P-selectin-dependent lung injury. *Nature 364*, 149-151.

268. MUNRO, J. R. & SCHACHTER, H. (1973): The presence of two GDP-L-fucose:glycoprotein fucosyltransferases in human serum. *Arch. Biochem. Biophys. 156*, 534-542.

269. MURAMATSU, H., KAMADA, Y. & MURAMATSU, T. (1986): Purification and properties of N-acetylglucosaminide α1→3 fucosyltransferase from embryonal carcinoma cells. *Eur. J. Biochem. 157*, 71-75.

270. MÜTHING, J., SPANBROEK, R., PETER-KATALINIC, J., HANISCH, F. G., HANSKI, C., HASEGAWA, A., UNLAND, F., LEHMANN, J., TSCHESCHE, H. & EGGE, H. (1996): Isolation and structural characterization of fucosylated gangliosides with linear poly-N-acetyllactosaminyl chains from human granulocytes. *Glycobiology 6*, 147-156.

271. NAKAMURA, J. I., MOGI, A., ASAO, T., NAGAMACHI, Y. & YAZAWA, S. (1997): Evidence that the aberrant α1→2 fucosyltransferase found in colorectal carcinoma may be encoded by Fuc-TIII (Le) gene. *Anticancer Res. 17*, 4563-4569.

272. NATSUKA, S., GERSTEN, K. M., ZENITA, K., KANNAGI, R. & LOWE, J. B. (1994): Molecular cloning of a cDNA encoding a novel human leukocyte α-1,3-fucosyltransferase capable of synthesizing the sialyl Lewis x determinant. *J. Biol. Chem. 269*, 16789-16794.

273. NGUYEN, A. T., HOLMES, E. H., WHITAKER, J. M., HO, S., SHETTERLY, S. & MACHER, B. A. (1998): Human alpha 1,3/4-fucosyltransferases. I. Identification of amino acids involved in acceptor

substrate binding by site-directed mutagenesis. *J. Biol. Chem. 273*, 25244-25249.

274. Nicholas, J. W., Jenkins, W. J. & Marsh, W. L. (1957): Human blood chimaeras. A study of surviving twins. *Brit. Med. J. i*, 1458-1460.

275. Nielsen, L. S., Eiberg, H. & Mohr, J. (1983): Another case of a lymphoytotoxic antibody with blood group A^1 Le^b and A Le^d associated specificity. *Tissue Antigens 21*, 177-183.

276. Nilsson, M. & Norberg, T. (1988): Synthesis of a dimeric Lewis X hexasaccharide derivative corresponding to a tumor-associated glycolipid. *Carbohydr. Res. 183*, 71-82.

277. Nishihara, S., Hiraga, T., Ikehara, Y., Iwasaki, H., Kudo, T., Yazawa, S., Morozumi, K., Suda, Y. & Narimatsu, H. (1999): Molecular behavior of mutant Lewis enzymes in vivo. *Glycobiology 4*, 373-382.

278. Nishihara, S., Narimatsu, H., Iwasaki, H., Yazawa, S., Akamatsu, S., Ando, T., Seno, T. & Narimatsu, I. (1994): Molecular genetic analysis of the human Lewis histo-blood group system. *J. Biol. Chem. 269*, 29271-29278.

279. Nishihara, S., Yazawa, S., Iwasaki, H., Nakazato, M., Kudo, T., Ando, T. & Narimatsu, H. (1993): α(1,3/1,4)Fucosyltransferase (FucT-III) gene is inactivated by a single amino acid substitution in Lewis histo-blood type negative individuals. *Biochem. Biophys. Res. Commun. 196*, 624-631.

280. Nomura, A., Stemmermann, G. N., Chyou, P. H., Kato, I., Perez-Perez, G. I. & Blaser, M. J. (1991): Helicobacter pylori infection and gastric carcinoma among Japanese Americans in Hawaii. *N. Engl. J. Med. 325*, 1132-1136.

281. Norgard, K. E., Moore, K. L., Diaz, S., Stults, N. L., Ushiyama, S., McEver, R. P., Cummings, R. D. & Varki, A. (1993): Characterization of a specific ligand for P-selectin on myeloid cells - a minor glycoprotein with sialylated O-linked oligosaccharides. *J. Biol. Chem. 268*, 12764-12774.

282. Nudelman, E., Fukushi, Y., Levery, S. B., Higuchi, T. & Hakomori, S. I. (1986): Novel fucolipids of human adenocarcinoma: disialosyl Le^a antigen (III^{4}FucIII^{6}NeuAcIV^{3}NeuAcLc$_4$) of human colonic adenocarcinoma and the monoclonal antibody (FH7) defining this structure. *J. Biol. Chem. 261*, 5487-5495.

283. Nudelman, E., Levery, S. B., Kaizu, T. & Hakomori, S. I. (1986): Novel fucolipids of human adenocarcinoma: characterization of the major Le^y antigen of human adenocarcinoma as trifucosylnonaosyl Le^y glycolipid (III^{3}FucV^{3}FucVI^{2}FucnLc$_6$). *J. Biol. Chem. 261*, 11247-11253.

284. Ohmori, K., Yoneda, T., Ishihara, G., Shigeta, K., Hirashima, K., Kanai, M., Itai, S., Sasaoki, T., Arii, S. & Arita, H. (1989): Sialyl SSEA-1 antigen as a carbohydrate marker of human natural killer cells and immature lymphoid cells. *Blood 74*, 255-261.

285. Oriol, R., Cartron, J. P., Cartron, J. & Mulet, C. (1980): Biosynthesis of ABH and Lewis antigens in normal and transplanted kidneys. *Transplantation 29*, 184-188.

286. Oriol, R., Danilovs, J., Lemieux, R. U., Terasaki, P. I. & Bernoco, D. (1980): Lymphocytotoxic definition of combined ABH and Lewis antigens and their transfer from sera to lymphocytes. *Hum. Immunol. 1*, 195-205.

287. Oriol, R., Le Pendu, J. & Mollicone, R. (1986): Genetics of ABO, H, Lewis, X, and related antigens. *Vox Sang. 51*, 161-171.

288. Ørntoft, T. F., Holmes, E. H., Johnson, P., Hakomori, S. & Clausen, H. (1991): Differential tissue expression of the Lewis blood group antigens: enzymatic, immunohistologic, and immunochemical evidence for Lewis a and b antigen expression in Le(a-b-) individuals. *Blood 77*, 1389-1396.

289. Ørntoft, T. F., Vestergaard, E. M., Holmes, E., Jakobsen, J. S., Grunnet, N., Mortensen, M., Johnson, P., Bross, P., Gregersen, N., Skorstengaard, K., Jensen, U. B., Bolund, L. & Wolf, H. (1996): Influence of Lewis α1-3/4-L-fucosyltransferase (FUT3) gene mutations on enzyme activity, erythrocyte phenotyping, and circulating tumor marker sialyl-Lewis a levels. *J. Biol. Chem. 271*, 32260-32268.

290. Oulmouden, A., Wierinckx, A., Petit, J. M., Costache, M., Palcic, M. M., Mollicone, R., Oriol, R. & Julien, R. (1997): Molecular cloning and expression of a bovine α(1,3)-fucosyltransferase gene homologous to a putative ancestor gene of the human FUT3-FUT5-FUT6 cluster. *J. Biol.*

Chem. 272, 8764-8773.

291. PACUSKA, T. & KOSCIELAK, J. (1974): α 1→2 Fucosyltransferase of human bone marrow. *FEBS Lett. 41*, 348-351.

292. PANG, H., KODA, Y., SOEJIMA, M. & KIMURA, H. (1999): Significance of each of three missense mutations, G484A, G667A, and G808A, present in an inactive allele of the human Lewis gene (FUT3) for α(1,3/1,4)fucosyltransferase inactivation. *Glycoconj. J. 15*, 961-967.

293. PANG, H., LIU, Y. H., KODA, Y., SOEJIMA, M., JIA, J. T., SCHLAPHOFF, T., DU TOIT, E. D. & KIMURA, H. (1998): Five novel missense mutations of the Lewis gene (FUT3) in African (Xhosa) and Caucasian populations in South Africa. *Hum. Genet. 102*, 675-680.

294. PARK, M. S., ORIOL, R., NAKATA, S., TERASAKI, P. I., FORD, R. & BERNOCO, D. (1979): ABH and Lewis antigens on lymphocytes: screening of pregnant women's sera with the B-cell cytotoxicity test. *Transplant. Proc. 11*, 1947-1949.

295. PARSONNET, J., FRIEDMAN, G. D., VANDERSTEEN, D. P., CHANG, Y., VOGELMAN, J. H., ORENTREICH, N. & SIBLEY, R. K. (1991): Helicobacter pylori infection and the risk of gastric carcinoma. *N. Engl. J. Med. 325*, 1127-1131.

296. PARSONNET, J., HANSEN, S., RODRIGUEZ, L., GELB, A. B., WARNKE, R. A., JELLUM, E., ORENTREICH, N., VOGELMAN, J. H. & FRIEDMAN, G. D. (1994): Helicobacter pylori infection and gastric lymphoma. *N. Engl. J. Med. 330*, 1267-1271.

297. POMPECKI, R., SHIVELY, J. E. & TODD, C. W. (1981): Demonstration of elevated anti-Lewis antibodies in sera of cancer patients using a carcinoembryonic antigen-polyethylene glycol immunoassay. *Cancer Res. 41*, 1910-1915.

298. POTAPOV, M. (1976): Production of immune-anti-Lewis sera in goats. *Vox Sang. 30*, 211-213.

299. POTAPOV, M. I. (1970): Dectection of the antigen of the Lewis system, characteristic of the erythrocytes of the secretory group Le(a-b-). *Probl. Haematol. (Moskow) 11*, 45-49.

300. PRIEELS, J. P., MONNOM, D., DOLMANS, M., BEYER, T. A. & HILL, R. L. (1981): Co-purification of the Lewis blood group N-acetylglucosaminide α1→4 fucosyltransferase and an N-acetylglucosaminide α1→3 fucosyltransferase from human milk. *J. Biol. Chem. 256*, 10456-10463.

301. PROHASKA, R., SCHENKEL-BRUNNER, H. & TUPPY, H. (1978): Enzymatic synthesis of blood-group Lewis-specific glycolipids. *Eur. J. Biochem. 84*, 161-166.

302. RACE, R. R. & SANGER, R. (1975): The Lewis groups. In: *Blood Groups in Man*. Blackwell Scientific Publications, Oxford, pp. 323-349.

303. RANA, S. S. & MATTA, K. L. (1983): A facile synthesis of 2-acetamido-2-deoxy-4-O-α-L-fucopyranosyl-3-O-β-D-galactopyranosyl-D-glucopyranose, the Lewis a blood-group antigenic determinant, and related compounds. *Carbohydr. Res. 117*, 101-112.

304. REGUIGNE, I., JAMES, M. R., RICHARD, C. W., MOLLICONE, R., SEAWRIGHT, A., LOWE, J. B., ORIOL, R. & COUILLIN, P. (1994): The gene encoding myeloid α-3-fucosyltransferase(FUT4) is located between D11S388 and D11S919 on 11q21. *Cytogenet. Cell Genet. 66*, 104-106.

305. REGUIGNE-ARNOULD, I., COUILLIN, P., MOLLICONE, R., FAURÉ, S., FLETCHER, A., KELLY, R. J., LOWE, J. B. & ORIOL, R. (1995): Relative positions of two clusters of human α-L- fucosyltransferases in 19q (FUT1-FUT2) and 19p (FUT6-FUT3- FUT5) within the microsatellite genetic map of chromosome 19. *Cytogenet. Cell Genet. 71*, 158-162.

306. REGUIGNE-ARNOULD, I., WOLFE, J., HORNIGOLD, N., FAURÉ, S., MOLLICONE, R., ORIOL, R. & COULLIN, P. (1996): Fucosyltransferase genes are dispersed in the genome: FUT7 is located on 9q34.3 distal to D9S1830. *Compt. Rend. Acad. Sci. Ser. III Sci. Vie 319*, 783-788.

307. RODECK, U., HERLYN, M., LEANDER, K., BORLINGHAUS, P. & KOPROWSKI, H. (1987): A mucin containing the X, Y, and H type 2 carbohydrate determinants is shed by carcinoma cells. *Hybridoma 6*, 389-401.

308. ROHR, T. E., SMITH, D. F., ZOPF, D. A. & GINSBURG, V. (1980): Le^b-active glycolipids in human plasma: measurement by radioimmunoassay. *Arch. Biochem. Biophys. 199*, 265-269.

309. ROSEN, S. D. & BERTOZZI, C. R. (1994): The selectins and their ligands. *Curr. Opin. Cell Biol. 6*, 663-673.

310. SAADI, A. T., BLACKWELL, C. C., RAZA, M. W., JAMES, V. S., STEWART, J., ELTON, R. A. & WEIR, D. M. (1993): Factors enhancing adherence of toxigenic Staphylococcus aureus to epithelial cells and their possible role in sudden infant death syndrome. *Epidemiol. Infect. 110*, 507-517.

311. SAADI, A. T., WEIR, D. M., POXTON, I. R., STEWART, J., ESSERY, S. D., BLACKWELL, C. C., RAZA, M. W. & BUSUTTIL, A. (1994): Isolation of an adhesin from Staphylococcus aureus that binds Lewis[a] blood group antigen and its relevance to sudden infant death syndrome. *FEMS Immunol. Med. Microbiol. 8*, 315-320.

312. SAKAMOTO, J., WATANABE, T., TOKUMARU, T., TAKAGI, H., NAKAZATO, H. & LLOYD, K. O. (1989): Expression of Lewis[a], Lewis[b], Lewis[x], Lewis[y], sialyl-Lewis[a], and sialyl-Lewis[x] blood group antigens in human gastric carcinoma and in normal gastric tissue. *Cancer Res. 49*, 745-752.

313. SAKAMOTO, J., YIN, B. W. T. & LLOYD, K. O. (1984): Analysis of the expression of H, Lewis, X, Y, and precursor blood group determinants in saliva and red cells using a panel of mouse monoclonal antibodies. *Mol. Immunol. 21*, 1093-1098.

314. SAKO, D., CHANG, X. J., BARONE, K. M., VACHINO, G., WHITE, H. M., SHAW, G., VELDMAN, G. M., BEAN, K. M., AHERN, T. J., FURIE, B. & ET AL. (1993): Expression cloning of a functional glycoprotein ligand for P-selectin. *Cell 75*, 1179-1186.

315. SARNESTO, A., KOHLIN, T., HINDSGAUL, O., VOGELE, K., BLASZCZYK-THURIN, M. & THURIN, J. (1992): Purification of the β-N-acetylglucosaminide α1→3-fucosyltransferase from human serum. *J. Biol. Chem. 267*, 2745-2752.

316. SASAKI, K., KURATA, K., FUNAYAMA, K., NAGATA, M., WATANABE, E., OHTA, S., HANAI, N. & NISHI, T. (1994): Expression cloning of a novel α1,3-fucosyltransferase that is involved in biosynthesis of the sialyl Lewis x carbohydrate determinants in leukocytes. *J. Biol. Chem. 269*, 14730-14737.

317. SATO, S., ITO, Y., NUKADA, T., NAKAHARA, Y. & OGAWA, T. (1987): Total synthesis of X hapten, III³Fucα-nLc₄Cer. *Carbohydr. Res. 167*, 197-210.

318. SATO, S., ITO, Y. & OGAWA, T. (1986): Stereo- and regio-controlled, total synthesis of the Le[b] antigen, III⁴FucIV²FucLcOse₄Cer. *Carbohydr. Res. 155*, C1-C5.

319. SCHENKEL-BRUNNER, H., CHESTER, M. A. & WATKINS, W. M. (1972): α-L-Fucosyltransferases in human serum from donors of different ABO, secretor, and Lewis blood group phenotypes. *Eur. J. Biochem. 30*, 269-277.

320. SCHENKEL-BRUNNER, H. & HANFLAND, P. (1981): Immunochemistry of the Lewis blood-group system. III. Studies on the molecular basis of the Le[x] property. *Vox Sang. 40*, 358-366.

321. SCUDDER, P. R., SHAILUBHAI, K., DUFFIN, K. L., STREETER, P. R. & JACOB, G. S. (1994): Enzymatic synthesis of a 6'-sulphated sialyl-Lewis[x] which is an inhibitor of L-selectin binding to peripheral addressin. *Glycobiology 4*, 929-932.

322. SEAMAN, M. J., CHALMERS, D. G. & FRANKS, D. (1968): Siedler: an antibody which reacts with A₁Le(a-b+) red cells. *Vox Sang. 15*, 25-30.

323. SHEN, L., GROLLMAN, E. F. & GINSBURG, V. (1968): An enzymatic basis for secretor status and blood group substance specificity in humans. *Proc. Natl. Acad. Sci. USA 59*, 224-230.

324. SHERWOOD, A. L., NGUYEN, A. T., WHITAKER, J. M., MACHER, B. A., STROUD, M. R. & HOLMES, E. H. (1998): Human α 1,3/4-fucosyltransferases - III. A Lys/Arg residue located within the α1,3-FucT motif is required for activity but not substrate binding. *J. Biol. Chem. 273*, 25256-25260.

325. SHIBATA, S., GOLDSTEIN, I. J. & BAKER, D. A. (1982): Isolation and characterization of a Lewis b-active lectin from Griffonia simplicifolia seeds. *J. Biol. Chem. 257*, 9324-9329.

326. SHINODA, K., MORISHITA, Y., SASAKI, K., MATSUDA, Y., TAKAHASHI, I. & NISHI, T. (1997): Enzymatic characterization of human α1,3-fucosyltransferase Fuc-TVII synthesized in a B cell lymphoma cell lines. *J. Biol. Chem. 272*, 31992-31997.

327. SHINOHARA, T. & YAMAMOTO, S. (1983): Partial characterization of blood group Le[a]- and Le[b]-active substances isolated from human urine. *Agric. Biol. Chem. 47*, 141-143.

328. SHINOHARA, T., YAMAMOTO, S. & ISEKI, S. (1977): Some immunochemical properties of Le[a]- and Le[b]-active substances in human urine. *J. Immunogenet. 4*, 159-165.

329. SIDDIQUI, B., WHITEHEAD, J. S. & KIM, Y. S. (1978): Glycosphingolipids in human colonic

adenocarcinoma. *J. Biol. Chem. 253*, 2168-2175.

330. SKACEL, P. O., EDWARDS, A. J., HARRISON, C. T. & WATKINS, W. M. (1991): Enzymic control of the expression of the X determinant (CD15) in human myeloid cells during maturation: the regulatory role of 6'- sialyltransferase. *Blood 78*, 1452-1460.

331. SKACEL, P. O. & WATKINS, W. M. (1987): Fucosyltransferase expression in human platelets and leucocytes. *Glycoconj. J. 4*, 267-272.

332. SLOMIANY, A. & SLOMIANY, B. L. (1975): Blood-group A active difucosyl glycolipid from hog gastric mucosa. *Biochim. Biophys. Acta 388*, 135-145.

333. SLOMIANY, B. L., SLOMIANY, A. & HOROWITZ, M. I. (1975): Characterization of three new fucolipids from hog gastric mucosa. *Eur. J. Biochem. 56*, 353-358.

334. SLOMIANY, B. L., ZDEBSKA, E. & SLOMIANY, A. (1984): Structural characterization of neutral oligosaccharides of human H⁺Le^{b+} gastric mucin. *J. Biol. Chem. 259*, 2863-2869.

335. SMITH, E. L., MCKIBBIN, J. M., BREIMER, M. E., KARLSSON, K. A., PASCHER, I. & SAMUELSSON, B. E. (1975): Identification of a novel heptaglycosylceramide with two fucose residues and a terminal hexosamine. *Biochim. Biophys. Acta 398*, 84-91.

336. SMITH, E. L., MCKIBBIN, J. M., KARLSSON, K. A., PASCHER, I., SAMUELSSON, B. E., LI, Y. T. & LI, S. C. (1975): Characterization of a human intestinal fucolipid with blood group Le^a activity. *J. Biol. Chem. 250*, 6059-6064.

337. SMITH, E. L., MCKIBBIN, J. M., KARLSSON, L. A., PASCHER & SAMUELSSON, B. E. (1975): Main structures of the Forssman glycolipid hapten and a Le^b-like glycolipid of dog small intestine, as revealed by mass spectrometry. Difference in ceramide structure related to tissue localization. *Biochim. Biophys. Acta 388*, 171-179.

338. SMITH, P. L., GERSTEN, K. M., PETRYNIAK, B., KELLY, R. J., ROGERS, C., NATSUKA, Y., ALFORD, J. A., SCHEIDEGGER, E. P., NATSUKA, S. & LOWE, J. B. (1996): Expression of the α(1,3)fucosyltransferase Fuc-TVII in lymphoid aggregate high endothelial venules correlates with expression of L-selectin ligands. *J. Biol. Chem. 271*, 8250-8259.

339. SNEATH, J. S. & SNEATH, P. H. A. (1955): Transformation of the Lewis groups of human red cells. *Nature 156*, 172.

340. SNEATH, J. S. & SNEATH, P. H. A. (1959): Adsorption of blood-group substances from serum on to red cells. *Brit. Med. Bull. 15*, 154-157.

341. SOLTER, D. & KNOWLES (1978): Monoclonal antibody defining a stage-specific mouse embryonic antigen (SSEA-1). *Proc. Natl. Acad. Sci. USA 75*, 5565-5569.

342. SPITALNIK, S., COWLES, J., COX, M. T. & BLUMBERG, N. (1985): Detection of IgG anti-Lewis^a antibodies in cord sera by kintic ELISA. *Vox Sang. 48*, 235-238.

343. SPOHR, U., HINDSGAUL, O. & LEMIEUX, R. U. (1985): Molecular recognition. II. The binding of the Lewis b and Y human blood group determinants by the lectin IV of Griffonia simplicifolia. *Can. J. Chem. 63*, 2644-2652.

344. SPOONCER, E., FUKUDA, M., KLOCK, J. C., OATES, J. E. & DELL, A. (1984): Isolation and characterization of polyfucosylated lactosaminoglycan from human granulocytes. *J. Biol. Chem. 259*, 4792-4801.

345. SRIVATSAN, J., SMITH, D. F. & CUMMINGS, R. D. (1992): The human blood fluke Schistosoma mansoni synthesizes glycoproteins containing the Lewis X antigen. *J. Biol. Chem. 267*, 20196-20203.

346. STAUDACHER, E. (1996): α1,3-Fucosyltransferases. *Trends Glycosci. Glycotechnol. 8*, 391-408.

347. STAUDACHER, E., DALIK, T., WAWRA, P., ALTMANN, F. & MÄRZ, L. (1995): Functional purification and characterization of a GDP-fucose: β-N-acetylglucosamine (Fuc to Asn linked GlcNAc) α1,3-fucosyltransferase from mung beans. *Glycoconj. J. 12*, 780-786.

348. STEEGMAIER, M., LEVINOVITZ, A., ISENMANN, S., BORGES, E., LENTER, M., KOCHER, H. P., KLEUSER, B. & VESTWEBER, D. (1995): The E-selectin-ligand ESL-1 is a variant of a receptor for fibroblast growth factor. *Nature 373*, 615-620.

349. STEPLEWSKI, Z., BLASZCZYK-THURIN, M., LUBECK, M., LOIBNER, H., SCHOLZ, D. & KOPROWSKI, H. (1990): Oligosaccharide Y specific monoclonal antibody and its isotype switch variants.

Hybridoma 9, 201-210.

350. STIGENDAL, L., OLSSON, R., RYDBERG, L. & SAMUELSSON, B. E. (1984): Blood group Lewis phenotype on erythrocytes and in saliva in alcoholic pancreatitis and chronic liver disease. *J. Clin. Pathol. 37*, 778-782.

351. STROUD, M. R., LEVERY, S. B., NUDELMAN, E. D., SALYAN, M. E. K., TOWELL, J. A., ROBERTS, C. E., WATANABE, M. & HAKOMORI, S. I. (1991): Extended type 1 chain glycosphingolipids: dimeric Le^a (III^{4}V^{4}Fuc$_2$Lc$_6$) as human tumor-associated antigen. *J. Biol. Chem. 266*, 8439-8446.

352. STROUD, M. R., LEVERY, S. B., SALYAN, M. E. K., ROBERTS, C. E. & HAKOMORI, S. (1992): Extended type-1 chain glycosphingolipid antigens. Isolation and characterization of trifucosyl-Le^b antigen (III^{4}V^{4}VI^{2}Fuc$_3$Lc$_6$). *Eur. J. Biochem. 203*, 577-586.

353. STURGEON, P. & ARCILLA, M. B. (1970): Studies on the secretion of blood group substances. I. Observations on the red cell phenotype Le(a+b+x+). *Vox Sang. 18*, 301-322.

354. SZULMAN, A. E. & MARCUS, D. M. (1973): The histologic distribution of the blod group substances in man as disclosed by immunofluorescence. VI. The Le^a and Le^b antigens during fetal development. *Lab. Invest. 28*, 565-574.

355. TAKADA, A., OHMORI, K., TAKAHASHI, N., TSUYUOKA, K., YAGO, A., ZENITA, K., HASEGAWA, A. & KANNAGI, R. (1991): Adhesion of human cancer cells to vascular endothelium mediated by a carbohydrate antigen, sialyl Lewis^a. *Biochem. Biophys. Res. Commun. 179*, 713-719.

356. TAKADA, A., OHMORI, K., YONEDA, T., TSUYUOKA, K., HASEGAWA, A., KISO, M. & KANNAGI, R. (1993): Contribution of carbohydrate antigens sialyl Lewis-A and sialyl Lewis-X to adhesion of human cancer cells to vascular endothelium. *Cancer Res. 53*, 354-361.

357. TAKI, T., TAKAMATSU, M., MYOGA, A., TANAKA, K., ANDO, S. & MATSUMOTO, M. (1988): Glycolipids of metastatic tissue in liver from colon cancer: appearance of sialylated Le^x and Le^x lipids. *J. Biochem. 103*, 998-1003.

358. TANAKA, M., DUBE, V. E. & ANDERSON, B. (1984): Structures of oligosaccharides cleaved by base-borohydride from an I, H, and Le^a active ovarian cyst glycoprotein. *Biochim. Biophys. Acta 798*, 283-290.

359. TAYLOR, R. A., RACHKEWICH, R. A., GARE, D. J., FALK, J. A., SHUMAK, K. H. & CROOKSTON, M. C. (1974): Effect of pregnancy of the reactivity of lymphocytes with cytotoxic antisera. *Transplantation 17*, 142-146.

360. TEGOLI, J., CORTEZ, M., JENSEN, L. & MARSH, W. L. (1971): A new antibody, anti-ILe^{bH}, specific for a determinant formed by the combined action of the I, Le, Se, and H gene products. *Vox Sang. 21*, 397-404.

361. TENEBERG, S., MILLER-PODRAZA, H., LAMPERT, H. C., EVANS, D. J., EVANS, D. G., DANIELSSON, D. & KARLSSON, K. A. (1997): Carbohydrate binding specificity of the neutrophil-activating protein of Helicobacter pylori. *J. Biol. Chem. 272*, 19067-19071.

362. TETTEROO, P. A. T., DE HEIJ, H. T., VAN DEN EIJNDEN, D. H., VISSER, F. J., SCHOENMAKER, E. & GEURTS VAN KESSEL, A. H. (1987): A GDP-fucose:[Galβ1→4]GlcNAc α1→3-fucosyltransferase activity is correlated with the presence of human chromosome 11 and the expression of the Le^x, Le^y, and sialyl-Le^x antigens in human-mouse cell hybrids. *J. Biol. Chem. 262*, 15984-15989.

363. TILLEY, C. A., CROOKSTON, M. C., BROWN, B. L. & WHERRETT, J. R. (1975): A and B and A$_1$Le^b substances in glycosphingolipid fractions of human serum. *Vox Sang. 28*, 25-33.

364. TSUBOI, S., ISOGAI, Y., HADA, N., KING, J. K., HINDSGAUL, O. & FUKUDA, M. (1996): 6'-Sulfo sialyl Le^x but not 6-sulfo sialyl Le^x expressed on the cell surface supports L-selectin-mediated adhesion. *J. Biol. Chem. 271*, 27213-27216.

365. VALLI, M., GALLANTI, A., BOZZARO, S. & TRINCHERA, M. (1998): β-1,3-galactosyltransferase and α-1,2-fucosyltransferase involved in the biosynthesis of type-1-chain carbohydrate antigens in human colon adenocarcinoma cell lines. *Eur. J. Biochem. 256*, 494-501.

366. VAN DAM, G. J., BERGWERFF, A. A., THOMAS-OATES, J. E., ROTMANS, J. P., KAMERLING, J. P., VLIEGENTHART, J. F. G. & DEELDER, A. M. (1994): The immunologically reactive O-linked polysaccharide chains derived from circulating cathodic antigen isolated from the human blood fluke Schistosoma mansoni have Lewis^x as repeating unit. *Eur. J. Biochem. 225*, 467-482.

367. VARKI, A. (1994): Selectin ligands. *Proc. Natl. Acad. Sci. USA 91*, 7390-7397.
368. VOS, G. H. & COMLEY, P. (1967): Red cell and saliva studies for the evaluation of ABH and Lewis factors among the Caucasians and Aboriginal populations of Western Australia. *Acta Genet. 17*, 495-510.
369. VOS, G. H., MOORES, P. P., DOWNING, H. J. & MIHIDEEN, A. F. C. (1976): Haemagglutination inhibition studies for the evaluation of blood group antigens in ethanol soluble substances (ESS) obtained from human, baboon, and vervet monkey red blood cells. *Transfusion 16*, 42-47.
370. WAGERS, A. J., LOWE, J. B. & KANSAS, G. S. (1996): An important role for the α 1,3 fucosyltransferase, FucT-VII, in leukocyte adhesion to E-selectin. *Blood 88*, 2125-2132.
371. WANG, J. W., AMBROS, R. A., WEBER, P. B. & ROSANO, T. G. (1995): Fucosyltransferase and α-L-fucosidase activities and fucose levels in normal and malignant endometrial tissue. *Cancer Res. 55*, 3654-3658.
372. WANG, W. T., LUNDGREN, T., LINDH, F., NILSSON, B., GRÖNBERG, G., BROWN, J. P., MENTZER-DIBERT, H. & ZOPF, D. (1992): Isolation of two novel sialyl-Lewis^x-active oligosaccharides by high-performance liquid affinity chromatography using monoclonal antibody Onc-M26. *Arch. Biochem. Biophys. 292*, 433-441.
373. WATKINS, W. M. (1972): Blood group specific substances. In: *Glycoproteins: Their Composition, Structure and Function* (A. Gottschalk, ed). Elsevier, Amsterdam, pp. 830-891.
374. WATKINS, W. M. (1980): Biochemistry and genetics of the ABO, Lewis and P blood group systems. In: *Advances in Human Genetics*. Vol. 10, pp. 1-136.
375. WESTON, B. W., NAIR, R. P., LARSEN, R. D. & LOWE, J. B. (1992): Isolation of a novel human α(1,3)fucosyltransferase gene and molecular comparison to the human Lewis blood group α(1,3/1,4)fucosyltransferase gene. Syntenic, homologous, nonallelic genes encoding enzymes with distinct acceptor substrate specificities. *J. Biol. Chem. 267*, 4152-4160.
376. WESTON, B. W., SMITH, P. L., KELLY, R. J. & LOWE, J. B. (1992): Molecular cloning of afourth member of a human α(1,3)fucosyltransferase gene family. Multiple homologous sequences that determine expression of the Lewis^x, sialyl Lewis^x, and difucosyl sialyl Lewis^x epitopes. *J. Biol. Chem. 267*, 24575-24584.
377. WU, J. T. & CHANG, J. (1992): Chromatographic characterization of CA-19-9 molecules from cystic fibrosis and pancreatic carcinoma. *J. Clin. Lab. Anal. 6*, 209-215.
378. WU, J. T., OLSON, J. & WALKER, K. (1992): Tumor markers CA 19-9 and CA 195 are also useful markers for cystic fibrosis. *J. Clin. Lab. Anal. 6*, 151-161.
379. WYATT, J. I. & DIXON, M. F. (1988): Chronic gastritis - a pathogenetic approach. *J. Pathol. 154*, 113-124.
380. XU, X. H., VO, L. & MACHER, B. A. (1996): Structure-function analysis of human α1,3-fucosyltransferase. Amino acids involved in acceptor substrate specificity. *J. Biol. Chem. 271*, 8818-8823.
381. YAGO, K., ZENITA, K., GINYA, H., SAWADA, M., OHMORI, K., OKUMA, M., KANNAGI R. & LOWE, J. B. (1993): Expression of α(1,3)-fucosyltransferases which synthesize sialyl Le^x and sialyl Le^a, the carbohydrate ligands for E-selectin and P-selectin, in human malignant cell lines. *Cancer Res. 53*, 5559-5565.
382. YAGO, T., TSUKUDA, M., YAMAZAKI, H., NISHI, T., AMANO, T. & MINAMI, M. (1995): Analysis of an initial step of T cell adhesion to endothelial monolayers under flow conditions. *J. Immunol. 154*, 1216-1222.
383. YAMAMOTO, S. (1982): Inhibitory activities of substances present in plant seeds and fruits against anti-Lewis agglutinins. *J. Immunogenet. 9*, 137-141.
384. YAZAWA, S. & FURUKAWA, K. (1980): α-L-Fucosyltransferases related to biosynthesis of blood group substances in human saliva. *J. Immunogenet. 7*, 137-148.
385. YAZAWA, S., MADIYALAKAN, R., IZAWA, H., ASAO, T., FURUKAWA, K. & MATTA, K. L. (1988): Cancer-associated elevation of α(1→3)-L-fucosyltransferase activity in human serum. *Cancer 62*, 516-520.
386. YAZAWA, S., NISHIHARA, S., IWASAKI, H., ASAO, T., NAGAMACHI, Y., MATTA, K. L. & NARIMATSU, H.

(1995): Genetic and enzymatic evidence for Lewis enzyme expression in Lewis-negative cancer patients. *Cancer Res. 55*, 1473-1478.

387. YOUNG, W. W., JOHNSON, H. S., TAMURA, Y., KARLSSON, K. A., LARSON, G., PARKER, J. M. R., KHARE, D. P., SPOHR, U., BAKER, D. A., HINDSGAUL, O. & LEMIEUX, R. U. (1983): Characterization of monoclonal antibodies specific for the Lewis[a] human blood group determinant. *J. Biol. Chem. 258*, 4890-4894.

388. YUEN, C. T., BEZOUSKA, K., O'BRIEN, J., STOLL, M., LEMOINE, R., LUBINEAU, A., KISO, M., HASEGAWA, A., BOCKOVICH, N. J., NICOLAOU, K. C. & FEIZI, T. (1994): Sulfated blood group Lewis[a] .A superior oligosaccharide ligand for human E-selectin. *J. Biol. Chem. 269*, 1595-1598.

389. YUEN, C. T., LAWSON, A. M., CHAI, W. G., LARKIN, M., STOLL, M. S., STUART, A. C., SULLIVAN, F. X., AHERN, T. J. & FEIZI, T. (1992): Novel sulfated ligands for the cell adhesion molecule E-selectin revealed by the neoglycolipid technology among O- linked oligosaccharides on an ovarian cystadenoma glycoprotein. *Biochemistry 31*, 9126-9131.

390. ZOLLNER, O. & VESTWEBER, D. (1996): The E-selectin ligand-1 is selectively activated in Chinese hamster ovary cells by the alpha(1,3)-fucosyltransferases IV and VII. *J. Biol. Chem. 271*, 33002-33008.

391. ZOPF, D. A., GINSBURG, A. & GINSBURG, V. (1975): Goat antibody directed against a human Le[b] blood group hapten, lacto-N-difucohexaose I. *J. Immunol. 115*, 1525-1529.

7 Antigens I and i

The antigens I and i[1] [144] are two genetically independent characters which are closely interrelated with each other chemically. Erythrocytes of adults normally show I activity; only in very few cases (in about 0.02% of Europids) is this antigen absent and the cells show i activity. The erythrocytes of newborns are always i specific.

These antigenic characters are not confined to erythrocytes, they are found on other blood and tissue cells as well as in various secretions and body fluids. I and i determinants have been detected throughout the whole animal kingdom [143].

The antigens I and i represent important receptors for various autoantibodies present in patients suffering from cold agglutinin syndrome. Interest in these antigens has increased with their significance as chemical markers for ontogenesis, onco-genesis, and cell differentiation in human and animal tissues.

7.1 Antisera

Most of the anti-I and anti-i reagents are cold agglutinins which react better at 4°C than at body temperature [115,118]. Though the majority belong to the IgM type, some IgG antibodies [13] and mixtures of IgG and IgM [59] have also been described.

The anti-I agglutinins found in most human sera are normally very weak (i.e. titer 64 and lower), and, because they do not react at body temperature, they are completely harmless [115]. Anti-I of higher titers are often found in connection with *Mycoplasma pneumoniae* infections [7,39], while anti-i is found in patients suffering from infectious mononucleosis [13,73,120]. Great quantities of anti-I and, occasionally, anti-i antibodies (with titers frequently much higher than 1000) are produced in patients with chronic cold agglutinin syndrome [116]; in these cases the antibodies are usually monoclonal and thus strictly monospecific. The serum of I-negative individuals, i.e. i subjects [19,74], or of newborns [1] often contains (polyclonal) allo-anti-I.

A series of 'compound-specific' antibodies have been described. These antibodies bind to epitopes composed of two different antigenic determinants and often occur

---

[1] ISBT Terminology (see *Chap. 1*): I = I1 or 207 001, i = I2 or 207 002.

together with 'normal' anti-I and anti-i [71]. The most frequent anti-**IH** antibodies agglutinate only I-positive erythrocytes exhibiting **H** activity (i.e. cells from **O** or **A₂** donors), and are inhibited by saliva of **ABH**-secretors of blood group **O** [88]. Further reports of compound-specific antibody types comprise anti-**IA** [7], anti-**IA₁** [133], anti-**IB** [30,131], anti-**I(A+B)** [29], and anti-**IP₁** [10,70].

Naturally occurring anti-I antibodies have been detected in the sera of cattle, sheep, and various marsupials [23]. Most samples of anti-**H** occurring in eel serum show anti-**HI** specificity [14]. Anti-I specific cold agglutinins [22,87] are produced by rabbits immunised either with heat-killed *Listeria monocytogenes* or *Mycoplasma pneumoniae* cells, or with erythrocytes coated with extracts of these bacteria.

More recently, hybridoma antibodies with anti-I(Ma) specificity (*LICR-LON-M18* and *LICR-LON-M39*, also referred to as *M18* and *M39* [60]) and anti-i specificity (*MCC-1004* [66], *GL-1* and *GL-2* [103]) have been prepared.

The different anti-I sera are extremely heterogeneous with regard to their serological specificities. Several groups [11,35,99] have suggested a classification of anti-I's on the basis of their reaction with erythrocytes of newborns (F = foetal, thus anti-I^F), of infants in transition stage from i to I (T = transient, thus anti-I^T), and of adults (D = developed, thus anti-I^D), or with water-soluble I substances (S = soluble, thus anti-I^S). A further criterion proposed by Feizi and Kabat [46] for the classification of the anti-I's involves the capacity of the antibodies to bind to defined I active glycoproteins. A more exact classification of anti-I sera is based on their receptor specificity and will be discussed in the following chapter.

It should be noted that the heavy chains of almost all pathogenic and most naturally occurring anti-I and anti-i antibodies are encoded by the $V_H4.21$ or a very closely related V_H4 heavy chain gene, whereas the light chains are preferentially κ chains encoded by different germline genes (see e.g. [72,108,123,126,127]).

7.2 Blood Group I and i Determinants

First evidence that carbohydrate structures determine I specificity was obtained by the group of Kabat [95]. The authors showed that I activity of erythrocytes is destroyed upon incubation with a mixture of β-galactosidase and β-N-acetylglucosaminidase. Analyses of I active substances from gastric mucosa and milk have in fact revealed a high content of galactose and N-acetylglucosamine. The first I structure

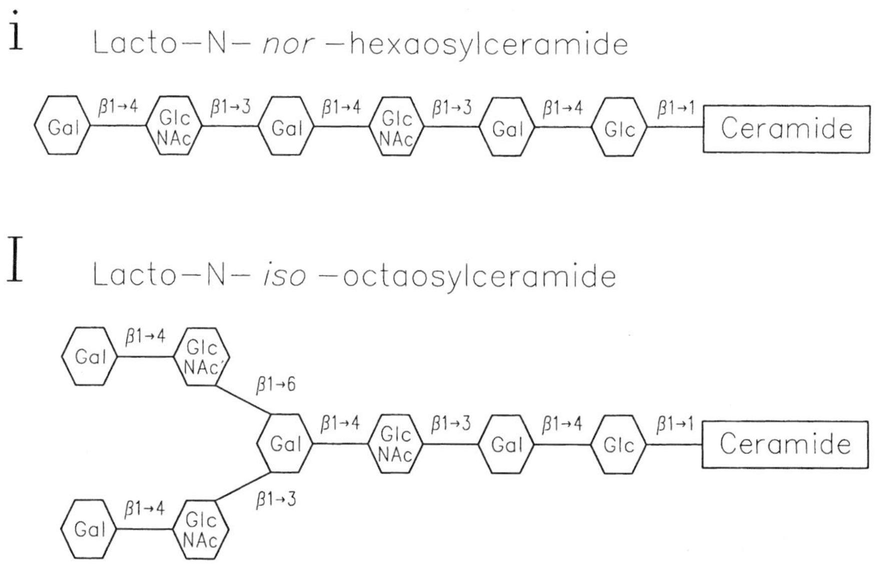

i Lacto−N− *nor* −hexaosylceramide

I Lacto−N− *iso* −octaosylceramide

Figure 7.1: Structures of the most basic blood group i and I active glycosphingolipids.

established was the disaccharide unit, **Galβ1→4GlcNAcβ1→6Gal**, which is the binding site for the type-1 antiserum, anti-I(Ma) [48]. Strong I activity exhibited by complex glycosphingolipids [142] and polyglycosyl-ceramides [57,80] of the erythrocyte membrane was found later (see *Sect. 5.3.2*).

In subsequent studies I and i active glycolipids isolated from cattle [141] and rabbit erythrocytes [38,63,64], and their degradation products, were used for closer characterisation of blood group I and i determinants. These investigations revealed the location of blood group I epitopes on highly branched oligosaccharide chains composed of galactose and N-acetylglucosamine. The simplest I active glyco-sphingolipid − i.e. the one which reacts with all anti-I sera − is lacto-N-*iso*-octaosyl-ceramide [141] (see *Fig. 7.1*). The i active structure is represented by linear, unbranched carbohydrate chains built of repeating N-acetyllactosamine units; the simplest i active glycosphingolipid has been identified as lacto-N-*nor*-hexaosyl-ceramide *(Fig. 7.1)* [104].

The variations in serological specificity among the anti-Is are due to the fact that the different anti-I antibodies may bind to different epitopes of the highly complex and

Water

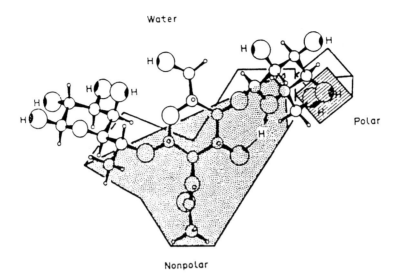

Polar

Nonpolar

Figure 7.2: Molecular model of the I(Ma) determinant structure.
The diagram presents the preferred conformation of the disaccharide structure. The antibody binds to the hydrophobic region (shaded) of the epitope.
Reproduced from Lemieux et al. [83] by permission of Pergamon Press Ltd., Oxford, UK (©1984).

variable carbohydrate units [56,61,141]. Two basic types of anti-I have been defined [61] on the basis of their towards partially degraded I-active glycolipids and oligosaccharides:

Type 1: This group of antibodies binds to the terminal disaccharide structure **Galβ1→4GlcNAcβ1→6(Gal)**- as represented by the short carbohydrate side chains attached to the poly-N-acetyllactosamine units [48]. Examples of anti-I type-1 antibodies are anti-I(Ma) [48], anti-I(Woj) [34], anti-I(Hy) [31], and the hybridoma antibodies *M18* and *M39* [60,61]. An exact characterisation of the anti-I(Ma) determinant structure revealed that the I(Ma) epitope comprises the **Gal–GlcNAc** disaccharide unit and the -OCH$_2$ group of the galactose residue at the branching point [48,76]. The three-dimensional structure of the hapten and the binding site of the I(Ma) antibody have been investigated by the groups of Kabat and Lemieux [75,76,83] (see *Fig. 7.2*).

Type 2: Anti-I's of this type react with complex, branched oligosaccharide structures. Some antibodies bind to a segment of a linear N-acetyllactosamine unit, i.e. $(\rightarrow 3Gal\beta 1\rightarrow 4GlcNAc\beta 1\rightarrow 3)_n$, which must carry a carbohydrate side chain for optimum serological activity; examples of antibodies belonging to this group are the anti-I sera *Step*, *Ver*, *Gra*, and *Ful*. Other antibodies of this type (e.g. the anti-I sera *Sch*, *Low*, *Da*, and *Phi*) bind directly to the branching sites of the I-active chains (see [42]).

Feizi and coworkers [45,130] report that preliminary investigations on anti-i sera suggest two types of anti-i antibodies: one group, represented by anti-i(Den), binds to the non-reducing end of the carbohydrate unit requiring a terminal $\beta 1\rightarrow 4$ galactose residue ('cavity type'): a second group, represented by anti-i(Tho), is able to bind to an inner segment of the polylactosamino chain ('groove type').

Further variability among the anti-I and anti-i sera is manifested mainly by the influence of substituents on the reactivity of the antibodies. Terminal $\alpha 2\rightarrow 3$ N-acetylneuraminic acid or $\alpha 1\rightarrow 3$ galactose residues effect in most cases partial or even complete masking of Ii determinants. Most substances carrying $\alpha 1\rightarrow 2$ fucose exhibit only reduced I and i activities; this finding is in accordance with the increased I and i reactivities of **'Bombay'** erythrocytes [33,91], and the fact that secretions of **ABH**-nonsecretors react better with anti-I sera than secretions of **ABH**-secretors [110] (see *Chap. 5*).

Additional differences in antibody specificity are based on the size of the antigenic binding site, which may range from a disaccharide to a hexa- [62] or heptasaccharide [61].

7.3 Blood Group I and i Substances

7.3.1 Blood Group I and i Substances of Erythrocytes

The blood group I and i determinants of human erythrocytes are located mainly on neutral glycosphingolipids, especially polyglycosyl-ceramides (see *Fig. 7.3*, refer also to *Sect. 5.3.2*). The carbohydrate chains of these substances are generally substituted with sialyl- and fucosyl residues [41,106].

Ii determinants have also been detected on polyglycosyl-peptides of the erythrocyte membrane [37] – the band-3 protein in particular (see *Sect. 4.1.2)* shows distinct I specificity [17,54]. It is widely agreed, however, that membrane glycoproteins contribute only little to the I and i activity of erythrocytes [107,148].

Figure 7.3: Structure of a blood group I active polyglycosylceramide.
Model proposed by Koscielak et al. [80].

Studies on the carbohydrate moieties of polyglycosyl-ceramides isolated from i erythrocytes revealed that these glycosphingolipids carry short and virtually unbranched oligosaccharide chains with an average length of 15 monosaccharide units, whereas polyglycosyl-ceramides isolated from I cells contain highly complex carbohydrate chains with 20 to 59 sugar residues and an average of five oligosaccharide side chains [81,148]. An analogous result was obtained for polyglycosyl-peptides – the band-3 protein obtained from the erythrocytes of I individuals carries complex carbohydrate units with approximately nine side branches, whereas band-3 protein of i subjects carries only short and linear poly-N-acetyl-lactosamine units [54,148] (see also *Sect. 7.4*).

Blood group I and i specific structures have also been detected on animal erythrocytes [143]: red cells of Cynamolgus monkeys show high i activity [73] and, as mentioned above, α-galactosyl and α-N-acetylneuraminyl derivatives of I and i active glycolipids have been isolated from cattle [141] and rabbit [25,26,63] erythrocytes.

7.3.2 Blood Group I and i Determinants on Other Cell Types

Thrombocytes [32], nucleated blood cells, as well as fibroblasts and other human or animal cells in tissue culture show blood group I and i activity [18,51,114,125,134]. In lymphocytes of adults and newborns the I and i determinants are localised on N-linked carbohydrate chains of high-molecular weight glycoproteins; the carrier molecules of these oligosaccharide units are different on T and B cells [16].

7.3.3 Water-Soluble Blood Group I and i Substances

I active substances have been detected in milk [47,98], saliva [36], gastric juice [110], amniotic fluid [20], and ovarian cyst fluid [49]. The expression of I in body secretions is independent of the I/i phenotype of the erythrocytes, and I activity also occurs in secretions of i individuals [36,97] and newborns [36].

A glycoprotein with strong i activity has been found in the serum of human blood group I individuals [21,28]; investigations performed by the group of Feizi [40] suggest that these i determinants are probably carried by orosomucoid (α_1-acid glycoprotein).

Water-soluble blood group I substances from non-human sources are milk and saliva of Rhesus monkeys [15], sheep gastric mucin [68,146], hydatid cyst fluid of sheep liver (when the cysts contain living proto-scolices of the tapeworm, *Echinococcus granulosus*) [36,47], and the AI active glycoprotein from the body fluid of the round worm, *Ascaris suum* [24].

All these water-soluble substances are glycoproteins containing oligosaccharide chains O-glycosidically linked to serine or threonine residues of the protein backbone [48]. The carbohydrate units normally carry fucose and/or neuraminic acid and show expectedly high variability in their reactions with different antisera.

7.3.4 Chemical Synthesis of I and i Determinant Structures

Linear and branched oligosaccharides have been prepared by chemical synthesis (e.g. [3,4,27,94,100,101]) and repeatedly used for characterisation of anti-I and anti-i sera (see [61,145]).

7.3.5 Conformation of i Determinant Structures

Theoretical calculations on the three-dimensional structure of an octameric poly-N-acetyllactosamine unit (**GlcNAcβ1→4Galβ1→3**)$_8$ indicated a high degree of flexibility in the carbohydrate chain – the low-energy conformations range from extended structures to spring-like coils ([117], *Fig. 7.4).*

7.4 Expression of I and i Activity during Embryonic Development and under Various Pathological Conditions

First investigations on mouse embryos found the I/i activity of cells changed in the course of embryonic development. I antigen was expressed in undifferentiated cells. The appearance of i was connected with the formation of the primary entoderm, the increase in the i antigen being associated with a decrease in I [77]. Other studies

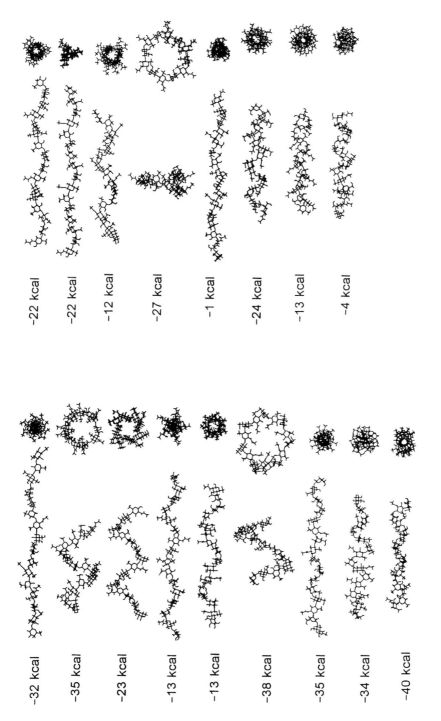

−32 kcal

−35 kcal

−23 kcal

−13 kcal

−13 kcal

−38 kcal

−35 kcal

−34 kcal

−40 kcal

−22 kcal

−22 kcal

−12 kcal

−27 kcal

−1 kcal

−24 kcal

−13 kcal

−4 kcal

Figure 7.4: Lowest energy conformations of the blood group i-active oligosaccharide structure (Galβ1→4GlcNAcβ1→3)₈.
Reproduced from Renouf & Hounsell [117] with permission from the authors and from Elsevier Science.

suggest that high expression of i may be a characteristics of dividing and differentiating cells from adults [132].

Further, the expression of I and i specificities on human erythrocytes changes in the course of embryonic development: fetal erythrocytes and red cells of newborns show strong i activity which gradually changes to I for I individuals; about 18 months after birth the I status of adults is reached [96]. However, body secretions of foetuses always exhibit I specificity [50].

Investigations on glycosphingolipids and glycoproteins of foetal erythrocytes revealed carbohydrate unit structures similar to those found in adult blood group i individuals (see above), the chains being short and virtually without branches [54,81]. Among the i active glycolipids, lacto-N-*nor*-hexaosylceramide (most of it sialylated) has been found in high concentration in foetal erythrocytes, whereas this same substance is found only in minute amounts in the cells of adult I individuals [106]. The oligosaccharide structures of the Ii-active band 3 proteins of erythrocytes from newborns and I adults [52,53] are presented in *Fig. 7.5*.

A considerable increase in i activity of the erythrocytes has been observed in a series of blood diseases, such as thalassaemia, hypoblastic anaemia, in some forms of acute leukaemia [58], and in sickle cell anaemia [6,93]; in the majority of these cases, however, I activity remains more or less unchanged [44]. Lymphocytes of patients suffering from chronic lymphocytic leukaemia often show reduced i activity [125]. A significant increase in I and i activity has also been observed in transformed tissue cells (e.g. [78,109,110]).

It must be assumed, however, that these changes are normally due to incomplete synthesis of oligosaccharide chains in transformed cells, and/or to disorders in membrane architecture, which influence the accessibility of the determinants. Thus the I/i antigens cannot be used as indicators for a specific pathological condition.

As discussed above a transient production of high titer anti-I antibodies following infection with *Mycoplasma pneumoniae* has been frequently observed. Subsequent investigations have identified the receptors for the infectious agent: $\alpha 2 \rightarrow 3$ sialylated I and i structures carried by glycoproteins and glycolipids [43,89,90]. It has therefore been proposed that the complexes formed between the oligosaccharide structures of the host and the lipid-rich *Mycoplasma* may serve as an adjuvant triggering antibody formation in the inflamed areas of the respiratory tissues.

Sialylated I and i structures have also been identified as receptors for Sendai [128] and Influenza virus [129]; it has recently been reported that the latter elicits cold agglutinins of anti-I specificity [139].

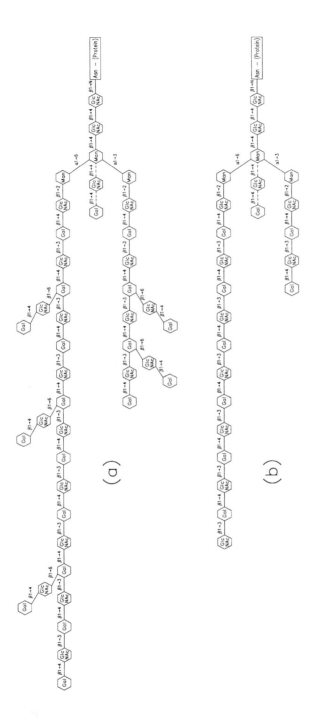

Figure 7.5: Basic structure of lactosaminoglycan chains N-glycosidically linked to band 3 protein of human erythrocyte membranes.
(a) from the erythrocytes of adults (blood group I) [53],
(b) from foetal erythrocytes [52].

Poly-N-acetyllactosamines are often modified to express differentiation antigens and functional oligosaccharides. Among those oligosaccharides, sialoyl-Le^x and its sulfated forms are ligands for E-, P-, and L-selectins [69,102,135]. Thus, branched polylactosamines on animal cell surfaces contribute to multivalent interactions in cell adhesion and cell signalling.

7.5 Biosynthesis of I and i Determinant Structures

The investigation on the chemical nature of Ii antigens showed that I and i determinant structures are common constituents of animal cell membranes. Consequently, it is generally accepted that the glycosyltransferases responsible for their biosynthesis are widely distributed in the animal kingdom.

The first I determinant to be synthesised *in vitro* was the I(Ma) specific structure [122]. De-fucosylated blood group H substance of hog gastric mucosa shows a distinct reaction with anti-I(Ma) sera, which is destroyed by treatment with β-galactosidase. The I(Ma) activity is restored upon transfer of galactosyl residues to this precursor substance by a galactosyltransferase isolated from human milk (*Fig. 7.6*). Enzyme/substrate competition experiments suggested that this enzyme is identical with the lactose-synthase. In the mammary gland, in the presence of α-lactalbumin, this enzyme is responsible for the formation of the disaccharide lactose; in other tissues, however, it is involved in the biosynthesis of glycoproteins and glycolipids.

More recent studies have investigated the biosynthesis of linear and branched oligosaccharide units (see also *Sect. 3.1*):

Linear oligosaccharide structures are formed by alternating action of β1,4-galactosyl- and β1,3-N-acetyl-glucosaminyltransferase.

β1,3-N-Acetylglucosaminyltransferases with respective properties have been found in serum and several human and animal tissues (e.g. [5,62,67,112,137,147]). Piller et al. [111] were able to synthesise blood group i active structures using these enzymes isolated from human serum.

The cDNA encoding a β1,3-N-acetylglucosaminyltransferase has been isolated by Sasaki et al. [121]. Nucleotide sequencing revealed a transmembrane type II protein of 415 amino acids (*Fig. 7.7*). Functional expression of this cDNA clearly showed that the encoded enzyme is essential for the formation of poly-N-acetyllactosamine chains. The transcript has been found ubiquitously expressed in adult tissues.

Zhou et al. [149] have succeeded in cloning another β1,3-N-acetylglucos-aminyltransferase, which showed a marked preference for **Galβ1→4Glc(NAc)**-based

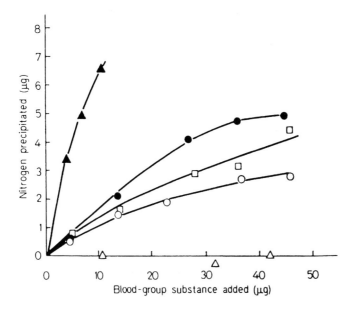

Figure 7.6: *In vitro* synthesis of the I(Ma) determinant using the β1→4 galactosyl-transferase from human milk.
Quantitative immunoprecipitation with anti-I(Ma) serum.
O : Blood group H substance from hog gastric mucosa.
□ : Defucosylated H substance (= 'I(Ma) substance').
Δ : 'I(Ma) substance' treated with β-galactosidase (= 'I(Ma) precursor').
● : 'I(Ma) precursor' after treatment with β1,4-galactosyltransferase and UDP-galactose.
▲ : Defined glycoprotein fraction (termed 'OG 20% from 10%' [50]) as a standard I substance.
Reproduced from Schenkel-Brunner et al. [122] by permission of Springer-Verlag, Heidelberg.

acceptors. The enzyme was found capable of both initiating and elongating poly-N-acetyllactosamine chains. Its primary structure, however, showed no similarity with the i-transferase described by Sasaki et al., but shares several amino acid motifs characteristic for β1,3-galactosyltransferases. In this case as well, the gene transcripts were detected in all human tissues examined (heart, brain, spleen, lung, liver, skeletal muscle, kidney, and testis).

Thus far four different β1,4-galactosyltransferases have been cloned [2]. Recently it has been demonstrated that β1,4-galactosyltransferase I (the lactose synthase, *Fig. 7.8)* acts most efficiently in adding galactose to linear and branched poly-N-acetyllactosamines [136].

```
  1 MQMSYAIRCA FYQLLLAALM LVAMLQLLYL SLLSGLHGQE EQDQYFEFFP PSPRSVDQVK   60

 61 AQLRTALASG GVLDASGDYR VYRGLLKTTM DPNDVILATH ASVDNLLHLS GLLERWEGPL  120

121 SVSVFAATKE EAQLATVLAY ALSSHCPDMR ARVAMHLVCP SRYEAAVPDP REPGEFALLR  180

181 SCQEVFDKLA RVAQPGINYA LGTNVSYPNN LLRNLAREGA NYALVIDVDM VPSEGLWRGL  240

241 REMLDQSNQW GGTALVVPAF EIRRARRMPM NKNELVQLYQ VGEVRPFYYG LCTPCQAPTN  300

301 YSRWVNLPEE SLLRPAYVVP WQDPWEPFYV AGGKVPTFDE RFRQYGFNRI SQACELHVAG  360

361 FDFEVLNEGF LVHKGFKEAL KFHPQKEAEN QHNKILYRQF KQELKAKYPN SPRRC
```

Figure 7.7: Amino acid sequence of human β1,3-N-acetylglucosaminyltransferase.
Underlined: membrane-spanning region, *N* : N-glycosylation sites.
According to Sasaki et al. [121]. The sequence data are deposited in the EMBL/GenBank data libraries (accession number AF029893).

```
  1 MRLREPLLSG AAMPGASLQR ACRLLVAVCA LHLGVTLVYY LAGRDLSRLP QLVGVSTPLQ   60

 61 GGSNSAAAIG QSSGELRTGG ARPPPPLGAS SQPRPGGDSS PVVDSGPGPA SNLTSVPVPH  120

121 TTALSLPACP EESPLLVGPM LIEFNMPVDL ELVAKQNPNV KMGGRYAPRD CVSPHKVAII  180

181 IPFRNRQEHL KYWLYYLHPV LQRQQLDYGI YGIYVINQAG DTIFNRAKLL NVGFQEALKD  240

241 YDYTCFVFSD VDLIPMNDHN AYRCFSQPRH ISVAMDKFGF SLPYVQYFGG VSALSKQQFL  300

301 TINGFPNNYW GWGGEDDDIF NRLVFRGMSI SRPNAVVGRC RMIRHSRDKK NEPNPQRFDR  360

361 IAHTKETMLS DGLNSLTYQV LDVQRYPLYT QITVDIGTPS
```

Figure 7.8: Amino acid sequence of human β1,4-galactosyltransferase I.
Underlined: membrane-spanning region, *N* : potential N-glycosylation site, *T* : N-terminus of the soluble form of the enzyme.
According to Masri et al. [Masri, 1988 #4152]. The sequence data are deposited in the EMBL/GenBank data libraries (accession number X14085)

The branched oligosaccharide structures are formed by a transferase which attaches N-acetylglucosaminyl residues to carbon 6 of galactose. These so-called 'branching enzymes' have been detected in serum and various tissue cells of human and mammalian origin (e.g. [5,12,62,65,79,85,113,119,138,150]).

More recent investigations revealed two types of β1,6-N-acetylglucosaminyl-transferases:

- the so-called 'distally acting type' transfers a β1,6-GlcNAc unit to the penultimate galactose residue at the growing end of a linear poly-N-acetyllactosamine chain [12,62,65,79,82,113,119];
- the 'centrally acting type' transfers β1,6-GlcNAc units to midchain galactoses of a linear (Galβ1→4GlcNAcβ1→3)$_n$ chain [62,85,92,105].

Investigations on the substrate specificity of a 'distally acting type' of β1,6-N-acetylglucosaminyltransferase isolated from hog gastric mucosa showed a strict requirement for terminal GlcNAcβ1→3Gal units; here, addition of a β1→4 galactose residue blocked the transfer reaction [12,113]. This finding indicated that in this case branching of lactosaminoglycan structures occurs exclusively during the chain elongation process: the transfer of N-acetylglucosamine to carbon 3 of a terminal galactose residue enables the 'branching enzyme' to add a second N-acetyl-glucosamine to the galactose at position 6. The subsequent galactosylation of the two N-acetylglucosamine residues also proceeds in a fairly ordered sequence, the β1→6 linked N-acetylglucosamine being galactosylated more easily than the β1→3 linked residue [9] (see *Fig. 7.9)*. Vilkman et al. [140] in investigating the subsequent step in chain elongation have shown that the β1,3-N-acetylglucosaminyltransferase of human serum acts equally well at both branches of the hexasaccharide LacNAcβ1→3' (LacNAcβ1→6')LacNAc. Thus, the pathway involving the 'distally acting' transferase is supposed to form not only poly-lactosaminoglycan units with short side chains but also fairly complex, highly branched structures [124].

In the case of the 'centrally acting type' a linear carbohydrate chain is formed by the alternating action of β1,3-N-acetylglucosaminyltransferase and β1,4-galactosyl-transferase. Elongated to its final size, N-acetylglucosamine side branches are attached to the oligosaccharide backbone by this type of β1,6-N-acetylglucosaminyl-transferase at different sites along the chain. In a subsequent step the N-acetyl-glucosaminyl branches are galactosylated.

According to Leppänen et al. [86] this pathway should be more likely to produce poly-lactosaminoglycan units with fairly uniform LacNAcβ1→6 side-branches along the entire primary oligosaccharide backbone (as found for example in human embryonal carcinoma cells [55] and erythrocyte band 3 protein from adults [53]).

The branch-forming action of the enzyme was completely inhibited at sites in the immediate neighbourhood of α1,3-fucosylated N-acetylglucosamine residues [84].

A cDNA encoding a β1,6-N-acetylglucosaminyltransferase shown to convert linear into branched poly-N-acetyllactosamine structures has been isolated from human teratocarcinoma cells [8]. Nucleotide sequencing of the cloned cDNA revealed an open

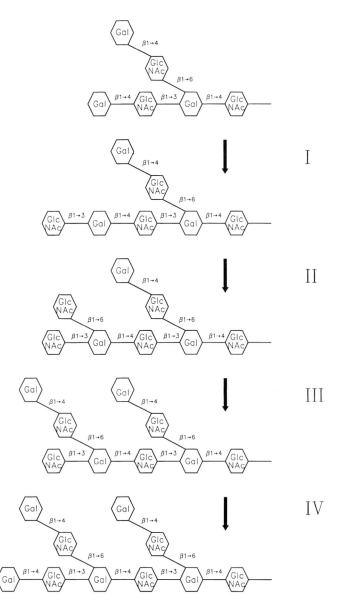

Figure 7.9: Proposed pathway for the biosynthesis of branched lactosaminoglycan structures by the 'distally acting type' of β1,6-N-acetylglucosaminyltransferase.
According to Piller et al. [113].

1 MPLSMR<u>YLFI ISVSSVIIFI VFSVF</u>NFGGD PSFQRL*N*ISD PLRLTQVCTS FINGKTRFLW 60

61 KNKLMIHEKS SCKEYLTQSH YITAPLSKEE ADFPLAYIMV IHHHFDTFAR LFRAIYMPQN 120

121 IYCVHVDEKA TTEFKDAVEQ LLSCFPNAFL ASKMEPVVYG GISRLQADLN CIRDLSAFEV 180

181 SWKYVINTCG QDFPLKTNKE IVQYLKGFKG K*N*ITPGVLPP AHAIGRTKYV HQEHLGKELS 240

241 YVIRTTALKP PPPH*N*LTIYF GSAYVALSRE FANFVLHDPR AVDLLQWSKD TFSPDEHFWV 300

301 TLNRIPGVPG SMP*N*ASWTGN LRAIKWSDME DRHGGCHGHY VHGICIYGNG DLKWLVNSPS 360

361 LFANKFELNT YPLTVECLEL RHRERTL*N*QS ETAIQPSWYF

Figure 7.10: Amino acid sequence of human β1,6-N-acetylglucosaminyltransferase.
Underlined: membrane-spanning region, *N* : N-glycosylation sites.
According to Bierhuizen et al. [8]. The sequence data are deposited in the EMBL/GenBank data libraries (accession number Z19550).

reading frame encoding a polypeptide of 400 amino acids (*Fig. 7.10*). Hydropathy analysis of the deduced protein predicted a membrane topology common to all mammalian glycosyltransferases thus far investigated: the transferase consists of a short cytoplasmic sequence of 6 amino acids, a hydrophobic transmembrane domain of 19 amino acids and a large C-terminal catalytic domain. Transfection experiments proved that the enzyme is able to synthesise I specific structures.

The gene encoding this 'branching enzyme' has been located on chromosome 9, band q21.

References

1. ADINOLFI, M. (1965): Anti-i antibodies in normal human newborn infants. *Immunology 9*, 43-52.
2. ALMEIDA, R., AMADO, M., DAVID, L., LEVERY, S. B., HOLMES, E. H., MERKX, G., VAN KESSEL, A. G., RYGAARD, E., HASSAN, H., BENNETT, E. & CLAUSEN, H. (1997): A family of human β4-galactosyl-transferases. Cloning and expression of two novel UDP-galactose: β-N-acetylglucosamine β1,4-galactosyltransferases, β4Gal-T2 and β4Gal-T3. *J. Biol. Chem. 272*, 31979-31991.
3. AUGÉ, C., DAVID, S. & VEYRIÈRES, A. (1977): Synthesis of a branched pentasaccharide: one of the core oligosaccharides of human blood group substances. *Chem. Commun. 132*, 449-450.
4. AUGÉ, C., MATHIEU, C. & MÉRIENNE, C. (1986): The use of an immobilized cyclic multi-enzyme system to synthesize branched penta- and hexasaccharides associated with blood-group I epitopes. *Carbohydr. Res. 151*, 147-156.
5. BASU, M. & BASU, S. (1984): Biosynthesis in vitro of Ii core glycosphingolipids from neolactotetraosylceramide by β1-3- and β1-6-N-acetylglucosaminyltransferases from mouse T-lymphoma. *J. Biol. Chem. 259*, 12557-12562.
6. BASU, M. K., LEE, M. M., MANIATIS, A. & BERTLES, J. F. (1984): Characteristics of I and i antigen receptors on the membrane of erythrocytes in sickle cell anemia. *J. Lab. Clin. Med. 103*, 712-719.

7. BELL, C. A., ZWICKER, H. & ROSENBAUM, D. L. (1973): Paroxysmal cold hemoglobinuria (P.C.H.) following mycoplasma infection: anti-I specificity of the biphasic hemolysin. *Transfusion 13*, 138-141.
8. BIERHUIZEN, M. F. A., MATTEI, M. G. & FUKUDA, M. (1993): Expression of the developmental I antigen by a cloned human cDNA encoding a member of a β-1,6-N-acetylglucosaminyltransferase gene family. *Genes Develop. 7*, 468-478.
9. BLANKEN, W. M., HOOGHWINKEL, G. J. M. & VAN DEN EIJNDEN, D. H. (1982): Biosynthesis of blood-group I and i substances. Specificity of bovine colostrum β-N-acetyl-D-glucosaminide β1→4 galactosyltransferase. *Eur. J. Biochem. 127*, 547-552.
10. BOOTH, P. B. (1970): Anti-$I^T P_1$: an antibody showing a further association between the I and P blood group systems. *Vox Sang. 19*, 85-90.
11. BOOTH, P. B., JENKINS, W. J. & MARSH, W. L. (1966): Anti-I^T: a new antibody of the I blood-group system occurring in certain Melanesian sera. *Br. J. Haematol. 12*, 341-344.
12. BROCKHAUSEN, I., MATTA, K. L., ORR, J., SCHACHTER, H., KOENDERMAN, A. H. L. & VAN DEN EIJNDEN, D. H. (1986): Mucin synthesis. Conversion of R_1-β1-3Gal-R_2 to R_1-β1-3(GlcNAcβ1-6)-Gal-R_2 and of R_1-β1-3GalNAc-R_2 to R_1-β1-3(GlcNAcβ1-6)GalNAc-R_2 by a β6-N-acetylglucosaminyltrans-ferase in pig gastric mucosa. *Eur. J. Biochem. 157*, 463-474.
13. CAPRA, J. D., DOWLING, P., COOK, S. & KUNKEL, H. G. (1969): An incomplete cold-reactive γG antibody with i specificity in infectious mononucleosis. *Vox Sang. 16*, 10-17.
14. CHESSIN, L. N. & McGINNISS, M. (1968): Further evidence for the serologic association of the O(H) and I blood groups. *Vox Sang. 14*, 194-201.
15. CHIEWSILP, P., COLLEDGE, K. E. & MARSH, W. L. (1971): Water soluble I blood group substance in the secretions of rhesus monkeys. *Vox Sang. 21*, 30-36.
16. CHILDS, R. A. & FEIZI, T. (1981): Differences in carbohydrate moieties of high molecular weight glycoproteins of human lymphocytes of T and B origins revealed by monoclonal autoantibodies with anti-I and anti-i specicities. *Biochem. Biophys. Res. Commun. 102*, 1158-1164.
17. CHILDS, R. A., FEIZI, T., FUKUDA, M. & HAKOMORI, S. I. (1978): Blood-group-I activity associated with band 3, the major intrinsic membrane protein of human erythrocytes. *Biochem. J. 173*, 333-336.
18. CHILDS, R. A., KAPADIA, A. & FEIZI, T. (1980): Expression of blood group I and i active carbohydrate sequences on cultured human and animal cell lines assessed by radioimmunoassays with monoclonal cold agglutinins. *Eur. J. Immunol. 10*, 379-384.
19. CLAFLIN, A. J. (1963): Three members of one family with the phenotype i; one with an anti-I antibody. *Transfusion 3*, 216-219.
20. COOPER, A. G. (1970): Soluble blood group I substance in human amniotic fluid. *Nature 227*, 508-509.
21. COOPER, A. G. & BROWN, M. C. (1973): Serum i antigen: a new human blood-group glycoprotein. *Biochem. Biophys. Res. Commun. 55*, 297-304.
22. COSTEA, N., YAKULIS, V. J. & HELLER, P. (1965): Experimental production of cold agglutinin in rabbits. *Blood 26*, 323-340.
23. CURTAIN, C. C. (1969): Anti-I agglutinins in non-human sera. *Vox Sang. 16*, 161-171.
24. CURTAIN, C. C. (1970): The occurrence of a substance with IA blood group activity in the body fluid of Ascaris. *Int. Arch. Allergy 38*, 449-456.
25. DABROWSKI, J., DABROWSKI, U., BERMEL, W., KORDOWICZ, M. & HANFLAND, P. (1988): Structure elucidation of the blood group B like and blood group I active octaantennary ceramide tetraconta-saccharide from rabbit erythrocyte membranes by two-dimensional ^{1}H NMR spectroscopy at 600 MHz. *Biochemistry 27*, 5149-5155.
26. DABROWSKI, U., HANFLAND, P., EGGE, H., KUHN, S. & DABROWSKI, J. (1984): Immunochemistry of I/i-active oligo- and polyglycosylceramides from rabbit erythrocyte membranes. Determination of branching patterns of a ceramide pentadecasaccharide by ^{1}H nuclear magnetic resonance. *J. Biol. Chem. 259*, 7648-7651.
27. DAVID, S. & VEYRIÈRES, A. (1975): The synthesis of 3,6-di-O-(2-acetamido-2-deoxy-β-D-gluco-pyranosyl)-D-galactose. A branched trisaccharide reported as a hydrolysis product of blood group

substances. *Carbohydr. Res. 40*, 23-29.

28. DE BOISSEZON, J. F., MARTY, Y., DUCOS, J. & ABBAL, M. (1970): Présence constante d'une substance inhibitrice de l'anticorps anti-i dans le sérum humain normal. *C.R. Acad. Sci. Paris 271*, 1448-1451.

29. DOINEL, C., ROPARS, C. & SALMON, C. (1974): Anti-I(A+B): an autoantibody detecting an antigenic determinant of I and a common part to A and B. *Vox Sang. 27*, 515-519.

30. DRACHMANN, O. (1968): An autoaggressive anti-BI(O) antibody. *Vox Sang. 14*, 185-193.

31. DUBE, V. E., KALLIO, P. & TANAKA, M. (1986): Specificity of the monoclonal anti-I antibody (Hy). *Mol. Immunol. 23*, 217-220.

32. DUNSTAN, R. A., SIMPSON, M. B. & ROSSE, W. F. (1984): The presence of the Ii blood group system on human platelets. *Am. J. Clin. Pathol. 82*, 74-77.

33. DZIERZKOWA-BORODEJ, W. (1971): HI and I^s fractions in the expression of H activity in human erythrocytes. *Ann. Immunol. 3*, 85-107.

34. DZIERZKOWA-BORODEJ, W., LISOWSKA, E. & SEYFRIEDOWA, H. (1970): The activity of glycoproteins from erythrocytes and protein fractions of human colostrum towards anti-I antibodies. *Life Sci. 9*, 111-120.

35. DZIERZKOWA-BORODEJ, W., SEYFRIED, H. & LISOWSKA, E. (1975): Serological classification of anti-I sera. *Vox Sang. 28*, 110-121.

36. DZIERZKOWA-BORODEJ, W., SEYFRIED, H., NICHOLS, M., REID, M. & MARSH, W. L. (1970): The recognition of water-soluble I blood group substance. *Vox Sang. 18*, 222-234.

37. EBERT, W., ROELCKE, D. & WEICKER, H. (1975): The I antigen of human red cell membrane. *Eur. J. Biochem. 53*, 505-515.

38. EGGE, H., KORDOWICZ, M., PETER-KATALINIC, J. & HANFLAND, P. (1985): Immunochemistry of I/i-active oligo- and polyglycosylceramides from rabbit erythrocyte membranes. Characterization of linear, di-, and triantennary neolactoglycosphingolipids. *J. Biol. Chem. 260*, 4927-4935.

39. FEIZI, T. (1967): Monotypic cold agglutinins in infection by Mycoplasma pneumoniae. *Nature 215*, 540-542.

40. FEIZI, T. (1980): Structural and biological aspects of blood group I and i antigens on glycolipids and glycoproteins. *Blood Transfus. Immunohaematol. 23*, 563-577.

41. FEIZI, T., CHILDS, R. A., HAKOMORI, S. I. & POWELL, M. E. (1978): Blood-group-Ii-active gangliosides of human erythrocyte membranes. *Biochem. J. 173*, 245-254.

42. FEIZI, T., CHILDS, R. A., WATANABE, K. & HAKOMORI, S. I. (1979): Three types of blood group I specificity among monoclonal anti-I autoantibodies revealed by analogues of a branched erythrocyte glycolipid. *J. Exp. Med. 149*, 975-980.

43. FEIZI, T., GOOI, H. C., LOOMES, L. M., SUZUKI, Y., SUZUKI, T. & MATSUMOTO, M. (1984): Cryptic I antigen activity and Mycoplasma pneumoniae-receptor activity associated with sialoglycoprotein GP-2 of bovine erythrocyte membranes. *Biosci. Rep. 4*, 743-749.

44. FEIZI, T. & HARDISTY, R. M. (1966): I antigen in leukaemic patients. *Nature 210*, 1066-1067.

45. FEIZI, T., HOUNSELL, E. F., ALAIS, J., VEYRIÈRES, A. & DAVID, S. (1992): Further definition of the size of the blood group-i antigenic determinant using a chemically synthesised octasaccharide of poly-N-acetyllactosamine type. *Carbohydr. Res. 228*, 289-297.

46. FEIZI, T. & KABAT, E. A. (1972): Immunochemical studies on blood groups. LIV. Classification of anti-I and anti-i sera into groups based on reactivity patterns with various antigens related to the blood group A, B, H, Le^a, Le^b, and precursor substances. *J. Exp. Med. 135*, 1247-1258.

47. FEIZI, T. & KABAT, E. A. (1974): Immunochemical studies on blood groups. LVI. Purification of glycoproteins with different I determinants from hydatid cyst fluid and from human milk on insoluble anti-I immunoadsorbents. *J. Immunol. 112*, 145-150.

48. FEIZI, T., KABAT, E. A., VICARI, G., ANDERSON, B. & MARSH, W. L. (1971): Immunochemical studies on blood groups. LXIX. The I antigen complex. Specificity differences among anti-I sera revealed by quantitative precipitin studies. Partial structure of the I determinant specific for one anti-I serum. *J. Immunol. 106*, 1578-1592.

49. FEIZI, T., KABAT, E. A., VICARI, G., ANDERSON, B. & MARSH, W. L. (1971): Immunochemical studies on blood groups. XLVII. The I antigen complex - precursors in the A, B, H, Le^a, and Le^b blood

group system – hemagglutination-inhibition studies. *J. Exp. Med. 133*, 39-52.

50. FEIZI, T. & MARSH, W. L. (1970): Demonstration of I- anti-I interaction in a precipitin system. *Vox Sang. 18*, 379-382.

51. FRANKS, D. (1966): Antigenic markers on cultured human cells I. Ii, Tjª, Donath-Landsteiner and "non-specific" autoantigens. *Vox Sang. 11*, 674-685.

52. FUKUDA, M., DELL, A. & FUKUDA, M. N. (1984): Structure of fetal lactosaminoglycan. The carbohydrate moiety of band 3 isolated from human umbilical cord erythrocytes. *J. Biol. Chem. 259*, 4782-4791.

53. FUKUDA, M., DELL, A., OATES, J. E. & FUKUDA, M. N. (1984): Structure of branched lactosaminoglycan, the carbohydrate moiety of band 3 isolated from adult human erythrocytes. *J. Biol. Chem. 259*, 8260-8273.

54. FUKUDA, M., FUKUDA, M. N. & HAKOMORI, S. I. (1979): Developmental change and genetic defect in the carbohydrate structure of band 3 glycoprotein of human erythrocyte membrane. *J. Biol. Chem. 254*, 3700-3703.

55. FUKUDA, M. N., DELL, A., OATES, J. E. & FUKUDA, M. (1985): Embryonal lactosaminoglycan. The structure of branched lactosaminoglycans with novel disialosyl (sialyl $\alpha2\rightarrow9$ sialyl) terminals isolated from PA1 human embryonal carcinoma cells. *J. Biol. Chem. 260*, 6623-6631.

56. GARDAS, A. (1976): Studies on the I-blood-group-active sites on macro-glycolipids from human erythrocytes. *Eur. J. Biochem. 68*, 185-191.

57. GARDAS, A. & KOSCIELAK, J. (1974): Megaloglycolipids - unusually complex glycosphingolipids of human erythrocyte membrane with A, B, H, and I blood group specificity. *FEBS Lett. 42*, 101-104.

58. GIBLETT, E. R. & CROOKSTON, M. C. (1964): Agglutinability of red cells by anti-i in patients with thlassaemia major and other haematological disorders. *Nature 201*, 1138-1139.

59. GOLDBERG, L. S. & BARNETT, E. V. (1967): Mixed γG-γM cold agglutinin. *J. Immunol. 99*, 803-809.

60. GOOI, H. C., UEMURA, K., EDWARDS, P. A. W., FOSTER, C. S., PICKERING, N. & FEIZI, T. (1983): Two mouse hybridoma antibodies against human milk-fat globules recognize the I(Ma) antigenic determinant β-D-Galp-(1$\rightarrow$4)-β-D-GlcpNAc-(1$\rightarrow$6). *Carbohydr. Res. 120*, 293-302.

61. GOOI, H. C., VEYRIÈRES, A., ALAIS, J., SCUDDER, P., HOUNSELL, E. F. & FEIZI, T. (1984): Further studies of the specificities of monoclonal anti-i and anti-I antibodies using chemically synthesized, linear oligosaccharides of the poly-N-acetyllactosamine series. *Mol. Immunol. 21*, 1099-1104.

62. GU, J., NISHIKAWA, A., FUJII, S., GASA, S. & TANIGUCHI, N. (1992): Biosynthesis of blood group I and i antigens in rat tissues. Identification of a novel β-1-6-N-acetylglucosaminyltransferase. *J. Biol. Chem. 267*, 2994-2999.

63. HANFLAND, P., EGGE, H., DABROWSKI, U., KUHN, S., ROELCKE, D. & DABROWSKI, J. (1981): Isolation and characterization of an I-active ceramide decasaccharide from rabbit erythrocyte membranes. *Biochemistry 20*, 5310-5319.

64. HANFLAND, P., KORDOWICZ, M., PETER-KATALINIC, J., EGGE, H., DABROWSKI, J. & DABROWSKI, U. (1988): Structure elucidation of blood group B-like and I-active ceramide eicosa- and pentacosa-saccharides from rabbit erythrocyte membranes by combined gas chromatography - mass spectrometry, electron-impact and fast-atom-bombardment mass spectrometry, and two-dimensional correlated, relayed-coherence transfer, and nuclear Overhauser effect 500-MHz ^{1}H -N.M.R. spectroscopy. *Carbohydr. Res. 178*, 1-21.

65. HELIN, J., PENTTILA, L., LEPPANEN, A., MAAHEIMO, H., LAURI, S., COSTELLO, C. E. & RENKONEN, O. (1997): The β1,6-GlcNAc transferase activity present in hog gastric mucosal microsomes catalyses site-specific branch formation on a long polylactosamine backbone. *FEBS Lett. 412*, 637-642.

66. HIROHASHI, S., CLAUSEN, H., NUDELMAN, E., INOUE, H., SHIMOSATO, Y. & HAKOMORI, S. I. (1986): A human monoclonal antibody directed to blood group i antigen: heterohybridoma between human lymphocytes from regional lymph nodes of a lung cancer patient and mouse myeloma. *J. Immunol. 136*, 4163-4168.

67. HOSOMI, O., TAKEYA, A. & KOGURE, T. (1984): Human serum contains N-acetyllactosamine: β1-3 N-acetylglucosaminyltransferase activity. *J. Biochem. 95*, 1655-1659.

68. HOUNSELL, E. F., WOOD, E., FEIZI, T., FUKUDA, M., POWELL, M. E. & HAKOMORI, S. I. (1981): Structural analysis of hexa- to octa-saccharide fractions isolated from sheep gastric-glycoproteins having

blood-group I and i activities. *Carbohydr. Res. 90*, 283-307.

69. IMAI, Y., LASKY, L. A. & ROSEN, S. D. (1993): Sulphation requirement for GlyCAM-1, an endothelial ligand for L-selectin. *Nature 361*, 555-557.

70. ISSITT, P. D., TEGOLI, J., JACKSON, V., SANDERS, C. W. & ALLEN, F. H. (1968): Anti-IP₁: antibodies that show an association between the I and P blood group systems. *Vox Sang. 14*, 1-8.

71. JACKSON, V. A., ISSITT, P. D., FRANCIS, B. J., GARRIS, M. L. & SANDERS, C. W. (1968): The simultaneous presence of anti-I and anti-i in sera. *Vox Sang. 15*, 133-141.

72. JEFFERIES, L. C., CARCHIDI, C. M. & SILBERSTEIN, L. E. (1993): Naturally occurring anti-i/I cold agglutinins may be encoded by different V$_H$3 genes as well as the V$_H$4.21 gene segment. *J. Clin. Invest. 92*, 2821-2833.

73. JENKINS, W. J., KOSTER, H. G., MARSH, W. L. & CARTER, R. L. (1965): Infectious mononucleosis: an unsuspected source of anti-i. *Br. J. Haematol. 11*, 480-483.

74. JENKINS, W. J., MARSH, W. L., NOADES, J., TIPPETT, P., SANGER, R. & RACE, R. R. (1960): The I antigen and antibody. *Vox Sang. 5*, 97-106.

75. KABAT, E. A., LIAO, J., BURZYNSKA, M. H., WONG, T. C., THOGERSEN, H. & LEMIEUX, R. U. (1981): Immunochemical studies on blood groups. LXIX. The conformation of the trisaccharide determinant in the combining site of anti-I Ma (group 1). *Mol. Immunol. 18*, 873-881.

76. KABAT, E. A., LIAO, J. & LEMIEUX, R. U. (1978): Immunochemical studies on blood groups. LXVIII. The combining site of anti-I Ma (group 1). *Immunochemistry 15*, 727-731.

77. KAPADIA, A., FEIZI, T. & EVANS, M. J. (1981): Changes in the expression and polarization of blood group I and i antigens in post-implantation embryos and teratocarcinomas of mouse associated with cell differentiation. *Exp. Cell Res. 131*, 185-195.

78. KAPADIA, A., FEIZI, T., JEWELL, D., KEELING, J. & SLAVIN, G. (1981): Immunocytochemical studies of blood group A, H, I, and i antigens in gastric mucosae of infants with normal gastric histology and of patients with gastric carcinoma and chronic benign peptic ulceration. *J. Clin. Pathol. 34*, 320-337.

79. KOENDERMAN, A. H. L., KOPPEN, P. L. & VAN DEN EIJNDEN, D. H. (1987): Biosynthesis of polylactosaminoglycans - Novikoff ascites tumor cells contain two UDP-GlcNAc:β-galactoside β1→6 N-acetylglucosaminyltransferase activities. *Eur. J. Biochem. 166*, 199-208.

80. KOSCIELAK, J., MILLER-PODRAZA, H., KRAUZE, R. & PIASEK, A. (1976): Isolation and characterization of poly(glycosyl)ceramides (megaloglycolipids) with A, H, and I blood group activities. *Eur. J. Biochem. 71*, 9-18.

81. KOSCIELAK, J., ZDEBSKA, E., WILCZYNSKA, Z., MILLER-PODRAZA, H. & DZIERZKOWA-BORODEJ, W. (1979): Immunochemistry of Ii-active glycosphingolipids of erythrocytes. *Eur. J. Biochem. 96*, 331-337.

82. KUHNS, W., RUTZ, V., PAULSEN, H., MATTA, K. L., BAKER, M. A., BARNER, M., GRANOVSKY, M. & BROCKHAUSEN, I. (1993): Processing of O-glycan core 1, Gal β1-3GalNAac α-R - specificities of core 2, UDP-GlcNAc:gal β1-3GalNAc-R (GlcNAc to GalNAc) β6-N-acetyl-glucosaminyltransferase and CMP-Sialic Acid - Gal β1-3GalNAc-R α3- sialyltransferase. *Glycoconj. J. 10*, 381-394.

83. LEMIEUX, R. U., WONG, T. C., LIAO, J. & KABAT, E. A. (1984): The combining site of anti-I Ma (group 1). *Mol. Immunol. 21*, 751-759.

84. LEPPÄNEN, A., NIEMELÄ, R. & RENKONEN, O. (1997): Enzymatic midchain branching of polylactosamine backbones is restricted in a site-specific manner in α1,3-fucosylated chains. *Biochemistry 36*, 13729-13735.

85. LEPPÄNEN, A., PENTTILÄ, L., NIEMELÄ, R., HELIN, J., SEPPO, A., LUSA, S. & RENKONEN, O. (1991): Human serum contains a novel β1,6-N-acetylglucosaminyltransferase activity that is involved in midchain branching of oligo(N-acetyllactosaminoglycans). *Biochemistry 30*, 9287-9296.

86. LEPPÄNEN, A., ZHU, Y., MAAHEIMO, H., HELIN, J., LEHTONEN, E. & RENKONEN, O. (1998): Biosynthesis of branched polylactosaminoglycans. Embryonal carcinoma cells express midchain β1,6-N-acetyl-glucosaminyltransferase activity that generates branches to preformed linear backbones. *J. Biol. Chem. 273*, 17399-17405.

87. LIND, K. (1973): Production of cold agglutinins in rabbits induced by Mycoplasma pneumoniae, Listeria monocytogenes or Streptococcus MG. *Acta Pathol. Microbiol. Scand. 81B*, 487-496.

88. Lodge, T. W. & Voak, D. (1968): An example of inhibitable anti-HI in a group B donor. *Vox Sang. 14*, 60-62.

89. Loomes, L. M., Uemura, K. & Feizi, T. (1985): Interaction of Mycoplasma pneumoniae with erythrocyte glycolipids of I and i antigen types. *Infect. Immun. 47*, 15-20.

90. Loomes, L. M., Uemura, K. I., Childs, R. A., Paulson, J. C., Rogers, G. N., Scudder, P. R., Michalski, J. C., Hounsell, E. F., Taylor-Robinson, D. & Feizi, T. (1984): Erythrocyte receptors for Mycoplasma pneumoniae are sialylated oligosaccharides of Ii antigen type. *Nature 307*, 560-563.

91. Lopez, M., Gerbal, A. & Salmon, C. (1972): Excés d'antigène I dans les érythrocytes de phénotypes O_h, A_h et B_h. *Rev. Fr. Transf. 15*, 187-193.

92. Maaheimo, H., Rabina, J. & Renkonen, O. (1997): H^1 and C^{13} NMR analysis of the pentasaccharide Gal β(1→4)GlcNAc β(1→3)-[GlcNAcβ(1→)] Galβ(→4)GlcNAc synthesized by the mid-chain beta-(1→6)-D-N-acetylglucosaminyltransferase of rat serum. *Carbohydr. Res. 297*, 145-151.

93. Maniatis, A., Frieman, B. & Bertles, J. F. (1977): Increased expression in erythrocyte Ii antigens in sickle cell disease and sickle cell trait. *Vox Sang. 33*, 29-36.

94. Maranduba, A. & Veyrières, A. (1986): Glycosylation of lactose: synthesis of branched oligosaccharides involved in the biosynthesis of glycolipids having blood-group I activitiy. *Carbohydr. Res. 151*, 105-119.

95. Marcus, D. M., Kabat, E. A. & Rosenfield, R. E. (1963): The action of enzymes from Clostridium tertium on the I antigenic determinant of human erythrocytes. *J. Exp. Med. 118*, 175-194.

96. Marsh, W. L. (1961): Anti-i: a cold antibody defining the Ii relationship in human red cells. *Br. J. Haematol. 7*, 200-209.

97. Marsh, W. L., Jensen, L., Decary, F. & Colledge, K. (1972): Water-soluble I blood group substance in the secretions of i adults. *Transfusion 12*, 222-226.

98. Marsh, W. L., Nichols, M. E. & Allen, F. H. (1970): Inhibition of anti-I sera by human milk. *Vox Sang. 18*, 149-154.

99. Marsh, W. L., Nichols, M. E. & Reid, M. E. (1971): The definition of two I antigen components. *Vox Sang. 20*, 209-217.

100. Matsuzaki, Y., Ito, Y. & Ogawa, T. (1992): Stereoselective total synthesis of the blood group I-active biantennary neolacto-glycodecaosyl ceramide. *Tetrahedron Lett. 33*, 4025-4028.

101. Matsuzaki, Y., Ito, Y. & Ogawa, T. (1992): Synthesis of triantennary blood group I antigens: neolactoglycopentadecaosyl ceramide. *Tetrahedron Lett. 33*, 6343-6346.

102. McEver, R. P., Mooore, K. L. & Cummings, R. D. (1995): Leukocyte trafficking mediated by selectin-carbohydrate interactions. *J. Biol. Chem. 270*, 11025-11028.

103. Nagatsuka, Y., Watarai, S., Yasuda, T., Higashi, H., Yamagata, T. & Ono, Y. (1995): Production of human monoclonal antibodies to i blood group by EBV-induced transformation: possible presence of a new glycolipid in cord red cell membranes and human hematopoietic cell lines. *Immunol. Lett. 46*, 93-100.

104. Niemann, H., Watanabe, K., Hakomori, S. I., Childs, R. A. & Feizi, T. (1978): Blood group i and I activities of "lacto-N-norhexaosylceramide" and its analogues: the structural requirements for i-specificities. *Biochem. Biophys. Res. Commun. 81*, 1286-1293.

105. Niemelä, R., Rabina, J., Leppanen, A., Maaheimo, H., Costello, C. E. & Renkonen, O. (1995): Site-directed enzymatic alpha-(1→3)-l-fucosylation of the tetrasaccharide Gal beta(→4)GlcNAc beta(1→3)Gal beta(1→4)GlcNAc at the distal N-acetyllactosamine unit. *Carbohydr. Res. 279*, 331-338.

106. Okada, Y., Kannagi, R., Levery, S. B. & Hakomori, S. I. (1984): Glycolipid antigens with blood group I and i specificities from human adult and umbilical cord erythrocytes. *J. Immunol. 133*, 835-842.

107. Oppenheim, J. D., Nachbar, M. S. & Blank, M. (1983): The distribution of and the biochemical and serological relationships between the I/i and ABH blood-group antigens of the human erythrocyte membrane as determined by immunoelectrophoretic techniques. *Electrophoresis 4*, 53-62.

108. Pascual, V., Kimberly, V., Lelsz, D., Spellerberg, M. B., Hamblin, T. J., Thompson, K. M.,

RANDEN, I., NATVIG, J., CAPRA, J. D. & STEVENSON, F. K. (1991): Nucleotide sequence analysis of the V regions of two IgM cold agglutinins. Evidence that the V_H4-21 gene segment is responsible for the major cross-reactive idiotype. *J. Immunol. 146*, 4385-4391.

109. PICARD, J., WALDRON-EDWARD, D. & FEIZI, T. (1978): Changes on the expression of the blood-group A, B, H, Le[a] and Le[b] antigens and the blood-group precursor associated I (Ma) antigen in glycoprotein-rich extracts of gastric carcinomas. *J. Clin. Lab. Immunol. 1*, 119-128.

110. PICARD, J. K. & FEIZI, T. (1983): Peanut lectin and anti-Ii antibodies reveal structural differences among human gastrointestinal glycoproteins. *Mol. Immunol. 20*, 1215-1220.

111. PILLER, F. & CARTRON, J. P. (1983): Biosynthesis of i-antigenic structures. In: *Red Cell Membrane Glycoconjugates and Related Genetic Markers* (J. P. Cartron, P. Rouger, and C. Salmon, eds.). Librairie Arnette, Paris, pp. 175-182.

112. PILLER, F. & CARTRON, J. P. (1983): UDP-GlcNAc:Galβ1-4Glc(NAc)β1-3 N-acetylglucosaminyl-transferase. Identification and characterization in human serum. *J. Biol. Chem. 258*, 12293-12299.

113. PILLER, F., CARTRON, J. P., MARANDUBA, A., VEYRIERES, A., LEROY, Y. & FOURNET, B. (1984): Biosynthesis of blood group I antigens. Identification of a UDP-GlcNAc: GlcNAcβ1-3Gal(-R) β1-6(GlcNAc to Gal) N-acetylglucosaminyltransferase in hog gastric mucosa. *J. Biol. Chem. 259*, 13385-13390.

114. PRUZANSKI, W., FARID, N., KEYSTONE, E. & ARMSTRONG, M. (1975): The influence of homogeneous cold agglutinins on polymorphonuclear and mononuclear phagocytes. *Clin. Immunol. Immunopathol. 4*, 277-285.

115. PRUZANSKI, W. & SHUMAK, K. H. (1977): Biological activity of cold-reacting autoantibodies (Part 1). *N. Engl. J. Med. 297*, 538-542.

116. RACE, R. R. & SANGER, R. (1975): The I and i antigens. In: *Blood Groups in Man*. Blackwell Scientific Publications, Oxford, pp. 447-462.

117. RENOUF, D. V. & HOUNSELL, E. F. (1993): Conformational studies of the backbone (poly-N-acetyllactosamine) and the core region sequences of O-linked carbohydrate chains. *Int. J. Biol. Macromol. 15*, 37-42.

118. ROELCKE, D. (1974): Cold agglutination. Antibodies and antigens. A review. *Clin. Immunol. Immunopathol. 2*, 266-280.

119. ROPP, P. A., LITTLE, M. R. & CHENG, P. W. (1991): Mucin biosynthesis: purification and characterization of a mucin β 6N-acetylglucosaminyltransferase. *J. Biol. Chem. 266*, 23863-23871.

120. ROSENFIELD, R. E., SCHMIDT, P. J., CALVO, R. C. & McGINNISS, M. H. (1965): Anti-i, a frequent cold agglutinin in infectious mononucleosis. *Vox Sang. 10*, 631-634.

121. SASAKI, K., KURATA-MIURA, K., UJITA, M., ANGATA, K., NAKAGAWA, S., SEKINE, S., NISHI, T. & FUKUDA, M. (1997): Expression cloning of cDNA encoding a human β-1,3-N-acetylglucos-aminyltransferase that is essential for poly-N-acetyllactosamine synthesis. *Proc. Natl. Acad. Sci. USA 94*, 14294-14299.

122. SCHENKEL-BRUNNER, H., KABAT, E. A. & LIAO, J. (1979): Biosynthesis of a blood-group-I determinant reacting with anti-I Ma serum (group 1). *Eur. J. Biochem. 98*, 573-575.

123. SCHUTTE, M. E., VAN ES, J. H., SILBERSTEIN, L. E. & LOGTENBERG, T. (1993): V_H4.21-encoded natural autoantibodies with anti-i specificity mirror those associated with cold hemagglutinin disease. *J. Immunol. 151*, 6569-9576.

124. SEPPO, A., PENTTILÄ, L., NIEMELÄ, R., MAAHEIMO, H., RENKONEN, O. & KEANE, A. (1995): Enzymatic synthesis of octadecameric saccharides of multiply branched blood group I-type, carrying four distal α1,3- galactose or β1,3-GlcNAc residues. *Biochemistry 34*, 4655-4661.

125. SHUMAK, K. H., RACHKEWICH, R. A. & GREAVES, M. F. (1975): I and i antigens on normal human T and B lymphocytes and on lymphocytes from patients with chronic lymphocytic leukemia. *Clin. Immunol. Immunopathol. 4*, 241-247.

126. SILBERSTEIN, L. E., JEFFERIES, L. C., GOLDMAN, J., FRIEDMAN, D., MOORE, J. S., NOWELL, P. C., ROELCKE, D., PRUZANSKI, W., ROUDIER, J. & SILVERMAN, G. J. (1991): Variable region gene analysis of pathologic human autoantibodies to the related i and I red blood cell antigens. *Blood 78*, 2372-2386.

127. Smith, G., Spellerberg, M., Boulton, F., Roelcke, D. & Stevenson, F. (1995): The immunoglobulin V_H gene, V_H4-21, specifically encodes autoanti-red cell antibodies against the I or i antigens. *Vox Sang. 68*, 231-235.

128. Suzuki, Y., Hirabayashi, Y., Suzuki, T. & Matsumoto, M. (1985): Occurrence of O-glycosidically peptide-linked oligosaccharides of poly-N- acetyllactosamine type (erythroglycan II) in the I-antigenically active Sendai virus receptor sialoglycoprotein GP-2. *J. Biochem. 98*, 1653-1659.

129. Suzuki, Y., Nagao, Y., Kato, H., Matsumoto, M., Nerome, K. & Nakajima, K. (1986): Human influenza A virus hemagglutinin distinguishes sialyloligosaccharides inmembrane-associated gangliosides as its receptor which mediates the adsorption and fusion processes of virus infection. *J. Biol. Chem. 261*, 17057-17061.

130. Tang, P. W., Scudder, P., Mehmet, H., Hounsell, E. F. & Feizi, T. (1986): Sulphate groups are involved in the antigenicity of keratan sulphate and mask i antigen expression on their poly-N-acetyllactosamine backbones. An immunochemical and chromatographic study of keratan sulphate oligosaccharides after desulphation or nitrosation. *Eur. J. Biochem. 160*, 537-545.

131. Tegoli, J., Harris, J. P., Issitt, P. D. & Sanders, C. W. (1967): Anti-IB, an expected 'new' antibody detecting a joint product of the I and B genes. *Vox Sang. 13*, 144-157.

132. Thomas, D. B. (1974): The i antigen complex: a new specificity unique to dividing human cells. *Eur. J. Immunol. 4*, 819-824.

133. Tippett, P., Noades, J., Sanger, R., Race, R. R., Sausais, L., Holman, C. A. & Buttimer, R. J. (1960): Further studies of the I antigen and antibody. *Vox Sang. 5*, 107-121.

134. Toh, B. H., Diggle, T. A. & Koh, S. H. (1979): Ii blood group antigens on fibroblast cell surfaces. *Clin. Immunol. Immunopathol. 12*, 177-182.

135. Tsuboi, S., Isogai, Y., Hada, N., King, J. K., Hindsgaul, O. & Fukuda, M. (1996): 6'-Sulfo sialyl Le^x but not 6-sulfo sialyl Le^x expressed on the cell surface supports L-selectin-mediated adhesion. *J. Biol. Chem. 271*, 27213-27216.

136. Ujita, M., McAuliffe, J., Suzuki, M., Hindsgaul, O., Clausen, H., Fukuda, M. N. & Fukuda, M. (1999): Regulation of I-branched poly-N-acetyllactosamine synthesis. Concerted actions by i-extension enzyme, I-branching enzyme, and β 1,4-galactosyltransferase I. *J. Biol. Chem. 274*, 9296-9304.

137. Van den Eijnden, D. H., Koenderman, A. H. L. & Schiphorst, W. E. C. M. (1988): Biosynthesis of blood group i-active polylactosaminoglycans. Partial purification and properties of an UDP-GlcNAc : N-acetyllactosaminide β1→3-N-acetylglucosaminyltransferase from Novikoff tumor cell ascites fluid. *J. Biol. Chem. 263*, 12461-12471.

138. Van den Eijnden, D. H., Winterwerp, H., Smeeman, P. & Schiphorst, W. E. C. M. (1983): Novikoff ascites tumour cells contain N-acetyllactosaminide β1-3 and β1-6 N-acetylglucos-aminyltransferase activity. *J. Biol. Chem. 258*, 3435-3437.

139. Vilalta-Castel, E., Guerra-Vales, J. M., Gonzalez-Gamarra, A., Abarca-Costalago, M., Alonso-Navas, F. & Lopez-Pascual, J. J. (1986): Hemolytic anemia caused by cryoagglutinins associated with an influenza A virus infection. *Rev. Clin. Esp. 179*, 22-25.

140. Vilkman, A., Niemelä, R., Penttilä, L., Helin, J., Leppänen, A., Seppo, A., Maaheimo, H., Lusa, S. & Renkonen, O. (1992): Elongation of both branches of biantennary backbones of oligo-(N-acetyllactosamino)glycans by human serum (1→3)-N-acetyl-β-D-glucosaminyltransferase. *Carbohydr. Res. 226*, 155-174.

141. Watanabe, K., Hakomori, S. I., Childs, R. A. & Feizi, T. (1979): Characterization of a blood group I-active ganglioside. Structural requirements for I and i specificities. *J. Biol. Chem. 254*, 3221-3228.

142. Watanabe, K., Laine, R. A. & Hakomori, S. I. (1975): On neutral fucoglycolipids having long, branched carbohydrate chains: H-active and I-active glycosphingolipids of human erythrocyte membranes. *Biochemistry 14*, 2725-2733.

143. Wiener, A. S., Moor-Jankowski, J., Gordon, E. B. & Davis, J. (1965): The blood factors I and i in primates including man, and in lower species. *Amer. J. Phys. Anthropol. 23*, 389-396.

144. Wiener, A. S., Unger, L. J., Cohen, L. & Feldman, J. (1956): Type-specific cold auto-antibodies as a cause of acquired hemolytic anamia and hemolytic transfusion reactions: biologic test with

bovine red cells. *Ann. Int. Med. 44*, 221-240.

145. WOOD, E. & FEIZI, T. (1979): Blood group I and i acitivities of straight chain and branched synthetic oligosaccharides related to the precursors of the major blood group antigens. *FEBS Lett. 104*, 135-140.

146. WOOD, E., HOUNSELL, E. F., LANGHORNE, J. & FEIZI, T. (1980): Sheep gastric mucins as a source of blood-group-I and -i antigens. *Biochem. J. 187*, 711-718.

147. YATES, A. D. & WATKINS, W. M. (1983): Enzymes involved in the biosynthesis of glycoconjugates. A UDP-2-acetamido-2-deoxy-D-glucose:β-D-galactopyranosyl-(1$\rightarrow$4)-saccharide (1$\rightarrow$ 3)-2-acetamido-2-deoxy-β-D-glucopyranosyltransferase in human serum. *Carbohydr. Res. 120*, 251-268.

148. ZDEBSKA, E., MAGNUSKA, A., KUSNIERZ, G., SEYFRIED, H. & KOSCIELAK, J. (1980): Both polyglycosylceramides and polyglycosylpeptides are unbranched in i erythrocytes. *FEBS Lett. 120*, 33-36.

149. ZHOU, D. P., DINTER, A., GUTIERREZ-GALLEGO, R. G., KAMERLING, J. P., VLIEGENTHART, J. F. G., BERGER, E. G. & HENNET, T. (1999): A β-1,3-N-acetylglucosaminyltransferase with poly-N-acetyllactosamine synthase activity is structurally related to β-1,3-galactosyltransferases. *Proc. Natl. Acad. Sci. USA 96*, 406-411.

150. ZIELENSKI, J. & KOSCIELAK, J. (1983): Sera of i subjects have the capacity to synthesize the branched GlcNAc(β1$\rightarrow$6)[GlcNAc(β1$\rightarrow$3)]Gal... structure. *FEBS Lett. 163*, 114-118.

8 P System

The first antigen of the blood group **P** system, P_1, was discovered in 1927 by Landsteiner and Levine [66]; the system was expanded in 1935 when the antigens **P** [103] and P^k [58,59,85] were discovered and identified as belonging to the **P** system. Five blood group **P** phenotypes can be defined according to the presence or absence of these three serological specificities on the erythrocytes, two of which, P_1 and P_2, are frequent and three, P_1^k, P_2^k, and **p**, very rare (*Tables 8.1* and *8.2*) [101]:

- P_1 erythrocytes are characterised by the presence of P_1 and **P** antigen,
- P_2 cells lack the **P** antigen,
- in P_1^k and P_2^k erythrocytes the **P** determinant is absent and the cells show strong P^k activity; the presence of P_1 antigen distinguishes the P_1^k phenotype from P_2^k,
- none of the three antigenic characters are detectable on **p** cells.

The high-incidence **LKE** (= **Luke**) character described by Tippett et al. [118] is absent from P^k and **p** cells and thus was originally believed to be another antigen of the **P** blood group system. More recent investigations, however, have shown that this relationship is of an indirect nature, and the **Luke** antigen therefore is no longer considered part of the **P** system.

Table 8.1: Blood group P phenotypes.

Phenotype	Original term	Frequency in Europids[a]	Isoagglutinins in serum
P_1	P+	75 %	–
P_2	P–	25 %	anti-P_1[b]
P_1^k		very rare	anti-P
P_2^k		very rare	anti-P
p	Tj(a–)	very rare	anti-PP$_1$P^k

[a] According to Race and Sanger [101]; the frequency of P_1 is race-dependent, varying from 30% in Japanese to 90% in some African and South American peoples [89]
[b] Normally but not always present

Table 8.2: Presence of antigens P₁, P, and Pᵏ on erythrocytes and fibroblasts from individuals of different P phenotypes.

Phenotype	Antigen on erythrocytes			Antigen on fibroblasts		
	P_1	P	P^k	P_1	P	P^k
P_1	+	+	−	+	+	+
P_2	−	+	−	−	+	+
P_1^k	+	−	+	+	−	+
P_2^k	−	−	+	−	−	+
p	−	−	−	−	−	−

The antigens of the **P** system are transmitted by two gene systems, the *P1* gene which has been localised on chromosome 22, position q11.3–qter [86] and the *P* gene, the assignment of which to chromosome 6 [33] has yet to be confirmed. Models explaining the genetic regulation of the biosynthesis of **P** system antigens will be presented in *Sect. 8.3.3*.

P_1 expression is inhibited by **In(Lu)**, the rare dominant inhibitor of Lutheran and other erythrocyte antigens (see *Chap. 19)* [22,24].

According to the 'ISBT Working Party on Terminology for Red Cell Surface Antigens' (see *Chap. 1)* only the **P1** antigen (ISBT Nr. 003001) meets the criteria for inclusion in the **P1** blood group system as defined above. The antigens **P** (ISBT Nr. 209001, or GLOBO1), **Pᵏ** (ISBT Nr. 209002, or GLOBO2), and **LKE** (ISBT Nr. 209003, or GLOBO3), which are derived from a common precursor but are produced in different biosynthetic pathways, have been combined in a 'Collection'.

8.1 Antisera and Lectins

Anti-P₁

Anti-P_1 isoagglutinins (described in the original paper by Landsteiner in 1927 [66] as anti-**P**) are commonly found in the sera of P_2 individuals [101]; in most cases the antibodies are only weakly active and react better at low temperatures than at 37°C. Strong anti-P_1 has been repeatedly detected in the serum of blood group P_2 subjects suffering from echinococcosis (hydatid cyst disease) [21] and fascioliasis (liver fluke disease) [10,11], and occasionally in pigeon breeders (induced by P_1 active substance found in the ovomucoid of turtledove eggs). Antibodies with this specificity also occur naturally in non-immune sera of several animal species (primarily in the horse, but also in the cat, the rabbit, the pig, and cattle) [67]; furthermore, specific anti-P_1 reagents have been produced in goats and rabbits by immunisation with P_1 substance from hydatid cysts (see below) [72].

Anti-PP$_1$P^k

Antisera with this specificity (also described as anti-Tj^a = 'anti-**Jay**') [71,124] are regularly found in blood group **p** individuals. The sera represent mixtures of anti-**P** and anti-P_1P^k antibodies – absorption with P_1^k erythrocytes yields relatively pure anti-**P** reagents, whereas absorption with P_2 cells [101] or with globoside [91] yields anti-P_1P^k specific serum.   An entire spectrum of antibodies can be found in the anti-P_1P^k component, with specificities ranging from anti-P^k to anti-P_1(see [91]). Due to the cross-reactivity between the determinants P_1 and P^k (see *Sect. 8.2.2*), treatment of anti-PP$_1$P^k sera with P_1 cells has yielded an anti-P^k only in rare cases, and it has not been possible to obtain a specific anti-P_1 by absorption with P_2^k erythrocytes [101].

The anti-P_1P^k and anti-**P** components of anti-Tj^a sera consist mainly of IgG_3 antibodies [91,102,107] which can penetrate the placenta. Some authors (see [79]) assume that the anti-P_1P^k type antibodies are responsible for the high rate of spontaneous abortion observed in women of blood group **p** [45,74]; in other cases, however, it is thought that abortion may be caused by the anti-**P** component of anti-Tj^a sera [41].

Glycolipid analysis on tissues of spontaneously aborted fetuses and on placenta from blood group **p** women revealed substantial amounts of globotetraosylceramide (P antigen) and globotriaosylceramide (P^k antigen) in the placental fraction; the fetuses, however, contained only trace amounts of these structures [41,77]. These results suggest that the primary target for the maternal anti-P_1PP^k antibodies is the placental tissue.

Anti-P

Anti-**P** antibodies occur as isoagglutinins in P^k individuals [101]; anti-**P** can also be obtained by absorption of anti-PP$_1$P^k sera with P_1^k erythrocytes (see above).

Furthermore, cold agglutinins[1] with this specificity ('Donath-Landsteiner antibodies' [27]) were detected in patients with paroxysmal cold haemoglobinuria [73,128].

Several authors [56,93,105] have noticed that anti-P sera frequently show a good reaction with the **Forssman** glycolipid *(Fig. 8.4)*. It has therefore been assumed that many anti-P isoagglutinins are naturally occurring anti-**Forssman** antibodies induced by **Forssman** or 'Forssman-like' antigens which are widely distributed in animal tissues and in bacteria.

Anti-P^k

This antibody is always accompanied by anti-P$_1$ and anti-P [101]. As mentioned above, specific anti-P^k can be obtained from some anti-PP$_1$P^k sera of **p** individuals by absorption with P$_1$ erythrocytes.

'Anti-p'

Antisera which agglutinate preferentially **p** erythrocytes have been described by several groups [28,47,87]. Closer investigations on the substrate specificity (see *Sect. 8.2.2)* show, however, that the 'anti-**p**' antibodies are not directed towards an antigen of the blood group **P** system but instead react with sialic acid-containing glycosphingolipids which occur in excessive quantities in **p** erythrocytes (see *Sect. 8.2.3)*.

Anti-LKE (= Luke)

Only a few examples of anti-**LKE** have been described to date [20,101,118]. The serological properties of this antibody are similar to those of anti-P: it agglutinates neither **p** erythrocytes nor P^k cells. Unlike anti-P, however, it does not react with ~ 2% of P$_1$ and P$_2$ cells. Furthermore, the reactivity of this antibody is correlated with the **ABO** group of the donor – A$_1$ and A$_1$B erythrocytes are distinctly more weakly agglutinated than cells of blood groups O, A$_2$, B, or A$_2$B.

Anti-IP$_1$

Anti-P antibodies serologically associated with **Ii** antigens are relatively common. Anti-**IP**$_1$ is an antibody which agglutinates only erythrocytes from **IP**$_1$ individuals and fails to react with **iP**$_1$ cells ([48], see also *Chap. 7)*. Examples of other 'compound-specific' antibodies detected in human sera are anti-**IP** [1] and anti-I^{T}P$_1$ [15].

---

[1] Actually 'biphasic haemolysins', i.e. antibodies which bind to erythrocytes at 0°C in the presence of complement and haemolyse the cells when subsequently warmed.

Hybridoma Antibodies

A series of hybridoma antibodies towards antigens of the blood group **P** system have been described:

- an IgM antibody towards Burkitt-Lymphoma cells (*38-13*, [126]) shows anti-**P**^{k} specificity [29,94],
- the murine IgM antibody *CLB-ery-2* prepared by immunisation with a bacterial immunogen [123] was identified as anti-**P**,
- both the so-called anti-**SSEA-3** serum (*MC631*) raised against embryonic cells of the mouse [53,117] and the 'anti-**P**-like' antibody *LM147/328* directed towards an *E. coli* immunogen [44] show good **P** specificity when tested with erythrocytes; haemagglutination inhibition tests, however, led to the assumption that the antibodies bind preferentially to β1,3-galactosyl-globoside, a derivative of the **P** substance,
- a hybridoma obtained from a mouse immunised with the **P**$_1$ substance of turtle dove ovomucoid produced a specific anti-**P**$_1$ antibody (*154-IX-B6* [7]),
- antibodies obtained by immunising mice with neo-glycoproteins containing the **P**$_1$ trisaccharide showed anti-**P**$_1$**P**^{k} specificity [17],
- the antibody *MC813-70* towards a human teratocarcinoma cell line, originally described as anti-**SSEA-4** [51], was found to have anti-**LKE** specificity [117].

Lectins

The lectins in female gonads and roe of fishes, e.g. roach (*Rutilus rutilus*) [104], herring (*Clupea harengus*) [4], and various *Salmonidae* species, such as trout (*Salmo trutta*), and salmon (*S. salar*) [4,122], agglutinate **P**$_1$, **P**$_2$, and **P**^{k} erythrocytes intensely. These lectins, however, also react readily with **B** erythrocytes. Since these specificities cannot be separated from each other it must be assumed that the lectins bind to α-galactosyl residues common to the determinants **P**$_1$, **P**^{k} and **B**.

Similiar galactose-binding lectins have been found in bacteria, such as *Pseudomonas aeruginosa*, *Streptococcus suis*, and *Escherichia coli* [12,13,40,110]

8.2 Blood Group P Substances

8.2.1 Occurrence of Blood Group P Active Substances

In humans the blood group **P** substances are constituents of cell membranes exclusively and do not occur in secretions. They are not confined to erythrocytes but have also been detected on different leukocyte populations [30], thrombocytes, megacaryocytes, endothelial cells [123], fibroblasts [32], and smooth muscle cells of the digestive tract and the urogenital system [55].

Certain tumour cells and malignant cell lines show an increased content of **P** and **P**^{k} glycolipids [14,17,54,75,94,123].

The antigens of the **P** system have also been detected in various animal species (see below), where the characters occur not only on erythrocytes and tissue cells but also on water-soluble glycoproteins of secretions and body fluids. Strong **P**$_1$(+**P**^{k}) activity has been found in the ovomucoid of turtledove eggs [35,36]. Another excellent source for **P**$_1$(+**P**^{k}) active material is the fluid in hydatid cysts taken from sheep liver containing live protoscolices of the tapeworm *Echinococcus granulosus* [21,108]. Further, **P**$_1$ specific substances were found in the extracts of earthworm *Lumbricus terrestris* [100] and the roundworm *Ascaris suum* [99].

P and **P**^{k} determinants have recently been identified in the lipo-oligosaccharides of the bacteria *Neisseria gonorrhoeae*, *N. meningitidis*, *N. lactamica*, and *Branhamella catarrhalis* [80,129].

8.2.2 Chemical Structures of the Blood Group P Substances

(a) Glycoproteins from Hydatid Cysts

The first investigations on the structure of the **P**$_1$ determinant were performed by Morgan and Watkins on material isolated from hydatid cysts [88,125]. The authors identified this substance as a glycoprotein with a carbohydrate moiety composed mainly of galactose and N-acetylglucosamine. Haemagglutination inhibition tests with different monosaccharides and their glycosides [125], as well as treatment with different glycosidases [5,132] indicated that the **P**$_1$ activity is determined by terminal α-galactosyl residues. In analogous experiments Voak et al. [121] showed that α-galactose is also an essential part of the immunodominant group of the **P**^{k} determinant.

In the course of these investigations the disaccharide **Galα1→4Gal** [125] was isolated as was later the trisaccharide **Galα1→4Galβ1→4GlcNAc** [23] after mild hydrolysis of the **P**$_1$ glycoprotein. The trisaccharide was able to inhibit the reaction of **P**$_1$ erythrocytes with the majority of anti-**P**$_1$ sera, whereas the disaccharide was only weakly **P**$_1$ active. Both oligosacchardes inhibited the agglutination of **P**^{k} cells with anti-**P**^{k} to approximately the same degree [131]; none of the fragments reacted with anti-**P**.

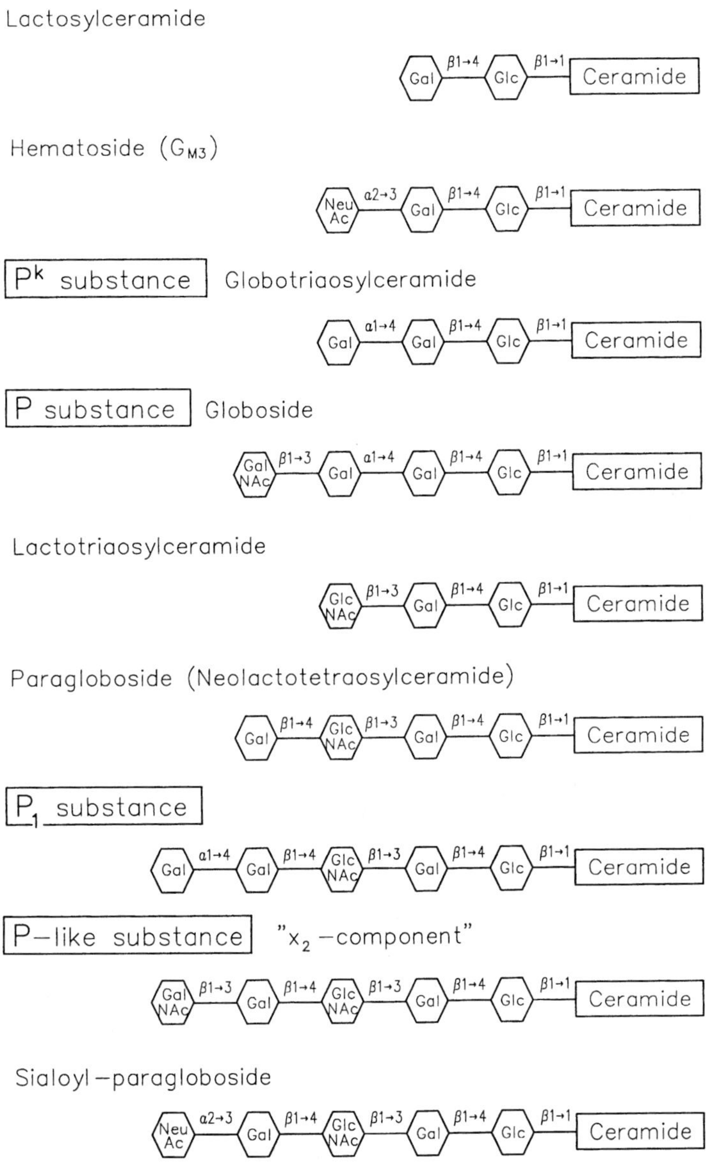

Lactosylceramide

Hematoside (G$_{M3}$)

P^k substance   Globotriaosylceramide

P substance   Globoside

Lactotriaosylceramide

Paragloboside (Neolactotetraosylceramide)

P$_1$ substance

P—like substance   "x$_2$ —component"

Sialoyl—paragloboside

Figure 8.1: Chemical structures of P^k, P, and P$_1$ substances obtained from human erythrocyte membranes.

These results showed that the determinants **P₁** and **Pᵏ** have closely similar structures, and thus confirmed the serological findings of significant cross-reactivity between these two antigens [121].

(b) Blood Group P Active Glycolipids of the Erythrocyte Membrane

Research on the blood group **P** active substances of the erythrocyte membrane has been conducted by Naiki and Marcus. Since **P** and **P₁** active material was found soluble in organic solvents [6,81], they suspected that the blood group **P** substances were already well-defined glycolipids of the red cell membrane. When various glycolipids of the erythrocyte membrane were tested for their inhibitory activity the authors were able to identify **globoside** (globotetraosylceramide) as **P** substance and **globotriaosylceramide** as **Pᵏ** substance [92]. Shortly thereafter a **P₁** active substance was isolated and characterised as α1,4-galactosyl-paragloboside [90]. The chemical structures of the **P**, **P₁**, and **Pᵏ** active glycosphingolipids are presented in *Fig. 8.1*.

Poppe et al. have studied the three-dimensional structures of the oligo-saccharide termini of **Pᵏ** [97] and **P** substances [98]. Analysis by NMR spectroscopy revealed that the globotriaosyl chain exists in two stable but rapidly interconverting conformations

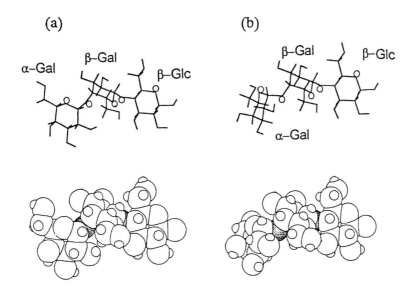

Figure 8.2: Three-dimensional structure of globotriaose (Pᵏ epitope).
(a) and **(b)** diagrams of two stable conformations.
Molecular models as based on the torsion angles published by Poppe et al. [97].

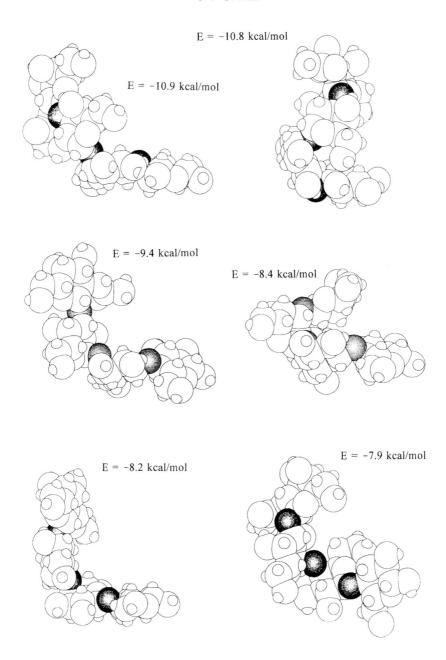

E = −10.8 kcal/mol

E = −10.9 kcal/mol

E = −9.4 kcal/mol

E = −8.4 kcal/mol

E = −8.2 kcal/mol

E = −7.9 kcal/mol

Figure 8.3: Three-dimensional structure of globotetraose (P epitope).
Examples of minimum energy conformations.
Molecular models as based on the torsion angles published by Poppe et al. [98].

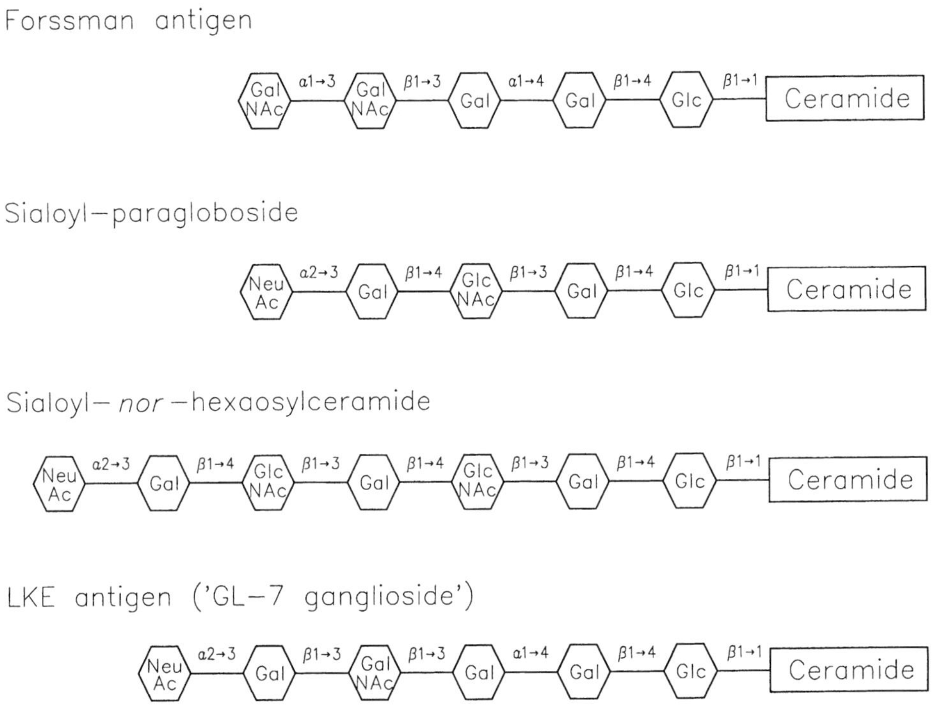

Forssman antigen

Sialoyl—paragloboside

Sialoyl— *nor* —hexaosylceramide

LKE antigen ('GL—7 ganglioside')

Figure 8.4: Chemical structures of Forssman antigen, sialoyl-paragloboside, sialoyl-*nor*-hexaosylceramide, and the LKE (= SSEA-4) antigen.

(Fig. 8.2). The higher flexibility of the globotetraosyl chain, however, permitted the oligosaccharide unit to adopt an increased number of different minimum energy conformations *(Fig. 8.3)*.

As mentioned above (see *Sect. 8.1*), the so-called 'anti-p' sera preferentially agglutinate **p** erythrocytes [28,47,87]. When the specificity of one of those sera, *'Föl'* [87], was studied [62,106] it was found that **p** cells treated with neuraminidase could no longer be agglutinated by this antibody, whereas protease treatment had only weak influence; furthermore, it was shown that the antibody is inhibited specifically by sialoyl-paragloboside and sialoyl-*nor*-hexaosyl-ceramide (see *Fig. 8.4* for formulae). On the basis of these results it becomes obvious that the 'anti-p' sera do not react with a proper 'p determinant'. They react instead with glycosphingolipids containing neuraminic acid — and may possibly also react with protease-resistant sialoglycoproteins occuring in increased amounts in **p** erythrocytes (see *Sect. 8.2.3)*. Thus, anti-**p** sera express a serological specificity similar to that of anti-**Gd** cold agglutinins (see *Chap. 12)*.

Investigations on hybridoma antibodies specific for antigens of embryonic cells showed that the antibody *MC813-70* originally described as anti-**SSEA-4** ('stage specific embryonic antigen 4') [51] had a specificity similar to anti-**Luke** sera [117]. In chromatographical binding tests [51] the anti-**SSEA-4** serum reacted with a derivative of globoside with additional α2,3- N-acetylneuraminyl and galactosyl residues (see 'LKE-Antigen' in *Fig. 8.4)*; this may explain the finding that anti-**Luke** sera are specific for **P(+)** erythrocytes. Final proof of the serological identity of anti-**Luke** and anti-**SSEA-4**, however, is still outstanding.

A series of additional glycosphingolipids of the erythrocyte membrane reacts with anti-globoside sera [52]. One of these substances was identified as a ceramide penta-saccharide of the lacto series with terminal β1,3-N-acetylgalactosamine ('x_2-component', see *Fig. 8.1)*; this glycolipid was also clearly inhibitory with an anti-**P** serum. As these membrane components occur only in minute quantities, their significant contribution to blood group **P** activity of red cells is questionable. It is very likely, however, that the repeatedly observed weak **P** reactivity of **p** erythrocytes [82] may be caused by such glycolipids cross-reacting with anti-**P** sera [52,116]. Glycosphingolipids carrying x_2 and sialoyl-x_2 structures are widely distributed in normal and malignant human tissues [54,116].

(c) P and P₁ Active Glycoproteins of the Erythrocyte Membrane

As the **P₁** active glycolipid occurs only in minute amounts in erythrocyte membranes it was suspected that the **P₁** determinant would be found not only on glycosphingolipids but also on glycoproteins of the red cell membrane.

Fractionation of the erythrocyte stroma in fact showed considerable **P₁** activity present in pronase-labile material; in SDS-polyacrylamide-gel electrophoresis the main **P₁** activity was found in band 4.5 [42]. The fact that membrane glycoproteins contain highly complex and branched carbohydrate chains which are responsible for blood group **Ii** specificity (see *Sect. 7.2)* suggests that membrane glycoproteins act as receptors for the above-mentioned anti-**IP₁** antibodies. This result, however, is in contrast to a more recent investigation, in which no **P₁** and **P^k** antigens could be detected in the glycoprotein fraction of human erythrocytes [130].

Strong evidence for the occurrence of **P** active glycoproteins in erythrocyte membranes also exists. Tonegawa and Hakomori [119] were able to show three protein fractions binding anti-globoside sera (the so-called 'globoproteins'). In SDS-polyacrylamide-gel electrophoresis the material with the highest activity showed an electrophoretic mobility similar to that of the PAS-2 band (see *Sect. 4.1.2)*; the substance, however, is not identical with glycophorin A. The reactivity of the globoproteins with anti-**P** sera has not yet been tested, but it is assumed that these glycoproteins carry **P**-like x_2 determinants.

8.2.3 Occurrence of Blood Group P Active Glycosphingolipids in Erythrocytes and Fibroblasts of Different P Phenotypes

As had been expected, chemical analyses of red cell membranes of various blood group **P** phenotypes [60,63,83] showed considerable variations in their glycosphingolipid content: the **P**-active globoside is the predominant glycolipid of **P**$_1$ and **P**$_2$ erythrocytes. In **P**^{k} cells this glycolipid is present only in traces or is completely absent, whereas the concentration of the blood group **P**^{k} active globotriaosylceramide is considerably increased compared to the level found in normal erythrocytes. Blood group **p** cells lack globoside and globotriaosylceramide (*Table 8.3*) but contain 2–5 times the normal amount of lactosylceramide, hematoside (= G_{M3}), paragloboside, sialoyl-paragloboside (*Fig. 8.1*), and several other more slowly migrating N-acetylglucosamine- and fucose-containing glycosphingolipids, which are probably various **ABH** active substances (see *Fig. 8.5*).

Table 8.3: Presence of neutral glycosphingolipids in the membranes of P$_1$ / P$_2$, P^k, and p erythrocytes.

P phenotype	Concentration [a] of			
	Lactosylceramide	Globotriaosylceramide	Globoside	Paragloboside
P$_1$/P$_2$[b]	2.3	6.3	18.1	1.2
P$_1$/P$_2$[b]	2.7	4.2	16.7	n.d.
P^k	–	19.7	–	6.5
p	14.1	–	–	22.0

[a] μmol glycosphingolipid per 100 ml of packed erythrocytes,
[b] P$_1$/P$_2$ phenotype not indicated.
From Kundu et al. [64] by permission of S. Karger AG, Basel.

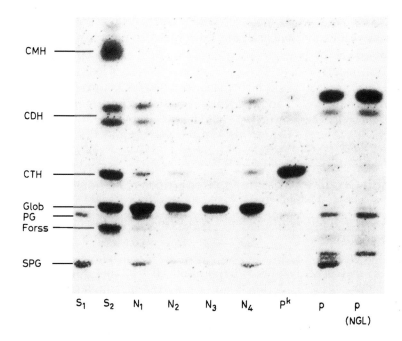

Figure8.5:Thin layer chromatogram of erythrocyte membrane glycosphingolipids from P₁/P₂, Pᵏ, and p donors.

In lanes **S₁** and **S₂** the following glycosphingolipids were used as standards: **CMH** = glucosylceramide, **CDH** = lactosylceramide, **CTH** = globotriaosylceramide, **Glob** = globoside, **PG** = paragloboside, **Forss** = Forssman antigen, **SPG** = sialoyl-paragloboside. **NGL** = neutral glycosphingolipids.
N₁ – **N₄**: four P₁/P₂ individuals.
Reproduced from Marcus et al. [84] by permission of S. Karger AG, Basel.

In erythrocyte membranes the **P₁** active glycolipid is present only in minute amounts; its occurrence in different **P** phenotypes has not yet been investigated.

In thin-layer chromatography the lactoxsylceramide of the erythrocyte membrane is resolved into two bands (see *Fig. 8.5*, 'CDH') which differ in the ceramide moieties: the band that migrates more rapidly contains primarily C_{22} and C_{24} fatty acids, whereas the more slowly moving band contains mostly C_{16} and C_{18} fatty acids [2]. In **p** erythrocytes only the faster moving component is present in increased quantity (*Fig. 8.5*, [83]). Since most erythrocyte membrane glycosphingolipids (e.g. globoside, paragloboside, sialoyl-paragloboside, and **ABH** active substances) contain predominantly C_{22} and C_{24} fatty acids [3] it can be assumed that lactosylceramide with long-chained fatty acids is the 'preferred' precursor in the biosynthesis of complex glycosphingolipids in erythrocytes (see also *Chap. 3*).

Analogous variations in glycosphingolipid content have been observed in gastric mucosal tissue [16] and fibroblasts [57]. However, in contrast to normal erythrocytes where globoside is the predominant glycosphingolipid and globotriaosylceramide is present only in minute amounts [3,83,120], an inverse ratio of these glycosphingolipids was found in fibroblasts. Here, globotriaosylceramide is the main glycosphingolipid [57], thus explaining why P_1 and P_2 fibroblasts react much more strongly with anti-P^k sera than P_1 and P_2 erythrocytes.

8.2.4 Blood Group P Antigens as Receptors for Bacteria and Viruses

The blood group P substances of tissue cells mediate the adhesion of a series of pathogenic bacteria:

– P_1 and P^k glycolipids are the preferential attachment site of pyelonephritogenic *Escherichia coli* to epithelial cells of the human urinary tract, the disaccharide motif Galα1→4Gal representing the receptor for the adhesin molecules of their fimbriae [13,50,69],
– several strains of *Streptococcus suis* which occasionally cause meningitis in humans bind to P^k and P_1 substances [39],
– the PA-I lectin of *Pseudomonas aeruginosa* mediating the adherence of the bacterium to human tissues bound to P^k, (B), and P_1 determinants [37],
– the Galα1→4Gal motif acts as a receptor for bacterial toxins, such as the verotoxins of some pathogenic *Escherichia coli* strains [12,78] and the Shiga toxin of *Shigella dysenteriae* [76,95].

Further, P (globoside) has been shown to be the cellular receptor for parvovirus B19, the cause of *Erythema infectiosum* ('fifth disease') in children [18], p individuals are resistant towards infection by this virus [19].

8.2.5 Chemical Synthesis of the P Determinants

Several groups have succeeded in synthesising P_1 [25,133], P^k [26,49], and P [70,96] determinants using organochemical methods. More recently the total synthesis of the SSEA-4 (= LKE) antigen was reported [68].

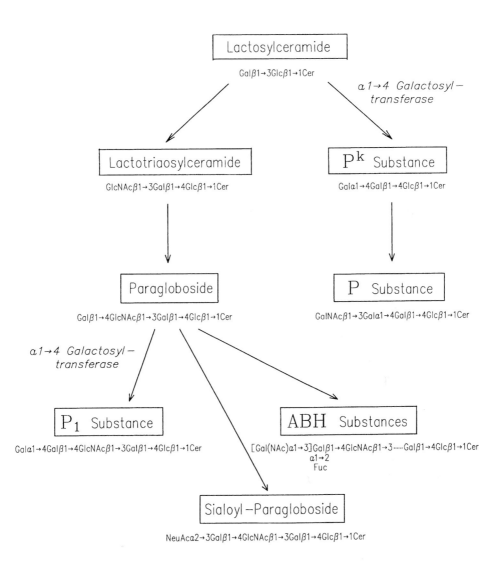

Figure 8.6: Pathways for the biosynthesis of the antigens P, P^k, and P$_1$.

The serologically active oligosaccharides synthesised have repeatedly been linked to various water-insoluble matrices (e.g. Synsorb® products offered by Chembiomed Ltd., Edmonton, Alberta, Canada) and used as immunosorbents for isolation of antibodies.

8.3 Biosynthesis of the Blood Group P Antigens

8.3.1 Biosynthetic Pathway

On the basis of the structures of the blood group **P** active glycosphingolipids a simple scheme for the biosynthesis of the antigens **P**, **P$_1$**, and **P^k** can be established *(Fig. 8.6)*:

In this pathway the lactosylceramide represents the common precursor of **P**, **P$_1$**, and **P^k** antigens. The **P^k** active globotriaosylceramide is produced by the transfer of an α-galactosyl residue to the terminal β-galactose of lactosylceramide, and the **P** antigen (globoside) by the addition of a β-N-acetylgalactosamine to the globotriaosylceramide. The **P$_1$** antigen is synthesised from lactosylceramide by sequential transfer of β-N-acetylglucosamine (→ lactotriaosylceramide), β-galactose (→ paragloboside), and α-galactose to the respective terminal sugar residue.

The characteristic distribution of glycosphingolipids in erythrocyte membranes *(Table 8.3)* shows that in **P$_1$** and **P$_2$** cells lactosylceramide is converted mainly to globoside and only to a much lesser extent to paragloboside. When the synthesis of globotriaosylceramide and globoside is blocked as in the case of **p** cells, the accumulation of lactosylceramide favours the formation of paragloboside and its derivatives, such as sialoyl-paragloboside and **ABH**-active glycolipids (*Fig. 8.7*, cf. *Sect. 8.2.3*).

Two additional **P** phenotypes have been defined by Kundu et al. on the basis of lipid analyses (see *Table 8.4)*. These phenotypes are not characterised by the absence of a transferase but rather can be considered variants with 'less efficient' enzymes: in the case of **F.L.** [64] the synthesis of **P^k** substance is impaired, while in the case of **N.W.** [61] it is the synthesis of globoside which is affected.

In a gastric carcinoma from a **p** individual, significant amounts of the **P** active 'x$_2$-glycolipid' mentioned above have been detected. The fact that in **p** subjects the biosynthesis of **P** substance, globoside, is blocked suggested that the x$_2$-glycolipid is synthesised according to an independent pathway. This assumption was further corroborated by recent investigations reported by Takeya et al. [111]: the authors found an N-acetylgalactosaminyltransferase in human plasma which transferred

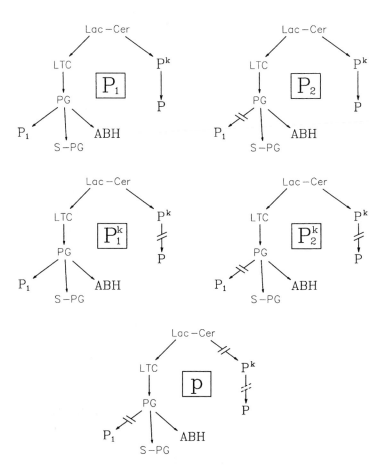

Figure 8.7 Schematic representation of the biosynthesis of antigens P, P^k, and P_1, in P_1, P_2, P_1^k, P_2^k, and p individuals.

specifically β1→3 N-acetylgalactosamine residues onto acceptor substrates with terminal **Galβ1→4GlcNAc** groups; the enzyme, however, was not able to synthesise globoside. Subsequent investigations provided evidence that this enzyme is a β1,3-N-acetylglucosaminyltransferase with a fairly high affinity for UDP-GalNAc [112].

Table 8.4: Presence of neutral glycosphingolipids in the membranes of erythrocytes from the subjects F.L. and N.W.

Phenotype	Concentration [a] of			
	Lactosyl-ceramide	Globotriaosyl-ceramide	Globoside	Paragloboside
P_1/P_2 [b]	2.3	6.3	18.1	1.2
F.L.	2.3	1.7	4.1	2.1
N.W.	4.8	26.5	10.2	n.d.

[a] µmol glycosphingolipid per 100 ml of packed erythrocytes
[b] Control cells, P_1/P_2 phenotype not indicated.
From Kundu et al. [61,64] by permission of S. Karger AG, Basel, and Blackwell Scientific Publications, Oxford, UK.

8.3.2 Investigations on Enzymes

One or probably two α1,4-galactosyltransferases (see *Sect. 8.3.3)* and one β1,3-N-acetylgalactosaminyltransferase are involved in the formation of the three determinants P^k, P_1, and P from their respective precursors. On the basis of the biosynthesis scheme it is generally assumed that the α1,4-galactosyltransferase which converts paragloboside into P_1 substance is absent in P_2 and P_2^k subjects. Similarly it can be suggested that the β1,3-N-acetylgalactosaminyltransferase which transfers the blood group P determinant, β-N-acetylgalactosamine, onto globotriaosylceramide will not be expressed in P_1^k and P_2^k individuals. Further, virtually no α1,4-galactosyl-transferase will be present in p individuals *(Fig. 8.7)*.

These assumptions were generally confirmed when homogenates of fibroblasts and Epstein-Barr virus (EBV)-transformed peripheral lymphocytes from donors of different P phenotypes were tested for glycosyltransferase content:

– β-N-acetylgalactosaminyltransferase synthesising globoside from globotriaosyl-ceramide ('globoside synthase' or P-transferase) was easily detectable in fibroblasts derived from P_2 individuals, but was absent in P_2^k cells [57] (see *Table 8.5)*,

Table 8.5: Activities of globotriaosylceramide (GTC) synthase and globoside synthase in fibroblasts from P_2, P_2^k, and p donors.

Enzyme	Blood group of fibroblast donor		
	P_2	P_2^k	p
GTC synthase	11.4	16.0	0.193
Globoside synthase	0.76	0.013	0.033

Enzyme activities in pmol monosaccharide transferred per mg protein per 2 h (averages of the values published by Kijimoto-Ochiai et al. [57]. Published by permission of the authors.

– an α-galactosyltransferase able to synthesise globotriaosylceramide from lactosylceramide ('GTC-synthase' or P^k-transferase) was present in fibroblasts and cells obtained from P_2 and P_2^k individuals, but could not be located in p cells [57]. An analogous result has been obtained with EBV-transformed lymphocytes [127]. Iizuka et al. [43] detected considerable amounts of P^k-transferase in EBV-transformed B-cells from a p individual, although the intact cells were not able to produce the P^k antigen. This may be based on a different genetic background of the p phenotype.

According to the biosynthetic pathway mentioned above *(Fig. 8.6)* and due to the high frequency of the P^p gene, all indications show that the globoside-synthase should be present in normal quantities in p cells. The scheme furthermore suggests that the lack of P substance is caused by the lack of the respective precursor, globotriaosylceramide. Contrary to expectations, p fibroblasts show only traces of this enzyme [57]. Kijimoto et al. propose that the expression of globoside-synthase is induced by its substrate globotriaosylceramide. This assumption is corroborated by fibroblast fusion experiments performed by Fellous et al. [30,31]: approximately 60% of the hybrids of P_2 and p cells (i.e. the P_2P_2 and P_2p recombinants) showed P activity immediately after cell fusion, but 30–40% of polykaryon cells obtained by fusion of P^k and p fibroblasts (i.e. the P^kp recombinants, in which one parent cell provides P^k substance and the other the gene for the globoside synthase) expressed P antigenicity three to four days after hybridisation.

In a more recent investigation by Wiels et al. [127] EBV-transformed B cell lines from p individuals were found to contain high P-transferase activity.

P^k-transferase (GTC-synthase) has also been detected in human placenta [65], and in liver [115], and kidney [109] of the rat. P-transferase (globoside synthase) has been found in human lymphoblastic cells [114], in dog spleen [113], and in guinea pig kidney [46].

Most recently Steffensen et al. [108a] have cloned an α1,4-galactosyltransferase which was able to synthesise the P^k epitope (*Fig. 8.8*). The gene encoding this enzyme (provisionally designated α1,4Gal-T1) has been located on chromosome 22 at position q13.2 and showed significant homology ($\sim$35%) to the gene coding for a human α1,4-N-acetylglucosaminyltransferase. Expression of full coding constructs of α1,4Gal-T1 in insect cells and transfection experiments using P^k negative Namalwa cells revealed that the encoded enzyme synthesises only the P^k antigen but showed no P_1 synthase activity.

Sequence analysis of the coding region of the α1,4-galactosyltransferase in P_1 and P_2 individuals did not reveal polymorphisms correlating with the P_1P_2 typing.

A single homozygous missense mutation, $M^{183} \to K$, was found in 6 Swedish individuals of the rare **p** type.

```
  1 MSKPPDLLLR LLRGAPRQRV CTLFIIGFKF TFFVSIMIYW HVVGEPKEKG QLYNLPAEIP  60

 61 CPTLTPPTPP SHGPTPGNIF FLETSDRTNP NFLFMCSVES AARTHPESHV LVLMKGLPGG 120

121 NASLPRHLGI SLLSCFPNVQ MLPLDLRELF RDTPLADWYA AVQGRWEPYL LPVLSDASRI 180

181 ALMWKFGGIY LDTDFIVLKN LRNLTNVLGT QSRYVLNGAF LAFERRHEFM ALCMRDFVDH 240

241 YNGWIWGHQG PQLLTRVFKK WCSIRSLAES RACRGVTTLP PEAFYPIPWQ DWKKYFEDIN 300

301 PEELPRLLSA TYAVHVWNKK SQGTRFEATS RALLAQLHAR YCPTTHEAMK MYL         353
```

Figure 8.8: Predicted amino acid sequence of human P^k α1,4-galactosyltransferase.
Underlined: the putative transmembrane domain; *N*: potential N-glycosylation site.
According to Steffensen et al. [108a]. The sequence data is deposited in the EMBL/GenBank data library (accession number AJ245581).

8.3.3 Genetic Regulation of the Biosynthesis of Blood Group P Substances

At present, two genetic models claiming to explain the genetic control of the biosynthesis of blood group **P** antigens are under discussion: one put forward by Naiki and Marcus [93] and the other by Graham and Williams [38].

In both schemes one gene locus with two alleles is postulated as controlling the formation of **P** substance: the active allele, P^2, responsible for the formation of the β1,3-N-acetylgalactosaminyltransferase (= globoside synthase) and the silent allele, P_2^o, which does not code for an active transferase.

The two models propose divergent forms of control for the synthesis of P_1 and P^k determinants. Since the terminal disaccharide structure **Galα1→4Gal** is common to both antigens, both groups of authors suggest that the respective α-galactosyl-transferases show similar specificities. A close relationship between the two transferases is also indicated by the simultaneous occurrence of the **P** of P^k antigens whenever the P_1 antigen is present; there is no known case in which P_1 is detected in the absence of P^k.

In the Naiki–Marcus model two different genes are postulated, P^k and P_1: the P^k gene codes for an α-galactosyltransferase able to form P^k substance by transferring galactosyl residues to lactosyl ceramide; the second gene, P_1, which is inherited independently of the former, encodes a regulatory protein (an enzyme subunit?) that enables the product of the P^k gene to utilise not only lactosyl ceramide but also paragloboside as galactose acceptors, thus forming both P^k and P_1 substance (*Fig. 8.8.a*)[2]. The P_2 phenotype would thus result from a mutation of the P_1 gene, the **p** phenotype from a mutation of the P^k gene impairing the expression of both globo-triaosylceramide and P_1 substance.

In contrast, the Graham–Williams model postulates one single gene locus with three alleles, P^k, P_1^k, and **p**. The authors assume the product of the P^k allele to catalyse only the formation of globotriaosylceramide (P^k substance) from lactosyl-ceramide; according to their hypothesis, the product of the P_1^k allele should convert lactosylceramide into P^k substance, as well as paragloboside into P_1 substance (see *Fig. 8.8.b*). The rare allele **p** represents the inactive form of the gene.

---

[2] A similar control mechanism is known from lactose synthase, the substrate specificity of which is changed by α-lactalbumin (see *Sect. 3.3*).

(a) Genetic Model according to Naiki & Marcus

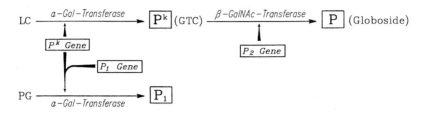

(b) Genetic Model according to Graham & Williams

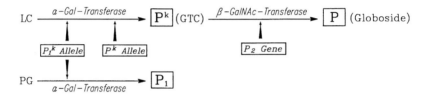

(c) Genetic Model according to Graham & Williams as modified by Fletcher et al.

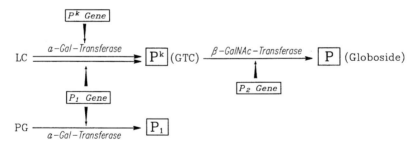

Figure 8.9 Models for the genetic control of the biosynthesis of antigens P, P_1, and P^k.
LC = lactosylceramide, **GTC** = globotriaosyl-ceramide, **PG** = paragloboside.
Following the proposals by Naiki and Marcus [93], Graham and Williams [38], and Fletcher et al. [34].
Reproduced by permission of S. Karger AG, Basel.

Results of family analyses and fibroblast fusion experiments favour the two-enzyme model rather than the one-enzyme model: following Naiki and Marcus, P_1 offspring should be expected from P_2 x p matings in which one parent supplies the P^k allele and the other the P_1 regulatory gene; likewise fusion of $P_2{}^k$ and p, or P_2 and p fibroblasts should produce a certain number of P_1 active cell populations. The investigation by Race and Sanger of five P_2 x p matings with a total of 17 children showed, as proposed in the Graham–Williams model, no P_1 offspring [101]. By analogy, Fellous et al. reported the occurrence of P but not of P_1 specific cells among P^k x p and P_2 x p fibroblast polykaryons [31].

Further, investigations on α1,4-galactosyltransferase activity in human kidney microsomes reported by Bailly et al. [8,9] showed that microsomal proteins from P_1 kidneys catalysed the synthesis of P_1 and P^k glycolipids, whereas microsomes from P_2 kidneys produced only P^k glycolipid. Competition experiments using mixtures of suitable acceptor substrates (i.e. lactosylceramide and neo-lactotetraosylceramide) clearly indicated the presence of two different α1,4-galactosyltransferases.

Table 8.6: Presence of glyco-sphingolipids in blood group P_1 and P_2 erythrocytes.

Glycosphingolipid	Concentration of glycolipid[a] in	
	P_1 cells	P_2 cells
Glucosyl-ceramide	0.37 ± 0.11	0.45 ± 0.19
Lactosyl-ceramide	1.09 ± 0.32	3.15 ± 0.73
Globotriaosyl-ceramide (P^k)	2.07 ± 0.78	1.38 ± 0.53
Globoside (P)	17.49 ± 2.25	17.73 ± 2.56
Ratio lactosyl-ceramide to globotriaosyl-ceramide	0.58 ± 0.21	2.45 ± 1.07

[a] μmol glycosphingolipid per 100 ml of packed erythrocytes.
From Fletcher et al. [34] by permission of the authors.

Additional evidence for the existence of two galactosyltransferases was obtained from the investigations by Fletcher et al. [34]. Analysis of the glycosphingolipid content of human erythrocytes showed that membranes of P_1 cells contain more globotriaosylceramide and less lactosyl ceramide than do membranes from P_2 cells (*Table 8.6*). On the basis of these results Fletcher et al. proposed a modification of the Graham–Williams model on the assumption that the two α1,4-galactosyltransferases are encoded by two independent genes, P_1 and P^k (*Fig. 8.8.c*). Following this model the increased content of globotriaosylceramide in P_1 erythrocytes is caused by the action of the P_1- and P^k-transferases, both of which are able to convert lactosylceramide into P^k substance; in P_2 individuals only the P^k enzyme is present.

References

1. ALLEN, F. H., MARSH, W. L., JENSEN, L. & FINK, J. (1974): Anti-IP: an antibody defining another product of interaction between the genes of the I and P blood group systems. *Vox Sang. 27*, 442-446.
2. ANDO, S., ISOBE, M. & NAGAI, Y. (1976): High performance preparative column chromatography of lipids using a new porous silica, iatrobeads. I. Separation of molecular species of sphingoglycolipids. *Biochim. Biophys. Acta 424*, 98-105.
3. ANDO, S. & YAMAKAWA, T. (1973): Separation of polar glycolipids from human red blood cells with special reference to blood group A activity. *J. Biochem. 73*, 387-396.
4. ANSTEE, D. J., HOLT, P. D. J. & PARDOE, G. I. (1973): Agglutinins from fish ova defining blood groups B and P. *Vox Sang. 25*, 347-360.
5. ANSTEE, D. J. & PARDOE, G. I. (1973): Structural aspects of heterologous blood-group-P_1 substance. *Eur. J. Biochem. 39*, 149-156.
6. ANSTEE, D. J. & TANNER, M. J. A. (1974): The distribution of blood group antigens on butanol extraction of human erythrocyte 'ghosts'. *Biochem. J. 138*, 381-386.
7. BAILLY, P., CHEVALÈYRE, J., SONDAG, D., FRANÇOIS-GÉRARD, C., PIQUET, Y., VEZON, G. & CARTRON, J. P. (1987): Characterization of a murine monoclonal antibody specific for the human P_1 blood group antigen. *Mol. Immunol. 24*, 171-176.
8. BAILLY, P., PILLER, F. & CARTRON, J. P. (1986): Identification of UDP-galactose:lactose (lactosylceramide) α-4 and β-3 galactosyltransferases in human kidney. *Biochem. Biophys. Res. Commun. 141*, 84-91.
9. BAILLY, P., PILLER, F., GILLARD, B., VEYRIÈRES, A., MARCUS, D. & CARTRON, J. P. (1992): Biosynthesis of the blood group-P^k and P_1 antigens by human kidney microsomes. *Carbohydr. Res. 228*, 277-287.
10. BEN-ISMAIL, R., ROUGER, P., CARME, B., GENTILINI, M. & SALMON, C. (1980): Comparative automated assay of anti-P_1 antibodies in acute hepatic distomiasis (fasciolasis) and in hydatidosis. *Vox Sang. 38*, 165-168.
11. BEVAN, B., HAMMOND, W. & CLARKE, R. L. (1970): Anti-P_1 associated with liver-fluke infection. *Vox Sang. 18*, 188-189.
12. BITZAN, M., RICHARDSON, S., HUANG, C., BOYD, B., PETRIC, M. & KARMALI, M. A. (1994): Evidence that verotoxins (Shiga-like toxins) from Escherichia coli bind to P blood group antigens of human erythrocytes in vitro. *Infect. Immun. 62*, 3337-3347.
13. BOCK, K., BREIMER, M. E., BRIGNOLE, A., HANSSON, G. C., KARLSSON, K. A., LARSON, G., LEFFLER, H., SAMUELSSON, B. E., STRÖMBERG, N., SVANBORG-EDÉN, C. & THURIN, J. (1985): Specificity of

binding of a strain of uropathogenic Escherichia coli to Galα1→4Gal-containing glycosphingolipids. *J. Biol. Chem. 260*, 8545-8551.

14. BONO, R., CARTRON, J. P., MULET, C., AVNER, P. & FELLOUS, M. (1981): Selective expression of blood group antigens on human teratocarcinoma cell lines. *Blood Transfus. Immunohaematol. 24*, 97-107.

15. BOOTH, P. B. (1970): Anti-I^{T}P$_1$: an antibody showing a further association between the I and P blood group systems. *Vox Sang. 19*, 85-90.

16. BREIMER, M. E., CEDERGREN, B., KARLSSON, K. A., NILSON, K. & SAMUELSSON, B. E. (1980): Glycolipid pattern of stomach tissue of a human with the rare blood group A,p. *FEBS Lett. 118*, 209-211.

17. BRODIN, N. T., DAHMÉN, J., NILSSON, B., MESSETER, L., MÅRTENSSON, S., HELDRUP, J., SJÖGREN, H. O. & LUNDBLAD, A. (1988): Monoclonal antibodies produced by immunization with neoglycoproteins containing Galα1-4Galβ1-4Glcβ-O and Galα1-4Galβ1-4GlcNAcβ-O residues: useful immuno-chemical and cytochemical reagents for blood group P antigens and a differentiation marker in Burkitt lymphoma and other B-cell malignancies. *Int. J. Cancer 42*, 185-194.

18. BROWN, K. E., ANDERSON, S. M. & YOUNG, N. S. (1993): Erythrocyte P antigen: cellular receptor for B19 parvovirus. *Science 262*, 114-117.

19. BROWN, K. E., HIBBS, J. R., GALLINELLA, G., ANDERSON, S. M., LEHMAN, E. D., McCARTHY, P. & YOUNG, N. S. (1994): Resistance to parvovirus B19 infection due to the lack of virus receptor (erythrocyte P antigen). *N. Engl. J. Med. 330*, 1192-1196.

20. BRUCE, M., WATT, A., GABRA, G. S., MITCHELL, R., LAKHESAR, D. & TIPPETT, P. (1988): LKE red cell antigen and its relationship to P$_1$ and P^k: serological study of a large family. *Vox Sang. 55*, 237-240.

21. CAMERON, G. L. & STAVELEY, J. M. (1957): Blood-group P substance in hydatid cyst fluid. *Nature 179*, 147-148.

22. CONTRERAS, M. & TIPPETT, P. (1974): The Lu(a-b-) syndrome and an apparent upset of P$_1$ inheritance. *Vox Sang. 27*, 369-371.

23. CORY, H. T., YATES, A. D., DONALD, A. S. R., WATKINS, W. M. & MORGAN, W. T. J. (1974): The nature of the human blood group P$_1$ determinant. *Biochem. Biophys. Res. Commun. 61*, 1289-1296.

24. CRAWFORD, M. N., TIPPETT, P. & SANGER, R. (1974): Antigens Au^a, i, and P$_1$ of cells of the dominant type of Lu(a-b-). *Vox Sang. 26*, 283-287.

25. DAHMÉN, J., FREJD, T., MAGNUSSON, G., NOORI, G. & CARLSTRÖM, A. S. (1984): Synthesis of spacer-arm, lipid, and ethyl glycosides of the terminal trisaccharide [α-D-Gal-(1→4)-β-D-Gal-(1→4)-β-D-GlcNAc] portion of the blood group P$_1$ antigen: preparation of neoglycoproteins. *Carbohydr. Res. 129*, 63-71.

26. DAHMÉN, J., FREJD, T., MAGNUSSON, G., NOORI, G. & CARLSTRÖM, A. S. (1984): Synthesis of spacer-arm, lipid, and ethyl glycosides of the trisaccharide portion [α-D-Gal-(1→4)-β-D-Gal-(1→4)-β-D-Glc] of the blood-group P^k antigen: preparation of neoglycoproteins. *Carbohydr. Res. 127*, 15-25.

27. DONATH, J. & LANDSTEINER, K. (1904): Über paroxysmale Hämoglobinurie. *Münch. Med. Wochenschr. 51*, 1590-1593.

28. ENGELFRIET, C. P., BECKERS, D., VON DEM BORNE, A. E. K. G., REYNIERSE, E. & VAN LOGHEM, J. J. (1971): Haemolysins probably recognizing the antigen p. *Vox Sang. 23*, 176-181.

29. FELLOUS, M., CARTRON, J. P., WIELS, J. & TURSZ, T. (1985): A monoclonal antibody against a Burkitt lymphoma associated antigen has an anti-P^k red blood cell specificity. *Brit. J. Haematol. 60*, 559-566.

30. FELLOUS, M., GERBAL, A., KAMOUN, M., CHEREAU, C., JULLIEN, A. & DUMONT, J. (1977): Studies on the P and ABO red blood cell systems using somatic cell genetics. In: *Human Blood Groups* (J. F. Mohn, R. W. Plunkett, R. K. Cunningham, and R. M. Lambert, eds.). S. Karger, Basel, pp. 417-425.

31. Fellous, M., Gerbal, A., Nobillot, G. & Wiels, J. (1977): Studies on the biosynthetic pathway of human P erythrocyte antigen using genetic complementation tests between fibroblasts from rare p and P^k phenotype donors. *Vox Sang. 32*, 262-268.

32. Fellous, M., Gerbal, A., Tessier, C., Frézal, J., Dausset, J. & Salmon, C. (1974): Studies on the biosynthetic pathway of human P erythrocyte antigens using somatic cells in culture. *Vox Sang. 26*, 518-536.

33. Fellous, M., Hors, M. C. & Rebourcet, R. (1977): The expression and relation of HLA, β_2-microglobulin, and receptor for marmoset red cells on man/mouse and man/Chinese hamster hybrid cells. *Eur. J. Immunol. 7*, 22-30.

34. Fletcher, K. S., Bremer, E. G. & Schwarting, G. A. (1979): P blood group regulation of glycosphingolipid levels in human erythrocytes. *J. Biol. Chem. 254*, 11196-11198.

35. François-Gérard, C., Brocteur, J. & Andre, A. (1980): Turtledove: a new source of P$_1$-like material cross-reacting with the human erythrocyte antigen. *Vox Sang. 39*, 141-148.

36. François-Gérard, C., Brocteur, J., André, A., Gerday, C., Pierce-Cretel, A., Montreuil, J. & Spik, G. (1980): Demonstration of the existence of a specific blood-group P$_1$ antigenic determinant in turtle-dove ovomucoid. *Blood Transfus. Immunohaematol. 23*, 579-588.

37. Gilboa-Garber, N., Sudakevitz, D., Sheffi, M., Sela, R. & Levene, C. (1994): PA-I and PA-II lectin interactions with the ABO(H) and P blood group glycosphingolipid antigens may contribute to the broad spectrum adherence of Pseudomonas aeruginosa to human tissues in secondary infections. *Glycoconj. J. 11*, 414-417.

38. Graham, H. A. & Williams, A. N. (1980): A genetic model for the inheritance of the P, P$_1$ and P^k antigens. *Immunol. Commun. 9*, 191-201.

39. Haataja, S., Tikkanen, K., Liukkonen, J., François-Gérard, C. & Finne, J. (1993): Characterization of a novel bacterial adhesion specificity of Streptococcus suis recognizing blood group P receptor oligosaccharides. *J. Biol. Chem. 268*, 4311-4317.

40. Haataja, S., Tikkanen, K., Nilsson, U., Magnusson, G., Karlsson, K. A. & Finne, J. (1994): Oligosaccharide-receptor interaction of the Galα1-4Gal binding adhesin of Streptococcus suis. Combining site architecture and characterization of two variant adhesin specificities. *J. Biol. Chem. 269*, 27466-27472.

41. Hansson, G. C., Wazniowska, K., Rock, J. A., Ness, P. M., Kickler, T. S., Shirey, R. S., Niebyl, J. R. & Zopf, D. (1988): The glycosphingolipid composition of the placenta of a blood group P fetus delivered by a blood group P$^k_1$ woman and analysis of the anti-globoside antibodies found in maternal serum. *Arch. Biochem. Biophys. 260*, 168-176.

42. Haselberger, C. & Schenkel-Brunner, H. (1982): Evidence for erythrocyte membrane glycoproteins of being the predominant carriers of blood group P$_1$ determinants. *FEBS Lett. 149*, 126-128.

43. Iizuka, S., Chen, S. H. & Yoshida, A. (1986): Studies on the human blood group P system: an existence of UDP-Gal:lactosylceramide α1→4galactosyltransferase in the small p type cells. *Biochem. Biophys. Res. Commun. 137*, 1187-1195.

44. Inglis, G., Fraser, R. H., Mitchell, A. A. B., Mackie, A., Allan, E. K. & Mitchell, R. (1987): Serological characterization of a mouse monoclonal anti-P-like antibody. *Vox Sang. 52*, 79-82.

45. Iseki, S., Masaki, S. & Levine, P. (1954): A remarkable family with the rare human isoantibody anti-Tj^a in four siblings; anti-Tj^a and habitual abortion. *Nature 173*, 1193-1194.

46. Ishibashi, T., Kijimoto, S. & Makita, A. (1974): Biosynthesis of globoside and Forssman hapten from trihexosylceramide and properties of β-N-acetylgalactosaminyltransferase of guinea pig kidney. *Biochim. Biophys. Acta 337*, 92-106.

47. Issitt, C. H., Duckett, J. B., Osborne, B. M., Gut, J. B. & Beasley, J. (1976): Another example of an antibody reacting optimally with p red cells. *Br. J. Haematol. 34*, 19-22.

48. Issitt, P. D., Tegoli, J., Jackson, V., Sanders, C. W. & Allen, F. H. (1968): Anti-IP$_1$: antibodies that show an association between the I and P blood group systems. *Vox Sang. 14*, 1-8.

49. Jacquinet, J. C. & Sinay, P. (1985): Chemical synthesis of the human P^k-antigenic determinant. *Carbohydr. Res. 143*, 143-150.

50. Källenius, G., Möllby, R., Svensson, S. B., Winberg, J., Lundblad, A., Svensson, S. & Cedergren, B. (1980): The P^k antigen as receptor for the hemagglutinin of pyelonephrotic E. coli. *FEMS Lett. 7*, 297-302.
51. Kannagi, R., Cochran, N. A., Ishigami, F., Hakomori, S. I., Andrews, P. W., Knowles, B. B. & Solter, D. (1983): Stage-specific embryonic antigens (SSEA-3 and -4) are epitopes of a unique globo-series ganglioside isolated from human teratocarcinoma cells. *EMBO J. 2*, 2355-2361.
52. Kannagi, R., Fukuda, M. N. & Hakomori, S. I. (1982): A new glycolipid antigen isolated from human erythrocyte membranes reacting with antibodies directed to globo-N-tetraosylceramide (globoside). *J. Biol. Chem. 257*, 4438-4442.
53. Kannagi, R., Levery, S. B., Ishigami, F., Hakomori, S. I., Shevinsky, L. H., Knowles, B. B. & Solter, D. (1983): New globoseries glycosphingolipids in human teratocarcinoma reactive with the monoclonal antibody directed to a developmentally regulated antigen, stage-specific embryonic antigen 3. *J. Biol. Chem. 258*, 8934-8942.
54. Kannagi, R., Levine, P., Watanabe, K. & Hakomori, S. I. (1982): Recent studies of glycolipid and glycoprotein profiles and characterization of the major glycolipid antigen in gastric cancer of a patient of blood group genotype pp (Tj^a) first studied in 1951. *Cancer Res. 42*, 5249-5254.
55. Kasai, K., Galton, J., Terasaki, P. I., Wakisaka, A., Kawahara, M., Root, T. & Hakomori, S. I. (1985): Tissue distribution of the P^k antigen as determined by a monoclonal antibody. *J. Immunogenet. 12*, 213-220.
56. Kato, M., Kubo, S. & Naiki, M. (1978): Complement fixation antibodies to glycosphingolipids in sera of rare blood group p and P^k phenotypes. *J. Immunogenet. 5*, 31-40.
57. Kijimoto-Ochiai, S., Naiki, M. & Makita, A. (1977): Defects of glycosyltransferase activities in human fibroblasts of P^k and p blood group phenotypes. *Proc. Natl. Acad. Sci. USA 74*, 5407-5410.
58. Kortekangas, A. E., Kaarsalo, E., Melartin, L., Tippett, P., Gavin, J., Noades, J., Sanger, R. & Race, R. R. (1965): The red cell antigen P^k and its relationship to the P system: the evidence of three more P^k families. *Vox Sang. 10*, 385-404.
59. Kortekangas, A. E., Noades, J., Tippett, P., Sanger, R. & Race, R. R. (1959): A second family with the red cell antigen P^k. *Vox Sang. 4*, 337-349.
60. Koscielak, J., Miller-Podraza, H., Krauze, R. & Cedergren, B. (1976): Glycolipid composition of blood group P erythrocytes. *FEBS Lett. 66*, 250-253.
61. Kundu, S. K., Evans, A., Rizvi, J., Glidden, H. & Marcus, D. M. (1980): A new P^k phenotype in the P blood group system. *J. Immunogenet. 7*, 431-439.
62. Kundu, S. K., Marcus, D. M. & Roelcke, D. (1982): Glycosphingolipid receptors for anti-Gd and anti-p cold agglutinins. *Immunol. Lett. 4*, 263-267.
63. Kundu, S. K., Suzuki, A., Sabo, B., McCreary, J., Niver, E., Harman, R. & Marcus, D. M. (1981): Erythrocyte glycosphingolipids of four siblings with the rare blood group p phenotype and their parents. *J. Immunogenet. 8*, 357-365.
64. Kundu, S. M., Steane, S. M., Bloom, J. E. C. & Marcus, D. M. (1978): Abnormal glycolipid composition of erythrocytes with a weak P antigen. *Vox Sang. 35*, 160-167.
65. Lampio, A., Airaksinen, A. & Maaheimo, H. (1993): UDP-galactose:lactosylceramide α-galactosyltransferase activity in human placenta. *Glycoconj. J. 10*, 165-169.
66. Landsteiner, K. & Levine, P. (1927): Further observations on individual differences of human blood. *Proc. Soc. Exp. Biol. Med. 24*, 941-942.
67. Landsteiner, K. & Levine, P. (1931): The differentiation of a type of human blood by means of normal animal serum. *J. Immunol. 20*, 179-185.
68. Lassaletta, J. M., Carlsson, K., Garegg, P. J. & Schmidt, R. R. (1996): Total synthesis of sialylgalactosylgloboside: stage-specific embryonic antigen 4. *J. Org. Chem. 61*, 6873-6880.
69. Leffler, H. & Svanborg-Eden, C. (1980): Chemical identification of a glycosphingolipid receptor for Escherichia coli attaching to human urinary tract epithelial cells and agglutinating human erythrocytes. *FEMS Microbiol. Lett 8*, 127-134.
70. Leontein, K., Nilsson, M. & Norberg, T. (1985): Synthesis of the methyl and 1-octyl glycosides

of the P-antigen tetrasaccharide (globotetraose). *Carbohydr. Res. 144*, 231-240.

71. LEVINE, P., BOBBITT, O. B., WALLER, R. K. & KUHMICHEL, A. (1951): Isoimmunization by a new blood factor in tumor cells. *Proc. Soc. Exp. Biol. Med. 77*, 403-405.

72. LEVINE, P., CELANO, M. & STAVELEY, J. M. (1958): The antigenicity of P substance in Echinococcus cyst fluid onto tanned red cells. *Vox Sang. 3*, 434-438.

73. LEVINE, P., CELANO, M. J. & FALKOWSKI, F. (1963): The specificity of the antibody in paroxysmal cold hemoglobinuria (P.C.H.). *Transfusion 3*, 278-280.

74. LEVINE, P. & KOCH, E. A. (1954): The rare human isoagglutinin anti-Tjª and habitual abortion. *Science 120*, 239-241.

75. LI, S. C., KUNDU, S. K., DEGASPERI, R. & LI, Y. T. (1986): Accumulation of globotriaosylceramide in a case of leiomyosarcoma. *Biochem. J. 240*, 925-927.

76. LINDBERG, A. A., BROWN, J. E., STRÖMBERG, N., WESTLING-RYD, M., SCHULTZ, J. E. & KARLSSON, K. A. (1987): Identification of the carbohydrate receptor for Shiga toxin produced by Shigella dysenteriae type 1. *J. Biol. Chem. 262*, 1779-1785.

77. LINDSTRÖM, K., VON DEM BORNE, A. E. G. K., BREIMER, M. E., CEDERGREN, B., OKUBO, Y., RYDBERG, L., TENEBERG, S. & SAMUELSSON, B. E. (1992): Glycosphingolipid expression in spontaneously aborted fetuses and placenta from blood group p women. Evidence for placenta being the primary target for anti-Tjª-antibodies. *Glycoconj. J. 9*, 325-329.

78. LINGWOOD, C. A., LAW, H., RICHARDSON, S., PETRIC, M., BRUNTON, J. L., DE GRANDIS, S. & KARMALI, M. (1987): Glycolipid binding of purified and recombinant Escherichia coli produced verotoxin in vitro. *J. Biol. Chem. 262*, 8834-8839.

79. LOPEZ, M., CARTRON, J., CARTRON, J. P., MARIOTTI, M., BONY, V., SALMON, C. & LEVENE, C. (1983): Cytotoxicity of anti-PP₁Pᵏ antibodies and possible relationship with early abortions of p mothers. *Clin. Immunol. Immunopathol. 28*, 296-303.

80. MANDRELL, R. E. (1992): Further antigenic similarities of Neisseria gonorrhoeae lipooligosaccharides and human glycosphingolipids. *Infect. Immun. 60*, 3017-3020.

81. MARCUS, D. M. (1971): Isolation of a substance with blood group P₁ activity from human erythrocyte stroma. *Transfusion 11*, 16-18.

82. MARCUS, D. M., KUNDU, S. K. & SUZUKI, A. (1981): The P blood group system: recent progress in immunochemistry and genetics. *Sem. Hematol. 18*, 63-71.

83. MARCUS, D. M., NAIKI, M. & KUNDU, S. K. (1976): Abnormalities in the glycosphingolipid content of human Pᵏ and p erythrocytes. *Proc. Natl. Acad. Sci. USA 73*, 3263-3267.

84. MARCUS, D. M., NAIKI, M., KUNDU, S. K. & SCHWARTING, G. A. (1977): Immunochemical studies of the human blood group P system. In: *Human Blood Groups* (J. F. Mohn, R. W. Plunkett, R. K. Cunningham, and R. M. Lambert, eds.). S. Karger, Basel, pp. 206-215.

85. MATSON, G. A., SWANSON, J., NOADES, J., SANGER, R. & RACE, R. R. (1959): A 'new' antigen and antibody belonging to the P blood group system. *Amer. J. Hum. Genet. 11*, 26-34.

86. MCALPINE, P. J., KAIA, H. & LEWIS, M. (1978): Is the Dia₁ locus linked to the P blood group locus? *Cytogenet. Cell Genet. 22*, 629-632.

87. METAXAS, M. J., METAXAS-BÜHLER, M. & TIPPETT, P. (1975): A 'new' antibody in the P blood group system. *XIVth Congress of the International Society of Blood Transfusion, Helsinki, 1975*, Finnish Red Cross Blood Transfusion Service, Helsinki. Abstracts p. 95.

88. MORGAN, W. T. J. & WATKINS, W. M. (1964): Blood group P₁ substance. I. Chemical properties. *Proceedings of the IXth Congress of the International Society of Blood Transfusion, Mexico City, 1962*, Karger, Basel, pp. 225-229.

89. MOURANT, A. E., KOPEC, A. C. & DOMANIEWSKA-SOBCZAK, K. (1976): *The distribution of the human blood groups and other polymorphisms*. 2nd edn., Oxford University Press, London 1976.

90. NAIKI, M., FONG, J., LEDEEN, R. & MARCUS, D. M. (1975): Structure of the human erythrocyte blood group P₁ glycosphingolipid. *Biochemistry 14*, 4831-4837.

91. NAIKI, M. & KATO, M. (1979): Immunological identification of blood group Pᵏ antigen on normal human erythrocytes and isolation of anti-Pᵏ with different affinity. *Vox Sang. 37*, 30-38.

92. NAIKI, M. & MARCUS, D. M. (1974): Human erythrocyte P and Pᵏ blood group antigens: identification

as glycosphingolipids. *Biochem. Biophys. Res. Commun. 60*, 1105-1111.

93. NAIKI, M. & MARCUS, D. M. (1975): An immunochemical study of the human blood group P_1, P, and P^k glycosphingolipid antigens. *Biochemistry 14*, 4837-4841.

94. NUDELMAN, E., KANNAGI, R., HAKOMORI, S. I., PARSONS, M., LIPINSKI, M., WIELS, J., FELLOUS, M. & TURSZ, T. (1983): A glycolipid antigen associated with Burkitt lymphoma defined by a monoclonal antibody. *Science 220*, 509-511.

95. OBRIG, T. G., LOUISE, C. B., LINGWOOD, C. A., BOYD, B., BARLEY-MALONEY, L. & DANIEL, T. O. (1993): Endothelial heterogeneity in Shiga toxin receptors and responses. *J. Biol. Chem. 268*, 15484-15488.

96. PAULSEN, H. & BÜNSCH, A. (1982): Synthese der Tetrasaccharid-Kette des P-Antigen-Globosids. Eine β-D-Glycosidsynthese für 2-Amino-2-Desoxyzucker. *Carbohydr. Res. 101*, 21-30.

97. POPPE, L., DABROWSKI, J., VON DER LIETH, C. W., KOIKE, K. & OGAWA, T. (1990): Three-dimensional structure of the oligosaccharide terminus of globotriaosylceramide and isoglobotriaosylceramide in solution. A rotating-frame NOE study using hydroxyl groups as long-range sensors in conformational analysis by ^{1}H-NMR spectroscopy. *Eur. J. Biochem. 189*, 313-325.

98. POPPE, L., VON DER LIETH, C. W. & DABROWSKI, J. (1990): Conformation of the glycolipid globoside head group in various solvents and in the micelle-bound state. *J. Amer. Chem. Soc. 112*, 7762-7771.

99. PROKOP, O. & SCHLESINGER, D. (1965): Über das Vorkommen von P_1-Blutgruppensubstanz in Ascaris suum. *Dtsch. Gesundh. Wes.*, 1584.

100. PROKOP, O. & SCHLESINGER, D. (1965): Über das Vorkommen von P_1-Blutgruppensubstanz oder einer 'P_1'- like-substance' bei Lumbricus terrestris. *Acta Biol. Med. Germ. 15*, 180-181.

101. RACE, R. R. & SANGER, R. (1975): The P-blood groups. In: *Blood Groups in Man.* Blackwell Scientific Publications, Oxford, pp. 139-177.

102. RYDBERG, L., CEDERGREN, B., BREIMER, M. E., LINDSTRÖM, K., NYHOLM, P. G. & SAMUELSSON, B. E. (1992): Serological and immunochemical characterization of anti-PP$_1$P^k (anti-Tj^a) antibodies in blood group little p individuals. Blood group A type 4 recognition due to internal binding. *Molec. Immunol. 29*, 1273-1286.

103. SANGER, R. (1955): An association between the P and Jay systems of blood groups. *Nature 176*, 1163-1164.

104. SCHNITZLER, S., MÜLLER, G. & PROKOP, O. (1967): Ein "neuer" Antikörper, Anti-P$_{rut}$, aufgefunden im Rogen von Rutilus rutilus. *Z. Immun.-Forsch. 134*, 45-53.

105. SCHWARTING, G. A., KUNDU, S. K. & MARCUS, D. M. (1979): Reaction of antibodies that cause paroxysmal cold hemoglobinuria (PCH) with globoside and Forssman glycosphingolipid. *Blood 53*, 186-192.

106. SCHWARTING, G. A., MARCUS, D. M. & METAXAS, M. (1977): Identification of sialosylparagloboside as an erythrocyte receptor for an 'anti-p' antibody. *Vox Sang. 32*, 257-261.

107. SÖDERSTRÖM, T., ENSKOG, A., SAMUELSSON, B. E. & CEDERGREN, B. (1985): Immunoglobulin subclass (IgG3) restriction of anti-P and anti-P^k antibodies in patients of the rare p blood group. *J. Immunol. 134*, 1-3.

108. STAVELEY, J. M. & CAMERON, G. L. (1958): The inhibiting action of hydatid cyst fluid on anti-Tj^a sera. *Vox Sang. 3*, 114-118.

108a. STEFFENSEN, R., CARLIER, K., WIELS, J., LEVERY, S.B., STROUD, M., CEDERGREN, B., NILSSON-SOJKA, B., BENNETT, E.P., JERSILD, C. & CLAUSEN, H. (2000): Cloning and expression of the histo-blood group P^k UDP-galactose: Galβ1-4Glcβ1-Cer α1,4-galactosyltransferase: molecular genetic basis of the p phenotype. *J. Biol. Chem.* (in press).

109. STOFFYN, A., STOFFYN, P. & HAUSER, G. (1974): Structure of trihexosylceramide biosynthesized in vitro by rat kidney galactosyltransferase. *Biochim. Biophys. Acta 360*, 174-178.

110. SUDAKEVITZ, D., LEVENE, C., SELA, R. & GILBOA-GARBER, N. (1996): Differentiation between human red cells of P^k and p blood types using Pseudomonas aeruginosa PA-I lectin. *Transfusion 36*, 113-116.

111. TAKEYA, A., HOSOMI, O., SHIMODA, N. & YAZAWA, S. (1992): Biosynthesis of the blood group P

antigen-like GalNAcβ1→3Galβ1→4GlcNAc/Glc structure: a novel acetylgalactosaminyltransferase in human blood plasma. *J. Biochem. 112*, 389-395.

112. TAKEYA, A., HOSOMI, O., YAZAWA, S. & KOGURE, T. (1993): Biosynthesis of the blood group P antigen-like GalNAcβ1→3Galβ1→4GlcNAc/Glc structure: kinetic evidence for the responsibility of N-acetylglucosaminyltransferase. *Jp. J. Med. Sci. Biol. 46*, 1-15.

113. TANIGUCHI, N. & MAKITA, A. (1984): Purification and characterization of UDP-N-acetylgalactos-amine: globotriaosylceramide β-3-N-acetylgalactosaminyltransferase, a synthase of human blood group P antigen, from canine spleen. *J. Biol. Chem. 259*, 5637-5642.

114. TANIGUCHI, N., YANAGISAWA, K., MAKITA, A., MIZUNO, F. & OSATO, T. (1985): Globoside and Forssman synthases in human lymphocytes exposed to Epstein-Barr virus and mitogens. *J. Natl. Cancer Inst. 74*, 563-568.

115. TANIGUCHI, N., YANAGISAWA, K., MAKITA, A. & NAIKI, M. (1985): Purification and properties of rat liver globotriaosylceramide synthase, UDP-galactose:lactosylceramideα1-4-galactosyltransferase. *J. Biol. Chem. 260*, 4908-4913.

116. THORN, J. J., LEVERY, S. B., SALYAN, M. E. K., STROUD, M. R., CEDERGREN, B., NILSSON, B., HAKOMORI, S. & CLAUSEN, H. (1992): Structural characterization of x_2 glycosphingolipid, its extended form, and its sialosyl derivatives: accumulation associated with the rare blood group p phenotype. *Biochemistry 31*, 6509-6517.

117. TIPPETT, P., ANDREWS, P. W., KNOWLES, B. B., SOLTER, D. & GOODFELLOW, P. N. (1986): Red cell antigens P (globoside) and Luke: identification by monoclonal antibodies defining the murine stage-specific embryonic antigens-3 and -4 (SSEA-3 and SSEA-4). *Vox Sang. 51*, 53-56.

118. TIPPETT, P., SANGER, R., RACE, R. R., SWANSON, J. & BUSH, S. (1965): An agglutinin associated with the P and the ABO blood group systems. *Vox Sang. 10*, 269-280.

119. TONEGAWA, Y. & HAKOMORI, S. I. (1977): "Ganglioprotein and globoprotein": the glycoproteins reacting with anti-ganglioside and anti-globoside antibodies and the ganglioprotein change associated with transformation. *Biochem. Biophys. Res. Commun. 76*, 9-17.

120. VANCE, D. E. & SWEELEY, C. C. (1967): Quantitative determination of the neutral glycosyl ceramides in human blood. *J. Lipid Res. 8*, 621-630.

121. VOAK, D., ANSTEE, D. J. & PARDOE, G. (1973): The α-galactose specificity of anti-P^k. *Vox Sang. 25*, 263-270.

122. VOAK, D., TODD, G. M. & PARDOE, G. I. (1974): A study of the serological behaviour and nature of the anti-B/P/P^k activity of Salmonidae roe protectin. *Vox Sang. 26*, 176-188.

123. VON DEM BORNE, A. E. G., BOS, M. J. E., JOUSTRA-MAAS, N., TROMP, J. F., VAN WIJNGAARDEN - DU BOIS, R. & TETTEROO, P. A. T. (1986): A murine monoclonal IgM antibody specific for blood group P antigen (globoside). *Brit. J. Haematol. 63*, 35-46.

124. WALSH, R. J. & KOOPTZOFF, O. (1954): The human blood group Tj^a. *Aust. J. Exp. Biol. Med. Sci. 32*, 387-392.

125. WATKINS, W. M. & MORGAN, W. T. J. (1964): Blood-group P_1 substance. II. Immunological properties. *Proceedings of the IXth Congress of the International Society of Blood Transfusion, Mexico City 1962*, Karger, Basel, pp. 230-234.

126. WIELS, J., FELLOUS, M. & TURSZ, T. (1981): Monoclonal antibody against a Burkitt lymphoma-associated antigen. *Proc. Natl. Acad. Sci. USA 78*, 6485-6488.

127. WIELS, J., TAGA, S., TÉTAUD, C., CEDERGREN, B., NILSSON, B. & CLAUSEN, H. (1996): Histo-blood group p: biosynthesis of globoseries glycolipids in EBV-transformed B cell lines. *Glycoconj. J. 13*, 529-535.

128. WORLLEDGE, S. M. & ROUSSO, C. (1965): Studies on the serology of paroxysmal cold haemoglobinuria (P.C.H.), with special reference to its relationship with the P blood group system. *Vox Sang. 10*, 293-298.

129. YAMASAKI, R., BACON, B. E., NASHOLDS, W., SCHNEIDER, H. & GRIFFISS, J. M. (1991): Structural determination of oligosaccharides derived from lipooligosaccharide of Neisseria gonorrhoeae F62 by chemical, enzymatic, and two-dimensional NMR methods. *Biochemistry 30*, 10566-10575.

130. YANG, Z. T., BERGSTRÖM, J. & KARLSSON, K. A. (1994): Glycoproteins with Galα4Gal are absent

from human erythrocyte membranes, indicating that glycolipids are the sole carriers of blood group P activities. *J. Biol. Chem. 269*, 14620-14624.

131. YATES, A. D., DONALD, A. S. R., WATKINS, W. M. & MORGAN, W. T. J. (1975): Immunochemical observations on the human blood group P system. *XIVth Congress of the International Society of Blood Transfusion, Helsinki, 1975*, Finnish Red Cross Blood Transfusion Service, Helsinki. Abstract p. 95.

132. YATES, A. D., MORGAN, W. T. J. & WATKINS, W. M. (1975): Linkage-specific α-D-galactosidases from Trichomonas foetus: characterization of the blood-group B-destroying enzyme as a 1,3-α-galactosidase and the blood group P$_1$-destroying enzyme as a 1,4-α-galactosidase. *FEBS Lett. 60*, 281-285.

133. ZOLLO, P. H. A., JACQUINET, J. C. & SINAY, P. (1983): Chemical synthesis of the human blood-group P$_1$-antigenic determinant. *Carbohydr. Res. 122*, 201-208.

9 MNS System

The **MNS** blood group system was discovered by Landsteiner & Levine in 1927. Originally represented by the allelic antigens **M** [192] and **N** [191] it was later expanded to include the newly discovered allelic characters **S** [265,312] and **s** [196], as well as **U** [319] (*Table 9.1*). In the course of further investigations a large number of low-frequency antigenic characters closely associated with the **MNS** system have also been described (*Table 9.2*).

The **MNS** antigens are generally considered characteristics of erythrocytes and their precursor cells in bone marrow. Although glycophorin A, the blood group **MN** substance, can be found on the earliest recognisable red cell precursor, significant **MN** activity does not begin to develop until the polychromatic normoblast stage [102,119,324]. The antigens of the **MNS** system can be found on mature erythrocytes in young human fetuses (e.g. [277]). To the extent already investigated, these antigens are fully developed at birth [249]. **MN** determinants have also been detected on erythroid leukemia cells [9,120]. Some reports suggest the occurrence of blood group

Table 9.1: MN and Ss phenotypes.

Phenotype	Frequency in Europids	Frequency in Negrids
M+ N−	28%	25%
M+ N+	50%	49%
M− N+	22%	26%
S+ s−	11%	5.9%
S+ s+	44%	24.5%
S− s+	45%	68.1%
S− s−	0%	1.5%

From Issitt [161].

Table 9.2: Antigens of the MNS system.

Antigen	Historical name	ISBT Nr.[a]	References
M		MNS1	[192]
N		MNS2	[191]
S		MNS3	[265,312]
s		MNS4	[196]
U		MNS5	[319]
He	Henshaw	MNS6	[157]
Mi^a	Miltenberger	MNS7	[197]
M^c		MNS8	[100]
Vw	Gr, Verweyst	MNS9	[304]
Mur	Murrell	MNS10	[161]
M^g		MNS11	[5]
Vr	Verdegaal	MNS12	[304]
M^e		MNS13	[317]
Mt^a	Martin	MNS14	[286]
St^a	Stones	MNS15	[49]
Ri^a	Ridley	MNS16	[49]
Cl^a	Caldwell	MNS17	[310]
Ny^a	Nyberg	MNS18	[237]
Hut	Hutchinson	MNS19	[50]
Hil	Hill	MNS20	[50]
M^v		MNS21	[124]
Far(Kam)		MNS22	[56]
s^D	Dreyer	MNS23	[271]
Mit	Mitchell	MNS24	[20]
Dantu		MNS25	[301,303]
Hop	Hopper	MNS26	[125]
Nob	Noble	MNS27	[125]
En^a		MNS28	[95,117]
ENKT	En^{a}KT	MNS29	
'N'		MNS30	[4,135]
Or	Orriss	MNS31	[19]
DANE		MNS32	[275]
TSEN		MNS33	[259]
MINY		MNS34	[260]
MUT		MNS35	[296]
SAT		MNS36	[152]
ERIK		MNS37	[94]
Os^a		MNS38	[93]
ENEP		MNS39	
ENEH		MNS40	[154,283]
HAG		MNS41	[244]
ENAV (AVIS)		MNS42	[169]
MARS		MNS43	[169]
M$_1$			[168]
U^x			[39]
U^z			[38]
Duclos			[129]
Hu	Hunter		[193]
Sul	Sullivan		[184]
Sj	Stenbar-James		[165]

[a] 002 001 to 002 040 in numerical notation, see *Chap. 1.*

M and N reactive material in renal capillary endothelium [132,133] and U reactive material in leukocytes and thrombocytes [214]. The blood group M and N active glycolipid material reported as occurring in meconium [55] was later identified as cholesterol-sulfate, which, obviously due to a characteristic distribution of electrical charge and hydrophobic regions, binds preferentially to anti-N and anti-M sera [123].

The MNS antigens are confined to humans and hominoid primates [223,256,316].

9.1 Genetics

The MNS locus (= GYP) consists of three closely linked genes, GPA, GPB, and GPE. Blood group M and N specificities are controlled by the GPA gene, S, s, and U specificities by the GPB gene; the GPE gene is not responsible for any MNS specificity of erythrocytes. Investigations on structure and organisation of the MNS locus are discussed in Sect. 9.4.

Most of the MNS variants and some erythrocyte phenotypes connected with the MNS system (see Sect. 9.6) can be attributed to alleles of MN or Ss, or to mutants deficient in one or both of these genes. Some low-incidence blood group characters are located on the products of MN /Ss hybrid genes which obviously are the result of chromosomal misalignment followed by unequal crossing-over events during meiosis. Only in rare cases is the MNS antigenicity influenced by genes which are independent of the MNS locus.

The MNS gene complex has been localised on the long arm of chromosome 4 in the area q28–q31 [53,251].

9.2 Antisera and Lectins

Most examples of anti-M and anti-N antibodies occurring in human sera are cold-reactive isoagglutinins (see [160]). They generally show a distinct dosage effect. In contrast, anti-S and anti-s are usually immune antibodies induced by transfer of incompatible blood in the course of blood transfusions or pregnancies. Various cases of autoimmune haemolytic anaemias have been reported as being caused by MNS specific auto-antibodies [161].

Strong anti-M, anti-N, and anti-s sera have repeatedly been prepared in rabbits by immunisation with erythrocytes of the respective blood group [191,192,248]. Anti-M is also frequently found as a 'naturally occurring' antibody in horse and cattle sera [249].

More recently a series of **hybridoma antibodies** with anti-**M** [14,25,109,204, 230,263] and anti-**N** specificities [8,14,24,25,107,110,111,263] have been described.

Some of the anti-**M**, anti-**N**, anti-**S,** and anti-**U** antibodies found in human serum are more reactive at pH 6.5 [22,23,195,257] or at decreased ionic strength [103,320]. Another type of unusual antibodies are the 'glucose-dependent' anti-**M** and anti-**N** which react only with erythrocytes pre-incubated in glucose-containing media. According to the investigations by Drzeniek et al. [96] these antibodies recognise a structure formed by the addition of glucose to the amino group of the N-terminal amino acid of the **M** or **N** antigen.

Naturally occurring anti-**En**[a] antisera (see 'Antigen Complex En[a]', *Sect. 9.6*), as well as various hybridoma antibodies [13,14,232] react with different epitopes on the glycophorin A molecule [122]. Though not connected to any **MNS** character, these antibodies have proved valuable tools for structural studies on blood group **MNS** substances and investigations on glycophorin A variants and hybrid molecules.

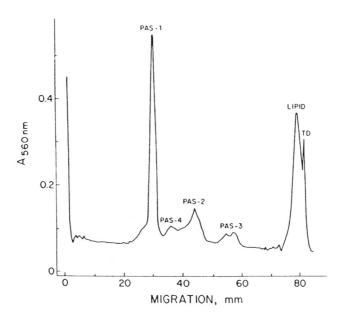

Figure 9.1: Polyacrylamide-gel electrophoresis of erythrocyte membrane sialo-glycoproteins.
Separation by SDS polyacrylamide-gel electrophoresis and densitometric scannning of the gels stained for carbohydrate with PAS reagent. **TD** = 'tracking dye'. For nomenclature of bands compare *Table 9.3*. Reproduced from Steck [284] by permission of the author and The Rockefeller University Press, New York (©1974).

The occurrence and specificity of antisera towards rare **MNS** variants are referred to in the sections on the respective antigens and **MNS** phenotypes (see *9.6*).

Lectins strongly agglutinating **N** but not (or only weakly) agglutinating native **M** erythrocytes have been detected in the seeds of *Vicia graminea* (anti-N_{Vg}) [238,247] (see *Sect. 9.5*) and *Vicia unijuga* [322]. Other sources of anti-**N** specific lectins are plants of the genus *Bauhinia* (e.g. *B. purpurea* and *B. variegata*) [40,108,158]. The lectin from *Moluccella laevis* shows anti-**(A+N)** specificity [26,99,198]. An anti-**M** lectin has been found in the seeds of *Iberis amara* [7].

Thus far no anti-**M** or anti-**N** specific lectins have been detected in animals.

9.3 Blood Group MNS Substances

The antigenic determinants of the **MNS** system are located on glycophorin A and glycophorin B, two sialoglycoproteins of the human erythrocyte membrane [131] (see *Sect. 4.1.2*). Upon separation of the membrane constituents by SDS-polyacryl-amidegel electrophoresis, these proteins are resolved in a series of bands which can easily be detected by periodic acid Schiff (= PAS) reagent [104] (*Fig. 9.1*). Note the discontinuous gel system by Lämmli, which shows a fairly complex pattern, since the molecules may appear as monomers as well as homo- or heterodimers [225]. A survey of the different PAS bands and a comparison of the different nomenclature systems found in the literature is presented in *Table 9.3*.

Glycophorin A:

Glycophorin A (α-sialoglycoprotein[1]) is the carrier of **M** and **N** specificities [82]. It is an intrinsic membrane protein with an apparent molecular mass of 37 kDa [60]. When erythrocyte membrane proteins are analysed by SDS-polyacrylamide-gel electrophoresis and stained with PAS reagent, glycophorin A appears as band PAS-2 in its monomeric form, as band PAS-1 in its dimeric form, and as band PAS-4 as a glycophorin A/B hybrid-dimer which may have been formed during the isolation procedure [82].

In situ (i.e. in the erythrocyte membrane), glycophorin A is sensitive to trypsin and only partially degraded by chymotrypsin [84,180]. The number of glycophorin A molecules per erythrocyte is of the order of 1 million [60,122].

---

[1] Other synonyms: GPA, **MN**-sialoglycoprotein, PAS-1 glycoprotein.

Table 9.3: Sialoglycoproteins of the erythrocyte membrane.
(Comparing different nomenclature systems).

Glycophorin	Designation of the band according to PAS nomenclature[a]	Anstee et al.[b]	% of PAS stained material[c]
	A		
Glycophorin A dimer	PAS-1	α_2	51
	F		
Glycophorin A/B dimer	PAS-4 or B	$\alpha\delta$	12
Glycophorin B dimer	C	δ_2	6
Glycophorin A monomer	PAS-2	α	17
Glycophorin C	PAS-2' or D	β	6
	L		
	K_1 and K_2	β_1	
		ε	
Glycophorin D[d]	E	γ	1
Glycophorin B monomer	PAS-3	δ	7
	J		

[a] According to Fairbanks et al. [104], Steck et al. [285], and Dahr et al. [86]
[b] According to Anstee et al. [16]
[c] According to Furthmayr [114]
[d] Designation proposed by Dahr [60].

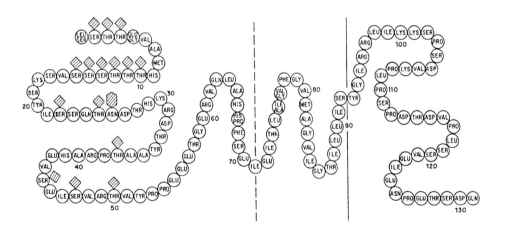

Figure 9.2: The glycophorin A molecule.
| |: Position of the erythrocyte membrane, ⬦: position of the O-linked oligosaccharide chains, ▨: position of the N-linked oligosaccharide chains.

Reproduced from Marchesi et al. [213]. With permission of the publisher.

The peptide moiety of the glycophorin A molecule is composed of 131 amino acids. Their sequence was established by Tomita and Marchesi [298,299] (*Figs. 9.2* and *9.4*; two positions have later been corrected by Dahr et al. [66]). The results of these investigations were subsequently confirmed by nucleotide sequencing of isolated glycophorin A-specific cDNA clones [251,272,291].

On the glycophorin A molecule three domains can be distinguished:

- a heavily glycosylated segment (residues 1–72) predominantly composed of hydrophilic amino acids and situated on the outside of the cell,
- a segment (residues 73–95) composed mainly of hydrophobic amino acids which spans the lipid bilayer of the erythrocyte membrane, and
- the C-terminal end (residues 96–131) which projects into the cytoplasm; this segment is characterised by its high content of acid amino acids and proline.

As mentioned in *Sect. 4.2* phosphatidyl-serine and phosphatidyl-inositide bind to the hydrophobic segment of glycophorin A. The formation of this complex presumably stabilises a special conformation of the molecule and thus might influence the expression of **M** and **N** antigenicity.

More recent investigations [45] have shown that amino acid residues 73–95 adopt a helical configuration and are oriented parallel to the acyl chains of the lipid bilayer, whereas residues 66–72 adopt a β-sheet conformation and are oriented parallel to the lipid-water interface.

Glycophorin A is a glycoprotein with approximately 60% carbohydrate. It represents 1.6% of the total protein mass and provides about 67% of the neuraminic acid content of normal erythrocytes [41,87]. The glycophorin A molecule carries ~17 carbohydrate side chains located in the section between amino acids 1 and 50, consisting of up to 16 copies of an alkali-labile tetrasaccharide chain O-linked to serine or threonine residues, and one alkali-stable chain N-linked to asparagine at position 26 ([242], see *Fig. 9.2*). It should be mentioned, however, that studies by Dahr et al. [77] have revealed a fairly heterogeneous glycosylation pattern of the serine and threonine residues especially at positions 33 – 50 of the peptide chain.

The structure of the alkali-labile tetrasaccharide chain has been determined by Thomas and Winzler [294] (*Fig. 9.3.a*); apart from this tetrasaccharide, trisaccharide units with only one neuraminyl residue have been detected – some 25% of the chains have terminal α2→3 neuraminic acid (*Fig. 9.3.b*) and about 8% have an α2→6 neuraminic acid attached to the N-acetylgalactosamine [202].

Besides these chain types small amounts of other alkali-labile oligosaccharide structures have been detected in human erythrocytes:

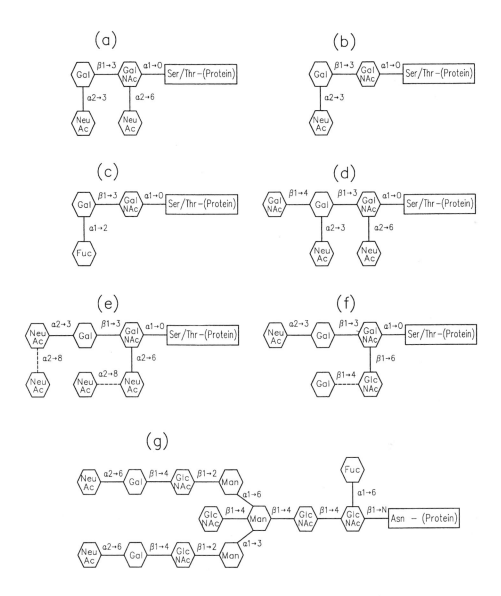

Figure 9.3: Structures of the oligosaccharide chains of glycophorin A.
(a) Disialotetrasaccharide [294], **(b)** monosialotrisaccharide [202], **(c)** blood group **H** active trisaccharide [287], **(d)** blood group **Cad** specific oligosaccharide [30], **(e)** oligosaccharide with disialoyl goups [112], **(f)** N-acetylglucosamine-containing oligosaccharide [2], **(g)** alkali-stable chain after Yoshima et al. [323].

- short fucose-containing oligosaccharide units with weak blood group **ABH** activity (*Fig. 9.3.c*); in the majority of these chains, however, the innermost N-acetyl-galactosamine residue is substituted by $\alpha 2 \rightarrow 6$ neuraminic acid [287],
- blood group **Cad** specific oligosaccharide units (see *Sect. 11.3*) characterised by a terminal N-acetylgalactosamine residue (*Fig. 9.3.d*),
- oligosaccharides with disialoyl groups (*Fig. 9.3.e*) [112],
- oligosaccharides in which the $\alpha 2 \rightarrow 6$ neuraminic acid is exchanged by N-acetylglucosamine or N-acetyllactosamine units [2,34] (*Fig. 9.3.f*). More complex chains of this type may show blood group **i** or **I** activity or may carry **ABH** determinants (see also [287]). These oligosaccharide units occur only rarely in Europids but are found fairly frequently in Negrids. There is some evidence that these chains are associated with rare **MNS** phenotypes, such as M_1, **Can**, **Tm**, and **Hunter**.

The exact structure of the alkali-stable oligosaccharide unit has not yet been firmly established. The investigations conducted by several groups revealed a complex-type carbohydrate chain [159,295,323]; a model of its structure as proposed by Yoshima et al. [323] is presented in *Fig. 9.3.g*. N-linked oligosaccharide chains with higher molecular weights detected by the same group probably differ from the basic structure by additional N-acetyllactosamine units.

The N-linked carbohydrate chain is not involved in any known **MNS** determinant; it is, however, responsible for the binding of various lectins (e.g. wheat germ agglutinin and the lectins of *Phaseolus vulgaris* and *Lens culinaris*) to glycophorin A [1,113,185].

Glycophorin B

Glycophorin B (δ-sialoglycoprotein[(2)]) is a minor red cell sialoglycoprotein with an apparent molecular mass of 24 kDa [60]. It is responsible for **S**, **s**, and **U** specificities of human erythrocytes [88,131]. When analysed by SDS polyacrylamide-gel electrophoresis glycophorin B appears as band PAS-3 in its monomeric form, as band C in its dimeric form, and as band PAS-4 as a hybrid-dimer with glycophorin A [82].

In contrast to glycophorin A, glycophorin B is not trypsin-sensitive *in situ*, but is easily cleaved by chymotrypsin [89]. The values determined for the average number of glycophorin B molecules per erythrocyte vary between 80,000 and 300,000 [58,122,217].

The polypeptide backbone of glycophorin B contains 72 amino acids. The N-terminal segment has been characterised by classical amino acid sequencing [33,

---

[(2)] Other synonyms: GPB, **Ss**-sialoglycoprotein, PAS-3 glycoprotein.

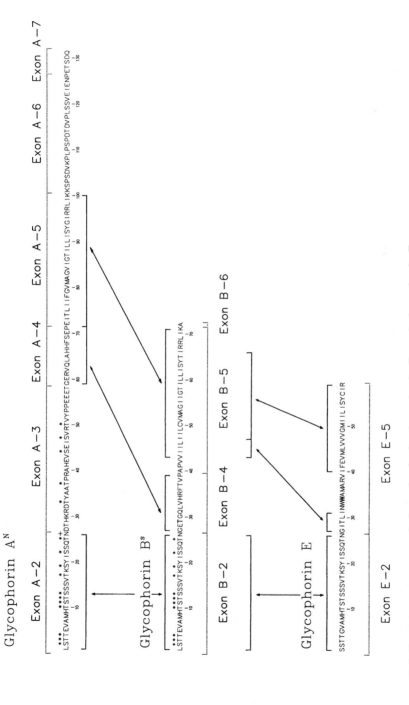

Figure 9.4: Peptide sequences of glycophorin A, glycophorin B, and glycophorin E.

*: O-linked carbohydrate chains, +: N-linked carbohydrate chain.

According to Kudo & Fukuda [187,188].

The sequence data of the glycophorin molecules are deposited in the EMBL/GenBank data library under the accession numbers M12857 and J02578 (glycophorin A), J02982 and X08055 (glycophorin B), and M 29610 (glycophorin E).

66,115]; more recently the structure of the complete polypeptide chain has been determined by cDNA analysis [273,291] (see *Fig. 9.4*). These studies revealed that glycophorin A and glycophorin B share highly homologous amino acid sequences: in particular, the sequence of the first 26 N-terminal residues in the extracellular domain of glycophorin B (residues 1–44) is completely identical to the corresponding sequence in group **N** glycophorin A. Further homologies can be observed between the amino acids at positions 27–35 of glycophorin B and those at positions 59–67 of glycophorin A. Moreover, the transmembrane domain (residues 45–64) is almost identical to that of glycophorin A. Glycophorin B, on the other hand, lacks sequences corresponding to segment 27–58 and to the C-terminal cytoplasmic domain of glycophorin A; it has a cytoplasmic domain of 8 amino acids.

Glycophorin B carries up to 11 alkali-labile tetrasaccharide chains which are identical to those found in glycophorin A; glycophorin B lacks an N-linked carbohydrate unit.

As mentioned in *Sect. 4.1.2*, glycophorin A and B act as receptors for myxoviruses [181,321]. A strain of pathogenic *Escherichia coli* (IH 11165) specifically binds to glycophorin A^M [177]. Moreover, various investigations showed a significant resistance towards infection with *Plasmodium falciparum* of glycophorin-deficient erythrocytes (e.g. **En(a–)**, **M^k**, **Dantu+**, and **S–s–U–** cells [105,239]). It was therefore suspected that glycophorins might represent an attachment site for this malaria parasite. Further studies, however, revealed that the invasion of the merozoite into the blood cells is controlled by several parameters and that different strains of *Plasmodium* probably attach to different receptors sites [130,222,269,274,276].

9.4 Structure and Organisation of the MNS Gene Complex

Molecular genetic investigations on genomic DNA [187] have provided direct proof for the view originally deduced from immunogenetic studies, which was that glycophorin A and glycophorin B are encoded by separate and distinct, but closely linked single-copy genes. In more recent studies [188,309] a further gene has been identified which encodes a third type of glycophorin designated as glycophorin E. The three genes are organised within a ~330-kb genomic segment in the order of 5'-*GPA-GPB-GPE*-3'. The genes have approximately the same size (each comprises >30 kb of DNA) and show a high degree of sequence homology (~95%); the distance between the genes is about equal [234].

EXON A-1/B-1/E-1

```
Glycophorin A: acactgacacttgcAGTGTCTTTGGTAGTTTTTTGCACTAACTTCAGGAACCAGCTCATGATCTCAGGATG TAT GGA AAA ATA ATC TTT GTA TTA CTA TTG TCA G / gtaagt.........
Glycophorin B: --------------------------------------------------------------g------- --- --- --- --- --- --- --- --- --- --- --- / -------
Glycophorin E: --------------------------------------------------------------g------- --- --- --- --- --- --- --- --- --- --- --- / -------

*******   *** ***
```

EXON A-2/B-2/E-2

```
cgtcttaatcccttctcaacttctatgttatacag / CA ATT GTG AGC ATA TCA GCA TCA AGT ACC ACT GGT GTG GCA ATG CAC ACT TCA ACC TCT TCT TCA GTC ACA AAG AGT TAC ATC
----------------------------------- / A-- --- --- --- --T- --- --AG --- --- --- --- --- --- --- --- --- --- --- --- --- --- --- --- --- ---
----------------------------------- / G-- --- --- --- --- --- --- --- --- --- --- --- --- --- --- --- --- --- --- --- --- --- --- --- ---

*******           *** ***
```

EXON A-3

```
TCA TCA CAG ACA AAT G / gtttgt.....tcattcttgacccctttctcaacttctctttatatgcag / AT ACG CAC AAA CGG GAC ACA TAT GCA GCC ACT CCT AGA GCT CAT GAA GTT TCA GAA
--- --- --- --- --- / -------- tt-- ------ --- --a- --- --- --a- --- --- --- --- c-- --- --a- --ct --- --- --a- --- ---
--- --- --- --- --- / -------- tt-- ------ --- --a- --- --- --a- --- --- --- --- c-- --- --a- --t a-- --- --t- --- c--

                      **** **           ****** 
```

EXON A-4/B-3

```
ATT TCT GTT AGA ACT GTT TAC CCT CCA GAA GAG GAA ACC G / gtatgt.....ttacttattggacttacattgaaatttgctttatag / GA GAA AGG GTA CAA CTT GCC CAT CAT TTC TCT
--- --- --- c-- --- --- --- --- --- --- --- --- --- / --t---- ------ a-- --- --- --C- --G- --- --T- --g- --- A--
-a- --- --c --- --- --- --- --- --- -at a-- / a------ ------ --- --- --c --- --- --g- --t- --g- --- c--

          ** *****                        (+)
```

EXON A-5/B-4/E-3

```
GAA CCA G / gtatgt.....ctttcataatttgctgctctctttat    ct cct gta g / AG ATA ACA CTC ATT ATT TTT GGG GTG ATG GCT GGT GTT ATT GGA ACG ATC CTC TTA ATT TCT
--T-- --- / -------- ------------------------g / CT CCT GTA G T-- --- --T- ---G T-T- --- A-- --- --- --- --- --- --- --- --- ---
--- g-- --- / a------ ------------------------c / ct cct gta g / G-- --- --- --|--- --- -A- --- CT- -T- --- G-- --- --T- --- A-- --- --- --- ---

        ** *****                   ***** **              (+)
```

EXON A-6

```
TAC GGT ATT CGC CGA CTG ATA AAG / gtgaga...aaaatgtgttattaatattttatggtattcttcatag / AAA AGC CCA TCT GAT GTA AAA CCT CTC CCC TCA CCT GAC ACA GAC GTG
--- AC- --- --- --- --- --- --- / ------- --- (+)
--- T-- --- --A T-- --- --- --- / ------- --- (++)

**** *** ****
```

EXON A-7

```
CCT TTA AGT TCT GTT GAA ATA GAA AAT CCA G / gttggtg.....tttcggtcttgtatttttttactataatcccttctag / AG ACA AGT GAT CAA TGA GAATCTGTTCA....AGCAAAATATTGTA
```

```
ATAAAGAAATCTTTCCtgtgaagatacccatgacccatg
```

(+) Insert in glycophorin E genome: AAT TGG TGG GCG ATG GCT CGT GTT

```
                                       **** **
```

EXON B-5/E-4

(++) Glycophorin B: gtgaga...ctgaagtggaaacttctgtctttttatcacag / GCA TGA GGATGTGG .... CTAAATAAAATAAAACAAAATACAAACCGTTtcatgtattagtca
(++) Glycophorin E: -------.ctgaagtggaaacttctgtctttttatcacag / --- -------- CTAAATAAAATAAAACAAAATACAAATTATTtgcatgtagtagtca

Figure 9.5: Nucleotide sequences of glycophorin A, glycophorin B, and glycophorin E genes.

The sequences are aligned, identical sequences being shown by dashes; the splicing branch points are indicated by asterisks.
According to Kudo and Fukuda [187,188] by permission of the authors and of the American Society for Biochemistry and Molecular Biology Inc., Baltimore, MD, USA.
The gene structure has also been determined by Vignal et al. [309].

The coding sequence of the **glycophorin A gene** is distributed over 7 exons (*Fig. 9.5*): exon A-1 and part of exon A-2 code for a leader peptide (codons −19 through −1), exons A-2, A-3, and A-4 for the extracellular domain (codons 1 through 71), exon A-5 for the transmembrane domain (codons 72 through 101), and exon A-6 and part of exon A-7 for the cytoplasmic domain (codons 102 through 131). Exons A-1 and A-2 are separated by a large intron of approximately 30 kb, exons A-2 to A-7 by short introns of 0.7–3.2 kb.

The **glycophorin B gene** comprises five transcribed exons[3]. The nucleotide sequence is very similar to that of the glycophorin A^N gene: except for a point mutation in exon B-2 resulting in an amino acid exchange (Ala → Glu) in the leader peptide at position −7, exon 1 and its 5' flanking region, intron 1 and exon 2 are identical in both genes. The sequences which follow exon B-2 are highly homologous to their counterparts in the glycophorin A gene (i.e. to exons A-3, A-4, and A-5). Due to a mutation at the 5' consensus splice signal, however, the region corresponding to exon A-3 (= pseudoexon ΨB3) is passed over, and exon B-2 is ligated to the exon corresponding to A-4. The next exon coding for the transmembrane domain (exon B-5) is almost identical to the respective exon of glycophorin A but contains three additional amino acid residues; this insert is a result of an upstream shift in the 3' splice site of intron B-4 due to a point mutation. The glycophorin B gene lacks an exon encoding a C-terminal cytoplasmic domain (corresponding to exon A-6), and exon B-5 is directly spliced to the exon B6 encoding the 3' untranslated region which is totally different from that of glycophorin A (*Figs. 9.4* and *9.5*).

The coding sequence of the **glycophorin E** gene contains only four transcribed exons. Analyses of genomic DNA, however, have revealed that organisation and nucleotide sequence of the glycophorin E gene closely resemble those of the glycophorin B gene. Nevertheless, several differences characteristic of glycophorin E have been observed (*Fig. 9.5*):

– first, exon E-2 contains a point mutation which changes the amino acid at position −7 in the (putative) leader peptide; moreover, the gene codes for the Ser^1/Gly^5 polymorphism typical for blood group **M** specific glycophorin A;
– second, due to an additional splice site mutation the sequences corresponding to exons A-3 and A-4 (pseudoexons ΨE-3 and ΨE-4) are not expressed;
– third, the next exon (E-5, which in the case of glycophorin B codes for the transmembrane domain) contains an in-frame 24-nucleotide insert encoding 8

---

[3] The numbering of the homologous exons, pseudoexons, and introns of glycophorin A, B, and E is maintained in order to point out the relationship of the exon–intron structures.

amino acids unique for the glycophorin E gene. This insert represents a duplication of the adjacent 3' coding sequence of the ancestral glycophorin B gene). Further, as the result of a point mutation, a stop codon is generated which shortens the reading frame by four codons in comparison to its B5 counterpart. Downstream from E-5, the nucleotide sequences of glycophorin E and glycophorin B genes are almost identical.

The nucleotide sequences of cDNA clones isolated from erythroblasts and erythroleukaemic K562 cells have confirmed that glycophorin E gene codes for a protein of 78 amino acids. Residues 1 through 19 correspond to the leader peptide in glycophorin A and glycophorin B. The sequence of the following 26 residues is identical to that of the N-terminal domain of blood group **M** specific glycophorin A. The C-terminal region, with the exception of an insert of 8 amino acids not found in glycophorin A or B, shares a high degree of homology with the transmembrane domain of glycophorin B. The proposed amino acid sequence shows that glycophorin E lacks a consensus sequence for N-glycosylation but may carry O-linked oligosaccharide chains.

The discovery of glycophorin E-specific mRNA in a human erythroleukemia cell line shows that the *GPE* gene is indeed transcribed [188]. Further, the fact that anti-**M** specific hybridoma antibodies bind to a 20 kDa glycoprotein revealed by polyacrylamidegel electrophoresis – regardless of the **M/N** group of the erythrocytes [11] – suggests that the glycophorin E gene product is expressed as a constituent of normal red cell membranes.

The glycophorin E gene (or at least exons E-2 to E-6) has been found in all human genomes thus far examined [143,309].

As discussed above, human erythrocytes contain 3–4 times more copies of glycophorin A compared to glycophorin B (10^6 *vs.* 2×10^5). Glycophorin E, however, is only poorly represented or absent. These quantitative differences reflect differences in stability of the transcripts rather than a transcriptional regulation; the functional activities of the glycophorin promoters and the transcription rates of the glycophorin genes were shown to be closely similiar [250].

A polymerase chain reaction-based assay of **M** and **N** blood group specificity has already been developed [54].

The high degree of sequence homology between the glycophorin genes in both coding and noncoding regions increases the chance of recombination due to unequal crossing-over or gene conversion events. In the majority of cases the glycophorin A and B genes are involved – the aberrant genes formed by unequal crossing-over

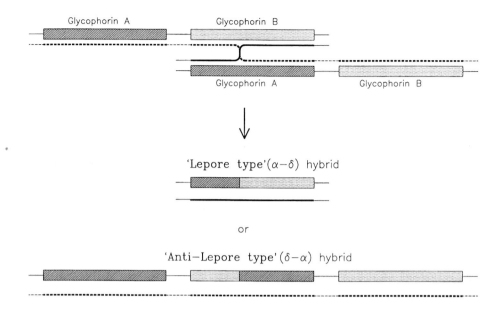

Figure 9.6: Schematic diagrams showing the formation of glycophorin A-B hybrid molecules by unequal crossing-over of the genes MN and Ss.
According to Anstee et al. [16] and Tanner et al. [289].

encode either glycophorin A-B (α-δ) hybrids or glycophorin B-A (δ-α) hybrids[4] (*Fig. 9.6*). Furthermore, unequal crossing-over seems to be the main reason for the glycophorin A and/or glycophorin B gene deletions observed in the rare **MNS** phenotypes **En(a–)**, **S–s–U–**, and **M^k**. In some cases, glycophorin A-B-A (**M^z**, **Mi.IX**) and glycophorin B-A-B (**Mi.III**, **Mi.VI**, **Mi.X**) hybrid molecules have been detected (see *9.6*); these gene variants are probably formed by gene conversion in which a DNA segment of a glycophorin A or B gene is replaced by the homologous sequence of a glycophorin B or A gene. The major 'hot spot' of these recombination events is located between the 3' end of intron 2 and the 5' end of exon 4, another 'hot spot' spans the intron 1 – exon 2 junction through the exon 2 – intron 2 junction.

---

[4] A similar phenomenon [44] has been described in haemoglobins – in analogy to these investigations the glycophorin A-B hybrid molecules are often designated as 'Lepore type' and the glycophorin B-A hybrid molecules as 'anti-Lepore type' hybrids.

As a consequence of re-arranged glycophorin A and B domains and/or inserted new peptide units these glycophorin variants contain novel polypeptide sequences which define epitopes not present in the original glycophorin molecules.

Evolution of the glycophorin genes

The observed high degree of sequence homology between *GPA*, *GPB*, and *GPE* genes confirmed the assumption that they developed from a common ancestral gene. Investigations on the glycophorin gene complex in hominoid primates revealed a *GPA* gene in all species studied (i.e. gibbon, orangutan, gorilla, chimpanzee, and bonobo). The *GPB* and *GPE* genes have been detected only in the genomes of gorilla, chimpanzee, and bonobo [254,256]. Thus, the glycophorin gene complex apparently evolved from an ancestral glycophorin A gene by two successive gene duplications and subsequent diversification: first and prior to gorilla divergence, the ancestral gene was duplicated, and one of the duplicated genes became a direct precursor to the *GPA* gene; then the other duplicated gene acquired a different 3' sequence through homologous recombination at *Alu* repeat sequences and was subsequently duplicated to generate *GPB* and *GPE* genes [234,235]. The 24-base pair insert found in the transmembrane exon of *GPE* gene was inserted into the ancestral *GPE* gene prior to gorilla divergence. It has further been suggested that the *GPE* gene acquired the sequence for the blood group **M** determinant from the *GPA^M* gene through gene conversion [189].

Genes significantly homologous to the human glycophorin A gene have also been found in Old World monkeys ([228], see also [208]). The glycophorin A-like sialoglycoproteins in other vertebrate species, i.e. mouse [293], horse [227], pig [136], cattle [226], and dog [229] are similar in molecular size and glycosylation; the primary structures, though consistent with the transmembrane orientation of glycophorins, show only little homology to human glycophorin A.

9.5 Chemical Structures of the Main Determinants of the MNS System

M (MNS1) and N (MNS2)

The blood group **M** and **N** specificities are located at the amino-terminal end of glycophorin A. Both antigens are trypsin-sensitive *in situ*. The smallest still serologically active unit is the terminal octapeptide with three alkali-labile tetrasaccharide chains attached [80,206]. **M** and **N** specificities are defined by an amino acid dimorphism in the N-terminal segment: blood group **M** specific glycophorin A has **serine** at position 1 and **glycine** at position 5, whereas the **N** specific form has **leucine** at position 1 and **glutamic acid** at position 5 [35,80,83,115,314][5] (*Fig. 9.7*).

Removal of the N-terminal amino acid by Edman degradation eliminates both **M** and **N** activity [200]. Similarly, alteration of the terminal amino group (e.g. acylation or attaching an additional amino acid) results in the loss of serological activity [201,205]; the effect is reversible and **M/N** activity can be restored by removing the attached groups.

These results suggest that the N-terminal amino acid plays the pivotal role in **M** and **N** specificity, whereas amino acid Nr. 5 is of only secondary importance [172,173]; however, antibodies have been described which bind to an epitope in which the fifth amino acid represents the immunodominant group [14,70,171,204] (see also anti-**M**[e] and anti-**M**[c] sera, below).

For the majority of anti-**M** and anti-**N** antibodies the presence of neuraminic acid in the **M** and **N** determinant group is essential for their serological activity [212,278]. Erythrocytes treated with neuraminidase are not agglutinated by these sera. Likewise, removal of neuraminic acid or elimination of the O-linked carbohydrate chains from glycophorin A by treatment with alkaline borohydride results in the total loss of **M** or **N** specificity [200]. Further evidence of the important role of neuraminic acid in the reactivity of many anti-**M** and anti-**N** antibodies is provided by the drastic reduction of **MN** activity in **T** and **Tn** transformed cells, both of which lack neuraminic acid (see *Sects. 10.2* and *10.6*).

Nevertheless, a series of antibodies which recognise the **M** and **N** antigens in the absence of neuraminic acid have been detected in humans [179] and have also been produced in rabbits [203]. Of the anti-**N** specific hybridoma antibodies thus far described, two show a reactivity independent of sialylation but are dependent on the presence of **Gal-GalNAc** disaccharide units, and one exhibits maximum reactivity with partially desialylated antigen [172].

Anti-**Can** sera are specific for blood group **M** but agglutinate only about one-third of **M** erythrocyte samples [178]. In analogy to this, anti-**Tm** is an **N** specific serum with only restricted reactivity towards **N** cells [164,165]. Both sera, however, agglutinate all neuraminidase-treated erythrocytes of the respective specificity [166,178].

---

[5] For a long time the exact role of the carbohydrate moiety in **MN** specificity was a matter of major controversy. It was initially suspected that the antigens of the **MNS** system were determined by carbohydrate structures. Furthermore, it was also assumed that the **N** determinant structure is the precursor substrate used for the biosynthesis of **M** antigen [279,280,282].

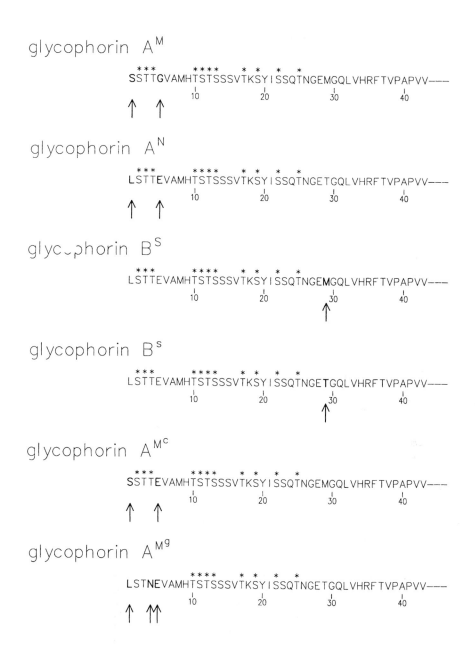

Figure 9.7: N-terminal amino acid sequences of glycophorin A and glycophorin B.
★ : O-linked carbohydrate chain, ✦ : N-linked carbohydrate chain.

Though both receptors are undoubtedly located in the N-terminal segment of glycophorin A, no difference in peptide sequence as compared to normal **M** or **N** can be detected [62,69]. However, it was found that **Tm**- as well as **Can**-positive erythrocytes have an increased content of the alkali-labile tetrasaccharide chains in which one of the neuraminyl residues is replaced by N-acetylglucosamine (*Fig. 9.3.f*) [69]. Since N-acetylglucosamine does not represent an essential component of the epitopes [167], it is assumed that the change in tertiary structure of the glycophorin A molecule induced by the different glycosylation favours the binding of anti-**Can** and anti-**Tm** to the **M** or **N** determinant group. The high variability in the serological reaction suggests that the degree of **Tm** and **Can** activity is determined by location and/or number of the altered tetrasaccharide chains.

S (MNS3) and s (MNS4)

The antigens **S** [265,267,312] and **s** [196] are located on the glycophorin B molecule [81,131]. They are not influenced *in situ* by trypsin treatment but are readily degraded by chymotrypsin.

The **S** gene appears to produce more glycophorin B than does **s**. On the basis of PAS staining intensity on polyacrylamide gels, Dahr et al. [88] estimated that the amount of glycophorin B^S found in **S+s−** erythrocytes is about 1.5 times higher than the amount of glycophorin B^s found in **S−s+** cells.

Structural analyses revealed that the antigens **S** and **s** are determined by an amino acid dimorphism at position 29 − **methionine**, characteristic for the **S** active glycophorin B, is replaced by **threonine** in the **s** active form ([66], see *Fig. 9.7*)[6]. Further investigations suggested that Glu^{28}, as well as the oligosaccharide unit attached to Thr^{25} may well be part of the **S** and **s** epitopes; removal of neuraminic acid has no influence whatsoever on **S** or **s** activity [67].

U (MNS5)

The **U** antigen [318,319] is a common character in Europids but is frequently absent in Negrids (in certain populations as much as 35% of the individuals are **U**-negative [207]). Moreover, **U** activity is not detectable in Rh_{null} erythrocytes of the regulator type (see *Sect. 13.3*) [72,268].

---

[6] Without affecting the **s** antigen, **S** of intact erythrocytes is destroyed by carboxymethylation, performic acid, hydrogen peroxide, cyanogen bromide, and low concentrations of sodium hypochlorite, and by reagents with selectively react with the **S** determinant methionine residue [67,264].

Thus far, anti-**U** has been detected only in human sera [249]: anti-**U** allo-antibodies occur in most **S–s–U–** subjects, whereas immune anti-**U** is produced occasionally in the course of pregnancies or following **U**-incompatible blood transfusions. Anti-**U** specific autoantibodies have been repeatedly reported in patients suffering from acquired autoimmune haemolytic anaemia.

In the course of their investigations, Dahr et al. [76] were able to locate the **U** character on the glycophorin B molecule. The finding that the **U** antigen is only slightly degraded by ficin, papain, and pronase, as well as the fact that the presence of lipid is necessary for optimum expression of serological activity in isolated glycophorin B, suggest that the **U** determinant is situated in the carbohydrate-free segment of the molecule somewhere between amino acids 33 and 39 (see *Fig. 9.8*).

'N' (MNS30)

Human erythrocytes (with the exception of **S–s–U+(var)** and **S–s–U–** cells) show low **N** activity, which is not destroyed by trypsin treatment [4,106,135]. As discussed above, the occurrence of this so-called 'cryptic **N** activity' or **'N'** (known as the term 'N quotes') [58] led to the erroneous assumption that **N** represents the biosynthetic precursor of **M**.

Investigations by Dahr [83] revealed that the **'N'** character is located on the N-terminal end of glycophorin B – the same end which is identical to that of glycophorin A^N (*Fig. 9.7*). Furthermore, it has been shown that the **'N'** epitope comprises amino acids 1–5 of the N-terminus of glycophorin B as well as portions of the carbohydrate chains attached to that region.

Despite the close structural similarity of **N** and **'N'**, normal anti-**N** sera are obviously able to distinguish between *MM* on the one hand, and *MN* and *NN* on the other, yet cells from *MM* individuals will adsorb anti-**N** antibodies to exhaustion, as has been shown in haemagglutination tests. It is generally known that anti-**N** sera differ in the amount of cross-reactivity with the **'N'** receptor: anti-**N** agglutinins from *MM* individuals show a fairly high specificity towards **N**, whereas anti-**N** sera from *MMuu* (= **S–s–U–**) individuals who lack the **'N'** antigen agglutinate erythrocytes from *MM* individuals almost as strongly as cells from *NN* subjects. Likewise, the absorption of rabbit anti-**N** sera with *MMUU* cells results in a fairly specific anti-**N**, and treatment with *MMuu* cells leads to an antiserum which agglutinates both *MM* and *NN* erythrocytes (see [58]).

These differences in reactivity may be determined by two parameters: first, due to the fact that glycophorin B is much shorter than glycophorin A, the **'N'** receptors in the intact membrane or in aggregates of extracted **Ss** glycoprotein are less accessible than the **N** antigen located on the glycophorin A molecule [58]. Second, it must be assumed that the number of copies of '**N**' on **MM** erythrocytes is far too small to sustain agglutination by ordinary anti-**N**; the cells, nevertheless, may be weakly agglutinated by more potent anti-**N** samples.

N_{Vg}

The lectin from *Vicia graminaea* agglutinates specifically native **N** erythrocytes [238] and therefore is widely used as an anti-**N** reagent ('anti-N_{Vg}'); after treatment of the erythrocytes with neuraminidase or trypsin, however, the lectin also agglutinates cells from **MM** individuals [97]. **T** transformed cells (see *Sect. 10.2*) which have a decreased level of neuraminic acid, show an increased N_{Vg} activity [199], **Cad** erythrocytes (see *Sect. 11.3*) in which the tetrasaccharide units are extended by an N-acetylgalactosamine residue are only weakly agglutinated by the lectin [29], and **Tn** cells which lack the $\alpha2{\rightarrow}3$ neuraminic acid and $\beta1{\rightarrow}3$ linked galactose (see *Sect. 10.6*) show no reaction with anti-N_{Vg} at all [27,29].

Inhibition tests with isolated glycophorin confirm these findings. It has been demonstrated that the *Vicia graminaea* lectin binds to both glycophorin A^N and glycophorin B ('**N**'!) [98,246], but shows only weak affinity towards glycophorin A^M. However, after removal of neuraminic acid, or modification of the N-terminal amino acid, the lectin reacts with both **M** and **N** specific glycophorin A molecules [201]. Edman degradation of the terminal amino acid totally eliminates N_{Vg} activity [98].

The results of the above-mentioned investigations suggest that the N_{Vg} epitope comprises at least one carbohydrate unit and a hydrophobic structure in position 1 for optimum binding of the *Vicia graminaea* lectin.

9.6 Rare Variants of the MNS Antigens, MNS-Deficient Phenotypes, and Antigens Associated with the MNS System

In addition to the determinants discussed above, a variety of erythrocyte characters closely associated with the **MNS** system have been defined (see *Table 9.2*). Most of these **MNS** variants and erythrocyte phenotypes are extremely rare in Europids (frequency of <0.1%, see [161]). Some of them, however, do occur fairly frequently in non-white populations.

Different factors are responsible for the occurrence of **MNS** variant phenotypes:

- hybrid genes which either originated from unequal crossing-over events between glycophorin A and B (e.g. **Dantu, St³, Mi.V**) or were formed through the mechanism of gene conversion (e.g. **Mi.III**),
- alleles of **MN** or **Ss** genes encoding glycophorin A or glycophorin B molecules with one or more amino acid substitutions (e.g. **M^c, M^g**),
- genomes in which the glycophorin A and/or glycophorin B genes are deleted (e.g. **En(a–), S^u**, and **M^k**),
- abnormal glycosylation of the glycophorin molecules (e.g. **M₁, Can, Tm**, and **Hu**),
- antigen site density influencing the expression of some **MNS** phenotypes (e.g. **U^x** and **U^z**).

Due to alterations in glycophorin A and glycophorin B structure many **MNS** variants differ in neuraminic acid content (see [10]).

M^c (MNS8)

The **M^c** antigen [100] is characterised by its reaction with most anti-**M** sera as well as with some anti-**N** samples, in which reactions it shows features of both **M** and **N** blood groups.

The serological properties of this antigen are reflected by a characteristic amino acid sequence of glycophorin A [70,116]: in the **M^c** specific sialoglycoprotein a **serine** residue specific for blood group **M** is found at position 1 and a **glutamic acid** residue characteristic for blood group **N** at position 5 (*Fig. 9.7*).

M^g (MNS11)

The **M^g** antigen [5] — to which neither anti-**M** nor anti-**N** bind [221] — is defined by its reaction with anti-**M^g** sera. Anti-**M^g** specific antibodies are fairly common in human sera [161] and are also produced in rabbits by immunisation with **M^g** cells [156]. The reactivity of **M^g** erythrocytes with anti-**M^g** is not dependent on neuraminic acid [281].

The trypsin-sensitivity of the **M^g** antigen, and the fact that **S** and **s** are expressed normally, suggested that the **M^g** determinant is also located on the glycophorin A molecule. Peptide analyses have revealed an amino acid polymorphism in the N-terminal region of the antigen: the blood group **M^g** specific glycophorin A has the same amino acids at positions 1 and 5 as glycophorin A^N (i.e. **leucine** and **glutamic acid**, respectively). In position 4, however, **threonine** is substituted by an **asparagine** residue [34,63,116] (*Fig. 9.7*). The amino acid exchange at position 4 prevents the glycosylation of the hydroxy amino acids at positions 2 and 3 [63,116]; the oligosaccharide chains on the other serine and threonine residues, however, are not affected.

The sequence Asn^4-Glu^5-Val^6, which is also part of the **DANE** epitope (see below), is probably responsible for the cross-reactivity of anti-**DANE** serum with M^g erythrocytes [275].

En(a–) Phenotype and Antigen Complex 'En$^{a'}$'

Erythrocytes of the rare **En(a–)** phenotype lack the high-incidence antigen designated **En**^{a} (**MNS28**) – the term 'En' was proposed to indicate a serological character of the erythrocyte envelope.

Two categories of **En(a–)** have thus far been described – the *Finnish* type ('**En(a–) Fin**' [117,118]) and the *English* type ('**En(a–) UK**' [95]). Both **En(a–)** categories are characterised by a deficiency in M/N antigens, whereas **S** and **s** are normally expressed. In contrast to **En(a–) Fin**, the *English* type of **En(a–)** lacks 'N' but has a weak, trypsin-resistant **M** activity denoted as (**M**) or, in analogy to 'N', as 'M'.

Analysis of the constituents of **En(a–)** membranes [87,117] showed a significant decrease in neuraminic acid level. Probably due to a reduction in zeta-potential, saline suspensions of **En(a–)** cells are strongly agglutinated by incomplete anti-**Rh** sera; further, their reactivity towards various lectins is greatly enhanced [117].

Serological and immunogenetic studies [89,219] suggest that the *Finnish* **En(a–)** variant is homozygous for a gene complex, *'En(Fin)'*, which does not express the **MN** antigens but contains a normal glycophorin B (*Fig. 9.9*). On the other hand, the **En(a–) UK** individuals described are of genotype *En(UK)* / *M^k* in which the *'En(UK)'* gene complex lacks the glycophorin A gene and codes for a variant of glycophorin B, whereas the *M^k* allele does not express **MN** or **Ss** antigens (see below).

SDS-polyacrylamide-gel electrophoresis of the membrane proteins [12,121,288], as well as immunochemical investigations with an antibody directed towards the C-terminal domain of glycophorin A [115], have revealed that **En(a–)** erythrocytes completely lack glycophorin A. In the cells of **En(a–) Fin** individuals glycophorin B is present to a normal extent [121]. In the case of the *English* type of **En(a–)**, however, an abnormal glycoprotein molecule has been detected [12]: it has an electrophoretic mobility identical to normal glycophorin B but has a serine residue at the N-terminal end [86]. The reaction of **En(a–) UK** erythrocytes with an **M**-specific hybridoma antibody (*6A7*) which binds to an epitope comprising Gly^1 and Glu^5 [24] clearly shows that this glycophorin B variant carries an **M** specific structure. Furthermore, the finding that two anti-glycophorin-A antibodies were non-reactive suggested that **En(a–)** erythrocytes of the *English* type carry a glycophorin A-B hybrid molecule [24].

In all cases of **En(a–)** the apparent molecular mass of the band-3 protein is significantly increased by ~5–10 kDa due to enhanced glycosylation [121,290].

Analyses of genomic DNA obtained from **En(a–)** individuals [251,253,292] corroborated the results of the serological and biochemical investigations mentioned above. In the case of **En(a–) Fin** the structural gene coding for glycophorin A was completely deleted, and no aberration detected in the gene coding for glycophorin B. In the case of the **En(a–) UK** type the investigations strongly support the view that the *En(UK)* gene codes for a glycophorin hybrid composed of a short N-terminal segment of glycophorin A^M and a large segment of the C-terminal portion of glycophorin B. Subsequent studies performed by Vignal et al. [309] suggested that the deletion of the glycophorin A gene in the genome of **En(a–) UK** individuals may be a consequence of an unequal crossing-over event between the 5' end of glycophorin A exon 2 (in the region of codons 2–5) and the corresponding region of exon B-2.

The sera of all **En(a–)** individuals contain so-called anti-En^a antibodies, which are often accompanied by anti-Wr^b antibodies [240]. Anti-En^a sera have also been produced by immunising rabbits with normal human erythrocytes [117]. Moreover, hybridoma antibodies with respective specificities have been prepared by Anstee et al. [13].

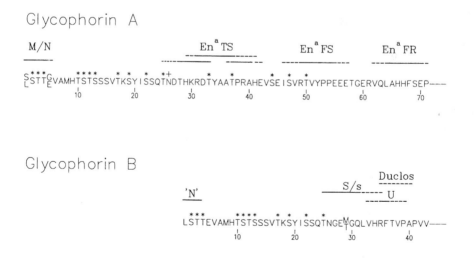

Figure 9.8: Location of the binding sites on anti-En^a sera on the glycophorin A molecule, and of the S/s, U, and Duclos epitopes on glycophorin B.
According to Dahr et al. [59,77].

However, anti-**En**[a] is not a homogeneous antibody type [240,305]: sera from **En(a–)** individuals tend to contain mixtures of antibodies directed towards different epitopes of the glycophorin A molecule, on the basis of which Issitt [163] proposed a nomenclature system for a characterisation of these antibodies. His system is based on the protease-sensitivity of the antigen:

- antibodies recognising Trypsin-Sensitive determinants located towards the N-terminal end of glycophorin A are designated as anti-**En**[a]**TS**,
- antibodies binding to Ficin/papain-Sensitive determinants are designated as anti-**En**[a]**FS**,
- and antibodies specific for Ficin-Resistant determinants located near the membrane bilayer are termed as anti-**En**[a]**FR**.

Dahr et al. [77,91] were able to localise the binding sites of the different types of anti-**En**[a] antibodies (see *Fig. 9.8*). Anti-**En**[a]**TS** recognise regions between amino acids 28 and 42. The binding site of the anti-**En**[a]**FS** antibodies is situated between amino acids 46 and 56; most anti-**En**[a]**FS** need a neuraminic acid residue at Thr^{50} for optimum reaction. The anti-**En**[a]**FR** antibodies bind to a segment located between amino acids 62 and 72 and are reactive only in the presence of lipid.

The Phenotypes S–s–U+(variant) and S–s–U–

Erythrocytes from individuals of the **S–s–U+(var)** phenotype [6,18] exhibit neither **S** nor **s** activity; they do show, however, weak and abnormal **U**, very weak '**N**', and a weak but significant reaction with anti-**Duclos** sera [59]. Cells from **S–s–U–** subjects totally lack the antigens **S**, **s**, **U**, and '**N**' [128,266]. Though anti-**U** sera which react directly with **S–s–U+(var)** cells have been described, absorption-elution tests are necessary to differentiate the two phenotypes [162]. Both variants are prevalent among Negrids [161].

Analysis of the membrane proteins from erythrocytes of **S–s–U+(var)** and **S–s–U–** phenotypes revealed no change in glycophorin A. It has been shown that **S–s–U+(var)** cells contain a minute amount of glycophorin B (<5% of the level found in normal erythrocytes) with a decreased electrophoretic mobility when compared with protein from normal cells [10,68]. Erythrocytes of the **S–s–U–** phenotype completely lack glycophorin B [68,81,144].

Molecular biological investigations have revealed that the **S/s/U** deficiency is based on two different genetic backgrounds:

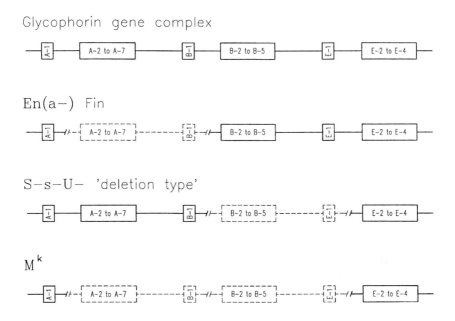

Figure 9.9: Proposed model for glycophorin A and/or B gene deletions in En(a–) Fin, S–s–U–, and Mk individuals.
According to Vignal et al. [309].

In the '**deletion type**' a large portion of the glycophorin B gene is absent from the genome. Analyses of genomic DNA suggested that the absence of the glycophorin B gene in **S–s–U–** individuals is a consequence of unequal crossing-over between intron 1 of glycophorin B and intron 1 of glycophorin E genes ([144,309], see *Fig. 9.9*).

The '**non-deletion type**' is characterised by an altered but grossly intact glycophorin B gene [147,252]. In one case the abnormal gene associated with the **S–s–U–** phenotype has been described in detail ([151]. The extracellular domain of this glycophorin B variant, termed *GPHe(P₂)* (see *Fig. 9.10*), is characterised by a glycophorin A insertion within exon 2 accompanied by multiple untemplated nucleotide replacements. These changes not only define the sequence for the **Henshaw (He)** epitope (see below) but also abolish the glycophorin B-associated '**N**' antigenicity. In addition, this same variant gene carries two splice site mutations which provoke the skipping of exon B-5 (the exon encoding the membrane-spanning domain of glycophorin B). Due to the shift in the open reading frame caused by the ligation of exon 4 to exon 6, the original transmembrane domain is displaced by an entirely different, highly hydrophobic transmembrane segment of 41 amino acids. Expression of the **He** antigen in the proband's erythrocytes shows that the abnormal glycophorin B molecule is in fact inserted into the red cell membrane. This remodeling obviously affects the surface presentation of **S**, **s**, and **U**, since the variant glycophorin molecule does not display these antigens although it still contains the intact sequences defining these epitopes.

In a more recent investigation it has been shown that this variant occurs fairly frequently in the non-deletion type of the **S–s–** phenotypes [261].

M^k

The **M^k** variant has been described by Metaxas & Metaxas-Bühler [218]. Erythrocytes from heterozygous **M^k** individuals show distinctly reduced **MN**, **Ss**, and **U** activities, whereas homozygous subjects (genotype **M^{k}M^k**) lack these characters completely [297]. These findings suggest that the **M^k** gene affects both the **MN** and the **Ss** locus.

The content of neuraminic acid is greatly reduced in **M^k** erythrocytes (homozygous individuals contain only 25–31% of the amount present in normal control cells [297]). As also found for **En(a–)** erythrocytes, saline suspensions of **M^k** cells are agglutinated by incomplete **Rh** antisera and show an increased agglutinability with *Sophora japonica* lectin [231]. Furthermore, band-3 and band-4.1 proteins show a molecular mass up to 5 kDa greater than expected, reflecting a higher degree of glycosylation [43,233,297]. The reduced sulfate transport activity of **M^{k}M^k** cells (~60% of the value of normal erythrocytes) is due to a decreased anion binding affinity of the band-3 protein [43].

First investigations on the membranes of erythrocytes obtained from heterozygous **M^k** individuals indicate a ~50% reduction in glycophorin A and glycophorin B content [17,85,115], whereas in the erythrocyte membranes from **M^{k}M^k** subjects [233,297] both sialoglycoproteins were totally absent. The fact that the viability of **M^k** erythrocytes is close to normal clearly indicates that glycophorin A and B play no role in maintaining membrane architecture.

Analysis of the genomic DNA derived from an **M^k** individual revealed deletions in the glycophorin A and B genes [292]. Subsequent investigations have shown that the **M^k** genome contained exon 1 of the glycophorin A gene and exons 2–4 of the glycophorin E gene [309]. It has therefore been suggested that the deletion of glycophorin A and glycophorin B genes in the **M^k** phenotype is due to unequal crossing-over between intron 1 of glycophorin A and intron 1 of glycophorin E genes (see *Fig. 9.9*).

He (Henshaw; MNS6)

The **He** receptor [157] occurs only very rarely in Europids but is particularly prevailing in Negrids [249] from malaria endemic regions.

The first anti-**He** antibodies were produced in rabbits [46], later they were detected in human sera as well [209], and monoclonal anti-**He** can be produced by immunising mice with **He+** red cells [258]. The reactivity of most anti-**He** antibodies is independent of neuraminic acid [71,179]. The so-called anti-**M**^{e} sera from humans [216] and rabbits [317] react with both **M** and **He** erythrocytes – in this case the specificities cannot be separated. The **He** antigen *in situ* is resistant to trypsin but destroyed by chymotrypsin [71].

The **He** character has been localised on glycophorin B. Structural analyses [71] revealed an amino acid polymorphism in this variant – the amino acids leucine, threonine and glutamic acid found in positions 1, 4, and 5 of normal glycophorin B are exchanged to *tryptophan, serine*, and *glycine*, respectively; the hydroxy amino acids at positions 2, 3, and 4 are glycosylated. The glycine residue in position 5 which is found also in glycophorin A^M, probably represents the immunodominant group for the anti-**M**^{e} sera, which, as already mentioned, react both with **He** and **M**; the tryptophan at position 1 is obviously immunodominant for anti-**He** sera.

Molecular biological investigations on the glycophorin genes carrying the **He**-defining sequence have shown that in all cases the extracellular domain is characterised by replacement of part of exon B-2 with the homologous segment of glycophorin A [146,151]. Further, this gene conversion is accompanied by six untemplated nucleotide replacements, which define the sequence for the Henshaw (**He**) epitope and abolish the glycophorin B-associated 'N' antigenicity. It is unlikely that these nucleotide exchanges arise from six independent spontaneous mutations. Rather they may originate from the heteroduplex DNA repair after the gene conversion event – not an error-free process [211].

The investigations further revealed that the **He**-specific glycophorins are a heterogeneous group of (B-A-B) hybrids. Thus far three different variant glycophorin B genes have been described:

(A) The **He**-active glycophorin variant '*GPHe(P$_2$)*' detected in an **S–s–U–** proband [151] has been described above. It is characterised by two additional mutations in the glycophorin B gene: a C $\rightarrow$ G exchange at the 3' end of exon B-5, which creates a cryptic acceptor splice site, and a G $\rightarrow$ T change at position +5 of intron 5, which alters the consensus donor splice site. In a coordinated mode, the two mutations cause a complete skipping of exon 5; thus, due to a frameshift a new hydrophobic membrane-

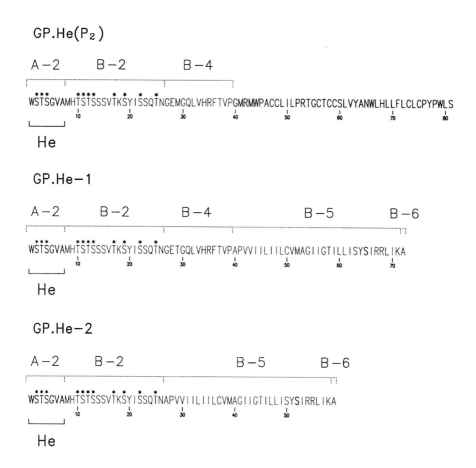

Figure 9.10: Peptide sequences of He-active glycophorin variants.
★ : O-linked carbohydrate chain, the new hydrophobic sequence in GP.He(P₂) caused by frameshift is printed in boldface.

spanning sequence is encoded (*Fig. 9.10*). Interestingly, the **He**-active protein product does not display **S** reactivity, although it contains the linear sequence for this antigen.

(B) The same working group has identified a further glycophorin **He** allele, 'GPHe(GL)', which gives rise to the formation of two **He**-positive protein isoforms in the erythrocyte membrane [141]. In addition to the nucleotide changes defining the epitopic sequence of **He**, this allele is characterised by two nucleotide changes – a T → G mutation at nucleotide –6 of the acceptor splice site for exon 4 and a C → G substitution in exon 5. The latter point mutation results not only in a Thr → Ser substitution at position 65 but concurrently creates a new acceptor splice site. Partial activation of the new acceptor splice site and partial inactivation of the normal splice site results in the production of four different cDNAs. The full-length transcript, *GPHe-1*, which is equivalent to glycophorin B in molecular size, contains a Thr → Ser substitution at position 65 and encodes **He**, **S**, and **U** antigens. Another transcript, *GPHe-2*, codes for a polypeptide with an intact transmembrane segment; due to its deficiency in exon 4 the encoded protein lacks the sequence defining the **S** and **U** antigens as well as the potential cleavage sites of α-chymotrypsin [60]. *GPHe-3* and *GPHe-4* are low-level transcripts with major deletions of internal sequences and premature chain termination. Truncation of an essential portion of their transmembrane segment should impede the insertion of their putative protein products into the plasma membrane, and presumably these isoforms are not expressed at the surface.

(C) Another glycophorin B variant of the human red blood cell membrane, *GPHe(St^a)*, exhibited **He** and **St^a** antigens [146]. This allele has originated from genetic recombination of a glycophorin hybrid carrying the **He** epitope and a glycophorin A gene. The unequal crossing-over event took place in the third intron at a recombination site identical to that observed in the glycophorin B-A hybrid (type A) encoding the **St^a** antigen (see below).

SAT (MNS36)

This low incidence antigen has been detected thus far in only two unrelated Japanese families [92].

First investigations suggested that the **SAT** antigen is associated with two different glycophorin isoforms:

In one family an isoform has been identified as a glycophorin A-B hybrid composed of exons 1–4 of the glycophorin A^N gene and exons 5 and 6 of the glycophorin B gene (**GP.TK**) (*Fig. 9.12*) [152]. Reticulocyte RNA polymerase chain reaction showed in the **SAT** homozygote individual transcripts of the hybrid gene and glycophorin E, but no glycophorin A and B transcripts; the hybrid thus shows an arrangement reciprocal to that found in the glycophorin B-A hybrid encoding the **Dantu** antigen (see above).

The **SAT**-specific glycophorin gene encodes a protein of 104 amino acid residues which lacks the sequence defining **S**, **s**, and **U** antigens. The hexapetide, ^{69}Ser-Glu-Pro-Ala-Pro-Val^{74}, encoded by the junction of exon A-4 and exon B-5 obviously represents the epitope site of the **SAT** antigen.

In the other family the **SAT** antigen is not associated with a glycophorin hybrid but rather with a glycophorin A variant. This variant is characterised by an insert between exons A-4 and A-5 of nine bases originating from the 5' end of glycophorin B exon 5. The tripeptide sequence Ala-Pro-Val thus inserted into the glycophorin A molecule created the **SAT**-specific sequence [300].

Dantu (MNS25)

Dantu-positive erythrocytes are characterised by a protease-resistant sialo-glycoprotein which carries 'N' and N_{Vg}, a qualitatively altered **s** antigen [52], but no **U**. *Phaseolus vulgaris* lectin does not bind to the glycoprotein, thus indicating the absence of an N-linked carbohydrate chain [289].

The aberrant glycophorin is not precipitated by the mouse hybridoma antibody *R18* which recognises the amino acid sequence 31–54 of glycophorin A. Instead it reacts with *I16*, a rabbit antibody specific for its C-terminal portion [215]. This clearly shows that the **Dantu** character is located on a glycophorin B-A (δ-α, 'anti-Lepore' type) hybrid molecule, in which glycophorin B represents the N-terminal part and glycophorin A the C-terminus[7].

The peptide sequence data obtained from glycophorin variant molecules isolated from **Dantu**-positive individuals [37,64,79] corresponded fully with these earlier findings. It has been shown that amino acid residues 1–39 of this aberrant glycophorin molecule are derived from glycophorin B, and that residues 40–99 correspond to amino acids 72–131 of glycophorin A *(Fig. 9.11)*; the hybrid gene thus consists of exons B-2, ΨB-3 and B-4, and exons A-5 to A-7. The glycosylation of the molecule is identical to that of glycophorin B. The **Dantu** character is presumably located within the region 31–40.

Thus far three types of **Dantu** phenotype have been described:

- in the case of the **M.D.** variant detected in Europids [79,241] the glycophorin B-A hybrid is flanked by glycophorin A and glycophorin B genes, suggesting that this type originated from a single unequal crossing-over event,
- the gene cluster of **N.E.** and **Ph** variants found in Negrids [52,289,302] contains a functional glycophorin A gene but lacks a glycophorin B gene [137]. It has been proposed that in this type of **Dantu** haplotype an S^u gene was involved in the

---

[7] *In situ* δ-α hybrid molecules are not degraded by proteases. Based on this protease-resistance the reaction of ficin-treated erythrocytes with anti-N_{Vg} has been used for the quick-screening of blood samples for δ-α hybrids [306].

gene recombination event. Further, dosage quantitation experiments suggest that in the case of **N.E.** the **Dantu** gene is duplicated and tandemly arranged [137]. This assumption is corroborated by the substantially increased amount of the hybrid protein present in **Dantu+** cells of the **N.E.** type (hybrid : glycophorin A ratio being 2.4 : 1 [37,75] as compared to 1 : 1 in the **Ph** type [289]),

– the erythrocytes of the heterozygote donor **J.O.** have only one-half the normal levels of glycophorin A, thus leading to the assumption [217] that the **J.O.**variant may contain a glycophorin A-B-A hybrid molecule and an unchanged glycophorin B gene rather than an unchanged glycophorin A gene and a glycophorin B-A hybrid.

In **Dantu+** erythrocytes the apparent molecular mass of band-3 protein is reduced by about 3000 Da due to shortening of the N-glycan chain [75].

St^a (Stones; MNS15)

The **St^a** antigen occurs at an unexpectedly high frequency in Chinese (1%) and Japanese (6.4%), it is, however, extremely rare in Europids [49,210].

In most cases **St^a** erythrocytes contain an aberrant sialoglycoprotein (**GP.Sch**) with properties similar to those of the **Dantu** variant glycophorin. This suggests that the **St^a** determinant is also located on a glycophorin B-A hybrid molecule [15,31,36], a fact later proved by peptide sequencing of a **St^a** active sialoglycoprotein [32]: the molecule is composed of amino acids 1–26 of glycophorin B and amino acids 59–131 of glycophorin A *(Fig. 9.11)*. The glycosylation corresponds to that of glycophorin B.

The structure of the **St^a** specific glycophorin **GP.Sch** was later confirmed by analyses of genomic DNA isolated from **St^a** individuals [138,142,255]. These investigations revealed that the hybrid gene arose from a single unequal crossing-over event between misaligned glycophorin B and A. This misalignment led to a gene in which exons B-1 and B-2 (and pseudoexon B-3) were linked to exons A-4 to A-7. The *St^a* gene is flanked by functional glycophorin A and glycophorin B genes. This hybrid gene is thus reciprocal to that found in the **Mi.V** variant (see below).

The **St^a** epitope is determined by the amino acid sequence of the junction of exon B-2 and exon A-4, i.e. by the heptapeptide (-Gln^{24}-Thr-Asn-Gly-Glu-Arg-Val^{30}-). For some antisera the carbohydrate unit attached to Thr^{25} is an essential part of the **St^a** epitope; Asn^{26} is not glycosylated [32].

Three distinct isoforms of the gene encoding the glycophorin variant **GP.Sch** have been detected. These isoforms produce identical proteins and differ from each

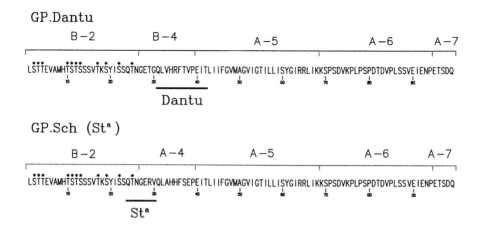

Figure 9.11: Peptide sequences of glycophorin B-A hybrid molecules.
★ : O-linked carbohydrate chain.

other only in the location of crossing-over sites [140]: in type A the crossing-over point is situated between nucleotides 134 and 270, in type B between 326 and 514, and in type C between 746 and 788. Types A and B have been found in Negrids, types B and C in Mongolids. A St^a donor studied by Rearden et al. [255] had a variant glycophorin A gene, in which substitution of A → G at nucleotide 117 resulted in a Lys → Arg substitution at position 39.

Some other glycophorin variants carrying the St^a character have been described:

In the case of the glycophorin hybrid carrying **He** and St^a antigens [146] the allele arose from recombination between a **He**-specific glycophorin hybrid allele and a normal glycophorin A gene. The unequal crossing-over event took place at the recombination site identical to that in type A isoform of **GP.Sch**.

Another St^a carrying glycophorin (**GP.EBH**) is associated with the low frequency **ERIK (MNS37)** antigen [94]. This glycophorin variant is characterised by a G → A transition at the terminal nucleotide of glycophorin A exon 3 [148]. The mutation not only created a Gly → Arg substitution at position 59 of the molecule, but also affected pre-mRNA splicing due to partial inactivation of the adjacent 5' donor splice site.

As a consequence of the alternative use of other constitutive splice sites several transcripts are produced by this glycophorin A mutant. The full-length transcript encodes a variant glycophorin A molecule with the Arg^{59} substitution defining the **ERIK** epitope. A second transcript in which exon A-3 is deleted specifies a shorter glycophorin carrying the **St**^{a} antigen. Further, two abnormally spliced mRNA species with deletions of exons A-2 and A-3 and exons A-2 to A-4, respectively, are generated by selection of 5' splice sites far distant from the mutated site: Although a correct translation frame is present, the predicted polypeptides are not inserted into the erythrocyte membrane, probably due to the severe truncation of the signal sequence required for targeting and translocation.

In an Australian family another glycophorin variant exhibiting **ERIK** and **St**^{a} specificity has been detected. The primary structure of the protein is identical to that of **GP.Sch**: The loss of exon A-3, however, resulted from its replacement with pseudoexon E-3 and the inactive splice site of intron E-3 [149].

The primary structure of another **St**^{a} specific glycophorin, **GP.Zan** (originally described as **M**^{z} [220]) is also identical to that of the **GP.Sch** hybrid molecule. The gene encoding this variant molecule, however, has been characterised as a glycophorin A-B-A hybrid [150]. In this case exon A-3 and the 5' end of intron A-3 are replaced by the homologous segment of glycophorin B, thus introducing the defective donor splice site of the pseudoexon into the glycophorin A gene.

Transcript analysis shows the presence in the **M**^{z} reticulocytes of two glycophorin cDNA species derived from the **M**^{z} gene: one species lacks exon A-3 and encodes a protein of 99 amino acids bearing the **M** and **St**^{a} antigens, whereas the other species lacks exons A-3 and A-4 and encodes an **M** active protein of 86 amino acids.

The Miltenberger Series

The **Miltenberger** series includes a group of low-frequency erythrocyte characters. Both the inheritance and the serological properties of these antigens clearly reveal a close association with the **MNS** system [311]. Originally, erythrocytes reactive with anti-**Mi**^{a} (= Miltenberger) serum [197] were classified into four groups on the basis of their different reactions with the antisera **Vw** (= Verweyst), **Mur** (= Murrell), and **Hil** (= Hill) [50]. In the course of further investigations the **Miltenberger** subsystem has been extended, and thus far eleven serological types are defined based on the reaction patterns of the erythrocytes with defined antisera (see *Table 9.4*).

Table 9.4: Serological classification of the Miltenberger variants.

Miltenberger class	Notation acc. to Tippett et al. [c]	Reaction of erythrocytes with antiserum[a][b]									
		Vw	Hut	MUT	Mur	Hil	Hop	Nob	DANE	TSEN	MINY
Mi.I	GP.Vw	+	−	−	−	−	−	−	−	−	−
Mi.II	GP.Hut	−	+	+	−	−	−	−	−	−	−
Mi.III	GP.Mur	−	−	+	+	+	−	−	−	−	+
Mi.IV	GP.Hop	−	−	+	+	−	+	−	−	+	+
Mi.V	GP.Hil	−	−	−	−	+	−	−	−	−	+
Mi.VI	GP.Bun	−	−	+	+	+	+	−	−	−	+
Mi.VII	GP.Nob	−	−	−	−	−	−	+	−	−	−
Mi.VIII	GP.Joh	−	−	−	−	−	+	+	−	n.t.	−
Mi.IX	GP.Dane	−	−	−	+	−	−	−	+	−	−
Mi.X	GP.HF	−	−	+	−	+	−	−	−	−	+
Mi.XI[d]	GP.JL	−	−	n.t.	−	−	−	−	−	+	+

[a] Vw = Verweyst, Hut = Hutchinson, Mur = Murrell, Hil = Hill, Hop = Hopper, Nob = Noble.
[b]The specificity of serum Anek [47] is similar to that of serum Hop, the specificities of the sera Raddon and Lane [315]are similar to that of serum Nob.
[c] [262, 296]
[d] Originally termed Mi.V–like or Mi.V^{J.L.} [175].

Since most of these sera are in extremely limited supply and often polyspecific, King et al. [182] have proposed an alternative classification of **Miltenberger** classes I to VI according to the reaction patterns of their abnormal erythrocyte sialoglycoproteins with murine hybridoma antibodies towards different glycophorin A epitopes.

More recently a notation has been suggested to replace the classification of the increasingly complicated **Miltenberger** subsystem [262,296]. In this notation the serologically specified phenotypes are defined by characteristic glycophorin variants: the symbol used is composed of 'GP' (for glycophorin) followed by a full stop and the abbreviated name of the propositus in which the variant had first been described, e.g. **Mi.V** becomes **GP.Hil**; the type of the hybrid is then designated as **GP(A-B)Hil**, and the respective gene *GP(A-B)Hil*.

All glycophorin variants are based on glycophorin hybrid molecules.

Glycophorin A-B hybrids

The **Mi.V** variant glycophorin, **GP.Hil**, is characterised by trypsin-sensitive **M** or **N**, strong trypsin-resistant **s**, and by the presence of the **Hil** and **MINY** antigens; **En°FR** is absent. In erythrocytes from homozygous **Mi.V** individuals the content of neuraminic acid is reduced to about one-half and the cells completely lack normal glycophorin A and B. Furthermore, probably due to a higher degree of glycosylation, the band 3 protein shows a greater molecular mass when compared to the protein of 'normal' erythrocytes.

When investigated by polyacrylamide gelelectrophoresis [16,74,215,305] the membranes of **Mi.V** erythrocytes were found to contain a glycophorin A-B hybrid. Molecular analyses revealed that the gene encoding the **GP.Hil** variant glycophorin is composed of exons A-1 to A-3 of the glycophorin A gene and exons B-4 to B-6 of the glycophorin B^s gene [138,307,308]. The **Mi.V** specific sialoglycoprotein thus contains amino acids 1–58 of glycophorin A and residues 27–72 of glycophorin B^s; its peptide sequence is shown in *Fig. 9.12*.

Mi.XI (**GP.JL**, formerly termed 'Mi.V-like' or **Mi.V**$^{J.L}$) is characterised by weak **M** and an **S** specific methionine residue at position 61, and further by the antigens **TSEN** and **MINY**. The variant protein is encoded by a glycophorin A-B^S hybrid gene analogous to that encoding the **Mi.V** (**GP.Hil**) sialoglycoprotein (*Fig. 9.12*) [138,175,186].

The **GP.JR** hybrid described by Langley et al. [194] is probably similar to the **GP.JL** variant.

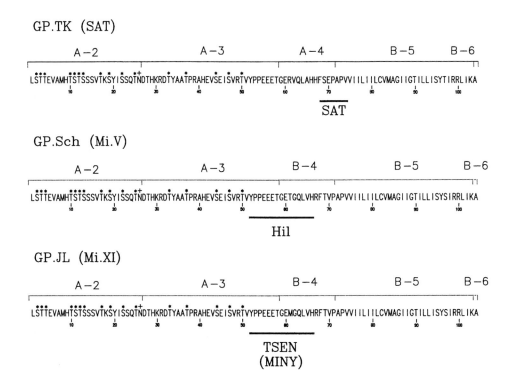

Figure 9.12: Peptide sequences of glycophorin A-B hybrid molecules.
★ : O-linked carbohydrate chain, + : N-linked carbohydrate chain.

The sequence data show that the antisera towards **Hil (MNS20)**, **TSEN (MNS33)**, and **MINY (MNS34)** antigens bind to epitopes formed by the peptide sequence at the junction of exon A-3 with exon B-4 (*Fig. 9.12*). The **Hil** epitope depends essentially on the presence of an **s** specific threonine residue present in **GP.Hil** [176], whereas the **TSEN** epitope depends on the **S** specific methionine residue present in **GP.JL** [259]. The **MINY** epitope is found on both hybrid molecules since the antibody binds to the glycophorin A–B junction regardless of whether the glycophorin B part carries **S** or **s** antigen [260].

Both **Mi.V** and **Mi.XI** genes contain a single copy of the glycophorin A-B hybrid residing in a chromosome lacking glycophorin A and B genes [138,186]. The genes are thus reciprocal to the gene encoding the **St^a** glycophorin variant.

Glycophorin A-B-A hybrids

First investigations on the erythrocyte membrane sialoglycoproteins have shown that the **Mi.I** and **Mi.II** characters are located on glycophorin A variants [28]. Structural analyses by Dahr et al. [78] and recent molecular genetic studies by Huang et al. [154] revealed an amino acid polymorphism at position 28: the threonine present in normal glycophorin A is changed in the case of **Mi.I** (**GP.Vw**) to *methionine*, and in the case of **Mi.II** (**GP.Hut**) to *lysine* (see *Fig. 9.13)*; the newly created determinants are recognised by the human antibodies towards **Vw** (**MNS9**) and **Hut** (**MNS19**) antigens, respectively. Due to the amino acid exchange at position 28 the **Asn-X-Thr/Ser** sequence obligatory for the addition of an N-linked carbohydrate chain is lost (see *Sect. 3.1.1*). The absence of the alkali-stable carbohydrate chain at position 26 (which accounts for the 3 kDa decrease in molecular mass) is further corroborated by the fact that the lectin from *Phaseolus vulgaris* does not bind to the changed glycoprotein [28].

As a probable origin of the mutant genes Huang et al. [154] have proposed a gene conversion event in which a small segment of glycophorin B (spanning the junction of intron 2 and pseudoexon 3) has replaced the homologous segment of glycophorin A; the authors assumed the amino acid polymorphism between **Mi.I** and **Mi.II** to be due to untemplated and templated nucleotide replacements (*Fig. 9.13*).

The glycophorin isolated from **Mi.VII** erythrocyte membranes shows normal PAS characteristics but exhibits **Nob** specificity (**GP.Nob**) [10]. First serological investigations suggested the location of the **Mi.VII** determinant on a glycophorin A variant with an amino acid substitution in the region between amino acids 46 and 56 [190].

Structural analysis of glycophorin A isolated from a homozygous **Mi.VII** individual [65] did in fact reveal two amino acid exchanges in that very region − at position 49 the arginine found in normal glycophorin A is substituted by glycosylated threonine, and the tyrosine at position 52 is replaced by serine (*Fig. 9.13*).

These amino acid substitutions can easily be explained: ten nucleotides of exon A-3 (nt 67–76) have been replaced by the corresponding sequence of the glycophorin pseudoexon, which contains the codons for Thr[49] and Ser[52] [153].

Serologically the **Mi.VIII** class [101] closely resembles **Mi.VII** − the glycophorin variant shows both **Hop** and **Nob** specificities. Investigations on the respective glycophorin A variant (**GP.Joh**) revealed that this class differs from **Mi.VII** by having only the arginine → threonine substitution at position 49 [90] (*Fig. 9.13*); also in this case the Thr[49] residue is glycosylated.

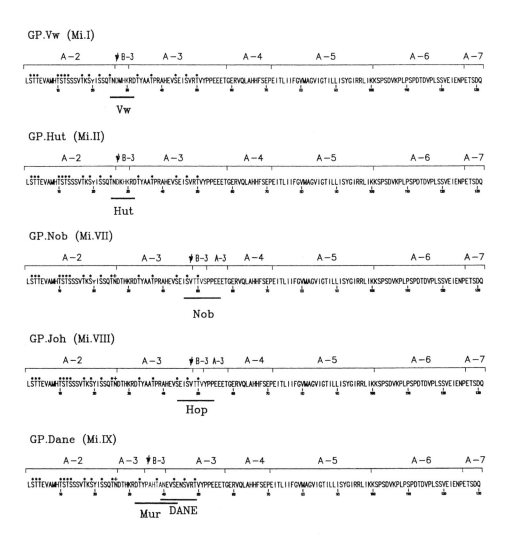

Figure 9.13: Peptide sequences of glycophorin A-B-A hybrid molecules.
★ : O-linked carbohydrate chain, ✚ : N-linked carbohydrate chain.

Inhibition experiments with synthetic peptides were able to locate the **Hop** antigenic site at residues 44–53 [176]. In contrast to the **Nob** antigen, which is probably located in the same region, Tyr^{52} is obviously an essential part of the **Hop** epitope. Both **Hop** and **Nob** antigens are sialidase-sensitive: they apparently depend on the glycosylation of Thr^{49}. The **Anek** serum, which has a specificity similar to that of anti-**Hop** has also been shown to bind to this region; the epitope also comprises sialic acid attached to O-glycosidically linked oligosaccharide [65].

The **Mi.IX** class is defined by the glycophorin variant **GP.Dane**. **Mi.IX** erythrocytes are agglutinated by anti-**Mur** (**MNS10**) and anti-**DANE** (**MNS32**) and are characterised by a trypsin-resistant and chymotrypsin-sensitive **M** antigen carried by an aberrant glycophorin A molecule with an apparent molecular mass ~1 kDa lower than that of normal glycophorin A [275].

Analysis of the gene encoding the **Mi.IX** variant revealed a glycophorin A-B-A hybrid molecule in which an internal segment of exon A-3 (codons 35–41) is replaced by the corresponding sequence of the glycophorin B pseudoexon [153]. As a consequence of this gene conversion event, the heptapeptide sequence of glycophorin A, ^{35}Ala–Ala–Thr–Pro–Arg–Ala–His^{41}, has been changed to the hexapeptide sequence ^{35}Pro–Ala–His–Thr–Ala–Asn^{40} (*Fig. 9.13*). The glycophorin hybrid contains an additional amino acid substitution (Ile^{46} → Asn^{45}) caused by an untemplated adenyl nucleotide mutation, probably as a consequence of a defective repair of the heteroduplex DNA during segment replacement.

The **Mur** (**MNS10**) epitope is defined by the expressed pseudoexon B sequence, which data obtained from inhibition tests confirms [176]. The Asn substitution at position 45 obviously is responsible for the **DANE** antigen expression; the presence of the same tripeptide, Thr^3-Asn-Glu^5, in the **M^g** specific glycophorin may explain the reaction of anti-**DANE** with **M^g** cells [127].

Glycophorin B-A-B hybrids

The **Miltenberger** variant, **Mi.III**, occurs at a remarkably high frequency (5–10%) in certain Mongolid populations [42,48,245]. **Mi.III** cells are **s+**, **Mur+**, **Hil+**, and **MINY+**, and are characterised by an elevated expression of '**N**' and a significantly increased level of sialic acid (in homozygotes by about 21% more than normal cells) [16,74,182].

The **Mi.III** variant glycophorin, **GP.Mur**, is based on a glycophorin B-A-B hybrid gene. Genome analyses [139] revealed that the **Mi.III** specific sialoglycoprotein is encoded by a *GPB^s* gene, in which a short segment (about 55 base pairs) of the pseudoexon containing the non-functional donor splice site has been replaced by its homologous counterpart of the glycophorin A gene (*Fig. 9.14*). Since this segment comprises a portion of both exon A-3 and intron A-3, which carries a functional 5' splicing signal, the rearrangement results in the expression of a normally unexpressed pseudoexon sequence of the glycophorin B gene. The *GP.Mur* haplotype carries a normal glycophorin A gene.

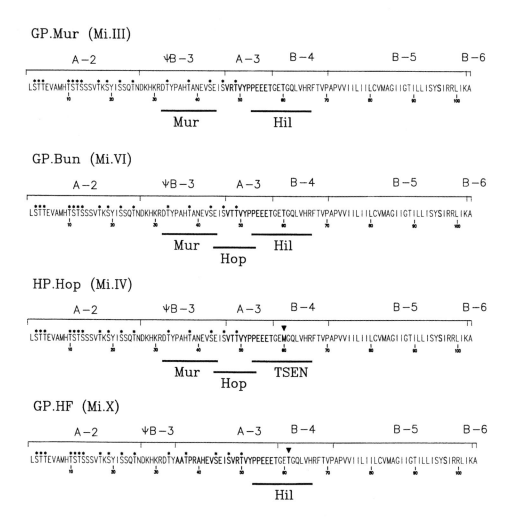

Figure 9.14: Peptide sequences of glycophorin B-A-B hybrid molecules.
★ : O-linked carbohydrate chain.

The structure of the **Mi.III** sialoglycoprotein deduced from genome analysis has been verified by peptide sequencing [174]. The **Mur** epitope has been located at residues 32–44 [176].

In one sample of **Mi.III** cells a significant increase in the amount of the tetrasaccharide chain containing N-acetyl-glucosamine (*Fig. 9.3.f*) has been detected

[34] (see also variant M_1). It has not yet been possible, however, to duplicate this result in a second **Mi.III** individual [2].

The **Mi.VI** variant glycophorin hybrid, **GP.Bun**, is almost identical to the **Mi.III** variant hybrid [139]. In contrast to the latter **Mi.VI** cells are **Hop+**. Though differing in the location of upstream (δ-α) and downstream (α-δ) breakpoints and in the length of sequence replaced (i.e. 131 bp), the **Mi.VI** gene differs from the **Mi.III** gene by only one nucleotide in the coding sequence. This results in an **Arg** ($\rightarrow$ **Mi.III**) / **Thr** ($\rightarrow$ **Mi.VI**) dimorphism at position 48 (see *Fig. 9.14*) and a few nucleotides in the noncoding sequence.

The **Mi.IV** specific glycophorin, **GP.Hop**, although **Hil**-negative and **TSEN**-positive, reacts in a manner closely similar to the **Mi.III** and **Mi.VI** variants. It is inherited with **N** and **S** [50]. The sequence of the **Mi.IV** glycophorin has not been investigated to date. However, the presence of **Mur**, **Hop**, and **TSEN** epitopes suggests that **GP.Hop**, with the exception of an **S** specific methionine residue at position 60, is identical to the **Mi.VI** variant glycophorin, **GP.Bun** (*Fig. 9.14*) [61,134,190].

Erythrocytes carrying the **Mi.X** variant (**GP.HF**) are characterised by **M** and unusually strong **s** activity, as well as by their reactivity with anti-**Hil** and anti-**MINY**.
Investigation [145] of the **Mi.X** specific glycophorin revealed a glycophorin B-A-B hybrid molecule similar to those detected in **Mi.III** and **Mi.VI** erythrocytes (*Fig. 9.14*). In this case a 98 bp insert from exon A-3 created a hybrid glycophorin B gene which encodes a peptide differing from **GP.Mur** by five amino acid residues.

ENEP, HAG, ENAV, and MARS

Recently a glycophorin A variant has been found in which the alanine residue at position 65 is replaced by proline [243]. This mutant lacks the high-incidence antigen **ENEP** (MNS39) and shows the low-incidence antigen **HAG** (MNS41), the **ENEP** antigen thus being dependent on Ala^{65} and **HAG** on Pro^{65}.

ENAV (MNS42), previously termed **AVIS**, is a high-incidence antigen and **MARS** (MNS43) its low-incidence antithetical antigen. A $Gln^{63} \rightarrow Lys$ substitution in glycophorin A results in absence of **ENAV** and presence of **MARS** [169,170].

The fact that both mutations markedly affect the expression of the Wr^b (= DI4) antigen shows that the peptide sequence around residue 65 is essentially involved in formation of the Wr^b epitope (see *Sect. 20.2.2)*.

Further MNS Variants

M_1 [168] is an antigen characteristic for Negrids – up to 25% of the **M** individuals among American Blacks and almost 50% among Bantus exhibit M_1 activity. The frequency of M_1 is less than 5% in Europids [126].

Anti-M_1 is found fairly frequently in the serum of human *NN* individuals, in most cases it occurs together with anti-**M** antibodies [126].

There is evidence that the M_1 determinant is located at the N-terminal end of glycophorin A^M; thus far, however, investigations have not revealed any amino acid substitution [62,69]. On the other hand, a raised level of N-acetylglucosamine has been observed in the membranes of M_1 erythrocytes. Because M_1 activity increases with glucosamine content, it has been suspected that the glucosamine-containing tetrasaccharide chains (see *Fig. 9.3.f)* may play a role in blood group M_1 specificity [21]. Ever since Issitt et al. [167] showed that the carbohydrate chain is not an essential part of the epitope, it has been assumed that the anti-M_1 sera bind to a glycophorin A^M chain, the sterical configuration of which is altered when the $\alpha2{\to}6$ neuraminic acid residue in the oligosaccharide units at positions 2, 3, and/or 4 is exchanged to N-acetylglucosamine.

In M^v (MNS21) erythrocytes [124] the **M** and **N** characters are normally expressed, but 'N' and N_{Vg} are absent [73]; the **s** antigen, when present, is detected by some but not all samples of anti-**s**.

Several antisera with anti-M^v specificities [57,313] and one example of an anti-(M^v+N) antiserum [124] have been described.

Investigations by Dahr et al. revealed that the M^v epitope is located in the N-terminal octapeptide of a trypsin-sensitive glycophorin B variant ([73,74], see also Mawby, cited in [10]). M^v erythrocytes lack the 'N' character, whereas isolated sialoglycoproteins show 'N' activity which is destroyed by a single cycle of Edman degradation. This finding led to the assumption that at least the N-terminal amino acid is changed; further, sensitivity of the receptor towards neuraminidase [73] shows that neuraminic acid must be part of the M^v epitope.

Molecular biological investigations revealed in two M^v individuals a $C^{65} \to G$ mutation resulting in a Thr $\to$ Ser substitution at position 3 of glycophorin B [285a].

The low incidence antigen Ri^a (**Ridley, MNS16**) [49,51] has been found in one family only. Ri^a is sensitive to trypsin but insensitive to papain or α-chymotrypsin.

The glycophorin A-specific cDNA clones from Ri^a-positive individuals showed a point mutation (G → A) at nucleotide 220 which changes Glu^{57} to Lys in exon 3 [285b].

Two other antigen specificities have been located on glycophorin A [285c] – the **Vr (Verdegaal, MNS12)** [304] antigen is caused by a C^{197} → A mutation which changes serine to tyrosine at position 47, and the Mt^a (**Martin, MNS14**) [183,286] antigen is based on a C^{230} → T mutation which changes threonine to isoleucein at position 58.

The **Mit (Mitchell, MNS24)** [20] and the s^D (**Dreyer, MNS23**) [271] antigens are both located on glycophorin B [285a] – the **Mit** antigen is determined by a G^{160} → A mutation which changes Arg to His at position 35, and the molecular basis of the s^D (**Dreyer, MNS23**) [271] antigen is a C^{173} → G mutation which causes a Pro → Arg substitution at position 39.

The **Duclos** antigen [129] is a common character of human erythrocytes; however, its activity is repressed in the rare Rh_{null} individuals of the 'regulator type' (see *Sect. 13.3*) [72] and in **S–s–U–** subjects.

There is strong evidence that the **Duclos** antigen is located on glycophorin B [76], the structure of the **Duclos** determinant, however, has not yet been established. The character is completely insensitive to protease treatment and needs the presence of lipid for optimum expression of serological activity. This fact leads to the assumption that the binding site of anti-**Duclos** sera must be situated near the cell membrane close to the blood group **U** determinant (i.e. between amino acids 34 and 36) [76] (see *Fig. 9.8*).

The characters U^x and U^z are defined by anti-U^x and anti-U^z antibodies which are extremely rare in Europids but are found fairly frequently in Melanesians [38,39]. Both sera react with all human erythrocytes except those of the **S–s–U–** phenotype. Significant variations in agglutinability suggest that the serological activity of both antibodies is highly dependent on antigenic site density [58].

Investigations on the protease-sensitivity of the two determinants [58], as well as the fact that anti-**U** or anti-**s** antibodies prevent the binding of anti-U^x and anti-U^z to erythrocytes, indicate that U^x and U^z are located on the glycophorin B molecule close to the **S/s** antigens. Inhibition tests showed that anti-U^x binds to glycophorin B, although a weak affinity towards glycophorin A has also been observed [58,66]. Since U^x activity is destroyed upon hydrolysis or modification of the neuraminic acid, it can be assumed that the anti-U^x antibodies bind to a receptor located in the glycosylated segment of glycophorin B near amino acid 26. The chymotrypsin sensitivity of the U^z

character suggests that the U^z receptor is situated close to the chymotrypsin cleavage site, i.e. near amino acid residues 32–35 [58].

The low-frequency antigen Os^a (**MNS38**) [270] resides on glycophorin A of normal electrophoretic mobility [93]. It is trypsin sensitive, but papain, ficin, and pronase resistant. MAIEA tests located the antigen in the region around residue 40 of glycophorin A.

Results of family investigations on the inheritance of the very low-incidence antigen **Orriss** (**Or, MNS31**) by Bacon et al. [19] confirm Issitt's assumption that this character is part of the **MNS** system [161]. This view is further corroborated by the finding that glycophorin A isolated from **Or(+)** erythrocytes is able to neutralise Anti-**Or**. The chemical basis of the **Orriss** antigen has not yet been investigated.

In the literature a variety of further antigenic characters closely connected with the **MNS** system have been described:
Cl^a (**Caldwell, MNS17**) [310], Ny^a (**Nyberg, MNS18**) [236,237], **Far** (**MNS22**) [56], as well as **Sj** (**Stenbar-James**) [165], **Hu** (**Hunter**) [46,193], **Sul** (**Sullivan**) [184], and the phenotypes S_2 [155] and N_2 [220]. Since the chemical basis of these characters is still unclear they will not be subject of discussion here.

References

1. ADAIR, W. L. & KORNFELD, S. (1974): Isolation of the receptor for wheat germ agglutinin and the Ricinus communis lectins from human erythrocytes using affinity chromatography. *J. Biol. Chem.* 249, 4696-4704.
2. ADAMANY, A. M., BLUMENFELD, O. O., SABO, B. & McCREARY, J. (1983): A carbohydrate structural variant of MM glycoprotein (glycophorin A). *J. Biol. Chem. 258*, 11537-11545.
3. ALBERTS, B., BRAY, D., LEWIS, J., RAFF, M., ROBERTS, K. & WATSON, J. D. (1989): *Molekularbiologie der Zelle.* VCH Verlagsgesellschaft mbH, Weinheim, Germany.
4. ALLEN, F. H., CORCORAN, P. A. & ELLIS, F. R. (1960): Some new observations on the MN system. *Vox Sang. 5*, 224-231.
5. ALLEN, F. H., CORCORAN, P. A., KENTON, H. B. & BREARE, N. (1958): M^g, a new blood group antigen in the MNS system. *Vox Sang. 3*, 81-91.
6. ALLEN, F. H., MADDEN, H. J. & KING, R. W. (1963): The MN gene MU, which produces M and U but no N, S, or s. *Vox Sang. 8*, 549-556.
7. ALLEN, N. K. & BRILLANTINE, L. (1959): A survey of hemagglutinins in various seeds. *J. Immunol.* 102, 1295-1304.
8. ALLEN, R. W., NUNLEY, N., KIMMETH, M. E., WALLHERMFECHTEL, M. & VENGELEN- TYLER, V. (1984): Isolation and serological characterization of a monoclonal antibody recognizing the N blood group antigen. *Transfusion 24*, 136-140.
9. ANDERSSON, L. C., GAHMBERG, C. G., TEERENHOVI, L. & VICOPIO, P. (1979): Glycophorin A as a cell surface marker of early erythroid differentiation in acute leukaemia. *Int. J. Cancer 23*, 717-723.
10. ANSTEE, D. J. (1980): Blood group MNSs-active sialoglycoproteins of the human erythrocyte

membrane. In: *Immunobiology of the Erythrocyte* (S. G. Sandler, J. Nusbacher, and M. S. Schanfield, eds.). Alan R. Liss Inc., New York, pp. 67-98.

11. ANSTEE, D. J. (1990): The nature and abundance of human red cell surface glycoproteins. *J. Immunogenet. 17*, 219-225.

12. ANSTEE, D. J., BARKER, D. M., JUDSON, P. A. & TANNER, M. J. A. (1977): Inherited sialoglycoprotein deficiencies in human erythrocytes of type En(a-). *Brit. J. Haematol. 35*, 309-320.

13. ANSTEE, D. J. & EDWARDS, P. A. W. (1982): Monoclonal antibodies to human erythrocytes. *Eur. J. Immunol. 12*, 228-232.

14. ANSTEE, D. J. & LISOWSKA, E. (1990): Monoclonal antibodies against glycophorins and other glycoproteins. *J. Immunogenet. 17*, 301-308.

15. ANSTEE, D. J., MAWBY, W. J., PARSONS, S. F., TANNER, M. J. A. & GILES, C. M. (1982): A novel hybrid sialoglycoprotein in St^a positive human erythrocytes. *J. Immunogenet. 9*, 51-55.

16. ANSTEE, D. J., MAWBY, W. J. & TANNER, M. J. A. (1979): Abnormal blood-group-Ss-active sialoglycoproteins in the membrane of Miltenberger class III, IV, and V human erythrocytes. *Biochem. J. 183*, 193-203.

17. ANSTEE, D. J. & TANNER, M. J. A. (1978): Genetic variants involving the major membrane sialoglycoprotein of human erythrocytes. Studies on erythrocytes of type M^k, Miltenberger class V and M^g. *Biochem. J. 175*, 149-157.

18. AUSTIN, R. J. & RICHES, G. (1978): Inherited lack of Ss antigens with weak expression of U in a Caucasian family. *Vox Sang. 34*, 343-346.

19. BACON, J. M., MACDONALD, E. B., YOUNG, S. G. & CONNELL, T. (1987): Evidence that the low frequency antigen Orriss is part of the MN blood group system. *Vox Sang. 52*, 330-334.

20. BATTISTA, N., STOUT, T. D., LEWIS, M. & KAITA, H. (1980): A new rare blood group antigen 'Mit'. Probable genetic relationship with the MNSs blood groups system. *Vox Sang. 39*, 331-334.

21. BAUMEISTER, G., DAHR, W., BEYREUTHER, K., MOULDS, J., JUDD, W. J., ISSITT, P. D. & KRÜGER, J. (1982): Untersuchungen zur Struktur der M$_1$, Tm und Can Antigene. In: *Forschungsergebnisse der Transfusionsmedizin und Immunohaematologie* (V. Nagel and W. Stangel, eds.). Medicus Verlag GmbH., Berlin, pp. 121-125.

22. BEATTIE, K. M. & ZUELZER, W. W. (1965): The frequency and properties of pH dependent anti-M. *Transfusion 5*, 322-326.

23. BELL, C. A. & ZWICKER, H. (1980): pH-dependent anti-U in autoimmune hemolytic anemia. *Transfusion 20*, 86-89.

24. BIGBEE, W. L., LANGLOIS, R. G., VANDERLAAN, M. & JENSEN, R. H. (1984): Binding specificities of eight monoclonal antibodies to human glycophorin A. Studies with M^{c}M and M^{k}En(UK) variant human erythrocytes and M- and MN^V-type chimpanzee erythrocytes. *J. Immunol. 133*, 3149-3155.

25. BIGBEE, W. L., VANDERLAAN, M., FONG, S. S. N. & JENSEN, R. H. (1983): Monoclonal antibodies specific for the M- and N-forms of human glycophorin A. *Mol. Immunol. 20*, 1353-1362.

26. BIRD, G. W. G. & WINGHAM, J. (1970): Agglutinins for antigens of two different human blood group systems in the seeds of Moluccella laevis. *Vox Sang. 18*, 235-239.

27. BIRD, G. W. G. & WINGHAM, J. (1974): The M, N, and N$_{Vg}$ receptors of Tn-erythrocytes. *Vox Sang. 26*, 171-175.

28. BLANCHARD, D., ASSERAF, A., PRIGENT, M. J. & CARTRON, J. P. (1983): Miltenberger class I and II erythrocytes carry a variant of glycophorin A. *Biochem. J. 213*, 399-404.

29. BLANCHARD, D., ASSERAF, A., PRIGENT, M. J., MOULDS, J. J., CHANDANAYINGYONG, D. & CARTRON, J. P. (1984): Interaction of Vicia graminaea anti-N lectin with cell surface glycoproteins from erythrocytes with rare blood group antigens. *Hoppe-Seyler's Z. Physiol. Ch. 365*, 469-478.

30. BLANCHARD, D., CARTRON, J. P., FOURNET, B., MOUNTREUIL, J., VAN HALBEEK, H. & VLIEGENTHART, J. F. G. (1983): Primary structure of the oligosaccharide determinant of blood group Cad specificity. *J. Biol. Chem. 258*, 7691-7695.

31. BLANCHARD, D., CARTRON, J. P., ROUGER, P. & SALMON, C. (1982): Pj variant; a new hybrid MNSs glycoprotein of the human red-cell membrane. *Biochem. J. 203*, 419-426.

32. BLANCHARD, D., DAHR, W., BEYREUTHER, K., MOULDS, J. & CARTRON, J. P. (1987): Hybrid

glycophorins from human erythrocyte membranes. Isolation and complete structural analysis of the novel sialoglycoprotein from St(a+) red cells. *Eur. J. Biochem. 167*, 361-366.

33. BLANCHARD, D., DAHR, W., HUMMEL, M., LATRON, F., BEYREUTHER, K. & CARTRON, J. P. (1987): Glycophorins B and C from human erythrocyte membranes. Purification and sequence analysis. *J. Biol. Chem. 262*, 5808-5811.

34. BLUMENFELD, O. O., ADAMANY, A. & PUGLIA, K. V. (1981): Amino acid and carbohydrate structural variants of glycoprotein products (M-N glycoproteins) of the M-N allelic locus. *Proc. Natl. Acad. Sci. USA 78*, 747-751.

35. BLUMENFELD, O. O. & ADAMANY, A. M. (1978): Structural polymorphism within the amino-terminal region of MM, NN, and MN glycoproteins (glycophorins) of the human erythrocyte membrane. *Proc. Natl. Acad. Sci. USA 75*, 2727-2731.

36. BLUMENFELD, O. O., ADAMANY, A. M., KIKUCHI, M., SABO, B. & MCCREARY, J. (1986): Membrane glycophorins in St^a blood group erythrocytes. *J. Biol. Chem. 261*, 5544-5552.

37. BLUMENFELD, O. O., SMITH, A. J. & MOULDS, J. J. (1987): Membrane glycophorins of Dantu blood group erythrocytes. *J. Biol. Chem. 262*, 11864-11870.

38. BOOTH, P. B. (1972): A 'new' blood group antigen associated with S and s. *Vox Sang. 22*, 524-528.

39. BOOTH, P. B. (1978): Two melanesian antisera reacting with SsU components. *Vox Sang. 34*, 212-220.

40. BOYD, W. C., EVERHART, D. L. & MCMASTER, M. H. (1958): The anti-N lectin of Bauhinina purpurea. *J. Immunol. 81*, 414-418.

41. BRETSCHER, M. S. (1971): Major human erythrocyte glycoprotein spans the cell membrane. *Nat. New Biol. 231*, 229-232.

42. BROADBERRY, R. E. & LIN, M. (1996): The distribution of the MiIII (GP.Mur) phenotype among the population of Taiwan. *Transfus. Med. 6*, 145-148.

43. BRUCE, L. J., GROVES, J. D., OKUBO, Y., THILAGANATHAN, B. & TANNER, M. J. A. (1994): Altered band 3 structure and function in glycophorin A- and B-deficient (M^{k}M^k) red blood cells. *Blood 84*, 916-922.

44. BUNN, H. F., FORGET, B. G. & RANNEY, H. M. (1977): *Human Haemoglobins.* Saunders, Philadelphia, pp. 151-154.

45. CHALLOU, N., GOORMAGHTIGH, E., CABIAUX, V., CONRATH, K. & RUYSSCHAERT, J. M. (1994): Sequence and structure of the membrane-associated peptide of glycophorin A. *Biochemistry 33*, 6902-6910.

46. CHALMERS, J. N. M., IKIN, E. W. & MOURANT, A. E. (1953): A study of two unusual blood-group antigens in West Africans. *Brit. Med. J. 2*, 175-177.

47. CHANDANAYINGYONG, D., PEJRACHANDRA, S. & POOLE, J. (1977): Three antibodies of the MNSs system and their association with the Miltenberger complex of antigens. I. Anek serum. *Vox Sang. 32*, 272-273.

48. CHANDANAYINGYONG, D. & PEJRACHANDRA, S. (1975): Studies on the Miltenberger complex frequency in Thailand and family studies. *Vox Sang. 28*, 152-155.

49. CLEGHORN, T. E. (1962): Two human blood group antigens, St^a (Stones) and Ri^a (Ridley), closely related to the MNSs system. *Nature 195*, 297-298.

50. CLEGHORN, T. E. (1966): A memorandum on the Miltenberger blood groups. *Vox Sang. 11*, 219-222.

51. CONTRERAS, M., ARMITAGE, S. E. & STEBBING, B. (1984): The MNSs antigen Ridley (Ri^a). *Vox Sang. 46*, 360-365.

52. CONTRERAS, M., GREEN, C., HUMPHREYS, J., TIPPETT, P., DANIELS, G., TEESDALE, P., ARMITAGE, S. & LUBENKO, A. (1984): Serology and genetics of the MNSs-associated antigen Dantu. *Vox Sang. 46*, 377-386.

53. COOK, P. J. L., LINDENBAUM, R. H., SALONEN, R., DE LA CHAPELLE, A., DAKER, M. G., BUCKTON, K. E., NOADES, J. E. & TIPPETT, P. (1981): The MNSs blood groups of families with chromosome 4 rearrangements. *Ann. Hum. Genet. 45*, 39-47.

54. CORFIELD, V. A., MOOLMAN, J. C., MARTELL, R. & BRINK, P. A. (1993): Polymerase chain reaction-based detection of MN blood group-specific sequences in the human genome.

Transfusion 33, 119-124.

55. Côté, R. H. (1970): Human sources of blood group substances. In: *Blood and Tissue Antigens* (D. Aminoff, ed.) Academic Press, New York, London, pp. 249-264.

56. Cregut, R., Liberge, G., Yvart, J., Brocteur, J. & Salmon, C. (1974): A new rare blood group antigen, 'FAR', probably linked to the MNSs system. *Vox Sang. 21*, 522-530.

57. Crossland, J. D., Pepper, M. D., Giles, C. M. & Ikin, E. W. (1970): A British family possessing two variants of the MNSs blood group system, M^v and a new class within the Miltenberger complex. *Vox Sang. 18*, 407-413.

58. Dahr, W. (1981): Serology, genetics, and chemistry of the MNSs blood group system. *Blood Transfus. Immunohaematol. 24*, 85-95.

59. Dahr, W. (1983): Biochemical studies of erythrocyte membrane glycoprotein variants. In: *Red Cell Membrane Glycoconjugates and Related Genetic Markers* (J. P. Cartron, P. Rouger, and C. Salmon, eds.). Librairie Arnette, Paris, pp. 27-36.

60. Dahr, W. (1986): Immunochemistry of sialoglycoproteins in human red blood cell membranes. In: *Recent Advances in Blood Group Biochemistry* (V. Vengelen-Tyler and W. J. Judd, eds.). American Association of Blood Banks, Arlington, VA, pp. 23-65.

61. Dahr, W. (1992): Miltenberger subsystem of the MNSs blood group system. Review and outlook. *Vox Sang. 62*, 129-135.

62. Dahr, W., Baumeister, G. & Beyreuther, K.(1982): Studies on structure of the M$_1$, Tm, and Can antigens (Abstract). *Transfusion 22*, 420.

63. Dahr, W., Beyreuther, K., Gallasch, E., Krüger, J. & Morel, P. (1981): The amino acid sequence of the blood group M^g-specific major human erythrocyte membrane sialoglycoprotein. *Hoppe-Seyler's Z. Physiol. Chem. 362*, 81-85.

64. Dahr, W., Beyreuther, K., Moulds, J. & Unger, P. (1987): Hybrid glycophorins from human erythrocyte membrane. I. Isolation and complete structural analysis of the hybrid sialoglycoprotein from Dantu-positive red cells of the N.E. variety. *Eur. J. Biochem. 166*, 31-36.

65. Dahr, W., Beyreuther, K. & Moulds, J. J. (1987): Structural analysis of the major human erythrocyte membrane sialoglycoprotein from Miltenberger VII cells. *Eur. J. Biochem. 166*, 27-30.

66. Dahr, W., Beyreuther, K., Steinbach, H., Gielen, W. & Krüger, J. (1980): Structure of the Ss blood group antigens. II. A methionine/ threonine polymorphism within the N-terminal sequence of the Ss-glycoprotein. *Hoppe-Seyler's Z. Physiol. Chem. 361*, 895-906.

67. Dahr, W., Gielen, W., Beyreuther, K. & Krüger, J. (1980): Structure of the Ss blood group antigens. I. Isolation of Ss-active glycopeptides and differentiation of the antigens by modification of methionine. *Hoppe-Seyler's Z. Physiol. Chem. 361*, 145-152.

68. Dahr, W., Issitt, P., Moulds, J. & Pavone, B. (1978): Further studies on the membrane glycoprotein defects of S-s- and En(a-) erythrocytes. *Hoppe-Seyler's Z. Physiol. Chem. 359*, 1217-1224.

69. Dahr, W., Knuppertz, G., Beyreuther, K., Moulds, J. J., Moulds, M., Wilkinson, S., Capon, C., Fournet, B. & Issitt, P. D. (1991): Studies on the structures of the Tm, Sj, M$_1$, Can, Sext, and Hu blood group antigens. *Biol. Chem. Hoppe-Seyler 372*, 573-584.

70. Dahr, W., Kordowicz, M., Beyreuther, K. & Krüger, J. (1981): The amino acid sequence of the M^c-specific major red cell membrane sialoglycoprotein - an intermediate of the blood group M- and N-active molecules. *Hoppe-Seyler's Z. Physiol. Chem. 362*, 363-366.

71. Dahr, W., Kordowicz, M., Judd, W. J., Moulds, J., Beyreuther, K. & Krüger, J. (1984): Structural analysis of the Ss sialoglycoprotein specific for Henshaw blood group from human erythrocyte membranes. *Eur. J. Biochem. 141*, 51-55.

72. Dahr, W., Kordowicz, M., Moulds, J., Gielen, W., Lebeck, L. & Krüger, J. (1987): Characterization of the Ss sialoglycoproteins and its antigens in Rh$_{null}$ erythrocytes. *Blut 54*, 13-24.

73. Dahr, W. & Longster, G. (1984): Studies on M^v red cells. II. Immunochemical investigations. *Blut 49*, 299-306.

74. Dahr, W., Longster, G., Uhlenbruck, G. & Schumacher, K. (1978): Studies on Miltenberger class

III, V, M^v , and M^k red cells. I. Sodium-dodecylsulfate polyacrylamide gel electrophoretic investigations. *Blut 37*, 129-138.

75. DAHR, W., MOULDS, J., UNGER, P. & KORDOWICZ, M. (1987): The Dantu erythrocyte phenotype of the NE variety. I. Dodecylsulfate polyacrylamide gel electrophoretic studies. *Blut 55*, 19-31.

76. DAHR, W. & MOULDS, J. J. (1987): High-frequency antigens of human erythrocyte membrane sialoglycoproteins. IV. Molecular properties of the U antigen. *Biol. Chem. Hoppe-Seyler 368*, 659-667.

77. DAHR, W., MÜLLER, T., MOULDS, J., BAUMEISTER, G., ISSITT, P. D., WILKINSON, S. & GARRATTY, G. (1985): High frequency antigens of human erythrocyte membrane sialoglycoproteins. I. En^a receptors in the glycosylated domain of the MN sialoglycoprotein. *Biol. Chem. Hoppe-Seyler 366*, 41-51.

78. DAHR, W., NEWMAN, R. A., CONTRERAS, M., KORDOWICZ, M., TEESDALE, P., BEYREUTHER, K. & KRÜGER, J. (1984): Structures of the Miltenberger class I and II specific major human erythrocyte membrane sialoglycoproteins. *Eur. J. Biochem. 138*, 259-265.

79. DAHR, W., PILKINGTON, P. M., REINKE, H., BLANCHARD, D. & BEYREUTHER, K. (1989): A novel variety of the Dantu gene complex (Dantu^{MD}) detected in a Caucasian. *Blut 58*, 247-253.

80. DAHR, W. & UHLENBRUCK, G. (1978): Structural properties of the human MN blood group antigen receptor sites. *Hoppe-Seyler's Z. Physiol. Chem. 359*, 835-843.

81. DAHR, W., UHLENBRUCK, G., ISSITT, P. D. & ALLEN, F. H. (1975): SDS-polyacrylamide gel electrophoretic analysis of the membrane glycoproteins from S-s-U- erythrocytes. *J. Immunogenet. 2*, 249-251.

82. DAHR, W., UHLENBRUCK, G., JANSSEN, E. & SCHMALISCH, R. (1976): Heterogeneity of human red cell membrane sialoglycoproteins. *Blut 32*, 171-184.

83. DAHR, W., UHLENBRUCK, G., JANSSEN, E. & SCHMALISCH, R. (1977): Different N-terminal amino acids in the MN-glycoprotein from MM and NN erythrocytes. *Hum. Genet. 35*, 335-343.

84. DAHR, W., UHLENBRUCK, G. & KNOTT, H. (1975): Immunochemical aspects of the MNSs-blood group system. *J. Immunogenet. 2*, 87-100.

85. DAHR, W., UHLENBRUCK, G. & KNOTT, H. (1977): The defect of M^k erythrocytes as revealed by sodium dodecylsulphate-polyacrylamide gel electrophoresis. *J. Immunogenet. 4*, 191-200.

86. DAHR, W., UHLENBRUCK, G., LEIKOLA, J. & WAGSTAFF, W. (1978): Studies on the membrane glycoprotein defect of En(a-) erythrocytes. III. N-terminal amino acids of sialoglycoproteins from normal and En(a-) red cells. *J. Immunogenet. 5*, 117-127.

87. DAHR, W., UHLENBRUCK, G., LEIKOLA, J., WAGSTAFF, W. & LANDFRIED, K. (1976): Studies on the membrane glycoprotein defect of En(a-) erythrocytes. I. Biochemical aspects. *J. Immunogenet. 3*, 329-346.

88. DAHR, W., UHLENBRUCK, G., SCHMALISCH, R. & JANSSEN, E. (1976): Ss-blood group associated PAS-staining polymorphism of glycoprotein 3 from human erythrocyte membranes. *Hum. Genet. 32*, 121-132.

89. DAHR, W., UHLENBRUCK, G., WAGSTAFF, W. & LEIKOLA, J. (1976): Studies on the membrane glycoprotein defect of En(a-) erythrocytes. II. MN antigenic properties of En(a-) erythocytes. *J. Immunogenet. 3*, 383-393.

90. DAHR, W., VENGELEN-TYLER, V., DYBKJAER, E. & BEYREUTHER, K. (1989): Structural analysis of glycophorin A from Miltenberger class VIII erythrocytes. *Biol. Chem. Hoppe-Seyler 370*, 855-859.

91. DAHR, W., WILKINSON, S., ISSITT, P. D., BEYREUTHER, K., HUMMEL, M. & MOIREL, P. (1986): High frequency antigens of human erythrocyte membrane sialoglycoproteins. III. Studies on the En^a FR, Wr^b, and Wr^a antigens. *Biol. Chem. Hoppe-Seyler 367*, 1033-1045.

92. DANIELS, G. L., GREEN, C. A., OKUBO, Y., SENO, T., YAMAGUCHI, H., OTA, S., TAGUCHI, T. & TOMONARI, Y. (1991): SAT, a 'new' low frequency blood group antigen, which may be associated with two different MNS variants. *Transfus. Med. 1*, 39-45.

93. DANIELS, G. L., GREEN, C. A., PETTY, A. C., GREEN F & OKUBO, Y. (1993): The low frequency antigen Os^a is located on glycophorin A. (Abstract). *Transfus. Med. 3 (Suppl. 1)*, 85.

94. DANIELS, G. L., GREEN, C. A., POOLE, J., JERNE, D., SMART, E., WILCOX, D. & YOUNG, S. (1993):

ERIK, a low-frequency red cell antigen of the MNS blood group system associated with St^a. *Transfus. Med. 3*, 129-135.

95. DARNBOROUGH, J., DUNSFORD, I. & WALLACE, J. A. (1969): The En^a antigen and antibody, a genetical modification of human red cells affecting their blood grouping reactions. *Vox Sang. 17*, 241-255.

96. DRZENIEK, Z., KUSNIERZ, G. & LISOWSKA, E. (1981): A human antiserum reacting with modified blood group M determinants. *Immunol. Commun. 10*, 185-197.

97. DUK, M. & LISOWSKA, E. (1981): Vicia graminaea anti-N lectin: partial characterization of the purified lectin and its binding to erythrocytes. *Eur. J. Biochem. 118*, 131-136.

98. DUK, M., LISOWSKA, E., KORDOWICZ, M. & WASNIOWSKA, K. (1982): Studies on the specificity of the binding site of Vicia graminaea anti-N lectin. *Eur. J. Biochem. 123*, 105-112.

99. DUK, M., MITRA, D., LISOWSKA, E., KABAT, E. A., SHARON, N. & LIS, H. (1992): Immunochemical studies on the combining site of the A + N blood type specific Moluccella laevis lectin. *Carbohydr. Res. 236*, 245-258.

100. DUNSFORD, I., IKIN, E. W. & MOURANT, A. E. (1953): A human blood group gene intermediate between M and N. *Nature 172*, 688-699.

101. DYBKJAER, E., POOLE, J. & GILES, C. M. (1981): A new Miltenberger class detected by a second example of Anek serum. *Vox Sang. 41*, 302-305.

102. EKBLOM, M., GAHMBERG, C. G. & ANDERSSON, L. C. (1985): Late expression of M and N antigens on glycophorin A during erythroid differentiation. *Blood 66*, 233-236.

103. ELLIOT, M., BOSSOM, E., DUPUY, M. E. & MASOUREDIS, S. P. (1964): Effect of ionic strength on the serologic behaviour of red cell isoantibodies. *Vox Sang. 9*, 396-411.

104. FAIRBANKS, G., STECK, T. L. & WALLACH, D. F. H. (1971): Electrophoretic analysis of the major polypeptides of the human erythrocyte membrane. *Biochemistry 10*, 2606-2617.

105. FIELD, S. P., HEMPELMANN, E., MENDELOW, B. V. & FLEMING, A. F. (1994): Glycophorin variants and Plasmodium falciparum: protective effect of the Dantu phenotype in vitro. *Hum. Genet. 93*, 148-150.

106. FIGUR, A. M. & ROSENFIELD, R. E. (1965): The crossreaction of anti-N with type M erythrocytes. *Vox Sang. 10*, 169-176.

107. FLETCHER, A. & HARBOUR, C. (1984): An interesting monoclonal anti-N produced following immunization with human group O, NN erythrocytes. *J. Immunogenet. 11*, 121-126.

108. FLETCHER, G. (1959): The anti-N phytagglutinin of Bauhinia variegata. *Aust. J. Sci. 22*, 67-70.

109. FRASER, R. H., INGLIS, G., MACKIE, A., MUNRO, A. C., ALLAN, E. K., MITCHELL, R., SONNEBORN, H. H. & UTHEMANN, H. (1985): Mouse monoclonal antibodies reacting with M blood group-related antigens. *Transfusion 25*, 261-266.

110. FRASER, R. H., MUNRO, A. C., WILLIAMSON, A. R., BARRIE, E. K., HAMILTON, E. A. & MITCHELL, R. (1982): Mouse monoclonal anti-N. I. Production and serological characterization. *J. Immunogenet. 9*, 295-301.

111. FRASER, R. H., MUNRO, A. C., WILLIAMSON, A. R., BARRIE, E. K., HAMILTON, E. A. & MITCHELL, R. (1982): Mouse monoclonal anti-N. II. Physicochemical characterization and assessment for routine blood grouping. *J. Immunogenet. 9*, 303-309.

112. FUKUDA, M., LAUFFENBURGER, M., SASAKI, H., ROGERS, M. E. & DELL, A. (1987): Structures of novel sialylated O-linked oligosaccharides isolated from human erythrocyte glycophorins. *J. Biol. Chem. 262*, 11952-11957.

113. FUKUDA, M. & OSAWA, T. (1973): Isolation and characterization of a glycoprotein from human group O erythrocyte membrane. *J. Biol. Chem. 248*, 5100-5105.

114. FURTHMAYR, H. (1978): Glycophorins A, B, and C: a familiy of sialoglycoproteins. Isolation and preliminary characterization of trypsin derived peptides. *J. Supramol. Strct. 9*, 79-95.

115. FURTHMAYR, H. (1978): Structural comparison of glycophorins and immunochemical analysis of genetic variants. *Nature 271*, 519-524.

116. FURTHMAYR, H., METAXAS, M. N. & METAXAS-BÜHLER, M. (1981): M^g and M^c: mutations within the amino-terminal region of glycophorin A. *Proc. Natl. Acad. Sci. USA 78*, 631-635.

117. FURUHJELM, U., MYLLYLÄ, G., NEVANLINNA, H. R., NORDLING, S., PIRKOLA, A., GAVIN, J., GOOCH, A.,

SANGER, R. & TIPPETT, P. (1969): The red cell phenotype En(a-) and anti-En^a: serological and physicochemical aspects. *Vox Sang. 17*, 256-278.

118. FURUHJELM, U., NEVANLINNA, H. R. & PIRKOLA, A. (1973): A second Finnish En(a-) propositus with anti-En^a. *Vox Sang. 24*, 545-549.

119. GAHMBERG, C. G., EKBLOM, M. & ANDERSSON, L. C. (1984): Differentiation of human erythroid cells is associated with increased O-glycosylation of the major sialoglycoprotein, glycophorin A. *Proc. Natl. Acad. Sci. USA 81*, 6752-6756.

120. GAHMBERG, C. G., JOKINEN, M. & ANDERSSON, L. C. (1979): Expression of the major red cell sialoglycoprotein, glycophorin A, in the human leukemic cell line K562. *J. Biol. Chem. 254*, 7442-7448.

121. GAHMBERG, C. G., MYLLYLÄ, G., LEIKOLA, J., PIRKOLA, A. & NORDLING, S. (1976): Absence of the major sialoglycoprotein in the membrane of human En(a-) erythrocytes and increased glycosylation of band 3. *J. Biol. Chem. 251*, 6108-6116.

122. GARDNER, B., PARSONS, S. F., MERRY, A. H. & ANSTEE, D. J. (1989): Epitopes on sialyglycoprotein α: evidence for heterogeneity in the molecule. *Immunology 68*, 283-289.

123. GAUSTER, D. & TUPPY, H. (1983): Cholesteryl sulfate in meconium: inhibitory action on rabbit anti-M and anti-N antisera. *Monatsh. Chem. 114*, 453-464.

124. GERSHOWITZ, H. & FRIED, K. (1966): Anti-M^v, a new antibody of the MNSs blood group system. 1. M^v, a new inherited variant of the M gene. *Amer. J. Hum. Genet. 18*, 264-281.

125. GILES, C. M. (1982): Serological activity of low frequency antigens of the MNSs system and reappraisal of the Miltenberger complex. *Vox Sang. 42*, 256-261.

126. GILES, C. M. & HOWELL, P. (1974): An antibody in the serum of an MN patient which reacts with the M$_1$ antigen. *Vox Sang. 27*, 43-51.

127. GREEN, C., DANIELS, G., SKOV, F. & TIPPETT, P. (1994): M^g + MNS blood group phenotype: further observations. *Vox Sang. 66*, 237-241.

128. GREENWALT, T. J., SASAKI, T., SANGER, R., SNEATH, J. & RACE, R. R. (1954): An allele of the S(s) blood group genes. *Proc. Natl. Acad. Sci. USA 40*, 1126-1129.

129. HABIBI, B., FOUILLADE, M. T., DUEDARI, N., ISSITT, P. D., TIPPETT, P. & SALMON, C. (1978): The antigen Duclos. A new high frequency red cell antigen related to Rh and U. *Vox Sang. 34*, 302-309.

130. HADLEY, T. J., KLOTZ, F. W., PASVOL, G., HAYNES, J. D., McGINNISS, M. H., OKUBO, Y. & MILLER, L. H. (1987): Falciparum malaria parasites invade erythrocytes that lack glycophorin A and B (M^k M^k). Strain differences indicate receptor heterogeneity and two pathways for invasion. *J. Clin. Invest. 80*, 1190-1193.

131. HAMAGUCHI, H. & CLEVE, H. (1972): Solubilization of human erythrocyte membrane glycoproteins and separation of the MN glycoprotein from a glycoprotein with I, S, and A activity. *Biochim. Biophys. Acta 278*, 271-280.

132. HARVEY, J., PARSONS, S. F., ANSTEE, D. J. & BRADLEY, B. A. (1988): Evidence for the occurrence of human erythrocyte membrane sialoglycoproteins in human kidney endothelial cells. *Vox Sang. 55*, 104-108.

133. HAWKINS, P., ANDERSON, S. E., McKENZIE, J. L., McLOUGHLIN, K., BEARD, M. E. J. & HART, D. N. J. (1985): Localization of MN blood group antigens in kidney. *Transplant. Proc. 17*, 1697-1700.

134. HEMMING, N. J. & REID, M. E. (1994): Evaluation of monoclonal anti-glycophorin B as an unusual anti-S. *Transfusion 34*, 333-336.

135. HIRSCH, W., MOORES, P., SANGER, R. & RACE, R. R. (1957): Notes on some reactions of human anti-M and anti-N sera. *Brit. J. Haematol. 3*, 134-142.

136. HONMA, K., TOMITA, M. & HAMADA, A. (1980): Amino acid sequence and attachment sites of oligosaccharide units of porcine erythrocyte glycophorin. *J. Biochem. 88*, 1679-1791.

137. HUANG, C. H. & BLUMENFELD, O. O. (1988): Characterization of a genomic hybrid specifying the human erythrocyte antigen Dantu: Dantu gene is duplicated and linked to a γ glycophorin gene deletion. *Proc. Natl. Acad. Sci. USA 85*, 9640-9544.

138. HUANG, C. H. & BLUMENFELD, O. O. (1991): Identification of recombination events resulting in three

hybrid genes encoding human MiV, MiV(J.L.), and St^a glycophorins. *Blood 77*, 1813-1820.

139. HUANG, C. H. & BLUMENFELD, O. O. (1991): Molecular genetics of human erythrocyte MiIII and MiVI glycophorins. Use of a pseudoexon in construction of two γ-α-γ hybrid genes resulting in antigenic diversification. *J. Biol. Chem. 266*, 7248-7255.

140. HUANG, C. H. & BLUMENFELD, O. O. (1991): Multiple origins of the human glycophorin St^a gene. Identification of hot spots for independent unequal homologous recombinations. *J. Biol. Chem. 266*, 23306-23314.

141. HUANG, C. H., BLUMENFELD, O. O., REID, M. E., CHEN, Y., DANIELS, G. L. & SMART, E. (1997): Alternative splicing of a novel glycophorin allele GPHe(GL) generates two protein isoforms in the human erythrocyte membrane. *Blood 90*, 391-397.

142. HUANG, C. H., GUIZZO, M. L., KIKUCHI, M. & BLUMENFELD, O. O. (1989): Molecular genetic analysis of a hybrid gene encoding St^a glycophorin of the human erythrocyte membrane. *Blood 74*, 836-843.

143. HUANG, C. H., GUIZZO, M. L., McCREARY, J., LEIGH, E. M. & BLUMENFELD, O. O. (1991): Typing of MNSs blood group specific sequences in the human genome and characterization of a restriction fragment tightly linked to S-s- alleles. *Blood 77*, 381-386.

144. HUANG, C. H., JOHE, K., MOULDS, J. J., SIEBERT, P. D., FUKUDA, M. & BLUMENFELD, O. O. (1987): γ glycophorin (glycophorin B) gene deletion in two individuals homozygous for the S-s-U- blood group phenotype. *Blood 70*, 1830-1835.

145. HUANG, C. H., KIKUCHI, M., McCREARY, J. & BLUMENFELD, O. O. (1992): Gene conversion confined to a direct repeat of the acceptor splice site generates allelic diversity at human glycophorin (GYP) locus. *J. Biol. Chem. 267*, 3336-3342.

146. HUANG, C. H., LOMAS, C., DANIELS, G. & BLUMENFELD, O. O. (1994): Glycophorin He(St^a) of the human red blood cell membrane is encoded by a complex hybrid gene resulting from two recombinational events. *Blood 83*, 3369-3376.

147. HUANG, C. H., LU, W. M., BOOTS, M., GUIZZO, M. L. & BLUMENFELD, O. O. (1989): Two types of glycophorin gene alterations in S-s-U- individuals. *Transfusion 29*, 35S.

148. HUANG, C. H., REID, M., DANIELS, G. & BLUMENFELD, O. O. (1993): Alteration of splice site selection by an exon mutation in the human glycophorin A gene. *J. Biol. Chem. 268*, 25902-25908.

149. HUANG, C. H., REID, M., DANIELS, G. & BLUMENFELD, O. O. (1994): Gene conversion between glycophorins A and E results in St^a glycophorin in a family exhibiting the ERIK/St^a blood group phenotype. (Abstract). *Blood 84 (Suppl. 1)*, 238a.

150. HUANG, C. H., REID, M. E. & BLUMENFELD, O. O. (1993): Exon skipping caused by DNA recombination that introduces a defective donor splice site into the human glycophorin A gene. *J. Biol. Chem. 268*, 4945-4952.

151. HUANG, C. H., REID, M. E. & BLUMENFELD, O. O. (1994): Remodeling of the transmembrane segment in human glycophorin by aberrant RNA splicing. *J. Biol. Chem. 269*, 10804-10812.

152. HUANG, C. H., REID, M. E., OKUBO, Y., DANIELS, G. L. & BLUMENFELD, O. O. (1995): Glycophorin SAT of the human erythrocyte membrane is specified by a hybrid gene reciprocal to glycophorin Dantu gene. *Blood 85*, 2222-2227.

153. HUANG, C. H., SKOV, F., DANIELS, G., TIPPETT, P. & BLUMENFELD, O. O. (1992): Molecular analysis of human glycophorin MiIX gene shows a silent segment transfer and untemplated mutation resulting from gene conversion via sequence repeats. *Blood 80*, 2379-2387.

154. HUANG, C. H., SPRUELL, P., MOULDS, J. J. & BLUMENFELD, O. O. (1992): Molecular basis for the human erythrocyte glycophorin specifying the Miltenberger class-I (Mil) phenotype. *Blood 80*, 257-263.

155. HURD, J. K., JAKOX, R. F., SWISHER, S. N., CLEGHORN, T. E., CARLIN, S. & ALLEN, F. H. (1964): S$_2$, a new phenotype in the MN blood group system. *Vox Sang 9*, 487-491.

156. IKIN, E. W. (1966): The production of anti-M^g in rabbits. *Vox Sang. 11*, 217-218.

157. IKIN, E. W. & MOURANT, A. E. (1951): A rare blood group antigen occurring in negroes. *Brit. Med. J. 1*, 456-457.

158. IRIMURA, T. & OSAWA, T. (1972): Studies on a hemagglutinin from Bauhinia purpurea alba seeds.

Arch. Biochem. Biophys. 151, 475-482.

159. IRIMURA, T., TSUJI, T., TAGAMI, S., YAMAMOTO, K. & OSAWA, T. (1981): Structure of a complex-type sugar chain of human glycophorin A. *Biochemistry 20*, 560-566.

160. ISSITT, P. D. (1981): *The MN Blood Group System*. Montgomery Scientific Publication, Cincinnati, Ohio.

161. ISSITT, P. D. (1985): The MN blood group system. In: *Applied Blood Group Serology*. Montgomery Scientific Publications, Miami, FL, pp. 316-374.

162. ISSITT, P. D. (1990): Heterogeneity of anti-U. *Vox Sang. 58*, 70-71.

163. ISSITT, P. D., DANIELS, G. & TIPPETT, P. (1981): Proposed new terminology for En^a. *Transfusion 21*, 473-474.

164. ISSITT, P. D., HABER, J. M. & ALLEN, F. H. (1965): Anti-Tm, an antibody defining a new antigenic determinant within the MN blood-group system. *Vox Sang. 10*, 742-773.

165. ISSITT, P. D., HABER, J. M. & ALLEN, F. H. (1968): Sj, a new antigen in the MN system, and further studies on Tm. *Vox Sang. 15*, 1-14.

166. ISSITT, P. D. & WILKINSON, S. L. (1981): Anti-Tm is anti-N polypeptide. *Transfusion 21*, 493-497.

167. ISSITT, P. D., WREN, M. R., MOORE, R. E. & ROY, R. B. (1986): The M$_1$ and Tm antigens require M and N gene-specified amino acids for expression. *Transfusion 26*, 413-418.

168. JACK, J. A., TIPPETT, P., NOADES, J., SANGER, R. & RACE, R. R. (1960): M$_1$, a subdivision of the human blood group antigen M. *Nature 186*, 642.

169. JAROLIM, P., MOULDS, J. M., MOULDS, J. J., RUBIN, H. L. & DAHR, W. (1996): MARS and AVIS blood group antigens: polymorphism of glycophorin A affects band 3 glycophorin A interaction. (Abstract). *Blood 88 (Suppl. 1)*, 182a.

170. JAROLIM, P., MOULDS, J. M., MOULDS, J. J., RUBIN, H. L. & DAHR, W. (1997): Molecular basis of the MARS and AVIS blood group antigens. *Transfusion 37*, 90S.

171. JASKIEWICZ, E., CZERWINSKI, M., SYPER, D. & LISOWSKA, E. (1994): Anti-M monoclonal antibodies cross-reacting with variant M^g antigen: an example of modulation of antigenic properties of peptide by its glycosylation. *Blood 84*, 2340-2345.

172. JASKIEWICZ, E., LISOWSKA, E. & LUNDBLAD, A. (1990): The role of carbohydrate in the blood group N-related epitopes recognized by three new monoclonal antibodies. *Glycoconj. J. 7*, 255-268.

173. JASKIEWICZ, E., MOULDS, J. J., KRAEMER, K., GOLDSTEIN, A. S. & LISOWSKA, E. (1990): Characterization of the epitope recognized by a monoclonal antibody highly specific for blood group M antigen. *Transfusion 30*, 230-235.

174. JOHE, K. K., SMITH, A. J. & BLUMENFELD, O. O. (1991): Amino acid sequence of MiIII glycophorin. Demonstration of γ-α and α-γ junction regions and expression of γ-pseudoexon by direct protein sequencing.Appendix. *J. Biol. Chem. 266*, 7256.

175. JOHE, K. K., SMITH, A. J., VENGELEN-TYLER, V. & BLUMENFELD, O. O. (1989): Amino acid sequence of an α-γ-glycophorin hybrid. A structure reciprocal to St^a γ-α-glycophorin hybrid. *J. Biol. Chem. 264*, 17486-17493.

176. JOHE, K. K., VENGELEN-TYLER, V., LEGER, R. & BLUMENFELD, O. O. (1991): Synthetic peptides homologous to human glycophorins of the Miltenberger complex of variants of MNSs blood group system specify the epitopes for Hil, S^{JL}, Hop, and Mur antisera. *Blood 78*, 2456-2461.

177. JOKINEN, M., EHNHOLM, C., VAISANEN-RHEN, V., KORHONEN, T., PIPKORN, R., KALKKINEN, N. & GAHMBERG, C. G. (1985): Identification of the major human sialoglycoprotein from red cells; glycophorin A^M, as the receptor for Escherichia coli IH 11165 and characterization of the recetor site. *Eur. J. Biochem. 157*, 47-52.

178. JUDD, W. J., ISSITT, P. D. & PAVONE, B. G. (1979): The Can serum: demonstrating further polymorphism of M and N blood group antigens. *Transfusion 19*, 7-11.

179. JUDD, W. J., ISSITT, P. D., PAVONE, B. G., ANDERSON, J. & AMINOFF, D. (1979): Antibodies that define NANA-independent MN-system antigens. *Transfusion 19*, 12-18.

180. JUDSON, P. A. & ANSTEE, D. J. (1977): Comparative effect of trypsin and chymotrypsin on blood group antigens. *Med. Lab. Sci. 34*, 1-6.

181. KATHAN, R. A., WINZLER, R. J. & JOHNSON, C. A. (1961): Preparation of an inhibitor of viral

haemagglutinin from human erythrocytes. *J. Exp. Med. 113*, 37-45.

182. KING, M. J., POOLE, J. & ANSTEE, D. J. (1989): An application of immunoblotting in the classification of the Miltenberger series of blood group antigens. *Transfusion 29*, 106-112.

183. KONUGRES, A. A., HUBERLIE, M. M., SWANSON, J. & MATSON, G. A. (1963): The production of anti-Mt^a in rabbits. *Vox Sang. 8*, 632-633.

184. KONUGRES, A. A. & WINTER, N. M. (1967): Sul, a new blood group antigen in the MN system. *Vox Sang. 12*, 221-224.

185. KORNFELD, R. & KORNFELD, S. (1970): The structure of a phytohaemagglutinin receptor site from human erythrocytes. *J. Biol. Chem. 245*, 2536-2542.

186. KUDO, S., CHAGNOVICH, D., REARDEN, A., MATTEI, M. G. & FUKUDA, M. (1990): Molecular analysis of a hybrid gene encoding human glycophorin variant Miltenberger V-like molecule. *J. Biol. Chem. 265*, 13825-13829.

187. KUDO, S. & FUKUDA, M. (1989): Structural organization of glycophorin A and B genes: glycophorin B gene evolved by homologous recombination at Alu repeat sequences. *Proc. Natl. Acad. Sci. USA 86*, 4619-4623.

188. KUDO, S. & FUKUDA, M. (1990): Identification of a novel human glycophorin, glycophorin E, by isolation of genomic clones and complementary DNA clones utilizing polymerase chain reaction. *J. Biol. Chem. 265*, 1102-1110.

189. KUDO, S. & FUKUDA, M. (1994): Contribution of gene conversion to the retention of the sequence for M blood group type determinant in glycophorin E gene. *J. Biol. Chem. 269*, 22969-22974.

190. LAIRD-FRYER, B., MOULDS, J. J., DAHR, W., MIN, Y. O. & CHANDANAYINGYONG, D. (1986): Anti-En^{a}FS detected in the serum of an Mi^{VII} homozygote. *Transfusion 26*, 51-56.

191. LANDSTEINER, K. & LEVINE, P. (1927): Further observations on individual differences of human blood. *Proc. Soc. Exp. Biol. Med. 24*, 941-942.

192. LANDSTEINER, K. & LEVINE, P. (1927): A new agglutinable factor differentiating individual human bloods. *Proc. Soc. Exp. Biol. Med. 24*, 600-602.

193. LANDSTEINER, K., STRUTTON, W. R. & CHASE, M. W. (1934): An agglutination reaction observed with some human bloods, chiefly among negroes. *J. Immunol. 27*, 469-472.

194. LANGLEY, J. W., ISSITT, P. D., ANSTEE, D. J., McMAHAN, M., SMITH, N., PAVONE, B. G., TESSEL, J. A. & CARLIN, M. A. (1981): Another individual (J.R.) whose red blood cells appear to carry a hybrid MNSs sialoglycoprotein. *Transfusion 21*, 15-24.

195. LEIGH, K. & DE RUITER, K. (1988): An example of anti-S enhanced at acid pH. *Transfusion 28*, 86.

196. LEVINE, P., KUHMICHEL, A. B., WIGOD, M. & KOCH, E. (1951): A new blood factor s, allelic to S. *Proc. Soc. Exp. Biol. Med. 78*, 218-220.

197. LEVINE, P., STOCK, A. H., KUHMICHEL, A. B. & BRONIKOVSKY, N. (1951): A new human blood factor of rare incidence in the general population. *Proc. Soc. Exp. Biol. Med. 77*, 402-403.

198. LIS, H., LATTER, H., ADAR, R. & SHARON, N. (1988): Isolation of two blood type A and N specific isolectins from Moluccella laevis seeds. *FEBS Lett. 233*, 191-195.

199. LISOWSKA, E. (1963): Reaction of erythrocyte mucoproteins with anti-N phytoagglutinin from Vicia graminaea seeds. *Nature 198*, 865-866.

200. LISOWSKA, E. (1981): Biochemistry of M and N blood group specificities. *Blood Transfus. Immunohaematol. 24*, 75-84.

201. LISOWSKA, E. & DUK, M. (1975): Modification of amino groups of human erythrocyte glycoproteins and the new concept on the structural basis of M and N blood group specificity. *Eur. J. Biochem. 54*, 469-474.

202. LISOWSKA, E., DUK, M. & DAHR, W. (1980): Comparison of alkali-labile oligosaccharide chains of M and N blood-group glycopeptides from human erythrocyte membrane. *Carbohydr. Res. 79*, 103-113.

203. LISOWSKA, E. & KORDOWICZ, M. (1977): Specific antibodies for desialyzed M and N blood group antigens. *Vox Sang. 33*, 164-169.

204. LISOWSKA, E., MESSETER, L., DUK, M., CZERWINSKI, M. & LUNDBLAD, A. (1987): A monoclonal anti-glycophorin A antibody recognizing the blood group M determinant: studies on the

subspecificity. *Mol. Immunol. 24*, 605-613.

205. LISOWSKA, E. & MORAWIECKI, A. (1967): The role of free amino groups in the blood group activity of M and N mucoids. *Eur. J. Biochem. 3*, 237-241.

206. LISOWSKA, E. & WASNIOWSKA, K. (1978): Immunochemical characterization of cyanogen bromide degradation products of M and N blood-group glycopeptides. *Eur. J. Biochem. 88*, 247-252.

207. LOWE, R. F. & MOORES, P. (1972): S-s-U- red cell factor in Africans of Rhodesia, Malawi, Mozambique, and Natal. *Hum. Hered. 22*, 344-350.

208. LU, Y. Q., LIU, J. F., SOCHA, W. W., NAGEL, R. L. & BLUMENFELD, O. O. (1987): Polymorphism of glycophorins in nonhuman primate erythrocytes. *Biochem. Genet. 25*, 477-491.

209. MACDONALD, K. A., NICHOLS, M. E., MARSH, W. L. & JENKINS, W. J. (1967): The first example of anti-Henshaw in human-serum. *Vox Sang. 13*, 346-348.

210. MADDEN, H. J., CLEGHORN, T. E., ALLEN, F. H., ROSENFIELD, R. E. & MACKEPRANG, M. (1964): A note on the relatively high frequency of St^a on the red blood cells of orientals, and report of a third example of anti-St^a. *Vox Sang. 9*, 502-504.

211. MAIZELS, N. (1989): Might gene conversion be the mechanism of somatic hypermutation of mammalian immunoglobulin genes? *Trends Genet. 5*, 4-8.

212. MÄKELÄ, O. & CANTELL, K. (1958): Destruction of M and N blood group receptors of human red cells by some influenza viruses. *Ann. Med. Exp. Biol. Fenn. 36*, 366-374.

213. MARCHESI, V. T., FURTHMAYR, H., TOMITA, M., SILVERBERG, M. & COTMORE, S. (1977): Molecular features of glycophorin A, the major sialoglycoprotein of the human red cell embrane. *In: Human Blood Groups - 5th International Convocation on Immunology, 1976, Buffalo* (J. F. Mohn, R. W. Plunkett, R. K. Cunningham, and R. M. Lambert, eds.). S. Karger, Basel, pp. 374-382.

214. MARSH, W. L., OYEN, R., NICHOLS, M. E. & CHARLES, H. (1974): Studies of MNSsU antigen activity on leucocytes and platelets. *Transfusion 14*, 462-466.

215. MAWBY, W. J., ANSTEE, D. J. & TANNER, M. J. A. (1981): Immunochemical evidence for hybrid sialoglycoproteins of human erythrocytes. *Nature 291*, 161-162.

216. MCDOUGALL, D. C. J. & JENKINS, W. J. (1981): The first human example of anti-M^e. *Vox Sang 40*, 412-415.

217. MERRY, A. H., HODSON, C., THOMSON, E., MALLINSON, G. & ANSTEE, D. J. (1986): The use of monoclonal antibodies to quantify the levels of sialoglycoproteins α and γ and variant sialoglycoproteins in human erythrocyte membranes. *Biochem. J. 233*, 93-98.

218. METAXAS, M. N. & METAXAS-BÜHLER, M. (1964): M^k: an apparently silent allele at the MN locus. *Nature 202*, 1123.

219. Metaxas, M. N. & Metaxas-Bühler, M. (1977): Rare genes of the MNSs system affecting the red cell membrane. In: *Human Blood Groups - 5th International Convocation on Immunology, 1976, Buffalo* (J. F. Mohn, R. W. Plunkett, R. K. Cunningham, and R. M. Lambert, eds.). S. Karger, Basel, pp. 344-352.

220. METAXAS, M. N., METAXAS-BÜHLER, M. & IKIN, E. W. (1968): Complexities of the MN locus. *Vox Sang. 15*, 102-117.

221. METAXAS-BÜHLER, M., CLEGHORN, T. E., ROMANSKI, J. & METAXAS, M. N. (1966): Studies on the blood group antigen M^g. II. Serology of M^g. *Vox Sang. 11*, 170-183.

222. MILLER, L. H., HAYNES, J. D., MCAULIFFE, F. M., SHIROISHI, T., DUROCHER, J. R. & MCGINNISS, M. H. (1977): Evidence for differences in erythrocyte surface receptors for the malarial parasite Plasmodium falciparum and P. knowlesi. *J. Exp. Med. 146*, 277-281.

223. MOOR-JANKOWSKI, J., WIENER, A. S. & ROGERS, C. M. (1964): Human blood group factors in non-human primates. *Nature 202*, 663-665.

224. MOULDS, J. J. & DAHR, W. (1989): MNSs and Gerbich blood group systems. In: *Human Immunogenetics. Basic Principles and Clinical Relevance* (S.D. Litwin ed.). Marcel Dekker Inc., New York, Basel, pp. 713-741.

225. MUELLER, T. J., DOW, A. W. & MORRISON, M. (1976): Heterogeneity of the sialoglycoproteins of the normal human erythrocyte membrane. *Biochem. Biophys. Res. Commun. 72*, 94-99.

226. MURAYAMA, J., TOMITA, M. & HAMADA, A. (1982): Glycophorins of bovine erythrocyte membranes.

Isolation and preliminary characterization of the major component. *J. Biochem. 91*, 1829-1836.

227. MURAYAMA, J. I., TOMITA, M. & HAMADA, A. (1982): Primary structure of horse erythrocyte glycophorin HA. Its amino acid sequence has a unique homology with those of human and porcine erythrocyte glycophorins. *J. Membr. Biol. 64*, 205-215.

228. MURAYAMA, J. I., UTSUMI, H. & HAMADA, A. (1989): Amino acid sequence of monkey erythrocyte glycophorin MK. Its amino acid sequence has a striking homology with that of human glycophorin A. *Biochim. Biophys. Acta 999*, 273-280.

229. MURAYAMA, J. I., YAMASHITA, T., TOMITA, M. & HAMADA, A. (1983): Amino acid sequence and oligosaccharide attachment sites of the glycosylated domain of dog erythrocyte glycophorin. *Biochim. Biophys. Acta 742*, 477-483.

230. NICHOLS, M. E., ROSENFIELD, R. E. & RUBINSTEIN, P. (1985): Two blood group M epitopes disclosed by monoclonal antibodies. *Vox Sang. 49*, 138-148.

231. NORDLING, S., SANGER, R., GAVIN, J., FURUHJELM, U., MYLLYLÄ, G. & METAXAS, M. N. (1969): M^k and M^g: some serological and physicochemical observations. *Vox Sang. 17*, 300-302.

232. OCHIAI, Y., FURTHMAYR, H. & MARCUS, D. M. (1983): Diverse specificites of five monoclonal antibodies reactive with glycophorin A of human erythrocytes. *J. Immunol. 131*, 864-868.

233. OKUBO, Y., DANIELS, G. L., PARSONS, S. F., ANSTEE, D. J., YAMAGUCHI, H., TOMITA, T. & SENO, T. (1988): A Japanese family with two sisters apparently homozygous for M^k. *Vox Sang. 54*, 107-111.

234. ONDA, M., KUDO, S. & FUKUDA, M. (1994): Genomic organization of glycophorin A gene family revealed by yeast artificial chromosomes containing human genomic DNA. *J. Biol. Chem. 269*, 13013-13020.

235. ONDA, M., KUDO, S., REARDEN, A., MATTEI, M. G. & FUKUDA, M. (1993): Identification of a precursor genomic segment that provided a sequence unique to glycophorin B and glycophorin E genes. *Proc. Natl. Acad. Sci. USA 90*, 7220-7224.

236. ÖRJASÆTER, H., KORNSTAD, L. & HEIER, A. M. (1964): Studies on the Ny^a blood group antigen and antibodies. *Vox Sang. 9*, 673-683.

237. ÖRJASÆTER, H., KORNSTAD, L., HEIER, A. M., VOGT, E., HAGEN, P. & HARTMANN, O. (1964): A human blood group antigen, Ny^a (Nyberg), segregating with the Ns gene complex of the MNSs system. *Nature 201*, 832.

238. OTTENSOOSER, P. & SILBERSCHMIDT, K. (1953): Haemagglutinin anti-N in plant seeds. *Nature 172*, 914.

239. PASVOL, G., WAINSCOAT, J. S. & WEATHERALL, D. J. (1982): Erythrocytes deficient in glycophorin resist invasion by the malarial parasite Plasmodium falciparum. *Nature 297*, 64-66.

240. PAVONE, B. G., BILLMAN, R., BRYANT, J., SNIECINSKI, I. & ISSITT, P. (1981): An auto-anti-En^a, inhibitable by MN sialoglycoprotein. *Transfusion 21*, 25-31.

241. PILKINGTON, P., DAHR, W., LACEY, P., MOULDS, J. J. & WHEELER, N. (1985): A Dantu positive Caucasian demonstrating a variant hybrid sialoglycoprotein. *Transfusion 25*, 464.

242. PISANO, A., REDMOND, J. W., WILLIAMS, K. L. & GOOLEY, A. A. (1993): Glycosylation sites identified by solid-phase Edman degradation: O-linked glycosylation motifs on human glycophorin-A. *Glycobiology 3*, 429-435.

243. POOLE, J., BANKS, J., BRUCE, L. J., RING, S. M., LEVENE, C., STERN, H., OVERBEEKE, M. A. M. & TANNER, M. J. A. (1999): Glycophorin A mutation Ala65 → Pro gives rise to a novel pair of MNS alleles ENEP (MNS39) and HAG (MNS41) and altered Wr^b expression: direct evidence for GPA/band 3 interaction necessary for normal Wr^b expression. *Transfus. Med. 9*, 167-174.

244. POOLE, J., BANKS, J. A., BRUCE, L. J., RING, S. M. & TANNER (1995): A novel glycophorin A polymorphism affecting Wr^b expression . (Abstract). *Transfusion 35 (Suppl.)*, 40S.

245. POOLE, J., KING, M. J., MAK, K. H., LIEW, Y. W., LEONG, S. & CHUA, K. M. (1991): The MiIII phenotype among Chinese donors in Hong Kong: immunochemical and serological studies. *Transfus. Med. 1*, 169-175.

246. PRIGENT, M. J., BLANCHARD, D. & CARTRON, J. P. (1983): Membrane receptors for Vicia graminaea anti-N lectin and its binding to native and neuraminidase-treated human erythrocytes. *Arch.*

Biochem. Biophys. 222, 231-244.

247. PRIGENT, M. J. & BOURILLON, R. (1976): Purification and characterization of a lectin (plant hemagglutinin) with N blood group specificity from Vicia graminaea seeds. *Biochim. Biophys. Acta 420*, 112-121.

248. PUNO, C. S. & ALLEN, F. H. (1969): Anti-s produced in rabbits. *Vox Sang. 16*, 155-156.

249. RACE, R. R. & SANGER, R. (1975): The MNSs Blood Groups. In: *Blood Groups in Man.* Blackwell Scientific Publiations, Oxford, pp. 92-138.

250. RAHUEL, C., ELOUET, J. F. & CARTRON, J. P. (1994): Post-transcriptional regulation of the cell surface expression of glycophorins A, B, and E. *J. Biol. Chem. 269*, 32752-32758.

251. RAHUEL, C., LONDON, J., D'AURIOL, L., MATTEI, M. G., TOURNAMILLE, C., SKRZYNIA, C., LEBOUC, Y., GALIBERT, F. & CARTRON, J. P. (1988): Characterization of cDNA clones for human glycophorin A. Use for gene localization and for analysis of normal of glycophorin A-deficient (Finnish type) genomic DNA. *Eur. J. Biochem. 172*, 147-153.

252. RAHUEL, C., LONDON, J., VIGNAL, A., BALLAS, S. K. & CARTRON, J. P. (1991): Erythrocyte glycophorin B deficiency may occur by two distinct gene. *Amer. J. Hematol. 37*, 57-58.

253. RAHUEL, C., LONDON, J., VIGNAL, A., CHÉRIF-ZAHAR, B., COLIN, Y., SIEBERT, P., FUKUDA, M. & CARTRON, J. P. (1988): Alteration of the genes for glycophorin A and B in glycophorin-A-deficient individuals. *Eur. J. Biochem. 177*, 605-614.

254. REARDEN, A., MAGNET, A., KUDO, S. & FUKUDA, M. (1993): Glycophorin-B and glycophorin-E genes arose from the glycophorin-A ancestral gene via 2 duplications during primate evolution. *J. Biol. Chem. 268*, 2260-2267.

255. REARDEN, A., PHAN, H., DUBNICOFF, T., KUDO, S. & FUKUDA, M. (1990): Identification of the crossing-over point of a hybrid gene encoding human glycophorin variant St^a. Similarity to the crossing-over point in haptoglobin-related genes. *J. Biol. Chem. 265*, 9259-9263.

256. REARDEN, A., PHAN, H., KUDO, S. & FUKUDA, M. (1990): Evolution of the glycophorin gene family in the hominoid primates. *Biochem. Genet. 28*, 209-222.

257. REID, M. E., ELLISOR, S. S. & BARKER, J. M. (1984): A human alloanti-N enhanced by acid media. *Transfusion 24*, 222-223.

258. REID, M. E., LOMAS-FRANCIS, C., DANIELS, G. L., CHEN, V., SHEN, J., HO, Y. C., HARE, V., BATTS, R., YACOB, M., SMART, E. & GREEN, C. A. (1995): Expression of the erythrocyte antigen Henshaw (He; MNS6): serological and immunochemical studies. *Vox Sang. 68*, 183-186.

259. REID, M. E., MOORE, B. P. L., POOLE, J., PARKER, N. J., ASENBRYL, E., VENGELEN-TYLER, V., LUBENKO, A. & GALLIGAN, B. (1992): TSEN: a novel MNS-related blood group antigen. *Vox Sang. 63*, 122-128.

260. REID, M. E., POOLE, J., GREEN, C., NEILL, G. & BANKS, J. (1992): MINY: a novel MNS-related blood group antigen. *Vox Sang. 63*, 129-132.

261. REID, M. E., STORRY, J. R., RALPH, H., BLUMENFELD, O. O. & HUANG, C. H. (1996): Expression and quantitative variation of the low-incidence blood group antigen He on some S-s- red cells. *Transfusion 36*, 719-724.

262. REID, M. E. & TIPPETT, P. (1993): Review of a terminology proposed to supersede Miltenberger. *Immunohematology 9*, 91-95.

263. RUBOCKI, R. & MILGROM, F. (1986): Reactions of murine monoclonal antibodies to blood group MN antigens. *Vox Sang. 51*, 217-225.

264. RYGIEL, S. A., ISSITT, C. H. & FRUITSTONE, M. J. (1985): Destruction of the S antigen by sodium hypochlorite. *Transfusion 25*, 274-277.

265. SANGER, R. & RACE, R. R. (1947): Subdivisions of the MN blood groups in man. *Nature 160*, 505.

266. SANGER, R., RACE, R. R., GREENWALT, T. J. & SASAKI, T. (1955): The S, s, and S^u blood group genes in Americal Negroes. *Vox Sang. 5*, 73-81.

267. SANGER, R., RACE, R. R., WALSH, R. J. & MONTGOMERY, C. (1948): An antibody which subdivides the human MN blood groups. *Heredity 2*, 131-139.

268. SCHMIDT, P. J., LOSTUMBO, M. M., ENGLISH, C. T. & HUNTER, O. B. (1967): Aberrant U blood group accompanying Rh_{null}. *Transfusion 7*, 33-34.

269. SCHULMAN, S., ROTH, E. F., CHENG, B., RYBICKI, A. C., SUSSMAN, I.I., WONG, M., WANG, W., RANNEY, H. M., NAGEL, R. L. & SCHWARTZ, R. S. (1990): Growth of Plasmodium falciparum in human erythrocytes containing abnormal membrane proteins. *Proc. Natl. Acad. Sci. USA 87*, 7339-7343.

270. SENO, T., YAMAGUCHI, H., OKUBO, Y., SUMI, R., GREEN, C. A. & TIPPETT, P. (1983): Os^a, a 'new' low-frequency red cell antigen. *Vox Sang. 45*, 60-61.

271. SHAPIRO, M. & LE ROUX, M. E. (1981): Serology and genetics of a "new" red cell antigen: s^D (the Dreyer antigen). (Abstract). *Transfusion 21*, 614.

272. SIEBERT, P. D. & FUKUDA, M. (1986): Isolation and characterization of human glycophorin A cDNA clones by a synthetic oligonucleotide approach: nucleotide sequence and mRNA structure. *Proc. Natl. Acad. Sci. USA 83*, 1665-1669.

273. SIEBERT, P. D. & FUKUDA, M. (1987): Molecular cloning of a human glycophorin B cDNA: Nucleotide sequence and genomic relationship to glycophorin A. *Proc. Natl. Acad. Sci. USA 84*, 6735-6739.

274. SIM, B. K. L., CHITNIS, C. E., WASNIOWSKA, K., HADLEY, T. J. & MILLER, L. H. (1994): Receptor and ligand domains for invasion of erythrocytes by Plasmodium falciparum. *Science 264*, 1941-1944.

275. SKOV, F., GREEN, C., DANIELS, G., KHALID, G. & TIPPETT, P. (1991): Miltenberger Class IX of the MNS blood group system. *Vox Sang. 61*, 130-136.

276. SOUBES, S. C., REID, M. E., KANEKO, O. & MILLER, L. H. (1999): Search for the sialic acid-independent receptor on red blood cells for invasion by Plasmodium falciparum. *Vox Sang. 76*, 107-114.

277. SPEISER, P. (1959): Über die bisher jüngste menschliche Frucht (27mm/22g) an der bereits die Erbmerkmale A, M, N, s, Fy(a+), C, c, D, E, e, Jk(a+?) im Blut festgestellt werden konnten. *Wien. Klin. Wochenschr. 71*, 549-551.

278. SPRINGER, G. F. & ANSELL, N. J. (1958): Inactivation of human erythrocyte agglutinogens M and N by influenza viruses and receptor-destroying enzyme. *Proc. Natl. Acad. Sci. USA 44*, 182-188.

279. SPRINGER, G. F. & DESAI, P. R. (1974): Common precursors of human blood group MN specificities. *Biochem. Biophys. Res. Commun. 61*, 470-475.

280. SPRINGER, G. F. & DESAI, P. R. (1976): Enzymatic synthesis of human blood group M-, N-, and T-specific structures. *Naturwissenschaften 63*, 488-489.

281. SPRINGER, G. F. & STALDER, K. (1961): Action of influenza viruses, receptor destroying enzyme, and proteases on blood group agglutinogen M^g. *Nature 191*, 187.

282. SPRINGER, G. F., TEGTMEYER, H. & HUPRIKAR, S. V. (1972): Anti-N reagents in elucidation of the genetical basis of human blood group MN specificities. *Vox Sang. 22*, 325-343.

283. SPRUELL, P., MOULDS, J. J., MARTIN, M., GILCHER, R. O., HOWARD, P. B. & BLUMENFELD, O. O. (1993): An anti-En^{a}TS detected in the serum of an Mi^i homozygote. *Transfusion 33*, 848-851.

284. STECK, T. L. (1974): The organization of proteins in the human red blood cell membrane. A review. *J. Cell Biol. 62*, 1-19.

285. STECK, T. L., FAIRBANKS, G. & WALLACH, D. F. H. (1971): Disposition of the major proteins in the isolated erythrocyte membrane. Proteolytic dissection. *Biochemistry 10*, 2617-2624.

285a. STORRY, J. R., NORTMAN, P. & REID, M. E. (2000): Molecular basis of three low incidence antigens on glycophorin B (Abstract). *Vox Sanguinis* (in press).

285b. STORRY, J. R. & REID, M. E. (2000): A point mutation in GypA exon 3 encodes the low incidence antigen MNS16 (Abstract). *Vox Sanguinis* (in press).

285c. STORRY, J. R., COGHLAN, G., POOLE, J., FIGUEROA, D. & REID, M. E. (2000): The MNS Blood Group Antigens, Vr (MNS12) and Mt^a (MNS14), Each Arise from an Amino Acid Substitution on Glycophorin A. *Vox Sang. 78*, 52-56.

286. SWANSON, J. & MATSON, G. A. (1962): Mt^a , a "new" antigen in the MNSs system. *Vox Sang. 7*, 585-590.

287. TAKASAKI, S., YAMASHITA, K. & KOBATA, A. (1978): The sugar chain structures of ABO blood group active glycoproteins obtained from human erythrocyte membrane. *J. Biol. Chem. 253*, 6086-6091.

288. TANNER, M. J. A. & ANSTEE, D. J. (1976): The membrane change in En(a-) human erythrocytes.

Absence of the major erythrocyte sialoglycoprotein. *Biochem. J. 153*, 271-277.

289. TANNER, M. J. A., ANSTEE, D. J. & MAWBY, W. J. (1980): A new human erythrocyte variant (Ph) containing an abnormal membrane sialoglycoprotein. *Biochem. J. 187*, 493-500.

290. TANNER, M. J. A., JENKINS, R. E., ANSTEE, D. J. & CLAMP, J. R. (1976): Abnormal carbohydrate composition of the major penetrating membrane protein of En(a-) human erythrocytes. *Biochem. J. 155*, 701-703.

291. TATE, C. G. & TANNER, M. J. A. (1988): Isolation of cDNA clones for human erythrocyte membrane sialoglycoproteins α and γ. *Biochem. J. 254*, 743-750.

292. TATE, C. G., TANNER, M. J. A., JUDSON, P. A. & ANSTEE, D. J. (1989): Studies on human red-cell membrane glycophorin A and glycophorin B genes in glycophorin-deficient individuals. *Biochem. J. 263*, 993-996.

293. TERAJIMA, M., MATSUI, Y., COPELAND, N. G., GILBERT, D. J., JENKINS, N. A. & OBINATA, M. (1994): Structural organization of the mouse glycophorin A gene. *J. Biochem. 116*, 1105-1110.

294. THOMAS, D. B. & WINZLER, R. J. (1969): Structural studies on human erythrocyte glycoprotein. Alkali-labile oligosaccharides. *J. Biol. Chem. 244*, 5943-5946.

295. THOMAS, D. B. & WINZLER, R. J. (1971): Structure of glycoproteins of human erythrocytes. Alkali-stable oligosaccharides. *Biochem. J. 124*, 55-59.

296. TIPPETT, P., REID, M. E., POOLE, J., GREEN, C. A., DANIELS, G. L. & ANSTEE, D. J. (1992): The Miltenberger subsystem: is it obsolescent? *Transfus. Med. Rev. 6*, 170-182.

297. TOKUNAGA, E., SASAKAWA, S., TAMAKA, K., KAWAMATA, H., GILES, C. M., IKIN, E. W., POOLE, J., ANSTEE, D. J., MAWBY, W. J. & TANNER, M. J. A. (1979): Two apparently healthy japanese individuals of type M^kM^k have erythrocytes which lack both the blood group MN and Ss-active sialoglycoproteins. *J. Immunogenet. 6*, 383-390.

298. TOMITA, M., FURTHMAYR, H. & MARCHESI, V. T. (1978): Primary structure of human erythrocyte glycophorin A. Isolation and characterization of peptides and complete amino acid sequence. *Biochemistry 17*, 4756-4770.

299. TOMITA, M. & MARCHESI, V. (1975): Amino-acid sequence and oligosaccharide attachment sites of human erythrocyte glycophorin. *Proc. Natl. Acad. Sci. USA 72*, 2964-2968.

300. UCHIKAWA, M., TSUNEYAMA, H., WANG, L., TOKUNAGO, K., JUJI, T., TADOKORO, K. & ENOMOTO, T. (1994): A novel amino acid sequence results in the expression of the MNS related private antigen, SAT. (Abstract). *Vox Sang. 34 (Suppl. 2)*, 116.

301. UNGER, P., ORLINA, A., DAHR, W., KORDOWICZ, M., MOULDS, J. & NEWMAN, R. A. (1981): Two new gene complexes in the MNSs blood group system. *Proceedings of the 9th International Congress of the Society for Forensic Haematogenetics*, Schmitt & Meyer, Würzburg, pp. 189-193.

302. UNGER, P., PROCTER, J. L., MOULDS, J. J., MOULDS, M., BLANCHARD, D., GUIZZO, M. L., McCALL, L. A., CARTRON, J. P. & DAHR, W. (1987): The Dantu erythrocyte phenotype of the NE variety. II. Serology; immunochemistry, genetics, and frequency. *Blut 55*, 33-43.

303. UNGER, P. J., ORLINA, A. R., DAHR, W., KORDOWICZ, M., MOULDS, J. J. & NEWMAN, R. (1981): Two new gene complexes in the MNSs blood group system. *Transfusion 21*, 614.

304. VAN DER HART, M., VAN DER VEER, M., VAN LOGHEM, J. J., SANGER, R. & RACE, R. R. (1958): Vr, an antigen belonging to the MNSs blood group system. *Vox Sang. 3*, 261-265.

305. VENGELEN-TYLER, V., ANSTEE, D. J., ISSITT, P. D., PAVONE, B. G., FERGUSON, S. J., MAWBY, W. J., TANNER, M. J. A., BLAJCHMAN, M. A. & LORQUE, P. (1981): Studies on the blood of an Mi^V homozygote. *Transfusion 21*, 1-14.

306. VENGELEN-TYLER, V. & MOGCK, N. (1986): A new test useful in identifying red cells with a (γ-α) hybrid sialoglycoprotein. *Transfusion 26*, 231-233.

307. VIGNAL, A., LONDON, J., RAHUEL, C. & CARTRON, J. P. (1990): Promoter sequence and chromosomal organization of the genes encoding glycophorins A, B, and E. *Gene 95*, 289-293.

308. VIGNAL, A., RAHUEL, C., EL MALIKI, B., LONDON, J., LE VAN KIM, C., BLANCHARD, D., ANDRE, C., D'AURIOL, L., GALIBERT, F., BLAJCHMAN, M. A. & CARTRON, J. P. (1989): Molecular analysis of glycophorin A and B gene structure and expression in homozygous Miltenberger class V (Mi.V) human erythrocytes. *Eur. J. Biochem. 184*, 337-344.

309. VIGNAL, A., RAHUEL, C., LONDON, J., CHÉRIF-ZAHAR, B., SCHAFF, F., HATTAB, OKUBO, Y. & CARTRON, J. P. (1990): A novel gene member of the human glycophorin A and B gene family. Molecular cloning and expression. *Eur. J. Biochem. 191*, 619-625.

310. WALLACE, J. & IZATT, M. M. (1963): The Clᵃ (Caldwell) antigen: a new and rare human blood group antigen related to the MNSs system. *Nature 200*, 689-690.

311. WALLACE, J., MILNE, G., MOHN, J., CAMBERT, R. M., ROSAMILIA, H. G., MOORES, P., SANGER, R. & RACE, R. R. (1957): Blood group antigens Miᵃ and Vw and their relation to the MNSs system. *Nature 179*, 478.

312. WALSH, R. J. & MONTGOMERY, C. M. (1947): A new human iso-agglutinin subdividing the MN group. *Nature 160*, 504-505.

313. WALSH, T. J., GILES, C. M. & POOLE, J. (1981): Anti-Mᵛ causing haemolytic disease of the newborn; serological considerations of the Mᵛ red cell determinant. *Clin. Lab. Haematol. 3*, 137-142.

314. WASNIOWSKA, K., DRZENIEK, Z. & LISOWSKA, E. (1977): The amino acids of M and N blood group glycopeptides are different. *Biochem. Biophys. Res. Commun. 76*, 385-390.

315. WEBB, A. J. & GILES, C. M. (1977): Three antibodies of the MNSs system and their association with the Miltenberger complex of antigens. II. Raddon and Lane sera. *Vox Sang. 32*, 274-276.

316. WIENER, A. S., GORDON, E. B., MOOR-JANKOWSKI, J. & SOCHA, W. W. (1972): Homolgoues of the human M-N blood types in gorillas and other nonhuman primates. *Haematologia 6*, 419-428.

317. WIENER, A. S. & ROSENFIELD, R. E. (1961): Mᵉ, a blood factor common to the antigenic properties M and He. *J. Immunol. 87*, 376-378.

318. WIENER, A. S., UNGER, L. J. & COHEN, L. (1954): Distribution and heredity of blood factor U. *Science 119*, 734-735.

319. WIENER, A. S., UNGER, L. J. & GORDON, E. D. (1953): Fatal hemolytic transfusion reation caused by sensitation to a new blood factor U. *J. Amer. Med. Assoc. 153*, 1444-1446.

320. WOJCICKI, R. E., HARDMAN, J. T., BECK, M. L., PLAPP, F. V., WRIGHT, T., BROWN, P. J. & WOOD, M. G. (1980): pH, temperature and ionic strength: significant variables in detection of auto anti-U. (Abstract). *Transfusion 20*, 828.

321. WYBENGA, L. E., EPAND, R. F., NIR, S., CHU, J. W. K., SHAROM, F. J., FLANAGAN, T. D. & EPAND, R. M. (1996): Glycophorin as a receptor for Sendai virus. *Biochemistry 35*, 9513-9518.

322. YANAGI, K., OHYAMA, K., YAMAKAWA, T., HASHIMOTO, K. & OHKUMA, S. (1990): Purification and characterization of anti-N lectin from Vicia unijuga leaves. *Int. J. Biochem. 22*, 43-52.

323. YOSHIMA, H., FURTHMAYR, H. & KOBATA, A. (1980): Structures of the asparagine-linked sugar chains of glycophorin A. *J. Biol. Chem. 255*, 9713-9718.xc

324. YURCHENKO, P. D. & FURTHMAYR, H. (1980): Expression of red cell membrane proteins in erythroid precursor cells. *J. Supramol. Struct. 13*, 255-269.

10 Polyagglutination

The phenomenon of polyagglutination is characterised by the agglutinability of erythrocytes and some other blood cells by almost all sera from normal human adults, independent of standard blood groups [9,62]. This unusual reactivity is due to the fact that antibodies normally present in human sera bind to secondarily altered erythrocyte membrane antigens.

The receptors recognised by these antibodies are essential constituents of oligosaccharide chains found in membrane glycoconjugates. They are normally hidden on human erythrocytes but become exposed in polyagglutinable cells.

This chapter will discuss the following types of polyagglutination[1]:

- forms of acquired polyagglutination (**T**, **Tk**, **Tx**, and **Th**) which develop as a result of bacterial or viral infections; the transformation is usually transient and disappears after recovery of the patient. In these cases the antigens are uncovered by various enzymes released in the blood stream by viruses or microorganisms. Because some bacteria strains produce several enzymes, the simultaneous appearance of two or three of these characteristics has been observed;
- **Tn**, another form of polyagglutination, in this case caused by somatic mutations in the haematopoietic stem cells. These mutations affect genes responsible for the biosynthesis of glycoconjugates, and the resulting disorder in glycosylation of the membrane constituents leads to a changed antigen pattern of the erythrocyte membrane,
- the characters **VA** and **NOR**, which have not yet been classified.

Polyagglutination is generally accompanied by haemolytic anaemia and other haematological disorders. These pathological conditions, however, are not caused by defects in membrane structure but are due to the accelerated degradation of blood cells induced by endogenous serum antibodies.

---

[1] '**Acquired B**' and **Cad** are often mentioned in connection with polyagglutinability; they are discussed in *Chaps. 5.5.5* and *11*, respectively. In rare cases, absorption of organic substances such as drugs or insecticides [10] or various bacterial antigens [84,116] may affect erythrocytes, causing them to become polyagglutinable.

Table 10.1: Reaction of human erythrocytes with AB serum and various lectins.

Agglutinin	Normal erythrocytes[a]		Polyagglutinable erythrocytes				
	native	RDE-tr.[b]	T	Tn	Tk	Th	Tx
AB serum	−	+	+	+	+	+	+
Arachis hypogaea	−	+	+	−	+[c]	+	+
Bandeiraea simplicifolia (BS II)	−	−	−	−	+	−	−
Dolichos biflorus [d]	−	−	+	−	−	−	−
Glycine soja	−	+	+/±	+	−	−	−
Salvia sclarea	−	−	−	+	−	−	−
Vicia cretica	−	+	+	−	−	+	−

[a] Independent of **ABO** group
[b] RDE-tr. = neuraminidase-treated
[c] Only in the presence of proteases
[d] Also reacts with A_1 and **Cad** erythrocytes

The antigen characters discussed in this chapter also occur in undifferentiated and malignantly transformed cells. Furthermore, **T** and **Tn** active structures are not confined to humans but are found wide-spread throughout the animal kingdom.

10.1 Antisera and Lectins

Antibodies reacting with polyagglutinable erythrocytes occur in virtually all sera from human adults. They are also found regularly as 'naturally occurring' antibodies in the sera of various animals and can be easily prepared by immunisation of rabbits with appropriate cells. These antisera represent mixtures of antibodies of different specificities, which in most cases can be separated from each other by absorption/elution using suitable erythrocytes.

More recently, hybridoma antibodies with anti-**T** [2,95], anti-**Th** [95], and anti-**Tn** specificity [2,54,85,104] have been described.

A series of lectins has also been detected in the course of investigations on polyagglutination. These lectins react with the antigen characters discussed in this chapter. Thanks to their high specificity some of these lectins have displaced the antisera and are now widely used for serological characterisation of the different forms of polyagglutination. Lectins have further become indispensable tools for structural investigations.

In *Table 10.1* the reactions of normal and polyagglutinable erythrocytes with normal human sera and some lectins are presented. Further lectins used for the identification of these characters are mentioned below or can be found in the review article written by Bird [11].

10.2 T Antigen

The appearance *in vitro* (in stored blood samples) of erythrocyte polyagglutination was first observed by Hübener [57] and Thomsen [106]. Later, Friedenreich [41] realised that this phenomenon was caused by bacterial contamination.

This so-called **Thomsen-Friedenreich antigen** appears also *in vivo* during infections with influenza viruses or bacteria (e.g. *Bacteroides fragilis*, *Diplococcus pneumoniae*, and various strains of *Streptococcus*, *Clostridium*, and *Corynebacterium* [12,46]); it normally disappears after the patient's recovery. Occasionally T-transformation is accompanied by **Tk**-transformation [60].

T-transformation is not confined to erythrocytes − it can also be detected on other blood cells, i.e. leukocytes and thrombocytes [58]. Strong **T** activity repeatedly has been observed in malignant breast epithelium, gastric and colonic carcinomas, leukaemic lymphocytes and teratocarcinoma cells; Springer discusses these findings in detail in his review article [98].

Anti-**T** antibodies are found in almost all sera of adults; they are absent, however, in newborns and infants [67,90]. Absorption/elution with **T**-transformed erythrocytes often yields highly specific anti-**T** sera from normal human sera (see also [56]). Anti-**T** also are found as 'naturally occurring' antibodies in various animal species − horse sera especially show strong anti-**T** activity [109]. Anti-**T** specific hybridoma antibodies have been described [2,95].

The agglutinin of peanuts (*Arachis hypogaea*) is widely used for the identification of **T**-transformed erythrocytes [7,75] (see *Table 10.1*). A Leu^{212} → Asn mutant of the peanut lectin has recently been prepared which shows an even higher specificity towards the **T** antigen than the 'wild-type' lectin [96]. Anti-**T** specific lectins have also been detected in *Vicia cretica* [19], *Salvia glutinosa* [16], *Amaranthus caudatus* [91], and *Artocarpus integrifolia* [76,94] seeds.

Gottschalk's investigations [43] revealed that the appearance of **T** antigen is caused by neuraminidases released in the blood stream by microorganisms or

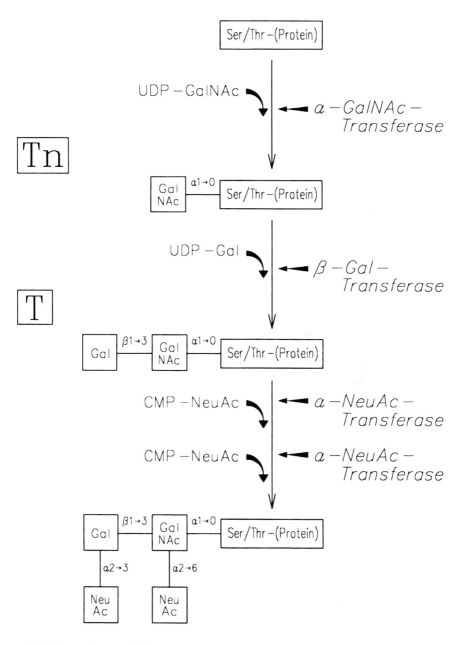

Figure 10.1: Biosynthesis of Tn and T determinants.

viruses[2]. **T**-transformation can also be induced *in vitro* by treating native erythrocytes with neuraminidase; this reaction is widely used in blood group serology for preparing test samples of **T** red cells [56].

Due to the action of neuraminidase, the level of neuraminic acid in **T**-transformed erythrocytes is significantly decreased [43]; as a consequence, the cells show a greatly reduced electrophoretic mobility and are not aggregated by polycations such as polybrene or protamine [62]. Furthermore, blood group **M** and **N** activities are greatly reduced, whereas reactivity towards the anti-**N** specific lectin of *Vicia graminaea* (**N**$_{Vg}$, see *Sect. 9.5)* is increased [74,92].

Inhibition tests with desialylated gangliosides and oligosaccharides of known structure have revealed that the **T** determinant is represented by the disaccharide unit **Galβ1→3GalNAc** [33,112]. This structure forms throughout the animal kingdom part of the carbohydrate chains of glycoconjugate constituents present in cellular membranes and various body fluids [112]. In erythrocytes this disaccharide represents the basic structure of the O-linked disialo-tetrasaccharide occurring in great quantities in glycophorins [105] (see *Sect. 9.3* and *Fig. 9.3.a)*. Thus the assumption that the **T** antigens of the red cells are localised mainly on these sialo-glycoproteins is justified.

Characterisation with synthetic antigens [55] has revealed that anti-**T** antisera usually represent heterogeneous mixtures of antibodies binding to different epitopes of the **Galβ1→3GalNAc** structure; the reactivity is variously influenced by the carrier molecule.

The biosynthetic scheme depicted in *Fig. 10.1* shows that the **T** determinant represents a precursor structure in the synthesis of the disialo-tetrasaccharide of the glycophorins referred to in [100]. Different groups have succeeded in performing the synthesis step from the precursor, **Tn** (see *Sect. 10.6)*, to **T** *in vitro* – **Tn** erythrocytes acquire **T** activity when incubated with normal human serum (the source of the 'T-transferase') and UDP-galactose (the sugar donor) [27,37,101].

10.3 Tk Antigen

This form of acquired polyagglutination has been described by Bird and Wingham [14]. **Tk** can occur either alone [14,20,61], or together with **T** [60] and/or **'acquired B'** [26].

---

[2] Because neuraminidases destroy the receptors for influenza viruses, these enzymes are widely designated as 'receptor-destroying enzymes' (= RDE).

Tk erythrocytes are agglutinated by most adult sera and by the 'BS-II lectin' of *Bandeiraea* (= *Griffonia*) *simplicifolia* [65] (see *Table 10.1)*; the repeatedly noticed reaction of **Tk** cells with the lectin of *Arachis hypogaea* [14] can be detected only in the presence of proteases [21,38,79] and probably has nothing to do with the **Tk**-transformation *per se*. In contrast to **T** the agglutination of **Tk** cells is enhanced by protease treatment [14,65]. Normally **Tk** transformed erythrocytes also show decreased **ABH** activity [61]. The cells have a normal content of neuraminic acid and thus are strongly aggregated by polybrene [14].

Soon after the detection of the **Tk** antigen it was suspected that **Tk**-transformation was caused by the action of bacterial enzymes [20,60]. Based on the observation that the **Tk** character occurs fairly frequently in patients suffering from *Bacteroides fragilis* septicaemia it has been assumed that this bacterium produces the 'transforming agent' which is responsible for the change in membrane antigenicity. In the course of further investigations **Tk**-transformation of erythrocytes was in fact induced by *in vitro* treatment of native erythrocytes with cultural filtrates from certain strains of *Bacteroides fragilis* [59]. This same reaction is now used for the *in vitro* preparation of **Tk** cells.

Studies by Doinel et al. [38,39] on the chemical basis of **Tk**-transformation suggest that this change of the antigen pattern of the erythrocyte membrane may be caused by the action of endo-β-galactosidases: the authors were able to induce **Tk**-transformation of erythrocytes *in vitro* by treatment with endo-β-galactosidase from *Escherichia freundii*. The cells incubated with the supernatant of *Bacteroides fragilis*

Table 10.2: *In vitro* **Tk transformation of human erythrocytes.**

Agglutinin	Untreated cells	Erythrocytes treated with B. fragilis supernatant	endo-β-Gal-ase
Anti-B	46	7	39
Anti-I	58	7	7
Anti-i[a]	0 (30)	0	0
AB serum (Anti-Tk)	0	13	15
BS II lectin	0	20	20

Serological reactivity (agglutination score) of I-positive erythrocytes before and after treatment with culture supernatant of *Bacteroides fragilis* or endo-β-galactosidase of *Escherichia freundii*.
[a] i activity of I-positive cells from adults is unmasked by papain treatment (agglutination score in brackets), i activity of foetal erythrocytes is totally destroyed by protease treatment [1,39].
From Doinel et al. [38] by permission of S. Karger AG, Basel.

cultures or the enzyme from *Escherichia freundii* [61] showed a decrease in **ABH** activity similar to that found in **Tk** cells transformed *in vivo* [1,38]; furthermore, a significant decrease in **I** activity and the loss of **i** has been observed [38] (*Table 10.2*).

The exact structure of the **Tk** determinant has not yet been established. However, due to the fact that oligosaccharide chains are cleaved by endo-β-galactosidases between galactose and N-acetylglucosamine [42] and considering the specificity of BS-II lectin which binds to terminal N-acetylglucosamine residues [65], it must be assumed that the immunodeterminant group of the **Tk** antigen is represented by a terminal **GlcNAcβ1→3Galβ1→4** unit.

The discovery that the endo-β-galactosidase from *E. freundii* acts on the carbohydrate chains of band 3 and band 4.5 proteins and splits the carbohydrate moiety of polyglycosyl-ceramides [78] suggests that the **Tk** determinants are localised on these components of the red cell membrane.

10.4 Th Antigen

Th polyagglutinability has been detected in patients with bacterial infections by Bird et al. [6]. **Th** activation of erythrocytes has also been observed in nearly 11% of pregnant women and newborns [115], as well as in children with congenital hypoblastic anaemias [52]. In these cases the cells reacted with *Arachis hypogaea* lectin, but were not polyagglutinable.

Th transformed erythrocytes are characterised by their reaction with the lectins of *Arachis hypogaea* [6,114], *Vicia cretica* [19,118], *Medicago disciformis* [97], and *Hibiscus sabdariffa* [119]; in contrast to **T** transformed cells, however, **Th** cells are not agglutinated by *Glycine soja* (= *G. max*) lectin (see *Table 10.1*). The preparation of an anti-**Th** specific hybridoma antibody has been reported [95].

Th erythrocytes show a normal content of neuraminic acid and are aggregated by polybrene. In contrast to the **T** determinant, which is destroyed only by high concentrations of proteases, mild treatment with papain is sufficient to remove the **Th** determinant from the erythrocytes.

Since the serological characteristics of the **Th** antigen are similar to those of the **T** antigen it was originally assumed that **Th**-transformation represented a weak expression or an early stage of **T**-transformation [6]. However, the fact that anti-**Th** antibodies occurring in normal adult sera can clearly be separated from anti-**T** specific antibodies [12] and the finding that **Th** transformed erythrocytes lack **T** activity [53] suggest that **Th** is a separate antigenic entity.

Because patients with **Th** transformed erythrocytes often suffer from bacterial infections [6,114], it has been assumed that the appearance of the **Th** character might also be due to the action of a bacterial enzyme.

In fact, *in vitro* **Th**-transformation has recently been achieved by incubating erythrocytes with a culture supernatant of *Corynebacterium aquaticum* isolated from samples of **Th** transformed blood (Shechter, cited in [97]). In subsequent investigations the transforming agent has been identified as a weakly active neuraminidase [97]. The studies furthermore revealed that **Th**-transformation is likewise induced by treatment with other neuraminidases, but only under extremely mild conditions: **Th** reactivity appeared upon release of a very small quantity of sialic acid (less than 20 µg per 10^{10} red blood cells), whereas hydrolysis of greater amounts (>15% of total) resulted in classical **T** polyagglutinability.

The exact chemical structure of the **Th** determinant has not yet been established. However, the known specificity of the *Arachis hypogaea* lectin indicates that the **Th** epitope may be represented by terminal galactose residues located on membrane glycoproteins.

10.5 Tx Antigen

Tx erythrocytes, like **T** and **Th** cells, are agglutinated by the lectin of *Arachis hypogaea* but show no reaction with *Vicia cretica* lectin (see *Table 10.1*). The **Tx** activity of erythrocytes is only slightly influenced by protease treatment.

Tx-transformation was first discovered by Bird et al. [23] on erythrocytes of children suffering from *Pneumococcus* infections. As in the cases of **T** and **Th** the appearance of the **Tx** character could be induced by *in vitro* treatment of normal erythrocytes with culture supernatants of the *Pneumococcus* strain which had been isolated from these patients.

The structure of the **Tx** antigen has, however, not yet been established.

10.6 Tn Antigen

In most cases **Tn**-transformation [36,77] is not accompanied by pathological symptoms, although mild forms of leukopenia, thrombocytopenia, or haemolytic anaemia have occasionally been observed [13,36]. In rare cases the occurrence of **Tn** is linked to leukaemic diseases [3,22,83,93]. In contrast to T-transformation, **Tn** is usually persistent; in some cases, however, it has disappeared spontaneously or after treatment with cytostatica [9,22].

Tn-transformation is not confined to erythrocytes; it is detected on all blood cells [4,25,72,86]. In each case only a limited percentage of the cells is affected, while the other cells show virtually normal characteristics [102,103]. In serological tests the presence of two sub-populations is manifested by mixed-field agglutination of the erythrocytes [13], with the number of agglutinated cells ranging between 5% and 95% [30].

Vainchenker et al. [110,111] have shown that in cultures of erythroids, granulocyte-macrophages, and megakaryocytes obtained from **Tn** individuals, the colonies were composed exclusively of either **Tn(+)** or **Tn(−)** cells. This result confirmed the view that **Tn**-transformation affects a pluripotent stem cell of the bone marrow (see [25,51,93]).

Tn receptors are also a characteristic feature of malignantly transformed cells. **Tn** activity has been detected in various carcinomas of mammary glands, lung, pancreas, and colon [104]; **Tn** active glycoproteins have also been isolated from cultured human breast carcinoma cells [99] and colonic cancer cells [82]. Further information on this topic can be obtained from the review article written by Springer [98].

The **Tn** antigen is marked by antibodies which occur regularly in sera of adults [103]; they are also found as 'naturally occurring' antibodies in goats, horses, sheep, rabbits, and chimpanzees [36,113], and have been induced in rabbits by immunisation with **Tn**-cells [48]. More recently **Tn** specific hybridoma antibodies have been prepared (e.g. [2,54,68,85,104]).

Anti-**Tn** specific lectins occur in plants of the genera *Salvia* [16,17] and *Lamium* [18], as well as in the seeds of *Dolichos biflorus* [13,48] and *Vicia villosa* [108]. Of these reagents only the lectins of *Salvia sclarea* (see also [87,88]) and *S. haematodes* have proved useful as anti-**Tn** reagents. Extracts from the seeds of *Salvia hormium*, *S. farinacea* and *Dolichos biflorus* also show anti-**Cad** activity; *Dolichos biflorus* lectin is furthermore used as a powerful anti-**A**$_1$ *(Sect. 5.2.2)*. Among the lectins of animal origin the anti-**A** agglutinin from *Helix pomatia* has frequently been used as anti-**Tn** reagent [30].

The serological characteristics of **Tn** erythrocytes can be found in *Table 10.1*.

Tn erythrocytes have a greatly reduced level of neuraminic acid (they average only 50–60% of the normal level); as a consequence, the cells show a very low electrophoretic mobility [47,80,102] and are not aggregated by polybrene [13]. This property has been used for the separation of **Tn** transformed ('polybrene-negative') erythrocytes from unchanged ('polybrene-positive') cells in blood samples ([13], see below). Furthermore, a series of antigen characteristics is altered in **Tn** cells, whereby

M, **N**, N_{Vg}, and **T** activities are significantly reduced [15,103]. The reactivity of the cells towards myxo- and paramyxoviruses, however, is not changed [13,73]. The **Tn** antigen is destroyed by treatment of the erythrocytes with papain or ficin [48,103].

A greatly reduced PAS reaction of glycophorin A and B was observed when membrane proteins of erythrocytes obtained from **Tn** transformed blood were investigated by SDS-polyacrylamide-gel electrophoresis [34]. Moreover, the different bands showed a slightly increased electrophoretic mobility when compared to that of glycophorins from normal cells [30].

Likewise, minor variations have been found in the membrane glycoproteins of thrombocytes and granulocytes isolated from **Tn** transformed blood: the fraction GP Ib in thrombocytes was affected [86], and in granulocytes several membrane proteins were also changed [28,66].

Chemical analyses of erythrocyte membranes and membrane glycoproteins showed a significantly lower content of galactose in **Tn** erythrocytes, in addition to the reduced level of neuraminic acid mentioned above [35]. Furthermore, removal of neuraminic acid and galactose from isolated erythrocyte membrane sialoglycoproteins (by using enzymatical or chemical methods) induces **Tn** specificity [32,100]. The results of serological investigations show that the **Tn** determinant is characterised by terminal N-acetylgalactosamine [13,80]; these residues are released as N-acetylgalactosaminitol during treatment of red cell membrane sialoglycoproteins with alkaline borohydride [35]. The findings clearly indicated that the **Tn** determinant of human erythrocytes is represented by unsubstituted O-linked α-N-acetylgalactosamine as carried by glycophorin A and B molecules [24,34,64] (see *Chapter 9.3)*. Serological investigations using an anti-**Tn** specific hybridoma antibody showed that the **Tn** epitope comprises a cluster of three vicinal **GalNAc-Ser/Thr** residues [81].

Recently the presence of the sialoyl-**Tn** structure (**NeuAc**α**2→6GalNAc** α**1→OSer/Thr**) has been reported in **Tn** polyagglutinable erythrocytes [24,70]. Only small amounts of sialoyl-**Tn** are found in normal human tissues; the same antigen, however, is strongly expressed in many adenocarcinomas [63,69,82,117].

As shown in *Fig. 10.1*, **Tn** and **T** represent precursor structures in the biosynthetic pathway leading to the alkali-labile disialo-tetrasaccharide units of glycophorin A and glycophorin B [100]. Based on this scheme it had been assumed that, due to a deficient β1,3-galactosyltransferase ('T-transferase'), unchanged **GalNAc**α**1→OSer/Thr** structures were accumulated in **Tn** transformed cells [102].

Confirmation of this fact was borne out by the findings of several groups working independently of each other, which were able to show that **Tn** transformed cells lack the T-transferase. The activity of this enzyme was greatly reduced in the membranes of erythrocytes [5,27,31], thrombocytes [29,30], and granulocytes [28] obtained from **Tn** transformed blood. When **Tn**-positive cells were separated from **Tn**-negative cells

Table 10.3: Change in 'T-transferase' activity at Tn transformation.

Donor	Activity of T-transferase[a] in						
	serum	erythrocytes			thrombocytes		
		total	Tn(−)	Tn(+)	total	Tn(−)	Tn(+)
normal	3.4	181			178		
Tn(Du)	3.5	36	144	4	58	125	17
Tn (Gon)		126	179	7			
Tn (Az)	3.15	10			46	203	16
Tn (Ba)	3.1	5.8			13		

[a] pMol Gal per mg protein per h; p-nitrophenyl-α-N-acetylgalactosaminide as substrate.
From Cartron et al. [30] by permission of Librairie Arnette, Paris.

(as in the case of erythrocytes by selective aggregation by polybrene [27], or, as in the case of thrombocytes, by affinity chromatography on *Helix pomatia* lectin columns [30]) this effect became even more significant − in **Tn**-positive cells the enzyme was virtually absent, whereas in **Tn**-negative, i.e. not transformed cells, normal values were found. A small amount of β1,3-galactosyltransferase detected in serum, however, was not influenced by **Tn**-transformation (see *Table 10.3*).

Investigations on cloned T-lymphocytes derived from a patient suffering from **Tn** syndrome and selected for T-transferase-deficiency showed that treatment of the cells with 5-azacytidine or sodium n-butyrate (both of which induce gene expression) resulted in re-expression of **T** antigen and T-transferase. These studies demonstrated that the absence of β1,3-galactosyltransferase in **Tn** cells is not due to a gene mutation or deletion, but rather due to a persistent but reversible repression of the gene encoding the T-transferase [40,107]. In accordance with this result are the small but significant amounts of disialotetrasaccharide found in **Tn** erythrocytes, suggesting that the T-transferase may not be totally inactive in **Tn** transformed cells [24].

As already mentioned, accumulation of **Tn** and sialoyl-**Tn** has been reported in many carcinomas [98]. In a preliminary investigation on the human T leukemia cell line 'Jurkat' [89] it was not possible to induce the β1,3-galactosyltransferase by 5-azacytidine or Na-butyrate [107]. This finding suggests that in transformed tissue other mechanisms affecting the expression of T-transferase may be of importance, e.g. irreversible repression, gene mutation, or even gene deletion.

Changes in the thrombocyte glycoproteins have been observed in some cases of congenital platelet disorders: cells from patients suffering from Glanzmann's Thrombasthenia show a decreased PAS reaction in fractions GP Ib and GP IIIa, in Bernard-Soulier Syndrome GP Ib/Is is affected [49]; in contrast to **Tn**-transformed cells, however, the respective carrier proteins are absent in those cases [30,49]. Enzyme tests also confirm that the mechanisms leading to these glycoprotein abnormalities are different from the mechanism causing the **Tn** syndrome [30,49].

10.7 VA and NOR

VA (Vienna)

This form of polyagglutination has been described in only two cases. In a patient with haemolytic anaemia it proved persistent [45], in the second case it appeared together with **Tk** and disappeared after some time (Beck et al., cited in [8]).

The **VA** character is clearly distinguished from all other forms of polyagglutination and is confined to erythrocytes. **VA** cells showed a significant reduction of **H** receptors and partial agglutination with *Helix pomatia* lectin (Anti-A_{HP}). When treated with fluorescence-labelled anti-A_{HP} the cells displayed an irregular and patchy distribution of fluorescence [44]. The **VA** antigen is resistant to protease treatment.

No chemical investigations on the structure of the **VA** determinant have been performed to date.

NOR

In contrast to the forms of acquired polyagglutination thus far described, **NOR** is inherited as an apparently dominant trait [50]. Characterisation of **NOR** erythrocytes by various plant and animal lectins showed only minute differences to normal cells, the agglutinability, however, is enhanced after proteolytic treatment, and the haemagglutination is inhibited by avian P_1 substance and hydatid cyst fluid (see *Chap. 8*).

Recent studies revealed that **NOR** is related to an altered neutral glycolipid pattern, the antigen responsible for **NOR** polyagglutination being characterised by a terminal α-linked galactosyl residue. In contrast to expectations, however, serological tests have shown unambiguously that the **NOR** antigen is distinct from the P_1 glycolipid, thus demonstrating that **NOR** polyagglutination is not the result of overexpression of P_1 [71].

375

References

1. ANDREU, G., DOINEL, C., CARTRON, J. P. & MATIVET, S. (1979): Induction of Tk polyagglutination by Bacteroides fragilis culture supernatants. *Blood Transfus. Immunohaematol. 22*, 551-561.
2. ANSTEE, D. J. & LISOWSKA, E. (1990): Monoclonal antibodies against glycophorins and other glycoproteins. *J. Immunogenet. 17*, 301-308.
3. BALDWIN, M. L., BARRASSO, C. & RIDOLFI, R. L. (1979): Tn polyagglutinability associated with acute myelomonocytic leukemia. *Amer. J. Clin. Pathol. 72*, 1024-1027.
4. BECK, M. L., HICKLIN, B. L., PIERCE, S. R. & EDWARDS, R. L. (1977): Observations on leucocytes and platelets in six cases of Tn polyagglutination. *Med. Lab. Sci. 34*, 325-332.
5. BERGER, E. G. & KOZDROWSKI, I. (1978): Permanent mixed-field polyagglutinable erythrocytes lack galactosyltransferase activity. *FEBS Lett. 93*, 105-108.
6. BIRD, G. W., WINGHAM, J., BECK, M. L., PIERCE, S. R., OATES, G. D. & POLLOCK, A. (1978): Th, a 'new' form of erythrocyte polyagglutination. *Lancet i*, 1215-1216.
7. BIRD, G. W. G. (1964): Anti-T in peanuts. *Vox Sang. 9*, 748-749.
8. BIRD, G. W. G. (1977): Compexity of erythrocyte polyagglutinability. *Human Blood Groups - 5th International Convocation on Immunology, Buffalo, 1976* (J.F. Mohn, Plunkett, R.W., Cunningham, R.K., and Lambert, R.M., eds). S. Karger, Basel, 335-343.
9. BIRD, G. W. G. (1977): Erythrocyte polyagglutination. In: *CRC Handbook Series in Clinical Laboratory Science. Section D, Blood Banking* (D. Seligson, T. J. Greenwalt, and E. A. Steane, eds.). CRC Press Inc., Cleveland, Ohio, pp. 443-454.
10. BIRD, G. W. G. (1977): Hemagglutination dependent on drugs or other chemical substances. In: *CRC Handbook Series in Clinical Laboratory Science. Section D, Blood Banking* (D. Seligson, T. J. Greenwalt, and E. A. Steane, eds.). CRC Press Inc., Cleveland, Ohio, pp. 431-438.
11. BIRD, G. W. G. (1977): Lectins. In: *CRC Handbook Series in Clinical Laboratory Science. Section D, Blood Banking* (D. Seligson, T. J. Greenwalt, and E. A. Steane, eds.). CRC Press Inc., Cleveland, Ohio, pp. 459-473.
12. BIRD, G. W. G. (1982): Clinical aspects of red blood cell polyagglutinability of microbial origin. In: *Blood Groups and Other Red Cell Surface Markers in Health and Disease* (C. Salmon, ed.). Masson Publishing USA, Inc., New York, pp. 55-64.
13. BIRD, G. W. G., SHINTON, N. K. & WINGHAM, J. (1971): Persistent mixed-field polyagglutination. *Brit. J. Haematol. 21*, 443-453.
14. BIRD, G. W. G. & WINGHAM, J. (1972): Tk: a new form of red cell polyagglutination. *Brit. J. Haematol. 23*, 759-763.
15. BIRD, G. W. G. & WINGHAM, J. (1974): The M, N, and N_{Vg} receptors of Tn-erythrocytes. *Vox Sang. 26*, 171-175.
16. BIRD, G. W. G. & WINGHAM, J. (1976): More Salvia agglutinins. *Vox Sang. 30*, 217-219.
17. BIRD, G. W. G. & WINGHAM, J. (1977): Yet more Salvia agglutinins. *Vox Sang. 32*, 121-122.
18. BIRD, G. W. G. & WINGHAM, J. (1981): Tn-specific lectins from Lamium. *Clin. Lab. Haematol. 3*, 169-171.
19. BIRD, G. W. G. & WINGHAM, J. (1981): Vicia cretica: a powerful lectin for T- and Th- but not Tk- or other polyagglutinable erythrocytes. *J. Clin. Pathol. 34*, 69-70.
20. BIRD, G. W. G., WINGHAM, J., INGLIS, G. & MITCHELL, A. A. B. (1975): Tk polyagglutination in Bacteroides fragilis septicaemia. *Lancet i*, 286-287.
21. BIRD, G. W. G., WINGHAM, J. & LIEW, Y. W. (1983): Reaction of peanut lectin with 'pure' Tk-cryptantigen. *Transfusion 23*, 271.
22. BIRD, G. W. G., WINGHAM, J., PIPPARD, M. J., HOULT, J. G. & MELIKIAN, V. (1976): Erythrocyte membrane modification in malignant diseases of myeloid and lymphoreticular tissues. I. Tn-polyagglutination in acute myelocytic leukaemia. *Brit. J. Haematol. 33*, 289-294.
23. BIRD, G. W. G., WINGHAM, J., SEGER, R. & KENNY, A. B. (1982): Tx, a "new" red cell cryptantigen

exposed by pneumococcal enzymes. *Blood Transfus. Immunohaematol. 25*, 215-216.

24. BLUMENFELD, O. O., LALEZARI, P., KHORSHIDI, M., PUGLIA, K. & FUKUDA, M. (1992): O-linked oligosaccharides of glycophorins A and B in erythrocytes of two individuals with the Tn polyagglutinability syndrome. *Blood 80*, 2388-2395.

25. BROUET, J. C., VAINCHENKER, W., BLANCHARD, D., TESTA, U. & CATRON, J. P. (1983): The origin of human B and T cells from multipotent stem cells. A study of the Tn syndrome. *Eur. J. Immunol. 13*, 350-357.

26. BYRNE, U., BROWN, A., ROPARS, C. & MOORE, B. P. L. (1979): Acquired B antigen, Tk activation and A_1 destroying enzyme activity in a patient with septicaemia. *Vox Sang. 36*, 208-212.

27. CARTRON, J. P., ANDREU, G., CARTRON, J., BIRD, G. W. G., SALMON, C. & GERBAL, A. (1978): Demonstration of T-transferase deficiency in Tn-polyagglutinable blood samples. *Eur. J. Biochem. 92*, 111-119.

28. CARTRON, J. P., BLANCHARD, D., NURDEN, A., CARTRON, J., RAHUEL, C., LEE, D., VAINCHENKER, W., TESTA, U. & ROCHANT, H. (1982): Tn syndrome: a disorder affecting red blood cell, platelet, and granulocyte cell surface components. In: *Blood Groups and Other Red Cell Surface Markers in Health and Disease* (C. Salmon, ed).Masson Publishing USA, Inc., New York, pp. 39-54.

29. CARTRON, J. P. & NURDEN, A. T. (1979): Galactosyltransferase and membrane glycoprotein abnormality in human platelets from Tn-syndrome donors. *Nature 282*, 621-623.

30. CARTRON, J. P., NURDEN, A. T., BLANCHARD, D., LEE, H., DUPUIS, D. & SALMON, C. (1980): The Tn receptors of human red cells and platelets. *Blood Transfus. Immunohaematol. 23*, 613-628.

31. DAHR, W., GIELEN, W., PIERCE, S. & SCHAPER, R. (1979): UDP-Gal:GalNAc-α-galactosyl transferase deficiency in Tn-syndrome. In: *Glycoconjugates - Proceedings of the 5th International Symposium, Kiel, Germany.* (R. Schauer, P. Boer, E. Buddecke, M. F. Kramer, J. F. G. Vliegenthart, and H. Wiegandt, eds.). Georg Thieme Publishers, Stuttgart, pp. 272-273.

32. DAHR, W., UHLENBRUCK, G. & BIRD, G. W. G. (1974): Cryptic A-like receptor sites in human erythrocyte glycoproteins: proposed nature of Tn-antigen. *Vox Sang. 27*, 29-42.

33. DAHR, W., UHLENBRUCK, G. & BIRD, G. W. G. (1975): Further characterization of some heterophile agglutinins reacting with alkali-labile carbohydrate chains of human erythrocyte glycoproteins. *Vox Sang. 28*, 133-148.

34. DAHR, W., UHLENBRUCK, G., GUNSON, H. H. & VAN DER HART, M. (1975): Molecular basis of Tn-polyagglutinability. *Vox Sang. 29*, 36-50.

35. DAHR, W., UHLENBRUCK, G., GUNSON, H. H. & VAN DER HART, M. (1975): Studies on glycoproteins and glycopeptides from Tn-polyagglutinable erythrocytes. *Vox Sang. 28*, 249-252.

36. DAUSSET, J., MOULLEC, J. & BERNARD, J. (1959): Acquired hemolytic anemia with polagglutinability of red blood cells due to a new factor present in normal human serum (anti-Tn). *Blood 14*, 1079-1093.

37. DESAI, P. R. & SPRINGER, G. F. (1979): Biosynthesis of human blood group T-, N-, and M-specific immunodeterminants on human erythrocyte antigens. *J. Immunogenet. 6*, 403-417.

38. DOINEL, C., ANDREU, G., CARTRON, J. P., SALMON, C. & FUKUDA, M. N. (1980): T_k polyagglutination produced in vitro by an endo-β-galactosidase. *Vox Sang. 38*, 94-98.

39. DOINEL, C., RUFIN, J. M. & ANDREU, G. (1981): The T_k antigenic determinant: studies of T_k activated red blood cells with endoglycosidases. *Blood Transfus. Immunohaematol. 24*, 109-116.

40. FELNER, K. M., DINTER, A., CARTRON, J. P. & BERGER, E. G. (1998): Repressed β-1,3-galactosyltransferase in the Tn syndrome. *Biochim. Biophys. Acta 1406*, 115-125.

41. FRIEDENREICH, V. (1930): *The Thomsen Haemagglutination Phenomenon.* Copenhagen, Lewis and Munksgaard.

42. FUKUDA, M. N. & MATSUMURA, G. (1976): Endo-β-galactosidase of E. freundii. Purification and endo-glycosidic action on keratan sulfates, oligosaccharides and blood group active glycoproteins. *J. Biol. Chem. 251*, 6218-6225.

43. GOTTSCHALK, A. (1960): *The Chemistry and Biology of Sialic Acids and Related Substances,* Cambridge University Press.

44. GRANINGER, W., POSCHMANN, A., FISCHER, K., SCHEDL-GIOVANNONI, I., HÖRANDNER, H. &

KLAUSHOFER, K. (1977): "VA" a new type of erythrocyte polyagglutination characterized by depressed H receptors and associated with hemolytic anaemia. II. Observations by immunofluorescence, electron microscopy, cell electrophoresis and biochemistry. *Vox Sang. 32*, 201-207.

45. GRANINGER, W., RAMEIS, H., FISCHER, K., POSCHMANN, A., BIRD, G. W. G., WINGHAM, J. & NEUMANN, E. (1977): "VA" a new type of erythrocyte polyagglutination characterized by depressed H receptors and associated with hemolytic anaemia. I. Serological and hematological observations. *Vox Sang. 32*, 195-200.

46. GRAY, J. M., BECK, M. L. & OBERMAN, H. A. (1972): Clostridial-induced type I polyagglutinability with associated intravascular haemolysis. *Vox Sang. 22*, 379-383.

47. GUNSON, H. H., BETTS, J. J. & NICHOLSON, J. T. (1971): The electrophoretic mobility of Tn polyagglutinable cells. *Vox Sang. 21*, 455-461.

48. GUNSON, H. H., STRATTON, F. & MULLARD, G. W. (1970): An example of polyagglutinability due to the Tn antigen. *Brit. J. Haematol. 18*, 309-316.

49. HAGEN, I., NURDEN, A., BJERRUM, O. J., SOLUM, N. O. & CAEN, J. (1980): Immunochemical evidence for protein abnormalities in platelets with Glanzmann's Thrombasthenia and the Bernard-Soulier syndrome. *J. Clin. Invest. 65*, 722-731.

50. HARRIS, P. A., ROMAN, G. K., MOULDS, J. J., BIRD, G. W. G. & SHAH, N. G. (1982): An inherited RBC characteristic, NOR, resulting in erythrocyte polyagglutination. *Vox Sang. 42*, 134-140.

51. HAYNES, C. R., DORNER, I., LEONARD, G. L., ARROWSMITH, W. R. & CHAPLIN, H. (1970): Persistent polyagglutinability in vivo unrelated to T-antigen activation. *Transfusion 10*, 43-51.

52. HERMAN, J. H., SHIREY, R. S., SMITH, B., KICKLER, T. S. & NESS, P. M. (1987): Th activation in congenital hypoplastic anemia. *Transfusion 27*, 253-256.

53. HERMAN, J. H., WHITEHEART, W., SHIREY, R. S., JOHNSON, R. J., KICKLER, T. S. & NESS, P. M. (1987): Red cell Th activation: biochemical studies. *Brit. J. Haematol. 65*, 205-209.

54. HIROHASHI, S., CLAUSEN, H., YAMADA, T., SHIMOSATO, Y. & HAKOMORI, S. I. (1985): Blood group A cross-reacting epitope defined by monoclonal antibodies NCC-LU-35 and -81 expressed in cancer of blood group O or B individuals: its identification as Tn antigen. *Proc. Natl. Acad. Sci. USA 82*, 7039-7043.

55. HÖPPNER, W., FISCHER, K., POSCHMANN, A. & PAULSEN, H. (1985): Use of synthetic antigens with the carbohydrate structure of asialoglycophorin A for the specification of Thomsen-Friedenreich antibodies. *Vox Sang. 48*, 246-253.

56. HOWARD, D. R. (1979): Expression of T-antigen on polyagglutinable erythrocytes and carcinoma cells: preparation of T-activated erythrocytes, anti-T lectin, anti-T absorbed human serum, and purified anti-T antibody. *Vox Sang. 37*, 107-110.

57. HÜBENER, G. (1926): Untersuchungen über die Iso-Agglutination mit besonderer Berücksichtung scheinbarer Abweichungen vom Gruppenschema. *Z. Immun.-Forsch. 45*, 223-248.

58. HYSELL, J. K., HYSELL, J. W., NICHOLS, M. E., LEONARDI, R. G. & MARSH, W. L. (1976): In vivo and in vitro activation of T-antigen receptors on leucocytes and platelets. *Vox Sang. 31 (Suppl. 1)*, 9-15.

59. INGLIS, G., BIRD, G. W. G., MITCHELL, A. A. B., MILNE, G. R. & WINGHAM, J. (1975): The effect of Bacteroides fragilis on the human erythrocyte membrane: pathogenesis of Tk polyagglutination. *J. Clin. Pathol. 28*, 964-968.

60. INGLIS, G., BIRD, G. W. G., MITCHELL, A. A. B., MILNE, G. R. & WINGHAM, J. (1975): Erythrocyte polyagglutination showing properties of both T and Tk, probably induced by Bacteroides fragilis infection. *Vox Sang. 28*, 314-317.

61. INGLIS, G., BIRD, G. W. G., MITCHELL, A. A. B. & WINGHAM, J. (1978): Tk polyagglutination associated with reduced A and H activity. *Vox Sang. 35*, 370-374.

62. ISSITT, P. D. (1985): Polyagglutination. In: *Applied Blood Group Serology*. Montgomery Scientific Publications, Miami, Florida, USA, pp. 455-476.

63. ITZKOWITZ, S. H., YUAN, M., MONTGOMERY, C. K., KJELDSEN, D., TAKAHASHI, H. K., BIGBEE, W. L. & KIM, Y. S. (1989): Expression of Tn, sialosyl-Tn, and T-antigens in human colon cancer. *Cancer*

Res. 49, 197-204.

64. JOKINEN, M. (1981): Characterization of glycophorin A and band 3 from Tn polyagglutinable erythrocytes. *Scand. J. Haematol. 26*, 272-280.

65. JUDD, W. J., BECK, M. L., HICKLIN, B. L., IYER, P. N. S. & GOLDSTEIN, I. J. (1977): BS II Lectin: a second hemagglutinin isolated from Bandeiraea simplicifolia seeds with affinity for type III polyagglutinable red cells. *Vox Sang. 33*, 246-251.

66. JUDSON, P. A., SPRING, F. A., TAYLOR, M. A. & ANSTEE, D. J. (1983): Evidence for carbohydrate-deficient forms of the major sialoglycoproteins of human platelets, granulocytes, and T lymphocytes in individuals with Tn syndrome. *Immunology 50*, 415-422.

67. KIM, Y. D. (1980): Immunochemical characteristics of human anti-T antibodies. *Vox Sang. 39*, 162-168.

68. KING, M. J., PARSONS, S. F., WU, A. M. & JONES, N. (1991): Immunochemical studies on the differential binding properties of two monoclonal antibodies reacting with Tn red cells. *Transfusion 31*, 142-149.

69. KJELDSEN, T., CLAUSEN, H., HIROHASHI, S., OGAWA, T., IIJIMA, H. & HAKOMORI, S. I. (1988): Preparation and characterization of monoclonal antibodies directed to the tumor-associated O-linked sialosyl-2→6 α-N-acetylgalactosaminyl (sialosyl-Tn) epitope. *Cancer Res. 48*, 2214-2220.

70. KJELDSEN, T., HAKOMORI, S. I., SPRINGER, G. F., DESAI, P., HARRIS, T. & CLAUSEN, H. (1989): Coexpression of sialosyl-Tn (NeuAcα2→6GalNAcα1→O-Ser/Thr) and Tn (GalNAcα→OSer/Thr) blood group antigens on Tn erythrocytes. *Vox Sang. 57*, 81-87.

71. KUSNIERZ-ALEJSKA, G., DUK, M., STORRY, J. R., REID, M. E., WIECEK, B., SEYFRIED, H. & LISOWSKA, E. (1999): NOR polyagglutination and Stª glycophorin in one family: relation of NOR polyagglutination to terminal α-galactose residues and abnormal glycolipids. *Transfusion 39*, 32-38.

72. LALEZARI, P. & AL-MONDHIRY, H. (1973): Sialic acid deficiency of human red blood cells associated with persistent red cell, leucocyte, and platelet polyagglutinability. *Brit. J. Haematol. 25*, 399-405.

73. LEE, L. T., FRANK, S., DE JONGH, D. S. & HOWE, C. (1981): Immunochemical studies on Tn erythrocyte glycoprotein. *Blood 58*, 1228-1231.

74. LISOWSKA, E. (1963): Reaction of erythrocyte mucoproteins with anti-N phytoagglutinin from Vicia graminaea seeds. *Nature 198*, 865-866.

75. LOTAN, R., SKUTELSKY, E., DANON, D. & SHARON, N. (1975): The purification, composition, and specificity of the anti-T lectin from peanut (Arachis hypogaea). *J. Biol. Chem. 250*, 8518-8523.

76. MAHANTA, S. K., SASTRY, M. V. K. & SUROLIA, A. (1990): Topography of the combining region of a Thomsen-Friedenreich-antigen-specific lectin jacalin (Artocarpus integrifolia agglutinin). A thermodynamic and circular-dichroism spectroscopic study. *Biochem. J. 265*, 831-840.

77. MOREAU, R., DAUSSET, J., BERNARD, J. & MOULLEC, J. (1957): Anémie hémolytique acquisé avec polyagglutinabilité des hématies par un nouveau facteur présent dans le sérum humain normal (anti-Tn). *Bull. Soc. Med. Hôp. (Paris) 73*, 569-587.

78. MUELLER, T. J., LI, Y. T. & MORRISON, M. (1979): Effect of endo-β-galactosidase on intact human erythrocytes. *J. Biol. Chem. 254*, 8103-8106.

79. MULLARD, G. W., HAWORTH, C. & LEE, D. (1978): A case of atypical polyagglutinability due to Tk-transformation. *Brit. J. Haematol. 40*, 571-582.

80. MYLLYLÄ, G., FURUHJELM, U., NORDLING, S., PIRKOLA, A., TIPPETT, P., GAVIN, J. & SANGER, R. (1971): Persistent mixed-field polyagglutinability. Electrokinetic and serological aspects. *Vox Sang. 20*, 7-23.

81. NAKADA, H., INOUE, M., NUMATA, Y., TANAKA, N., FUNAKOSHI, I., FUKUI, S., MELLORS, A. & YAMASHINA, I. (1993): Epitopic structure of Tn glycophorin A for an anti-Tn antibody (MLS 128). *Proc. Natl. Acad. Sci. USA 90*, 2495-2499.

82. NAKADA, H., INOUE, M., NUMATA, Y., TANAKA, N., FUNAKOSHI, I., FUKUI, S. & YAMASHINA, I. (1992): Cancer-associated glycoproteins defined by a monoclonal antibody, MLS 128, recognizing the Tn antigen. *Biochem. Biophys. Res. Commun. 187*, 217-224.

83. NESS, P. M., GARRATY, G., MOREL, P. A. & PERKINS, H. A. (1979): Tn polyagglutination preceding

acute leukemia. *Blood 54*, 30-34.
84. NETER, E. (1956): Bacterial hemagglutination and hemolysis. *Bacteriol. Rev. 20*, 166-188.
85. NUMATA, Y., NAKADA, H., FUKUI, S., KITAGAWA, H., OZAKI, K., INOUE, M., KAWASAKI, H., FUNAKOSHI, I. & YAMASHINA, I. (1990): A monoclonal antibody directed to Tn antigen. *Biochem. Biophys. Res. Commun. 170*, 981-985.
86. NURDEN, A. T., DUPUIS, D., PIDARD, D., KIEFFER, N., KUNICKI, T. J. & CARTRON, J. P. (1982): Surface modifications in the platelets of a patient with α-N-acetyl-D-galactosamine residues, the Tn syndrome. *J. Clin. Invest. 70*, 1281-1291.
87. PILLER, V., PILLER, F. & CARTRON, J. P. (1986): Isolation and characterization of an N-acetylgalactosamine specific lectin from Salvia sclarea seeds. *J. Biol. Chem. 261*, 14069-14075.
88. PILLER, V., PILLER, F. & CARTRON, J. P. (1990): Comparison of the carbohydrate-binding specificities of seven N-acetyl-D-galactosamine-recognizing lectins. *Eur. J. Biochem. 191*, 461-466.
89. PILLER, V., PILLER, F. & FUKUDA, M. (1990): Biosynthesis of truncated O-glycans in the T cell line Jurkat. Localization of O-glycan initiation. *J. Biol. Chem. 265*, 9264-9271.
90. RACE, R. R. & SANGER, R. (1975): Polyagglutinability. In: *Blood Groups in Man*. Blackwell Scientific Publications, Oxford, pp. 486-496.
91. RINDERLE, S. J., GOLDSTEIN, I. J., MATTA, K. L. & RATCLIFFE, R. M. (1989): Isolation and characterization of amaranthin, a lectin present in the seeds of Amaranthus caudatus, that recognizes the T- (or cryptic T)-antigen. *J. Biol. Chem. 264*, 16123-16131.
92. ROMANOWSKA, E. (1964): Reactions of M and N blood-group substances natural and degraded with specific reagents of human and plant origin. *Vox Sang. 9*, 578-588.
93. ROXBY, D. J., MORLEY, A. A. & BURPEE, M. (1987): Detection of the Tn antigen in leukaemia using monoclonal anti-Tn antibody and immunohistochemistry. *Brit. J. Haematol. 67*, 153-156.
94. SASTRY, M. V. K., BANARJEE, P., PATANJALI, S. R., SWAMY, M. J., SWARNALATHA, G. V. & SUROLIA, A. (1986): Analysis of saccharide binding to Artocarpus integrifolia lectin reveals specific recognition of T-antigen (β-D-Gal(1→3)D-GalNAc). *J. Biol. Chem. 261*, 11726-11733.
95. SEITZ, R., FISCHER, K. & POSCHMANN, A. (1983): Differentiation of red cell membrane abnormalities causing T-polyagglutination by use of monoclonal antibodies. *Rev. Fr. Transfus. Immunohematol. 26*, 420.
96. SHARMA, V., VIJAYAN, M. & SUROLIA, A. (1996): Imparting exquisite specificity to peanut agglutinin for the tumor-associated Thomsen-Friedenreich antigen by redesign of its combining site. *J. Biol. Chem. 271*, 21209-21213.
97. SONDAG-THULL, D., LEVENE, N. A., LEVENE, C., MANNY, N., LIEW, Y. W., BIRD, G. W. G., SCHECHTER, Y., FRANÇOIS-GÉRARD, C., HUET, M. & BLANCHARD, D. (1989): Characterization of a neuraminidase from Corynebacterium aquaticum responsible for Th polyagglutination. *Vox Sang. 57*, 193-198.
98. SPRINGER, G. F. (1984): T and Tn, general carcinoma autoantigens. *Science 224*, 1198-1206.
99. SPRINGER, G. F., CHANDRASEKARAN, E. V., DESAI, P. R. & TEGTMEYER, H. (1988): Blood group Tn-active macromolecules from human carcinomas and erythrocytes: characterization of and specific reactivity with mono- and poly-clonal anti-Tn antibodies induced by various immunogens. *Carbohydr. Res. 178*, 271-292.
100. SPRINGER, G. F. & DESAI, P. R. (1974): Common precursors of human blood group MN specificities. *Biochem. Biophys. Res. Commun. 61*, 470-475.
101. SPRINGER, G. F. & DESAI, P. R. (1976): Enzymatic synthesis of human blood group M-, N-, and T-specific structures. *Naturwissenschaften 63*, 488-489.
102. STURGEON, P., LUNER, S. J. & McQUISTON, D. T. (1973): Permanent mixed-field polyagglutinability (PMFP): II. Haematological, biophysical, and biochemical observations. *Vox Sang. 25*, 498-512.
103. STURGEON, P., McQUISTON, D. T., TASWELL, H. F. & ALLAN, C. J. (1973): Permanent mixed-field polyagglutinability (PMFP): I. Serological observations. *Vox Sang. 25*, 481-497.
104. TAKAHASHI, H. K., METOKI, R. & HAKOMORI, S. I. (1988): Immunoglobulin G3 monoclonal antibody directed to Tn antigen (tumor-associated α-N-acetylgalactosaminyl epitope) that does not cross-react with blood group A antigen. *Cancer Res. 48*, 4361-4367.

105. THOMAS, D. B. & WINZLER, R. J. (1969): Structural studies on human erythrocyte glycoprotein. Alkali-labile oligosaccharides. *J. Biol. Chem. 244*, 5943-5946.
106. THOMSEN, O. (1927): Ein vermehrungsfähiges Agens als Veränderer des iso-agglutinatorischen Verhaltens der roten Blutkörperchen, ein bisher unbekannter Fall der Fehlbestimmung. *Z. Immun.-Forsch. 52*, 85-107.
107. THURNHER, M., RUSCONI, S. & BERGER, E. G. (1993): Persistent repression of a functional allele can be responsible for galactosyltransferase deficiency in Tn syndrome. *J. Clin. Invest. 91*, 2103-2110.
108. TOLLEFSEN, S. E. & KORNFELD, R. (1983): The B$_4$ lectin from Vicia villosa seeds interacts with N-acetylgalactosamine residues α-linked to serine of threonine residues in cell surface glycoproteins. *J. Biol. Chem. 258*, 5172-5176.
109. UHLENBRUCK, G., PARDOE, G. I. & BIRD, G. W. G. (1969): On the specificity of lectins with a broad agglutination spectrum. II. Studies of the nature of the T-antigen and the specific receptors for the lectin of Arachis hypogaea (ground-nut). *Z. Immun.-Forsch. 138*, 423-433.
110. VAINCHENKER, W., TESTA, U., DESCHAMPS, J. F., HENRI, A., TITEUX, M., BRETON-GORIUS, J., ROCHANT, H., LEE, D. & CARTRON, J. P. (1982): Clonal expression of Tn antigen in erythroid and granulocyte colonies and its application to determination of the clonality of the human megacaryocyte colony assay. *J. Clin. Invest. 69*, 1081-1091.
111. VAINCHENKER, W., VINCI, G., TESTA, U., HENRI, A., TABILIO, A., FACHE, M. P., ROCHANT, H. & CARTRON, J. P. (1985): Presence of the Tn antigen on hematopoietic progenitors from patients with the Tn syndrome. *J. Clin. Invest. 75*, 541-546.
112. VAITH, P. & UHLENBRUCK, G. (1978): The Thompson agglutination phenomenon: a discovery revisited 50 years later. *Z. Immun.-Forsch. 154*, 1-14.
113. VAN DER HART, M., MOES, M., VAN LOGHEM, J. J., ENNEKING, J. H. J. & LEEKSMA, C. H. W. (1961): A second example of red cell polyagglutinability caused by the Tn antigen. *Vox Sang. 6*, 358-361.
114. VENEZIANO, G., RASORE-QUARTINO, A. & SANSONE, G. (1978): Th erythrocyte polyagglutination. *Lancet ii*, 483.
115. WAHL, C. M., HERMAN, J. H., SHIREY, R. S., KICKLER, T. S. & NESS, P. M. (1989): Th activation of maternal and cord blood. *Transfusion 29*, 635-637.
116. WEEDEN, A. R., DATTA, N. & MOLLISON, P. L. (1960): Adsorption of bacteria on to red cells leading to positive antiglobulin reactions. *Vox Sang. 5*, 523-531.
117. WERTHER, J. L., RIVERA-MACMURRAY, S., BRUCKNER, H., TATEMATSU, M. & ITZKOWITZ, S. H. (1994): Mucin-associated sialosyl-Tn antigen expression in gastric cancer correlates with an adverse outcome. *Brit. J. Cancer 69*, 613-616.
118. YANG, E., MOORE, B. P. L., LIEW, Y. W. & BIRD, G. W. G. (1986): Further observations on the Vicia cretica lectin. *Transfusion 26*, 306-307.
119. YANG, E. K. L., SPENCE, L. R., HARDING, R. Y. & MOORE, B. P. L. (1982): A "new" lectin for detection of T, Tn, and Th polyagglutination. *Transfusion 22*, 338-339.

11 Sid and Cad

The **Sid** antigen **Sd^a** (ISBT Nr. 901 012) is inherited as a dominant autosomal character [38,55]. It is not confined to red cell membranes but occurs also in different tissues and body secretions. In about 96% of humans this specificity is present on erythrocytes and/or in secretions as well; 4% lack **Sd^a** activity and contain strong anti-**Sd^a** in their serum [44].

There is considerable variation in the strength of **Sd^a** antigen in different individuals [38,55]: only 1% of the red cell samples are strongly agglutinated by anti-**Sd^a** antibodies, 80% show a moderate reaction with partial agglutination; in approximately 20% of the **Sd(a+)** individuals the **Sd^a** antigen is scarcely detectable on the erythrocytes, and the **Sid** phenotype can be determined only by testing body secretions for **Sd^a** substance.

Studies of tissues and secretions [44,45] show that high **Sd^a** activity can be found in kidneys, stomach, and colon, whereas this specificity is only weakly expressed or totally absent in other tissues. The main sources of water-soluble **Sd^a** active material are urine and meconium; saliva and gastric juice show only low **Sd^a** activity.

The occurrence of **Sd^a** in the human organism is age-dependent. The antigen cannot be detected on the erythrocytes of infants under ten weeks [38]. In foetuses **Sd^a** is equally distributed along all segments of the intestinal tract, in adults, however, it is found only in gastric and colonic mucosa [51]. Saliva of newborn infants contains about four times the **Sd^a** activity present in the saliva of adults [44].

In pregnancy the **Sd^a** activity of erythrocytes is significantly reduced – about 75% of all pregnant women are **Sd(a–)**, whereas the **Sd^a** activity of urine is not influenced [51].

Further, the expression of **Sd^a** antigen dramatically decreases in malignantly transformed cells [21,53].

Sd^a active material is not confined to humans but is found widely distributed in tissues and body fluids of various species of mammals; the urine of guinea pigs in particular has proved an excellent source of water-soluble **Sd^a** substance [44]. In birds **Sd^a** antigen has not been detected to date.

Cad [14,37] is a rare human red cell antigen which is also inherited as an autosomal dominant character [75]. It is characterised by its strong reactivity with the

N-acetylgalactosamine-specific lectin of *Dolichos biflorus*. Since **Cad** erythrocytes are agglutinated by the majority of normal human sera this serological property is often considered a form of polyagglutinability. Further investigations show, however, that this phenomenon is caused by the weak anti-**Sd**^{a} antibodies frequently found in human sera [54]. Sanger et al. [57] have already proposed that **Cad** may be a very intense form of **Sd**^{a} ($\rightarrow$ **Sd(a++)** or 'Super-Sid', see also [54]). Indeed, structural and biochemical investigations on **Sd**^{a} and **Cad** determinants confirm a close similarity between these antigens (see below), although their distinct serological differences cannot yet be explained.

It should also be mentioned here that during *in vitro* experiments a remarkable resistance of **Cad** erythrocytes towards infection by the malaria parasite *Plasmodium falciparum* was noted [13].

Two examples of a cold agglutinin unrelated to the **I**, **Pr**, and **Gd** blood groups were described by Marsh et al. [42]. As the antibodies were inhibited by urine from **Sd(a+)** persons, a relationship to the **Sd**^{a} blood group was assumed and the autoantibodies were named anti-**Sd**^{x}.

Subsequent investigations by Bass et al. [2], however, showed that the inhibition was a nonspecific effect caused by charged molecules, and no direct association with the **Sid** blood group could be found. The authors therefore recommend that the antibody be renamed anti-**Rx** (after the initials of the original proposita).

11.1 Antisera and Lectins

Anti-**Sd**^{a} is a relatively common antibody found in small quantities in nearly all human sera [54]; it reacts preferentially with **Sd(a++)** (= **Cad**) erythrocytes. The only useful form of anti-**Sd**^{a} is found in the sera of **Sd(a–)** individuals [55].

Two hybridoma antibodies directed towards murine cytotoxic T cells have turned out to be anti-**Sd**^{a}/**Cad** specific [16]. More recently two **Cad**-specific hybridoma antibodies (*2A3D3* and *2D11E2*) have been produced using gangliosides from human hepatocellular carcinoma cells as immunogens [34].

Specific, naturally occurring anti-**Cad** has been detected in chicken serum [7]; it can also be produced by immunisation of the animals with **Cad** erythrocytes.

Anti-**Cad** specific lectins occur in *Dolichos biflorus* [3,57], *Vicia villosa* [71], *Salvia hormium* and *Salvia farinacea* [4,6]. Extracts from *Salvia* seeds also contain an anti-**Tn** agglutinin (see *Sect. 10.6*), which in most cases can be removed by absorption with **Tn** erythrocytes; the anti-**A**$_1$ specificity of the *Dolichos biflorus* lectin (see *Sect. 5.2.1*), however, cannot be separated from its anti-**Cad** specificity [54].

The anti-**A** reagents from the albumen glands of *Helix pomatia* and *Helix aspersa* have also proved excellent anti-**Cad** reagents [73].

11.2 Sd^a Active Substances

Morton and Terry were the first to provide evidence that a carbohydrate structure determines the **Sd^a** character [46]: the authors were able to show that the **Sd^a** activity in urine is heat stable but can be destroyed by treatment with acids, bases or periodate. Human anti-**Sd^a** sera cannot be inhibited by monosaccharides or their glycosides. However, the close serological relationship between **Sd^a** and **Cad** led to the assumption that the immunodominant sugar is the same for both antigens. Because the agglutination of **Cad** erythrocytes by the lectin of *Dolichos biflorus* is specifically inhibited by N-acetylgalactosamine it was suspected that this sugar represents an essential part of the immunodeterminant group of the **Sd^a** antigen [5,57].

In subsequent investigations on the **Sd^a** active material from human urine by Morgan et al. [43] the **Tamm-Horsfall glycoprotein** [69,70] was recognised as the main carrier of the **Sd^a** character. It is produced in considerable quantities (up to 85 mg a day [35]) by the kidney tubular epithelium, namely by the cells of the thick ascending limb of Henle's loop, and the early distal convoluted tubule segments of the nephron [63].

The Tamm-Horsfall glycoprotein has a molecular weight of 76–82 kDa and contains about 30% carbohydrate (7–10% of which is sialic acid) [27]. The molecule tends to form aggregates of very high molecular mass. It is a phosphatidylinositol-linked membrane protein, which is released into the urine after the loss of its hydrophobic anchor, presumably by the action of phospholipase or protease [56].

The primary structure of the peptide moiety of the Tamm-Horsfall glycoprotein was determined by nucleic acid sequencing of the respective cDNA [33,50]. The predicted peptide contains 640 amino acids including a 25 amino acid leader sequence (*Fig. 11.1*). The occurrence of 48 cysteine residues suggests a secondary and tertiary structure highly stabilised by disulfide bridges; there are eight potential N-glycosylation sites, the carbohydrate chains being of the 'complex' (mainly tetra-antennary) type [1,31,60,74] (see *Sect. 2.2* and *Fig. 11.2*).

Several *in vivo* functions of the Tamm-Horsfall glycoprotein have been suggested:

```
  1 MGQPSLTWML  MVVVASWFIT  TAATDTSEAR  WCSECHSNAT  CTEDEAVTTC  TCQEGFTGDG   60

 61 LTCVDLDECA  IPGAHNCSAN  SSCVNTPGSF  SCVCPEGFRL  SPGLGCTDVD  ECAEPGLSHC  120

121 HALATCVNVV  GSYLCVCPAG  YRGDGWHCEC  SPGSCGPGLD  CVPEGDALVC  ADPCQAHRTL  180

181 DEYWRSTEYG  EGYACDTDLR  GWYRFVGQGG  ARMAETCVPV  LRCNTAAPMW  LNGTHPSSDE  240

241 GIVSRKACAH  WSGHCCLWDA  SVQVKACAGG  YYVYNLTAPP  ECHLAYCTDP  SSVEGTCEEC  300

301 SIDEDCKSNN  GRWHCQCKQD  FNITDISLLE  HRLECGANDM  KVSLGKCQLK  SLGFDKVFMY  360

361 LSDSRCSGFN  DRDNRDWVSV  VTPARDGPCG  TVLTRNETHA  TYSNTLYLAD  EIIIRDLNIK  420

421 INFACSYPLD  MKVSLKTALQ  PMVSALNIRV  GGTGMFTVRM  ALFQTPSYTQ  PYQGSSVTLS  480

481 TEAFLYVGTM  LDGGDLSRFA  LLMTNCYATP  SSNATDPLKY  FIIQDRCPHT  RDSTIQVVEN  540

541 GESSQGRFSV  QMFRFAGNYD  LVYLHCEVYL  CDTMNEKCKP  TCSGTRFRSG  SVIDQSRVLN  600

601 LGPITRKGVQ  ATVSRAFSSL  GLLKVWLPLL  LSATLTLTFQ
```

Figure 11.1: Amino acid sequence of the Tamm-Horsfall glycoprotein as deduced from cDNA analysis.

N: potential N-glycosylation sites.

According to Hession et al. [33] and Pennica et al. [50]. The sequence data are deposited in the EMBL/GenBank data library (accession numbers M15881 and M17778).

— it has been shown that the Tamm-Horsfall glycoprotein is a specific ligand for neutrophils, probably through an arginine-glycine-aspartate sequence characteristic for cell adhesion molecules [72]. Therefore, this glycoprotein is potentially important in leukocyte migration across renal epithelium and in the pathogenesis of tubulo-interstitial nephritis;

— the molecule (also termed as uromodulin) acts as an immunosuppressive agent which suppresses T-cell and monocyte reactivity *in vitro* by inhibiting the action of interleukin 1 [33,47,48];

— the fact that Tamm-Horsfall glycoprotein can inhibit the binding of *Escherichia coli* S fimbriae to epithelial cells of the urinary tract also suggests a protective role of the glycoprotein in preventing infection by N-acetylgalactosamine-binding bacteria [49].

As had been suspected, analysis of the carbohydrate moiety of the Tamm-Horsfall glycoprotein showed that the content of N-acetylgalactosamine depends exclusively on the **Sd^a** phenotype (*Table 11.1*), whereas the content of the other monosaccharides and the amino acids remains constant [66].

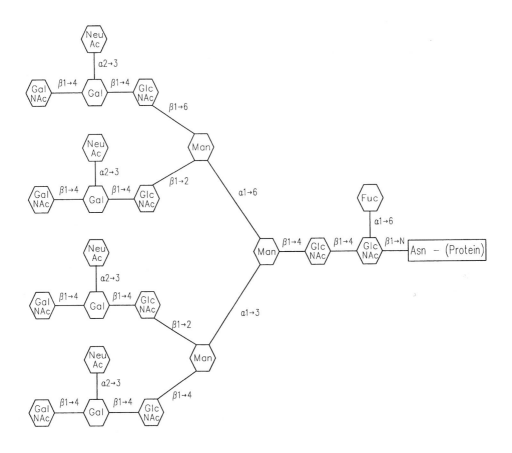

Figure 11.2: N-linked carbohydrate chains of human Tamm-Horsfall glycoprotein.
Structure of the largest N-acetylgalactosamine-containing compound.
According to Hård et al. [31].

Hydrazinolysis and mild acid hydrolysis of Tamm-Horsfall glycoprotein isolated from the urine of **Sd^a** individuals yielded a disaccharide characterised as **GalNAc** β1→4**Gal** [25]. This substance was not able to inhibit human anti-**Sd^a** sera, but showed a distinct reaction with the *Dolichos biflorus* lectin. This result was initially unexpected since the *Dolichos* lectin reacts preferentially with terminal α-N-acetyl-galactosamine residues [30]. In the course of further investigations a sialoyl-penta-saccharide containing this disaccharide structure was isolated by treatment of the

Table 11.1: Relationship between the Sd[a] phenotype of donors and the N-acetyl-galactosamine content of their Tamm-Horsfall glycoprotein.

Sd[a] phenotype of the donor erythrocytes	Anti-Sd[a] in donor serum	Tamm-Horsfall glycoprotein	
		Sd[a] activity[a]	GalNAc content[b]
Sd(a+)	−	0.04–0.08	1.21 %
Sd(a−)	−	250–500	0.21 %
Sd(a−)	+	>1000	0.06 %

[a] Minimum amount of substance (µg/ml) giving complete inhibition of agglutination of **Sd(a++)** cells by human anti-**Sd**[a] serum.
[b] Mean values from the published data.
From Soh et al. [66] by permission of the authors and of Academic Press, Inc., San Diego, CA., USA.

Tamm-Horsfall with endo-β-galactosidases [23,26] (*Fig. 11.3.a*); based on the distinct reaction of this oligosaccharide with anti-**Sd**[a] sera it became evident that this structure represented the **Sd**[a] determinant.

It can be assumed that a pentasaccharide with this structure is also responsible for the **Sd**[a] activity of human erythrocytes [76]; the 'proper' carrier of the **Sd**[a] determinant, however, has not yet been identified.

The group led by Cartron isolated another **Sd**[a] active substance from human urine [12]. In contrast to the Tamm-Horsfall glycoprotein this substance is a glycoprotein of the mucin type (i.e. the oligosaccharide chains are O-glycosidically linked to the protein backbone); it has a molecular mass of approximately 340 kDa. The structure of oligo-saccharide units of this glycoprotein, however, has not yet been determined.

In the stomach **Sd**[a] determinant is carried by glycolipid(s) (see *Fig. 11.4.a*), whereas in colonic mucosa the **Sd**[a] determinant is carried exclusively by glycoprotein(s) [22].

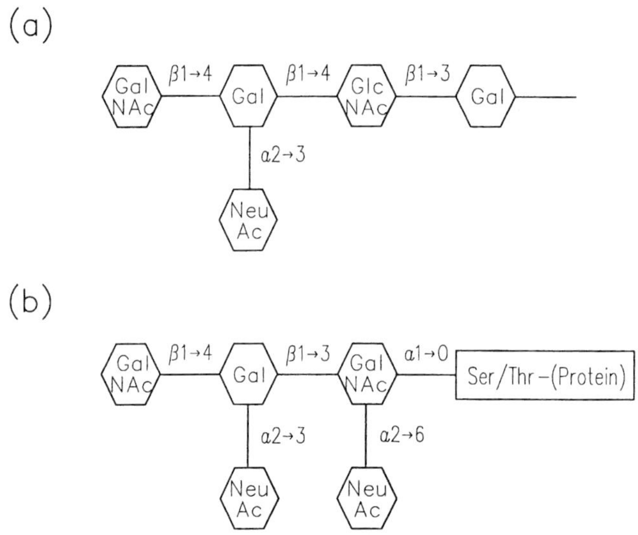

(a)

(b)

Figure 11.3: Chemical structures of Sd^a and Cad determinants.
(a) **Sd^a** determinant of human Tamm-Horsfall glycoprotein [26].
(b) **Cad** specific carbohydrate chain of glycophorin A obtained from human **Cad** erythrocytes [9].

11.3 Cad Active Substances

In their attempt to identify the carrier of the **Cad** determinant of human erythrocytes, Cartron and Blanchard [11] observed a decreased electrophoretic mobility of the main sialoglycoproteins, glycophorin A and glycophorin B (see *Sect. 9.3)* in **Cad** cells. This suggested an increase in apparent molecular mass of 3 and 2 kDa, respectively. When the carbohydrate content of glycophorin A isolated from **Cad** erythrocytes was compared to that isolated from **Sd(a+)** control cells, a distinct increase in N-acetylgalactosamine was noted (*Table 11.2).*

The predominant product released after alkaline-borohydride treatment of the glycophorin A of **Cad** erythrocytes was a sialoyl pentasaccharide [9,32]. This oligosaccharide strongly inhibited not only the agglutination of **Cad** erythrocytes by the *Dolichos* lectin, but also the agglutination of **Sd(a+)** cells by anti-Sd^a sera [32]. The structure was found to be the classic disialotetrasaccharide described by Thomas and Winzler (see *Sect. 9.3)* extended by an N-acetylgalactosamine unit β1→4 linked to the galactose residue (*Fig. 11.3.b).*

Table 11.2: Sugar composition of glycophorin A preparations obtained from Cad erythrocytes and Sd(a+) control cells.

Glycophorin A from	Carbohydrate content[a]					NeuAc / GalNAc ratio
	Gal	Man	GalNAc	GlcNAc	NeuAc	
Cad cells	577	115	911	165	910	1.0
Control cells	660	156	541	206	1080	2.0

[a] nMol/mg glycophorin A.
From Cartron & Blanchard [11] by permission of Portland Press, London.

Analysis of glycophorin from a donor with high **Cad** activity (donor *Cad.*) revealed about 12 chains of this pentasaccharide per molecule, whereas in the glycophorin fraction from donors with lower **Cad** activity (*Bui.* and *Des.*) only 2–3 chains were found. In **Sd(a+)** and **Sd(a–)** cells this **Cad** specific oligosaccharide could not be detected at all [8].

In subsequent investigations a glycosphingolipid which reacts with the lectin of *Helix pomatia* as well as with human anti-Sd^a sera has been isolated from the membranes of **Cad** erythrocytes [10,29]. The glycolipid represents a sialoyl-paragloboside with the carbohydrate chain extended by a β-N-acetylgalactosamine residue ($IV^4GalNAcIV^3NeuAcnLc_4Cer$) (*Fig. 11.4.a*). This glycolipid has also been found in human gastric mucosa, it is absent, however, from intestinal mucosa [21]; the lacto-isomer ($IV^4GalNAcIV^3NeuAcLc_4Cer$) was detected in meconium [28]. Though the structure is obviously identical to the Sd^a determinant, this glycosphingolipid is present in **Sd(a+)** erythrocytes only in trace amounts; it can thus be excluded as 'the' Sd^a substance of the erythrocytes.

The anti-**Cad** specific hybridoma antibodies, *2A3D2* and *2D11E2* [34], react with glycosphingolipids and glycoproteins of a cultured human hepatocellular carcinoma cell line (*PLC/PRF/5*). The major glycolipid has been characterised as $IV^4GalNAcβ-G_{D1a}$ (*Fig. 11.4.b*), a ganglioside which has been detected in human brain [67] and peripheral nerves [36]. The glycoproteins reacting with these antibodies appeared as triplet bands with molecular masses of 92 kDa, 75 kDa, and 61 kDa when analysed in SDS-polyacrylamide gelelectrophoresis under either reducing or nonreducing conditions.

389

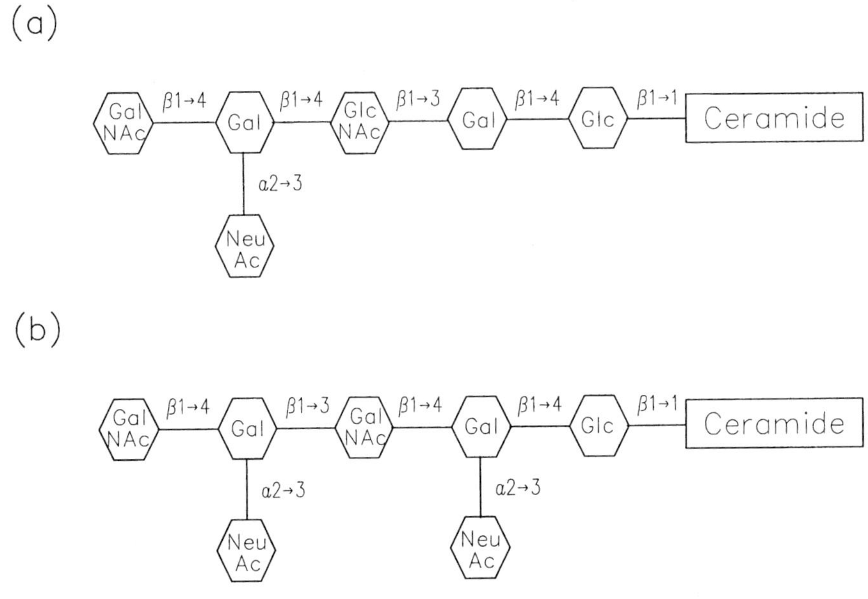

(a)

(b)

Figure 11.4: Chemical structures of Cad active glycosphingolipids.
(a) Isolated from the membranes of human **Cad** erythrocytes [29] (also found in human gastric mucosa [21]. **(b)** Isolated from a human hepatocellular carcinoma cell line [34].

Virtually the same protein triplet has been observed in several other samples of hepatocellular carcinoma tissue; normal and cirrhotic liver tissues contained no appreciable amount of these glycoproteins. These results provide additional evidence that the expression of **Cad** or **Cad**-like antigens may be associated with certain human cancers.

11.4 Investigations on the Biosynthesis of Sdᵃ and Cad Determinants

The product of the **SID** gene is a β1,4-N-acetylgalactosaminyltransferase, which links galactosaminyl residues to terminal **NeuAcα2→3Galβ1→4** disaccharide structures.

Enzymes of this specificity have been found in human serum [68] and urine [24,61], as well as in the microsomes of human gastric [19,20] and large intestinal mucosa [39], human colon carcinoma CaCo-2 cells [62], human [52] and guinea pig kidneys [58,59,65], and pig large intestine [40]. Moreover, transferases with a respective specificity have been detected in the large intestine and in cytotoxic T lymphocytes of the mouse [15].

The transferases investigated showed very similar characteristics probably due to a close phylogenetic relationship: all the enzymes are able to transfer N-acetyl-galactosamine residues to oligosaccharides and various glycoproteins containing terminal **NeuAcα2$\rightarrow$3Galβ1$\rightarrow$4** groups; the presence of α2$\rightarrow$3 neuraminic acid proved essential for the transfer reaction. The N-acetylgalactosamine units attach preferentially to the sialylated terminus of N-linked oligosaccharides (as found in Tamm-Horsfall glycoprotein, fetuin, or orosomucoid) [52,59,61,62,68]. The O-linked oligosaccharide chains of native glycophorin A proved virtually inactive, whereas tryptic peptides of this sialoglycoprotein were fairly good acceptor substrates. The enzymes were also able to use sialoyl-paragloboside as acceptor substrates, but were incapable of transferring an N-acetylgalactosamine residue to G$_{M3}$[1].

Moreover, strong evidence has been obtained that these transferases are responsible for the biosynthesis of the **Sd^a** determinant:

– the β1,4-N-acetylgalactosaminyltransferase is detectable only in the urine of human **Sd(a+)** individuals and is absent in **Sd(a–)** subjects [24],
– the Tamm-Horsfall glycoprotein from **Sd(a–)** individuals proved a better substrate than that obtained from **Sd(a+)** subjects [58,61],
– when N-acetylgalactosamine residues were transferred by a microsomal preparation from human kidney onto sialoyl-paragloboside [10], a product was obtained that was chromatographically indistinguishable from the **Sd^a/Cad** active glyco-sphingolipid isolated from human **Cad** erythrocytes,
– when N-acetylgalactosamine residues were transferred to 3'-sialoyl-N-acetyllactos-amine using the transferase from guinea pig kidney, the obtained tetrasaccharide was highly active in **Sd^a** serological tests [65].

More recent investigations on the expression of the β1,4-N-acetylgalactos-aminyltransferase in the course of embryonic development and in malignantly transformed tissue revealed that the occurrence of the transferase corresponds exactly to the occurrence of the **Sd^a** antigen in the organism (see above). In the kidney and

---

[1] The G$_{M2}$ ganglioside, though having the same terminal trisaccharide sequence as the **Sd^a**-specific glycolipid, does not react with anti-**Sd^a** sera [65]. It is synthesised from G$_{M3}$ by a different transferase [22] and is accumulated in gastric carcinoma [19].

```
  1  MTSSVSFASF  RFPWLLKTFV  LMVGLATVAF  MVRKVSLTTD  FSTFKPKFPE  PARVDPVLKL   60

 61  LPEEHLRKLF  TYSDIWLFPK  NQCDCNSGKL  RMKYKFQDAY  NQKDLPAVNA  RRQAEFEHFQ  120

121  RREGLPRPPP  LLAPPNLPFG  YPVHGVEVMP  LHTILIPGLQ  YEGPDAPVYE  VILKASLGTL  180

181  NTLADVPDDE  VQGRGQRQLT  ISTRHRKVLN  FILQHVTYTS  TEYYLHKVDT  VSMEYESSVA  240

241  KFPVTIKQQT  VPKLYDPGPE  RKIRNLVTIA  TKTFLRPHKL  KILLQSIRKY  YPDITVIVAD  300

301  DSKEPLEIND  DYVEYYTMPF  GKGWFAGRNL  AISQVTTKYV  LWVDDDFLFS  DKTKIEVLVD  360

361  VLEKTELDVV  GGSVQGNTYQ  FRLLYEQTKN  GSCLHQRWGS  FQALDGFPGC  TLTSGVVNFF  420

421  LAHTEQLRRV  GFDPILQRVA  HGEFFIDGLG  RLLVGSCPGV  IINHQVRTPP  KDPKLAALEK  480

481  TYDKYRANTN  SVIQFKVALQ  YFKNHLYCST
```

Figure 11.5: Peptide sequence of murine β1,4-N-acetylgalactosaminyltransferase.
Underlined: the putative transmembrane domain, **N**: the potential N-glycosylation site.
According to Smith and Lowe [64]. The sequence data are deposited in the EMBL/GenBank data library
(accession number L30104).

large intestine of rats and guinea pigs the enzyme is virtually absent at birth; its activity increases slowly in the first days of life and the transferase is clearly expressed only after weaning [17,18]. In adults the **Sid**-transferase and the **Sd**[a] antigen appear restricted to intestine and kidney [39,52]. Very high enzyme activity has been found in epithelial cells of the large intestine with a proximal-distal gradient of enzyme expression [39]. Studies on human carcinomas showed that the expression of the β1,4-N-acetylgalactosaminyltransferase in gastric mucosa [19,20] and colon epithelial cells [39,41] is greatly reduced in malignantly transformed tissue.

Smith & Lowe [64] have determined the peptide sequence of a murine β1,4-N-acetylgalactosaminyltransferase, which by its substrate specificity is probably homologous to the human **Sid**-transferase (*Fig. 11.5*). It is a 510-amino acid type II transmembrane protein with a short N-terminal cytosolic domain (17 residues), a 15-amino acid transmembrane domain flanked by basic residues, and a 478-amino acid C-terminal domain with a single potential N-glycosylation site.

More recently a 390-bp cDNA segment has been isolated from the total RNA fraction of human gastric mucosa using primers according to the cDNA sequence of this murine enzyme [22]. This cDNA segment shared 85% nucleotide identity with the murine transferase.

In PCR tests, mRNA specific for this β1,4-N-acetylgalactosaminyltransferase has been detected in all samples of normal stomach and small intestine examined and in the majority of normal colonic specimens; the mRNA, however, was absent in most samples of gastric and colonic cancer cells [22]. These results correlate with the β1,4-N-acetylgalactosaminyltransferase activity measured in the same samples and thus confirm that the level of expression of **Sd**[a] epitopes in gastrointestinal mucosa is determined by the transcriptional level of the **Sid**-transferase.

Thus far the β1,4-N-acetylgalactosaminyltransferase from a human **Cad** individual has not yet been investigated.

References

1. AFONSO, A. M. M., CHARLWOOD, P. A. & MARSHALL, R. D. (1981): Isolation and characterization of glycopeptides from digests of Tamm-Horsfall glycoprotein. *Carbohydr. Res. 89*, 309-319.
2. BASS, L. S., RAO, A. H., GOLDSTEIN, J. & MARSH, W. L. (1983): The Sd[x] antigen and antibody: biochemical studies on the inhibitory property of human urine. *Vox Sang. 44*, 191-196.
3. BIRD, G. W. G. (1970): Comparative serological studies of the T, Tn, and Cad receptors. *Blut 21*, 366-370.
4. BIRD, G. W. G. & WINGHAM, J. (1971): Some serological properties of the Cad receptor. *Vox Sang. 20*, 55-61.
5. BIRD, G. W. G. & WINGHAM, J. (1972): Cad and Sid. *Vox Sang. 22*, 362-363.
6. BIRD, G. W. G. & WINGHAM, J. (1974): Haemagglutinins from Salvia. *Vox Sang. 26*, 163-166.
7. BIZOT, M. & CAYLA, J. P. (1972): Hétéroanticorps anti-Cad du poulet. *Rev. Fr. Transfus. 15*, 195-202.
8. BLANCHARD, D., CAPON, C., LEROY, Y., CARTRON, J. P. & FOURNET, B. (1985): Comparative study of glycophorin A derived O-glycans from human Cad, Sd(a+), and Sd(a-) erythrocytes. *Biochem. J. 232*, 813-818.
9. BLANCHARD, D., CARTRON, J. P., FOURNET, B., MOUNTREUIL, J., VAN HALBEEK, H. & VLIEGENTHART, J. F. G. (1983): Primary structure of the oligosaccharide determinant of blood group Cad specificity. *J. Biol. Chem. 258*, 7691-7695.
10. BLANCHARD, D., PILLER, F., GILLARD, B., MARCUS, D. & CARTRON, J. P. (1985): Identification of a novel ganglioside on erythrocytes with blood group Cad specificity. *J. Biol. Chem. 260*, 7813-7816.
11. CARTRON, J. P. & BLANCHARD, D. (1982): Association of human erythrocyte membrane glycoproteins with blood-group Cad specificity. *Biochem. J. 207*, 497-504.
12. CARTRON, J. P., KORNPROBST, M., LEMONNIER, M., LAMBIN, P., PILLER, F. & SALMON, C. (1982): Isolation from human urines of a mucin with blood group Sd[a] activity. *Biochem. Biophys. Res. Commun. 106*, 331-337.
13. CARTRON, J. P., PROU, O., LUILIER, M. & SOULIER, M. (1983): Susceptibility to invasion by Plasmodium falciparum of some human erythrocytes carrying rare blood group antigens. *Brit. J. Haematol. 55*, 639-647.
14. CAZAL, R., MONIS, M., CAUBEL, J. & BRIVES, J. (1968): Polyagglutinabilité héréditaire dominante: antigène privé (Cad) correspondant à un anticorps public et à une lectine de Dolichos biflorus. *Rev. Fr. Transfus. 11*, 209-221.
15. CONZELMANN, A. & BRON, C. (1987): Expression of UDP-N-acetylgalactosamine:β-galactose β1,4-N-acetylgalactosaminyltransferase in functionally defined T-cell clones. *Biochem. J. 242*, 817-824.

16. CONZELMANN, A. & LEFRANÇOIS, L. (1988): Monoclonal antibodies specific for T cell-associated carbohydrate determinants react with human blood group antigens Cad and Sd[a]. *J. Exp. Med.* *167*, 119-131.

17. DALL'OLIO, F., MALAGOLINI, N., DI STEFANO, G., CIAMBELLA, M. & SERAFINI-CESSI, F. (1990): Postnatal development of rat colon epithelial cells is associated with changes in the expression of the β1,4-N-acetylgalactosaminyltransferase involved in the synthesis of Sd[a] antigen and of α2,6-sialyltransferase activity towards N-acetyl-lactosamine. *Biochem. J. 270*, 519-524.

18. DALL'OLIO, F., MALAGOLINI, N. & SERAFINI-CESSI, F. (1987): Tissue distribution and age-dependent expression of β-4-N-acetylgalactosaminyltransferase in guinea-pig. *Biosci. Rep. 7*, 925-932.

19. DOHI, T., HANAI, N., YAMAGUCHI, K. & OSHIMA, M. (1991): Localization of UDP-GalNAc: NeuAcα2,3Gal-R β1,4(GalNAc to Gal) N-acetylgalactosaminyltransferase in human stomach. Enzymatic synthesis of a fundic gland-specific ganglioside and G_{M2}. *J. Biol. Chem. 266*, 24038-24043.

20. DOHI, T., NISHIKAWA, A., ISHIZUKA, I., TOTANI, M., YAMAGUCHI, K., NAKAGAWA, K., SAITOH, O., OHSHIBA, S. & OSHIMA, M. (1992): Substrate specificity and distribution of UDP-GalNAc : sialylparagloboside N-acetylgalactosaminyltransferase in the human stomach. *Biochem. J. 288*, 161-165.

21. DOHI, T., OHTA, S., HANAI, N., YAMAGUCHI, K. & OSHIMA, M. (1990): Sialylpentaosylceramide detected with anti-G_{M2} monoclonal antibody. Structural characterization and complementary expression with G_{M2} in gastric cancer and normal gastric mucosa. *J. Biol. Chem. 265*, 7880-8885.

22. DOHI, T., YUYAMA, Y., NATORI, Y., SMITH, P. L., LOWE, J. B. & OSHIMA, M. (1996): Detection of N-acetylgalactosaminyltransferase mRNA which determines expression of Sd[a] blood group carbohydrate structure in human gastrointestinal mucosa and cancer. *Int. J. Cancer 67*, 626-631.

23. DONALD, A. S. R. & FEENEY, J. (1986): Oligosaccharides obtained from a blood-group-Sd(a+) Tamm-Horsfall glycoprotein. An n.m.r. study. *Biochem. J. 236*, 821-828.

24. DONALD, A. S. R., SOH, C. P. C., FEENEY, J. & WATKINS, W. M. (1987): Genetic and enzymic basis of the Sd[a] negative phenotype. In: *Glycoconjugates, Proceedings of the 6th Int. Symposium on Glycoconjugates, Lille 1987*, Abstract F 66.

25. DONALD, A. S. R., SOH, C. P. C., WATKINS, W. M. & MORGAN, W. T. J. (1982): N-Acetyl-D-galactosaminyl-β-(1$\rightarrow$4)-D-galactose: a terminal non-reducing structure in human blood-group Sd[a] active Tamm-Horsfall urinary glycoprotein. *Biochem. Biophys. Res. Commun. 104*, 58-65.

26. DONALD, A. S. R., YATES, A. D., SOH, C. P. C., MORGAN, W. T. J. & WATKINS, W. M. (1983): A blood group Sd[a]-active pentasaccharide isolated from Tamm- Horsfall urinary glycoprotein. *Biochem. Biophys. Res. Commun. 115*, 625-631.

27. FLETCHER, A. P. (1972): The Tamm and Horsfall glycoprotein. In: *Glycoproteins. Their Composition, Structure, and Function* (A. Gottschalk, ed.). Elsevier, Amsterdam, pp. 892-908.

28. FREDMAN, P., MÅNSSON, J. E., WIKSTRAND, C. J., VRIONIS, F. D., RYNMARK, B. M., BIGNER, D. D. & SVENNERHOLM, L. (1989): A new ganglioside of the lactotetraose series, GalNAc-3'-isoL$_{M1}$, detected in human meconium. *J. Biol. Chem. 264*, 12122-12125.

29. GILLARD, B. K., BLANCHARD, D., BOUHOURS, J. F., CARTRON, J. P., VAN KUIK, J. A., KAMERLING, J. P., VLIEGENTHART, J. F. G. & MARCUS, D. M. (1988): Structure of a ganglioside with Cad blood group antigen activity. *Biochemistry 27*, 4601-4606.

30. HAMMARSTRÖM, S., MURPHY, L. A., GOLDSTEIN, I. J. & ETZLER, M. E. (1977): Carbohydrate binding specificity of four N-acetyl-D-galactosamine- "specific" lectins: Helix pomatia A hemagglutinin, soy bean agglutinin, lima bean lectin, and Dolichos biflorus lectin. *Biochemistry 16*, 2750-2755.

31. HARD, K., VAN ZADELHOFF, G., MOONEN, P., KAMERLING, J. P. & VLIEGENTHART, J. F. G. (1992): The Asn-linked carbohydrate chains of human Tamm-Horsfall glycoprotein of one male. Novel sulfated and novel N-acetylgalactosamine-containing N-linked carbohydrate chains. *Eur. J. Biochem. 209*, 895-915.

32. HERKT, F., PARENTE, J. P., LEROY, Y., FOURNET, B., BLANCHARD, D., CARTRON, J. P., VAN HALBEEK, H. & VLIEGENTHART, J. F. G. (1985): Structure determination of oligosaccharides isolated from

Cad erythrocyte membranes by permethylation analysis and 500-MHz ^{1}H-NMR spectroscopy. *Eur. J. Biochem. 146*, 125-129.

33. HESSION, C., DECKER, J. M., SHERBLOM, A. P., KUMAR, S., YUE, S. S., MATTALIANO, R. J., TIZARD, R., KAWASHIMA, E., SCHMEISSNER, U., HELETKY, S., CHOW, E. P., BURNE, C. A., SHAW, A. & MUCHMORE, A. V. (1987): Uromodulin (Tamm-Horsfall glycoprotein): a renal ligand for lymphokines. *Science 237*, 1479-1484.

34. HIRAIWA, N., TSUYUOKA, K., LI, Y. T., TANAKA, M., SENO, T., OKUBO, Y., FUKUDA, Y., IMURA, H. & KANNAGI, R. (1990): Gangliosides and sialoglycoproteins carrying a rare blood group antigen determinant, Cad, associated with human cancers as detected by specific monoclonal antibodies. *Cancer Res. 50*, 5497-5503.

35. HUNT, J. S., MCGIVEN, A. R., GROUFSKY, A., LYNN, K. L. & TAYLOR, M. C. (1985): Affinity-purified antibodies of defined specificity for use in a solid-phase microplate radioimmunoassay of human Tamm-Horsfall glycoprotein in urine. *Biochem. J. 227*, 957-963.

36. ILYAS, A. A., LI, S. C., CHOU, D. K. H., LI, Y. T., JUNGALWALA, F. B., DALAKAS, M. C. & QUARLES, R. H. (1988): Gangliosides G_{M2}, IV^{4}GalNAcG$_{M1b}$, and IV^{4}GalNAcG$_{D1a}$ as antigens for monoclonal immunoglobulin M in neuropathy associated with gammopathy. *J. Biol. Chem. 263*, 4369-4373.

37. LOPEZ, M., GERBAL, A., BONY, V. & SALMON, C. (1975): Cad antigen: comparative studies of 50 samples. *Vox Sang. 28*, 305-313.

38. MACVIE, S. I., MORTON, J. A. & PICKLES, M. M. (1967): The reactions and inheritance of a new blood group antigen Sd^a. *Vox Sang. 13*, 485-492.

39. MALAGOLINI, N., DALL'OLIO, F., DI STEFANO, G., MINNI, F., MARRANO, D. & SERAFINI-CESSI, F. (1989): Expression of UDP-GalNAc:NeuAc-α-2,3Gal-β-R β-1,4(GalNAc to Gal) N-acetylgalactosaminyl-transferase involved in the synthesis of Sd^a antigen in human large intestine and colorectal carcinomas. *Cancer Res. 49*, 6466-6470.

40. MALAGOLINI, N., DALL'OLIO, F., GUERRINI, S. & SERAFINI-CESSI, F. (1994): Identification and characterization of the Sd^a β1,4,N-acetylgalactosaminyltransferase from pig large intestine. *Glycocon. J. 11*, 89-95.

41. MALAGOLINI, N., DALL'OLIO, F. & SERAFINI-CESSI, F. (1991): UDP-GalNAc:NeuAcα2,3Galβ-R (GalNAc to Gal) β1,4-N-acetylgalactosaminyltransferase responsible for the Sd^a specificity in human colon carcinoma CaCo-2 cell line. *Biochem. Biophys. Res. Commun. 180*, 681-686.

42. MARSH, W. L., JOHNSON, C. L., ØYEN, R., NICHOLS, M. E., DINAPOLI, J., YOUNG, H., BRASSEL, J., CUSAMANO, I., BAZAZ, G. R., HABER, J. M. & WOLF, C. F. W. (1980): Anti-Sd^x: a "new" auto-agglutinin related to the Sd^a blood group. *Transfusion 20*, 1-8.

43. MORGAN, W. T. J., SOH, C. & WATKINS, W. M. (1979): Blood group Sd^a specificity as a possible genetic marker on Tamm and Horsfall urinary glycoprotein. In: *Glycoconjugates* (R. Schauer, P. Broer, E. Buddecke, M. F. Kramer, J. F. G. Vliegenthart, and H. Wiegandt, eds.), Thieme, Stuttgart, pp. 582-583.

44. MORTON, J. A., PICKLES, M. M. & TERRY, A. M. (1970): The Sd^a blood group antigens in tissues and body fluids. *Vox Sang. 19*, 472-482.

45. MORTON, J. A., PICKLES, M. M. & VAN HEGAN, R. I. (1988): The Sd^a antigen in the human kidney and colon. *Immunol. Invest. 17*, 217-224.

46. MORTON, J. A. & TERRY, A. M. (1970): The Sd^a blood group antigen. Biochemical properties of urinary Sd^a. *Vox Sang. 19*, 151-161.

47. MUCHMORE, A. V. & DECKER, J. M. (1985): Uromodulin: a unique 85-kilodalton immunosuppressive glycoprotein isolated from urine of pregnant women. *Science 229*, 479-481.

48. MUCHMORE, A. V. & DECKER, J. M. (1986): Uromodulin. An immunosuppressive 85-kilodalton glycoprotein isolated from human pregnancy urine is a high affinity ligand for recombinant interleukin 1α. *J. Biol. Chem. 261*, 13404-13407.

49. PARKKINEN, J., VIRKOLA, R. & KORHONEN, T. K. (1988): Identification of factors in human urine that inhibit the binding of Escherichia coli adhesins. *Infect. Immun. 56*, 2623-2630.

50. PENNICA, D., KOHR, W. J., KUANG, W. J., GLAISTER, D., AGGARWAL, B. B., CHEN, E. Y. & GOEDDEL, D. V. (1987): Identification of human uromodulin as the Tamm-Horsfall urinary glycoprotein.

Science 236, 83-88.

51. PICKLES, M. M. & MORTON, J. A. (1977): The Sd^a blood group. In: *Human Blood Groups* (J. F. Mohn, R. W. Plunkett, R. K. Cunningham, and R. M. Lambert, eds.). S. Karger, Basel, pp. 277-286.

52. PILLER, F., BLANCHARD, D., HUET, M. & CARTRON, J. P. (1986): Identification of a α-NeuAc-(2→3)-β-D-galactopyranosyl N-acetyl-β-D-galactosaminyltransferase in human kidney. *Carbohydr. Res. 149*, 171-184.

53. PILLER, F., CARTRON, J. P. & TUPPY, H. (1980): Increase of blood group A and loss of blood group Sd^a activity in the mucus from human neoplastic colon. *Blood Transfus. Immunohaematol. 23*, 599-611.

54. RACE, R. R. & SANGER, R. (1975): The Sid groups. In: *Blood Groups in Man*. Blackwell Scientific Publications, Oxford, pp. 395-405.

55. RENTON, P. H., HOWELL, P., IKIN, E. W., GILES, C. M. & GOLDSMITH, K. L. G. (1967): Anti-Sd^a, a new blood group antibody. *Vox Sang. 13*, 493-501.

56. RINDLER, M. J., NAIK, S. S., LI, N., HOOPS, T. C. & PERALDI, M. N. (1990): Uromodulin (Tamm-Horsfall glycoprotein / uromucoid) is a phosphatidylinositol-linked membrane protein. *J. Biol. Chem. 265*, 20784-20789.

57. SANGER, R., GAVIN, J., TIPPETT, P., TEESDALE, P. & ELDON, K. (1971): Plant agglutinin for another human blood group. *Lancet i*, 1130.

58. SERAFINI-CESSI, F. & DALL'OLIO, F. (1983): Guinea pig kidney β-N-acetylgalactosaminyltransferase towards Tamm-Horsfall glycoprotein. Requirement of sialic acid in the acceptor for transferase activity. *Biochem. J. 215*, 483-489.

59. SERAFINI-CESSI, F., DALL'OLIO, F. & MALAGOLINI, N. (1986): Characterization of N-acetyl-β-D-galactosaminyltransferase from guinea-pig kidney involved in the biosynthesis of Sd^a antigen associated with Tamm-Horsfall glycoprotein. *Carbohydr. Res. 151*, 65-76.

60. SERAFINI-CESSI, F., MALAGOLINI, N. & DALL'OLIO, F. (1984): A tetraantennary glycopeptide from human Tamm-Horsfall glycoprotein inhibits agglutination of desialylated erythrocytes induced by leucoagglutinin. *Biosci. Rep. 4*, 973-978.

61. SERAFINI-CESSI, F., MALAGOLINI, N. & DALL'OLIO, F. (1988): Characterization and partial purification of β-N-acetylgalactosaminyltransferase from urine of Sd(a+) individuals. *Arch. Biochem. Biophys. 266*, 573-582.

62. SERAFINI-CESSI, F., MALAGOLINI, N., GUERRINI, S. & TURRINI, I. (1995): A soluble form of Sd^a-β1,4-N-acetylgalactosaminyltransferase is released by differentiated human colon carcinoma CaCo-2 cells. *Glycoconj. J. 12*, 773-779.

63. SIKRI, K. L., FOSTER, C. L., BLOOMFIELD, F. J. & MARSHALL, R. D. (1979): Localization by immunofluorescence and by light- and electron-microscopic immunoperoxidase techniques of Tamm-Horsfall glycoprotein in adult hamster kidney. *Biochem. J. 181*, 525-532.

64. SMITH, P. L. & LOWE, J. B. (1994): Molecular cloning of a murine N-acetylgalactosamine transferase cDNA that determines expression of the T lymphocyte-specific CT oligosaccharide differentiation antigen. *J. Biol. Chem. 269*, 15162-15171.

65. SOH, C. P. C., DONALD, A. S. R., FEENEY, J., MORGAN, W. T. J. & WATKINS, W. M. (1989): Enzymic synthesis, chemical characterization and Sd^a activity of GalNAcβ1-4[NeuAcα2-3]Galβ1-4GlcNAc and GalNAcβ1-4[NeuAcα2-3]Galβ1-4Glc. *Glycoconj. J. 6*, 319-332.

66. SOH, C. P. C., MORGAN, W. T. J., WATKINS, W. M. & DONALD, A. S. R. (1980): The relationship between the N-acetylgalactosamine content and the blood-group Sd^a activity of Tamm and Horsfall urinary glycoprotein. *Biochem. Biophys. Res. Commun. 93*, 1132-1139.

67. SVENNERHOLM, L., MÅNSSON, J. E. & LI, Y. T. (1973): Isolation and structural determination of a novel ganglioside, a disialosylpentahexosylceramide from human brain. *J. Biol. Chem. 248*, 740-742.

68. TAKEYA, A., HOSOMI, O. & KOGURE, T. (1987): Identification and characterization of UDP-GalNAc: NeuAcα2-3Galβ1-4Glc(NAc) β1-4(GalNAc to Gal) N-acetylgalactosaminyltransferase in human blood plasma. *J. Biochem. 101*, 251-259.

69. TAMM, I. & HORSFALL, F. L. (1950): Characterization and separation of an inhibitor of viral

hemagglutination present in urine. *Proc. Soc. Exp. Biol. Med. 74*, 108-114.

70. Tᴀᴍᴍ, I. & Hᴏʀꜱꜰᴀʟʟ, F. L. (1952): A mucoprotein derived from human urine which reacts with influenza, mumps, and newcastle disease virus. *J. Exp. Med. 95*, 71-97.

71. Tᴏʟʟᴇꜰꜱᴇɴ, S. E. & Kᴏʀɴꜰᴇʟᴅ, R. (1984): The B_4 lectin from Vicia villosa seeds interacts with N-acetylgalactosamine residues on erythrocytes with blood group Cad specificity. *Biochem. Biophys. Res. Commun. 123*, 1099-1106.

72. Tᴏᴍᴀ, G., Bᴀᴛᴇꜱ, J. M. & Kᴜᴍᴀʀ, S. (1994): Uromodulin (Tamm-Horsfall Protein) is a leukocyte adhesion molecule. *Biochem. Biophys. Res. Commun. 200*, 275-282.

73. Uʜʟᴇɴʙʀᴜᴄᴋ, G. (1971): Diagnosis of the "Cad" blood group with agglutinins from snails and plants. *Z. Immun.-Forsch. 141*, 290-291.

74. Wɪʟʟɪᴀᴍꜱ, J., Mᴀʀꜱʜᴀʟʟ, R. D., ᴠᴀɴ Hᴀʟʙᴇᴇᴋ, H. & Vʟɪᴇɢᴇɴᴛʜᴀʀᴛ, J. F. G. (1984): Structural analysis of the carbohydrate moieties of human Tamm-Horsfall glycoprotein. *Carbohydr. Res. 134*, 141-155.

75. Yᴀᴍᴀɢᴜᴄʜɪ, H., Oᴋᴜʙᴏ, Y., Oɢᴀᴡᴀ, Y. & Tᴀɴᴀᴋᴀ, M. (1973): Japanese families with group O and B red cells agglutinable by Dolichos biflorus extracts. *Vox Sang. 25*, 361-369.

76. Yᴀᴛᴇꜱ, A. D., Dᴏɴᴀʟᴅ, A. S. R., Fᴇᴇɴᴇʏ, J. & Wᴀᴛᴋɪɴꜱ, W. M. (1987): Blood group Sd^a-active structures on erythrocytes. In: *Proceedings of the IXth Int. Symp. on Glycoconjugates, Lille 1987*, Abstract F 27.

12 Sialic Acid-Containing Receptors for Cold Agglutinins

Various cold agglutinins which react with sialic acid-dependent antigens of the red cell membrane have been described in human sera. Seven specificities have been defined to date: **Pr** [49], **Gd (Sia-lb1)** [48], **Sa** [47], **Fl (Sia-b1)** [30], **Lud** [31], **Vo (Sia-l1)** [46], and **Li** [34].

The antigens are characterised as follows (see *Table 12.1*):

- all these serological properties are destroyed by neuraminidase, thus confirming the role of neuraminic acid as the immunodominant component;
- the **Pr** antigen is completely destroyed by protease-treatment of the erythrocytes, whereas **Sa** and **Lud** are only partially degraded, and **Gd**, **Fl**, **Vo**, and **Li** are protease-resistant;
- the occurrence of **Pr**, **Gd**, and **Sa** is independent of age: the antigens **Fl** and **Lud** are fully expressed only on the red cells of adults, whereas **Vo** and **Li** are present primarily on erythrocytes of newborns and can also be found on i_{adult} cells.

12.1 Antisera

All antibodies with the above-mentioned specificities are without exception cold agglutinins. They are found in patients with lymphoreticular leukaemia or B-cell lymphoma; occasionally they occur together with anti-**I** and anti-**i** [20,50]. These antibodies are generally of the IgM type; in rare cases IgA [8,29,42,55] and IgG antibodies [4] have also been detected. Anti-**Pr** specific antibodies may also appear following infection with rubella [9,19] and varicella virus [12], and anti-**Fl** and anti-**Gd** specific antibodies may be induced by *Mycoplasma pneumoniae* infection [20].

The antibodies which cause chronic cold agglutinin disease are monoclonal and thus strictly monospecific. As expected, however, a distinct variability is observed among the individual antibodies, especially the Anti-**Pr** antibodies. Postinfectious cold agglutinins are commonly oligoclonal.

A hybridoma antibody with Anti-**Pr** specificity has been described [2].

Table 12.1: Characterisation of various cold agglutinins.

Antiserum	Reaction[a] of native erythrocytes			Reaction[b] of erythrocytes after treatment with	
	adult	foetal	i_{ad}	neuraminidase	protease
Anti-Pr$_{1-3}$	+	+	+	↓	↓
Anti-Pr$_a$	+	+	−	↓	
Anti-Gd	+	+	+	↓	−
Anti-Sa	+	+	+	↓	(↓)
Anti-Fl	+	(+)	(+)	↓	−
Anti-Lud	+	(+)	+	↓	(↓)
Anti-Vo	(+)	+	+	(↓)	−
Anti-Li	(+)	+	+	↓	−
Anti-I^F	+	(+)	(+)	−	−
Anti-I^D	+	−	−	↑	↑
Anti-I^T	(+)	+	−		
Anti-i	(+)	+	+	↑	↑

[a] **+** strong agglutination; **(+)** weak agglutination; − no reaction.
[b] − no effect; ↑ activated; ↓ inactivated; (↓) degree of inactivation depending on the enzyme used.
From Roelcke et al. [36,48] by permission of S. Karger AG, Basel.

12.2 Structures of Antigenic Determinants

The investigations of the specificity of the antibodies treated in this chapter show that they bind to carbohydrate units with terminal α2→3 N-acetylneuraminyl residues occurring in glycoproteins and glycolipids[1]. The antibodies of the various specificities and sub-specificities differ in their binding-topography, in the size of the antigenic site

---

[1] More recently anti-**Pr** antibodies have been found which bind to oligosaccharide chains with terminal α2→6 N-acetylneuraminic acid [41].

to be recognised, and in their degree of cross-reactivity. The binding specificity is thus greatly influenced by the environment of the epitope (i.e. the structure of the oligosaccharide chain), by the carrier molecule, as well as by the sterical conditions provided by the membrane architecture.

Pr

The **Pr** determinant [49] (originally described as **Sp**$_1$ [24] or **HD** [27]) is protease sensitive. Anti-**Pr** antibodies bind preferentially to the glycophorins of the erythrocyte membrane [3]. Cells of the rare **En(a–)** phenotype lacking in glycophorin A exhibit only weak **Pr** activity, and cells from *M*^{k}*M*^{k} individuals in which glycophorin A as well as glycophorin B are absent (see *Sect. 9.6)* are **Pr** negative [1]. Inhibition experiments with isolated sialo-glycoproteins and glycopeptides from normal and **MNS** deficient erythrocytes [5,17,40] show that the anti-**Pr** antibodies bind to the sialic acid-containing tetrasaccharides which are O-glycosidically linked to serine or threonine residues of the glycophorin A and B molecules (see *Fig. 9.3.a*). This assumption is corroborated by the discovery of an antibody which reacts at 4°C as anti-**Pr** and at 25□C as anti-**M** (= anti-**Pr**^{M} [38]); this 'compound-specific' antibody is neutralised by glycophorin A^M.

Three types of anti-**Pr** sera have been distinguished on the basis of their reaction with chemically modified erythrocyte membrane sialoglycoproteins: anti-**Pr**$_1$, anti-**Pr**$_2$, and anti-**Pr**$_3$ [6,23,39] (see *Table 12.2)*. The reactivity of isolated sialoglycoproteins towards anti-**Pr**$_1$ and anti-**Pr**$_3$ is abolished and that towards anti-**Pr**$_2$ is enhanced following treatment with periodate (periodate oxidation cleaves the side chain of neuraminic acid [54]). The **Pr**$_1$ and **Pr**$_2$ determinants are destroyed, whereas **Pr**$_3$ activity is increased by treating the sialoglycoproteins with carbodiimide. This reagent causes intra- and intermolecular coupling of the sialic acid residues to nucleophilic centres of the backbone protein via the carboxylic group [13]. Different activities towards erythrocyte glycoprotein fractions [40] suggest that the **Pr** subspecificities are independent antigen characters.

The anti-**Pr**$_1$ and anti-**Pr**$_3$ sera have been further differentiated into those which react only with human erythrocytes (anti-**Pr**$_{1h}$ and anti-**Pr**$_{3h}$) and those which are able to agglutinate canine red cells as well (anti-**Pr**$_{1d}$ and anti-**Pr**$_{3d}$) [28,44].

Anti-**Pr**$_2$ has also been shown to react with gangliosides and type 2 chain-based sialo-glycosphingolipids [56].

Cold auto-agglutinins which have been described as anti-**Pr**$_a$ [35,44] bind to a protease-sensitive erythrocyte membrane antigen which is not destroyed by neuraminidase. Serological tests with native and modified sialoglycoproteins have revealed that anti-**Pr**$_a$ antibodies differ markedly in their binding specificity – one antibody was inhibited by periodate- or carbodiimide-treated glycophorins, whereas

Table 12.2: Classification of anti-Pr sera.

Sialo-glycoprotein used	Reaction with anti-Pr$_1$	anti-Pr$_2$	anti-Pr$_3$
Native	+	+	+
Periodate-oxidised	−	++	−
Carbodiimide-treated	−	−	++

The reaction of the antisera with native and chemically modified sialoglycoproteins of the erythrocyte membrane: **+** : normal inhibition, **−** : inactivation, **++** : activation.
From Roelcke et al. [40], with permission of S. Karger AG, Basel.

another antibody showed no reaction whatsoever (periodate and carbodiimide are thought to affect sialic acid exclusively, as mentioned above). The exact antigenic specificity of the anti-**Pr$_a$** antibodies therefore calls for further investigations.

Gd (Sia-lb1)

The **Gd** determinant [48] is resistant to proteases. Thus, its localisation on erythrocyte membrane glycolipids has been proposed (→ '**G**lycolipid-**d**ependent'). Following this assumption, Kundu et al. [21] found that anti-**Gd** sera are inhibited by two erythrocyte membrane glycosphingolipids, namely sialoyl-neolactotetraosyl-ceramide and, more potently, sialoyl-lacto-N-nor-hexaosylceramide (*Fig. 12.1*). In a subsequent investigation by Uemura et al. [56] erythrocyte membrane glycolipids were separated on thin-layer chromatograms and tested for **Gd** activity. Chromatogram binding assays confirmed the above results and showed that anti-**Gd** antibodies bind to a series of linear and branched, long-chained glycosphingolipids of the neolacto-series containing the **NeuAcα2→3Gal** sequence. More recently it has been found that anti-**Gd** antibodies bind to both sialoyl-**Le^x** and sialoyl-**Le^a** determinants present on nucleated cells, cancer cells, and cancer-related mucins [7].

Based on inhibition tests, two types of anti-**Gd** antisera have been defined – anti-**Gd1** (= anti-**Sia-lb1**), which binds preferentially to the terminal α-sialic acid residue, and anti-**Gd2** (= anti-**Sia-lb2**), which, in addition, requires a subterminal galactose [36].

Anti-**Gd** sera show properties strongly similar to those of the so-called 'anti-**p**' sera, which bind to sialic acid-containing glycosphingolipids with linear oligosaccharide

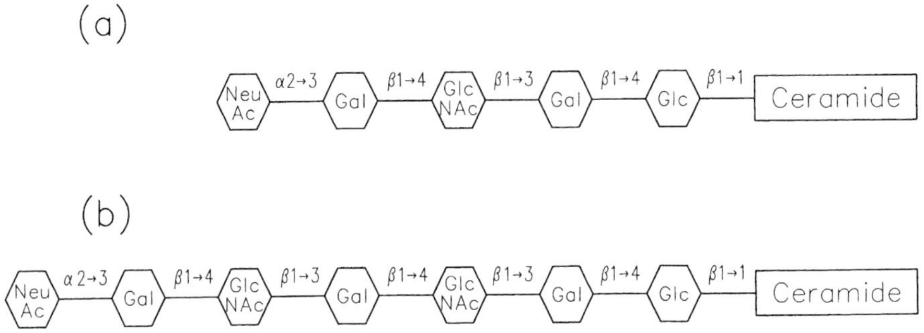

(a)

(b)

Figure 12.1: Chemical structures of sialoyl-neolactotetraosylceramide (a) and sialoyl-lacto-N-*nor*-hexaosylceramide (b).

chains [53] (see *Sect. 8.1)*. Not all anti-**Gd** sera react with **p** erythrocytes, however, as Roelcke et al. [33] have noticed.

Sa

Unlike the above antigens, the antigen **Sa** [47] is only partially inactivated by proteases. It can thus be assumed that the **Sa** determinant is located on glycoproteins and glycolipids of the erythrocyte membrane. Preliminary experiments revealed that anti-**Sa** sera are weakly inhibited by $\alpha 2,3$-sialoyllactose [37] and by the glycophorins of the erythrocyte membrane [3]. Subsequent investigations were able to locate the **Sa** epitope on the internal part of glycophorin A in the region of residues 40–61 [3]. In chromatogram binding tests anti-**Sa** sera showed a binding pattern towards glycosphingolipids similar to that of anti-**Gd** sera [56]. In haemagglutination inhibition tests, however, neither sialoyl-neolactotetraosylceramide nor sialoyl-lacto-N-*nor*-hexaosyl-ceramide proved to be inhibitors [21].

Fl (Sia-b1)

In contrast to anti-**Pr**, anti-**Gd**, and anti-**Sa** sera, anti-**Fl** [30,50] reacts more strongly with **I** erythrocytes from adults than with **i** cells from cord blood or adult **i** individuals. The preference of the antibody for the **I** structure and the protease resistance of the antigen suggests that anti-**Fl** binds mainly to glycolipids with branched oligosaccharide chains.

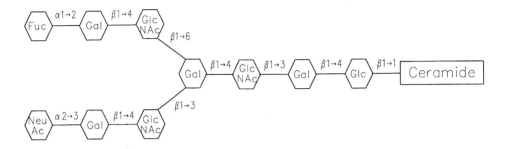

Figure 12.2: Blood group FI specific glycosphingolipid isolated from human erythrocyte membranes.
According to Kannagi et al. [17].

An **FI** reactive glycosphingolipid has been isolated from human erythrocyte membranes by Kannagi et al. [16]. It was identified as a derivative of lacto-N-*iso*-octaosylceramide (described earlier by Watanabe et al. [57]) which contains a blood group **H** determinant α1→2 fucose at the C-6 branch and a terminal α2→3 N-acetylneuraminic acid at the C-3 branch *(Fig. 12.2)*; **A** and **B** determinants at the C-6 branch did not influence the binding capacity. **Bombay (O_h)** erythrocytes show a strongly reduced activity with anti-**FI** [45], thus providing additional evidence for the view that the antibody recognises an epitope containing the α-sialoyl group at one chain-end and the α-fucosyl group at the other. In a subsequent paper, Uemura et al. [56] verified the preferential binding of anti-**FI** to branched glycosphingolipids of the erythrocyte membrane; a certain affinity of the antibody to glycolipids of the neolacto-series with longer linear chains was, however, also noted.

More recently an **I/FI** active glycoprotein has been isolated from human red cell membranes [43].

Several sera of patients suffering from *Mycoplasma pneumoniae* infections contained mixtures of anti-I and anti-**FI** [18,20]. These anti-**FI** antibodies, however, had a slightly different specificity – their epitope is partially destroyed by endo-β-galactosidase and the antibodies show a strong reaction with **Bombay (O_h)** erythrocytes. Further investigations indicated that **FI** antigens act as receptors for *M. pneumoniae* [11]. The binding of the pathogen obviously renders the host receptor immunogenic, thereby inducing the production of cold agglutinins towards the basic structure (anti-I) and its sialyl derivative (anti-**FI**).

Lud

The **Lud** antigen [31] is only partially released from human erythrocytes by protease treatment. Anti-**Lud** sera agglutinate preferentially **I** and **i** erythrocytes from adults and are only weakly reactive with cord cells.

Recently, Kajii et al. [15] have been able to localise the **Lud** determinant on a 43 kDa protein of the erythrocyte membrane.

Vo (Sia-l1)

Only one serum with anti-**Vo** specificity has been found to date [46]. It agglutinates preferentially erythrocytes from cord blood and from **i** adults. Agglutination is markedly enhanced by protease treatment of the red cells. On native erythrocytes the **Vo** determinant is only weakly sensitive to neuraminidase; the antigen, however, is totally destroyed when protease-treated cells are incubated with this enzyme.

In accordance with the protease resistance of the antigen, its susceptibility to endo-β-galactosidase [43], and the specific reaction of the antibody with **i** erythrocytes, it is generally assumed that glycolipids with sialylated linear type-2 oligosaccharide chains act as receptors for the anti-**Vo** antibody.

Li

The characteristics of the **Li** antigen [34] are nearly identical to those of **Vo**; in contrast to **Vo**, however, **Li** is completely destroyed by neuraminidase on intact cells and is not degraded *in situ* by endo-β-galactosidase [43]. The structure of the **Li** determinant has not yet been established.

12.3 Occurrence of Pr, Gd, and Sa Antigens on Other Cell Types

The antigens **Pr**, **Gd**, and **Sa** are not confined to erythrocytes; they have also been found on B and T lymphocytes and various phagocytic cells [25,26]; the epidermis shows strong Pr_1 activity, and thrombocytes also react distinctly with anti-**Gd** sera (Conte et al. and Riesen, respectively, cited in [32]). All three specificities can be detected in the human kidney; they are, however, localised in different regions of the organ [22] *(Fig. 12.3)*.

Pr and **Gd** antigens have also been found in kidney, liver, stomach, pancreas, lung, and brain of various animals as the dog, the guinea pig, the rat, and the rabbit [51].

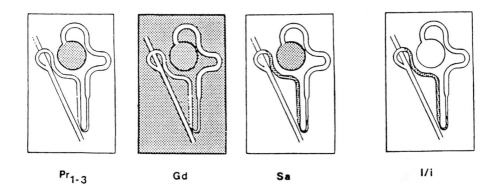

Figure 12.3: Schematic distribution pattern of the antigens Pr$_{1-3}$, Gd, and Sa in human kidneys.
In comparison: distribution of I/i antigens.
Reproduced from Roelcke [33] by permission of Librairie Arnette, Paris.

12.4 Other Receptors for Cold Agglutinins

Ju

The cold autoantibody anti-**Ju** [10] agglutinates erythrocytes from adults as well as cord cells; its reactivity is somewhat reduced by treatment of the cells with protease, and is significantly reduced by treatment with neuraminidase.

Me

The anti-**Me** cold agglutinin described by Salama et al. [52] agglutinates both cord erythrocytes and red blood cells from adults; its reactivity is enhanced by treatment of the cells with protease and neuraminidase. Its haemagglutinating activity against erythrocytes is markedly increased in the presence of preheated human milk.

Om

The serological properties of the anti-**Om** cold agglutinin [14] closely resemble anti-**Me**. It reacted strongly with papainised and neuraminidase-treated erythrocytes from newborn and adults. Unlike anti-**Me**, however, this antibody is markedly inhibited by galactose and by human milk.

References

1. Anstee, D. J. (1980): Blood group MNSs-active sialoglycoproteins of the human erythrocyte membrane. In: *Immunobiology of the Erythrocyte* (S.G. Sandler, J. Nusbacher, and M. S. Schanfield, eds.). Alan R. Liss Inc., New York, pp. 67-98.
2. Anstee, D. J. & Edwards, P. A. W. (1982): Monoclonal antibodies to human erythrocytes. *Eur. J. Immunol. 12*, 228-232.
3. Dahr, W., Lichthardt, D. & Roelcke, D. (1981): Studies on the receptor sites of the monoclonal anti-Pr and -Sa cold agglutinins. *Protides Biol. Fluids 29*, 365-368.
4. Dellagi, K., Brouet, J. C., Schenmetzler, C. & Praloran, V. (1981): Chronic hemolytic anemia due to a monoclonal IgG cold agglutinin with anti-Pr specificity. *Blood 57*, 189-191.
5. Ebert, W., Fey, J., Gärtner, C., Geisen, H. P., Rautenberg, U., Roelcke, D. & Weicker, H. (1979): Isolation and partial characterization of the Pr autoantigen determinants. *Molec. Immunol. 16*, 413-419.
6. Ebert, W., Metz, J. & Roelcke, D. (1972): Modifications of N-acetylneuraminic acid and their influence on the antigen activity of erythrocyte glycoproteins. *Eur. J. Biochem. 27*, 470-472.
7. Gallart, T., Roelcke, D., Blay, M., Pereira, A., Martinez, A., Massó, O., Viñas, O., Cid, M., Esparza, J., Molina, R. & Barceló, J. (1997): Anti-Sia-lb (anti-Gd) cold agglutinins bind the domain NeuNAcα2-3Gal in sialyl Lewis[x], sialyl Lewis[a], and related carbohydrates on nucleated cells and in soluble cancer-associated mucins. *Blood 90*, 1576-1587.
8. Garratty, G., Petz, L. D., Brodsky, I. & Fudenberg, H. H. (1973): An IgA high-titer cold agglutinin with an unusual blood group specificity within the Pr complex. *Vox Sang. 25*, 32-38.
9. Geisen, H. P., Roelcke, D., Rhen, K. & Konrad, G. (1975): Hochtitrige Kälteagglutinine der Spezifität Anti-Pr nach Rötelinfektion. *Klin. Wschr. 53*, 767-772.
10. Göttsche, B., Salama, A. & Mueller-Eckhardt, C. (1990): Autoimmune hemolytic anemia caused by a cold agglutinin with a new specificity (anti-Ju). *Transfus. 30*, 261-262.
11. Hengge, U. R., Kirschfink, M., König, A. L., Nicklas, W. & Roelcke, D. (1992): Characterization of I/FI glycoprotein as a receptor for Mycoplasma pneumoniae. *Infect. Immun. 60*, 79-83.
12. Herron, B., Roelcke, D., Orson, G., Myint, H. & Boulton, F. E. (1993): Cold autoagglutinins with anti-Pr specificity associated with fresh Varicella infection. *Vox Sang. 65*, 239-242.
13. Hoare, D. G. & Koshland, D. E. (1967): A method for the quantitative modification and estimation of carboxylic acid groups in proteins. *J. Biol. Chem. 242*, 2447-2453.
14. Kajii, E. & Ikemoto, S. (1989): A cold agglutinin: Om. *Vox Sang. 56*, 104-106.
15. Kajii, E., Ikemoto, S. & Miura, Y. (1988): Localization of the Lud antigen by immunoblotting. *Vox Sang. 54*, 248.
16. Kannagi, R., Roelcke, D., Peterson, K. A., Okada, Y., Levery, S. B. & Hakomori, S. I. (1983): Characterization of epitope (determinant) structure in a developmentally regulated glycolipid antigen defined by a cold agglutinin FI, Recognition of α-sialosyl and α-L-fucosyl groups in a branched structure. *Carbohydr. Res. 120*, 143-157.
17. Kewitz, S., Groß, H. J., Kosa, R. & Roelcke, D. (1995): Anti-Pr cold agglutinins recognize immunodominant α2,3- or α2,6-sialyl groups on glycophorins. *Glycocon. J. 12*, 714-720.
18. König, A. L., Kather, H. & Roelcke, D. (1984): Autoimmune hemolytic anemia by coexisting anti-I and anti-FI cold agglutinins. *Blut 49*, 363-368.
19. König, A. L., Keller, H. E., Braun, R. W. & Roelcke, D. (1992): Cold agglutinins of anti-Pr specificity in Rubella embryopathy. *Ann. Hematol. 64*, 277-280.
20. König, A. L., Kreft, H., Hengge, U., Braun, R. W. & Roelcke, D. (1988): Coexisting anti-I and anti-FI/Gd cold agglutinins in infections by Mycoplasma pneumoniae. *Vox Sang. 55*, 176-180.
21. Kundu, S. K., Marcus, D. M. & Roelcke, D. (1982): Glycosphingolipid receptors for anti-Gd and anti-p cold agglutinins. *Immunol. Lett. 4*, 263-267.
22. Lenhard, V., Seelig, H. P., Geisen, H. P. & Roelcke, D. (1978): Identification of I, i, Pr$_{1-3}$ and Gd antigens in the human kidney: possible relevance to hyperacute graft rejection induced by cold agglutinins. *Clin. Exp. Immunol. 33*, 276-282.

23. LISOWSKA, E. & ROELCKE, D. (1973): Differentiation of anti-Pr_1 and anti-Pr_2 - sera with periodate-oxidized erythrocyte glycoproteins. *Blut 26*, 339-341.
24. MARSH, W. L. & JENKINS, W. J. (1968): Anti-Sp_1: the recognition of a new cold auto-antibody. *Vox Sang. 15*, 177-186.
25. PRUZANSKI, W., ARMSTRONG, M. & ROELCKE, D. (1981): New antigenic determinant (Sa) on human lymphocytes and phagocytes. *Blut 43*, 307-313.
26. PRUZANSKI, W., ROELCKE, D., ARMSTRONG, M. & MANLY, M. S. (1980): Pr and Gd antigens on human B and T lymphocytes and phagocytes. *Clin. Immunol. Immunopathol. 15*, 631-641.
27. ROELCKE, D. (1969): A new serological specificity in cold antibodies of high titre: anti-HD. *Vox Sang. 16*, 76-79.
28. Roelcke, D. (1973): Serological studies on the Pr_1/Pr_2 antigens using dog erythrocytes. Differentiation of Pr_2 from Pr_1 and detection of a Pr_1 heterogeneity: Pr_{1h} / Pr_{1d}. *Vox Sang. 24*, 354-361.
29. ROELCKE, D. (1973): Specificity of IgA cold agglutinins: anti-Pr_1. *Eur. J. Immunol. 3*, 206-212.
30. ROELCKE, D. (1981): A further cold agglutinin, Fl, recognizing a N-acetylneuraminic acid-determined antigen. *Vox Sang. 41*, 98-101.
31. ROELCKE, D. (1981): The Lud cold agglutinin: a further antibody recognizing N-acetylneuraminic acid-determined antigens not fully expressed at birth. *Vox Sang. 41*, 316-318.
32. ROELCKE, D. (1981): Pr and Gd antigens. *Blood Transfus. Immunohaematol. 24*, 27-36.
33. ROELCKE, D. (1984): Reaction of anti-Gd, anti-Fl, and anti-Sa cold agglutinins with p erythrocytes. *Vox Sang. 46*, 161-164.
34. ROELCKE, D. (1985): Li cold agglutinin: a further antibody recognizing sialic acid-dependent antigens fully expressed on newborn erythrocytes. *Vox Sang. 48*, 181-183.
35. ROELCKE, D., ANSTEE, D. J., JUNGFER, H., NUTZENADEL, W. & WEBB, A. J. (1971): IgG-type cold agglutinins in children and corresponding antigens. Detection of a new Pr antigen: Pr_a. *Vox Sang. 20*, 218-229.
36. ROELCKE, D. & BROSSMER, R. (1984): Different fine specificities of human monoclonal anti-Gd cold agglutinins. *Protides Biol. Fluids 31*, 1075-1082.
37. ROELCKE, D., BROSSMER, R. & EBERT, W. (1982): Anti-P1, -Gd, and related cold agglutinins: human monoclonal antibodies against neuraminyl groups. *Protides Biol. Fluids 29*, 619-622.
38. ROELCKE, D., DAHR, W. & KALDEN, J. R. (1986): A human monoclonal IgMκ cold agglutinin recognizing oligosaccharides with immunodominant sialyl groups preferentially at the blood group M-specific peptide backbone of glycophorins: anti-Pr^M. *Vox Sang. 51*, 207-211.
39. ROELCKE, D., EBERT, W. & GEISEN, H. P. (1976): Anti-Pr_3: serological and immunochemical identification of a new anti-Pr subspecificity. *Vox Sang. 30*, 122-133.
40. ROELCKE, D., EBERT, W., METZ, J. & WEICKER, H. (1971): I-, MN-, and Pr_1/Pr_2-activity of human erythrocyte glycoprotein fractions obtained by ficin treatment. *Vox Sang. 21*, 352-361.
41. ROELCKE, D., HACK, H., KREFT, H. & GROSS, H. J. (1998): α2,3-specific desialylation of human red cells: effect on the autoantigens of the Pr, Sa and Sia-l1, -b1, -lb1 series. *Vox Sang. 74*, 109-112.
42. ROELCKE, D., HACK, H., KREFT, H., MACDONALD, B., PEREIRA, A. & HABIBI, B. (1993): IgA cold agglutinins recognize Pr and Sa antigens expressed on glycophorins. *Transfus. 33*, 472-475.
43. ROELCKE, D., HENGGE, U. & KIRSCHFINK, M. (1990): Neolacto (type-2 chain) sialoautoantigens recognized by human cold agglutinins. *Vox Sang. 59*, 235-239.
44. ROELCKE, D. & KREFT, H. (1984): Characterization of various anti-Pr cold agglutinins. *Transfus. 24*, 210-213.
45. ROELCKE, D., KREFT, H., NORTHOFF, H. & GALLASCH, E. (1991): Sia-b1 and I antigens recognized by Mycoplasma pneumoniae-induced cold agglutinins. *Transfus. 31*, 627-631.
46. ROELCKE, D., KREFT, H. & PFISTER, A. M. (1984): Cold agglutinin Vo. An IgMλ monoclonal human antibody recognizing a sialic acid determined antigen fully expressed on newborn erythrocytes. *Vox Sang. 47*, 236-241.
47. ROELCKE, D., PRUZANSKI, W., EBERT, W., RÖMER, W., FISCHER, E., LENHARD, V. & RAUTERBERG, E. (1980): A new human monoclonal cold agglutinin Sa recognizing terminal N-acetylneuraminyl

groups on the cell surface. *Blood 55*, 677-681.

48. ROELCKE, D., RIESEN, W., GEISEN, H. P. & EBERT, W. (1977): Serological identification of the new cold agglutinin specificity anti-Gd. *Vox Sang. 33*, 304-306.

49. ROELCKE, D. & UHLENBRUCK, G. (1970): Letter to the editor. *Vox Sang. 18*, 478-479.

50. ROELCKE, D. & WEBER, M. T. (1984): Simultaneous occurrence of anti-Fl and anti-I cold agglutinins in a patient's serum. *Vox Sang. 47*, 122-124.

51. RÖMER, W., SEELIG, H. P., LENHARD, V. & ROELCKE, D. (1979): The distribution of I/i, Pr, and Gd antigens in mammalian tissues. *Invest. Cell Pathol. 2*, 157-165.

52. SALAMA, A., PRALLE, H. & MUELLER-ECKHARDT, C. (1985): A new red blood cell cold autoantibody (anti-Me). *Vox Sang. 49*, 277-284.

53. SCHWARTING, G. A., MARCUS, D. M. & METAXAS, M. (1977): Identification of sialosylparagloboside as an erythrocyte receptor for an 'anti-p' antibody. *Vox Sang. 32*, 257-261.

54. SUTTAJIT, M. & WINZLER, R. J. (1971): Effect of modification of N-acetylneuraminic acid on the binding of glycoproteins to influenza virus and on susceptibility to cleavage by neuraminidase. *J. Biol. Chem. 246*, 3398-3404.

55. TONTHAT, H., ROCHANT, H., HENRY, A., LEPORRIER, M. & DREYFUS, B. (1976): A new case of monoclonal IgA kappa cold agglutinin with anti-Pr₁d specificity in a patient with persistent HB antigen cirrosis. *Vox Sang. 30*, 464-468.

56. UEMURA, K. I., ROELCKE, D., NAGAI, Y. & FEIZI, T. (1984): The reactivities of human erythrocyte autoantibodies anti-Pr₂, anti-Gd, Fl, and Sa with gangliosides in a chromatogram binding assay. *Biochem. J. 219*, 865-874.

57. WATANABE, K., POWELL, M. & HAKOMORI, S. I. (1978): Isolation and characterization of a novel fucoganglioside of human erythrocyte membranes. *J. Biol. Chem. 253*, 8962-8967.

13 Rh System

In 1939, Levine and Stetson [160] first assumed that serological incompatibility between mother and child was the cause of erythroblastosis fetalis, also known as haemolytic disease of the newborn (HDN). The authors found that a maternal antibody entering the foetal circulation led to destruction of fetal erythrocytes. One year later, Wiener and Peters [277] were able to identify the same antibody in the serum of individuals showing transfusion reactions after **ABO**-compatible blood transfusions. In 1941, Levine et al. [156] reported that the antibody responsible for this disease had the same specificity as the anti-**Rhesus** antibody obtained by Landsteiner and Wiener [147,148] after injecting rabbits and guinea pigs with erythrocytes of rhesus monkeys.

However, in the course of further investigations it became obvious that the human antibodies responsible for haemolytic disease of the newborn, and the 'anti-**Rhesus**' antibody produced in animals by Landsteiner and Wiener recognised different antigens. Therefore, the term anti-**Rh** was chosen to designate the clinically significant human alloantibody, whereas the heteroantibody was renamed anti-'**LW**' in honor of its discoverers, Landsteiner and Wiener [205] (see *Chap. 14).*

Though 45 different **Rh** characters have been described thus far, very few of them are actually of clinical importance. The major **Rh** antigen is **RhD**; it is highly immunogenic in humans and is therefore the principal antigen involved in immune haemolytic anaemias. According to the presence or absence of the **RhD** antigen, human erythrocytes are commonly subdivided into **Rh**-positive and **Rh**-negative cells.

In a few cases the **Rhc** and **RhE** antigens were found responsible for the haemolytic disease of the newborn and various transfusion reactions. All other antigens are either extremely rare or only weakly immunogenic; though only of minor clinical importance they are, nevertheless, of great interest in family studies and anthropological investigations.

The **Rh** antigens are found only in humans and a few non-human primates [187,188,276]. The blood group active material is exclusively membrane-bound and confined to erythrocytes and their precursor cells [80], that is to say, the **Rh** antigens are detected on erythroid burst-forming units (BFU-E), and their expression increases during erythroid differentiation and erythrocyte ageing [17,80,233]. All **Rh** determinants are fully developed at birth [128].

Table 13.1: The Rh antigens.

Antigen term acc. to Fisher & Race	Wiener	Numerical[1]	Frequency (%) in Europids	References	Antigen term acc. to Fisher & Race	Wiener	Numerical[1]	Frequency (%) in Europids	References
D	Rh_o	RH1	85	[147,160]	RH ('total Rh')		RH29	100	[4]
C	rh'	RH2	70	[274]	D^{Cor}(Go^a, Gonzales)		RH30	<1	[231]
E	rh''	RH3	30	[209,278]		hr^B	RH31	98	[55]
c	hr'	RH4	80	[157]			RH32	<1	[90]
e	hr''	RH5	98	[189]			RH33	<1	[91]]
ce (f)	hr	RH6	64	[227]	Bastiaan	Hr^B	RH34	100	[70]
Ce	rh_i	RH7	70	[217]			RH35	<1	[61]
C^w	rh^{w1}	RH8	2	[36]	Be^a (Berrens)		RH36	<1	[129]
C^x	rh^x	RH9	<1	[240]	Evans		RH37	<1	[163]
V (ce^s)	hr^v	RH10	(2)	[73]	C-like		RH39	>99	[246]
E^w	rh^{w2}	RH11	<1	[100]	Tar (Targett)		RH40	<1	[58]
G	rh^G	RH12	85	[3]	Ce-like,		RH41	70	[101]
	Rh^A	RH13 (3)	85	[260,275]	Ce^s	hr^H-like	RH42	(2)	[72]
	Rh^B	RH14 (3)	85	[261]	Crawford		RH43	<1	[150]
	Rh^C	RH15 (3)	85	[259]	Nou		RH44		[65]
	Rh^D	RH16 (3)	85	[225]	Riv		RH45	<1	[170,203]
	Hr$_o$	RH17	100	[206]	Sec		RH46		[177]
	Hr (Hr^s)	RH18	100	[229]	Dav		RH47		[168]
	hr^s	RH19	98	[229]	JAL		RH48	<1	[236]
VS (e^s)		RH20	(2)	[226]	STEM		RH49	<1	[92]
C^G		RH21	70	[159]	FPTT		RH50	<1	
CE		RH22	<1	[74]	MAR		RH51	>99	
D^w (Wiel)		RH23 (3)	<1	[57]	BARC		RH52		
E^T		RH24 (3)	30	[265]	**Antigens serologically related to the Rh System:**				
c-like		RH26	80	[119]	HOFM		700050	<1	[105]
cE		RH27	30	[138]	LOCR		700053	<1	[59]
	hr^H	RH28	(2)	[230]	Ol^a (Oldeide)			<1	[142]

According to Issitt [128] by permission of the author; additional data are taken from Lewis et al. [161] and Daniels et al. [67,69].
(1) The six-digit identification numbers (see *Chapter 1*) are 004 001 - 004 052, (2) rare among Europids, however fairly frequent in Negrids (20 - 25%),
(3) these antigens have been declared obsolete as the reagents essential for their definition are no longer available.

Any discussion of the vast number of investigations on **Rh** serology would be beyond the scope of this monograph, which is dedicated to the chemical and biochemical aspects of the **Rh** antigens. For more extensive information on the serological investigations of the **Rh** system the reader should refer to the detailed monographs published by Issitt [127] and by the Americal Association of Blood Banks [262] as well as to Race and Sanger's 'Blood Groups in Man' [205] and Issitt's 'Applied Blood Group Serology' [128].

13.1 Genetics

The **Rh** system is one of the most complex blood group systems in humans, comprising as it does the 50 different antigenic characters thus far defined (see *Table 13.1*). The **Rh** antigens are encoded by a complex *RH* locus which has been located on the short arm of chromosome 1 in the region between 1p34.3 and 1p36.13 [49,174,179].

There are two classical models to explain the transmission of the **Rh** characters:

– according to the theory advanced by Fisher and Race [205] the **Rh** antigens are transmitted *en bloc* by a set of three closely linked genes, *Cc*, *Dd* and *Ee*, which encode three independent proteins. With the exception of *d*, which is considered amorph, all alleles of the *RH* gene complex code for distinct antigenic characters. The occurrence of rare **Rh** antigens is explained by a variety of further variants of these genes;
– according to another theory proposed by Wiener [128], the expression of all **Rh** characters is controlled by a single multiallelic gene which encodes a single protein carrying a variable number of **Rh** epitopes.

Molecular biological investigations, however, have shown that the *RH* locus of the human genome consists of two closely linked and highly homologous structural genes, *RHD* and *RHCE*, which transmit the **RhD** and **RhCcEe** characters respectively ([60], see also [252]). Based on serological data a *D-CE* order of the *RH* gene loci on the chromosome has been proposed [205]. This assumption has recently been confirmed by molecular biological studies [198].

Investigations on the *RHCE* gene structure have revealed that the coding region is composed of 10 exons spread over 75 kb of DNA [46]. The *RHD* gene has not yet been investigated, but preliminary studies (see [190]) suggest an organisation similar to that of the *RHCE* gene. The expression of the genes as examined by Northern blot analysis is restricted to erythroid tissues and cell lines [45].

Serological studies on the **Rh** blood groups in anthropoid apes [272,276] suggested that the *RH* gene complex originated from a single 'ancestral' gene which encoded the **c** antigen. This primordial gene duplicated and differentiated during evolution, and subsequently formed the human *RH* locus (see [273]). The **Ee** antigens which are not found in anthropoid apes presumably arose from mutations within the *Cc* gene.

More recent investigations detected **Rh**-related sequences in the genome and **Rh**-like antigens in the erythrocytes of monkeys and many non-primate mammalians [6,194,223,270,272].

13.2. Rh Antisera

The antibodies towards the different **Rh** antigens are almost exclusively immune antibodies of the IgG class which had been induced by **Rh**-incompatible blood transfusions or pregnancies. Most of them are incomplete antibodies which agglutinate erythrocytes only in the presence of 'supplement' (usually serum albumin), after enzyme treatment of the erythrocytes, or by using Coombs technique. Details on origin and specificity of the different anti-**Rh** sera are found in the monographs published by Race and Sanger [205] and Issitt [128].

More recently a number of **Rh** specific hybridoma antibodies have been prepared, e.g. anti-**D** [33,83,144,171,210,249], anti-**C** [248], anti-**c** [210,248], anti-**E** [248], anti-**e** [84,248], anti-**G** [32,82,248], anti-**C**ʷ [251], and anti-**Rh17**, anti-**Rh29**, and anti-**Rh46** [218] (see also [254]). Since the specificity is in many cases equal to that of the human **Rh** antisera some of these antibodies are already in use in routine blood group testing (see [250]), or represent promising agents in immunoprophylaxis, where they can be used to prevent haemolytic disease of the newborn [143].

13.3 Chemical Properties of Rh Antigens

Investigations on the chemical nature of **Rh** substances revealed that the **Rh** antigens are carried by integral membrane proteins which penetrate the erythrocyte membrane and are associated with the membrane skeleton [87,201,214]. In SDS-polyacrylmide gelelectrophoresis the **Rh** proteins show an apparent molecular mass of 30–32 kDa [1,86,186]. They are not phosphorylated or glycosylated; *in situ* (i.e. in the membrane of intact erythrocytes) they resist proteolytic degradation [86]. The finding that both **RhD** and **Rhc** activity are destroyed by N-etylmaleinimide, p-chloromercuribenzoate, and 5,5'-dithiobis-(2-nitrobenzoic acid) suggests that thiol groups are required for the expression of these **Rh** antigens [93,94,98,228,244]. Recent findings also show that cysteine residues located adjacent to the cytoplasmic leaflet of the lipid bilayer are acylated by fatty acids (generally palmitic acid [71,102]).

The **Rh** proteins are highly hydrophobic and, *in situ*, form fairly strong aggregates with other hydrophobic membrane proteins [102]. Various findings show a close association with band-3 protein, glycophorin B, **LW**, **Duffy** protein, and the so-called 'Rh-related glycoproteins', **RhAG** (formerly termed **Rh50**) and **CD47** (see *Sect. 13.5)*.

The serological activity of the **Rh** antigens in the intact erythrocyte membrane is highly influenced by lipid. Green [95,96] was able to show that erythrocyte membranes lose their **Rh** activity upon treatment with n-butanol; after incubation of the treated stroma with the lipid extract the activity was fully regenerated. Lipid is obviously necessary for the maintenance of optimum sterical conformation and suitable orientation of the molecule within the erythrocyte membrane. Further studies revealed that this effect is mainly caused by lecithin [97]. The finding that erythrocytes treated with phospholipase A_2, an enzyme which cleaves unsaturated fatty acids from carbon 2 of the glycerol residue destroying mainly lecithin, phosphatidyl-ethanolamine, and phosphatidyl-serine in isolated membranes, significantly reduced **c**, **D**, and **e** activity of the cells [99,123,202] further corroborated this assumption[1].

The formation of these protein – lipid aggregates is essential for biological activity of the membrane. Lack of **Rh** polypeptides (see **Rh$_{null}$** syndrome) causes anomalies in ion transport, as well as the changes in lipid organisation of the membrane mentioned above. Association with **Rh** substances obviously has a significant influence on the conformation of certain membrane proteins or their position within the membrane. Moreover, lack or modification of **Rh** substances evidently inhibits a set of membrane components from being incorporated into the membrane.

13.4 Number of Rh Sites

The density of the antigens **D**, **E**, and **e** amounts to ~20,000 determinants per cell, that of **C** to ~40,000; in most cases a significant gene-dose effect is observed (see *Table 13.2)*.

---

[1] **Rh** activity of the erythrocytes is greatly influenced by membrane fluidity [18,31,232]: a high cholesterol content (cholesterol to phospholipid ratio 1.55) increases membrane viscosity and enhances **RhD** antigenic reactivity, whereas a low cholesterol content (cholesterol to phospholipid ratio 0.55) decreases membrane viscosity and reduces **RhD** activity.

Table 13.2: Number of Rh sites per erythrocyte.

Haplotype	D-sites	C-sites	c-sites	E-sites	e-sites
DD	22,900	–	–	–	–
Dd	14,600	–	–	–	–
CC	–	51,500	–	–	–
Cc	–	31,600	45,000	–	–
cc	–	–	65,000	–	–
EE	–	–	–	25,500	–
Ee	–	–	–	n.t.	14,000
ee	–	–	–	–	21,300
D--	156,000	–	–	–	–

Average values compiled from several publications [28,38,121,122,180,215]

The antigen density shows also great variations in **Rh** variant phenotypes. For example: erythrocytes of the rare **D--/D--** phenotype contain 110,000–202,000 **D** sites [121], whereas in some 'partial **D**' phenotypes the density drops to less than 2000 and to 500–1000 in **D^u** cells [21]. It can be assumed that these variations are mainly caused by changes in **Rh** protein conformation due to single residue substitution or exon exchange. In other cases the reduction of antigen site density is caused by 'weakly active' *RH* alleles (e.g. **D^{el}** or **D^u**).

13.5 Molecular Biological Investigations

13.5.1 Rh Proteins

cDNA clones encoding the **RhcE** [15,45] and **RhD** [8,135,153] proteins have been isolated from a human bone marrow library by using oligonucleotide primers based on the N-terminal amino acid sequence of the **Rh** proteins [14,27,224]. The predicted translation products of the **RhcE** and **RhD** cDNAs are polypeptides consisting of 417 amino acids, most of which are hydrophobic *(Fig. 13.1)*. The *RHD*

```
                                         50
cE   MSSKYPRSVR RCLPLWALTL EAALILLFYF FTHYDASLED QKGLVASYQV GQDLTVMAAL
D    ---------- ---------- ---------- ---------- ---------- ---------I


                                        100
cE   GLGFLTSNFR RHSWSSVAFN LFMLALGVQW AILLDGFLSQ FPPGKVVITL FSIRLATMSA
D    -------S-- ---------- ---------- ---------- --S------- ----------


                                   150
cE   MSVLISAGAV LGKVNLAQLV VMVLVEVTAL GTLRMVISNI FNTDYHMNLR HFYVFAAYFG
D    L-----VD-- ---------- ---------- -N-------- --------MM -I--------


               200
cE   LTVAWCLPKP LPKGTEDNDQ RATIPSLSAM LGALFLWMFW PSVNSPLLRS PIQRKNAMFN
D    -S-------- --E----K-- T--------- -------I-- --F--A---- --E----V--


          250                                                     300
cE   TYYALAVSVV TAISGSSLAH PQRKISMTYV HSAVLAGGVA VGTSCHLIPS PWLAMVLGLV
D    -----V---- ---------- --G---K--- ---------- ---------- ----------


                                             350
cE   AGLISIGGAK CLPVCCNRVL GIRRISVMHS IFSLLGLLGE ITYIVLLVLR TVWNGNGMIG
D    -----V---- Y--G------ --P-S-I-GY N--------- -I-------D --GA------


                                   400
cE   FQVLLSIGEL SLAIVIALTS GLLTGLLLNL KIWKAPHVAK YFDDQVFWKF PHLAVGF
D    ---------- ---------- ---------- -------E-- ---------- -------
```

Figure 13.1: Amino acid sequences deduced from CcEe and D gene transcripts.
Identical amino acids are shown by dashes.
According to Mouro et al. [190] by permission of Nature Publishing Company. The sequence data are deposited in the EMBL / GenBank data library (accession numbers M34015, X54534, and L08429)

and **RhCE** gene encoded proteins are 92% identical at the protein level, differing by only 35 amino acid substitutions spread along the whole molecule.

The highly similar primary structure suggests that both polypeptides closely resemble each other in their three-dimensional structure: hydropathy analyses predict 12 membrane-spanning domains connected by very short segments extending into the surrounding medium; the greatest part of the polypeptide thus resides within the phospholipid bilayer (Fig. 13.2). Both the N-terminus and the C-terminus of the **Rh** proteins were found intracellular, extending into the cytoplasm [15,104]. Five of the six cysteine residues of the **RhCcEe** polypeptide are located at the same positions

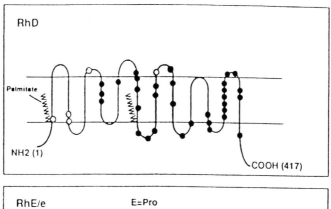

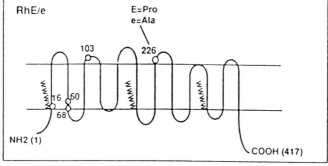

Figure 13.2: Polypeptides encoded by RhD and RhCcEe genes – postulated organisation within the erythrocyte membrane.
Black circles depict amino acid residues characteristic for the RhD protein, open circles indicate residues responsible for Cc or Ee antigenic expression.
Reproduced from Hopkinson [106] by permission of Y. Colin and Nature Publishing Company, New York.

in the **RhD** protein. It is predicted that the cysteine at position 285 lies in an extracellular loop of the **Rh** proteins or very near the outer leaflet of the lipid bilayer, thus representing the sulfhydryl group which influences **RhD** and **c** antigenicity. The cysteines at positions 12, 186, 311, 315, and 316 are situated near the cytoplasmic leaflet of the bilayer and probably represent palmitoylation sites of the **Rh** proteins (Cys^{311} of the **RhcE** protein is replaced by tyrosine in RhD). Since the partial Cys–His–Leu–Ile–Pro sequence (positions 285–289) is common to the **D** and **CcEe** proteins, this motif is presumably associated with all **D**, **Cc**, and **Ee** epitopes. Further evidence for this assumption is provided by the findings that histidine-reactive compounds partially reduce all **Rh** antigenic activities [263].

Surprisingly, the calculated molecular mass of the deduced proteins is 45.5 kDa – considerably higher than that determined for the **Rh** polypeptides (i.e. 30–32 kDa). On the basis of a tryptic peptide, the amino acid sequence of which corresponds to residues 400–406 of the predicted **RhD** protein [243], a post-translational proteolytic modification is highly unlikely. It is generally assumed that the higher electrophoretic mobility of the **Rh** polypeptide is due to an increased binding of sodium dodecyl sulfate.

The high degree of homology between *RHD* and *RHCE* genes facilitates the production of hybrid genes by unequal crossing-over events or by gene conversion. The hybrid proteins encoded by these genes are in many cases characterised by 'new' **Rh** epitopes defined by the junctions of *RHD* and *RHCE* specific amino acids.

Further **Rh**-related cDNA clones proved to be splicing isoforms transcribed from the *RHCE* gene [151]. The deduced proteins are derived from the **RhcE** polypeptide described above differing by peptide deletions, by the number of cysteine residues, and, in two cases, by changed C-terminal domains as a result of frameshift. The boundaries of the deletions found in the cDNAs all correspond to intron/exon transitions ([46,256]. Alternatively spliced mRNA's derived from the *RHD* gene have been detected more recently. These isoforms lack sequences corresponding to exons 7, 8, and/or 9 [51,117,245,271].

The functional significance of these splicing isoforms, however, remains elusive, as no corresponding protein products have been shown to be expressed in the erythrocyte membranes.

13.5.2 Rh-Associated Glycoproteins

As mentioned above, the **Rh** proteins *in situ* are associated with other membrane components, viz. glycophorin B, **LW** protein, **Duffy** protein (see *Fig. 13.3* and *Chaps. 9, 14* and *16*, respectively), and the so-called **Rh**-related glycoproteins, **RhAG** and **CD47** [40].

Peptide analysis of the **RhAG glycoprotein**[2] showed that its N-terminal region is similar to that of the **Rh** polypeptides [14]. By using primers based on the known

---

[2] This glycoprotein was originally referred to as **Rh50**. It has been renamed to **RhAG** (= 'Rh-associated glycoprotein') in order to avoid confusion with the **Rh50** antigen **FPTT** [66].

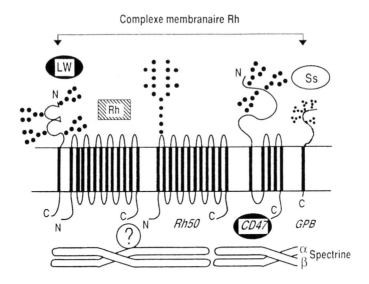

Complexe membranaire Rh

Figure 13.3: 'Rh membrane complex'.
Reproduced from Cartron [40a] by permission of the author and Editions Elsevier Paris.

```
  1 MRFTFPLMAI VLEIAMIVLF GLFVEYETDQ TVLEQLNITK PTDMGIFFEL YPLFQDVHVM  60
 61 IFVGFGFLMT FLKKYGFSSV GINLLVAALG LQWGTIVQGI LQSQGQKFNI GIKNMINADF 120
121 SAATVLISFG AVLGKTSPTQ MLIMTILEIV FFAHNEYLVS EIFKASDIGA SMTIHAFGAY 180
181 FGLAVAGILY RSGLRKGHEN EESAYYSDLF AMIGTLFLWM FWPSFNSAIA EPGDKQCRAI 240
241 VNTYFSLAAC VLTAFAFSSL VEHRGKLNMV HIQNATLAGG VAVGTCADMA IHPFGSMIIG 300
301 SIAGMVSVLG YKFLTPLFTT KLRIHDTCGV HNLHGLPGVV GGLAGIVAVA MGASNTSMAM 360
361 QAAALGSSIG TAVVGGLMTG LILKLPLWGQ PSDQNCYDDS VYWKVPKTR
```

Figure 13.4: Primary structure of Rh-related glycoprotein RhAG.
Underlined: hydrophobic membrane-spanning domains, *N* : Putative N-glycosylation site.
From Ridgwell et al. [213]. The sequence data are deposited in the EMBL / GenBank data library (accession numbers X64594 and AF031548).

partial peptide sequence, it has been possible to isolate from a human bone marrow library a cDNA clone encoding the **RhAG** glycoprotein [213]. DNA sequencing predicted a polypeptide of 409 mainly hydrophobic amino acid residues containing 12 transmembrane domains. Both the C- and N-termini are situated within the cytoplasm [75]. The protein contains one N-glycosidically linked carbohydrate chain at Asn^{37} [75], which carries blood group **ABH** antigenic structures [185]. Though independent of the *RH* gene (the gene encoding the **RhAG** glycoprotein, *RHAG*, has been mapped to chromosome 6 at position p11–21.1 [52]), the amino acid sequence of the deduced protein was clearly related to that of the **Rh** polypeptide, with about 36% amino acid identity *(Fig. 13.4)*. It has thus been assumed that both proteins adopt a similar overall structure and an analogous topology within the erythrocyte membrane.

Likewise, the organization of the human *RHAG* gene closely resembles that of the *Rh* genes [109,181]. Both gene types are composed of 10 exons, their size and the exon/intron structures being nearly identical. Like the *RH* genes, *RHAG* is expressed only in erythroid cells [130,211].

The transmembrane topology of the **RhAG** glycoprotein with 12 membrane-spanning domains strongly resembles that of multispanning membrane proteins associated with transport or channel functions. It has recently been found that **RhAG** and, to a lesser extent, the **Rh** proteins share significant sequence similarity with the Mep/Amt family of NH_4^+ transporters found in plants and microorganisms [178].

This significant homology suggests that *RHAG* and *the RH* genes indeed belong to the same gene family, having been formed as two separate genetic loci from a common ancestor via a trans-chromosomal insertion event.

CD47 is a glycoprotein of 50 kDa with wide cell and tissue distribution (see [182]). It proved identical to previously described membrane components, viz. the ovarian tumour antigen (**OA3**), the integrin-associated protein, which plays a role in cell adhesion in non-erythroid cells, and the Ca^{++} channel of endothelial cells [34,37,166,182]. Its expression is reduced by 75% on **Rh$_{null}$** erythrocytes [166].

The nucleotide sequence of cloned **CD47**-specific cDNA [37,165] predicted a 323-amino acid polypeptide with six putative N-glycosylation sites *(Fig. 13.5)*. The protein shows five C-terminal transmembrane helices and an N-terminal extracellular domain with significant homology to the v regions of the immunoglobulin gene superfamily, particularly to the IgV-related sequences from cartilage link proteins [200] and the poliovirus receptor [183]. The gene encoding the **CD47** protein has been mapped to chromosome 3q13.1–q13.2 [166].

```
  1 MWPLVAALLL GSACCGSAQL LFNKTKSVEF TFCNDTVVIP CFVTNMEAQN TTEVYVKWKF  60

 61 KGRDIYTFDG ALNKSTVPTD FSSAKIEVSQ LLKGDASLKM DKSDAVSHTG NYTCEVTELT 120

121 REGETIIELK YRVVSWFSPN ENILIVIFPI FAILLFWGQF GIKTLKYRSG GMDEKTIALL 180

181 VAGLVITVIV IVGAILFVPG EYSLKNATGL GLIVTSTGIL ILLHYYVFST AIGLTSFVIA 240

241 ILVIQVIAYI LAVVGLSLCI AACIPMHGPL LISGLSILAL AQLLGLVYMK FVASNQKTIQ 300

301 PPRKAVEEPL NAFKESKGMM NDE
```

Figure 13.5: Primary structure of CD47 (= OA3) protein.
Residues 1–18 represent a presumed leader peptide; N : potential N-glycosylation sites (there is evidence that only N^{34}, N^{73}, and N^{111} are glycosylated); underlined: putative transmembrane domains. According to Campbell et al. [37]. The peptide sequence of the protein is deposited in EMBL / GenBank data library (accession number X69398).

Studies with **Rh**-specific hybridoma antibodies have identified those **Rh**-related glycoproteins which co-precipitate with the **Rh** polypeptides [14,185]. The fact that the **Rh**-related glycoproteins, together with glycophorin B, **LW** protein, and **Duffy** protein, are absent or severely reduced in **Rh**-deficient cells further indicates that these membrane constituents form a protein cluster within the erythrocyte membrane.

The central core of this protein complex is presumably composed of a tetramer made of two **Rh** and two **RhAG** protein subunits [75]. Earlier investigations suggested that this multi-subunit complex is closely associated with band 3 protein [264].

Investigation of **RhAG** glycoprotein fragments which co-immunoprecipitate with the **Rh** polypeptides has suggested an interaction between the N-terminal half of the **Rh** polypeptides and amino acids 35–196 of **RhAG** glycoprotein. This region encompasses predicted transmembrane helices 2–6 of the **RhAG** molecule [75]. More recent investigations [48,109], however, suggested a participation of the C-terminus in the interaction as well.

The membrane abnormalities detected in **Rh**-deficient erythrocytes (to be discussed below) clearly show that the multi-subunit protein complex is essential to erythrocyte membrane integrity. Further, the interaction between **Rh** and **RhAG** polypeptides is thought crucial to the correct assembly and transport of the complex to the cell surface. It also seems likely that the **Rh** proteins require the stabilising effect of associated **RhAG** glycoproteins for a correct presentation of the **Rh** determinants.

13.5.3 Diagnostic Application of Molecular Biological Investigations

In order to determine the *RHD* phenotype and in particuoar to manage pregnancies in sensitised **RhD**-negative mothers, various molecular genetic methods for **Rh** genotyping have been developed. They are based on either Southern blot or polymerase chain reaction (PCR).

The Southern blot technique uses genetic markers or restriction fragment length polymorphism [60,118,125].

PCR-based methods are easier to perform. The procedures applied take advantage of the sequence differences in the 3' non-coding region [25] or in exon 7 of the *RHD* and *RHCE* genes [164,279], others focus on the different sizes of intron 4 [8,39,234,235]. Moreover, allele-specific PCR methods have been developed for genotyping of **c** and **E** [152] and **E/e** [79] or simultaneous genotyping of **D** and **C/c** [204].

The *RHD* genotypes determined by these methods were not concordant in all cases with the serologically established phenotypes. 'Weak **D**' and 'partial **D**' phenotypes in particular showed discordances between standard serological typing and the PCR assay. Therefore, procedures have been developed for the use in routine applications, whereby at least two PCR assays directed towards different regions of the *RHD* gene areconsidered [13,220]. The most reliable results in determining the *RHD* genotype, however, were obtained with multiplex PCR assays focussing on exons 3–7 and 9 [88,155,173]. These methods also made it possible to find new **D** variants, but – the only fairly harmless disadvantage – will type **D**-negative individuals who have an intact but non-expressed *RHD* gene as **D**-positive.

13.6 Rh Antigens and Phenotypes

Rh is a highly polymorphic blood group system. In contrast to other systems, only few alleles are formed by single amino acid substitutions. The majority of **Rh** alleles is based on hybrid genes originating from exon exchange driven by genetic recombination at the *RH* locus via segment transfer or gene conversion. In some cases unequal crossing-over events can be assumed.

As mentioned above, **Rh** antigens are conformation dependent structures on highly hydrophobic polypeptides of the erythrocyte membrane. Small changes in primary structure of the protein could therefore alter its mode of membrane insertion exposing or concealing amino acid sequences located within or outside the membrane bilayer in 'normal' *RH* alleles. In addition, the novel exon junctions in the hybrid genes, define a series of additional **Rh** epitopes.

Table 13.3: Frequencies of the main Rh gene complexes in England.

Nomenclature as proposed by Fisher-Race	Wiener	Frequency (%)*
CDe	R¹	40.76
cde	r	38.86
cDE	R²	14.11
cDe	R°	2.57
CᵂDe	R¹ʷ	1.29
cdE	r″	1.19
Cde	r′	0.98
CDE	Rᶻ	0.24

* according to Race and Sanger [205].

It should be noted here that great variations in occurrence and distribution of alleles have been observed at the population level.

Three different nomenclature systems are in use to designate the different **Rh** antigens and **Rh** phenotypes [128,205]:

- the **CDE** system by Fisher and Race,
- the **Rh-Hr** system by Wiener,
- the numerical terminology proposed by Rosenfield et al. [216] in which the antigen specificities are provided with successive numbers. This terminology has later been adopted by the ISBT commission for blood group nomenclature (see *Chap. 1*).

The **Rh** antigens thus far described, their designation in the three nomenclature systems and their frequencies are presented in *Table 13.1*; the most common **Rh** phenotypes are listed in *Table 13.3*.

RhD (Rh1)

The **D** antigen present on the erythrocytes of genotype *D/D* or *D/d* (= 'Rh-positive') individuals is the only **Rh** antigen of clinical relevance: it is mainly anti-**D** antibodies which cause haemolytic anaemia of the newborn when induced in **Rh**-negative mothers by an **Rh**-positive embryo.

As mentioned above, the **RhD** and **RhcE** proteins are 92% identical at the protein level, differing by only 32-amino-acid substitutions spread along the whole molecule. Which of them are essential for **D** antigen expression is thus far unknown — only eight of these substitutions (positions 169, 170, 172, 233, 238, 350, 353, and 354) occur on the predicted extracellular hydrophilic loops and are thus available for binding of anti-**D** antibodies on intact erythrocytes. However, substitutions at intramembraneous or intracytoplasmic locations may also be important in that they modulate protein conformation over long distances.

Evidence for the necessity of *RHD* exons 4 and 5 for the formation of the **D** epitopes is provided by the hybrid *RHD* allele of **D**^{VI} variant type I (see below). In this variant, which is characterised by extremely weak **D** expression, exons *RHD* 4 and 5 are replaced by the corresponding exons of *RHCE*.

In **D-negative** individuals the *RHD* gene is not expressed. This explains why no **d** isoform of the **D** antigen nor anti-**d** antibodies have ever been identified.

Molecular biological investigations have shown that the **D**-negative phenotype is heterogeneous:

In the majority of **D**– individuals the *RHD* gen is completely absent, probably as a result of gene deletion [8,60,126,153,235].

In rare cases in Europids and Negrids, but fairly frequently in Mongolids, apparently intact or only partially deleted *RHD* genes have been detected [13,126, 197,257]. The presence of *RHD* (-like?) genes is restricted to individuals carrying the **C** and/or **E** antigens — in all *dce* haplotypes complete *RHD* deletion has been found; in *dCe* and *dcE* haplotypes a significant proportion of these individuals carried portions of the *RHD* gene, an altered, or a seemingly unaltered but non-functional *RHD* gene.

Thus far only two **D**-negative cases have been investigated in which the *RHD* gene was not deleted:

– in one *dCCee* individual a full-length *RHD* gene has been detected which had a point mutation (C → T) at nucleotide 121 altering the Gln codon 41 to a stop codon [13]. Three other mutations were found in both *RHD* transcripts of this proband, i.e. at codons 215 (TTC → CTC, Phe → Leu), 216 (TTG → CTG, Leu → Leu), and 330 (TAC → CAC, Tyr → His);

– in another **dCCee** individual a four-nucleotide deletion at the splice junction at the intron-3/exon-4 boundary has been found [5]. This mutation alters the reading frame of the otherwise normal **RHD** gene and introduces a premature stop codon at position 166.

Weak-D (D^u) phenotype

The **weak-D** or 'low-grade' **D^u** phenotype [2,239] is characterised by a severe reduction of **D** antigen expression on red blood cells[3]. The average number of **D** antigenic sites is decreased to less than 5–10% of the value obtained for 'normal' **D**-positive erythrocytes [19,35,247]; the dissociation constants between anti-**D** and the two phenotypes, however, are indistinguishable [63]. Under normal conditions, the **RhD** antigen can be detected only by the indirect antiglobulin test.

Erythrocytes of the **weak-D** phenotype never induce antibody production in pregnancy or transfusion incompatibilities [184].

Genetical studies have suggested that the **weak-D** variant is transmitted by an allele at the **RHD** locus which produces a weak **D** antigen [141]. The phenotype occurs in 0.2% to 1% of Europids.

The first molecular biological analyses of **weak-D** variants disclosed a typical **RHD** mRNA sequence with a normal promoter region. However, comparative polymerase chain reaction analysis showed a decreased amount of **RHD**-specific transcripts in **weak-D** reticulocytes when compared with normal D-positive controls. These results suggested that the reduced expression of the **D** antigen is not the consequence of rearrangement or mutation in the coding sequence of the **RHD** gene, but may result from an altered transcription activity of the **RHD** gene, of changed pre-mRNA processing, or of reduced mRNA stability [19,220].

Evidence obtained by Wagner et al. [266], however, indicates a heterogeneous basis of this phenotype. In their investigation, all samples of genomic DNA obtained from **weak-D** probands encoded altered **RhD** proteins. The authors have defined 16 different types of **weak-D**. As shown in *Table 13.4* the majority of alleles is characterised by single base exchanges; two alleles (type 4 and 14) are probably derived from gene conversion, by which parts of exons 4 and 5 were substituted by the corresponding sequences of the **RHCE** gene. All amino acid substitutions of **weak-D** types are located in intracellular and transmembraneous protein segments

---

[3] Other phenotypes with a depression in the apparent numbers of **D** antigen sites are caused by the inhibition of **D** by a C antigen in *cis* or *trans* [43,54] or are based on hybrid **RH** alleles (see 'partial D').

Table 13.4: Molecular basis of RHD alleles encoding weak-D phenotypes.

Weak-D type	Nucleotide change	Nucleotide	Amino acid substitution	Position	Exon	Membrane position
Type 1	T → G	809	Val → Gly	270	6	TM
Type 2	G → C	1154	Gly → Ala	385	9	TM
Type 3	C → G	8	Ser → Cys	3	1	IC
	C → G	602	Thr → Arg	201	4	IC
Type 4	T → G	667	Phe → Val	223	5	TM
	G → A	819	–	–	–	–
Type 5	C → A	446	Ala → Asp	149	3	TM
Type 6	G → A	29	Arg → Gln	10	1	IC
Type 7	G → A	1016	Gly → Glu	339	7	TM
Type 8	G → A	919	Gly → Arg	307	6	IC
Type 9	G → C	880	Ala → Pro	294	6	TM
Type 10	T → C	1177	Trp → Arg	393	9	IC
Type 11	G → T	885	Met → Ile	295	6	TM
Type 12	G → A	830	Gly → Glu	277	6	TM
Type 13	G → C	826	Ala → Pro	276	6	TM
	T → A	544	Ser → Thr	182	4	TM
Type 14	A → T	594	Lys → Asn	198	4	IC
	C → G	602	Thr → Arg	201	4	IC
Type 15	G → A	845	Gly → Asp	282	6	TM
Type 16	T → C	658	Trp → Arg	220	5	TM

TM: transmembraneous, IC: intracellular.
From Wagner et al. [266].

and are clustered in four regions of the protein (amino acid positions 2 to 13, around 149, 179 to 225, and 267 to 397). It can thus be assumed that these mutations will not alter the exofacial domain structures which define **RhD** specificity, but may affect the membrane integration of the **RhD** protein or the formation of the **RhD/RhAG** complex.

D_{el} Phenotype

D_{el} ('D-elute') has been defined as an **RhD** variant with very weak **D** antigen [196]. No agglutination is detectable by the indirect antiglobulin test, but the erythrocytes are able to adsorb trace amounts of anti-**D** antibodies. This phenotype has thus far been found only in Mongolids.

It has been suggested that the **D**$_{el}$ phenotype is a **D**^{u} variant gene which is suppressed by the inhibitory effect of **Cde** [103]. Further, it also possible that the **D**-negative individuals of Mongolid ancestry in whom an apparently intact but suppressed **RHD** gene has been detected (see above) are at least in part **D**$_{el}$ variant individuals.

Analysis of the **RHD** gene of **D**$_{el}$ probands revealed a 1013 bp deletion between introns 8 and 9 and including the whole exon 9. The transcript maintains an open reading frame and encodes a 463 amino acid protein with a new C-terminus extending from codon 384 [44].

C/c and E/e

Investigations on cDNA transcripts and genomic DNA isolated from individuals of different **Rh** phenotypes [190] showed that **RhE** and **Rhe** differ by a proline → alanine substitution at position 226 (exon 5); the **C** and **c** alleles differ by three amino acid substitutions in exon 2, viz. isoleucine → leucine (position 60), serine → asparagine (position 68), and serine → proline (position 103) (see *Fig. 13.2*). **Rh** proteins expressing **c** specificity may contain cysteine or tryptophane at position 16 (exon 1) [280], but **C** antigen expression seems to require a cystein residue at this position (see [191]).

In accordance with the predicted secondary structure of the **Rh** polypeptides within the cell membrane, the residue characteristic for the **E/e** dimorphism (amino acid 226) is located on the fourth extracellular loop. Among the four positions determining the **C/c** dimorphism, however, only amino acid 103 is exofacial (second loop), suggesting that the amino acid dimorphisms at positions 103 and 226 are crucial for **C/c** and **E/e** antigenicities, respectively. Whether the intracellular or intramembraneous **C/c** polymorphic positions participate indirectly in the **C/c** antigenicity by conformational changes, remains a matter of speculation. Erythrocytes from **D-- /D--** individuals (see below) do not react with anti-**C**, anti-**c** or anti-**e** sera [205]; this finding demonstrates that the **D** protein does not express **C** or **c** and **e** antigens although the **D** polypeptide exhibits the **e** associated amino acid at position 226, and shows a sequence which is typical in part for **C** (positions 68 and 103) and **c** (positions 16 and 60). Interaction with the third extracellular loop encoded by the **RHCE** gene is probably required for the formation of the **C** epitope.

On the basis of peptide mapping experiments it has been proposed that the **C/c**, and **E/e** epitopes are expressed on different polypeptides of approximately equal molecular masses [27,120,190]; the proteins were found highly homologous but not identical [26]. A further strong argument in favour of the expression of **Cc** and **Ee** on two different polypeptides is the number of **Cc** antigen sites, of which there are almost twice as many as of the **Ee** antigen sites (see *Table 13.2*).

Since there is convincing evidence that **C/c** and **E/e** are encoded by one single gene [41], it has been suggested that alternatively spliced transcripts of the *RhCE* gene might account for the presence of the **C/c** and **E/e** epitopes on distinct polypeptides. According to this theory **RhCc** specific transcripts reside on shortened **Rh** polypeptides lacking exon 5, where residues critical for **E/e** antigens are found [151,190]; the **E/e** antigens are carried by a full-length **Rh** polypeptide, which, due to a different membrane topology expresses **C** and **c** only weakly.

Results of further studies cast doubt on the validity of this hypothesis; and the following facts refulte this theory:

- **Rh** polypeptides immunoprecipitated by anti-**c** and anti-**E** are apparently identical in size,
- analysis of immunopurified **RhCc** proteins has shown that these antigens are expressed on full-length, not trucated **Rh** polypeptide chains [12], and
- the product of the **RhC** isoform lacking exon 5 has not been found,
- a single full-length *RhcE* transcript, when expressed in K562 cells, resulted in the expression of both **c** and **E** antigens on a single 416-amino acid polypeptide chain [237],

It also seems unlikely that shortened isoforms of the *RHCE* gene would encode stable proteins, which are readily incorporated into the tightly packed multi-subunit **Rh** membrane complex of the red cell membrane. The hypothesis suggesting that **C/c** antigens are expressed on a truncated **Rh** polypeptide created by alternate splicing deserved reconsideration.

Nevertheless, Umenishi et al. [257,258] have published findings which they believe are best explained by the existence of three *RH* genes, each of which encodes full length transcripts of **Rh** bearing proteins.

C^W (**Rh8**) and C^X (**Rh9**)

Analysis of **Rh** transcripts and gene fragments from donors with the C^W and C^X phenotypes indicated that the expression of the C^W (**Rh8**) and C^X (**Rh9**) antigens are associated with point mutations in exon 1 of the the *RHCE* gene [192]. As compared with the common transcripts of the *RHCE* gene, the C^W-specific cDNA exhibited an $A^{122} \rightarrow G$ transition resulting in a $Gln^{41} \rightarrow Arg$ exchange in the C^W polypeptide, whereas the C^X cDNA was characterised by a $G^{106} \rightarrow A$ transition resulting in an $Ala^{36} \rightarrow Thr$ exchange in the C^X polypeptide. Both substitutions are located at the first extracellular loop of the **Rh** polypeptide. Therefore, although serologically the C^W and C^X specificities behave as if they were allelic, they cannot not be considered products of antithetical allelic forms of the *RHCE* gene.

It was originally suggested that the high-frequency antigen **MAR (Rh51)** has an antithetical relationship to both antigens, which are expressed only at the surface of erythrocytes which are hetero- or homozygous for *RHCE* alleles not expressing either C^W or C^X [236]. The **Rh51** epitope, therefore, is most likely associated with the presence of Ala^{36} and Gln^{41}.

RhG (RH12)

RhG expression is usually confined to erythrocytes carrying **D** and/or **C** antigens, whereas **D**- and/or **C**-negative cells only rarely express this antigen ($r^G r$ phenotype [3].

Investigations on genomic DNA and cDNA of two **G**-negative donors (genotype *ccDEe*) revealed a $T \rightarrow C$ substitution at nucleotide 307 in exon 2 of the *RHD* gene; this causes a serine $\rightarrow$ proline exchange at position 103 of the **RhD** polypeptide [76]. The fact that the **G**-specific Ser^{103} residue is also characteristic for exon 2 of the *RHC* allele reflects the expression of **G** with **D** and **C** antigens. These results suggest that both epitopes as recognised by anti-**C** and anti-**G** contain serine at position 103, but anti-**C** depends on the conformation of the protein and does not recognise Ser^{103} within a *RHD*-encoded environment; anti-**G**, however, is much less dependent on the conformation of the protein and recognises Ser^{103} in proteins encoded by both *RHD* or *RHCE* genes.

Further evidence that the **RhG** epitope is located on exon 2 is provided by the **G**-negative D^{IIIb} variant (see below), where exon 2 of **D** is replaced by exon 2 of the **Rhc** allele [222]. The D^{IIIc} variant, where exon 3 of *RHD* is replaced by the corresponding exon of *RHCE*, however, is **G**-positive [20].

It has been proven that an $r^G r$ donor (*S.F.*) who reacted with anti-**RhG** without expressing **RhD** and/or **RhC** (genotype *ccEe*) carries the complete exon 2 of **RHD** [76]. Sequence analysis revealed a hybrid gene between *RHCE* and *RHD* in which exons 4–7 (or 8) of *RHD* were replaced by exons 4–7 (or 8) of *RHCE* *(Fig.13.6)*.

In another $r^G r$ donor a hybrid gene has been detected; it arose either from a segmental DNA exchange between part of exon 2 of the *RHce* gene and the equivalent region of the *RHCE* or *RHD* genes, or from a crossing over between nucleotides 150 and 178 of the *RHce* and *RHCe* genes [191]. The predicted protein encoded by the hybrid r^G gene (*c–C–e* or *c–D–e*) is characterised by Ile^{60}, Ser^{68}, and Ser^{103} as found in **C** and **D** polypeptides *(Fig. 13.6)*.

Rh26

The **Rh26** epitope [119] is closely related to **Rhc**: red cells expressing **Rhc** usually express **Rh26**, and the rare **c+ Rh:−26** phenotype is associated with a weak expression of **Rhc** and, in some cases, with reduced expression of **Rhf** (= **Rhce**) as well.

The analysis of cDNA obtained from **c+ Rh:−26** individuals [78] revealed a single-point mutation ($G^{286} \rightarrow A$) in exon 2 of the *ce* allele, predicting a glycine → serine substitution at position 96 in the **ce**-carrying polypeptide.

According to the structure of the **Rh** polypeptide deduced from hydropathy analysis, residue 96 is located in a transmembrane segment close to a membrane boundary. Substitution of glycin by serine apparently induces extracellular conformation changes, leading to a loss of the **Rh26** epitope.

However, the expression of **Rh26** is not solely dependent on the presence of Gly^{96}. The **Rh26** epitope is restricted to the polypeptide encoded by the *c* allele, though a glycine residue at position 96 is normally present on the **C** and **D**-carrying polypeptides which do not express this antigen. It can therefore be assumed that additional parts of the peptide chain (presumably Trp^{16}, Leu^{60}, Asn^{68}, and Pro^{103}) are involved in the formation of the **Rh26** epitope.

R^N

The R^N variant is characterised by the presence of the low-incidence antigen **Rh32**, by severely decreased **C** and **e**, and lack of **Rh46** antigen. R^N is phenotypically expressed only in the absence of **D** in both *cis* and *trans* configurations.

Molecular genetic investigations [220] on three donors revealed that the R^N variant is associated with a hybrid *RHCe* gene in which exon 4 (in one case part of exon 3 as well) is replaced by the equivalent parts of *RHD*. The R^N phenotype may thus be due to at least two independent genomic alterations *(Fig. 13.6)*.

The hybrid R^N polypeptide differs from the *RHCe*-encoded protein by 7 (or 8) residues specific for the RHD protein. Membrane topology predicts that only the three amino acids Met^{169}, Met^{170}, and Ile^{172} are located on an extracellular loop, and that they may be responsible for the presence of **Rh32** and the absence of **Rh46**.

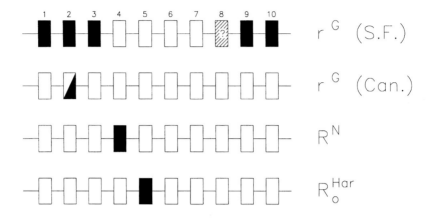

Figure 13.6: Structure of *RHCE* variant genes.
■ : *RHD* exons, ▯ :*RHCE* exons.
Redrawn according to Huang [108a] by permission of the author.

R₀^{Har}

R_o^{Har}, an **Rh** phenotype of low frequency [90], is characterized by normal **c**, weak expression of **D, e, f,** and **Rh17**, absence of **G, Rh18,** and **Rh19**, and the presence of the low-incidence antigens **Rh33** and **FPTT (Rh50)**.

Molecular biological investigations revealed that the R_o^{Har} phenotype is characterised by an *RH* haplotype which lacks the *RHD* gene and contains an *RHce* gene in which exon 5 is replaced by the corresponding exon of *RHD* [23,24] *(Fig.13.6)*.

VS and V

The **VS** (e^S, **Rh20**) character [226] is rare in Europids but fairly common in Negrids (in some populations up to 50% [68]). The presence of **VS** on red cells is associated with weak expression of **e** (= e^S) and with depression of **hr**^{B} (**Rh31**), which is present on most **e+** red cells [231]. **V (Rh10)** is characteristic for *ce*ˢ haplotypes [73].

The identification of a C → G transversion at nucleotide 733 in the *RHCE* gene of **VS** probands suggests that a Leu²⁴⁵ → Val substitution in exon 5 is associated with the expression of the **VS** antigen [77,110,238].
Since the valine residue at position 245 may – as predicted – lie within the red cell membrane bilayer, formation of the **VS** epitope is obviously due to a conformational change in the protein. It can therefore be assumed that the amino acid exchange at position 245 affects the **e** specific alanine residue at position 226 in such a way that **e**ˢ is expressed and **Rh31** is depressed. Val²⁴⁵ is normally present in the RhD protein and obviously leads to **VS** antigenicity only when present in vicinity to *RHCE*-encoded sequences.

The **(C)ce**ˢ (**r'**ˢ) complex is characterised by expression of **c** and **VS**, weak expression of **C** and absence of **V** and **D**. In these cases hybrid *RHD–CE–D* transcripts were found, which are composed of exons 1, 2, part of exon 3, and exons 9 and 10 of the *RHD* gene, and exons 4–7 (and probably 8) from the *RHCE* gene [29,77]. The hybrid gene of all probands tested contained Val²⁴⁵ in exon 5 [68]. The weak and variant form of **C** characteristic for the **r'**ˢ is obviously due to the Ser¹⁰³ residue encoded by the *RHD*-derived exon 2, which, however, is located in a *RHCE* environment.

In two unrelated probands the **VS** character was associated with the **D** category phenotype **D**ᴵᴵᴵᵃ [110]. In these cases an *RHD* gene has been found which, in addition to the valine residue at position 245, contained four amino acid substitutions scattered in three exons, viz. Asn¹⁵² → Thr, Thr²⁰¹ → Arg, Ile²¹⁸ → Met, and Phe²²³ → Val (see 'RhD Category Phenotypes', below). Since the nucleotides identified in this *RHD* variant gene are characteristic for the *RHCE* gene, these substitutions probably originated by templated micro-conversion events.

Investigations on the **V** character by Daniels et al. [68] suggested that both anti-**VS** and anti-**V** recognise the conformational changes created by Val²⁴⁵, but that anti-**V** is abolished by an additional conformational change created by a glycine → cystein substitution at position 336 caused by a G → T transversion at nucleotide 1006.
Amino acid position 336 is located in the 11th membrane-spanning domain. Thus, it is most likely that Cys³³⁶ affects the conformation of the 6th extracellular loop. The cystein residue may introduce an additional palmitoylation site.

RhD Category Phenotypes ('Partial D')

The detection of rare **D**-positive individuals who produce anti-**D** antibodies in response to immunisation with **D**-positive blood (by transfusion or pregnancy) suggested that the **D** antigen is a 'mosaic' structure. Investigations of these 'D category phenotypes' using a set of human monoclonal antibodies led to the identification of discrete epitopes on the **D** antigen. There are two current classification systems, one which uses a nine epitope pattern (*epD1 – epD9*) [169,253] *(Table 13.5)*, and a more recent model which proposes 30 different epitopes [132].

Table 13.5: Distribution of epitopes epD1–9 on RhD category erythrocytes.

RhD category cell	Epitopes present							
	epD1	epD2	epD3	epD4	epD5	epD6/7	epD8	epD9
D^{II}	+	+	+	–	+	+	+	–
$D^{III[1]}$	+	+	+	+	+	+	+	+
D^{IVa}	–	–	–	+	+	+	+	–
D^{IVb}	–	–	–	–	+	+	+	–
D^{Va}	–	+	+	+	–	+	+	+
D^{Vc}	–	–	–	–	+	+	+	–
D^{VI}	–	–	+	+	–	–	–	+
D^{VII}	+	+	+	+	+	+	–	+
DFR	+[2]	+/–[3]	+	+	+/–	+/–	–	+
DBT	–	–	–	–	–	+/–	+	–
DNU	+	+	+	–	+	+	+	–
D^{HMi}	–	+/–	+/–	+	+/–	+	–	+/–
D^{HMii}	+	+	+	+	+/–	+/–	+	+
R_o^{Har}	–	–	–	–	+/–	+/–	–	–

[1] D^{III} can be further sub-divided into D^{IIIa}, D^{IIIb}, and D^{IIIc}
[2] Positive with papain-treated cells
[3] Negative with some antibody samples
According to Lomas et al. [168] by permission of Blackwell Science, Oxford, UK.

Partial D variants are thus defined by characteristic **D** antigen anomalies, i.e. lack of expression of some **D** epitopes at the surface of their erythrocytes, and, in some cases, association with low-incidence antigens [40]. Moreover, in most variants the number of **D** sites per cell is significantly reduced [81,134]).

The loss of one or several **RhD** epitopes in 'D category phenotypes' is mainly the result of replacement of sections of **RHD** by the corresponding sections of **RHCE**. The DNA segments are transferred via gene conversion or unequal crossing-over events between homologous intronic regions of the two highly related genes at the **RH** locus (Fig. 13.7). In some cases, however, gene deletions and point mutations have been detected. The proteins encoded by these hybrid genes are characterised by new junctions between **D** and **CE** specific amino acids responsible for the formation of the low-incidence **Rh** antigens.

The **D** category phenotypes described to date are: D^{II}, D^{IIIa}, D^{IIIb}, D^{IIIc}, D^{IVa}, D^{IVb}, D^{Va}, D^{VI}, D^{VII}, D^{DFR} [168], D^{DHR} [133], D^{DBT}, D^{DNU}, D^{HMi} [131], and D^{HMii} [131] (D^{I} is obsolete).

D^{II}

Sequence analysis of exon 7 of the **RHD** gene derived from a D^{II} propositus indicates a single point mutation in this exon resulting in an amino acid change ($Ala^{354} \rightarrow Asp$) [10].

D^{IIIa}

Nucleotide sequencing of the **RHD** transcripts of two unrelated probands revealed four nucleotide changes distributed among three exons: $A^{455} \rightarrow C$ (exon 3), $C^{602} \rightarrow G$ (exon 4), $C^{654} \rightarrow G$ (exon 5), and $T^{667} \rightarrow G$ (exon 5). These variations resulted in amino acid substitutions characteristic of the **RhCE** encoded polypeptide: $Asn^{152} \rightarrow Thr$, $Thr^{201} \rightarrow Arg$, $Ile^{218} \rightarrow Met$, and $Phe^{223} \rightarrow Val$ [110]. Since none of the amino acids are located in an exofacial loop (they are either embedded in the membrane or reside at the cytoplasmic side), they obviously affect the positioning of their nearby surface loops and thus alter the conformation of given **D** epitopes.

The origin of this altered **RHD** gene is still not known. Gene conversion involving exons 3–5 has been excluded, since the changed nucleotides are flanked by **RHD** specific nucleotides. A templated micro-conversion event involving patches of **RHCE** sequence may be an explanation, and has therefore been a subject of discussion [136,175].

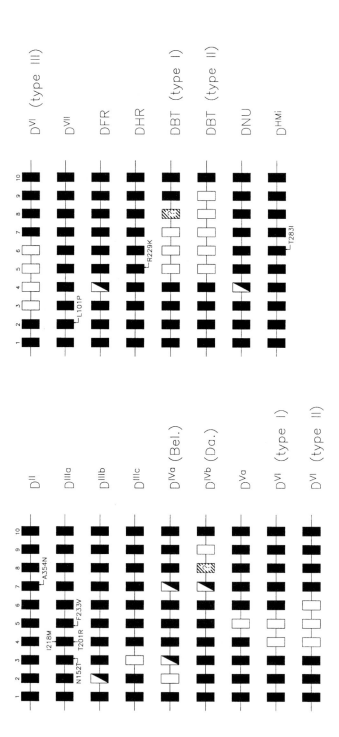

Figure 13.7: Schematic representation of *RHD* gene structures as found in 'D category' cells.
RHD and *RHCE* exons are represented by closed and open boxes, respectively.
Redrawn according to Huang [108a] by permission of the author.

D^{IIIb}

The **D^{IIIb}** category phenotype is characterized by the presence of most of the **D** epitopes and the total absence of the antigen **G** [255]; this variant occurs mainly in Negrids.

mRNA sequencing and Southern blot analysis of two unrelated samples indicated that the **D^{IIIb}** category phenotype is associated with a segmental DNA exchange between exon 2 of the **RHD** and **RHCE** genes. The resulting three amino acid substitutions (Ile^{60} → Leu, Ser^{68} → Asn and Ser^{103} → Pro) are characteristic for the **Rhc** allele [222].

D^{IIIc}

In this phenotype a hybrid **RHD-CE-D** allele has been found in which exon 3 of **RHD** is replaced by exon 3 of **RHCE** [20]. In the **D^{IIIc}** polypeptide four **D**-specific amino acids are thus changed to the corresponding **RHCE**-encoded residues, viz. Leu^{121} → Met, Val^{127} → Ala, Asp^{128} → Gly, and Asn^{152} → Thr. According to hydropathic analysis all four residues should reside in the transmembrane domain. However, since Thr^{152} lies among a cluster of hydrophilic residues (Glu^{146}, Arg^{154}, and Asp^{164}), Thr^{152} is possibly located in the third external loop of the **D^{IIIc}** polypeptide.

D^{IVa}

The **D^{IVa}** category is associated with **Go^a (Rh30)** and **Riv (Rh45)** antigens [72,162]. This **D** variant is probably the result of two independent recombination events in which exon 2 and part of exons 3 and 7 of the **RHD** gene were replaced by the corresponding sections of the **RHCE** gene [219]. This rearrangement is associated with the loss of the epitopes D1, 2, 3, and 9 due to a single amino acid exchange on the exofacial surface (His^{350}) and two in intramembraneous domains (Phe^{62} and Thr^{152}).

D^{IVb}

The **D^{IVb}** variant is based on a **RHD-CE-D** hybrid gene in which exons 7–9 (or parts thereof) are replaced by the corresponding sequences of the **RhCE** gene [219].

D^{Va}

This variant is associated with the **D^w (Rh23)** antigen [57] and is caused by an insertion of **RHCE** exon 5 or, in another case, a portion thereof into the corresponding sequence of the **RHD** gene [199,219]. This gene conversion led to two amino acid substitutions, viz. Phe^{223} → Val and Glu^{233} → Gln. Another **D^{Va}** variant ('**D^{Va}-like**') was characterised by two additional amino acid exchanges (Val^{238} → Met and Val^{245} → Leu) [199]. Residues 233 and 238 are located exofacially, while residues 233 and 245 are situated in the intramembrane domain.

D^{VI}

The **RhD** category VI (**D^{VI}**) phenotype is the most common of the partial **D** phenotypes (~0.02% in Europids, comprising about 5% of all weak **D** samples [154]). It is the clinically most significant partial **D**, since **D^{VI}** erythrocytes type as **RhD** when polyclonal anti-**D** reagents are used, but also lead to alloimmunisation. As a consequence, severe cases of HDN have occurred in **RhD** positive babies born to mothers with anti-**D** [146].

The **D^{VI}** phenotype is heterogeneous, having originated from at least three independent molecular events all leading to **RHD–RHCE–RHD** hybrid alleles [11,108,172,267]:

-   **D^{VI} type I** associated with the **cD^{VI}E** haplotype, is encoded by an **RHD** gene where exons 4 and 5 are replaced by their **RHCE** equivalents [9, 108];
-   **D^{VI} type II** associated with the **CD^{VI}e** haplotype, is encoded by an **RHD** gene where exons 4–6 are replaced by their **RHCE** equivalents [193,219]. This type is characterised by the presence of the low frequency antigen **BARC** (**RH52**) [253];
-   **D^{VI} type III** also associated with the **CD^{VI}e** haplotype, is encoded by an **RHD** gene where exons 3 to 6 are replaced by their **RHCE** equivalents [267][4].

D^{VII}

D-category VII is associated with the low-frequency antigen **Rh40** (**Tar**) [168 a]. This variant is due to a single point mutation, $T^{329} \rightarrow C$, in exon 2 of the **RHD** gene, resulting in a leucine → proline substitution at position 110 (i.e. in the second extracellular loop) of the **RhD** polypeptide [221,268].

DFR

The partial **D** phenotype, DFR [168], is associated with the low-incidence antigen **FPTT** (**Rh50**, formerly ISBT number 700 048).

The **DFR** variant is characterised by a **RHD–CE–D** hybrid gene in which exon 4 of **RHD** (or a part thereof) is replaced by the **RHCE** equivalent [219]. Further investigations [13] suggest that the **DFR** variant may occur on two different genetic backgrounds, one proband being **RHD** intron 4 positive, the other negative.

DHR

In the **DHR** variant a single point mutation (G → A at nucleotide 686) in exon 5 of the **RHD** gene has been found; this causes an Arg → Lys substitution at position

---

[4] This hybrid protein is complementary to the **RHCE-D-Ce** hybrid protein detected in the **D..** propositus *A.T.* [50,111].

229 of the **RhD** protein [133]. This residue is localised on the fourth predicted exofacial loop of the **RhD** polypeptide.

DBT

The partial **D** variant **DBT** is associated with the **Rh32** antigen [269]. In a Moroccan family a hybrid **RHD** gene has been found in which exons 5, 6 and 7 (and possibly the identical exon 8) were replaced by the corresponding exons of the **RHCE** gene [22] (**DBT-1**); the hybrid detected in a Japanese family was characterised by conversion of **RHD** exons 5-9 into the **RHCE** gene [112] (**DBT-2**).

It is worth noting that the **Rh32** antigen is also found on R^N erythrocytes. The R^N variant is based on a hybrid **RHCe** gene containing exon 4 (and in one case part of exon 3 as well) of **RHD** [220]. It can thus be assumed that the **Rh32** epitope is connected with the novel amino acid pattern at the linkage point between **RHD** exon 4 and **RHCE** exon 5 encoded sequences (see R^N).

DNU

The D^{DNU} phenotype arises from a single point mutation in the **RHD** gene, resulting in a glycine $\rightarrow$ arginine substitution at position 353 [10].

D^{HMi} and D^{HMii}

These **D** category phenotypes are characterised by their reactivity with anti-**D** hybridoma antibodies [131]. Family studies suggest that the variant **D** in both D^{HMi} and D^{HMii} is inherited as a **cDE** gene complex.

First investigations revealed that the D^{HMi} variant is caused by a threonine isoleucine substitution at position 283 [167]. The D^{HMii} variant has not yet been investigated.

Mapping of the D Epitopes

A preliminary epitope map of the **RhD** protein can be predicted from the sequence data of the different **RhD** category phenotypes. According to the model advanced by Cartron et al. [42], all **D** epitopes depend mainly on residues located on extracellular loops 3,4, and 6, although loops 2 and 5 may occasionally be involved. *Fig. 13.8* shows epitopes epD3, 4, D6/7, and 9 are located in one loop of the **RhD** protein, whereas epitopes epD1, D2, D5, and D8 are obviously conformation-dependent, comprising as they do amino acid sequences located on different loops.

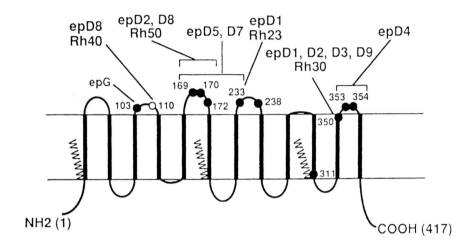

Figure 13.8: Proposed localisation of D epitopes on RhD protein.
Reproduced from Cartron [40a] by permission of the author and of Editions Elsevier, Paris.

RHCE-deficient phenotypes

These phenotypes are characterised by total or partial absence of *RHCE* gene-encoded antigens. The *CE*-deficient phenotypes thus far described are **D--** and **D..**, both of which lack all **CcEe** characters, and **Dc-**, **DC^w-**, and **D^IV(C)-**, which lack the **Ee** characters only.

The majority of these phenotypes are based on hybrid genes in which fairly large portions of the *RHD* gene (ranging from 5 to 9 exons) are inserted into the *RHCE* gene with concomitant loss of *RHCE* homologous sequences. This increase in **D** specific sequences reflects the significant increase in **D** activity in these phenotypes; and the formation of such hybrid genes also provides an explanation for the loss of **CcEe** expression, or the separation of **Cc** antigens from **Ee** antigens.

D-- and D..

Erythrocytes of the **D--** phenotype [206,208] express the **G** antigen and **Rh29** but lack all antigens of the **C** and **E** series. They are characterised by a high density of **D** antigenic sites – erythrocytes of individuals homozygote for the *D--* haplotype (= 'super-D') show a **RhD** activity which is up to six-fold higher than that of the cells from 'normal' **RhD** subjects (*cDE*/*cDE* genotype) (see *Table 13.2*). Due to their increased antigen density, **D--** erythrocytes are directly agglutinated by most incomplete anti-**RhD** antibodies; the cells are otherwise normal.

The molecular basis for the **D--** phenotype is heterogenous, the alterations having occurred on different genetic backgrounds:

Contrary to expectations, a series of **D--** probands contained a complete *RH* gene complex with no recognisable alteration of *RHCE* [51,60]. This apparently normal *RH* locus organisation suggests that in these cases the deficiency in **Cc** and **Ee** is due to impaired *RHCE* transcription caused by a thus far unknown defect in the regulatory mechanism for *RHCE* expression. The increased activity of **D** could be explained by an overexpression of **D**, which compensates the lack of the *RHCE* encoded protein in the **Rh** protein complex of the erythrocyte membrane.

In other cases of **D--** abnormal *RHCE* genes have been detected:

- in one proband the occurrence of a *RHCE* gene deletion has been proposed [107];
- in a **D--** haplotype found in an Icelandic family exons 2–10 of *RHCE* were absent [30]. In two other **D--** samples the *RHCE* gene lacked exons 2–8 [117]; analysis of *RH* cDNAs showed that no functional mRNA was produced from the altered genes. These forms were initially thought to result from internal deletions within the *RHCE* gene.

More recent studies, however, show that in most cases the missing parts of *RHCE* are replaced by the homologous sequences of the *RHD* gene *(Fig. 13.8)*:

- re-investigation of one **D--** proband (*L.M.*) previously studied by Huang et al. [117] showed that this **D--** gene complex resulted from a homologous recombination event between *RHD* and *RHCE* genes resulting in a hybrid gene composed of exons 1–9 of the *RHD* and exon 10 of the *RHCE* gene [50]; the haplotype of this proband also contained a hybrid *RHD* gene, in which exons 1 and 9 were replaced by the corresponding exons of the *RHCE* gene;
- five of six homozygous individuals with the **D--** phenotype were found to have rearranged *RHCE* genes in which internal sequences were replaced by the corresponding sequences from *RHD* [139,140]. The recombination hot spot in the *RH* genes was located within a small region around exon 2. This region is characterized by an exceptionally high degree of sequence homology between *RHCE* and *RHD*, a high density of dispersed repetitive elements, and the presence of an alternating purine-pyrimidine copolymer tract.

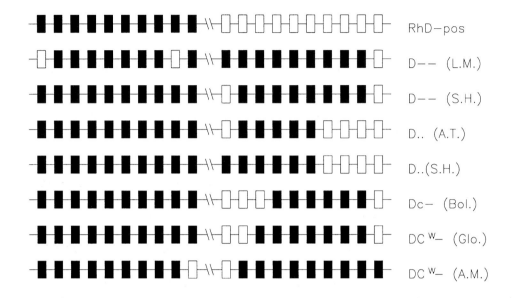

Figure 13.9: Structure and arrangement of *RH* gene complex in *RHCE*-deficient phenotypes.
■ :*RHD* exons, ☐ :*RHCE* exons.
Redrawn according to Huang [108a] by permission of the author.

Erythrocytes of the **D..** phenotype, like **D--** cells, lack all antigens of the **C** and **E** series but carry the low-incidence antigen **Evans (Rh37)** [128].

Here as well the phenotype occurs on different genetic backgrounds (*Fig. 13.8*):

- **D..** *(A.T.)* contained an **RHCE–D–CE** hybrid in which exons 2–6 of the **RHCE** gene were replaced by the homologous counterpart of the **RHD** gene [111]. The **Evans** antigen is probably specified by the novel sequence formed at the junction of **D** exon 6 – **CE** exon 7;
- **D..** *(Dav.)* contained an intact **RHD** gene and an **RHD-CE** hybrid in which exons 1–6 of **RHD** were fused with exons 7–10 of **RHCE** [50],
- a further **D..** *(H.D.)* complex may contain an **RHCE-D-CE** hybrid in which exons 2–8 (or 3–8) were derived from **RHD** [140].

440

Dc- and DC^w-

The **Dc-** and **DC^w-** phenotypes are characterised by lack of **Ee** antigens [128] and **D** activity of the erythrocytes significantly higher than that of normal **RhD**-positive cells [205].

In both gene complexes gene rearrangements leading to **RHCE–D–CE** genes have been found.

Thus far only one **Dc-** proband has has been studied (Bol.). In the hybrid **RHCE–D–CE** gene, exons 4 to 9 of **RHCE** were substituted by the equivalent exons of **RHD** [51] *(Fig. 13.8)*.

In the hybrid gene detected in one **DC^w-** phenotype *(Glo.)* exons 3–9 of **RHCE** have been substituted by the corresponding region of **RHD** [51] *(Fig. 13.8)*. The hybrid gene carries the Glu^{41} → Arg substitution characteristic for **C^w** specificity (see above).

In another case of **DC^w-** *(A.M.)* the hybrid **RHCE** gene was formed by fusion of exon 1 of **C^w**-specific **RHCE** with **RHD** exons 2–10 [107]. In this proband an altered **RHD** gene has been detected in which exon 10 was replaced by exon 10 of **RHCE**. The author assumed that the genome of this individual is composed of two different haplotypes: one characterised by the **RHD–CE** hybrid and a deleted **RHCE** gene (**D--**), and the other by the **RHCE-D** hybrid and a deleted **RHD** gene (**C^w**).

All hybrid genes thus far described lack **RHCE** exon 5, which is where the Pro^{226} → Ala dimorphism characterising **Ee** antigenicity is located. The encoded proteins, therefore, cannot produce proteins with **Ee** antigenicity.

Rh-deficient phenotypes

Rh-deficient erythrocytes are characterised by the lack of all **Rh** determinants [53,128,205,212,242].

Congenital deficiency of **Rh** structures is accompanied by morphological and functional anomalies of the erythrocytes, which cause mild to moderate chronic haemolytic anaemia ('**Rh$_{null}$** syndrome' or 'Rh-deficiency syndrome' [195]). The cells themselves have a decreased viability, increased osmotic fragility, abnormal cation permeability [16,149], and often show slight stomatocytosis and/or sphaerocytosis [241]; it has also been noticed that the membranes contain less cholesterol than normal, and that the phospholipid asymmetry of the erythrocyte membrane (see *Chap. 4.2)* is disturbed [89,145]. Further, **Rh$_{null}$** erythrocytes not only lack the products of the *RH* genes but are partially or totally deficient in the proteins carrying the **LW** and **Duffy** antigens and in the 'Rh-related glycoproteins': in all cases of **Rh**-deficiency, **CD47** was found severely reduced (~10%). In erythrocytes of the **Rh**$_{mod}$ phenotype **RhAG** was absent, whereas in the case of **Rh**$_{null}$ (amorph) most cells expressed low but significant amounts of **RhAG**[5] [52]. In addition, the glycophorin B content is reduced, reaching only approximately 30% of the normal level [64]; this results in low expression of **S**, **s**, **U**, and **Duclos** antigens in **Rh**$_{null}$ cells. These findings indicate that the protein products of the *RH* genes are essential for normal red cell membrane integrity.

The **Rh**-deficient phenotype may originate from two different genetic backgrounds:

- **Rh**$_{null}$ ('**Rh**$_{null}$ amorph' type) is based on the presence of silent alleles at the *RH* locus ($\bar{r}$) which do not encode **Rh** antigens;
- **Rh**$_{mod}$ ($\rightarrow$ '**Rh**$_{null}$ regulator' type), the more common **Rh**-deficient phenotype, is caused by the presence of an autosomal recessive suppressor gene (originally termed *X°r* [56,158]) which is genetically independent of the *RH* locus; when present on both chromosomes, this gene impairs the expression of the **Rh** antigens on the erythrocytes.

Thus far only two different mutations causing the **Rh**$_{null}$ (amorph) phenotype have been found (in both cases it was found that the *RHD* gene was deleted):

- a G $\rightarrow$ T mutation at the 5' donor splice site in intron 4 activating three cryptic splice-sites, two within intron 4 and one in exon 4; the aberrant transcripts derived are characterised by inserted amino acid sequences, exon skipping, and premature termination codons [48], and
- a non-contiguous deletion of two nucleotides in two adjacent codons (322 and 323) of *RHCE* exon 7 resulting in a TCA $\rightarrow$ C nucleotide change [48,113]. Due to frameshift and premature termination the variant gene encodes a shortened protein with 398 amino acids with only 10 transmembrane domains and a different C-terminus. Change in conformation of the encoded polypeptide (the **RhCe**-like transcript is readily transcribed) obviously affects the interaction of the **RHCe**-like protein with **RhAG**, thus impairing the correct assembly of the **Rh** complex in the erythrocyte membrane.

---

[5] A low amount of the **RhAG** glycoprotein is found in **Rh**$_{null}$ **U+** cells, which react weakly with anti-**U** antibodies; **RhAG**, however, is absent in **Rh**$_{null}$ **U–** cells [176].

In contrast to **Rh**$_{null}$ individuals, **Rh**$_{mod}$ subjects contain functional **RH** genes [53]. Family studies indicate that **Rh**$_{mod}$ (regulator type) individuals, though unable to express **Rh** antigens, are able to transmit intact **RH** genes from one generation to the next [205].

The investigations by Chérif-Zahar et al. [52] have shown that mutant alleles of **RHAG**, which impair the structure or synthesis of **RhAG** glycoprotein, represent the suppressor of the **RH** locus accounting for **Rh**-deficiency of the regulator type. Thus, the regulation in this case is not transcriptional but is related to defective assembly or transport of the **Rh** membrane complex to the cell surface.

Analysis of the genes and transcripts encoding the **RhAG** glycoprotein has revealed various alterations, such as frameshift, nucleotide mutations, defective splice site, or failure of amplification:

(1) Two nucleotide changes and a 2-bp-deletion (CCTC→GA), which introduce a frameshift after the codon for Tyr^{51} and result in a premature stop codon at nucleotides 323–325 [52].

(2) A single base deletion (adenine) at nucleotide 1086 introduces a frameshift after the codon for Ala^{362} and results in a premature stop codon at nucleotides 1130–132 [52].

(3) A Ser^{79} → Ala substitution in the second hydrophilic segment of the **RhAG** protein caused by a G → A transition at nucleotide 236 [52]; this mutant protein is poorly expressed at the cell surface (20% of its normal level).

(4) A single base exchange (G → A) at nucleotide 836 causing a Gly^{279} → Glu substitution at position 279 [116,124]. This amino acid is situated at the centre of the 9th predicted transmembrane domain of the **RhAG** polypeptide. Since in this case no **RhAG** protein could be detected at the erythrocyte surface, it can be assumed that the negatively charged Glu^{279} residue disrupts the continuity of the hydrophobic segment and, upon inducing a change in conformation, impairs the interaction of the mutant protein with the **Rh** polypeptides.

(5) A G → A mutation in the invariant GT element of 5' donor splice site caused complete skipping of exon 7 from the mature form of **RhAG** mRNA and resulted in a frameshift after the codon for Thr^{315} and a premature chain termination after the codon for Ile^{351} [109]; the deduced **RhAG** mutant protein contains only 351 amino acids, including a stretch of 36 new residues at the C-terminus.

(6) A single base substitution (G → T) at the consensus sequence of the 3' acceptor splice site of intron 6 caused a deletion of 122 bp (nucl. 946–1067) [47,137]; skipping of exon 7 and premature termination due to a frameshift effected a major structural change after the membrane-spanning domain 11.

(7) Splicing was completely inactivated in one mutant – characterised by a defective donor splice site caused by a G → A exchange in the invariant GT element of intron 1 [47,62,116].

(8) A G → T transversion in the ATG codon for translation initiation has been found in the **RHAG** gene of one family [115]. This mutation does not alter the gene expression at the level of transcription or splicing; in *in vitro* translation experiments it led to alternative translation initiation at adjacent ATGs (codons 8 and 16). Traces of **RhAG** protein and **Rh** antigens detected in the erythrocyte membranes of the proband suggest that this 'leaky' translation also occurs *in vivo*.

(9) In Rh_{null}(H.T.) two G → A transitions have been detected in **RHAG** exon 6, which cause Val^{270} → Ile and Gly^{280} → Arg substitutions [114]. These missense mutations are predicted to reside in endoloop 5 and transmembrane segment 9, respectively.

(10) Rh_{null}(W.O.) is characterised by a single G → T transversion in exon 9, which causes a Gly^{380} → Val missense change in transmembrane segment 12 [114].

References

1. AGRE, P., SABOORI, A. M., ASIMOS, A. & SMITH, B. L. (1987): Purification and partial characterization of the M, 30,000 integral membrane protein associated with the erythrocyte Rh(D) antigen. *J. Biol. Chem. 262*, 17497-17503.

2. AGRE, P. C., DAVIES, D. M., ISSITT, P. D., LAMY, B. M., SCHMIDT, P. J. & TREACY (1992): A proposal to standardize terminology for weak D antigen. *Transfusion 32*, 86-87.

3. ALLEN, F. H. & TIPPETT, P. A. (1958): A new Rh blood type which reveals the Rh antigen G. *Vox Sang. 3*, 321-330.

4. ALTER, A. A., GELB, A. G., CHOWN, B., ROSENFIELD, R. E. & CLEGHORN, T. E. (1967): Gonzales (Go^a), a new blood group character. *Transfusion 7*, 88-91.

5. ANDREWS, K. T., WOLTER, L. C., SAUL, A. & HYLAND, C. A. (1998): The RhD˙ trait in a white patient with the RhCCee phenotype attributed to a four-nucleotide deletion in the RHD gene. *Blood 92*, 1839-1840.

6. APOIL, P. A. & BLANCHER, A. (1999): Sequences and evolution of mammalian RH gene transcripts and proteins. *Immunogenetics 49*, 15-25.

7. ARASZKIEWICZ, P. & SZYMANSKI, I. O. (1987): Quantitative studies on the Rh-antigen D. Effect of the C gene. *Transfusion 27*, 257-261.

8. ARCE, M. A., THOMPSON, E. S., WAGNER, S., COYNE, K. E., FERDMAN, B. A. & LUBLIN, D. M. (1993): Molecular cloning of RhD cDNA derived from a gene present in RhD-positive, but not RhD-negative individuals. *Blood 82*, 651-655.

9. AVENT, N. D., JONES, J. W., LIU, W., L., S. M., VOAK, D., PISACKA, M., WATT, J. & FLETCHER, A. (1996): Molecular basis of the D^{VI} variant phenotype: evidence that a RHD gene deletion event does not generate all cD^{VI}E haplotypes. (Abstract). *Transfus. Clin. Biol. 3*, 34s.

10. AVENT, N. D., JONES, J. W., LIU, W., SCOTT, M. L., VOAK, D., FLEGEL, W. A., WAGNER, F. F. & GREEN, C. (1997): Molecular basis of the D variant phenotypes DNU and D^{II} allows localization of critical amino acids required for expression of RhD epitopes epD3, 4, and 9 to the sixth external domain of the RhD protein. *Br. J. Haematol. 97*, 366-371.

11. AVENT, N. D., LIU, W., JONES, J. W., SCOTT, M. L., VOAK, D., PISACKA, M., WATT, J. & FLETCHER, A. (1997): Molecular analysis of Rh transcripts and polypeptides from individuals expressing the D^{VI}

variant phenotype: an RHD gene deletion event does not generate all D^{VI}ccEe phenotypes. *Blood 89*, 1779-1786.

12. AVENT, N. D., LIU, W., WARNER, K. M., MAWBY, W. J., JONES, J. W., RIDGWELL, K. & TANNER, M. J. A. (1996): Immunochemical analysis of the human erythrocyte Rh polypeptides. *J. Biol. Chem. 271*, 14233-14239.

13. AVENT, N. D., MARTIN, P. G., ARMSTRONG-FISHER, S. S., LIU, W., FINNING, K. M., MADDOCKS, D. & URBANIAK, S. J. (1997): Evidence of genetic diversity underlying RhD⁻, weak D (D^u), and partial D phenotypes as determined by multiplex polymerase chain reaction analysis of the RHD gene. *Blood 89*, 2568-2577.

14. AVENT, N. D., RIDGWELL, K., MAWBY, W. J., TANNER, M. J. A., ANSTEE, D. J. & KUMPEL, B. (1988): Protein-sequence studies on Rh-related polypeptides suggest the presence of at least two groups of proteins which associate in the human red-cell membrane. *Biochem. J. 256*, 1043-1046.

15. AVENT, N. D., RIDGWELL, K., TANNER, M. J. A. & ANSTEE, D. J. (1990): cDNA cloning of a 30 kDa erythrocyte membrane protein associated with Rh (Rhesus)-blood group-antigen expression. *Biochem. J. 271*, 821-825.

16. BALLAS, S. K., CLARK, M. R., MOHANDAS, N., COLFER, H. F., CASWELL, M. S., BERGREN, M. O., PERKINS, H. A. & SHOHET, S. B. (1984): Red cell membrane and cation deficiency in Rh null syndrome. *Blood 63*, 1046-1055.

17. BALLAS, S. K., FLYNN, J. C., PAULINE, L. A. & MURPHY, D. L. (1986): Erythrocyte Rh antigens increase with red cell age. *Am. J. Hematol. 23*, 19-24.

18. BASU, M. K., FLAMM, M., SCHACHTER, D., BERTLES, J. F. & MANIATIS, A. (1980): Effects of modulating erythrocyte membrane cholesterol on Rh$_o$(D) antigen expression. *Biochem. Biophys. Res. Commun. 95*, 887-893.

19. BECKERS, E. A. M., FAAS, B. H. W., LIGTHART, P., OVERBEEKE, M. A. M., VON DEM BORNE, A. E. G. K., VAN DER SCHOOT, C. E. & VAN RHENEN, D. J. (1997): Lower antigen site density and weak D immunogenicity cannot be explained by structural genomic abnormalities or regulatory defects of the RHD gene. *Transfusion 37*, 616-623.

20. BECKERS, E. A. M., FAAS, B. H. W., LIGTHART, P., SIMSEK, S., OVERBEEKE, M. A. M., VON DEM BORNE, A. E. G. K., VAN RHENEN, D. J. & VAN DER SCHOOT, C. E. (1996): Characterization of the hybrid RHD gene leading to the partial D category IIIc phenotype. *Transfusion 36*, 567-574.

21. BECKERS, E. A. M., FAAS, B. H. W. & OVERBEEKE, M. A. M. (1995): Molecular aspects of the weak-D phenotype (abstract). *Transfusion 35 (Suppl. 10S)*, 50S.

22. BECKERS, E. A. M., FAAS, B. H. W., SIMSEK, S., OVERBEEKE, M. A. M., VAN RHENEN, D. J., WALLACE, M., VON DEM BORNE, A. E. G. K. & VAN DER SCHOOT, C. E. (1996): The genetic basis of a new partial D antigen: D^{DBT}. *Br. J. Haematol. 93*, 720-727.

23. BECKERS, E. A. M., FAAS, B. H. W., VON DEM BORNE, A. E. G. K., OVERBEEKE, M. A. M., VAN RHENEN, D. J. & VAN DER SCHOOT, C. E. (1996): The R$_o$^{Har} Rh:33 phenotype results from substitution of exon 5 of the RHCE gene by the corresponding exon of the RHD gene. *Br. J. Haematol. 92*, 751-757.

24. BECKERS, E. A. M., PORCELIJN, L., LIGTHART, P., VERMEY, H., VON DEM BORNE, A. E. G. K., OVERBEEKE, M. A. M. & VAN RHENEN, D. J. (1996): The R$_o$^{Har} antigenic complex is associated with a limited number of D epitopes and alloanti-D production: a study of three unrelated persons and their families. *Transfusion 36*, 104-108.

25. BENNETT, P. R., LE VAN KIM, C., COLIN, Y., WARWICK, R. M., CHERIF-ZAHAR, B., M., F. N. & CARTRON, J. P. (1993): Prenatal determination of fetal Rh D type by DNA amplification. *N. Engl. J. Med. 329*, 607-610.

26. BLANCHARD, D., BLOY, C., HERMAND, P., CARTRON, J. P., SABOORI, A. M., SMITH, B. L. & AGRE, P. (1988): Two-dimensional iodopeptide mapping demonstrates that erythrocyte Rh D, c, and E polypeptides are structurally homologous but nonidentical. *Blood 72*, 1424-1427.

27. BLOY, C., BLANCHARD, D., DAHR, W., BEYREUTHER, K., SALMON, C. & CARTRON, J. P. (1988): Determination of the N-terminal sequence of human red cell Rh(D) polypeptide and

demonstration the the Rh(D), (c), and (E) antigens are carried by distinct polypeptide chains. *Blood 72,* 661-666.

28. BLOY, C., BLANCHARD, D., LAMBIN, P., GOOSSENS, D., ROUGER, P., SALMON, C., MASOUREDIS, S. P. & CARTRON, J. P. (1988): Characterization of the D, c, E, and G antigens of the Rh blood group system with human monoclonal antibodies. *Mol. Immunol. 25,* 925-930.

29. BLUNT, T., DANIELS, G. & CARRITT, B. (1994): Serotype switching in a partially deleted RHD gene. *Vox Sang. 67,* 397-401.

30. BLUNT, T., STEERS, F., DANIELS, G. & CARRITT, B. (1994): Lack of Rh C/E expression in the Rhesus D-- phenotype is the result of a gene deletion. *Ann. Hum. Genet. 58,* 19-24.

31. BOROCHOV, H., ABBOTT, R. E., SCHACHTER, D. & SHINITZKY, M. (1979): Modulation of erythrocyte membrane proteins by membrane cholesterol and lipid fluidity. *Biochemistry 18,* 251-255.

32. BOUREL, D., LECOINTRE, M., GENETET, N., GUEGUEN-DUCHESNE, M. & GENETET, B. (1987): Murine monoclonal antibody suitable for use as an Rh reagent, anti-e. *Vox Sang. 52,* 85-88.

33. BRON, D., FEINBERG, M. B., TENG, N. N. H. & KAPLAN, H. S. (1984): Production of human monoclonal IgG antibodies against Rhesus (D) antigen. *Proc. Natl. Acad. Sci. USA 81,* 3214-3217.

34. BROWN, E., HOOPER, L., HO, T. & GRESHAM, H. (1990): Integrin-associated protein: a 50-kDa plasma membrane antigen physically and functionally associated with integrins. *J. Cell Biol. 111,* 2785-2794.

35. BUSH, M., SABO, B., STROUP, M. & MASOUREDIS, S. P. (1974): Red cell D antigen sites and titration scores in a family with weak and normal Du phenotypes. *Transfusion 14,* 433-439.

36. CALLENDER, S. T. & RACE, R. R. (1946): A serological and genetical study of multiple antibodies formed in response to blood transfusion by a patient with lupus erythematosus diffusus. *Ann. Eugen. 13,* 102-117.

37. CAMPBELL, I. G., FREEMONT, P. S., FOULKES, W. & TROWSDALE, J. (1992): An ovarian tumor marker with homology to vaccinia virus contains as IgV-like region and multiple transmembrane domains. *Cancer Res. 52,* 5416-5420.

38. CAREN, L. D., BELLAVANCE, R. & GRUMET, F. C. (1982): Demonstration of gene dosage effects on antigens in the Duffy, Ss, and Rh systems using an enzyme-linked immunosorbent assay. *Transfusion 22,* 475-478.

39. CARRITT, B., STEERS, F. J. & AVENT, N. D. (1994): Prenatal determination of fetal RhD type. *Lancet 344,* 205-206.

40. CARTRON, J. P. (1994): Defining the Rh blood group antigens - Biochemistry and molecular genetics. *Blood Rev. 8,* 199-212.

40a. CARTRON, J.P. (1996): Vers une approche moléculaire de la structure, du polymorphisme et de la fonction des groupes sanguins. *Transfus. Clin. Biol. 3,* 181-210.

41. CARTRON, J. P., LE VAN KIM, C., CHERIF-ZAHAR, B., MOURO, I., ROUILLAC, C. & COLIN, Y. (1995): The two-gene model of the RH blood-group locus. *Biochem. J. 306,* 877-878.

42. CARTRON, J. P., ROUILLAC, C., LE VAN KIM, C., MOURO, I. & COLIN, Y. (1996): Tentative model for the mapping of D epitopes on the RhD polypeptide. *Transfus. Clin. Biol. 3,* 497-503.

43. CEPPELLINI, R., DUNN, L. C. & TURRI, M. (1955): An interaction between alleles at the Rh locus in man which weakens the reactivity of the Rh_o factor (Dᵘ). *Proc. Natl. Acad. Sci. USA 41,* 283-288.

44. CHANG, J. G., WANG, J. C., YANG, T. Y., TSAN, K. W. & SHIH, M. C. (1998): Human RhDᵉˡ is caused by a deletion of 1,013 bp between introns 8 and 9 including exon 9 of RHD gene. *Blood 92,* 2602-2604.

45. CHÉRIF-ZAHAR, B., BLOY, C., LE VAN KIM, C., BLANCHARD, D., BAILLY, P., HERMAND, P., SALMON, C., CARTRON, J. P. & COLIN, Y. (1990): Molecular cloning and protein structure of a human blood group Rh polypeptide. *Proc. Natl. Acad. Sci. USA 87,* 6243-6247.

46. CHÉRIF-ZAHAR, B., LE VAN KIM, C., ROUILLAC, C., RAYNAL, V., CARTRON, J. P. & COLIN, Y. (1994): Organization of the gene (RhCE) encoding the human blood group RhCcEe antigens and characterization of the promoter region. *Genomics 19,* 68-74.

47. CHÉRIF-ZAHAR, B., MATASSI, G., RAYNAL, V., GANE, P., DELAUNAY, J., ARRIZABALAGA, B. & CARTRON, J. P. (1998): Rh-deficiency of the regulator type caused by splicing mutations in the human RH50

gene. *Blood 92*, 2535-2540.

48. CHÉRIF-ZAHAR, B., MATASSI, G., RAYNAL, V., GANE, P., MEMPEL, W., PEREZ, C. & CARTRON, J. P. (1998): Molecular defects of the RHCE gene in Rh-deficient individuals of the amorph type. *Blood 92*, 639-646.

49. CHÉRIF-ZAHAR, B., MATTEI, M. G., LE VAN KIM, C., BAILLY, P., CARTRON, J. P. & COLIN, Y. (1991): Localization of the human Rh blood group gene structure to chromosome region 1p34.3-1p36.1 by in situ hybridization. *Hum. Genet. 86*, 398-400.

50. CHÉRIF-ZAHAR, B., RAYNAL, V. & CARTRON, J. P. (1996): Lack of RHCE-encoded proteins in the D-- phenotype may result from homologous recombination between the two RH genes. *Blood 88*, 1518-1520.

51. CHÉRIF-ZAHAR, B., RAYNAL, V., D'AMBROSIO, A. M., CARTRON, J. P. & COLIN, Y. (1994): Molecular analysis of the structure and expression of the RH locus in individuals with D--, Dc-, and DC^w-gene complexes. *Blood 84*, 4354-4360.

52. CHÉRIF-ZAHAR, B., RAYNAL, V., GANE, P., MATTEI, M. G., BAILLY, P., GIBBS, B., COLIN, Y. & CARTRON, J. P. (1996): Candidate gene acting as a suppressor of the RH locus in most cases of Rh-deficiency. *Nat. Genet. 12*, 168-173.

53. CHÉRIF-ZAHAR, B., RAYNAL, V., LE VAN KIM, C., D'AMBROSIO, A. M., BAILLY, P., CARTRON, J. P. & COLIN, Y. (1993): Structure and expression of the RH locus in the Rh-deficiency syndrome. *Blood 82*, 656-662.

54. CHOWN, B. & LEWIS, M. (1957): Occurrence of D^u type of reaction when CDe or cDE is partnered with Cde. *Am. J. Hum. Genet. 22*, 58-64.

55. CHOWN, B., LEWIS, M. & KAITA, H. (1971): The Rh system. An anomaly of inheritance, probably due to mutation. *Vox Sang. 21*, 385-396.

56. CHOWN, B., LEWIS, M., KAITA, H. & LOWEN, B. (1972): An unlinked modifier of Rh blood groups: effects when heterozygous and when homozygous. *Am. J. Hum. Genet. 24*, 623-637.

57. CHOWN, B., LEWIS, M., KAITA, H. & PHILLIPS, S. (1964): The Rh antigen D^w (Wiel). *Transfusion 4*, 169-172.

58. COBB, M. L. (1980): Crawford: investigation of a new low frequency red cell antigen (Abstract). *Transfusion 20*, 631.

59. COGHLAN, G., McCREARY, J., UNDERWOOD, V. & ZELINSKI, T. (1994): A 'new' low-incidence red cell antigen, LOCR, associated with altered expression of Rh antigens. *Transfusion 34*, 492-495.

60. COLIN, Y., CHÉRIF-ZAHAR, B., LE VAN KIM, C., RAYNAL, V., VAN HUFFEL, V. & CARTRON, J. P. (1991): Genetic basis of the RhD-positive and RhD-negative blood group polymorphism as determined by Southern analysis. *Blood 78*, 2747-2752.

61. CONTRERAS, M., STEBBING, B., BLESSING, M. & GAVIN, J. (1978): The Rh antigen Evans. *Vox Sang. 34*, 208-211.

62. COWLEY, N. M., SAUL, A., CARTRON, J. P. & HYLAND, C. A. (1999): A single point mutation at a splice site generates a silent RH50 gene in a composite heterozygous Rh$_{null}$ blood donor. *Vox Sang. 76*, 247-248.

63. CUNNINGHAM, M. A., ZOLA, A. P., HUI, H. L., TAYLOR, L. M. & GREEN F.A (1985): Binding characteristics of anti-Rh$_o$(D) antibodies to Rh$_o$(D)-positive and D^u red cells. *Blood 65*, 765-769.

64. DAHR, W., KORDOWICZ, M., MOULDS, J., GIELEN, W., LEBECK, L. & KRÜGER, J. (1987): Characterization of the Ss sialoglycoproteins and its antigens in Rh$_{null}$ erythrocytes. *Blut 54*, 13-24.

65. DANIELS, G. L. (1982): An investigation of the immune response of homozygotes for the Rh haplotype -D- and related haplotypes. *Blood Transfus. Immunohaematol. 25*, 185-197.

66. DANIELS, G. L., ANSTEE, D. J., CARTRON, J. P., DAHR, W., GARRATTY, G., HENRY, S., JÖRGENSEN, J., JUDD, W. J., KORNSTAD, L., LEVENE, C., LOMAS-FRANCIS, C., LUBENKO, A., MOULDS, J. J., MOULDS, J. M., MOULDS, M., OVERBEEKE, M., REID, M. E., ROUGER, P., SCOTT, M., SEIDL, S., SISTONEN, P., TANI, Y., WENDEL, S. & ZELINSKI, T. (1999): Terminology for red cell surface antigens - ISBT Working Party Oslo Report. *Vox Sang 77*, 52-57.

67. DANIELS, G. L., ANSTEE, D. J., CARTRON, J. P., DAHR, W., ISSITT, P. D., JORGENSEN, J., KORNSTAD, L., LEVENE, C., LOMAS-FRANCIS, C., LUBENKO, A., MALLORY, D., MOULDS, J. J., OKUBO, Y., OVERBEEKE,

M., REID, M. E., ROUGER, P., SEIDL, S., SISTONEN, P., WENDEL, S., WOODFIELD, G. & ZELINSKI, T. (1995): Blood group terminology 1995 - ISBT working party on terminology for red cell surface antigens. *Vox Sang. 69*, 265-279.

68. DANIELS, G. L., FAAS, B. H. W., GREEN, C. A., SMART, E., MAASKANT VAN WIJK, P. A., AVENT, N. D., ZONDERVAN, H. A., VON DEM BORNE, A. E. G. K. & VAN DER SCHOOT, C. E. (1998): The VS and V blood group polymorphisms in Africans: a serologic and molecular analysis. *Transfusion 38*, 951-958.

69. DANIELS, G. L., MOULDS, J. J., ANSTEE, D. J., BIRD, G. W. G., BRODHEIM, E., CARTRON, J. P., DAHR, W., ENGELFRIET, C. P., ISSITT, P. D., JORGENSEN, J., KORNSTAD, L., LEWIS, M., LEVENE, C., LUBENKO, A., MALLORY, D., MOREL, P., NORDHAGEN, R., OKUBO, Y., REID, M., ROUGER, P., SALMON, C., SEIDL, S., SISTONEN, P., WENDEL, S., WOODFIELD, G. & ZELINSKI, T. (1993): ISBT Working Party on Terminology for Red Cell Surface Antigens - Sao Paulo report. *Vox Sang. 65*, 77-80.

70. DAVIDSOHN, I., STERN, K., STRAUSER, E. R. & SPURRIER, W. (1953): Be, a new 'private' blood factor. *Blood 8*, 747-754.

71. DE VETTEN, M. P. & AGRE, P. (1988): The Rh polypeptide is a major fatty acid-acylated erythrocyte membrane protein. *J. Biol. Chem. 263*, 18193-18196.

72. DELEHANTY, C. L., WILKINSON, S. L., ISSITT, P. D., TIPPETT, P., LOMAS, C. & WEILAND, D. (1983): Riv: a new low incidence Rh antigen (Abstract). *Transfusion 23*, 410.

73. DENATALE, A., CAHAN, A., A., J. J., RACE, R. R. & SANGER, R. (1955): V: a 'new' Rh antigen, common in Negroes, rare in white people. *J. Am. Med. Assoc. 159*, 247-250.

74. DUNSFORD, I. (1961): A new Rh anti-body - anti-CE. *Proceedings 8th Europ. Soc Haematol. Vienna 1961, paper No. 491.*

75. EYERS, S. A. C., RIDGWELL, K., MAWBY, W. J. & TANNER, M. J. A. (1994): Topology and organization of human Rh (Rhesus) blood group-related polypeptides. *J. Biol. Chem. 269*, 6417-6423.

76. FAAS, B. H. W., BECKERS, E. A. M., SIMSEK, S., OVERBEEKE, M. A. M., PEPPER, R., VAN RHENEN, D. J., VON DEM BORNE, A. E. G. K. & VAN DER SCHOOT, C. E. (1996): Involvement of Ser103 of the Rh polypeptides in G epitope formation. *Transfusion 36*, 506-511.

77. FAAS, B. H. W., BECKERS, E. A. M., WILDOER, P., LIGTHART, P. C., OVERBEEKE, M. A., ZONDERVAN, H. A., VON DEM BORNE, A. E. & VAN DER SCHOOT, C. E. (1997): Molecular background of VS and weak C expression in blacks. *Transfusion 37*, 38-44.

78. FAAS, B. H. W., LIGTHART, P. C., LOMAS-FRANCIS, C., OVERBEEKE, M. A. M., VON DEM BORNE, A. E. G. K. & VAN DER SCHOOT, C. E. (1997): Involvement of Gly96 in the formation of the Rh26 epitope. *Transfusion 37*, 1123-1130.

79. FAAS, B. H. W., SIMSEK, S., BLEEKER, P. M. M., OVERBEEKE, M. A. M., CUIJPERS, H. T. M., VON DEM BORNE, A. E. G. K. & VAN DER SCHOOT, C. E. (1995): Rh E/e genotyping by allele-specific primer amplification. *Blood 85*, 829-832.

80. FALKENBURG, J. H. F., FIBBE, W. E., VAN DER VAART-DUINKERKEN, N., NICHOLS, M. E., RUBINSTEIN, P. & JANSEN, J. (1985): Human erythroid progenitor cells express Rhesus antigens. *Blood 66*, 660-663.

81. FLEGEL, W. A. & WAGNER, F. F. (1996): RHD epitope density profiles of RHD variant red cells analzed by flow cytometry. *Transfus. Clin. Biol. 3*, 429-433.

82. FOUNG, S. K. H., BLUNT, J., PERKINS, S., WINN, L. & GRUMET, F. C. (1986): A human monoclonal antibody to rh^G. *Vox Sang. 50*, 160-163.

83. FOUNG, S. K. H., BLUNT, J. A., WU, P. S., AHEARN, P., WINN, L. C., ENGLEMAN, E. G. & GRUMET, F. C. (1987): Human monoclonal antibodies to Rh$_o$(D). *Vox Sang. 53*, 44-47.

84. FRASER, R. H., INGLIS, G., ALLAN, J. C., MURPHY, M. T., ALLAN, E. K., MACKIE, A. & MITCHELL, R. (1990): Murine monoclonal antibody with anti-e-like specificity: Suitability for screening for e-negative cells. *Transfusion 30*, 226-229.

85. FUKUMORI, Y., HORI, Y., OHNOKI, S., NAGAO, N., SHIBATA, H., OKUBO, Y. & YAMAGUCHI, H. (1997): Further analysis of D$_{el}$ (D-elute) using polymerase chain reaction (PCR) with RHD gene-specific primers. *Transfus. Med. 7*, 227-231.

86. GAHMBERG, C. G. (1983): Molecular characterization of the human red cell Rh$_o$(D) antigen. *EMBO*

J. 2, 223-227.

87. GAHMBERG, C. G. & KARHI, K. K. (1984): Association of Rh$_o$(D) polypeptides with the membrane skeleton in Rh$_o$(D)-positive human red cells. *J. Immunol. 133*, 334-337.

88. GASSNER, C., SCHMARDA, A., KILGA-NOGLER, S., JENNY-FELDKIRCHER, B., RAINER, E., MÜLLER, T. H., WAGNER, F. F., FLEGEL, W. A. & SCHÖNITZER, D. (1997): RHD/CE typing by polymerase chain reaction using sequence-specific primers. *Transfusion 37*, 1020-1026.

89. GELDWERTH, D., CHERIF-ZAHAR, B., HELLEY, D., GANE, P., FREYSSINET, J. M., COLIN, Y., DEVAUX, P. F. & CARTRON, J. P. (1997): Phosphatidylserine exposure and aminophospholipid translocase activity in Rh-deficient erythrocytes. *Mol. Membr. Biol. 14*, 125-132.

90. GILES, C. M., CROSSLAND, J. D., HAGGAS, W. K. & LONGSTER, G. (1971): An Rh gene complex which results in a 'new' antigen detectable by a specific antibody, anti-Rh33. *Vox Sang. 21*, 289-301.

91. GILES, C. M. & SKOV, F. (1971): The CDe rhesus gene complex; some considerations reveled by a study of a Danish family with an antigen of the rhesus gene complex (C)D(e) defined by a 'new' antibody. *Vox Sang. 20*, 328-334.

92. GREEN, C. A., LOMAS-FRANCIS, C., WALLACE, M., GOOCH, A. & SIMONSEN, A. C. (1996): Family evidence confirms that the low frequency antigen BARC is an Rh antigen (abstract). *Transfus. Med. 6 (Suppl. 2)*, 26.

93. GREEN, F. A. (1966): Sulfhydryl reagents and Rh activity of erythrocyte stroma. *Nature 211*, 852-853.

94. GREEN, F. A. (1967): Erythrocyte membrane sulfhydryl groups and Rh antigen activity. *Immunochemistry 4*, 247-257.

95. GREEN, F. A. (1968): Phospholipid requirement for Rh antigenic activity. *J. Biol. Chem. 243*, 5519-5524.

96. GREEN, F. A. (1968): Rh antigenicity: an essential component soluble in butanol. *Nature 219*, 86-87.

97. GREEN, F. A. (1972): Erythrocyte membrane lipids and Rh antigen activity. *J. Biol. Chem. 247*, 881-887.

98. GREEN, F. A. (1983): The mode of attenuation of erythrocyte membrane Rh$_o$(D) antigen activity by 5,5'-dithiobis-(2-nitrobenzoic acid) and protection against loss of activity by bound anti-Rh$_o$(D) antibody. *Mol. Immunol. 20*, 769-775.

99. GREEN, F. A., HUI, H. L., GREEN, L. A. D., HEUBUSCH, P. & PUDLAK, W. (1984): The phospholipid requirement for Rh$_o$(D) antigen activity: mode of inactivation by phospholipases and of protection by anti-Rh$_o$(D) antibody. *Mol. Immunol. 21*, 433-438.

100. GREENWALT, T. J. & SANGER, R. (1955): The Rh antigen E^w. *Br. J. Haematol. 1*, 52-54.

101. HABIBI, B., PERRIER, P. & SALMON, C. (1981): Antigen Nou. *Blood Transfus. Immunohaematol. 24*, 117-120.

102. HARTEL-SCHENK, S. & AGRE, P. (1992): Mammalian red cell membrane Rh polypeptides are selectively palmitoylated subunits of a macromolecular complex. *J. Biol. Chem. 267*, 5569-5574.

103. HASEKURA, H., OTA, M., ITO, S., HASEGAWA, Y., ICHINOSE, A., FUKUSHIMA, H. & OGATA, H. (1990): Flow cytometric studies of the D antigen of various Rh phenotypes with particular reference to D^u and D^{el}. *Transfusion 30*, 236-238.

104. HERMAND, P., MOURO, I., HUET, M., BLOY, C., SUYAMA, K., GOLDSTEIN, J., CARTRON, J. P. & BAILLY, P. (1993): Immunochemical characterization of Rhesus proteins with antibodies raised against synthetic peptides. *Blood 82*, 669-676.

105. HOFFMANN, J. J. M. L., OVERBEEKE, M. A. M., KAITA, H. & LOOMANS, A. A. H. (1990): A new, low-incidence red cell antigen (HOFM), associated with depressed C-antigen. *Vox Sang. 59*, 240-243.

106. HOPKINSON, D. A. (1993): The long [E/e] and the short [C/c] of the rhesus polymorphism. *Nat. Genet. 5*, 6-7.

107. HUANG, C. H. (1996): Alteration of RH gene structure and expression in human dCCee and DC^w- red blood cells: phenotypic homozygosity versus genotypic heterozygosity. *Blood 88*, 2326-2333.

108. HUANG, C. H. (1997): Human D^{VI} category erythrocytes: correlation of the phenotype with a novel hybrid RhD-CE-D gene but not an internally deleted RhD gene. *Blood 89*, 1834-1835.

108a. HUANG, C.H. (1997): Molecular insights into the Rh protein family and associated antigens. *Curr. Opin. Hematol. 4*, 94-103.

109. HUANG, C. H. (1998): The human Rh50 glycoprotein gene. Structural organization and associated splicing defect resulting in Rh$_{null}$ disease. *J. Biol. Chem. 273*, 2207-2213.

110. HUANG, C. H., CHEN, Y. & REID, M. (1997): Human D^{IIIa} erythrocytes: RhD protein is associated with multiple dispersed amino acid variations. *Am. J. Hematol. 55*, 139-145.

111. HUANG, C. H., CHEN, Y., REID, M. & GHOSH, S. (1996): Genetic recombination at the human RH locus: a family study of the red-cell Evans phenotype reveals a transfer of exons 2-6 from the RHD to the RHCE gene. *Am. J. Hum. Genet. 59*, 825-833.

112. HUANG, C. H., CHEN, Y., REID, M. E. & OKUBO, Y. (1999): Evidence for a separate genetic origin of the partial D phenotype DBT in a Japanese family. *Transfusion 39*, 1259-1265.

113. HUANG, C. H., CHEN, Y., REID, M. E. & SEIDL, C. (1998): Rh$_{null}$ disease: the amorph type results from a novel double mutation in RhCe gene on D-negative background. *Blood 92*, 664-671.

114. HUANG, C. H., CHENG, G. J., LIU, Z., CHEN, Y., REID, M. E., HALVERSON, G. & OKUBO, Y. (1999): Molecular basis for Rh-null syndrome: Identification of three new missense mutations in the Rh50 glycoprotein gene. *Am. J. Hematol. 62*, 25-32.

115. HUANG, C. H., CHENG, G. J., REID, M. E. & CHEN, Y. (1999): Rh$_{mod}$ syndrome: a family study of the translation-initiator mutation in the Rh50 glycoprotein gene. *Am. J. Hum. Genet. 64*, 108-117.

116. HUANG, C. H., LIU, Z., CHENG, G. J. & CHEN, Y. (1998): Rh50 glycoprotein gene and Rh$_{null}$ disease: a silent splice donor is trans to a Gly$_{279}$→Glu missense mutation in the conserved transmembrane segment. *Blood 92*, 1776-1784.

117. HUANG, C. H., REID, M. E. & CHEN, Y. (1995): Identification of a partial internal deletion in the RH locus causing the human erythrocyte D-- phenotype. *Blood 86*, 784-790.

118. HUANG, C. H., REID, M. E., CHEN, Y., COGHLAN, G. & OKUBO, Y. (1996): Molecular definition of red cell Rh haplotypes by tightly linked SphI RFLPs. *Am. J. Hum. Genet. 58*, 133-142.

119. HUESTIS, D. W., CANTINO, M. L. & BUSCH, S. A. (1964): A 'new' Rh antibody (anti-Rh26) which detects a factor usually accompanying hr'. *Transfusion 4*, 414-418.

120. HUGHES-JONES, N. C., BLOY, C., GORICK, B., BLANCHARD, D., DOINEL, C., ROUGER, P. & CARTRON, J. P. (1988): Evidence that the c, D, and E epitopes of the human Rh blood group system are on separate polypeptide molecules. *Mol. Immunol. 25*, 931-936.

121. HUGHES-JONES, N. C., GARDNER, B. & LINCOLN, P. J. (1971): Observations of the number of available c, D, and E antigen sites on red cells. *Vox Sang. 21*, 210-216.

122. HUGHES-JONES, N. C., GARDNER, B. & TELFORD, R. (1962): The kinetics of the reaction between the blood-group antibody anti-c and erythrocytes. *Biochem. J. 85*, 466-474.

123. HUGHES-JONES, N. C., GREEN, E. J. & HUNT, V. A. N. (1975): Loss of Rh antigen activity following the action of phospholipase A$_2$ on red cell stroma. *Vox Sang. 29*, 184-191.

124. HYLAND, C. A., CHÉRIF-ZAHAR, B., COWLEY, N., RAYNAL, V., PARKES, J., SAUL A & CARTRON, J. P. (1998): A novel single missense mutation identified along the RH50 gene in a composite heterozygous Rh$_{null}$ blood donor of the regulator type. *Blood 91*, 1458-1463.

125. HYLAND, C. A., WOLTER, L. C., LIEW, Y. W. & SAUL, A. (1994): A Southern analysis of Rh blood group genes: association between restriction fragment length polymorphism patterns and Rh serotypes. *Blood 83*, 566-572.

126. HYLAND, C. A., WOLTER, L. C. & SAUL, A. (1994): Three unrelated Rh D gene polymorphisms identified among blood donors with rhesus CCee (r'r') phenotypes. *Blood 84*, 321-324.

127. ISSITT, P. D. (1979): *Serology and Genetics of the Rhesus Blood Group System*. Montgomery Scientific Publications, Cincinnati, Ohio.

128. ISSITT, P. D. (1985): The Rh blood group system. In: *Applied Blood Group Serology*. Montgomery Scientific Publications, Miami, Florida, USA, pp. 219-277.

129. ISSITT, P. D., PAVONE, B. G. & SHAPIRO, M. (1979): Anti-Rh39 – a 'new' specificity Rh system antibody. *Transfusion 19*, 389-397.

130. IWAMOTO, S., OMI, T., YAMASAKI, M., OKUDA, H., KAWANO, M. & KAJII, E. (1998): Identification of 5' flanking sequence of RH50 gene and the core region for erythroid-specific expression. *Biochem.*

Biophys. Res. Commun. 243, 233-240.

131. JONES, J. (1995): Identification of two new D variants, D^HMi and D^HMii using monoclonal anti-D. *Vox Sang. 69*, 236-241.

132. JONES, J., SCOTT, M. L. & VOAK, D. (1995): Monoclonal anti-D specificity and Rh D structure: criteria for selection of monoclonal anti-D reagents for routine typing of patients and donors. *Transfus. Med. 5*, 171-184.

133. JONES, J. W., FINNING, K., MATTOCK, R., WILLIAMS, M., VOAK, D., SCOTT, M. L. & AVENT, N. D. (1997): The serological profile and molecular basis of a new partial D phenotype, DHR. *Vox Sang. 73*, 252-256.

134. JONES, J. W., LLOYD-EVANS, P. & KUMPEL, B. M. (1996): Quantitation of Rh D antigen sites on weak D and D variant red cells by flow cytometry. *Vox Sang. 71*, 176-183.

135. KAJII, E., UMENISHI, F., IWAMOTO, S. & IKEMOTO, S. (1993): Isolation of a new cDNA clone encoding an Rh polypeptide associated with the Rh blood group system. *Hum. Genet. 91*, 157-162.

136. KAPPES, D. & STROMINGER, J. L. (1988): Human class II major histocompatibility complex genes and proteins. *Annu. Rev. Biochem. 57*, 991-1128.

137. KAWANO, M., IWAMOTO, S., OKUDA, H., FUKUDA, S., HASEGAWA, N. & KAJII, E. (1998): A splicing mutation of the RHAG gene associated with the Rh_null phenotype. *Ann. Hum. Genet. 62*, 107-113.

138. KEITH, P., CORCORAN, P. A., CASPERSEN, K. & ALLEN, F. H. (1965): A new antibody, anti-Rh27 (cE) in the Rh blood group system. *Vox Sang. 10*, 528-535.

139. KEMP (1996): A recombination hot spot in the Rh genes revealed by analysis of unrelated donors with the Rare D - phenotype. *Am. J. Hum. Genet. 59,* 1066-1073.

140. KEMP, T. J., POULTER, M. & CARRITT, B. (1996): A recombination hot spot in the Rh genes revealed by analysis of unrelated donors with the rare D-- phenotype. *Am. J. Hum. Genet. 59*, 1066-1073.

141. KONUGRES, A. A., POLESKY, H. F. & WALKER, R. H. (1982): Rh immune globulin and the Rh-positive, D^u variant, mother. *Transfusion 22*, 76-77.

142. KORNSTAD, L. (1986): A rare blood group antigen, Ol^a (Oldeide), associated with weak Rh antigens. *Vox Sang 50*, 235-239.

143. KUMPEL, B. M., GOODRICK, M. J., PAMPHILON, D. H., FRASER, I. D., POOLE, G. D., MORSE, C., STANDEN, G. R., CHAPMAN, G. E., THOMAS, D. P. & ANSTEE, D. J. (1995): Human Rh D monoclonal antibodies (BRAD-3 and BRAD-5) cause accelerated clearance of Rh D+ red blood cells and suppression of Rh D immunization in Rh D- volunteers. *Blood 86*, 1701-1709.

144. KUMPEL, B. M., POOLE, G. D. & BRADLEY, B. A. (1989): Human monoclonal anti-D antibodies. I. Their production, serology, quantitation, and potential use as blood grouping reagents. *Br. J. Haematol. 71*, 125-139.

145. KUYPERS, F., VAN LINDE-SIBENIUS-TRIP, M., ROELOFSON, B., TANNER, M. J. A., ANSTEE, D. J. & OP DEN KAMP, J. A. F. (1984): Rh_null human erythrocytes have an abnormal membrane phospholipid organization. *Biochem. J. 221*, 931-934.

146. LACEY, P. A., CASKEY, C. R., WERNER, D. J. & MOULDS, J. J. (1983): Fatal hemolytic disease of a newborn due to anti-D in an Rh-positive D^u variant mother. *Transfusion 23*, 91-93.

147. LANDSTEINER, K. & WIENER, A. S. (1940): An agglutinable factor in human blood recognized by immune sera for rhesus blood. *Proc. Soc. Exp. Biol. N.Y. 43*, 223.

148. LANDSTEINER, K. & WIENER, A. S. (1941): Studies on an agglutinogen (Rh) in human blood reacting with anti-rhesus sera and with human isoantibodies. *J. Exp. Med. 74*, 309-320.

149. LAUF, P. K. & JOINER, C. H. (1976): Increased potassium transport and ouabain binding in human Rh_null red blood cells. *Blood 48*, 457-468.

150. LE PENNEC, P. Y., ROUGER, P., KLEIN, M. T., KORNPROBST, M., BROSSARD, Y., BOIZARD, B. & SALMON, C. (1989): A serologic study of red cells and sera from 18 Rh:32,-46 (R^N/R^N persons). *Transfusion 29*, 798-802.

151. LE VAN KIM, C., CHÉRIF-ZAHAR, B., RAYNAL, V., MOURO, I., LOPEZ, M., CARTRON, J. P. & COLIN, Y. (1992): Multiple Rh messenger RNA isoforms are produced by alternative splicing. *Blood 80*, 1074-1078.

152. LE VAN KIM, C., MOURO, I., BROSSARD, Y., CHAVINIE, J., CARTRON, J. P. & COLIN, Y. (1994): PCR-based determination of Rhc and RhE status of fetuses at risk of Rhc and RhE haemolytic disease. *Br. J. Haematol. 88*, 193-195.
153. LE VAN KIM, C., MOURO, I., CHÉRIF-ZAHAR, B., RAYNAL, V., CHERRIER, C., CARTRON, J. P. & COLIN, Y. (1992): Molecular cloning and primary structure of the human blood group RhD polypeptide. *Proc. Natl. Acad. Sci. USA 89*, 10925-10929.
154. LEADER, K. A., KUMPEL, B. M., POOLE, G. D., KIRKWOOD, J. T., MERRY, A. H. & BRADLEY, B. A. (1990): Human monoclonal anti-D with reactivity against category D^{VI} cells used in blood grouping and determination of the incidence of the category D^{VI} phenotype in the D^U population. *Vox Sang. 58*, 106-111.
155. LEGLER, T. J., MAAS, J. H., BLASCHKE, V., MALEKAN, M., OHTO, H., LYNEN, R., BUSTAMI, N., SCHWARTZ, D. W. M., MAYR, W. R., KOHLER, M. & PANZER, S. (1998): RHD genotyping in weak D phenotypes by multiple polymerase chain reactions. *Transfusion 38*, 434-440.
156. LEVINE, P., BURNHAM, L., KATZIN, E. M. & VOGEL, P. (1941): The role of isoimmunization in the pathogenesis of erythroblastosis fetalis. *Am. J. Obstet. Gynec. 42*, 925-937.
158. LEVINE, P., CELAMJ, VOS, G. H. & MORRISON, J. (1963): The first human blood; ---/---, which lacks the 'D-like' antigen. *Nature 194*, 304-305.
159. LEVINE, P. & ROSENFIELD, R. E. (1961): The first example of the Rh phenotype r^{G}r^G. *Am. J. Hum. Genet. 13*, 299-305.
160. LEVINE, P. & STETSON, R. E. (1939): An unusual case of intragroup agglutination. *J. Am. Med. Assoc. 113*, 126-127.
161. LEWIS, M., ANSTEE, D. J., BRID, G. W. G., BRODHEIM, E., CARTRON, J. P., CONTRERAS, M., CROOKSTON, M. C., DAHR, W., DANIELS, G. L., ENGELFRIET, C. P., GILES, C. M., ISSITT, P. D., JØRGENSEN, J., KORNSTAD, L., LUBENKO, A., MARSH, W. L., MCCREARY, J., MOORE, B. P. L., MOREL, P., MOULDS, J. J., NEVANLINNA, H., NORDHAGEN, R., OKUBO, Y., ROSENFIELD, R. E., ROUGER, P., RUBINSTEIN, P., SALMON, C., SEIDL, S., SISTONEN, P., TIPPETT, P., WALKER, R. H., WOODFIELD, G. & YOUNG, S. (1990): Blood group terminology 1990. *Vox Sang. 58*, 152-169.
162. LEWIS, M., B., C., KAITA, H., HAHN, D., KANGELOS, M., SHEPARD, W. L. & SHACKLETON, K. (1967): Blood group antigen Go^a and the Rh system. *Transfusion 7*, 440-441.
163. LEWIS, M., KAITA, H., ALLDERDICE, P. W., BARTLETT, S., SQUIRES, W. G. & HUNTSMAN, R. G. (1979): Assignment of the red cell antigen Targett (Rh40) to the Rh blood group system. *Am. J. Hum. Genet. 31*, 630-633.
164. LIGHTEN, A. D., OVERTON, T. G., SEPULVEDA, W., WARWICK, R. M., FISK, N. M. & BENNETT, P. R. (1995): Accuracy of prenatal determination of RhD type status by polymerase chain reaction with amniotic cells. *Am. J. Obstet. Gynec. 173*, 1182-1185.
165. LINDBERG, F. P., GRESHAM, H. D., SCHWARZ, E. & BROWN, E. J. (1993): Molecular cloning of integrin-associated protein: an immunoglobulin family member with multiple membrane-spanning domains implicated in $\alpha_v\beta_3$-dependent ligand binding. *J. Cell Biol. 123*, 485-496.
166. LINDBERG, F. P., LUBLIN, D. M., TELEN, M. J., VEILE, R. A., MILLER, Y. E., DONIS-KELLER, H. & BROWN, E. J. (1994): Rh-related antigen CD47 is the signal-transducer integrin-associated protein. *J. Biol. Chem. 269*, 1567-1570.
167. LIU, W., JONES, J. W., SCOTT, M. L., VOAK, D. & AVENT, N. D. (1996): Molecular analysis of two D-variants, D^{HMi} and D^{HMii}. *Transfus. Med. 6 (Suppl. 2)*, O28.
168. LOMAS, C., GRÄSSMANN, W., FORD, D., WATT, J., GOOCH, A., JONES, J., BEOLET, M., STERN, D., WALLACE, M. & TIPPETT, P. (1994): FPTT is a low-incidence Rh antigen associated with a 'new' partial Rh D phenotype, DFR. *Transfusion 34*, 612-616.
168a. LOMAS, C., BRUCE, M., WATT, A., GABRA, G. S., MULLER, S. & TIPPETT P. (1986): Tar+ individuals with anti-D: a new category DVII (Abstract). *Transfusion 26*, 560.
169. LOMAS, C., MCCOLL, K. & TIPPETT, P. (1993): Further complexities of the Rh antigen D disclosed by testing category D^{II} cells with monoclonal anti-D. *Transfus. Med. 3*, 67-69.
170. LOMAS, C., POOLE, J., SALARU, N., REDMAN, M., KIRKLEY, K., MOULDS, M., MCCREARY, J., NICHOLSON, G. S., HUSTINX, H. & GREEN, C. (1990): A low-incidence red cell antigen JAL associated with two

unusual Rh gene complexes. *Vox Sang. 59*, 39-43.

171. Lowe, A. D., Green, S. M., Voak, D., Gibson, D. & Lennox, E. S. (1986): A human-human monoclonal anti-D by direct fusion with a lymphoblastoid cell line. *Vox Sang. 51*, 212-216.

172. Maaskant- van Wijk, P. A., Beckers, E. A. M., van Rhenen, D. J., Mouro, I., Colin, Y., Cartron, J. P., Faas, B. H. W., van der Schoot, C. E., Apoil, P. A., Blancher, A. & von dem Borne, A. E. G. K. (1997): Evidence that the RHD^VI deletion genotype does not exist. *Blood 90*, 1709-1711.

173. Maaskant-van Wijk, P. A., Faas, B. H. W., de Ruijter, J. A. M., Overbeeke, M. A. M., von dem Borne, A. E. G. K., van Rhenen, D. J. & van der Schoot, C. E. (1998): Genotyping of RHD by multiplex polymerase chain reaction analysis of six RHD-specific exons. *Transfusion 38*, 1015-1021.

174. MacGeoch, C., Mitchell, C. J., Carritt, B., Avent, N. D., Ridgwell, K., Tanner, M. J. A. & Spurr, N. K. (1992): Assignment of the chromosomal locus of the human 30-kDal Rh (Rhesus) blood group-antigen-related protein (Rh30A) to chromosome region 1p36.13→p34. *Cytogenet. Cell Genet. 59*, 261-263.

175. Maizels, N. (1989): Might gene conversion be the mechanism of somatic hypermutation of mammalian immunoglobulin genes? *Trends Genet. 5*, 4-8.

176. Mallison, G., Anstee, D. J., Avent, N. D., Ridgwell, K., Tanner, M. J. A., Daniels, G. L., Tippett, P. & Von Dem Borne, A. E. G. (1990): Murine monoclonal antibody MB-2D10 recognizes Rh-related glycoproteins in the human red cell membrane. *Transfusion 30*, 222-225.

177. Marais, I., Moores, P., Smart, E. & Martell, R. (1993): STEM, a new low-frequency Rh antigen associated with the e-variant phenotypes hr^s- (Rh: -18, -19) and hr^B- (Rh: -31, -34). *Transfus. Med. 3*, 35-41.

178. Marini, A. M., Urrestarazu, A., Beauwens, R. & André, B. (1997): The Rh (Rhesus) blood group polypeptides are related to NH_4^+ transporters. *Trends Biochem. Sci. 22*, 460-461.

179. Marsh, W. L., Chaganti, R. S. K., Gardner, F. H., Mayer, K., Nowell, P. C. & German, J. (1974): Mapping human autosomes: evidence supporting the assigment of Rhesus to the short arm of chromosome No. 1. *Science 183*, 966-968.

180. Masouredis, S. P., Sudora, E. J., Mahan, L. & Victoria, E. J. (1976): Antigen site densities and ultrastructural distribution patterns of red cell Rh antigens. *Transfusion 16*, 94-106.

181. Matassi, G., Chérif-Zahar, B., Raynal, V., Rouger, P. & Cartron, J. P. (1998): Organization of the human RH50A gene (RHAG) and evolution of base composition of the RH gene family. *Genomics 47*, 286-293.

182. Mawby, W. J., Holmes, C. H., Anstee, D. J., Spring, F. A. & Tanner, M. J. A. (1994): Isolation and characterization of CD47 glycoprotein: a multispanning membrane protein which is the same as integrin-associated protein (IAP) and the ovarian tumour marker OA3. *Biochem. J. 304*, 525-530.

183. Mendelsohn, C. L., Wimmer, E. & Racaniello, V. R. (1989): Cellular receptor for poliovirus: molecular cloning, nucleotide sequence, and expression of a new member of the immunoglobulin superfamily. *Cell 56*, 855-865.

184. Mollison, P. L., Engelfriet, P. & Contreras, M. (1992): *Blood Transfusion in Clinical Medicine.* 9th ed, Blackwell, Oxford, UK, .

185. Moore, S. & Green, C. (1987): The identification of specific Rhesus-polypeptide-blood-group-ABH-active-glycoprotein complexes in the human red-cell membrane. *Biochem. J. 244*, 735-741.

186. Moore, S., Woodrow, C. F. & McClelland, D. B. L. (1982): Isolation of membrane components associated with human red cell antigens Rh(D), (c), (E), and Fy^a. *Nature 295*, 529-531.

187. Moor-Jankowski, J., Wiener, A. S. & Rogers, C. M. (1964): Human blood group factors in non-human primates. *Nature 202*, 663-665.

188. Moor-Jankowski, J., Wiener, A. S., Socha, W. W., Gordon, E. B. & Kaczera, Z. (1973): Blood-group homologues in orangutans and gorillas of the human Rh-Hr and the chimpanzee C-E-F-systems. *Fol. Primatol. 19*, 360-367.

189. Mourant, A. E. (1945): A new rhesus antibody. *Nature 155*, 542.

190. Mouro, I., Colin, Y., Chérif-Zahar, B., Cartron, J. P. & Le Van Kim, C. (1993): Molecular genetic basis of the human Rhesus blood group system. *Nat. Genet. 5*, 62-65.

191. MOURO, I., COLIN, Y., GANE, P., COLLEC, E., ZELINSKI, T., CARTRON, J. P. & LE VAN KIM, C. (1996): Molecular analysis of blood group Rh transcripts from a r^Gr variant. *Br. J. Haematol. 93*, 472-474.

192. MOURO, I., COLIN, Y., SISTONEN, P., LEPENNEC, P. Y., CARTRON, J. P. & LE VAN KIM, C. (1995): Molecular basis of the RhC^w (Rh8) and RhC^x (Rh9) blood group specificities. *Blood 86*, 1196-1201.

193. MOURO, I., LE VAN KIM, C., ROUILLAC, C., VAN RHENEN, D. J., LE PENNEC, P. Y., BAILLY, P., CARTRON, J. P. & COLIN, Y. (1994): Rearrangements of the blood group RhD gene associated with the D^{VI} category phenotype. *Blood 83*, 1129-1135.

194. MOURO, I., LE VAN KIM, C., CHERIF-ZAHAR, B., SALVIGNOL, I., BLANCHER, A., CARTRON, J. P. & COLIN, Y. (1994): Molecular characterization of the Rh-like locus and gene transcripts from the rhesus monkey (Macaca mulatta). *J. Mol. Evolut. 38*, 169-176.

195. NASH, R. & SHOJANIA, A. M. (1987): Hematological aspect of Rh deficiency syndrome: a case report and a review of the literature. *Am. J. Hematol. 24*, 267-276.

196. OKUBO, Y., YAMAGUCHI, H., TOMITA, T. & NAGAO, N. (1984): A D variant, D^{el}? *Transfusion 24*, 542.

197. OKUDA, H., KAWANO, M., IWAMOTO, S., TANAKA, M., SENO, T., OKUBO, Y. & KAJII, E. (1997): The RHD gene is highly detectable in RhD-negative Japanese donors. *J. Clin. Invest. 100*, 373-379.

198. OKUDA, H., SUGANUMA, H., TSUDO, N., OMI, T., IWAMOTO, S. & KAJII, E. (1999): Sequence analysis of the spacer region between the RHD and RHCE genes. *Biochem. Biophys. Res. Commun. 263*, 378-383.

199. OMI, T., TAKAHASHI, J., TSUDO, N., OKUDA, H., IWAMOTO, S., TANAKA, M., SENO, T., TANI, Y. & KAJII, E. (1999): The genomic organization of the partial D category D^{Va}: the presence of a new partial D associated with the D^{Va} phenotype. *Biochem. Biophys. Res. Commun. 254*, 786-794.

200. OSBORNE-LAWRENCE, S. I., SINCLAIR, A. K., HICKS, R. C., LACEY, S. W., EDDY, R. L., BYERS, M. G., SHOWS, T. B. & DUBY, A. D. (1990): Complete amino acid sequence of human cartilage link protein (CRTL1) deduced from cDNA clones and chromosomal assigment of the gene. *Genomics 8*, 562-567.

201. PARADIS, G., BAZIN, R. & LEMIEUX, R. (1986): Protective effect of the membrane skeleton on the immunologic reactivity of the human red cell Rh$_o$(D) antigen. *J. Immunol. 137*, 240-244.

202. PLAPP, F. V., KOWALSKI, M. M., EVANS, J. P., TILZER, L. L. & CHIGA, M. (1980): The role of membrane phospholipid in expression of erythrocyte Rh$_o$(D) antigen activity. *Proc. Soc. Exp. Biol. Med. 164*, 561-568.

203. POOLE, J., H., H., GERBER, H., LOMAS, C., LIEW, Y. W. & TIPPETT, P. (1990): The red cell antigen JAL in the Swiss population: family studies showing that JAL is an Rh antigen (RH48). *Vox Sang. 59*, 44-47.

204. POULTER, M., KEMP, T. J. & CARRITT, B. (1996): DNA-based Rhesus typing: simultaneous determination of RHC and RHD status using the polymerase chain reaction. *Vox Sang. 70*, 164-168.

205. RACE, R. R. & SANGER, R. (1975): The Rhesus blood groups. In: *Blood Groups in Man*. Blackwell Scientific Publications, Oxford, pp. 178-260.

206. RACE, R. R., SANGER, R. & SELWYN, J. G. (1950): A probable deletion in a human Rh chromosome. *Nature 166*, 520.

208. RACE, R. R., SANGER, R. & SELWYN, J. G. (1951): A possible deletion in a human Rh chromosome: a serological and genetical study. *Br. J. Exp. Pathol. 32*, 124-135.

209. RACE, R. R., TAYLOR, G. L., BOORMAN, K. E. & DODD, B. E. (1943): Recognition of Rh genotypes in man. *Nature 152*, 563.

210. RAPAILLE, A., FRANÇOIS-GÉRARD, C., DONNAY, D. & SONDAG-THULL, D. (1993): Production of stable human-mouse hybridomas secreting monoclonal antibodies against Rh D antigens and c antigens. *Vox Sang. 64*, 161-166.

211. RIDGWELL, K., EYERS, S. A. C., MAWBY, W. J., ANSTEE, D. J. & TANNER, M. J. A. (1994): Studies on the glycoprotein associated with Rh (Rhesus) blood group antigen expression in the human red blood cell membrane. *J. Biol. Chem. 269*, 6410-6416.

454

212. Ridgwell, K., Roberts, S. J., Tanner, M. J. A. & Anstee, D. J. (1983): Absence of two membrane proteins containing extracellular thiol groups in Rh_null human erythrocytes. *Biochem. J. 213*, 267-269.

213. Ridgwell, K., Spurr, N. K., Laguda, B., Macgeoch, C., Avent, N. D. & Tanner, M. J. A. (1992): Isolation of cDNA clones for a 50-kDa glycoprotein of the human erythrocyte membrane associated with Rh (Rhesus) blood-group antigen expression. *Biochem. J. 287*, 223-228.

214. Ridgwell, K., Tanner, M. J. A. & Anstee, D. J. (1984): The Rhesus (D) polypeptide is linked to the human erythrocyte cytoskeleton. *FEBS Lett. 174*, 7-10.

215. Rochna, E. & Hughes-Jones, N. C. (1965): The use of purified ^{125}J labelled anti-γ globulin in the determination of the number of D antigen sites on red cells of different phenotypes. *Vox Sang. 10*, 675-686.

216. Rosenfield, R. E., Allen, F. H., Swisher, S. N. & Kochwa, S. (1972): A review of Rh serology and presentation of a new terminology. *Transfusion 2*, 287-312.

217. Rosenfield, R. E. & Haber, G. V. (1958): An Rh blood factor, rh_i (Ce), and its relationship to hr (ce). *Am. J. Hum. Genet. 10*, 474-480.

218. Rouger, P. & Edelman, L. (1988): Murine monoclonal antibodies associated with Rh17, Rh29, and Rh46 antigens. *Transfusion 28*, 52-55.

219. Rouillac, C., Colin, Y., Hughes-Jones, N. C., Beolet, M., D'Ambrosio, A. M., Cartron, J. P. & Le Van Kim, C. (1995): Transcript analysis of D category phenotypes predicts hybrid Rh D-CE-D proteins associated with alteration of D epitopes. *Blood 85*, 2937-2944.

220. Rouillac, C., Gane, P., Cartron, J., Le Pennec, P. Y., Cartron, J. P. & Colin, Y. (1996): Molecular basis of the altered antigenic expression of RhD in weak D (D^u) and RhC/e in R^N phenotypes. *Blood 87*, 4853-4861.

221. Rouillac, C., Le Van Kim, C., Beolet, M., Cartron, J. P. & Colin, Y. (1995): Leu110Pro substitution in the RhD polypeptide is responsible for the D^VII category blood group phenotype. *Am. J. Hematol. 49*, 87-88.

222. Rouillac, C., Le Van Kim, C., Blancher, A., Roubinet, F., Cartron, J. P. & Colin, Y. (1995): Lack of G blood group antigen in D^IIIb erythrocytes is associated with segmental DNA exchange between RH genes. *Br. J. Haematol. 89*, 424-426.

223. Saboori, A. M., Denker, B. M. & Agre, P. (1989): Isolation of proteins related to the Rh polypeptides from nonhuman erythrocytes. *J. Clin. Invest. 83*, 187-191.

224. Saboori, A. M., Smith, B. B. L. & Agre, P. (1988): Polymorphism in the M_r 32,000 Rh protein purified from Rh(D)-positive and -negative erythrocytes. *Proc. Natl. Acad. Sci. USA 85*, 4042-4045.

225. Sacks, M. S., Wiener, A. S., Jahn, E. F., Spurling, C. L. & Unger, L. J. (1959): Isosensitization to a new blood factor Rh^D with special reference to its clinical importance. *Ann. Int. Med. 51*, 740-747.

226. Sanger, R., Noades, J., Tippett, P., Race, R. R., Jack, J. A. & Cunningham, C. A. (1960): An Rh antibody specific for V an R'^s. *Nature 186*, 171.

227. Sanger, R., Race, R. R., Rosenfield, R. E., Vogel, P. & Gibbel, N. (1953): Anti-f and the 'new' Rh antigen it defines. *Proc. Natl. Acad. Sci. USA 39*, 824-834.

228. Schmitz, G., Sonneborn, H. H., Ernst, M., Blanchard, D., Gielen, W. & Dahr, W. (1996): The effect of cysteine modification and proteinases on the major antigens (D, C, c, E and e) of the Rh blood group system. *Vox Sang. 70*, 34-39.

229. Shapiro, M. (1960): Serology and genetics of a new blood factor: hr^s. *J. Forens. Med. 7*, 96-105.

230. Shapiro, M. (1964): Serology and genetics of a 'new' blood factor: hr^H. *J. Forens. Med. 11*, 52-66.

231. Shapiro, M., le Roux, M. & Brink, S. (1972): Serology and genetics of a new blood factor: hr^B. *Haematologia 6*, 121-128.

232. Shinitzky, M. & Souroujon, M. (1979): Passive modulation of blood group antigens. *Proc. Natl. Acad. Sci. USA 76*, 4438-4440.

233. Sieff, C., Bicknell, D., Caine, G., Robinson, J., Lam, G. & Greaves, M. F. (1982): Changes in cell surface antigen expression during hemopoietic differentiation. *Blood 60*, 703-713.

234. SIMSEK, S., BLECKER, P. M. N. & VON DEM BORNE, A. E. G. K. (1994): Prenatal determination of fetal RhD type. *N. Engl. J. Med. 330*, 795-796.
235. SIMSEK, S., FAAS, B. H. W., BLEEKER, P. M. M., OVERBEEKE, M. A. M., CUIJPERS, H. T. M., VAN DER SCHOOT, C. E. & VON DEM BORNE, A. E. G. K. (1995): Rapid Rh D genotyping by polymerase chain reaction-based amplification of DNA. *Blood 85*, 2975-2980.
236. SISTONEN, P., SARENEVA, H., PIRKOLA, A. & EKLUND, J. (1994): MAR, a novel high-incidence Rh antigen revealing the existence of an allelic sub-system including C^w (Rh8) and C^x (Rh9) with exceptional distribution in the Finnish population. *Vox Sang. 66*, 287-292.
237. SMYTHE, J. S., AVENT, N. D., JUDSON, P. A., PARSONS, S. F., MARTIN, P. G. & ANSTEE, D. J. (1996): Expression of RHD and RHCE gene products using retroviral transduction of K562 cells establishes the molecular basis of Rh blood group antigens. *Blood 87*, 2968-2973.
238. STEERS, F., WALLACE, M., JOHNSON, P., CARRITT, B. & DANIELS, G. (1996): Denaturing gradient gel electrophoresis: A novel method for determining Rh phenotype from genomic DNA. *Br. J. Haematol. 94*, 417-421.
239. STRATTON, F. (1946): A new Rh allelomorph. *Nature 158*, 25-26.
240. STRATTON, F. & RENTON, P. H. (1954): Haemolytic disease of the newborn caused by a new Rh antibody, anti-C^x. *Br. Med. J. i*, 962-965.
241. STURGEON, P. (1970): Hematological observations on the anaemia associated with blood type Rh_{null}. *Blood 36*, 310-320.
242. SUYAMA, K. & GOLDSTEIN, J. (1988): Antibody produced against isolated Rh(D) polypeptide reacts with other Rh-related antigens. *Blood 72*, 1622-1626.
243. SUYAMA, K., GOLDSTEIN, J., AEBEWRSOLD, R. & KENT, S. (1991): Regarding the size of Rh proteins. *Blood 77*, 411.
244. SUYAMA, K., LUNN, R. & GOLDSTEIN, J. (1995): Red Cell surface cysteine residue (285) of D polypeptide is not essential for D antigenicity. *Transfusion 35*, 653-659.
245. SUYAMA, K., LUNN, R., HALLER, S. & GOLDSTEIN, J. (1994): Rh(D) antigen expression and isolation of a new Rh(D) cDNA isoform in human erythroleukemic K562 cells. *Blood 84*, 1975-1981.
246. SVOBODA, R. K., VAN WEST, B. & GRUMET, F. C. (1981): Anti-Rh41, a new Rh antibody found in association with an abnormal expression of chromosome 1 genetic markers. *Transfusion 21*, 150-156.
247. SZYMANSKI, I. O. & ARASZKIEWICZ, P. (1989): Quantitative studies on the D antigen of red cells with the D^u phenotype. *Transfusion 29*, 103-105.
248. THOMPSON, K., BARDEN, G., SUTHERLAND, J., BELDON, I. & MELAMED, M. (1990): Human monoclonal antibodies to C, c, E, e, and G antigens of the Rh system. *Immunology 71*, 323-327.
249. THOMPSON, K. M., MELAMED, M. D., EAGLE, K., GORICK, B. D., GIBSON, T., HOLBURN, A. M. & HUGHES-JONES, N. C. (1986): Production of human monoclonal IgG and IgM antibodies with anti-D (Rhesus) specificity using heterohybridomas. *Immunology 58*, 157-160.
250. THOMPSON, K. M., SUTHERLAND, J., BARDEN, G., MELAMED, M. D., WRIGHT, M. G., BAILEY, S. & THORPE, S. J. (1992): Human monoclonal antibodies specific for blood group antigens demonstrate multispecific properties characteristic of natural autoantibodies. *Immunology 76*, 146-157.
251. THORPE, S. J., BOULT, C. E. & THOMPSON, K. M. (1997): Immunochemical characterization of the Rh C^w antigen using human monoclonal antibodies. *Vox Sang. 73*, 174-181.
252. TIPPETT, P. (1986): A speculative model for the Rh blood groups. *Ann. Hum. Genet. 50*, 241-247.
253. TIPPETT, P., LOMAS-FRANCIS, C. & WALLACE, M. (1996): The Rh antigen D: partial D antigens and associated low incidence antigens. *Vox Sang. 70*, 123-131.
254. TIPPETT, P. & MOORE, S. (1990): Monoclonal antibodies against Rh and Rh related antigens. *J. Immunogenet. 17*, 309-319.
255. TIPPETT, P. & SANGER, R. (1977): Further observations on subdivisions of the Rh antigen, D. *Ärztl. Lab. 23*, 476-480.
256. UMENISHI, F., KAJII, E. & IKEMOTO, S. (1994): Identification of two Rh mRNA isoforms expressed in immature erythroblasts. *Biochem. Biophys. Res. Commun. 198*, 1135-1142.
257. UMENISHI, F., KAJII, E. & IKEMOTO, S. (1994): Molecular analysis of Rh polypeptides in a family with

RhD- positive and RhD-negative phenotypes. *Biochem. J. 299*, 207-211.

258. UMENISHI, F., KAJII, E. & IKEMOTO, S. (1994): A new genomic polymorphism of Rh-polypeptide genes. *Biochem. J. 299*, 203-206.

259. UNGER, L. J. & WIENER, A. S. (1959): A 'new' antibody, anti-Rh^C, resulting from isosensitization by pregnancy with special reference to the heredity of a new Rh-Hr agglutinogen Rh$^c_2$. *J. Lab. Clin. Med. 54*, 835-842.

260. UNGER, L. J. & WIENER, A. S. (1959): Some observations on the blood factor Rh^A of the Rh-Hr blood group system. *Acta Genet. Med. Gemellol. (2nd Suppl).*, 13-25.

261. UNGER, L. J., WIENER, A. S. & WIENER, L. (1959): New antibody (anti-Rh^B) resulting from blood transfusion in an Rh-positive patient. *J. Am. Med. Assoc. 170*, 1380-1383.

262. VENGELEN-TYLER, V. & PIERCE, S. R. (1987): *Blood Group Systems: Rh*. American Association of Blood Banks, Arlington, Virginia.

263. VICTORIA, E. J., BRANKS, M. J. & MASOUREDIS, S. P. (1986): Rh antigen immunoreactivity after histidine modification. *Mol. Immunol. 23*, 1039-1044.

264. VICTORIA, E. J., MAHAN, L. C. & MASOUREDIS, S. P. (1981): Anti-Rh$_o$(D) IgG binds to band 3 glycoprotein of the human erythrocyte membrane. *Proc. Natl. Acad. Sci. USA 78*, 2898-2902.

265. VOS, G. H. & KIRK, R. L. (1962): A 'naturally-occurring' anti-E which distinguishes a variant of the E antigen in Australian aborigines. *Vox Sang. 7*, 22-32.

266. WAGNER, F. F., GASSNER, C., MÜLLER, T. H., SCHÖNITZER, D., SCHUNTER, F. & FLEGEL, W. A. (1999): Molecular basis of weak D phenotypes. *Blood 93*, 385-393.

267. WAGNER, F. F., GASSNER, C., MÜLLER, T. H., SCHÖNITZER, D., SCHUNTER, F. & FLEGEL, W. A. (1998): Three molecular structures cause Rhesus D category VI phenotypes with distinct immunohematologic features. *Blood 91*, 2157-2168.

268. WAGNER, F. F., HILLESHEIM, B. & FLEGEL, W. A. (1997): D-category VII depends on amino acid substitution Leu(110)Pro. *Beiträge zur Infusionstherapie und Transfusionsmedizin 34*, 220-223.

269. WALLACE, M., LOMAS-FRANCIS, C., BECKERS, E., BRUCE, M., CAMPBELL, G., CHATFIELD, S., NAGAO, N., OKUBO, Y., OPALKA, A., OVERBEEKE, M., SCOTT, M. & VOAK, D. (1997): DBT: a partial D phenotype associated with the low-incidence antigen Rh32. *Transfus. Med. 7*, 233-238.

270. WESTHOFF, C. M., SCHULTZE, A., FROM, A., WYLIE, D. E. & SILBERSTEIN, L. E. (1999): Characterization of the mouse Rh blood group gene. *Genomics 57*, 451-454.

271. WESTHOFF, C. M. & WYLIE, D. E. (1994): Identification of a new RhD-specific mRNA from K562 cells. *Blood 83*, 3098-3100.

272. WESTHOFF, C. M. & WYLIE, D. E. (1994): Investigation of the human Rh blood group system in nonhuman primates and other species with serologic and southern blot analysis. *J. Mol. Evolut. 39*, 87-92.

273. WESTHOFF, C. M. & WYLIE, D. E. (1996): Investigation of the RH locus in gorillas and chimpanzees. *J. Mol. Evolut. 42*, 658-668.

274. WIENER, A. S. (1941): Hemolytic reaction following transfusion of blood of the homologous group. II. *Arch. Pathol. 32*, 227-250.

275. WIENER, A. S., GEIGER, J. & GORDON, E. B. (1957): Mosaic structure of the Rh$_o$ factor of human blood. *Exp. Med. Surg. 15*, 75-82.

276. WIENER, A. S., MOOR-JANKOWSKI, J. & GORDON, E. B. (1964): Blood groups of apes and monkeys. IV. The Rh-Hr blood types of anthropoid apes. *Am. J. Hum. Genet. 16*, 246-253.

277. WIENER, A. S. & PETERS, H. R. (1940): Hemolytic reactions following transfusions of blood of the homologous group, with three cases in which the same agglutinogen was responsible. *Ann. Int. Med. 13*, 2306-2322.

278. WIENER, A. S. & SONN, E. B. (1943): Additional variants of the Rh type demonstrable with a special human anti-Rh serum. *J. Immunol. 47*, 461-465.

279. WOLTER, L. C., HYLAND, C. A. & SAUL, A. (1993): Rhesus-D genotyping using polymerase chain reaction. *Blood 82*, 1682-1683.

280. WOLTER, L. C., HYLAND, C. A. & SAUL, A. (1994): Refining the DNA polymorphisms that associate with the Rhesus c phenotype. *Blood 84*, 985-986.

14 Landsteiner-Wiener System

In 1940 Landsteiner and Wiener [18] immunized rabbits and guinea pigs with the red cells of rhesus monkeys. The immune sera ('anti-**Rh**') obtained by this stimulus reacted with 85% of adult human blood samples. At that time the 'Rhesus' antigen has been considered the animal equivalent of the human antigen responsible for *erythroblastosis foetalis* (HDN), because the frequency of reactivity of rabbit sera with human erythrocytes matched that of a human antibody. This antibody had been obtained from a woman who had delivered a stillborn fetus which had died *in utero* of HDN [20]. Later investigations, however, proved conclusively that the human and animal antibodies described two different specificities (cf. [36]), and the '**Rh**' antigen was subsequently renamed **LW** in honor of Landsteiner and Wiener.

Two antithetical alleles have been described so far in the **LW** blood group system: **LW**^{a} and **LW**^{b} and a third character **LW**^{ab} which is defined by an epitope common to **LW**^{a} and **LW**^{b} [1] [15]. Thus a total of four phenotypes had been defined: **LW(a+b−)**, **LW(a−b+)**, **LW(a+b+)**, as well as **LW(a−b−)** in individuals homozygous for the extremely rare silent *LW* allele.

LW^a is a common antigen among Europids, whereas **LW**^{b} is found in fewer than 1% of the population. However, **LW**^{b} shows an increased frequency among Finns and other populations of Baltic origin (~5%) [29].

The *LW* gene has been assigned to chromosome l9pl3.3 by in situ hybridization [13,28].

Though **LW** characters are genetically independent of the **Rh** system, their expression is phenotypically closely associated with the expression of the **Rh** antigens: the fact that those **Rh**$_{null}$ erythrocytes which do not express any **Rh** antigens also lack the **LW** antigens (see [15]) indicatesshows that the **LW** protein is part of the '**Rh** protein cluster' (see *Chap. 13*). Moreover, it has been found that the number of **LW** determinants per erythrocyte depends on the presence of the **RhD** antigen (4,400 **LW** sites have been estimated in **RhD**-positive erythrocytes, but only 2,800 in **RhD**-negative cells [22]).

---

[1] The six-digit ISBT identification numbers are 016 005 for **LW**^{a} and 016 007 for **LW**^{b}; the **LW**^{ab} character is numbered 016 006. **LW**^{b} has originally been termed **Ne**^{a} by Sistonen et al. [30,31]

Occasionally, transient weakening or loss of **LW** antigens together with simultaneous **LW** antibody production has been observed during pregnancy [10,11] and in some immunological disorders, such as Hodgkin's disease [26] and autoimmune thrombocytopenic purpura [27]. A reduced expression of **LW** antigens has also been noticed in cases of congenital dyserythropoietic anemia [25] and in malignant lymphoma [16]. After delivery or recovery of the patients the depressed antigens re-express, and the anti-**LW** antibodies disappear.

LW antigens are not confined to erythrocytes − they are also found on 30–35% of T and B lymphocytes [24]. The antigen is expressed in equal amounts on each cell population.

The **LW** antigen has also been detected on the red cells of nonhuman primates, such as rhesus monkeys and baboons; chimpanzee red cells were found negative for the antigen [19,32].

Antibodies

Anti-**LW** antibodies are present as alloantibodies in genetically **LW**-negative individuals [9,37]. Autoantibodies with this specificity have been also described in patients suffering from autoimmune haemolytic anaemia [8,38] and transiently produced in patients with transient depression of **LW** antigen [10,11]. In one case anti-**LW** has been produced as a drug-induced antibody [3].

Two hybridoma antibodies with anti-**LW**[ab] specificity have also been described [33].

LW Antigens

The **LW** antigens in situ are resistant to trypsin; they are only slightly affected by papain, ficin, trypsin treatment, but are readily degraded by pronase [21]. Destruction by dithiothreitol and similar reducing agents suggests that the antigens require intramolecular disulphide bonds for serologic reactivity [17]. Research has shown that **LW** activity of the erythrocytes is lost upon treatment of the cells with EDTA, but is restored by addition of Mg^{++}: this suggests that the **LW** polypeptide contains a cation-binding domain in close proximity to the **LW** epitope, and may also indicate that Mg^{++} plays a role in the association of the **LW** protein to the **Rh** protein cluster [6]. In contrast to the **Rh**-deficient **Rh**$_{null}$ phenotype, which is characterised by abnormal red cells and reduced red cell survival, however, the absence of the **LW** antigens in **LW(a–b–)** individuals is not associated with any membrane disorder.

−29   MGSLFPLSL LFFLRPPTRE LGARWDAGLR −1

  1 AQSPKGSPLA PSGTSVPFWV RMSPEFVAVQ PGKSVQL*N*CS NSCPQPQ*N*SS LRTPLRQGKT 60

 61 LRGPGWVSYQ LLDVRAWSSL AHCLVTCAGK TRWATSRITA YKPPHSVILE PPVLKGRKYT 120

121 LRCHVTQVFP VGYLVVTLRH GSRVIYSESL ERFTGLDLA*N* VTLTYEFAAG PRDFWQPVIC 180

181 HARLNLDGLV VR*N*SSAPITL MLAWSPAP<u>TA LASGSIAALV GILLTVGAA</u>Y LCKCLAMKSQ 240

241 A

Figure 14.1: Amino acid sequence of the LW protein.
Underlined: hydrophobic transmembrane domain, *N* : potential N-glyosylation sites; −29 to −1 : leader peptide.
From Bailly et al. [1]. The sequence data are deposited in the EMBL/GenBank data library (accession number L27671).

Immunoblotting experiments using anti-**LW**[ab] specific hybridoma antibodies were able to locate the **LW** antigens on a 37–47-kDa glycoprotein of the erythrocyte membrane [22]. Subsequent studies revealed that the **LW** substance carries several N-glycosidically linked carbohydrate chains which are not part of the **LW** epitope [5,6].

Bailly et al. [1][2] haver determined the structure of the **LW** protein. By screening a human bone marrow cDNA library with a **LW**-specific PCR-amplified DNA fragment, the authors could identify a cDNA clone encoding a 271 amino acid polypeptide. Nucleotide sequence analysis predicted a transmembrane protein consisting of a signal peptide (30 residues), an N-terminal extracellular domain (208 residues), a single hydrophobic transmembrane region (21 residues), and a short C-terminal cytoplasmic segment (12 residues) *(Fig. 14.1)*. The protein contains four potential N-glycosylation sites (at positions 38, 48, 160, and 193).

The same authors identified a second cDNA which encoded a shortened version of the **LW** protein. This 236-amino acid polypeptide shares the same N-terminal sequence with the **LW** membrane protein but lacks the transmembrane and cytoplasmic domains. On the basis of their preliminary studies the authors suggest that

---

[2] The amino acid sequence of the leader peptide was later corrected by Hermand et al. [13].

this secreted isoform might be the result of a premature termination codon formed by aberrant splicing.

More recently, Hermand et al. [13] have shown that the **LW**^{a}/**LW**^{b} polymorphism is based on a single base mutation (A^{308} → G) that results in a Gln → Arg substitution at position 70 of the **LW** protein, glutamic acid being characteristic for **LW**^{a} and arginine for **LW**^{b}.

Individuals of the rare **LW(a–b–)** phenotype lack the **LW** protein [14]. A deletion of 10 bp in exon 1 of the *LW*^{a} allele was identified in the genome of an **LW(a–b–)** individual (*Big.*) of a normal **Rh** phenotype [14]. This deletion generates a premature stop codon at the beginning of exon 2 and encodes a truncated protein lacking the transmembrane and cytoplasmic domains. In another **LW(a-b-)** individual (*Nic.*) no detectable abnormality of the *LW* gene or its transcript could be found; this fact suggests a heterogeneity of the **LW(a–b–)** phenotype.

In contrast to **Rh**$_{null}$ phenotype – characterised by abnormal red cells and shortened red cell survival – absence of the **LW** antigens is not associated with a similar membrane disorder.

Studies by Southern blot analysis indicated that the *LW* gene is composed of a single gene organised into three exons spanning approximately 2.65 kb of DNA [14]. Exon 1 encodes the signal peptide and the first Ig-like domain, exon 2 the second Ig-like domain, and exon 3 the transmembrane and the cytoplasmic domains.

Sequence analysis has further shown that the **LW** protein belongs to the immunoglobulin superfamily (see e.g. **Lutheran** system and **CEA**) and is structurally related to the intercellular adhesion molecules (ICAMs), in particular to ICAM-2 [35].

It has been suggested [4] that the extracellular part of the **LW** protein consists of two immunoglobulin-like domains stabilised by disulfide links (between cystein residues 39 and 83, 132 and 180, and residues 43 and 87); this would correspond to the predicted ICAM structure. The critical residues of ICAM-2 involved in the binding of LFA-1 are partially conserved in **LW**.

The ICAMs are expressed on leukocytes, lymphocytes, and endothelial cells, but are absent from erythrocytes. They are involved in various adhesion processes in which they act as receptors for the 'lymphocyte function-associated antigens' (= LFA-1, Mac-1, and p150/95), also described as leukocyte-specific integrins (CDII/CD18), cf. [34].

The sequence corresponding with the ICAM's suggests that the **LW** protein may also be involved in cell adhesion processes. Human erythrocytes and **LW** protein purified from red cells were indeed found to bind to various leukocytes, as well as to isolated CD11/CD18 leukocyte integrins in a concentration-dependent manner [2]. The fact that *LW*, and the *ICAM* genes co-localise on the same chromosome (19p13) [7] suggests that these genes may have evolved from a common ancestral gene[3].

If the **LW** protein, like ICAM-1, acts as a receptor for pathogens such as rhinoviruses [12,23] or as a cytoadhesion receptor for *Plasmodium falciparum*-infected erythrocytes [4,23] is still under discussion.

References

1. BAILLY, P., HERMAND, P., CALLEBAUT, I., SONNEBORN, H. H., KHAMLICHI, S., MORNON, J. P. & CARTRON, J. P. (1994): The LW blood group glycoprotein is homologous to intercellular adhesion molecules. *Proc. Natl. Acad. Sci. USA 91*, 5306-5310.
2. BAILLY, P., TONTTI, E., HERMAND, P., CARTRON, J. P. & GAHMBERG, C. G. (1995): The red cell LW blood group protein is an intercellular adhesion molecule which binds to CD11/CD18 leukocyte integrins. *Eur. J. Immunol. 25*, 3316-3320.
3. BARTLETT, A. N., HOFFBRAND, A. V. & KONTOGHIORGHES, G. J. (1990): Long-term trial with the oral iron chelator 1,2-dimethyl-3-hydroxypyrid-4-one (L1). II. Clinical observations. *Brit. J. Haematol. 76*, 301-304.
4. BERENDT, A. R., MCDOWALL, A., CRAIG, A. G., BATES, P. A., STERNBERG, M. J., MARSH, K., NEWBOLD, C. I. & HOGG, N. (1992): The binding site on ICAM-1 for Plasmodium falciparum-infected erythrocytes overlaps, but is distinct from, the LFA-1-binding site. *Cell 68*, 71-81.
5. BLOY, C., BLANCHARD, D., HERMAND, P., KORDOWICZ, M., SONNEBORN, H. H. & CARTRON, J. P. (1989): Properties of the blood group LW glycoprotein and preliminary comparison with Rh proteins. *Molec. Immunol. 26*, 1013-1019.
6. BLOY, C., HERMAND, P., BLANCHARD, D., CHÉRIF-ZAHAR, B., GOOSSENS, D. & CARTRON, J. P. (1990): Surface orientation and antigen properties of Rh and LW polypeptides of the human erythrocyte membrane. *J. Biol. Chem. 265*, 21482-21487.
7. BOSSY, D., MATTEI, M. G. & SIMMONS, D. L. (1994): The human intercellular adhesion molecule 3 (ICAM3) gene is located in the 19p13.2-p13.3 region, close to the ICAM1 gene. *Genomics 23*, 712-713.
8. CELANO, M. & LEVINE, P. (1967): Anti-LW specificity in autoimmune acquired hemolytic anemia. *Transfusion 7*, 265-268.
9. CHAPLIN, H., HUNTER, V. L., ROSCHE, M. E. & SHIREY, R. S. (1985): Long-term in vivo survival of Rh(D)-negative donor red cells in a patient with anti-LW. *Transfusion 25*, 39-33.
10. CHOWN, B., KAITA, H., LOWEN, B. & LEWIS, M. (1971): Transient production of anti-LW by LW-positive people. *Transfusion 11*, 220-222.
11. GILES, C. M. & LUNDSGAAARD, A. (1967): A complex serological investigation involving LW. *Vox Sang. 13*, 406-416.
12. GREVE, J. M., DAVIS, G., MEYER, A. M., FORTE, C. P., YOST, S. C., MARLOR, C. W., KAMARCK, M. E. & MCCLELLAND, A. (1989): The major human rhinovirus receptor is ICAM-1. *Cell 56*, 839-847.
13. HERMAND, P., GANE, P., MATTEI, M. G., SISTONEN, P., CARTRON, J. P. & BAILLY, P. (1995): Molecular basis and expression of the LWᵃ/LWᵇ blood group polymorphism. *Blood 86*, 1590-1594.
14. HERMAND, P., LE PENNEC, P. Y., ROUGER, P., CARTRON, J. P. & BAILLY, P. (1996): Characterization of the gene encoding the human LW blood group protein in LW⁺ and LW⁻ phenotypes. *Blood 87*, 2962-2967.

---

[3] Accordingly, the designation **ICAM-4** has been suggested for the **LW** protein [2].

15. ISSITT, P. D. (1985): The Rh blood group system. In: *Applied Blood Group Serology*. Montgomery Scientific Publications, Miami, Florida, USA, pp. 219-277.
16. KOMATSU, F. & KAJIWARA, M. (1996): Transient depression of LW[a] antigen with coincident production of anti-LW[a] repeated in relapses of malignant lymphoma. *Transfus. Med. 6*, 139-143.
17. KONIGHAUS, G. J. & HOLLAND, T. I. (1984): The effect of dithiothreitol on the LW antigen. *Transfusion 24*, 536-537.
18. LANDSTEINER, K. & WIENER, A. S. (1940): An agglutinable factor in human blood recognized by immune sera for rhesus blood. *Proc. Soc. Exp. Biol. 43*, 223.
19. LEVINE, P. & CELANO, M. J. (1962): Presence of "D-like" antigens on various monkey red blood cells. *Nature 193*, 184-185.
20. LEVINE, P. & STETSON, R. E. (1939): An unusual case of intragroup agglutination. *J. Amer. Med. Assoc. 113*, 126-127.
21. LOMAS, C. G. & TIPPETT, P. (1985): Use of enzymes in distinguishing anti-LW[a] and anti-LW[b] from anti-D. *Med. Lab. Sci. 42*, 88-89.
22. MALLINSON, G., MARTIN, P. G., ANSTEE, D. J., TANNER, M. J. A., MERRY, A. H., TILLS, D. & SONNEBORN, H. H. (1986): Identification and partial characterization of the human erythrocyte membrane component(s) that express the antigens of the LW blood-group system. *Biochem. J. 234*, 649-652.
23. OCKENHOUSE, C. F., BETAGERI, R., SPRINGER, T. A. & STAUNTON, D. E. (1992): Plasmodium falciparum-infected erythrocytes bind ICAM-1 at a site distinct from LFA-1, Mac-1, and human rhinovirus. *Cell 68*, 63-69.
24. OLIVEIRA, O. L. P., THOMAS, D. B., LOMAS, C. G. & TIPPETT, P. (1984): Restricted expression of LW antigen on subsets of human B and T lymphocytes. *J. Immunogenet. 11*, 297-303.
25. PARSONS, S. F., JONES, J., ANSTEE, D. J., JUDSON, P. A., GARDNER, B., WIENER, E., POOLE, J., ILLUM, N. & WICKRAMASINGHE, S. N. (1994): A novel form of congenital dyserythropoietic anemia associated with deficiency of erythroid CD44 and a unique blood group phenotype [In(a-b-), Co(a-b-)]. *Blood 83*, 860-868.
26. PERKINS, H. A., McILROY, M., SWANSON, J. & KADIN, M. (1977): Transient LW-negative red blood cells and anti-LW in a patient with Hodgkin's disease. *Vox Sang. 33*, 299-303.
27. POOLE, J., WILLIAMSON, L. M., CLARK, N. & BLACK, A. (1991): Loss of Lutheran and LW antigen expression in a patient with autoimmune thrombocytopenic purpura (ITP). (Abstract). *Transfusion 31*, 38S.
28. SISTONEN, P. (1984): Linkage of the LW blood group locus with the complement C3 and Lutheran blood group loci. *Ann. Hum. Genet. 48*, 239-242.
29. SISTONEN, P., GREEN, C. A., LOMAS, C. G. & TIPPETT, P. (1983): Genetic polymorphism of the LW blood group system. *Ann. Hum. Genet. 47*, 277-284.
30. SISTONEN, P., NEVANLINNA, H. R., VIRTARANTA-KNOWLES, K., PIRKOLA, A., LEIKOLA, J., KEKOMÄKI, R., GAVIN, J. & TIPPETT, P. (1981): Ne[a], a new blood group antigen in Finland. *Vox Sang. 40*, 352-357.
31. SISTONEN, P. & TIPPETT, P. (1982): A 'new' allele giving further insight into the LW blood group system. *Vox Sang. 42*, 253-255.
32. SOCHA, W. W. & RUFFIE, J. (1990): Monoclonal antibodies directed against human Rh antigens in tests with the red cells of nonhuman primates. *Rev. Fr. Transfus. 33*, 39-48.
33. SONNEBORN, H. H., UTHEMANN, H., TILLS, D., LOMAS, C. G., SHAW, M. A. & TIPPETT, P. (1984): Monoclonal anti-LW[ab]. *Biotest Bull. 2*, 145-148.
34. SPRINGER, T. A. (1990): Adhesion receptors of the immune system. *Nature 346*, 425-434.
35. STAUNTON, D. E., DUSTIN, M. L. & SPRINGER, T. A. (1989): Functional cloning of ICAM-2, a cell adhesion ligand for LFA-1 homologous to ICAM-1. *Nature 339*, 61-64.
36. STORRY, J. R. (1992): Review: the LW blood group system. *Immunohematology 8*, 87-93.
37. SWANSON, J. & MATSON, G. A. (1964): The third example of a human 'D-like'antibody or anti-LW. *Transfusion 4*, 257-261.
38. VOS, G. H., PETZ, L. D., GARRATTY, G. & FUDENBERG, H. H. (1973): Autoantibodies in acquired hemolytic anemia with special reference to the LW system. *Blood 42*, 445-453.

15 Chido / Rodgers System

The character groups **Chido (Ch)** [28] and **Rodgers (Rg)** [33] are confined to humans. The antigens are found on erythrocytes [28,33] and in plasma [33,37,39,40], but are absent from tissue cells or secretions [17].

Thus far six **Chido** specificities, **Ch1** (originally **Ch**^{a}), **Ch2**, **Ch3**, **Ch4**, **Ch5**, and **Ch6** [18,19], and two **Rodgers** specificities, **Rg1** (originally **Rg**^{a}) and **Rg2** [18] have been defined; the **WH** determinant is a 'hybrid antigen' which is associated with **Ch6** and **Rg1** [25][1]. The frequencies of the **Ch/Rg** phenotypes thus far characterised are listed in *Table 15.1*.

The **Chido/Rodgers** epitopes are located on the fourth component of complement, C4 [42]. The characters are inherited by two closely linked and highly homologous genes, *C4A* and *C4B*, both of which encode isotypes of C4. The *C4* gene locus is situated next to genes for the second complement component C2, factor B (BF), and steroid 21-hydroxylases, *21-OHA* and *21-OHB*, in the major histocompatibility complex between the loci for **HLA** class I and class II molecules, i.e. on the short arm of chromosome 6 (region 6p21.2) [11,15,24,38].

Table 15.1: Incidence of Ch and Rg phenotypes.

Ch Phenotype	Frequency[1]	Rg Phenotype	Frequency[1]
Ch:1,2,3	88.2%	Rg:1,2	95.0%
Ch:1,−2,3	4.9%	Rg:1,−2	2.5%
Ch:1,2,−3	3.1%	Rg:−1,−2	2.5%
Ch:−1,−2,−3	3.8%		
Ch:−1,2,3	rare		
Ch:1,−2,−3	rare		

[1] Frequencies in Europids, from Giles [20] by permission of S. Karger AG, Basel.

---

[1] In ISBT terminology (see *Chap. 1*): **Ch1** – **Ch6** = **CH/RG1** – **CH/RG6** (or 017 001 – 017 006), **Rg1** = **CH/RG11** (or 017 011), **Rg2** = **CH/RG12** (or 017 012), and **WH** = **CH/RG7** (or 017 007).

C4A and *C4B* consist of 41 exons spanning approximately 22 kb; a 16 kb isotype of *C4B* lacks intron 9. Both genes encode transcripts for the pre-pro-C4 precursor protein of 1744 amino acid residues [59].

All anti-**Ch** and anti-**Rg** antibodies thus far described are IgG alloantibodies induced in most cases by blood transfusion in **Chido**- or **Rodgers**-negative individuals. The reaction of these antibodies with erythrocytes of the respective blood group is normally very weak and 'nebulous' and can often be detected only by the antiglobulin test [28,33]. The **Ch/Rg** activity in plasma is assayed by the haemagglutination inhibition test [20]. Closer serological investigations showed that most anti-**Ch** and anti-**Rg** sera contain mixtures of antibodies with different specificities: anti-**Rg1** is virtually always accompanied by anti-**Rg2**; anti-**Chido** sera, in addition to their predominant specificity — anti-**Ch1**, frequently contain antibodies towards **Ch4** (75%) and **Ch2** (25%), and less commonly, antibodies towards **Ch5** (16%) and **Ch3** (10%) (see [20]). Only in a few cases was it possible to separate the different anti-**Ch** specificities by serum adsorption [18].

A few hybridoma antibodies fairly specific for the determinants **Ch1** and **Rg1** have been described [22,23].

More recently PCR-based typing methods assaying the **Chido/Rodgers** antigenic determinants [4,58] have been developed.

The carrier of the **Ch/Rg** epitopes, the C4 molecule, is a glycoprotein with 7% carbohydrate content and a molecular mass of approximately 200 kDa [1,47]. It is composed of three disulfide-linked polypeptide chains, α, β, and γ (*Fig. 15.1*) and is synthesised by hepatocytes and macrophages as a pro-protein which is subsequently processed intracellularly to yield the secreted three-subunit molecule. C4 carries four carbohydrate units — on the α-chain three fucosylated oligosaccharides of the complex-type, and on the β-chain one oligosaccharide of the high-mannose type [14].

In the course of the classical pathway of complement activation (see [1]) the small peptide fragment C4a is split from the α-chain of the C4 molecule by the action of activated C1. This exposes the reactive thiolester group, which enables the residual molecule, C4b, to bind covalently to hydroxyl and amino groups of suitable receptors, such as cell membranes or different pathogens. During regulation of complement activity in organism the C4b molecule is split by the C4b-inactivator (or, *in vitro*, by trypsin or other proteases) into two fragments, C4c and C4d. C4d, which contains the **Ch/Rg** determinants [54], remains attached to the cell membrane.

The investigations by Tilley et al. [54] clearly showed that **Ch/Rg** antigens of red cells are acquired from the plasma: when erythrocytes were coated *in vitro* with C4 during incubation with sera in low ionic strength medium, their reaction with anti-**Ch**

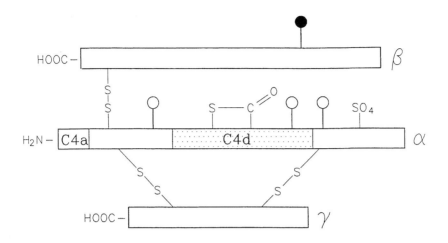

Figure 15.1: Schematic representation of the complement component C4.
● : high mannose type oligosaccharide chain, ♀ : biantennary complex type oligosaccharide chain.
According to Atkinson et al. [1].

and anti-**Rg** was significantly increased, occasionally even resulting in direct agglutination; the **Ch/Rg** phenotype acquired by the erythrocytes corresponded to the **Ch/Rg** phenotype of the serum used.

As mentioned above, two isotypes of the complement component C4 are present in human serum, **C4A** (formerly C4F [41]) and **C4B** (formerly C4S [41]). The isotypes are more than 99% identical in amino acid sequence, but are highly polymorphic in the C4d region: due to their electrophoretic mobility it has been possible to define 18 C4A and 21 C4B variants thus far in Europids [52,53]; some of these variants may be further subdivided on the basis of their **Ch/Rg** phenotype [21].

The two isotypes differ markedly in haemolytic activity and in chemical reactivity [31,32]: C4B exhibits an approximately four times greater haemolytic activity than C4A; C4B has a higher affinity to hydroxyl groups of carbohydrate antigens and binds preferentially to antibody-coated red cells, whereas C4A has a higher affinity to amino groups and binds more efficiently to immune complexes of soluble antigens.

Further, various studies [2,21] demonstrated that the expression of the **Ch/Rg** properties is associated with the C4 isotype – **Chido** specificities are usually found

Table 15.2: Ch/Rg phenotypes expressed by C4A and C4B allotypes.

Ch/Rg phenotype	C4A allotype						C4B allotype					
	1	2	3	4	5	6	1	2	3	4	5	6
Ch:1,2,3	0	0	0	0	n.d.	0	113	3	27	0	1	0
Ch:1,–2,3	12	0	0	0	n.d.	0	0	91	3	1	9	4
Ch:1,2,–3	0	0	0	0	n.d.	0	72	2	3	0	0	0
Ch:–1,–2,–3	0	5	129	8	n.d.	3	0	0	0	0	8	0
Rg:1,2	2	19	77	43	3	18	0	0	0	n.d.	1	n.d.
Rg:1,–2	0	1	28	0	1	0	0	0	0	n.d.	0	n.d.
Rg:–1,–2	2	0	0	0	0	0	52	13	1	n.d.	0	n.d.

A count taken from 325 families; from Giles et al.[27] by permission of Springer-Verlag, Heidelberg. As the families were selected for specific **Ch/Rg** phenotypes, the data do not give an estimate on frequency.

Table 15.3: Characteristic amino acid sequences of C4 allotypes of known Ch/Rg antigenic status.

C4 allotype	Ch/Rg expression	Amino acid sequence in the polymorphic Site[a]:			
		(I)	(II)	(III)	(IV)
C4A3	**Ch:–1,–2,–3,–4,–5,–6;Rg:1,2**	D	PCPVLD	N	VDLL
C4A1	**Ch:1,–2,3,–4,5,6;Rg:–1,–2**	G	PCPVLD	S	ADLR
C4B3	**Ch:1,2,3,4,5,6;Rg:–1,–2**	G	LSPVIH	S	ADLR
C4B1	**Ch:1,2,–3,4,5,–6[b];Rg:–1,–2[b]**	G	LSPVIH	N	ADLR
C4B2	**Ch:1,–2,3,4,–5,6[b];Rg:–1,–2[b]**	D	LSPVIH	S	ADLR
C4B5	**Ch:–1,–2,–3,4,–5,–6,WH;Rg:1,–2**	D	LSPVIH	S	VDLL

[a] (I) – (IV) represent four polymorphic regions of the C4d fragment within the α-chain: (I) at position 1054, (II) at positions 1101–1106, (III) at position 1157, and (IV) at positions 1188–1191; the numbers represent the location of the amino acid in the continuous sequence of pro-C4.
[b] Assumed **Ch/Rg** phenotype.
From Giles et al. [20] by permission of S. Karger AG, Basel.

on C4B chains, whereas **Rodgers** specificities are generally found on C4A chains; only in rare cases have **Ch** antigens been detected on C4A chains and **Rg** antigens on C4B chains [50]. More extensive investigations revealed a close connection between the **Ch/Rg** phenotype and the C4 allotype [19,22,27] (see *Table 15.2)*: the studies show that all C4B molecules express **Ch4**, and that **Ch1** and **Rg1** are obviously alternative determinants; moreover, the presence of **Ch1** and **Ch6** is essential for the expression of **Ch3**, that of **Ch4** and **Ch5** for the expression of **Ch2**, and that of **Ch6** and **Rg1** for the expression of **WH**. The same studies also recognise a close connection between **Rg2** and **Rg1**.

During investigations on **Ch/Rg** blood group substances the amino acid sequences of many C4 allotypes have been established by analysing tryptic C4 fragments (e.g. [13,30,34]), as well as by sequencing cDNA clones [5,6] or cloned C4 genes of different C4A and C4B allotypes [58,60].

These studies revealed that the amino acid residues determinant for the **Ch/Rg** phenotype are located in four discrete sectors of the C4d segment of the α-chain[2]:

> Region I: position 1054,
> Region II: positions 1101–1106,
> Region III: position 1157,
> Region IV: positions 1188–1191.

In these polymorphic regions a strong correlation between the **Ch/Rg** phenotype and the amino acid sequence of the respective C4 allotype has been found (*Table 15.3)*. Based on these studies a structural model has been proposed, which supports the interrelationships of the eight **Ch/Rg** determinants as deduced from serological studies [61] (*Fig. 15.2)*:

Following this model the C4 isotype is determined by the amino acid sequence in region II (positions 1101–1106), i.e. **Pro-Cys**-Pro-Val-**Leu-Asp** is characteristic for C4A and **Leu-Ser**-Pro-Val-**Ile-His** for C4B. Since all C4B chains carry **Ch4** it must be assumed that the Leu-Ser-Pro-Val-Ile-His structure also represents the **Ch4** epitope. Accordingly, **Ch1** is determined by **Ala-Asp-Leu-Arg** in region IV and **Rg1** by the alternative sequence **Val-Asp-Leu-Leu**. Furthermore, Gly^{1054} (region I) is characteristic for **Ch5**, and Ser^{1157} (region III) for **Ch6**. The hypothetical epitope **Rg3** defined by Giles et al. [27] is characterised by the asparagine at position 1157 alternating the **Ch6** specific glycine residue.

---

[2] The numbers indicated represent the position of the amino acid in the continuous sequence of the C4 pro-protein – C4d comprises residues 938 to 1317.

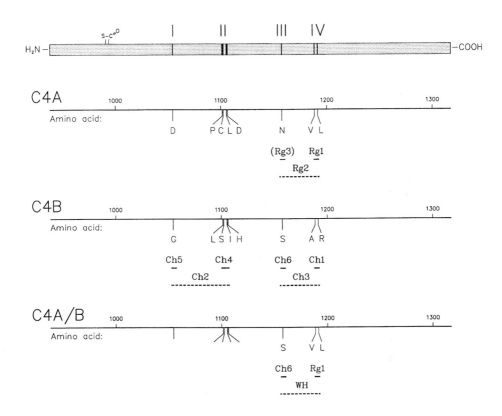

Figure 15.2: Ch/Rg antigenic determinants on the C4d fragment of C4.
According to Giles et al. [27] and Yu et al. [61]. Redrawn by permission of Springer-Verlag, Heidelberg.

The expression of **Ch2**, **Ch3**, **Rg2**, and **WH**, which depends closely on the presence of other **Ch** or **Rg** characters, suggests that these determinants are conformational epitopes involving amino acid residues from two separate regions. Thus it has been assumed that the **Ch2** epitope includes amino acids of regions I and II, whereas **Ch3**, **Rg2**, and **WH** comprise sequences of regions III and IV.

Thus far, no antisera have been detected that recognise either the C4A isotypic polymorphic site (Pro-Cys-Pro-Val-Leu-Asp, region II) or the epitopes that alternate with **Ch5** (region I) and **Ch6** (region III).

469

Null alleles of C4 (**C4Q0**) defined by the absence of C4 protein in plasma are present in the normal population at unusually high frequencies of up to 20% (e.g. [29,45,55]).

Lack of C4 proteins is associated with an increased susceptibility to systemic lupus erythaematosus [16,48], but also to other autoimmune or immune complex diseases, such as subacute sclerosing panencephalitis [49], chronic active hepatitis [57], and type 1 (insulin-dependent) diabetes [7,35,36].

These C4 null alleles are based on different genetic backgrounds.

(1) A proportion of the C4 haplotypes carrying a null allele are due to **gene deletions**. Usually, deletion of an entire **C4** gene also comprises a flanking steroid 21-hydroxylase gene [12,51,56].

(2) Point mutations as well as base insertions or deletions may produce **defective C4 genes** [8,44]. For example, in the majority of cases studied by Barba et al. [3] a 2 bp insertion in exon 29 was found, which led to nonexpression of the gene due to the creation of a premature termination codon at the beginning of exon 30.

(3) Fairly frequently the observed absence of an expected gene product is caused by **gene duplication**. Single unequal crossing-over or gene conversion events may lead to the expression of one C4 isotype at both C4 loci. In the majority of probands the C4B gene is replaced by a C4A gene isotype, in some cases C4A by a C4B gene isotype [3,9,26,43,46,58]. These C4 null individuals express either identical allotypes ('homoexpression') or different allotypes of the same isotype ('iso-expression') at both C4 loci.

References

1. ATKINSON, J. P., CHAN, A. C., KARP, D. R., KILLION, C. C., BROWN, R., SPINELLA, D., SHREFFLER, D. C. & LEVINE, R. P. (1988): Origin of the fourth component of complement related to Chido and Rodgers blood group antigens. *Complement 5*, 65-76.
2. AWDEH, Z. L. & ALPER, C. A. (1980): Inherited structural polymorphism of the fourth component of human complement. *Proc. Natl. Acad. Sci. USA 77*, 3576-3580.
3. BARBA, G., RITTNER, C. & SCHNEIDER, P. M. (1993): Genetic basis of human complement C4A deficiency. Detection of a point mutation leading to nonexpression. *J. Clin. Invest. 91*, 1681-1686.
4. BARBA, G. M. R., BRAUN-HEIMER, L., RITTNER, C. & SCHNEIDER, P. M. (1994): A new PCR-based typing of the Rodgers and Chido antigenic determinants of the fourth component of human complement. *Eur. J. Immunogenet. 21*, 325-339.
5. BELT, K. T., CARROLL, M. C. & PORTER, R. R. (1984): The structural basis of the multiple forms of human complement component C4. *Cell 36*, 907-914.
6. BELT, K. T., YU, C. Y., CARROLL, M. C. & PORTER, R. R. (1985): Polymorphism of human

complement component C4. *Immunogenetics 21*, 173-180.

7. Bertrams, J., Hintzen, U., Schlicht, V., Schoeps, S., Gries, F. A., Louton, T. K. & Baur, M. P. (1984): Gene and haplotype frequencies of the fourth component of complement (C4) in type 1 diabetics and normal controls. *Immunobiology 166*, 335-344.

8. Braun, L., Schneider, P. M., Giles, C. M., Bertrams, J. & Rittner, C. (1990): Null alleles of human complement C4. Evidence for pseudogenes at the C4A locus and for gene conversion at the C4B locus. *J. Exp. Med. 171*, 129-140.

9. Bruun-Petersen, G., Lamm, L. U., Jacobsen, B. K. & Kristensen, T. (1982): Genetics of complement C4. Two homoduplication haplotypes C4S C4S and C4F C4F in a family. *Hum. Genet. 61*, 36-38.

10. Campbell, R. D., Dunham, I., Kendall, E. & Sargent, C. A. (1990): Polymorphism of the human complement component C4. *Exp. Clin. Immunogenet. 7*, 69-84.

11. Carroll, M. C., Campbell, R. D., Bentley, D. R. & Porter, R. R. (1984): A molecular map of the human major histocompatibility complex class III region linking complement genes C4, C2, and factor B. *Nature 307*, 237-241.

12. Carroll, M. C., Palsdottir, A., Belt, K. T. & Porter, R. R. (1985): Deletion of complement C4 and 21-hydroxylase genes in the HLA class III region. *EMBO J. 4*, 2547-2552.

13. Chakravarti, D. N., Campbell, R. D. & Gagnon, J. (1983): Amino acid sequence of a polymorphic segment from fragment C4d of human complement component C4. *FEBS Lett. 154*, 387-390.

14. Chan, A. C. & Atkinson, J. P. (1985): Oligosaccharide structure of human C4. *J. Immunol. 134*, 1790-1798.

15. Dunham, I., Sargent, C. A., Trowsdale, J. & Campbell, R. D. (1987): Molecular mapping of the human major histocompatibility complex by pulsed-field gel electrophoresis. *Proc. Natl. Acad. Sci. USA 84*, 7237-7241.

16. Fielder, A. H., Walport, M. J., Batchelor, J. R., Rynes, R. I., Black, C. M., Dodi, I. A. & Hughes, G. R. (1983): Family study of the major histocompatibility complex in patients with systemic lupus erythematosus: importance of null alleles of C4A and C4B in determining disease. *Brit. Med. J. 286*, 425-428.

17. Giles, C. M. (1977): Serologically difficult red cell antibodies with special reference to Chido and Rodgers blood groups. In: *Human Blood Groups* (J. F. Mohn, R. W. Plunkett, R. K. Cunningham, and R. M. Lambert, eds.). S. Karger, Basel, pp. 268-276.

18. Giles, C. M. (1985): 'Partial inhibition' of anti-Rg and anti-Ch reagents. II. Demonstration of separable antibodies for different determinants. *Vox Sang. 48*, 167-173.

19. Giles, C. M. (1987): Three Chido determinants detected on the B5Rg allotype of human C4; their expression in Ch-typed donors and families. *Hum. Immunol. 8*, 111-122.

20. Giles, C. M. (1988): Antigenic determinants of human C4, Rodgers and Chido. *Exp. Clin. Immunogenet. 5*, 99-114.

21. Giles, C. M., Batchelor, J. R., Dodi, I. A., Fielder, A. H. L., Rittner, C., Mauff, G., Bender, K., Levene, C., Schreuder, G. M. T. & Well, L. J. (1984): C4 and HLA haplotypes associated with partial inhibition of anti-Rg and anti-Ch. *J. Immunogenet. 11*, 305-317.

22. Giles, C. M., Fielder, A. H. L., Lord, D. K., Robson, T. & O'Neill, G. J. (1987): Two monoclonal anti-C4d reagents react with epitopes closely related to Rg:1 and Ch:1. *Immunogenetics 26*, 309-312.

23. Giles, C. M. & Ford, D. S. (1986): A monoclonal anti-C4d that demonstrates a specificity related to anti-Ch. *Transfusion 26*, 370-374.

24. Giles, C. M., Gedde-Dahl, T., Robson, E. B., Thorsby, E., Olaisen, B., Aenason, A., Kissmeyer-Nielsen, F. & Schreuder, I. (1976): Rgᵃ (Rodgers) and the HLA region: linkage and associations. *Tissue Antigens 8*, 143-149.

25. Giles, C. M. & Jones, J. W. (1987): A new antigenic determinant for C4 of relatively low frequency. *Immunogenetics 26*, 392-394.

26. Giles, C. M., Uring-Lambert, B., Boksch, W., Braun, M., Goetz, J., Neumann, R., Mauff, G. & Hauptmann, G. (1987): The study of a French family with two duplicated C4A haplotypes. *Hum.*

Genet. 77, 359-365.

27. Giles, C. M., Uring-Lambert, B., Goetz, J., Hauptmann, G., Fielder, A. H. L., Ollier, W., Rittner, C. & Robson, T. (1988): Antigenic determinants expressed by human C4 allotypes; a study of 325 families provides evidence for the structural antigenic model. *Immunogenetics 27*, 442-448.

28. Harris, J. P., Tegoli, J., Swanson, J., Fisher, N., Gavin, J. & Noades, J. (1967): A nebulous antibody responsible for cross-matching difficulties (Chido). *Vox Sang. 12*, 140-142.

29. Hauptmann, G., Tappeiner, G. & Schifferli, J. A. (1988): Inherited deficiency of the fourth component of human complement. *Immunodeficiency Rev. 1*, 3-22.

30. Hellman, U., Eggertsen, G., Lundwall, A., Engström, A. & Sjöquist, J. (1984): Primary sequence differences between Chido and Rodgers variants of tryptic C4d of the human complement system. *FEBS Lett. 170*, 254-258.

31. Isenman, D. E. & Young, J. R. (1984): The molecular basis for the difference in immune hemolysis activity of the Chido and Rodgers isotypes of human complement component C4. *J. Immunol. 132*, 3019-3027.

32. Law, S. K. A., Dodds, A. W. & Porter, R. R. (1984): A comparison of the properties of two classes, C4A and C4B, of the human complement component C4. *EMBO J. 3*, 1819-1823.

33. Longster, G. & Giles, C. M. (1976): A new antibody specificity, anti-Rg[a], reacting with a red cell and serum antigen. *Vox Sang. 30*, 175-180.

34. Lundwall, A., Hellmann, U., Eggertsen, G. & Sjöquist, J. (1982): Isolation of tryptic fragments of human C4 expressing Chido and Rodgers antigens. *Mol. Immunol. 19*, 1655-1665.

35. Marcelli-Barge, A., Poirier, J. C., Schmid, M., Deschamps, I., Lestradet, H., Prevost, P. & Hors, J. (1984): Genetic polymorphism of the fourth component of complement and type 1 (insulin-dependent) diabetes. *Diabetologia*, 116-117.

36. McCluskey, J., McCann, V. J., Kay, P. H., Zilko, P. J., Christiansen, F. T., O'Neill, G. J. & Dawkins, R. L. (1983): HLA and complement allotypes in type 1 (insulin-dependent) diabetes. *Diabetologia 24*, 162-165.

37. Middleton, J. & Crookston, M. C. (1972): Chido substance in plasma. *Vox Sang. 23*, 256-261.

38. Middleton, J., Crookston, M. C., Falk, J. A., Robson, E. B., Cook, P. J. L., Batchelor, J. R., Bodmer, J., Ferrara, G. B., Festenstein, H., Harris, R., Kissmeyer-Nielsen, F., Lawler, S. D., Sachs, J. A. & Wolf, E. (1974): Linkage of Chido and HL-A. *Tissue Antigens 4*, 366-373.

39. Nordhagen, R., Olaisen, B., Teisberg, P. & Gedde-Dahl, T. (1980): Association between the electrophoretically-determined C4M haplotype product and partial inhibition of anti-Ch[a]. *J. Immunogenet. 7*, 301-306.

40. Nordhagen, R., Olaisen, B., Teisberg, P., Gedde-Dahl, T. & Thorsby, E. (1981): C4 haplotype products and partial inhibition of anti-Rodgers sera. *J. Immunogenet. 8*, 485-491.

41. O'Neill, G. J., Yang, S. Y. & Dupont, B. (1978): Two HLA-linked loci controlling the fourth component of human complement. *Proc. Natl. Acad. Sci. USA 75*, 5165-5169.

42. O'Neill, G. J., Yang, S. Y., Tegoli, J., Berger, R. & Dupont, B. (1978): Chido and Rodgers blood groups are distinct antigenic components of human complement C4. *Nature 273*, 668-670.

43. Palsdottir, A., Arnasson, A., Fossdal, R. & Jensson, O. (1987): Gene organization of haplotypes expressing two different C4A allotypes. *Hum. Genet. 76*, 220-224.

44. Partanen, J. & Campbell, R. D. (1989): Restriction fragment analysis of non-deleted complement C4 null genes suggests point mutations in C4 null alleles, but gene conversions in C4B null alleles. *Immunogenetics 30*, 520-523.

45. Partanen, J. & Koskimies, S. (1986): Human MHC class III genes, Bf, and C4. Polymorphism complotypes, and association with MHC class I genes in the Finnish population. *Hum. Hered. 36*, 269-275.

46. Raum, D., Awdeh, Z., Anderson, J., Strong, L., Garandos, J., Teran, L., Giblett, E., Yunis, E. J. & Alper, C. A. (1984): Human C4 haplotypes with duplicated C4A or C4B. *Amer. J. Hum. Genet. 36*, 72-79.

47. Reid, K. B. M. & Porter, R. R. (1981): The proteolytic activation systems of complement. *Annu. Rev. Biochem. 50*, 433-464.

48. Reveille, J. D., Arnett, F. C., Wilson, R. W., Bias, W. B. & McLean, R. H. (1985): Null alleles of the fourth component of complement and HLA haplotypes in familial systemic lupus erythematosus. *Immunogenetics 21*, 299-311.

49. Rittner, C., Meier, E. M., Stradmann, B., Giles, C. M., Kochling, R., Mollenhauer, E. & Kreth, H. W. (1984): Partial C4 deficiency in subacute sclerosing panencephalitis. *Immunogenetics 20*, 407-415.

50. Roos, M. H., Giles, C. M., Demant, P., Mollenhauer, E. & Rittner, C. (1984): Rodgers (Rg) and Chido (Ch) determinants on human C4: characterization of two C4 B5 subtypes, one of which contains Rg and Ch determinants. *J. Immunol. 133*, 2634-2640.

51. Schneider, P. M., Caroll, M. C., Alper, C. A., Rittner, C., Whitehead, A. S., Yunis, E. J. & Colten, H. R. (1986): Polymorphism of the human complement C4 and steroid 21-hydroxylase genes. Restriction fragment length polymorphisms revealing structural deletions, homo-duplications, and size variants. *J. Clin. Invest. 78*, 650-657.

52. Schneider, P. M., Stradmann-Bellinghausen, B. & Rittner, C. (1996): Genetic polymorphism of the fourth component of human complement: population study and proposal for a revised nomenclature based on genomic PCR typing of Rodgers and Chido determinants. *Eur. J. Immunogenet. 23*, 335-344.

53. WHO-IUIS Nomenclature Sub-Committee (1993): Revised nomenclature for human complement component C4. *J. Immunol. Methods 163*, 3-7.

54. Tilley, C. A., Romans, D. G. & Crookston, M. C. (1978): Localisation of Chido and Rodgers determinants to the C4d fragment of human C4. *Nature 276*, 713-715.

55. Tokunaga, K., Omoto, K., Akaza, T., Akiyama, N., Amemiya, H., Naito, S., Sasazuki, T., Satoh, H. & Juji, T. (1985): Haplotype study on C4 polymorphism in Japanese. Association with MHC alleles, complotypes, and HLA-complement haplotypes. *Immunogenetics 22*, 359-365.

56. Uring-Lambert, B., Goetz, J., Tongio, M. M., Mayer, S. & Hauptmann, G. (1984): C4 haplotypes with duplications at the C4A or C4B loci: frequency and associations with BF, C2, and HLA-A, B, C, DR alleles. *Tissue Antigens 24*, 70-72.

57. Vergani, D., Wells, L., Larcher, V. F., Nasaruddin, B. A., Davies, E. T., Mieli-Vergani, G. & Mowat, A. P. (1985): Genetically determined low C4: a predisposing factor to autoimmune chronic active hepatitis. *Lancet 2*, 294-298.

58. Yu, C. Y. & Campbell, R. D. (1987): Definitive RFLPs to distinguish between the human complement C4A/C4B isotypes and the major Rodgers/Chido determinants: application to the study of C4 null alleles. *Immunogenetics 25*, 383-390.

59. Yu, C. Y. (1991): The complete exon-intron structure of a human complement component C4A gene. DNA sequences, polymorphism, and linkage to the 21-hydroxylase gene. *J. Immunol. 146*, 1057-1066.

60. Yu, C. Y., Belt, K. T., Giles, C. M., Campbell, R. D. & Porter, R. R. (1986): Structural basis of the polymorphism of human complement components C4A and C4B: gene size, reactivity, and antigenicity. *EMBO J. 5*, 2873-2881.

61. Yu, C. Y., Campbell, R. D. & Porter, R. R. (1988): A structural model for the location of the Rodgers and the Chido antigenic determinants and their correlation with the human complement component C4A/C4B isotypes. *Immunogenetics 27*, 399-405.

16 Duffy System

Duffy is a polymorphic blood group system of which six characters are defined to date (*Table 16.1*)[(1)]:

- the two principal antigens, **Fy**[a] (= **FY1**) [16,17] and **Fy**[b] (= **FY2**) [26]. Based on the presence or absence of these antigens four phenotypes have been defined – **Fy(a+b–), Fy(a–b+), Fy(a+b+)** (with frequencies in Europids of 0.195, 0.33, and 0.475, respectively), and **Fy(a–b–)**. This last phenotype is the predominant phenotype in Negrids but exceedingly rare outside the black population;
- the characters **Fy3** (= **FY3**) [2], **Fy5** (= **FY5**) [14], and **Fy6** (= **FY6**) [45,51], being linked to the expression of **Fy**[a] and **Fy**[b] antigens, are present on the red cells of all persons except those of the **Fy(a–b–)** phenotype. These antigens thus represent epitopes localised on discrete domains of the **Duffy** blood group substance;
- **Fy4** (= **FY4**) [4] is present on almost all samples of **Fy(a–b–)** erythrocytes and some **Fy(a+b–)** and **Fy(a–b+)** cells from Negrids; the status of this antigenic character within the **Duffy** system is not yet clear.

It has been postulated that five allelic genes transmit the different **Duffy** characters, viz. *Fy*A*, *Fy*B*, *Fy*X*, *Fy*4*, and the silent allele *Fy*Fy* [27]. *FY*A* and *FY*B* (with respective gene frequencies in Europids of 0.425 and 0.557 [50]) are responsible for the inheritance of the respective specificities. *Fy*[x] [6,13,32] is a 'weak' *Fy*B* variant and does not encode a proper **Duffy** character: thus, erythrocytes of *Fy*[a]*Fy*[x] individuals can absorb all antibodies with **Fy**[b] specificity, but are agglutinated only by selected anti-**Fy**[b] sera (= **Fy(a+b±)** cells). *FY*Fy* represents a silent allele at the **Duffy** locus. *Fy*4* is an allele characteristic for Negrids. It is worth noting that the **Fy(a–b–)** phenotype reported to be common in blacks is in most cases based on the genotype *Fy4Fy4* rather than on the genotype *FyFy*.

---

(1) The six-digit ISBT identification numbers (see *Chap. 1*) are 008 001 to 008 006.

Table 16.1: Antigens of the Duffy blood group system.

ISBT nr.[1]	Common term	Frequency (%)[2] in in Europids	Negrids	References
FY1	Fy^a	42.5	8	[17]
FY2	Fy^b	55.7	92	[26]
FY3	Fy3	>99.9		[2]
FY4	Fy4	>99.9		[4]
FY5	Fy5	>99.9		[14]
FY6	Fy6	>99.9		[51]

[1] The six-digit ISBT identification numbers are 008 001 to 008 006 (see *Chap. 1*).
[2] According to Issitt [27].

The **Duffy** locus (*DARC*$^{(2)}$) is situated on chromosome 1 at position q22–23 [15,38].

The anti-**Duffy** sera [27,50] are in most cases immune sera induced by **Duffy**-incompatible erythrocytes acquired through blood transfusions or transplacental passage. The majority of the antisera do not agglutinate in saline and are usually detected by antiglobulin tests. Anti-**Fy**^{a} is found fairly frequently; anti-**Fy**^{b}, however, is a rare antibody usually occurring together with antibodies of other specificities. Antisera towards other **Duffy** characters have been described only in a few specimens. Anti-**Fy3** [2], anti-**Fy5** [14], and the hybridoma antibody anti-**Fy6** [45,51] react only with erythrocytes carrying **Fy**^{a} and/or **Fy**^{b} determinants, and show no reaction with **Fy(a–b–)** cells. This suggests that these antibodies bind to domains of the blood group **Duffy** substance common to both allelic forms. Anti-**Fy5** shows properties similar to those of anti-**Fy3**, but differs from this antibody in being dependent on the **Rh** phenotype (**Rh**$_{null}$ cells are not agglutinated at all, **D--** cells agglutinate only weakly). In contrast to anti-**Fy3**, anti-**Fy6** binds to a protease-sensitive epitope. Anti-**Fy4** [4] reacts with almost all **Fy(a–b–)** and some **Fy(a+b–)** and **Fy(a–b+)** erythrocytes of Negrids.

(2) Due to the function of the **Duffy** protein as a chemokine receptor the designation *DARC* (= Duffy Antigen Receptor for Chemokines) has been proposed for the *Duffy* gene [34].

The antigens of the **Duffy** system can be detected in very young foetuses and are already well developed at birth [2,14,16,50,53].

Further, the **Duffy** substance is not only expressed in human erythrocytes, but has also been found in endothelial cells of postcapillary venules throughout the body, as well as on endothelial cells lining the sinusoids of spleen, bone marrow, and choroid plexus, and in the Purkinje cells of the cerebellum [20,21].

Duffy antigens are also found in a few non-human primates [47].

Investigations on the density of **Duffy** antigens on erythrocytes of **Fy**^{a} or **Fy**^{b} homozygous individuals yielded an average value of 13,500 sites per cell [37].

The antigen characters **Fy**^{a}, **Fy**^{b}, and **Fy6** are destroyed by chymotrypsin, ficin, pronase, and papain. **Fy3**, **Fy4**, and **Fy5**, however, are resistant towards these proteases. Purified trypsin has no influence on the **Duffy** antigens [19,27,33]. Further, neuraminidase [5] and sulfhydryl reagents [51] have no effect on **Fy**^{a} or **Fy**^{b} activities.

The substance carrying the **Duffy** antigens is an integral membrane protein not (or only loosely) associated with the membrane skeleton [51]. It can be solubilised by detergents [19,33], and is released into the surrounding medium upon storage [59].

Treatment with endo-glycosidases indicated that the carrier of the **Duffy** antigens is a glycoprotein [51,52]. The **Duffy** substance moves in polyacrylamidegel electrophoresis as a fairly broad band corresponding to 35–43 kDa [19,42]; this variation in apparent molecular mass is probably due to variable glycosylation.

The **Duffy** protein when isolated by immunoprecipitation technique is highly hydrophobic and tends to spontaneous aggregation [33,42]. The fact that isolated **Duffy** active material is associated with about six other membrane proteins of molecular masses ranging from 21 to 68 kDa suggests that the **Duffy** substance *in situ* is a component of complex protein clusters [10].

The primary structure of the protein as deduced from the nucleotide sequence of **Duffy** protein encoding cDNA revealed a highly hydrophobic polypeptide composed of 338 amino acids [9] (*Fig. 16.1*). In a more recent investigation it has been found that the major product of the **Duffy** gene in the erythroid lineage is a spliceoform encoding 336 amino acids [28]. Hydropathy analysis of the predicted polypeptide suggested a 62 amino acid hydrophilic domain at the N-terminus, most probably seven membrane-spanning hydrophobic domains with short hydrophilic connecting segments, and a 28 amino acid hydrophilic domain at the C-terminus extending into the cytoplasm (*Fig. 16.2*) [44]. The protein contains two potential N-glycosylation sites.

```
  1 MGNCLHRAEL SPSTENSSQL DFEDVWNSSY GVNDSFPDGD YDANLEAAAP CHSCNLLDDS  60

 61 ALPFFILTSV LGILASSTVL FMLFRPLFRW QLCPGWPVLA QLAVGSALFS IVVPVLAPGL 120

121 GSTRSSALCS LGYCVWYGSA FAQALLLGCH ASLGHRLGAG QVPGLTLGLT VGIWGVAALL 180

181 TLPVTLASGA SGGLCTLIYS TELKALQATH TVACLAIFVL LPLGLFGAKG LKKALGMGPG 240

241 PWMNILWAWF IFWWPHGVVL GLDFLVRSKL LLLSTCLAQQ ALDLLLNLAE ALAILHCVAT 300

301 PLLLALFCHQ ATRTLLPSLP LPEGWSSHLD TLGSKS
```

Figure 16.1: Primary structure of the Duffy protein (= 'gpD' or 'gpFy').
The peptide sequence presented is the major product of the **Duffy** gene in the erythroid lineage as proposed by Iwamoto et al. [28]. In the sequence originally published by Chaudhuri et al. [9] and Neote et al. [44] the N-terminal heptapeptide MGNCLHR is replaced by the nonapeptide MASSGYVLQ. Underlined: hydrophobic segments representing the seven putative transmembrane domains as proposed by Neote et al. [44]; **N** : potential N-glycosylation sites.
The nucleotide sequence of the cDNA is deposited in the EMBL/GenBank data library (accession number U01839).

The peptide sequence of the **Duffy** protein (= gpFy[3]) shows significant homology to interleukin-8 receptors. Experimental evidence that the **Duffy** characters are located on the chemokine receptor of the erythrocyte membrane has been provided by the finding that several chemotactic and proinflammatory peptides, such as interleukin-8 (IL-8), melanoma growth-stimulating activity (MSGA), and monocyte chemotactic protein 1 (MCP-1), bind only to **Duffy**-positive erythrocytes [22,24]. This observation, as well as ligand binding and displacement experiments on transfected cells [44], strongly supports the idea that the **Duffy** glycoprotein represents an additional class of chemokine[4] receptors [11,25].

Southern blot analysis revealed that mRNA encoding the **Duffy** glycoprotein is not only present in bone marrow but can also be found in non-erythroid tissues, e.g. lung, muscle, spleen, heart, brain, pancreas, kidney, colon, and fetal liver (but not in liver from adults) [9,44]. Subsequent investigations confirmed previous serological

---

[3] The glycoprotein has originally been termed 'pD' [10], 'glycoprotein D' or gpD [9]. Later, in order to avoid confusion with **RhD**-specific protein, the term gpFy has been proposed [8].

[4] Chemokines are involved in immunoregulatory and inflammatory processes which specifically chemoattract and activate leukocytes [3].

477

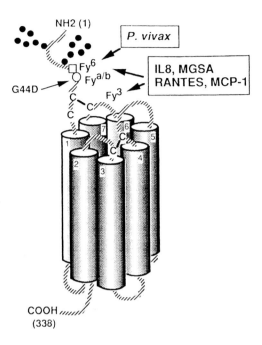

Figure 16.2: Model of the proposed structure of the Duffy glycoprotein.
From Cartron [6a] by permission of the author and of Editions Elsevier, Paris.

studies (see above) showing that the **DARC** gene is expressed mainly in endothelial cells and in the epithelial cells of the kidney collecting ducts and lung alveoli [7,20,49].

More recent investigations have revealed the molecular basis of some of the **Duffy** antigens:

The two codominant alleles **FY*A** and **FY*B** were shown to differ by a single G → A change at nucleotide 159 producing a glycine (**Fy^a**) ↔ aspartic acid (**Fy^b**) dimorphism at position 42[5] [8,30,56]. Comparison of the antigenicity of human **Duffy** glycoprotein with that of non-human primates suggested that residues 38, 39, 50, 55, and 58, in addition, are essential for anti-**Fy^b** specificity [8].

---

[5] The numbering of the amino acid residues according to the peptide sequence proposed by Iwamoto et al. [28].

As mentioned above, the *FY*X* allele is a 'weak' variant of the *Fy*2* allele which is associated with reduced expression of **Fy3, Fy5, Fy6** antigens. Further, erythrocytes from homozygous **Fy^x** individuals bind only 20–30% of the amount of chemokines as compared to homozygous **Fy^b** persons.

Investigations by Tournamille et al. [57] revealed that the *DARC* gene from **Fy(a–b+)** and **Fy(a-b^{weak})** individuals differ by a C → T exchange resulting in an Arg → Cys amino acid substitution at position 89. Transfection experiments have shown a reduced binding of anti-**Fy** antibodies and chemokines to the mutant protein. It has further been found that this substitution results in very low membrane expression of the **Duffy** antigen. The fact that the mutation is located in the first cytoplasmic loop of the protein suggests that the Arg → Cys exchange alters the conformation or impairs the proper membrane insertion of the protein.

Some *FY*X* alleles contained both the Arg^{89} → Cys substitution and an Ala → Thr exchange at position 100 [46,48]. The fact that this mutation is also found in normal **Fy^b** probands [35,44,46] shows that the substitution at position 100 does not affect **Fy^b** expression.

In another **Fy^x** donor a variant *FY*B* gene has been found which was characterised by a mutation in the binding site of a transcription factor [43]. Deletion of a single cytosine base in the region around nucleotide -89 which disrupts the Sp1 binding site obviously impairs the expression of the *Duffy* gene, allowing only a minimum amount of the protein to be produced.

As discussed above, **Fy(a–b–)** is the predominant phenotype in Negrid populations. When **Duffy**-negative individuals of Negrid ancestry were investigated by RNA blot analysis [10], no **Duffy**-specific cDNA clone could be detected in the bone marrow [8,9]. However, the finding that non-haematopoietic tissues contained **Duffy**-specific mRNA of the same size, yet in less quantity when compared to **Duffy**-positive individuals. These findings showed that **Duffy**-negative individuals have an intact *Duffy* gene, which, however, is not expressed in bone marrow. The expression of **Duffy** protein in endothelial cells of **Duffy**-negative individuals has been confirmed by biochemical and immunohistochemical studies [49].

Subsequent molecular biological investigations on the *Duffy* gene revealed that **Duffy**-negative Negrids are homozygous for a *FY*B* allele[6] characterised by a single T → C substitution in the proximal GATA motif. This mutation affects the promoter activity in erythroid cells by disrupting a binding site for the GATA-1 erythroid transcription factor. However, it does not alter the normal expression of the **Duffy** antigen in non-erythroid cells. These findings thus provide an explanation for the erythroid-specific repression of the *Duffy* gene in **Duffy**-negative Negrids [29,43,54].

---

[6] Accordingly, **Fy(a–b–)** Negrids produce anti-**Fy^a** and not anti-**Fy^b** [31].

A defective *FY* gene characterised by a 14 base-pair deletion (nucleotides 287–301) has been detected in one of the extremely rare **Fy(a–b–)** individuals of Europid origin [35]. The resulting frameshift introduced a stop codon and produced a putative truncated 116 amino acid protein.

The occurrence of this mutation in an apparently healthy individual questions the functional importance of the **Duffy** glycoprotein on normal erythrocytes but also in non-haematopoietic tissue.

Analyses of the *FY* gene in non-human primates revealed that the respective gene of chimpanzees differs from the human *Fy* gene only by a Val → Ile exchange at position 115 [8]. High homology was also found in squirrel and rhesus monkeys (94%), and aotus monkey (93%) as compared to the human *FY* gene. Thus far only the *FY*B* type allele has been found in non-human primates.

Two other **Duffy** determinants have been localised by antibody binding experiments:

Anti-**Fy6** antibodies were shown to bind to overlapping linear epitopes within residues Q^{19} and W^{25} which are located between the two glycosylation sites on the first extracellular domain of the **Duffy** protein [58]; subsequent investigations have identified the pivotal motif as being Phe^{22}-Glu^{23}-Asp^{24} [55].

The chemokine molecule binds to a pocket formed by sequences located in the first and fourth extracellular domains; they are brought into close proximity by an intrachain disulfide bridge between cysteines 51 and 276. Inhibition experiments suggest that the actual chemokine binding site is virtually identical to the **Fy6** epitope [55,60].

The **Fy3** epitope has been located on the fourth extracellular domain of the **Duffy** antigen [34].

Duffy Antigens and Malaria

In 1992 Miller et al. [39] found that **Fy(a–b–)** individuals are resistant to *Plasmodium vivax* infection; this suggests that the **Duffy** antigens play an important role in malaria transmission. Only recently has it become possible to cultivate *P. vivax in vitro* [18], therefore experiments were performed with *Plasmodium knowlesi*, a closely related monkey parasite which invades human erythrocytes but causes only mild infections in man. These studies [36,40,41] revealed that the malaria merozoites are able to bind to **Fy(a–b–)** erythrocytes but are unable to invade the cells. When the

Duffy antigens are destroyed by chymotrypsin treatment or blocked by the respective anti-**Duffy** sera, it then becomes possible to induce this resistance in **Duffy**-positive cells.

Later the ligand in *Plasmodium knowlesi* was identified as a 135 kDa protein [1,23] showing a high affinity towards the **Duffy** glycoprotein isolated from **Fy**[a] and **Fy**[b] erythrocytes. However, the finding that **Duffy** negative erythrocytes become infected by *P. knowlesi* after treatment with neuraminidase or trypsin [36] indicates that additional factors are involved in the invasion of the erythrocytes, at least by this particular malaria parasite.

These investigations reveal that at least two receptor sites are involved in the infection of a red cell by *Plasmodium vivax* and *P. knowlesi*: the binding site responsible for attachment of the parasite is present on all red cells irrespective of the **Duffy** phenotype, whereas the invasion site depends on the presence of **Duffy** antigens. More recent studies revealed that the site critical for *Plasmodium vivax* and *P. knowlesi* merozoite invasion of red blood cells is located within the N-terminal domain of the human **Duffy** substance between residues 20 and 38 [8,12].

On the basis of these findings it must be assumed that the **Duffy**-negative phenotype became predominant in Africans through natural selection as a consequence of adaptive evolution for an advantageous mutational change. Thus, lack of the **Duffy** antigens is one of the few known examples of selective advantage granted by a blood group phenotype.

References

1. ADAMS, J. H., HUDSON, D. E., TORII, M., WARD, G. E., WELLEMS, T. E., AIKAWA, M. & MILLER, L. H. (1990): The Duffy receptor family of Plasmodium knowlesi is located within the micronemes of invasive malaria merozoites. *Cell 63*, 141-153.
2. ALBREY, J. A., VINCENT, E. E. R., HUTCHINSON, J., MARSH, W. L., ALLEN, F. H., GAVIN, J. & SANGER, R. (1971): A new antibody, anti-Fy3, in the Duffy blood group system. *Vox Sang. 20*, 29-35.
3. BAGGIOLINI, M., DEWALD, B. & MOSER, B. (1994): Interleukin-8 and related chemotactic cytokines--CXC and CC chemokines. *Adv. Immunol. 55*, 97-179.
4. BEHZAD, O., LEE, C. L., GAVIN, J. & MARSH, W. L. (1973): A new anti-erythrocyte antibody in the Duffy system: anti-Fy4. *Vox Sang. 24*, 337-342.
5. BIRD, G. W. G. & WINGHAM, J. (1970): N-acetylneuraminic (sialic) acid and human blood group antigen structure. *Vox Sang. 18*, 240-243.
6. CEDERGREN, B. & GILES, C. M. (1973): An Fy[x] Fy[x] individual found in Northern Sweden. *Vox Sang. 24*, 264-266.
6a. CARTRON, J. P. (1996): Vers une approche moleculaire de la structure, du polymorphisme et de la fonction des groupes sanguins. *Transfus. Clin. Biol. 3*, 181-210.
7. CHAUDHURI, A., NIELSEN, S., ELKJAER, M. L., ZBRZEZNA, V., FANG, F. & POGO, O. (1997): Detection of Duffy antigen in the plasma membrane and caveolae of vascular endothelial and epithelial cells

of nonerythroid organs. *Blood 89*, 701-712.

8. CHAUDHURI, A., POLYAKOVA, J., ZBRZEZNA, V. & POGO, A. O. (1995): The coding sequence of Duffy blood group gene in humans and simians: restriction fragment length polymorphism, antibody and malarial parasite specificities, and expression in nonerythroid tissues in Duffy-negative individuals. *Blood 85*, 615-621.

9. CHAUDHURI, A., POLYAKOVA, J., ZBRZEZNA, V., WILLIAMS, K., GULATI, S. & POGO, A. O. (1993): Cloning of glycoprotein D cDNA, which encodes the major subunit of the Duffy blood group system and the receptor for the Plasmodium vivax malaria parasite. *Proc. Natl. Acad. Sci. USA 90*, 10793-10797.

10. CHAUDHURI, A., ZBRZEZNA, V., JOHNSON, C., NICHOLS, M., RUBINSTEIN, P., MARSH, W. L. & POGO, A. O. (1989): Purification and characterization of an erythrocyte membrane protein complex carrying Duffy blood group antigenicity. Possible receptor for Plasmodium vivax and Plasmodium knowlesi malaria parasite. *J. Biol. Chem. 264*, 13770-13774.

11. CHAUDHURI, A., ZBRZEZNA, V., POLYAKOVA, J., POGO, A. O., HESSELGESSER, J. & HORUK, R. (1994): Expression of the Duffy antigen in K562 cells. Evidence that it is the human erythrocyte chemokine receptor. *J. Biol. Chem. 269*, 7835-7835.

12. CHITNIS, C. E., CHAUDHURI, A., HORUK, R., POGO, A. O. & MILLER, L. H. (1996): The domain on the Duffy blood group antigen for binding Plasmodium vivax and P. knowlesi malarial parasites to erythrocytes. *J. Exp. Med. 184*, 1531-1536.

13. CHOWN, B., LEWIS, M. & KAITA, H. (1965): The Duffy blood group system in Caucasians: evidence for a new allele. *Am. J. Hum. Genet. 17*, 384-389.

14. COLLEDGE, K. I., PEZZULICH, M. & MARSH, W. L. (1973): Anti-Fy5, an antibody disclosing a probable association between the Rhesus and Duffy blood group genes. *Vox Sang. 24*, 193-199.

15. COOK, P. J. L., PAGE, B. M., JOHNSTON, A. W., STANFORD, W. K. & GAVIN, J. (1978): Four further families informative for 1q and the Duffy blood group. Human gene mapping 4. *Cytogenet. Cell Genet. 22*, 378-380.

16. CUTBUSH, M. & MOLLISON, P. L. (1950): The Duffy blood group system. *Heredity 4*, 383-389.

17. CUTBUSH, M., MOLLISON, P. L. & PARKIN, D. M. (1950): A new human blood group. *Nature 165*, 188.

18. GOLENDA, C. F., LI, J. & ROSENBERG, R. (1997): Continuous in vitro propagation of the malaria parasite Plasmodium vivax. *Proc. Natl. Acad. Sci. USA 94*, 6786-6791.

19. HADLEY, T. J., DAVID, P. H., MCGINNISS, M. H. & MILLER, L. H. (1984): Identification of an erythrocyte component carrying the Duffy blood group Fy[a] antigen. *Science 223*, 597-599.

20. HADLEY, T. J., LU, Z. H., WASNIOWSKA, K., MARTIN, A. W., PEIPER, S. C., HESSELGESSER, J. & HORUK, R. (1994): Postcapillary venule endothelial cells in kidney express a multispecific chemokine receptor that is structurally and functionally identical to the erythroid isoform, which is the Duffy blood group antigen. *J. Clin. Invest. 94*, 985-991.

21. HADLEY, T. J. & PEIPER, S. C. (1997): From malaria to chemokine receptor: the emerging physiologic role of the Duffy blood group antigen. *Blood 89*, 3077-3091.

22. HAUSMAN, E., DZIK, W. & BLANCHARD, D. (1996): The red cell chemokine receptor is distinct from the Fy6 epitope. *Transfusion 36*, 421-425.

23. HAYNES, J. D., DALTON, J. P., KLOTZ, F. W., MCGINNISS, M. H., HADLEY, T. J., HUDSON, D. E. & MILLER, L. H. (1988): Receptor-like specificity of a Plasmodium knowlesi malarial protein that binds to Duffy antigen ligands on erythrocytes. *J. Exp. Med. 167*, 1873-1881.

24. HORUK, R., CHITNIS, C. E., DARBONNE, W. C., COLBY, T. J., RYBICKI, A., HADLEY, T. J. & MILLER, L. H. (1993): A receptor for the malarial parasite Plasmodium vivax: the erythrocyte chemokine receptor. *Science 261*, 1182-1184.

25. HORUK, R., ZI-XUAN, W., PEIPER, S. C. & HESSELGESSER, J. (1994): Identification and characterization of a promiscuous chemokine-binding protein in a human erythroleukemic cell line. *J. Biol. Chem. 269*, 17730-17733.

26. IKIN, E. W., MOURANT, A. C., PETTENKOFER, H. J. & BLUMENTHAL, G. (1951): Discovery of the expected haemagglutinin, anti-Fy[b]. *Nature 168*, 1077.

27. ISSITT, P. D. (1985): The Duffy blood group system. In: *Applied Blood Group Serology*. Montgomery

Scientific Publications, Miami, Florida, USA, pp. 278-288.

28. IWAMOTO, S., LI, J., OMI, T., IKEMOTO, S. & KAJII, E. (1996): Identification of a novel exon and spliced form of Duffy mRNA that is the predominant transcript in both erythroid and postcapillary venule endothelium. *Blood 87*, 378-385.

29. IWAMOTO, S., LI, J., SUGIMOTO, N., OKUDA, H. & KAJII, E. (1996): Characterization of the Duffy gene promoter: evidence for tissue-specific abolishment of expression in Fy(a-b-) of black individuals. *Biochem. Biophys. Res. Commun. 222*, 852-859.

30. IWAMOTO, S., OMI, T., KAJII, E. & IKEMOTO, S. (1995): Genomic organization of the glycoprotein D gene: Duffy blood group Fy^a/Fy^b alloantigen system is associated with a polymorphism at the 44-amino acid residue. *Blood 85*, 622-626.

31. LE PENNEC, P. Y., ROUGER, P., KLEIN, M. T., ROBERT, N. & SALMON, C. (1987): Study of anti-Fy^a in five black Fy(a-b-) patients. *Vox Sang. 52*, 246-249.

32. LEWIS, M., KAITA, H. & CHOWN, B. (1972): The Duffy blood group system in Caucasians. A further population sample. *Vox Sang. 23*, 523-527.

33. LISOWSKA, E., DUK, M. & WASNIOWSKA, K. (1983): Duffy antigens: observations on biochemical properties. In: *Red Cell Membrane Glycoconjugates and Related Genetic Markers* (J. P. Cartron, P. Rouger, and C. Salmon, eds.). Librairie Arnette, Paris, pp. 87-96.

34. LU, Z. H., WANG, Z. X., HORUK, R., HESSELGESSER, J., LOU, Y. C., HADLEY, T. J. & PEIPER, S. C. (1995): The promiscuous chemokine binding profile of the Duffy antigen/receptor for chemokines is primarily localized to sequences in the amino-terminal domain. *J. Biol. Chem. 270*, 26239-26245.

35. MALLINSON, G., SOO, K. S., SCHALL, T. J., PISACKA, M. & ANSTEE, D. J. (1995): Mutations in the erythrocyte chemokine receptor (Duffy) gene: the molecular basis of the Fy^a/Fy^b antigens and identification of a deletion in the Duffy gene of an apparently healthy individual with the Fy(a-b-) phenotype. *Br. J. Haematol. 90*, 823-829.

36. MASON, S. J., MILLER, L. H., SHIROISHI, T., DVORAK, J. A. & MCGINNISS, M. H. (1977): The Duffy blood group determinants: their role in the susceptibility of human and animal erythrocytes to Plasmodium knowlesi malaria. *Br. J. Haematol. 36*, 327-335.

37. MASOUREDIS, S. P., SUDORA, E., MAHAN, L. & VICTORIA, E. L. (1980): Quantitative immunoferritin microscopy of Fy^a, Fy^b, Jk^a, U, and Di^b antigen site numbers on human red cells. *Blood 56*, 969-977.

38. MATHEW, S., CHAUDHURI, A., MURTY, V. V. & POGO, A. O. (1994): Confirmation of Duffy blood group antigen locus (FY) at 1q22→q23 by fluorescence in situ hybridization. *Cytogenet. Cell Genet. 67*, 68.

39. MILLER, L. H., MASON, S. J., CLYDE, D. F. & MCGINNISS, M. H. (1976): The resistance factor to Plasmodium vivax in blacks. The Duffy blood group genotype, FyFy. *N. Engl. J. Med. 295*, 302-304.

40. MILLER, L. H., MASON, S. J., DVORAK, J. A., MCGINNISS, M. H. & ROTHMANN, J. K. (1975): Erythrocyte receptors for (Plasmodium knowlesi) malaria: Duffy blood group determinants. *Science 189*, 561-563.

41. MILLER, L. H., SHIROISHI, T., DVORAK, J. A., DUROCHER, J. R. & SCHRIER, B. K. (1975): Enzymatic modification of the erythrocyte membrane surface and its effect on malarial merozoite invasion. *J. Mol. Med. 1*, 55-63.

42. MOORE, S., WOODROW, C. F. & MCCLELLAND, D. B. L. (1982): Isolation of membrane components associated with human red cell antigens Rh(D), (c), (E), and Fy^a. *Nature 295*, 529-531.

43. MOULDS, J. M., HAYES, S. & WELLS, T. D. (1998): DNA analysis of Duffy genes in American blacks. *Vox Sang. 74*, 248-252.

44. NEOTE, K., MAK, J. Y., KOLAKOWSKI, L. F. & SCHALL, T. J. (1994): Functional and biochemical analysis of the cloned Duffy antigen: identity with the red blood cell chemokine receptor. *Blood 84*, 44-52.

45. NICHOLS, M. E., RUBINSTEIN, P., BARNWELL, J., RODRIGUEZ DE CORDOBA, S. & ROSENFIELD, R. E. (1987): A new human Duffy blood group specificity defined by a murine monoclonal antibody.

Immunogenetics and association with susceptibility to Plamodium vivax. *J. Exp. Med. 166*, 776-785.

46. Olsson, M. L., Smythe, J. S., Hansson, C., Poole, J., Mallinson, G., Jones, J., Avent, N. D. & Daniels, G. (1998): The Fy^x phenotype is associated with a missense mutation in the Fy^b allele predicting Arg89Cys in the Duffy glycoprotein. *Br. J. Haematol. 103*, 1184-1191.

47. Palatnik, M. & Rowe, A. W. (1984): Duffy and Duffy-related human antigens in primates. *J. Hum. Evolut. 13*, 173-175.

48. Parasol, N., Reid, M., Rios, M., Castilho, L., Harari, I. & Kosower, N. S. (1998): A novel mutation in the coding sequence of the FY*B allele of the Duffy chemokine receptor gene is associated with an altered erythrocyte phenotype. *Blood 92*, 2237-2243.

49. Peiper, S. C. M., Wang, Z. X., Neote, K., Martin, A. W., Showell, H. J., Conklyn, M. J., Ogborne, K., Hadley, T. J., Lu, Z. H., Hesselgesser, J. & Horuk, R. (1995): The Duffy antigen/receptor for chemokines (DARC) is expressed in endothelial cells of Duffy negative individuals who lack the erythrocyte receptor. *J. Exp. Med. 181*, 1311-1317.

50. Race, R. R. & Sanger, R. (1975): The Duffy blood group. In: *Blood Groups in Man*. Blackwell Scientific Publications, Oxford, pp. 350-363.

51. Riwom, S., Janvier, D., Navenot, J. M., Benbunan, M., Muller, J. Y. & Blanchard, D. (1994): Production of a new murine monoclonal antibody with Fy6 specificity and characterization of the immunopurified N-glycosylated Duffy-active molecule. *Vox Sang. 66*, 61-67.

52. Tanner, M. J. A., Anstee, D. J., Mallinson, G., Ridgwell, K., Martin, P. G., Avent, N. D. & Parsons, S. F. (1988): Effect of endoglycosidase F-peptidyl N-glycosidase F preparations on the surface components of the human erythrocyte. *Carbohydr. Res. 178*, 203-212.

53. Toivanen, P. & Hirvonen, T. (1973): Antigens Duffy, Kell, Kidd, Lutheran, and Xg^a on fetal red cells. *Vox Sang. 24*, 372-376.

54. Tournamille, C., Colin, Y., Cartron, J. P. & Le Van Kim, C. (1995): Disruption of a GATA motif in the Duffy gene promoter abolishes erythroid gene expression in Duffy-negative individuals. *Nat. Genet. 10*, 224-228.

55. Tournamille, C., Le Van Kim, C., Gane, P., Blanchard, D., Proudfoot, A. E., Cartron, J. P. & Colin, Y. (1997): Close association of the first and fourth extracellular domains of the Duffy antigen/receptor for chemokines by a disulfide bond is required for ligand binding. *J. Biol. Chem. 272*, 16274-16280.

56. Tournamille, C., Le Van Kim, C., Gane, P., Cartron, J. P. & Colin, Y. (1995): Molecular basis and PCR-DNA typing of the Fya/Fyb blood group polymorphism. *Hum. Genet. 95*, 407-410.

57. Tournamille, C., Le Van Kim, C., Gane, P., Le Pennec, P. Y., Roubinet, F., Babinet, J., Cartron, J. P. & Colin, Y. (1998): Arg89Cys substitution results ins very low membrane expression of the Duffy antigen/receptor for chemokines in Fy^x individuals. *Blood 92*, 2147-2156.

58. Wasniowska, K., Blanchard, D., Janvier, D., Wang, Z. X., Peiper, S. C., Hadley, T. J. & Lisowska, E. (1996): Identification of the Fy6 epitope recognized by two monoclonal antibodies in the N-terminal extracellular portion of the Duffy antigen receptor for chemokines. *Mol. Immunol. 33*, 917-923.

59. Williams, D., Johnson, C. L. & Marsh, W. L. (1981): Duffy antigen changes on red blood cells stored at low temperature. *Transfusion 21*, 357-359.

60. Zhao-Hai, L., Zi-Xuan, W., Horuk, R., Hesselgesser, J., Yang-Chun, L., Hadley, T. J. & Peiper, S. C. (1995): The promiscuous chemokine binding profile of the Duffy antigen/receptor for chemokines is primarily localized to sequences in the amino-terminal domain. *J. Biol. Chem. 270*, 26239-26245.

17 Kell System

Kell ('Kelleher') is a complex blood group system of human erythrocytes, 23 antigens of which have been defined to date (see *Table 17.1*). The **Kell** phenotype of human erythrocytes is controlled by two proteins encoded by two independent genes: the **Kell** antigens *per se* are produced by alleles of an autosomal gene, *KEL*, which has been located on chromosome 7 at position q33–35 distal to the cystic fibrosis locus [56,71,78,109]. The expression of the **Kell** antigens on the red cell is controlled by another protein, **Kx**. This protein is encoded by the *XK* gene found on the short arm of the **X** chromosome at Xp21, between the Duchenne muscular dystrophy and chronic granulomatous disease loci [9,34].

A series of **Kell** characters are encoded in five sets of antithetical alleles expressing low- and high-frequency antigens: **KEL2/KEL1**, **KEL4/KEL3/KEL21**, **KEL7/KEL6**, **KEL11/KEL17**, and **KEL14/KEL24**. The other antigen specificities are independently expressed and have no known antithetical partner; these so-called para-**Kell** characters are absent from erythrocytes of the **K₀** phenotype and are expressed only weakly on **McLeod** phenotype cells [62]. Thus far no family studies have been performed which could have proved that these antigens are controlled by genes at the *KEL* locus. In subsequent investigations, however, these characters have been located on the **Kell** protein of the human erythrocyte membrane [53,77,80] – evidence that the para-**Kell** antigens are indeed part of the **Kell** system.

There is strong evidence that the high-incidence antigenic character **RAZ** [18] belongs to the **Kell** blood group system. It has, however, not yet been given a number in the **Kell** system.

The **Kell** antigenic characters are already well expressed on foetal erythrocytes and by birth are fully developed [99]. In adults the antigens develop early in red cell maturation and can be detected on cells of the early erythroblast stage [62].

Some **Kell** characters show a different frequency in certain ethnic groups: **KEL1** is a characteristic of Europids, it is less frequent in Negrids and extremely rare in Mongolids [68]. **KEL6** is present in 19.5% of Negrids and **KEL10** occurs in 2.6% of Finns; both characters occur in less than 1% of Europids [79].

Table 17.1: Notations and frequencies of antigens in the Kell blood group system.

ISBT Term[a]	Antigen	First case	Frequency (%)[b]	References
KEL1	K	Kelleher	9.0	[16]
KEL2	k	Cellano	99.8	[58]
KEL3	Kp^a	Penney	2.0	[3]
KEL4	Kp^b	Rautenberg	99.9	[4]
KEL5	Ku	Peltz (K$_0$)	>99.9	[17]
KEL6	Js^a	Sutter	<1.0[c]	[27]
KEL7	Js^b	Matthews	>99.9	[102,103]
KEL10	Ul^a	Karhula	>0.1[d]	[25]
KEL11		Côté	>99.9	[29]
KEL12[e]		Bockman	>99.9	[64]
KEL13[e]		Sgro.	>99.9	[59]
KEL14		Santini	>99.9	[104]
KEL16	'k-like'		99.8	[61]
KEL17	Wk^a	Weeks	0.3	[95]
KEL18[e]		Marshall (V.M.)	>99.9	[8]
KEL19[e]		Sublett	>99.9	[89]
KEL20	Km		>99.9	[100]
KEL21	Kp^c	Levay	<0.1	[26,107]
KEL22[e]		Ikar (N.I.)	>99.9	[7]
KEL23[e]			<0.1	[63]
KEL24			<2.0	[22]
KEL25	VLAN		<0.1	[41]
KEL26	TOU		>99.9	[39]
	RAZ[f]		>99.9	[18]

[a] The six-digit ISBT identification numners are 006 001 to 006 025 (see *Chap. 1*)
[b] In Europids
[c] 19.5% in Negrids
[d] 2.6% in Finns
[e] 'Para-Kell' antigens
[f] The antigen has not been given a **Kell** system number, though it is absent from cells of the K$_0$ phenotype and has been shown to be present on the **Kell** glycoprotein by MAIEA analysis [18]. From Issitt [36] by permission of the author, with additional data from Marsh et al. [62] and Lee et al. [53].

Kell antigens are not confined to humans but have also been found on the red cells of non-human primates. Mouse hybridoma antibodies of anti-**KEL2** specificity reacted with the erythrocytes of all primate species thus far tested; the **KEL14**

character, however, was expressed only on the red cells of the great apes (chimpanzee, gorilla, and gibbon) [72].

Subsequent studies showed that almost all common **Kell** group antigens found in humans are also found in the chimpanzee (*Pan troglodytes*) [82]. Further, the **Kell** protein of chimpanzee erythrocytes showed a close homology to that of human red blood cells, whereby its slightly greater molecular mass (97 kDa) being probably due to increased glycosylation [62].

Thus far no **Kell** protein has been found on the red cells of non-primate mammalian [38].

A **KEL1**-like antigen has also been detected in a series of pathogenic bacteria, e.g. *E. coli* (strain *O125:B15*) [60,90], *Mycobacterium* sp. [44], *Streptococcus faecium* [66], *Morganella morganii* [75], or *Enterococcus faecalis* [20].

17.1 Antibodies

Most antibodies [91] in the **Kell** system are IgG alloantibodies induced by **Kell** incompatible blood transfusions or pregnancies. Particularly antibodies towards the **KEL1** antigen are of great clinical interest since this antigen − a strong immunogen second in importance only to **RhD** − may cause severe reactions to incompatible blood transfusions and has been known to induce haemolytic disease of the newborn in sensitised mothers (see e.g. [10,65]. Antibodies with other specificities are encountered less commonly and are of less importance.

In some cases anti-**KEL1** antibodies are produced through immunisation by the **KEL1**-like antigens of bacteria (see above).

More recently hybridoma antibodies with anti-**KEL1** [37,38], anti-**KEL2**, and anti-**KEL14** [72,74] specificity have been prepared. Researchers have also produced a number of murine hybridoma antibodies towards purified **Kell** protein which react with all red cells except those of the K_o phenotype, e.g. *BRIC18*, *BRIC68*, and *BRIC203* [74].

Anti-**Kx** antibodies are produced in **McLeod** individuals and patients with chronic granulomatous disease when transfused with blood of common **Kell** type [28,61,100], and autoantibody with anti-**Kx** specificity has already been described [96].

17.2 Kell-Deficient Phenotypes

In the **Kell** system two rare **Kell**-deficient phenotypes have been described, **McLeod** and **K$_o$** ('**Kell** null'):

In erythrocytes of **McLeod** subjects [2] the **Kell** antigens are significantly depressed, and the **Kx** antigen is not detectable[1]. The red cells are predominantly of acanthocytic shape and show reduced *in vivo* survival. Furthermore, **McLeod** individuals express a series of neurological and muscular tissue disorders (= '**McLeod** syndrome' [62,87]). This rare phenotype either results from a deletion of part of the X chromosome or is inherited through a defective allele at the **XK** locus (see below). The **McLeod** phenotype is often associated with X-linked chronic granulomatous disease, Duchenne muscular dystrophy, or retinitis pigmentosa. As can be expected, all cases of this phenotype have been described in males.

K$_o$ erythrocytes lack all known **Kell** characters but show enhanced activity of **Kx** antigen. The red cells have normal morphology and *in vivo* survival and show no defect in red cell viability or function [62]. This phenotype is obviously inherited through a silent allele at the *Kell* locus. When immunised by transfusion or during pregnancy, **K$_o$** individuals produce an antibody (anti-**KEL5**) which reacts with all of the **Kell** antigens [83].

Some rare **Kell** phenotypes are characterised by a permanent weak expression of all **Kell** antigens and are presumably inherited through a variant allele at the *KEL* locus (= **K$_{mod}$** [83]). Erythrocytes of this phenotype show a higher level of **Kx**. Further, some **K$_{mod}$** individuals produce an antibody which resembles anti-**KEL5** but does not react with **K$_{mod}$** cells [62].

There is wide variability in the expression of **Kell** characters, ranging from a barely detectable decrease to the virtual absence of **Kell** antigens on the red cell membrane. Therefore, **K$_{mod}$** and **K$_o$** phenotypes should rather be grouped in a common **Kell** phenotype which is characterised by gradually decreasing expression of **Kell** antigens [81].

The **Kell** antigens are also weakly expressed on the erythrocytes of glycophorin C- and D-deficient individuals of **Leach** (**Ge:−2,−3,−4**) and **Ge** (**Ge:−2,−3,4**) phenotypes [36] (see *Sect. 18.2*): e.g. the site density of the **KEL1**

---

[1] Branch et al. [13], however, detected small amounts of **Kx** antigen on **McLeod**-type erythrocytes after treatment of the cells with dithiothreitol.

antigen on **Ge:–2,–3** cells was found to be about one-third of that present on **KEL1/KEL1** cells of the common **Gerbich** phenotype [37]. The fact that **Ge:–,3 (Yus** phenotype) cells show no weakened expression of the **Kell** antigens suggests that the **Kell** protein may interact with that part of the glycophorin C molecule which is encoded by exon 3.

Transient depression of **Kell** antigens is sometimes observed in connection with the production of autoimmune **Kell** antibodies towards the patient's own red cells (e.g. [14,92,101,106]).

17.3 Kell Glycoprotein

The **Kell** antigens are inactivated *in situ* by sulfhydryl reagents, suggesting that the **Kell** epitopes represent 'conformation-dependent determinants' stabilised by intrachain disulphide bonds. Treatment of erythrocytes with 6% 2-amino-ethylisothiouronium bromide (= AET) destroys most of the **Kell** antigens [1]. The majority of the **Kell** characters are inactivated by dithiothreitol at a concentration of 100–200 mM. The **KEL6** and **KEL7** antigens, however, are inactivated at a concentration as low as 2 mM [11]. Other disulfide-reducing reagents, like mercaptoethanol are less efficient. Since in all these cases the expression of **Kx** is enhanced (see e.g. [13]) sulfhydryl reagents – AET in particular – are used to produce artificial **K$_o$** cells (see above).

Single proteases have little or no effect on the expression of **Kell** antigens on erythrocytes, whereas simultaneous treatment of the cells with trypsin and α-chymotrypsin destroys the activity of most **Kell** and para-**Kell** antigens [42]. The **Kell** antigens are also denatured by treatment with a mixture of dithiothreitol and cysteine-activated papain (ZZAP reagent) [12].

The finding that preparations of membrane skeleton obtained by Triton X-100 treatment of erythrocytes retained most of the **Kell** protein [37] indicated that the **Kell** substance is a transmembrane protein which, with its cytoplasmic portion, interacts with the underlying membrane skeleton.

Under non-reducing conditions, the **Kell** glycoprotein is co-immunoprecipitated with the **Kx** protein [45], which indicates that *in situ* the **Kell** glycoprotein is covalently associated with the **Kx** protein via a disulfide bond [13].

```
  1 MEGGDQSEEE PRERSQAGGM GTLWSQESTP EERLPVEGSR PWAVARRVLT AILILGLLLC  60

 61 FSVLLFYNFQ NCGPRPCETS VCLDLRDHYL ASGNTSVAPC TDFFSFACGR AKETNNSFQE 120

121 LATKNKNRLR RILEVQNSWH PGSGEEKAFQ FYNSCMDTLA IEAAGTGPLR QVIEELGGWR 180

181 ISGKWTSLNF NRTLRLLMSQ YGHFPFFRAY LGPHPASPHT PVIQIDQPEF DVPLKQDQEQ 240

241 KIYAQIFREY LTYLNQLGTL LGGDPSKVQE HSSLSISITS RLFQFLRPLE QRRAQGKLFQ 300

301 MVTIDQLKEM APAIDWLSCL QATFTPMSLS PSQSLVVHDV EYLKNMSQLV EEMLLKQRDF 360

361 LQSHMILGLV VTLSPALDSQ FQEARRKLSQ KLRELTEQPP MPARPRWMKC VEETGTFFEP 420

421 TLAALFVREA FGPSTRSAAM KLFTAIRDAL ITRLRNLPWM NEETQNMAQD KVAQLQVEMG 480

481 ASEWALKPEL ARQEYNDIQL GSSFLQSVLS CVRSLRARIV QSFLQPHPQH RWKVSPWDVN 540

541 AYYSVSDHVV VFPAGLLQPP FFHPGYPRAV NFGAAGSIMA HELLHIFYQL LLPGGCLACD 600

601 NHALQEAHLC LKRHYAAFPL PSRTSFNDSL TFLENAADVG GLAIALQAYS KRLLRHHGET 660

661 VLPSLDLSPQ QIFFRSYAQV MCRKPSPQDS HDTHSPPHLR VHGPLSSTPA FARYFRCARG 720

721 ALLNPSSRCQ LW
```

Figure 17.1: Amino acid sequence of the Kell protein as derived from cDNA sequencing.
Underlined: hydrophobic segment representing the putative transmembrane domain; double underlined: cysteine residues; *N* : putative N-glycosylation sites.
From Lee et al. [54,55]. The sequence data are deposited in the EMBL/GenBank data library (accession numbers M64934 and S76819).

The membrane constituent carrying the blood group **Kell** characters is a 93-kDa glycoprotein [80,84,105] containing ~12% carbohydrate by weight [37]. Treatment with N-glycanase reduced the molecular mass to about 79 kDa, whereas treatment with O-glycanase had no effect whatsoever [82]. Carbohydrate analysis suggested that the **Kell** protein carries five or six N-linked oligosaccharide units on the extracellular domain, but lacks O-linked chains; further, the molar ratio of the sugars seems to propose oligosaccharide chains of the biantennary type [37].

Quantitative estimates of the **Kell** protein yielded values between 2,500 and 6,000 molecules per red cell [35,37,67,74]; site numbers in the range 4000–18000 were obtained using Fab fragments prepared from the hybridoma antibodies [74].

The primary structure of the blood group **Kell** protein was deduced from its cDNA sequence by Lee et al. [54,55]. Hydropathy analysis of the predicted a 732-amino-acid polypeptide (*Fig. 17.1)* indicated a type II membrane protein with a single membrane-spanning domain (residues 48–67). The N-terminal 47 amino acids are mainly hydrophilic; they contain no putative N-glycosylation sites, and are oriented into the cell cytoplasm. The 665-amino acid C-terminal segment, which is located extra-cellularly, contains five possible N-glycosylation sites (at positions 93, 115, 191, 345, and 627). Moreover, 15 of the 16 cysteine residues present in the deduced amino acid sequence are located in the extracellular domain; the cysteine residues are mainly located in two clusters – close to the membrane-spanning segment and near the C-terminus. This distribution of cysteine residues is in keeping with the secondary structure of the **Kell** protein proposed by Redman and Lee [81]. According to this study the extracellular portion consists of two heavily folded domains each of which is stabilised by disulfide bonds and separated from each other by a linear peptide segment. More recent literature suggests that the cysteine residue at position 72 forms a linkage to Cys^{347} of the **Kx** protein [88].

It has been noted [49,55] that the **Kell** protein showed both structural and sequence homology to a family of zinc metalloproteins with neutral endopeptidase activity, particularly to the 'common acute lymphoblastic leukaemia antigen' (= CALLA [93]). It also shares the pentapeptide sequence, H.E.X.X.H (at amino acid positions 581–595), which is characteristic for zinc-dependent endopeptidases found in nearly all organisms [40].

These endopeptidases show a broad tissue and species distribution, and are involved mainly in processing and inactivating peptide hormones, such as the enkaphalines, neurotensin, angiotensins, ocytocin, and bradykinin. However, in contrast to these endopeptidases which are found in many cell types, the **Kell** protein itself is expressed only in erythroid tissues [56]. The physiologic role of Kell on erythrocyte membranes has not yet been elucidated.

The coding sequence of the *KEL* gene is organised in 19 exons spanning approximately 21.5 kb. Exon 1 contains the 5' untranslated region and the initiation amino acid methionine; the membrane-spanning region is encoded in exon 3 and the putative zinc endo-peptidase active site is in exon 16 [54].

Transcripts of **Kell** mRNA were detected in human bone marrow and fetal liver, but not in brain, kidney, lung, or adult liver [56]. Further, no **Kell** antigen or **Kell** protein could be found in platelets, lymphocytes, monocytes or granulocytes [21,38]. These findings indicate that **Kell** expression is restricted to erythroid cells.

Table. 17.2: Point mutations in the Kell protein causing different phenotypes.

Allotype	Exon	Codon	Triplet	Amino acid	References
Antithetical alleles:					
KEL2	6	193	ACG	Thr	[55]
KEL1	6	193	ATG	Met	[55]
KEL4	8	281	CGG	Arg	[51]
KEL3	8	281	TGG	Trp	[51]
KEL21	8	281	CAG	Gln	[51]
KEL7	17	597	CTC	Leu	[50]
KEL6	17	597	CCC	Pro	[50]
KEL11	8	302	GTC	Val	[51]
KEL17	8	302	GCC	Ala	[51]
KEL14	6	180	CGC	Arg	[48]
KEL24	6	180	CCC	Pro	[48]
Low-incidence antigens:					
KEL10	13	494	GAA→GTA	Glu→Val	[51]
KEL23	10	382	CAG→CGG	Gln→Arg	[46]
Low-incidence phenotypes:					
KEL:-12	15	548	CAT→CGT	His→Arg	[49]
KEL:-18	4	130	CGG→TGG	Arg→Trp	[49]
KEL:-18	4	130	CGG→CAG	Arg→Gln	[49]
KEL:-19	13	492	CGA→CAA	Arg→Gln	[49]
KEL:-22	9	322	GCG→GTG	Ala→Val	[49]
KEL:-26	11	406	CGA→CAA	Arg→Gln	[49]
Weak expression:					
weak KEL2	11	423	GCG→GTG	Ala→Val	[46]
weak KEL2	6	193		Thr→Arg	[99a]

17.4 Kell Alleles

The chemical basis of most of the **Kell** characters has already been determined: all alleles thus far characterised result from single base mutations in the *KEL* gene causing amino acid substitutions in the **Kell** glycoprotein (*Table 17.2*):

- the **KEL2 / KEL1** dimorphism is based on a C → T base substitution in exon 6 (nucleotide 698) which results in an amino acid change at position 193, threonine being characteristic for **KEL2** and methionine for **KEL1** [57]. The methionine residue in the **KEL1** epitope disrupts the consensus N-glycosylation motif at Gln[191] thus preventing N-glycosylation at this site;
- **KEL7** and **KEL6** are distinguished by a T → C base substitution in exon 17 (nucleotide 1910) which gives rise to an amino acid change at position 597 of the **Kell** protein, leucine being determinant for **KEL7** and proline for **KEL6** [52][(2)]. A silent mutation at nucleotide 2019 (A → G) found in all the **KEL6** alleles investigated retains leucine at position 633;
- **KEL4**, **KEL3**, and **KEL21** are distinguished by point mutations at the same codon in exon 8, encoding amino acid residue 281 [53]: the high-prevalence antigen **KEL4** has the codon CGG for arginine, a C → T substitution at nucleotide 961 in the case of **KEL3** results in the TGG codon for tryptophan, and a G → A substitution at nucleotide 962 in the case of **KEL21** produces the CAG codon for glutamine;
- **KEL11** and **KEL17** are associated with a T → C base substitution in exon 8 (nucleotide 1025) changing the valine at position 302 in the case of **KEL11** to alanine in the case of **KEL17** [53];
- the **KEL14** and **KEL24** set is based on a G → C mutation in exon 6 (nucleotide 659) indicating an Arg (**KEL14**) ↔ Pro (**KEL24**) dimorphism at position 180 [50],
- **KEL10** is characterised by an A → T point mutation in exon 13 (nucleotide 1601), encoding a valine at residue 494 instead of glutamic acid [53];
- the low-frequency character **KEL23** is determined by a Gln → Arg substitution at position 382 [51].

More recent studies have identified the serologically determinant amino acid residues for a series of high-incidence antigens of the **Kell** blood group system [51]. Sequencing cDNAs obtained from probands lacking these antigens revealed single point mutations leading to amino acid substitutions: His[548] → Arg in **KEL:–12**,

---

[(2)] Lee et al. have also shown that **Kell** glycoprotein of chimpanzees who carry the **KEL:6,–7** phenotype [82] is characterised by the same base substitution as noted in human **KEL6** [52].

Arg^{492} → Gln in **KEL:–19**, Ala^{322} → Val in **KEL:–22**, and Arg^{406} → Gln in **KEL:–26** individuals; the **KEL:–18** phenotype is caused by two different point mutations in the same codon replacing Arg with either Trp or Gln.

It is therefore assumed that His^{548} is essential for the expression of **KEL12**, Arg^{130} for **KEL18**, Arg^{492} for **KEL19**, Ala^{322} for **KEL22**, and Arg^{406} for **KEL26**.

In four unrelated individuals with **K$_{mod}$** phenotype (**K+ ^{w}k–**, with weak expression of **Kp^a**, **Js^a**, **Ku**, **K14**, and **K22**) a **KEL2** allele has been detected which is characterised by a C^{698} → G point mutation in exon 6 resulting in a Thr → Arg substitution at position 193 [99a]. Immunoblotting experiments revealed a reduced quantity of the **Kell** protein at the cell surface.

Most of these single-base mutations lead to either the loss or the formation of a restriction endonuclease site. Subsequently, PCR-based diagnostic assays have been established which allow the determination of the *KEL1/KEL2* genotype from DNA samples [5,31,48,57,69,70]. Such genotyping procedures promise to be favourable for prenatal diagnosis of **KEL1**-related haemolytic disease of the newborn as compared to traditional serologic testing.

The spatial relationships between the different epitopes on the **Kell** protein was investigated by the so-called 'MAIEA' (= monoclonal antibody-specific immobilisation of erythrocyte antigens') technique [76]. This method is based on competition between human allo-antibodies and four mouse hybridoma antibodies which bind to different domains of the **Kell** glycoprotein (i.e. *BRIC18*, *BRIC107*, *BRIC68*, and *BRIC203*) [39,52,77,86]. The results of these tests indicated that the epitopes **KEL7/6** and **KEL12** are situated in close proximity to the **KEL2/1** epitope, even though the antigen sites are separated by more than 350 amino acid residues. Similarly, the point mutations which distinguish **KEL10** and **KEL14/24** are grouped by MAIEA but are located far apart on the peptide chain (positions 494 and 180, respectively). On the other hand, **KEL4/3/21** which is separated by only 88 amino acid residues from **KEL2/1** is obviously located on a different site of the molecule, and **KEL10** and **KEL19** although only two residues apart, do not interact at all in the MAIEA assay.

In one proband who expressed **'weak KEL2'**, a C^{1388} → T mutation has been detected which leads to an Ala to Val substitution at residue 423 [47]. The fact that this change in exon 11 influences and diminishes the expression of **KEL2** in exon 6 indicates that at least spatially the **KEL2** epitope is in the vicinity of the point of mutation.

The **KEL3** character is associated with a weakened expression of other **Kell** antigens in *cis* position (= **Kp(a+)** phenotype [98]). Recent investigations revealed that the depression of **Kell** antigens is due to a reduced amount of **Kell** protein at the cell

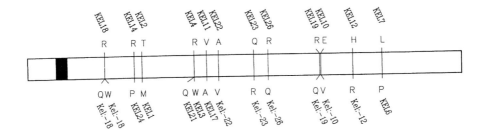

Figure 17.2: Kell protein.
Schematic representation of the protein showing point mutations characterising different **Kell** phenotypes.

surface. It has been demonstrated that this depression is effected by retention of the **Kell** glycoprotein in the pre-Golgi compartment due to differential processing of the protein chain [108].

In accordance with the above-mentioned serological data, immuno-precipitation experiments revealed that membranes of K_o erythrocytes lack the entire **Kell** protein molecule and yield about twice the amount of the **Kx** protein found in red cells of common **Kell** type [38].

Preliminary investigations on the DNA sequence of unrelated K_o individuals have shown that different mechanisms are responsible for the K_o phenotype. In one case a C → T mutation at nucleotide 502 introduced a premature stop codon in exon 4 leading to the expression of a truncated protein [47]. In other cases of K_o obviously 'normal' **Kell** mRNA has been detected [47,81]; it is not known yet whether this transcript is translated or whether the *KEL* allele is transcribed only in small amounts.

17.5 Kx Antigen

The **Kx** antigen (ISBT Nr. 019 001 or **XK1**, originally described as **K15** of the **Kell** blood group system) is genetically independent of the **Kell** system but plays an essential role in **Kell** antigen expression on the erythrocyte surface.

The **Kx** protein is a non-glycosylated 37-kDa erythrocyte membrane constituent [45,85]. *In situ* the protein is not sensitive to disulfide-reducing agents. In fact, quite to contrary: upon treatment with sulfhydryl reagents the erythrocytes show enhanced expression of **Kx** and thus react similarly to **K$_o$** cells [1,13]. *In situ* (i.e. within the membrane compound) the **Kx** protein is covalently associated with the **Kell** glycoprotein via a disulphide bond, as noted above.

Nucleotide sequence analysis of isolated cDNA predicted a membrane-associated protein of 444 amino acids (*Fig. 17.3.a*) with 10 putative transmembrane domains [33]. The **Kx** protein carries 16 cysteine residues: nine of which are located in the putative transmembrane segments, some of them are palmitoylated [15]. Only one cysteine (at position 347) is extracellular and most likely establishes the disulphide bridge to a Cys^{72} of the **Kell** protein [88]. Both the N-terminus and the C-terminus are intracellular. A topology of the **Kx** protein (*Fig. 17.3.b*) has been proposed which shows the structural characteristics of membrane transport proteins, such as the rabbit Na$^+$-dependent glutamate transporter [43] and the human noradrenaline transporter [73].

More recently two mutations of the **Kx** protein have been detected in Japanese donors, viz. C $\rightarrow$ G at codon 204 and G $\rightarrow$ C at codon 205, leading to Lys $\rightarrow$ Asp and Pro $\rightarrow$ Arg exchanges, respectively [94]. The amino acid substitutions are located at the 6th transmembrane segment of the **Kx** protein. Both **Kx** variants seem to represent a normal polymorphism frequently found in Asian ethnic groups.

The coding sequence of the *XK* gene is distributed over 3 exons. Northern blot analysis with different cDNA clones revealed widespread expression of the *XK* gene in human tissues, the expression is at its highest in skeletal muscle, brain, heart, and pancreas. This distribution is well correlated with the neurological and muscular defects in the **Kx**-deficient **McLeod** syndrome. Furthermore, the expression of the gene appears to be developmentally regulated in liver and spleen (which participate in erythropoiesis during fetal life), with high levels in the fetal stage and very low levels in the adult stage [33].

As mentioned above, the **McLeod** phenotype often results from gene deletions in the X chromosome [9,19,23,24]; in these cases the myopathy characteristic for the **McLeod** phenotype is frequently associated with X-linked chronic granulomatous disease and retinitis pigmentosa.

In other **McLeod** type probands, mutations of the *XK* gene have been detected. Molecular biological investigations on two probands have revealed point mutations at invariant residues of intron 2 – in one case a GT-to-AT transition at the 5' donor splice site, in the other case an AG to AA transition at the 3' acceptor splice site [33]. Such splice site mutations are known either to abolish splicing, resulting in aberrant or reduced normal splicing, or to lead to exon skipping. In another **McLeod** proband a single nucleotide deletion at codon 90 (exon 2) which creates a frameshift and leads

(a)

```
  1 MKFPASVLAS VFLFVAETTA ALSLSSTYRS GGDRMWQALT LLFSLLPCAL VQLTLLFVHR  60

 61 DLSRDRPLVL LLHLLQLGPL FRCFEVFCIY FQSGNNEEPY VSITKKRQMP KNGLSEEIEK 120

121 EVGQAEGKLI THRSAFSRAS VIQAFLGSAP QLTLQLYISV MQQDVTVGRS LLMTISLLSI 180

181 VYGALRCNIL AIKIKYDEYE VKVNRLAYVC IFLWRSFEIA TRVVVLVLFT SVLKTWVVVI 240

241 ILINFFSFFL YPWILFWCSG SPFPENIEKA LSRVGTTIVL CFLTLLYTGI NMFCWSAVQL 300

301 KIDSPDLISK SHNWYQLLVY YMIRFIENAI LLLLWYLFKT DIYMYVCAPL LVLQLLIGYC 360

361 TAILFMLVFY QFFHPCKKLF SSSVSEGFQR WLRCFCWACR QQKPCEPIGK EDLQSSRDRD 420

421 ETPSSSKTSP EPGQFLNAED LCSA
```

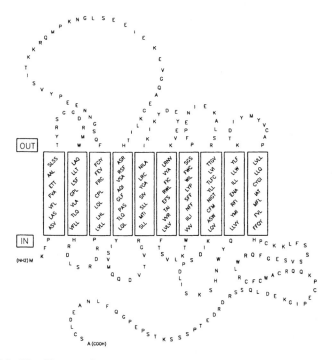

Figure 17.3: The Kx protein.
(a) Amino acid sequence of the Kx protein as derived from cDNA sequencing. Underlined: hydrophobic segments representing the putative transmembrane domains, double underlined: potential phosphorylation sites. **(b)** Proposed topology in the plasma membrane [33].

From Ho et al. [33] by permission of Cell Press. The sequence data are deposited in the EMBL/GenBank data library (accession number Z32684).

to premature termination at codon 128 has been found [32]. The resultant 38-amino acid peptide lacks seven of the 10 putative transmembrane segments present in a normal **XK** protein. More recently a further mutant has been detected in a Japanese family – a single base deletion (T) at nucleotide 1095 caused a frameshift in translation leading to a premature stop codon at position 408 [30].

These results confirm earlier findings which have shown that the **McLeod** phenotype is based on different genetic backgrounds – in some cases the expression of the **Kx** antigen is highly reduced (see [13]), in other cases the *XK* gene of **McLeod** type individuals either is not transcribed or encodes a defective protein which then is not incorporated into the cell membrane.

The exact nature of the abnormality which causes alterations in red cell morphology in **McLeod** erythrocytes is at present unknown. The composition of membrane proteins and lipids and the levels of various intracellular enzymes are virtually normal, **McLeod** cells, however, do show an increased phosphorylation of membrane proteins and membrane phospholipids [97]. Evidence that the lack of **Kx** protein impairs the organisation of the membrane lipid bilayer of **McLeod** cells [46] already exists: the investigations by Ballas et al. [6] showed a decreased cell deformability due to an increased mechanical stability of the cell membrane. Thus, **Kx** protein is obviously involved in maintaining the proper lipid bilayer balance and possibly also plays a role in stabilising the red cell membrane skeletal structure.

The fact that **McLeod** type red cells not only lack the **Kx** protein but contain a significantly reduced amount of the 93-kDa **Kell** protein [85] provides further evidence for a close interaction between the two proteins within the erythrocyte membrane.

The highly interesting hypothesis advanced by Ho et al. [33] according to which **Kx** may be responsible for the transport of peptide hormones to the **Kell** protein for processing will certainly be the focus of further investigation in the next years.

References

1. ADVANI, H., ZAMOR, J., JUDD, W. J., JOHNSON, C. L. & MARSH, W. L. (1982): Inactivation of Kell blood group antigens by 2-aminoethylisothiouronium bromide. *Brit. J. Haematol. 51*, 107-115.
2. ALLEN, F. H., KRABBE, S. M. R. & CORCORAN, P. A. (1961): A new phenotype (McLeod) in the Kell blood group system. *Vox Sang. 6*, 555-560.
3. ALLEN, F. H. & LEWIS, S. J. (1957): Kpª (Penney), a new antigen in the Kell blood group system. *Vox Sang. 2*, 81-87.
4. ALLEN, F. H., LEWIS, S. J. & FUDENBURG, H. (1958): Studies of anti-Kpᵇ, a new antibody in the Kell blood group system. *Vox Sang. 3*, 1-13.
5. AVENT, N. D. & MARTIN, P. G. (1996): Kell typing by allele-specific PCR (ASP). *Brit. J. Haematol. 93*, 728-730.
6. BALLAS, S. K., BATOR, S. M., AUBUCHON, J. P., MARSH, W. L., SHARP, D. E. & TOY, E. M. (1990):

Abnormal membrane physical properties of red cells in McLeod syndrome. *Transfusion 30*, 722-727.

7. BAR SHANY, S., BEN PORATH, D., LEVENE, C., SELA, R. & DANIELS, G. L. (1982): K22, a 'new' para-Kell antigen of high frequency. *Vox Sang. 42*, 87-90.

8. BARRASSO, C., ESKA, P., GRINDON, A. J., ØYEN, R. & MARSH, W. L. (1975): Anti-K18: an antibody defining another high-frequency antigen related to the Kell blood group system. *Vox Sang. 29*, 124-127.

9. BERTELSON, C. J., POGO, A. O., CHAUDHURI, A., MARSH, W. L., REDMAN, C. M., BANERJEE, D., SYMMANS, W. A., SIMON, T., FREY, D. & KUNKEL, L. M. (1988): Localization of the McLeod locus (XK) within Xp21 by deletion analysis. *Amer. J. Hum. Genet. 42*, 703-711.

10. BOWMAN, J. M., POLLOCK, J. M., MANNING, F. A., HARMAN, C. R. & MENTICOGLOU, S. (1992): Maternal Kell blood group alloimmunization. *Obstet. Gynecol. 79*, 239-244.

11. BRANCH, D. R., MUENSCH, H. A., SY SIOK HAN, A. L. & PETZ, L. D. (1983): Disulfide bonds are a requirement for Kell and Cartwright (Yt^a) blood group antigen integrity. *Brit. J. Haematol. 54*, 573-578.

12. BRANCH, D. R. & PETZ, L. D. (1982): A new reagent (ZZAP) having multiple appplications in immunohematology. *Amer. J. Clin. Pathol. 7*, 161-167.

13. BRANCH, D. R., SY SIOK HIAN, A. L. & PETZ, L. D. (1985): Unmasking of Kx antigen by reduction of disulphide bonds on normal and McLeod red cells. *Brit. J. Haematol. 59*, 505-512.

14. BRENDEL, W. L., ISSITT, P. D., MOORE, R. E., LENES, B. A., ZELLER, D. J., TSE, T. P. & ATKIN, N. (1985): Temporary reduction of red cell Kell system antigen expression and transient production of anti-Kp^b in a surgical patient. *Biotest Bull. 2*, 201-206.

15. CARBONNET, F., HATTAB, C., CALLEBAUT, I., COCHET, S., BLANCHER, A., CARTRON, J. P. & BERTRAND, O. (1998): Kx, a quantitatively minor protein from human erythrocytes, is palmitoylated in vivo. *Biochem. Biophys. Res. Commun. 250*, 569-574.

16. COOMBS, R. R. A., MOURANT, A. E. & RACE, R. R. (1946): In-vivo isosensitization of red cells in babies with haemolytic disease. *Lancet i*, 264-266.

17. CORCORAN, P. A., ALLEN, F. H., LEWIS, M. & CHOWN, B. (1961): A new antibody, anti-Ku (anti-Peltz), in the Kell blood group system. *Transfusion 1*, 181-183.

18. DANIELS, G. L., PETTY, A. C., REID, M., MOULDS, M., DEVENISH, A. & HALIL, O. (1994): Demonstration by the monoclonal antibody-specific immobilization of erythrocyte antigens assay that a new red cell antigen belongs to the Kell blood group system. *Transfusion 34*, 818-820.

19. DE SAINT BASILE, G., BOHLER, M. C., FISCHER, A., CARTRON, J., DUFIER, J. L., GRISCELLI, C. & ORKIN, S. H. (1988): Xp21 DNA microdeletion in a patient with chronic granulomatous disease, retinitis pigmentosa, and McLeod phenotype. *Hum. Genet. 80*, 85-89.

20. DOELMAN, C. J. A., WESTERMANN, W. F., VAN VOORST TOT VOORST, E. & MIEDEMA, K. (1992): An anti-K apparently induced by Enterococcus faecalis in a 30-year-old man. *Transfusion 32*, 790.

21. DUNSTAN, R. A. (1986): Status of major red cell blood group antigens on neutrophils, lymphocytes and monocytes. *Brit. J. Haematol. 62*, 301-309.

22. EICHER, C., KIRKLEY, K., PORTER, M. & KAO, Y. (1985): A new low frequency antigen in the Kell system: K24. *Transfusion 27*, 36-40.

23. FRANCKE, U., OCHS, H. D., DE MARTINVILLE, B., GIACALONE, J., LINDGREN, V., DISTECHE, C., PAGAN, R. A., HOFKER, M. H., VAN OMMEN, G. J., PEARSON, P. L. & WEDGEWOOD, R. J. (1985): Minor Xp21 chromosome deletion in a male associated with expression of Duchenne muscular dystrophy, chronic granulomatous disease, retinitis pigmentosa, and McLeod syndrome. *Amer. J. Hum. Genet. 37*, 250-267.

24. FREY, D., MACHLER, M., SEGER, R., SCHMID, W. & ORKIN, S. H. (1988): Gene deletion in a patient with chronic granulomatous disease and McLeod syndrome: fine mapping of the Xk gene locus. *Blood 71*, 252-255.

25. FURUHJELM, U., NEVANLINNA, H. R., NURKKA, R., GAVIN, J., TIPPETT, P., GOOCH, A. & SANGER, R. (1968): The blood group antigen Ul^a (Karhula). *Vox Sang. 15*, 118-124.

26. GAVIN, J., DANIELS, G. L., YAMAGUCHI, H., OKUBO, Y. & SENO, T. (1979): The red cell antigen once

called Levay is the antigen Kp^c of the Kell system. *Vox Sang. 36*, 31-33.

27. GIBLETT, E. R. & CHASE, J. (1959): Js^a, a 'new' red cell antigen found in Negroes; evidence for an eleventh blood goup system. *Brit. J. Haematol. 5*, 319-326.

28. GIBLETT, E. R., KLEBANOFF, S. J. & PINCUS, S. H. (1971): Kell phenotypes in chronic granulomatous disease: a potential transfusion hazard. *Lancet i*, 1235-1236.

29. GUÉVIN, R. M., TALIANO, V. & WALDMANN, O. (1976): The Côté serum (anti-K11), an antibody defining a new variant in the Kell system. *Vox Sang. 31 (Suppl. 1)*, 96-100.

30. HANAOKA, N., YOSHIDA, K., NAKAMURA, A., FURIHATA, K., SEO, T., TANI, Y., TAKAHASHI, J., IKEDA, S. & HANYU, N. (1999): A novel frameshift mutation in the McLeod syndrome gene in a Japanese family. *J. Neurol. Sci. 165*, 6-9.

31. HESSNER, M. J., McFARLAND, J. G. & ENDEAN, D. J. (1996): Genotyping of KEL1 and KEL2 of the human Kell blood group system by the polymerase chain reaction with sequence-specific primers. *Transfusion 36*, 495-499.

32. HO, M. F., CHALMER, R. M., DAVIS, M. B., HARDING, A. E. & MONACO, A. P. (1996): A novel point mutation in the McLeod syndrome gene in neuroacanthocytosis. *Ann. Neurol. 39*, 672-675.

33. HO, M. F., CHELLY, J., CARTER, N., DANEK, A., CROCKER, P. & MONACO, A. P. (1994): Isolation of the gene for McLeod syndrome that encodes a novel membrane transport protein. *Cell 77*, 869-880.

34. HO, M. F., MONACO, A. P., BLONDEN, L. A. J., VAN OMMEN, G. J. B., AFFARA, N. A., FERGUSON-SMITH, M. A. & LEHRACH, H. (1992): Fine mapping of the McLeod locus (XK) to a 150-380-kb region in Xp21. *Amer. J. Hum. Genet. 50*, 317-330.

35. HUGHES-JONES, N. C. & GARDNER, B. (1971): The Kell system studied with radioactively-labelled anti-K. *Vox Sang. 21*, 154-158.

36. ISSITT, P. (1985): The Kell blood group system. In: *Applied Blood Group Serology*. Montgomery Scientific Publications, Miami, Florida, pp. 289-307.

37. JABER, A., BLANCHARD, D., GOOSSENS, D., BLOY, C., LAMBIN, P., ROUGER, P., SALMON, C. & CARTRON, J. P. (1989): Characterization of the blood group Kell (K1) antigen with a human monoclonal antibody. *Blood 73*, 1597-1602.

38. JABER, A., LOIRAT, M. J., WILLEM, C., BLOY, C., CARTRON, J. P. & BLANCHARD, D. (1991): Characterization of murine monoclonal antibodies directed against the Kell blood group glycoprotein. *Brit. J. Haematol. 79*, 311-315.

39. JONES, J., REID, M. E., OYEN, R., HARRIS, T., MOSCARELLI, S., CO, S., LEGER, R., BEAL, C. & CARDILLO, K. (1995): A novel common Kell antigen, TOU, and its spatial relationship to other Kell antigens. *Vox Sang. 69*, 53-60.

40. JONGENEEL, C. V., BOUVIER, J. & BAIROCH, A. (1989): A unique signature identifies a family of zinc-dependent metallopeptidases. *FEBS Lett. 242*, 211-214.

41. JONGERIUS, J. M., DANIELS, G. L., OVERBEEKE, M. A. M., PETTY, A. C., REID, M., OYEN, R., RIJKSEN, H. & VANLEEUWEN, E. F. (1996): A new low-incidence antigen in the Kell blood group system: VLAN (KEL25). *Vox Sang. 71*, 43-47.

42. JUDSON, P. A. & ANSTEE, D. J. (1977): Comparative effect of trypsin and chymotrypsin on blood group antigens. *Med. Lab. Sci. 34*, 1-6.

43. KANAI, Y. & HEDIGER, M. A. (1992): Primary structure and functional characterization of a high-affinity glutamate transporter. *Nature 360*, 467-471.

44. KANEL, G. C., DAVIS, I. & BOWMAN, J. E. (1978): 'Naturally-occurring' anti-K1: possible association with Mycobacterium infection. *Transfusion 18*, 472-473.

45. KHAMLICHI, S., BAILLY, P., BLANCHARD, D., GOOSSENS, D., CARTRON, J. P. & BERTRAND, O. (1995): Purification and partial characterization of the erythrocyte Kx protein deficient in McLeod patients. *Eur. J. Biochem. 228*, 931-934.

46. KUYPERS, F. A., VAN LINDE-SIBENIUS, T. M., ROELOFSEN, B., OP DEN KAMP, J. A., TANNER, M. J. A. & ANSTEE, D. J. (1985): The phospholipid organisation in the membranes of McLeod and Leach phenotype erythrocytes. *FEBS Lett. 184*, 20-24.

47. LEE, S. (1997): Molecular basis of Kell blood group phenotypes. *Vox Sang. 73*, 1-11.

48. LEE, S., BENNETT, P. R., OVERTON, T., WARWICK, R., WU, X. & REDMAN, C. M. (1996): Prenatal diagnosis of Kell blood group genotypes: KEL1 and KEL2. *Amer. J. Obstet. Gynecol. 175*, 455-459.

49. LEE, S., LIN, M., MELE, A., CAO, Y., FARMAR, J., RUSSO, D. & REDMAN, C. (1999): Proteolytic processing of big endothelin-3 by the Kell blood group protein. *Blood 94*, 1440-1450.

50. LEE, S., NAIME, D., REID, M. & REDMAN, C. (1997): The KEL24 and KEL14 alleles of the Kell blood group system. *Transfusion 37*, 1035-1038.

51. LEE, S., NAIME, D. S., REID, M. E. & REDMAN, C. M. (1997): Molecular basis for the high-incidence antigens of the Kell blood group system. *Transfusion 37*, 1117-1122.

52. LEE, S., WU, X., REID, M. & REDMAN, C. (1995): Molecular basis of the K:6,-7 [Js(a+b-)] phenotype in the Kell blood group system. *Transfusion 35*, 822-825.

53. LEE, S., WU, X., SON, S., NAIME, D., REID, M., OKUBO, Y., SISTONEN, P. & REDMAN, C. (1996): Point mutations characterize KEL10, the KEL3, KEL4, and KEL21 alleles, and the KEL17 and KEL11 alleles. *Transfusion 36*, 490-494.

54. LEE, S., ZAMBAS, E., GREEN, E. D. & REDMAN, C. (1995): Organization of the gene encoding the human Kell blood group protein. *Blood 85*, 1364-1370.

55. LEE, S., ZAMBAS, E. D., MARSH, W. L. & REDMAN, C. M. (1991): Molecular cloning and primary structure of Kell blood group protein. *Proc. Natl. Acad. Sci. USA 88*, 6353-6357.

56. LEE, S., ZAMBAS, E. D., MARSH, W. L. & REDMAN, C. M. (1993): The human Kell blood group gene maps to chromosome 7q33 and its expression is restricted to erythroid cells. *Blood 81*, 2804-2809.

57. LEE, S., WU, X., REID, M., ZELINSKI, T. & REDMAN, C. (1995): Molecular basis of the Kell (K1) phenotype. *Blood 85*, 912-916.

58. LEVINE, P., BACKER, M., WIGOD, M. & PONDER, R. (1949): A new human hereditary blood property (Cellano) present in 99.8 per cent of all bloods. *Science 109*, 464-467.

59. MARSH, W. L., JENSEN, L., ØYEN, R., STROUP, M., GELLERMAN, M., McMAHON, F. J. & TSITSERA, H. (1974): Anti-K13 and the K:-13 phenotype: a blood-grouo variant related to the Kell system. *Vox Sang. 26*, 34-40.

60. MARSH, W. L., NICHOLS, M. E., ØYEN, R., THAYER, R. S., DEERE, W. L., FREED, P. J. & SCHMELTER, S. E. (1978): Naturally occurring anti-Kell stimulated by E. coli enterocolitis in a 20-day old child. *Transfusion 18*, 149-154.

61. MARSH, W. L., ØYEN, R., NICHOLS, M. E. & ALLEN, F. H. (1975): Chronic granulomatous disease and the Kell blood groups. *Brit. J. Haematol. 29*, 247-262.

62. MARSH, W. L. & REDMAN, C. M. (1990): The Kell blood group system: a review. *Transfusion 30*, 158-167.

63. MARSH, W. L., REDMAN, C. M., KESSLER, L. A., DiNAPOLI, J., SCARBOROUGH, A. L. & PHILIPPS, A. G. (1987): K23: a low-incidence antigen in the Kell blood group system identified by biochemical characterisation. *Transfusion 27*, 36-40.

64. MARSH, W. L., STROUP, M., MACILROY, M., ØYEN, R., REID, M. E. & HEISTO, H. (1973): A new antibody, anti-K12, associated with the Kell blood group system. *Vox Sang. 24*, 200-205.

65. MAYNE, K. M., BOWELL, P. J. & PRATT, G. A. (1990): The significance of anti-Kell sensitization in pregnancy. *Clin. Lab. Haematol. 12*, 379-385.

66. McGINNISS, M. H., MacLOWRY, J. D. & HOLLAND, P. V. (1984): Acquisition of K:1-like antigen during terminal sepsis. *Transfusion 24*, 28-30.

67. MERRY, A. H., THOMSON, E. E., ANSTEE, D. J. & STRATTON, F. (1984): The quantification of erythrocyte antigen sites with monoclonal antibodies. *Immunology 51*, 793-800.

68. MOURANT, A. E. (1983): *Blood Relations - Blood Groups and Anthropology*. University Press, London.

69. MURPHY, M. T. & FRASER, R. H. (1997): Detection of Kell blood groups: molecular methods in the diagnostic laboratory. *Blood Rev. 11*, 8-15.

70. MURPHY, M. T., FRASER, R. H. & GODDARD, J. P. (1996): Development of a PCR-based diagnostic assay for the determination of KEL genotype in donor blood samples. *Transfus. Med. 6*, 133-137.

71. MURPHY, M. T., MORRISON, N., MILES, J. S., FRASER, R. H., SPURR, N. K. & BOYD, E. (1993): Regional chromosomal assignment of the Kell blood group locus (KEL) to chromosome 7q33-q35 by fluorescence in situ hybridization:evidence for the polypeptide nature of antigenic variation. *Hum. Genet. 91*, 585-588.

72. NICHOLS, M. E., ROSENFIELD, R. E. & RUBINSTEIN, P. (1987): Monoclonal anti-K14 and anti-K2. *Vox Sang. 52*, 231-235.

73. PACHOLCZYK, T., BLAKELY, R. D. & AMARA, S. G. (1991): Expression cloning of a cocaine- and antidepressant-sensitive human noradrenaline transporter. *Nature 350*, 350-354.

74. PARSONS, S. F., GARDNER, B. & ANSTEE, D. J. (1993): Monoclonal antibodies against Kell glycoprotein: serology, immunochemistry and quantification of antigen sites. *Transfus. Med. 3*, 137-142.

75. PEREIRA, A., MONTEAGUDO, J. & ROVIRA, M. (1989): Anti-K1 of the IgA class associated with Morganella morganii infection. *Transfusion 29*, 549-551.

76. PETTY, A. C. (1993): Monoclonal antibody-specific immobilisation of erythrocyte antigens (MAIEA). A new technique to selectively determine antigenic sites on red cell membranes. *J. Immunol. Methods 161*, 91-95.

77. PETTY, A. C., DANIELS, G. L. & TIPPETT, P. (1994): Application of the MAIEA assay to the Kell blood group system. *Vox Sang. 66*, 216-224.

78. PUROHIT, K. R., WEBER, J. L., WARD, L. J. & KEATS, B. J. B. (1992): The Kell blood group locus is close to the cystic fibrosis locus on chromosome 7. *Hum. Genet. 89*, 457-458.

79. RACE, R. R. & SANGER, R. (1975): The Kell blood groups. In: *Blood Groups in Man*. Blackwell Scientific Publications, Oxford, pp. 283-310.

80. REDMAN, C. M., AVELLINO, G., PFEFFER, S. R., MUKHERJEE, T. K., NICHOLS, M., RUBINSTEIN, P. & MARSH, W. L. (1986): Kell blood group antigens are part of a 93,000-dalton red cell membrane protein. *J. Biol. Chem. 261*, 9521-9525.

81. REDMAN, C. M. & LEE, S. (1995): Kell blood group system and the McLeod syndrome. In: *Blood Cell Biochemistry. Molecular Basis of Human Blood Group* (J.P.Cartron and P. Rouger, eds.). Plenum Press, New York and London, pp. 227-242.

82. REDMAN, C. M., LEE, S., HUININK, T. B., RABIN, B. I., JOHNSON, C. L., ØYEN, R. & MARSH, W. L. (1989): Comparison of human and chimpanzee Kell blood group systems. *Transfusion 29*, 486-490.

83. REDMAN, C. M. & MARSH, W. L. (1993): The Kell blood group system and the Mcleod phenotype. *Sem. Hematol. 30*, 209-218.

84. REDMAN, C. M., MARSH, W. L., MUELLER, K. A., AVELLINO, G. P. & JOHNSON, C. L. (1984): Isolation of Kell-active protein from the red cell membrane. *Transfusion 24*, 176-178.

85. REDMAN, C. M., MARSH, W. L., SCARBOROUGH, A., JOHNSON, C. L., RABIN, B. I. & OVERBEEKE, M. (1988): Biochemical studies on McLeod phenotype red cells and isolation of Kx antigen. *Brit. J. Haematol. 68*, 131-136.

86. REID, M. E., ØYEN, R., REDMAN, C. M., GILLESPIE, G., JONES, J. & ECKRICH, R. (1995): K12 is located on the Kell blood group protein in proximity to K/k and Js^a/Js^b. *Vox Sang. 68*, 40-45.

87. ROUGER, P. (1990): Defects of McLeod red blood cells and association with disease. In: *Blood Group Systems: Kell* (J. Laird-Fryer, J. Levitt, and G. Daniels, eds.). American Association of Blood Banks, Arlington, Virginia, pp. 77-88.

88. RUSSO, D., REDMAN, C. & LEE, S. (1998): Association of XK and Kell blood group proteins. *J. Biol. Chem. 273*, 13950-13956.

89. SABO, B., MCCREARY, J., STROUP, M., SMITH, D. E. & WEIDNER, J. G. (1979): Another Kell-related antibody, anti-Kl9. *Vox Sang. 36*, 97-102.

90. SAVALONIS, J. M., KALISH, R. I., CUMMINGS, E. A., RYAN, R. W. & ALOISI, R. (1988): Kell blood group activity of gram-negative bacteria. *Transfusion 28*, 229-232.

91. SCHULTZ, M. H. (1990): Serology and clinical significance of Kell blood group system antibodies. In: *Blood Group Systems: Kell* (J. Laird-Fryer, J. Levitt, and G. Daniels, eds.). American Association of Blood Banks, Arlington, Virginia, pp. 37-64.

92. SEYFRIED, G. B., GÓRSKA, S., MAJ, S., SYLWESTROWICZ, M. T., GILES, C. M. & GOLDSMITH, K. L. G. (1972): Apparent depression of antigens of the Kell blood group system associated with autoimmune acquired haemolytic anaemia. *Vox Sang. 23*, 528-536.

93. SHIPP, M. A., VIJAYARAGHAVAN, J., SCHMIDT, E. V., MASTELLER, E. L., D'ADAMIO, L., HERSH, L. B. & REINHERZ, E. L. (1989): Common acute lymphoblastic leukemia antigen (CALLA) is active neutral endopeptidase 24.11 ("enkephalinase"): direct evidence by cDNA transfection analysis. *Proc. Natl. Acad. Sci. USA 86*, 297-301.

94. SHIZUKA, M., WATANABE, M., AOKI, M., IKEDA, Y., MIZUSHIMA, K., OKAMOTO, K., ITOYAMA, Y., ABE, K. & SHOJI, M. (1997): Analysis of the McLeod syndrome gene in three patients with neuroacanthocytosis. *J. Neurol. Sci. 150*, 133-135.

95. STRANGE, J. J., KENWORTHY, R. J., WEBB, A. J. & GILES, C. M. (1974): Wk^a (Weeks), a new antigen in the Kell blood group system. *Vox Sang. 27*, 81-86.

96. SULLIVAN, C. M., KLINE, W. E., RABIN, B. I., JOHNSON, C. L. & MARSH, W. L. (1987): The first example of autoanti-Kx. *Transfusion 27*, 322-324.

97. TANG, L. L., REDMAN, C. M., WILLIAMS, D. & MARSH, W. L. (1981): Biochemical studies on McLeod phenotype erythrocytes. *Vox Sang. 40*, 17-26.

98. TIPPETT, P. (1976): Some recent developments in the Kell and Lutheran systems. In: *Human Blood Groups (5th International Convocation on Immunology, Buffalo 1976*, (J.F. Mohn, R.W. Plunkett, R.K. Cunningham, and R.M. Lambert, eds.). S. Karger, Basel, pp. 401-409.

99. TOIVANEN, P. & HIRVONEN, T. (1973): Antigens Duffy, Kell, Kidd, Lutheran, and Xg^a on fetal red cells. *Vox Sang. 24*, 372-376.

99a. UCHIKAWA, M., ONODERA, T., TSUNEYAMA, H., ENOMOTO, T., ISHIJIMA, A., YUASA, S., MURATA, S., TADOKORO, K., NAKAJIMA, K. & JUJI, T. (2000): Molecular basis of unusual K$_{mod}$ phenotype with K+ ^{w}k– (Abstract). *Vox Sang.* (in press).

100. VAN DER HART, M., SZALOKY, A. & VAN LOGHEM, J. J. (1968): A 'new' antibody associated with the Kell blood group system. *Vox Sang. 15*, 456-458.

101. VENGELEN-TYLER, V., GONZALEZ, B., GARRATTY, G., KRUPPE, C., JOHNSON, C. L., MUELLER, K. A. & MARSH, W. L. (1987): Acquired loss of red cell Kell antigens. *Brit. J. Haematol. 65*, 231-234.

102. WALKER, R. H., ARGALL, C. I., STEANE, E. A., SASAKI, E. A. & GREENWALT, T. J. (1963): Anti-Js^b, the expected antithetical antibody of the Sutter blood group system. *Nature*, 295-296.

103. WALKER, R. H., ARGALL, C. I., STEANE, E. A., SASAKI, T. T. & GREENWALT, T. J. (1963): Js^b of the Sutter blood group system. *Transfusion 3*, 94-99.

104. WALLACE, M. E., BOUYSOU, C., DE JONGH, D. S., MANN, J. M., TEESDALE, P., ØYEN, R. & MARSH, W. L. (1976): Anti-K14: an antibody specificity associated with the Kell blood group system. *Vox Sang. 30*, 300-304.

105. WALLAS, C., SIMON, R., SHARPE, M. A. & BYLER, C. (1986): Isolation of a Kell-reactive protein from red cell membranes. *Transfusion 26*, 173-176.

106. WILLIAMSON, L. M., POOLE, J., REDMAN, C., CLARK, N., LIEW, Y. M., RUSSO, D. C., LEE, S., REID, M. E. & BLACK, A. J. (1994): Transient loss of proteins carrying Kell and Lutheran red cell antigens during consecutive relapses of autoimmune thrombocytopenia. *Brit. J. Haematol. 87*, 805-812.

107. YAMAGUCHI, H., OKUBO, Y., SENO, T., MATSUSHITA, K. & DANIELS, G. L. (1979): A 'new' allele, Kp^c, at the Kell complex locus. *Vox Sang. 36*, 29-30.

108. YAZDANBAKHSH, K., LEE, S., YU, Q. & REID, M. E. (1999): Identification of a defect in the intracellular trafficking of a Kell blood group variant. *Blood 94*, 310-318.

109. ZELINSKI, T., COGHLAN, G., MYAL, Y., WHITE, L. J. & PHILIPPS, S. E. (1991): Assignment of the Kell blood group locus to chromosome 7q. *Cytogenet. Cell Genet. 58*, 1927.

18 Gerbich System

The high-incidence antigens of the **Gerbich** group comprise thus far seven non-allelic characters – three high-frequency antigens, **Ge2**, **Ge3** [8,45], and **Ge4** [49], and four rare antigens, **Ls**ᵃ (= **Lewis II**) [13], **Wb** (= **Webb**) [62], **Dh**ᵃ (= **Duch**) [35], and **An**ᵃ (= **Ahonen**) [29][1].

The antigens presented in this chapter are all located on the same membrane constituents, glycophorin C and/or glycophorin D (see *Sect. 4.1*).

Isoantibodies with different anti-**Gerbich** specificities have been found in the sera of **Gerbich**-deficient subjects. They are also produced as a response to blood transfusions or pregnancies [8,9,60].

Furthermore, rare cases of **Gerbich**-specific autoantibodies have been described [5,52,58,59,61].

Hybridoma antibodies towards **Ge2** [34], **Ge3** [42] and **Ge4** antigen [41] have been produced. In addition, a series of hybridoma antibodies directed towards extracellular [63] and intracellular epitopes of glycophorin C (e.g. [3,19,24,41]) have been used for various studies on the blood group **Gerbich** substances.

18.1 Glycophorin C and Glycophorin D

Glycophorin C (synonyms: GPC, glycoconnectin, band-D protein, or β-sialo-glycoprotein) is an intrinsic protein of human erythrocyte membrane. It consists of a single polypeptide chain of 128 amino acids and contains one N-glycosidically linked and an average of 12 O-glycosidically linked carbohydrate side chains attached to its N-terminal domain; the N-linked chain is of the complex type without repeating lactosamine units [53], the O-linked oligosaccharides have the same structures as those found in glycophorin A [16] (see *Fig. 9.3*).

---

[1] ISBT numbers (see *Chap. 1*): **Ge2** = 020 002, **Ge3** = 020 003, **Ge4** = 020 004, **Wb** = 020 005, **Ls**ᵃ = 020 006, **An**ᵃ = 020 007, and **Dh**ᵃ = 020 008, or, alternatively, GE2 to GE8.
The antigen specificity **Ge1** originally described by Booth [9] is considered obsolete, since anti-**Ge1** is no longer available.

In polyacrylamidegel electrophoresis the glycoprotein has an apparent molecular weight of ± 32 kDa [10]. A partial amino acid sequence has been determined by peptide analysis [6,17,18]; the complete structure has been deduced by cDNA sequencing [15,31] (see *Fig. 18.1)*. These studies reveal that the glycophorin C molecule has the same basic characteristics as glycophorin A – those of a highly glycosylated extracellular segment (amino acids 1–58) followed by a hydrophobic region situated within the erythrocyte membrane (amino acids 59–81) and a cytoplasmic segment (residues 82–128). Glycophorin C *in situ* is degraded by trypsin.

Glycophorin D (synonyms: GPD, band-E protein, or γ-sialoglycoprotein) is an intrinsic membrane glycoprotein with an apparent molecular weight of about 23 kDa [10]. Preliminary investigations reveal great similarities to glycophorin C, the most obvious difference between the two glycoproteins being their N-terminal regions [28,37]: the N-terminus of glycophorin D is significantly smaller than that of glycophorin C and does not react with hybridoma antibodies directed towards the N-terminal region of glycophorin C; furthermore, glycophorin D is less glycosylated as it contains only six O-linked tetrasaccharide units and lacks the N-linked oligosaccharide chain. According to recent investigations [28] the N-terminus is blocked by a to date unidentified substituent. The C-terminal regions of glycophorin C and D, however, are immunologically [28,53] and structurally [7] indistinguishable.

The gene encoding glycophorin C (***GYPC***) has been located on chromosome 2, position q14–q21 [48]. It includes about 13.5 kilobase pairs of DNA and is organised in four exons separated by three intervening sequences. Exon 1 encodes amino acid residues 1–16, exon 2 residues 17–35, exon 3 residues 36–63, and exon 4 residues 64–128. Exons 1–3 encode the extracellular domain, exon 4 the membrane-spanning and cytoplasmic domains. In contrast to glycophorin A and B (see *Sect. 9.3)* the *GYPC* gene does not code for a signal peptide [31]. Except for a 9-amino acid insert at the 3' end, exon 3 shows a high degree of homology to exon 2, the respective 5'- and 3'-flanking intronic DNA sequences of the two exons being almost identical. The appearance in tandem of these two closely related 3.4 kbp domains probably derived from a recent duplication of an ancestral domain [14,32].

The glycophorin C gene is expressed not only in erythroid tissues but also in a wide variety of tissue cells and non-erythroid cell lines, though at a considerably lower level. However, the level of glycosylation of the mature glycophorin C differs in erythroid and non-erythroid cells [38].

Originally it was suspected that glycophorin C and glycophorin D were encoded by two adjacent and closely related genes [21]. However, genomic analyses with cDNA probes revealed that a unique glycophorin C gene was responsible for the

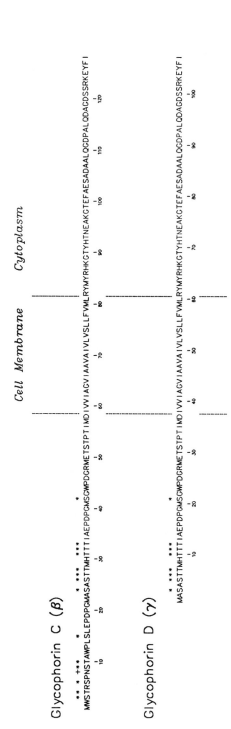

Cell Membrane   Cytoplasm

Glycophorin C (β)

```
 * *  * +++  *        * *** ***                                    *
MWSTRSPNSTAWPLSLEPDPGMASASTTMHTTTIAEPDPGMSGWPDGRMETSTPTIMDIVVIAGVIAAVAIVLSLLFVMLRYMYRHKGTYHTNEAKGTEFAESADAALQGDPALQDAGDSSRKEYFI
         -         -         -         -         -         -         -         -         -         -         -         -
        10        20        30        40        50        60        70        80        90       100       110       120
```

Glycophorin D (γ)

```
 * *** ***                                    *
MASASTTMHTTTIAEPDPGMSGWPDGRMETSTPTIMDIVVIAGVIAAVAIVLSLLFVMLRYMYRHKGTYHTNEAKGTEFAESADAALQGDPALQDAGDSSRKEYFI
         -         -         -         -         -         -         -         -         -         -         -
        10        20        30        40        50        60        70        80        90       100
```

Figure18.1: The amino acid sequence of glycophorin C and glycophorin D.
The glycophorin D sequence is based on the assumption that the peptide synthesis of this glycoprotein is initiated at Met[22] of glycophorin C [14,66].
★: potential O-glycosylation sites, +: potential N-glycosylation site
The peptide sequence of glycophorin C is deposited in the EMBL/GenBank data library (accession numbers M11802 and M36284).

synthesis of both sialoglycoproteins [37]. The close structural relationship between glycophorin C and D further led to the assumption that glycophorin D was a truncated version of glycophorin C. It has therefore been suggested that deletion of the first 21 N-terminal amino acids in glycophorin D arises from 'leaky' scanning of the mRNA, i.e. from alternative use of two in-phase AUG initiator codons at positions 1 and 22 [40,66] (see *Fig. 18.1*).

Quantitative binding assays using Fab fragments of glycophorin C-specific hybridoma antibodies yielded 143,000 molecules of glycophorin C and 82,000 molecules of glycophorin D per single erythrocyte [63].

Analysis of **Gerbich**-deficient variants indicated that glycophorin C and D play a pivotal role in regulating the mechanical stability and the deformability of the erythrocyte membrane (see also *Sect. 4.2)* [2,46,54]. The two sialoglycoproteins together with the palmitoylated erythrocyte membrane phosphoprotein p55 and the cytoskeleton protein 4.1 bind to ankyrin, which interacts with the N-terminal cytoplasmic domain of the band 3 protein [47]. Erythrocytes deficient in 4.1 protein exhibit a severe reduction of glycophorin C and D content [57], and erythrocytes which lack glycophorin C and D (= **Leach** phenotype, see below) show a distinct decrease of band 4.1 protein [1,30,64].

The invasion of the malaria parasite *Plasmodium falciparum* into erythrocytes deficient in band 4.1 membrane protein is significantly reduced [12,51], a fact which continues to be a focal point in malaria research.

18.2 Antigen Characters Ge1, Ge2, Ge3, and Ge4 and the Ls^a Determinant

In Europids and Negrids the **Gerbich** antigens are common red cell characteristics; phenotypes lacking one, two or more of these antigens, however, occur fairly frequently in certain Melanesian populations of Papua New Guinea [8,9]. To date four **Gerbich**-deficient phenotypes have been described: the **Melanesian** type, (**Ge:–1,2,3,4**), the **Yus** (**Yussef** or **Ge–Yus+**) type (**Ge:–2,3,4**), the **Gerbich** (or **Ge**) type (**Ge:–2,–3,4**), and the **Leach** type (**Ge:–2,–3,–4**) [3,33,49] (see *Table 18.1*).

As mentioned above, the **Gerbich** antigens are situated on glycophorin C and/or glycophorin D:

The **Ge2** epitope is found in the trypsin-sensitive domain of glycophorin D, encoded by exon 2; the determinant is also destroyed by papain but is resistant

Table 18.1: Definitions of Gerbich phenotypes to date.

Phenotype	Gerbich antigens
'normal'	Ge:1,2,3,4
Melanesian	Ge:−1,2,3,4
Yussef (Yus)	Ge:−1,−2,3,4
Gerbich (Ge)	Ge:−1,−2,−3,4
Leach	Ge:−1,−2,−3,−4

towards chymotrypsin and pronase. For most of the anti-**Ge2** antibodies N-acetylneuraminic acid attached to Ser/Thr-linked oligosaccharide(s) is essential for serological specificity [20]. On glycophorin C the **Ge2** epitope is a cryptantigen which, obviously, is expressed only when located at the amino terminus. A **Ge2** active tryptic glycopeptide from glycophorin D comprising 20- to 30-amino-acid residues has thus far not been characterised.

The **Ge3** epitope has been localised in the glycosylated segments of glycophorin C and glycophorin D, encoded by exon 3 [20]; the determinant is destroyed by trypsin but is resistant towards chymotrypsin, pronase and papain [50]. Thus, in the case of glycophorin C the **Ge3** epitope must be situated in the region surrounding the trypsin cleavage site at position 47/48, presumably comprising amino acids 42–50 (the respective position in glycophorin D is between residues 21 and 49). Further characterisation of the epitope suggested that methionine, aspartic acid or glutamic acid, tryptophan and/or arginine, as well as neuraminic acid as part of a carbohydrate unit attached to Ser^{42}, are directly involved in forming the **Ge3** epitope.

Ge4 is destroyed by trypsin, pronase, papain, and neuraminidase [49]. Following binding tests with hybridoma antibodies the epitope should be located within the first 21 amino acids of glycophorin C [22,41].

Gerbich-deficient individuals ('**Yus**', '**Ge**', and '**Leach**' phenotypes) do not express any form of glycophorin C or D on their erythrocytes [3,21,67][(2)].

Diffusely migrating sialoglycoprotein components have been detected in both **Yus** and **Ge** phenotypes. The glycoprotein from **Yus** phenotype erythrocytes migrates as

---

[(2)] It must be mentioned here that expression of **Kell** antigens, **K11** in particular, is generally weakened in **Ge** and **Leach** phenotype erythrocytes [23].

a broad band in the region of 32.5 to 36.5 kDa, which is cleaved *in situ* by trypsin, whereas the 30.5–34.5 kDa band found in **Ge** phenotype cells resists trypsin treatment [4]. Both components react with several hybridoma antibodies previously shown to bind to glycophorin C of normal erythrocytes. This finding indicates a high degree of immunologic homology between normal glycophorin C and the sialo-glycoproteins of the **Ge**-, and **Yus**-type [4,53]. It has furthermore been shown that both these sialoglycoproteins contain N-linked poly-N-acetyllactosamine chains which are considerably more complex than those found in normal glycophorin C [53].

The aberrant sialoglycoproteins occurring in the erythrocytes of **Gerbich** deficient individuals were originally suspected to be glycophorin C/D (β/γ) hybrid molecules analogous to haemoglobins of the Lepore type (see *Chapter 9.4)* [4,21]. Analyses of genomic DNA, however, showed that the sialoglycoproteins found in **Gerbich**-negative cells of **Ge** and **Yus** phenotypes are encoded by glycophorin C gene variants with internal gene deletions of three to four kilobases [37]. Further investigations [11,14,32,66] revealed that in the case of **Ge:–2,3 (Yus** phenotype) exon 2 is deleted (= β^{Yus}), while in the case of **Ge:–2,–3 (Ge** phenotype), it is exon 3 which is deleted (= β^{Ge}) (see *Fig. 18.2)*.

Surprisingly, the isoform of glycophorin C lacking exon 2 was found as a minor component in tissue cells [39].

Individuals of the **Ge** phenotype do not produce the respective glycophorin D, though the DNA sequence of the **Ge** variant glycophorin C has the potential alternative AUG initiation site [37]. In the case of the **Yus** phenotype the second methionine code is missing in the variant glycophorin C gene and glycophorin D cannot be produced.

In membranes of **Leach** phenotype erythrocytes no glycophorin C/D-like sialo-glycoproteins have been detected [3]. Here, genome analysis showed that a defective glycophorin C gene in which both exons 3 and 4 were either deleted or markedly altered is the most common molecular basis of this phenotype [32,66,68,70]. It should be noted that the mRNA encoded by this mutant gene has been detected in circulating reticulocytes from **Ge:-2-3-4** individuals [70]; however, the resulting polypeptide lacks both the transmembrane and the cytoplasmic domains and cannot be inserted into the membrane.

In one case of **Leach** phenotype (Patient *L.N.*) a virtually intact glycophorin C gene has been found which was characterised by a G $\rightarrow$ T substitution at nucleotide 131, thus changing Trp[44] to Leu, and a nucleotide deletion at position 134 [68]. The frameshift in the mRNA caused by this deletion effects a change in peptide sequence and forms a premature stop codon at the 56th triplet.

As discussed above, glycophorin C and glycophorin D interact with the membrane skeleton and contribute to maintaining the characteristic discoid shape of the normal

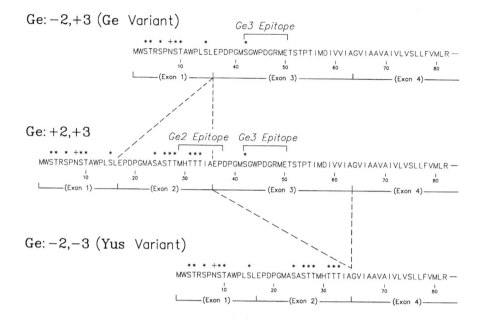

Figure18.2: The amino acid deletions in the glycophorin C molecule characteristic for the phenotypes Ge:-2,3 (Yus) and Ge:-2,-3 (Ge).

★: potential O-glycosylation sites, + : potential N-glycosylation site.
According to Colin et al. [14] by permission of the authors and the American Society for Biochemistry andMolecular Biology, Inc., Baltimore, MD, USA.

erythrocyte. The absence of glycophorin C and D, as is the case in the **Leach** phenotype, is associated with elliptocytosis [3,26,27]; erythrocytes containing the abnormal sialoglycoproteins encoded by the *Yus* and *Gerbich* genes, however, are of normal shape.

Glycophorin C and glycophorin D variants – both approximately 5.5 kDa larger than the normal sialoglycoproteins [43] – have been detected in the membranes of erythrocytes with the low frequency antigen **Ls**[a] (**Lewis II**) [13]. The **Ls**[a] antigen is destroyed by trypsin but resistant towards ficin; neuraminidase enhances the reaction with anti-**Ls**[a] [43]. Investigations on the genomic DNA in **Ls**[a] individuals revealed the presence of an aberrant glycophorin C gene containing one copy of exon 2 and two copies of exon 3 [32,55]. The glycophorin C and D molecules thus contain two **Ge3** determinants; the **Ls**[a] epitope is located at the exon 3-duplicated exon 3 boundary [55].

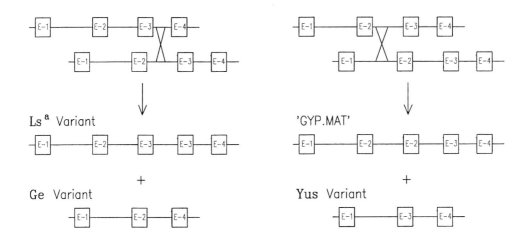

Figure 18.3: Possible unequal crossing-over between homologous domains within the glycophorin C gene.
(a) Recombination leading to a chromosome carrying two copies of exon 3.
(b) Recombination leading to a chromosome carrying two copies of exon 2.
According to Colin et al. [14] by permission of the authors and the American Society for Biochemistry and Molecular Biology, Inc., Baltimore, MD, USA.

Defective glycophorin C genes found in the genomes of **Gerbich**-deficient individuals arise from unequal crossing over due to non-homologous recombination between the two homologous domains of the normal glycophorin C gene [14,32]; breakpoints have been localised within introns 2 and 3 [14]. This mechanism explains not only the formation of aberrant genes lacking exon 2 or exon 3 (**Yus** or **Ge** phenotype, respectively), but also the generation of reciprocal genes containing an additional copy of exon 2 or exon 3. The latter variant is characteristic for **Ls^a**, whereas the glycophorin C variant with a duplication of exon 2 is not expected to result in a novel blood group antigen. The respective mutation ('GPC.MAT') which occurs with a frequency of ~0.02% in the Japanese population, however, has been identified by molecular biological methods [69].

A schematic representation of possible unequal crossing over between homologous domains within the glycophorin C gene is given in *Fig. 18.3*.

18.3 Wb (Webb)

The **Wb** [62] blood group character is a low-incidence antigen which is destroyed by trypsin and neuraminidase but is resistant towards chymotrypsin [44].

A study on the erythrocyte membrane proteins from **Wb**-positive individuals using SDS-polyacrylamide-gel electrophoresis revealed a decrease in glycophorin C content along with the appearance of an unusual sialoglycoprotein. The molecular mass of this membrane constituent is approximately 2.7 kDa less than that of normal glycophorin C [44,56] and is not changed after treatment with Endo F (an endo-N-acetyl-glucosaminidase cleaving N-linked oligosaccharide chains from glycoproteins). Furthermore, two mouse hybridoma antibodies (*BRIC 4* and *BRIC 10*) specific for normal glycophorin C react with this sialoglycoprotein. These findings suggested that blood group **Wb** is associated with an altered form of glycophorin C, which lacks the N-linked oligosaccharide chain normally found on this membrane glycoprotein.

Subsequent investigations by Chang et al. [11] proved the above assumption. Here, cDNA was prepared from glycophorin C mRNA and showed in the case of the **Webb** variant an A $\rightarrow$ G exchange at nucleotide 23. This mutation results in substitution of asparagine by serine at position 8 in the sialoglycoprotein molecule (*Fig. 18.4*) and thus in the loss of the single N-glycosylation site in glycophorin C.

18.4 Dh^a (Duch)

Dh^a is a low incidence erythrocyte character which, thus far, has been detected in only one family [35]. The antigen is degraded *in situ* by papain, trypsin, ficin, and pronase, and is susceptible to treatment with neuraminidase [65].

The **Dh^a** epitope has been located at the N-terminal portion of a variant glycophorin C which has the same molecular mass as normal glycophorin C [65]; the determinant is absent from glycophorin D. Sequencing of cDNA prepared from mRNA fractions isolated from the reticulocytes of **Dh(a+)** individuals [36] revealed a point mutation in the glycophorin C gene – the **Dh^a** variant differs from the normal glycophorin C molecule by a single base exchange (C $\rightarrow$ T) at nucleotide 40 resulting in a leucine $\rightarrow$ phenylalanine substitution at position 14 (*Fig. 18.4*).

18.5 An^a (Ahonen)

The rare blood group antigen **An^a** described by Furuhjelm et al. [29] is expressed exclusively on glycophorin D [22,25]. The presence of neuraminic acid is pivotal for its serological activity; the antigen is destroyed by trypsin, chymotrypsin, and papain [22].

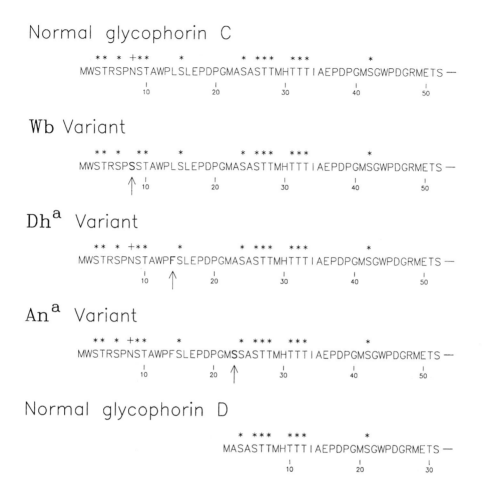

Figure 18.4: Peptide sequences of glycophorin C variant molecules and glycophorin D
★: potential O-glycosylation sites, + : potential N-glycosylation site
The peptide sequences of the variant glycophorin C molecules are deposited in the Swiss-Prot data bank (accession number P04921).

Analysis of cDNA isolated from **An(a+)** blood donors revealed that the expression of the **An**[a] antigen is associated with a single point mutation in exon 2 of the glycophorin C gene. A G → T substitution at nucleotide 67 of the coding sequence results in substitution of alanine by serine at amino acid residue 23 of glycophorin C

513

and, if the current view on glycophorin D formation is correct, at residue 2 of glycophorin D [22]. These results show that the **An**^{a} epitope is a cryptantigen on glycophorin C and is expressed only on glycophorin D where it is located in close proximity to the amino terminus (*Fig. 18.4*, see also **Ge2** antigen).

References

1. ALLOISIO, N., MORLÉ, L., BACHIR, D., GUETARNI, D., COLONNA, P. & DELAUNAY, J. (1985): Red cell membrane sialoglycoprotein β in homozygous and heterozygous 4.1(-) hereditary elliptocytosis. *Biochim. Biophys. Acta 816*, 57-62.
2. ALLOISIO, N., VENEZIA, N. D., RANA, A., ANDRABI, K., TEXIER, P., GILSANZ, F., CARTRON, J. P., DELAUNAY, J. & CHISHTI, A. H. (1993): Evidence that red blood cell protein-p55 may participate in the skeleton-membrane linkage that involves protein-4.1 and glycophorin-C. *Blood 82*, 1323-1327.
3. ANSTEE, D. J., PARSONS, S. F., RIDGWELL, K., TANNER, M. J. A., MERRY, A. H., THOMSON, E. E., JUDSON, P. A., JOHNSON, P., BATES, S. & FRASER, I. D. (1984): Two individuals with elliptocytic red cells apparently lack three minor erythrocyte membrane sialoglycoproteins. *Biochem. J. 218*, 615-619.
4. ANSTEE, D. J., RIDGWELL, K., TANNER, M. J. A., DANIELS, G. L. & PARSONS, S. F. (1984): Individuals lacking the Gerbich blood-group antigen have alterations in the human erythrocyte membrane sialoglycoproteins β and γ. *Biochem. J. 221*, 97-104.
5. BEATTIE, K. M. & SIGMUND, K. E. (1987): A Ge-like autoantibody in the serum of a patient receiving gold therapy for rheumatoid arthritis. *Transfusion 27*, 54-57.
6. BLANCHARD, D., DAHR, W., HUMMEL, M., LATRON, F., BEYREUTHER, K. & CARTRON, J. P. (1987): Glycophorins B and C from human erythrocyte membranes. Purification and sequence analysis. *J. Biol. Chem. 262*, 5808-5811.
7. BLANCHARD, D., EL-MALIKI, B., HERMAND, P., DAHR, W. & CARTRON, J. P. (1987): Structural homology between glycophorins C and D, two minor glycoproteins of the human erythrocyte membrane carrying blood group Gerbich antigens. *Proceedings of the 9th International Symposium on Glycoconjugates, Lille 1987*, Abstract F 54.
8. BOOTH, P. B., ALBREY, J. A., WHITTAKER, J. & SANGER, R. (1970): Gerbich blood group system: a useful genetic marker in certain Melanesians of Papua and New Guinea. *Nature 228*, 462.
9. BOOTH, P. B. & MCLOUGHLIN, K. (1972): The Gerbich blood group system, especially in Melanesians. *Vox Sang. 22*, 73-84.
10. CARTRON, J. P., LE VAN KIM, C. & COLIN, Y. (1993): Glycophorin C and related glycoproteins: structure, function, and regulation. *Sem. Hematol. 30*, 152-168.
11. CHANG, S., REID, M. E., CONBOY, J., KAN, Y. W. & MOHANDAS, N. (1991): Molecular characterization of erythrocyte glycophorin C variants. *Blood 77*, 644-648.
12. CHISHTI, A. H., PALEK, J., FISHER, D., MAALOUF, G. J. & LIU, S. C. (1996): Reduced invasion and growth of Plasmodium falciparum into elliptocytic red blood cells with a combined deficiency of protein 4.1, glycophorin C, and p55. *Blood 87*, 3462-3469.
13. CLEGHORN, T. E. & DUNSFORD, I. (1963): Unpublished results. Cited in *Race & Sanger, Blood Groups in Man*, p. 434.
14. COLIN, Y., LE VAN KIM, C., TSAPIS, A., CLERGET, M., D'AURIOL, L., LONDON, J., GALIBERT, F. & CARTRON, J. P. (1989): Human erythrocyte glycophorin C. Gene structure and rearrangement in genetic variants. *J. Biol. Chem. 264*, 3773-3780.
15. COLIN, Y., RAHUEL, C., LONDON, J., ROMEO, P. H., D'AURIOL, C., GALIBERT, F. & CARTRON, J. P. (1986): Isolation of cDNA clones and complete amino acid sequence of human erythrocyte glycophorin C. *J. Biol. Chem. 261*, 229-233.
16. DAHR, W. (1986): Immunochemistry of sialoglycoproteins in human red blood cell membranes. In:

Recent Advances in Blood Group Biochemistry (V. Vengelen-Tyler and W. J. Judd, eds.). American Association of Blood Banks, Arlington, VA,pp. 23-65.

17. DAHR, W. & BEYREUTHER, K. (1985): A revision of the N-terminal structure of sialoglycoprotein D (glycophorin C) from human erythrocyte membranes. *Biol. Chem. Hoppe-Seyler 366*, 1067-1070.

18. DAHR, W., BEYREUTHER, K., KORDOWICZ, M. & KRÜGER, J. (1982): N-terminal amino acid sequence of sialoglycoprotein D (glycophorin C) from human erythrocyte membranes. *Eur. J. Biochem. 125*, 57-62.

19. DAHR, W., BLANCHARD, D., KIEDROWSKI, S., POSCHMANN, A., CARTRON, J. P. & MOULDS, J. J. (1989): High-frequency antigens of human erythrocyte membrane sialoglycoproteins. VI. Monoclonal antibodies reacting with the N-terminal domain of glycophorin C. *Biol. Chem. Hoppe-Seyler 370*, 849-854.

20. DAHR, W., KIEDROWSKI, S., BLANCHARD, D., HERMAND, P., MOULDS, J. J. & CARTRON, J. P. (1987): High frequency antigens of human erythrocyte membrane sialoglycoproteins. V. Characterization of the Gerbich blood group antigens: Ge2 and Ge3. *Biol. Chem. Hoppe-Seyler 368*, 1375-1383.

21. DAHR, W., MOULDS, J., BAUMEISTER, G., MOULDS, M., KIEDROWSKI, S. & HUMMEL, M. (1985): Altered membrane sialoglycoproteins in human erythrocytes lacking the Gerbich blood group antigens. *Biol. Chem. Hoppe-Seyler 366*, 201-211.

22. DANIELS, G., KING, M. J., AVENT, N. D., KHALID, G., REID, M., MALLINSON, G., SYMTHE, J. & CEDERGREN, B. (1993): A point mutation in the GYPC gene results in the expression of the blood group Anᵃ antigen on glycophorin D but not on glycophorin C: further evidence that glycophorin D is a product of the GYPC gene. *Blood 82*, 3198-3203.

23. DANIELS, G. L. (1982): Studies on Gerbich negative phenotypes and Gerbich antibodies. (Abstract). *Transfusion 22*, 405.

24. DANIELS, G. L., BANTING, G. & GOODFELLOW, P. (1983): A monoclonal antibody related to the human blood group Gerbich. *J. Immunogenet. 10*, 103-105.

25. DANIELS, G. L., KHALID, G. & CEDERGREN, B. (1990): The low frequency red cell antigen Anᵃ is located on glycophorin D. (Abstract). *ISBT/AABB, Los Angeles*, Proceedings S775.

26. DANIELS, G. L., REID, M. E., ANSTEE, D. J., BEATTIE, K. M. & JUDD, W. J. (1988): Transient reduction in erythrocyte membrane sialoglycoprotein beta associated with the presence of elliptocytosis. *Brit. J. Haematol. 70*, 477-481.

27. DANIELS, G. L., SHAW, M. A., JUDSON, P. A., REID, M. E., ANSTEE, D. J., COLPITTS, P., CORNWALL, S., MOORE, B. P. L. & LEE, S. (1986): A family demonstrating inheritance of the Leach phenotype: a Gerbich-negative phenotype associated with elliptocytosis. *Vox Sang. 50*, 117-121.

28. EL MALIKI, B., BLANCHARD, D., DAHR, W., BEYREUTHER, K. & CARTRON, J. P. (1989): Structural homology between glycophorin C and D of human erythrocytes. *Eur. J. Biochem. 183*, 639-643.

29. FURUHJELM, U., NEVANLINNA, H. R., GAVIN, J. & SANGER, R. (1972): A rare blood group antigen Anᵃ. *J. Med. Genet. 9*, 385-391.

30. HEMMING, N. J., ANSTEE, D. J., MAWBY, W. J., REID, M. E. & TANNER, M. J. A. (1994): Localization of the protein 4.1-binding site on human erythrocyte glycophorins C and D. *Biochem. J. 299*, 191-196.

31. HIGH, S. & TANNER, M. J. A. (1987): Human erythrocyte membrane sialoglycoprotein β.The cDNA sequence suggests the absence of a cleaved N-terminal signal sequence. *Biochem. J. 243*, 277-280.

32. HIGH, S., TANNER, M. J. A., MACDONALD, E. B. & ANSTEE, D. J. (1989): Rearrangements of the red-cell membrane glycophorin C (sialoglycoprotein β) gene. A further study of alterations in the glycophorin C gene. *Biochem. J. 262*, 47-54.

33. ISSITT, P. (1985): Some high incidence antigens that may represent independent blood group systems. In: *Applied Blood Group Serology*. Montgomery Scientific Publications. Miami, Florida, USA, pp. 396-408.

34. JANVIER, D., VEAUX, S. & BENBUNAN, M. (1998): New murine monoclonal antibodies directed against glycophorins C and D, have anti-Ge2 specificity. *Vox Sang. 74*, 101-105.

35. JØRGENSEN, J., DRACHMANN, O. & GAVIN, J. (1982): Duch, Dhᵃ, a low frequency red cell antigen.

Hum. Hered. 32, 73-75.

36. KING, M. J., AVENT, N. D., MALLINSON, G. & REID, M. E. (1992): Point mutation in the glycophorin C gene results in the expression of the blood group antigen Dhª. *Vox Sang. 63,* 56-58.

37. LE VAN KIM, C., COLIN, Y., BLANCHARD, D., DAHR, W., LONDON, J. & CARTRON, J. P. (1987): Gerbich blood group deficiency of the Ge:-1,-2,-3 and Ge:-1,-2,3 types. Immunochemical study and genomic analysis with cDNA probes. *Eur. J. Biochem. 165,* 571-579.

38. LE VAN KIM, C., COLIN, Y., MITJAVILA, M. T., CLERGET, M., DUBART, A., NAKAZAWA, M., VAINCHENKER, W. & CARTRON, J. P. (1989): Structure of the promoter region and tissue specificity of the human glycophorin C gene. *J. Biol. Chem. 264,* 20407-20414.

39. LE VAN KIM, C., MITJAVILA, M. T., CLERGET, M., CARTRON, J. P. & COLIN, Y. (1990): An ubiquitous isoform of glycophorin C is produced by alternative splicing. *Nucleic. Acids Res. 18,* 3076.

40. LE VAN KIM, C., PILLER, V., CARTRON, J. P. & COLIN, Y. (1996): Glycophorin C and D are generated by the use of alternative translation initiation sites. *Blood 88,* 2364-2365.

41. LOIRAT, M. J., DAHR, W., MULLER, J. Y. & BLANCHARD, D. (1994): Characterization of new murine monoclonal antibodies directed against glycophorin-C and glycophorin-D. *Transfus. Med. 4,* 147-155.

42. LOIRAT, M. J., GOURBIL, A., FRIOUX, Y., MULLER, J. Y. & BLANCHARD, D. (1992): A murine monoclonal antibody directed against the Gerbich 3 blood group antigen. *Vox Sang. 62,* 45-48.

43. MACDONALD, E. B., CONDON, J., FORD, D., FISHER, B. & GERNS, L. M. (1990): Abnormal beta and gamma sialoglycoproteins associated with the low-frequency antigen Lsª. *Vox Sang. 58,* 300-304.

44. MACDONALD, E. B. & GERNS, L. M. (1986): An unusual sialoglycoprotein associated with the Webb-positive phenotype. *Vox Sang. 50,* 112-116.

45. MACGREGOR, A. & BOOTH, P. B. (1973): A second example of anti-Ge, and some observations on Gerbich subgroups. *Vox Sang. 25,* 474-478.

46. MARFATIA, S. M., LUE, R. A., BRANTON, D. & CHISHTI, A. H. (1994): In vitro binding studies suggest a membrane-associated complex between erythroid p55, protein 4.1, and glycophorin C. *J. Biol. Chem. 269,* 8631-8634.

47. MARFATIA, S. M., MORAIS-CABRAL, J. H., KIM, A. C., BYRON, O. & CHISHTI, A. H. (1997): The PDZ domain of human erythrocyte p55 mediates its binding to the cytoplasmic carboxyl terminus of glycophorin C. Analysis of the binding interface by in vitro mutagenesis. *J. Biol. Chem. 272,* 24191-24197.

48. MATTEI, M. G., COLIN, Y., LE VAN KIM, C., MATTEI, J. F. & CARTRON, J. P. (1986): Localization of the gene for human erythrocyte glycophorin C to chromosome 2, q14-q21. *Hum. Genet. 74,* 420-422.

49. MCSHANE, K. & CHUNG, A. (1989): A novel human alloantibody in the Gerbich system. *Vox Sang. 57,* 205-209.

50. MOHAMMED, M. T., O'DAY, T. & SUGASAWARA, E. (1986): Gerbich (Ge) antibody classification using enzyme-treated red cells. *Transfusion 26,* 120.

51. PASVOL, G., ANSTEE, D. J. & TANNER, M. J. A. (1984): Glycophorin C and the invasion of red cells by Plasmodium falciparum. *Lancet i,* 907-908.

52. POOLE, J., REID, M. E., BANKS, J., LIEW, Y. W., ADDY, J. & LONGSTER, G. (1990): Serological and immunochemical specificity of a human autoanti-Gerbich-like antibody. *Vox Sang. 58,* 287-291.

53. REID, M. E., ANSTEE, D. J., TANNER, M. J. A., RIDGWELL, K. & NURSE, G. T. (1987): Structural relationships between human erythrocyte sialoglycoproteins β and γ and abnormal sialoglycoproteins found in certain rare human erythrocyte variants lacking the Gerbich blood-group antigen(s). *Biochem. J. 244,* 123-128.

54. REID, M. E., CHASIS, J. A. & MOHANDAS, N. (1987): Identification of a functional role for human erythrocyte sialoglycoproteins β and γ. *Blood 69,* 1068-1072.

55. REID, M. E., MAWBY, W., KING, M. J. & SISTONEN, P. (1994): Duplication of exon 3 in the glycophorin C gene gives rise to the Lsª blood group antigen. *Transfusion 34,* 966-969.

56. REID, M. E., SHAW, M. A., ROWE, G., ANSTEE, D. J. & TANNER, M. J. A. (1985): Abnormal minor human erythrocyte membrane sialoglycoprotein (β) in association with the rare blood-group antigen Webb (Wb). *Biochem. J. 232,* 289-291.

57. REID, M. E., TAKAKUWA, Y., CONBOY, J., TCHERNIA, G. & MOHANDAS, N. (1990): Glycophorin C

content of human erythrocyte membrane is regulated by protein 4.1. *Blood 75*, 2229-2234.

58. REID, M. E., VENGELEN-TYLER, V., SHULMAN, I. & REYNOLDS, M. V. (1988): Immunochemical specificity of autoanti-Gerbich from two patients with autoimmune haemolytic anaemia and concomitant alteration in the red cell membrane sialoglycoprotein β. *Brit. J. Haematol. 69*, 61-66.

59. REYNOLDS, M. V., VENGELEN-TYLER, V. & MOREL, P. A. (1981): Autoimmune hemolytic anemia associated with autoanti-Ge. *Vox Sang. 41*, 61-67.

60. ROSENFIELD, R. E., HABER, G. V., KISSMEYER-NIELSEN, F., JACK, J. A., SANGER, R. & RACE, R. R. (1960): Ge, a very common red-cell antigen. *Brit. J. Haematol. 6*, 344-349.

61. SHULMAN, I. A., VENGELEN-TYLER, V., THOMPSON, J. C., NELSON, J. M. & CHEN, D. C. T. (1990): Autoanti-Ge associated with severe autoimmune hemolytic anemia. *Vox Sang. 59*, 232-234.

62. SIMMONS, R. T. & ALBREY, J. A. (1963): A 'new' blood group antigen Webb (Wb) of low frequency found in two Australian families. *Med. J. Aust. i*, 8-10.

63. SMYTHE, J., GARDNER, B. & ANSTEE, D. J. (1994): Quantitation of the number of molecules of glycophorins C and D on normal red blood cells using radioiodinated Fab fragments of monoclonal antibodies. *Blood 83*, 1668-1672.

64. SONDAG, D., ALLOISIO, N., BLANCHARD, D., DUCLUZEAU, M. T., COLONNA, P., BACHIR, D., BLOY, C., CARTRON, J. P. & DELAUNAY, J. (1987): Gerbich reactivity in 4.1(-) hereditary elliptocytosis and protein 4.1 level in blood group Gerbich deficiency. *Brit. J. Haematol. 65*, 43-50.

65. SPRING, F. A. (1991): Immunochemical characterisation of the low-incidence antigen, Dh^a. *Vox Sang. 61*, 65-68.

66. TANNER, M. J. A., HIGH, S., MARTIN, P. G., ANSTEE, D. J., JUDSON, P. A. & JONES, T. J. (1988): Genetic variants of human red-cell membrane sialoglycoprotein β. Study of the alterations occurring in the sialoglycoprotein-β gene. *Biochem. J. 250*, 407-414.

67. TELEN, M. J. & BOLK, T. (1987): Human red cell antigens. IV. The abnormal siaglycoprotein of Gerbich-negative red cells. *Transfusion 27*, 309-314.

68. TELEN, M. J., LE VAN KIM, C., CHUNG, A., CARTRON, J. P. & COLIN, Y. (1991): Molecular basis for elliptocytosis associated with glycophorin C and D deficiency in the Leach phenotype. *Blood 78*, 1603-1606.

69. UCHIKAWA, M., TSUNEYAMA, H., ONODERA, T., MURATA, S. & JUJI, T. (1997): A new high-molecular-weight glycophorin C variant with a duplication of exon 2 in the glycophorin C gene. *Transfus. Med. 7*, 305-309.

70. WINARDI, R., REID, M., CONBOY, J. & MOHANDAS, N. (1993): Molecular analysis of glycophorin C deficiency in human erythrocytes. *Blood 81*, 2799-2803.

19 Lutheran System

The **Lutheran** blood group system comprises 18 antigen characters to date (see *Table 19.1*), eight of which (i.e. **Lu^a – Lu^b, Lu6 – Lu9, Lu8 – Lu14,** and **Au^a – Au^b**) are products of antithetical alleles. Genetic studies are not yet able to show if the high-frequency para-**Lutheran** antigens are controlled by the **Lutheran** locus. Absence of these antigens from **Lutheran**-null cells and localisation on the **Lutheran** glycoproteins (see below) does, however, clearly show their close association with the **Lutheran** system.

The *LU* gene responsible for the transmission of the **Lutheran** antigens has been assigned to chromosome 19 at position q13.2–13.3 [40,54].

Three types of **Lutheran**-null phenotypes lacking all **Lutheran** antigenic characters have thus far been described [5]:

– **Lu(a–b–)** in which the expression of **Lutheran** antigens is impaired by the dominant suppressor gene, *In(Lu)* [1] [45];
– **X-linked Lu$_{null}$** presumably caused by an *X*-borne inhibitor gene ('*XS2*') [36]; and
– a recessive form caused by the occurrence of an amorph allele at the *LU* complex locus.

Anti-**Lutheran** antibodies are rare and usually show only very weak activity; in most cases they are of immune origin [20]. More recently a fairly specific anti-**Lu^b** hybridoma antibody (*BRIC 108*) [39] used in structural analyses of the **Lu^b** antigen has been described.

The **Lutheran** antigens are released from the red cell membrane by low concentrations of the non-ionic detergent Triton X-100 [7]. The antigens *in situ* are sensitive to trypsin and chymotrypsin; papain has only little effect [8]. Further, the

---

[1] The inhibitor gene *In(Lu)* also impairs the expression of the two otherwise unrelated red cell antigens **P$_1$, i**, and the antigens of the **Indian** system (see [8]). The proposal by Marsh et al. [30] to rename the *In(Lu)* locus *SYN* (*SYN-1A* being the common allele permitting normal biosynthesis, *SYN-1B* the rare dominant gene preventing normal biosynthesis of these red cell determinants) is not generally accepted.

20 Diego System

The **Diego** blood group system comprises to date 21 antigen specificities − the antithetical **Diego** antigens, **Di**[a] (= **DI1**) [43] and **Di**[b] (= **DI2**) [72], the antithetical **Wright** antigens, **Wr**[a] (= **DI3**) [24] and **Wr**[b] (= **DI4**) [1], as well as 17 low-incidence antigens (see *Table 20.1*).

The **Diego** antigens are of substantial importance for anthropological studies, as **Di**[a] is the only blood group character typical for the Mongolid peoples. Its frequency varies between 2 and 12% in South East Asians and reaches almost 70% in some South American Indians, in other populations, however, it is rarely found [50]; the **Di**[b] antigen, in contrast, is a common erythrocyte character in all populations tested so far.

Wr[a] is a low-incidence antigen (~0.1% in Europids), whereas the **Wr**[b] character is a common blood group character in most populations tested; **Wr**[b] is not expressed, however, in erythrocytes of some rare **MNS** variants (see this chapter below).

20.1 Antibodies

Most anti-**Di**[a] and anti-**Di**[b] antibodies are immune antibodies induced by incompatible blood transfusions or pregnancies [56].

Anti-**Wr**[a] is a not uncommon antibody type in normal donors; some samples agglutinate directly, but most samples are only detected by an antiglobulin test. Anti-**Wr**[b] is found in **Wr(a+b−)** individuals [1] and occurs regularly in **En(a−)** individuals (see *Sect. 9.6*), the accompanying anti-**En**[a] antibodies can easily be removed by serum absorption [30,52]. Autoantibodies with anti-**Wr**[a] (Cleghorn cit. in [63]) and anti-**Wr**[b] specificity are frequently detected in the serum of patients suffering from autoimmune haemolytic anaemia [22,29].

Hybridoma antibodies with anti-**Wr**[a] [63] and anti-**Wr**[b] specificity [2,20,61] have also been described.

Table 20.1: Antigens of the Diego blood group system.

ISBT nr.[1]	Common term	Historical term	References
DI1	Di^a	Diego a	[43]
DI2	Di^b	Diego b	[72]
DI3	Wr^a	Wright a	[24]
DI4	Wr^b	Wright b	[1]
DI5	Wd^a	Waldner	[44]
DI6	Rb^a	Redelberger	[14]
DI7	WARR	Warrior	[12]
DI8	ELO		[13]
DI9	Wu	Wulfsberg	[41]
DI10	Bp^a	Bishop	[57]
DI11	Mo^a	Moen	[40]
DI12	Hg^a	Hughes	[65]
DI13	Vg^a	van Vugt	[78]
DI14	Sw^a	Swann	[11]
DI15	BOW	Bowyer	[10]
DI16	NFLD		[39]
DI17	Jn^a	Nunhart	[42]
DI18	KREP		[54]
DI19	Tr^a	Traversu	[57]
DI20	Fra	Froese	[45]
DI21	SW1		

[1] The six-digit ISBT identification numbers are 010 001 to 010 018 (see *Chap. 1*).

20.2 Antigens of the Diego System

The **Diego** and **Wright** characters *in situ* are not affected by proteinases, neuraminidase, and sulfhydryl reagents (2-aminoethylisothiouronium bromide) [16,63]. The **Waldner** [7], **Tr**^{a} [34], and **Warrior** [32] antigens, however, are destroyed by chymotrypsin.

```
  1 MEELQDDYED MMEENLEQEE YEDPDIPESQ MEEPAAHDTE ATATDYHTTS HPGTHKVYVE  60

 61 LQELVMDEKN QELRWMEAAR WVQLEENLGE NGAWGRPHLS HLTFWSLLEL RRVFTKGTVL 120

121 LDLQETSLAG VANQLLDRFI FEDQIRPQDR EELLRALLLK HSHAGELEAL GGVKPAVLTR 180

181 SGDPSQPLLP QHSSLETQLF CEQGDGGTEG HSPSGILEKI PPDSEATLVL VGRADFLEQP 240

241 VLGFVRLQEA AELEAVELPV PIRFLFVLLG PEAPHIDYTQ LGRAAATLMS ERVFRIDAYM 300

301 AQSRGELLHS LEGFLDCSLV LPPTDAPSEQ ALLSLVPVQR ELLRRRYQSS PAKPDSSFYK 360

361 GLDLNGGPDD PLQQTGQLFG GLVRDIRRRY PYYLSDITDA FSPQVLAAVI FIYFAALSPA 420

421 ITFGGLLGEK TRNQMGVSEL LISTAVQGIL FALLGAQPLL VVGFSGPLLV FEEAFFSFCE 480

481 TNGLEYIVGR VWIGFWLILL VVLVVAFEGS FLVRFISRYT QEIFSFLISL IFIYETFSKL 540

541 IKIFQDHPLQ KTYNYNVLMV PKPQGPLPNT ALLSLVLMAG TFFFAMMLRK FKNSSYFPGK 600

601 LRRVIGDFGV PISILIMVLV DFFIQDTYTQ KLSVPDGFKV SNSSARGWVI HPLGLRSEFP 660

661 IWMMFASALP ALLVFILIFL ESQITTLIVS KPERKMVKGS GFHLDLLLVV GMGGVAALFG 720

721 MPWLSATTVR SVTHANALTV MGKASTPGAA AQIQEVKEQR ISGLLVAVLV GLSILMEPIL 780

781 SRIPLAVLFG IFLYMGVTSL SGIQLFDRIL LLFKPPKYHP DVPYVKRVKT WRMHLFTGIQ 840

841 IICLAVLWVV KSTPASLALP FVLILTVPLR RVLLPLIFRN VELQCLDADD AKATFDEEEG 900

901 RDEYDEVAMP V
```

Figure 20.1: Amino acid sequence of the band 3 protein of human erythrocyte membranes.
N : N-glycosylation site.
According to Lux et al. [48]. The sequence data are deposited in the EMBL/GenBank data library (accession number M27819).

The antigens of the **Diego** blood group system reside on the band 3 protein, the anion exchanger of the erythrocyte membrane.

The band 3 protein is the most abundant integral protein of the erythrocyte membrane, being present in over 1 million copies per cell. A homodimer of this 95 kDa membrane constituent [8] is responsible for the exchange of bicarbonate and chloride ions across the erythrocyte membrane; further, the band 3 protein mediates the attachment of the red cell membrane to the cytoskeleton (see *Sect. 4.1.2*).

The gene encoding the band 3 protein (***AE1*** = 'anion exchanger protein 1') has been assigned to the long arm of chromosome 17, position q12–q21 [67,79]. It consists of 20 exons covering 18 kb of DNA [66].

Sequencing of PCR amplified band 3 cDNA predicted a 911 amino acid protein (*Fig. 20.1* [48,70]). The polypeptide itself consists of three main domains – a hydrophilic cytoplasmic domain at the N-terminus (403 residues) which provides the attachment sites for the membrane skeleton (via ankyrin) and several cytoplasmic proteins, a hydrophobic transmembrane domain (479 residues) responsible for the anion transport and spanning the membrane with up to fourteen transmembrane segments (*Fig. 20.2* [60,76]), and a highly acidic C-terminal domain (29 residues) which is exposed to the inner surface of the membrane and is involved in the exchange of anions across the cell membrane. The asparagine residue at position 642 is glycosylated – the N-linked oligosaccharide chain of the complex type carries **ABH** and **Ii** determinants ([18,74], see also *Fig. 5.10*).

20.2.1 Diego Antigens

Spring et al. [68] located the **Di**^{a} antigen on the 'Memphis variant 2' of the band 3 protein; normal band 3 protein as well as the 'Memphis variant 1' protein were shown to carry the **Di**^{b} character.

The Memphis variant of the human erythrocyte anion transporter was first described by Mueller and Morrison [51] who found the variant protein characterised by a reduced electrophoretic mobility of the N-terminal fragment obtained by pronase or chymotrypsin treatment of red cells.

Band 3 Memphis is a fairly common polymorphism found in all ethnic groups; the frequency varies from 6–7% in Europids to more than 20% in American Indians and Japanese [27,58]. In contrast to other band 3 variants which cause changes in erythrocyte shape (e.g. [35,47], see also *Sect. 4.1.2*), the Memphis variants are not associated with any clinical, haematological, or biochemical abnormality except for a slightly reduced phosphoenolpyruvate transport activity [27].

More recently Hsu and Morrison [25] distinguished two types of the band 3 Memphis variant differing in their affinity towards a specific anion transport inhibitor – variant 2 reacted more readily with 4,4'-diisothiocyanato-1,2-diphenyl-ethane-2,2'-disulphonate (H$_2$DIDS) than did normal band 3 or Memphis variant 1. H$_2$DIDS covalently binds to a lysine residue at position 539 which is located close to the outer surface of the erythrocyte membrane [69].

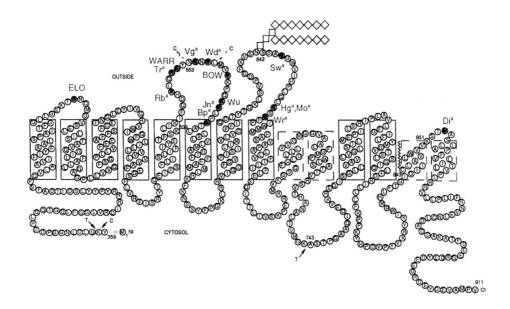

Figure 20.2: Proposed model for the orientation of the band 3 protein within the erythrocyte membrane and position of the antigens of the Diego blood group system on the protein.
Reproduced from Wang et al. [76] by permission of Oxford University Press.

Peptide sequencing and cDNA analysis have shown that the band 3 Memphis variant protein is characterised by a A → G exchange at nucleotide 166 causing a lysine → glutamic acid substitution at residue 56 [38,77]; variants 1 and 2 differ by a proline (= **Di**^{b}) or leucine (= **Di**^{a}) residue at position 854 caused by a C ↔ T dimorphism at nucleotide 2561 [4].

20.2.2 Wright Antigens

The **Wright** antigens are determined by a G ↔ A dimorphism at nucleotide 1972 which is responsible for an amino acid exchange at position 658 of the band 3 protein – lysine being characteristic of **Wr**^{a} and glutamic acid characteristic of **Wr**^{b} [6]. This amino acid dimorphism located within the fourth extracellular loop of band 3 is situated close to the membrane surface (see *Fig. 20.2*).

The **Wr^b** antigen was originally thought to reside on glycophorin A, thus making it part of the **MNS** system. This assumption was based on the fact that erythrocytes of glycophorin A-deficient individuals type as **Wr(a–b–)**, **En(a–)** and homozygous **M^k** probands [28,30,73], as well as red cells carrying glycophorin A hybrids lacking part of the glycophorin A molecule (e.g. **Mi V** and **Dantu**) fall into this category [62] (see *Sect. 9.6*). The finding that lipid [59] (see also the characters **En^{a}FS, U**, and **Duclos**) and band 3 protein [71] were necessary for optimum serological activity of **Wr^b** suggested the location of the antigen near the hydrophobic membrane domain of glycophorin A [15,62]. This assumption was further supported by a low inhibitory activity of anti-**Wr^b** haemagglutination by purified fragments of glycophorin A [15].

A series of studies have revealed a close association within the red cell membrane of band 3 protein with glycophorin A:

– glycophorin A and band 3 protein were co-precipitated by anti-**Wr^b** antibodies [15,64,71] and by a hybridoma antibody towards the cytoplasmic C-terminal domain of band 3 [75]. However, anti-glycophorin A sera precipitated only glycophorin A and anti-band 3 sera precipitated only band 3 [64,71];
– Telen and Chasis [71] found that monoclonal anti-**Wr^b** does not bind to human cell lines which do not express both glycophorin A and band 3;
– the increased size of the N-glycan chain of band 3 and the altered anion transport activity in glycophorin A-deficient erythrocytes (e.g. **En(a–)** or **M^k/M^k** cells, see *Sect. 9.6)* confirms a role of glycophorin A in biosynthesis and/or processing of band 3 protein [5,19].

Based on the close association of the two erythrocyte membrane proteins and the above-mentioned serological results it can be assumed that the **Wr^b** antigenic structure is formed by both glycophorin A and band 3 protein sequences. It has been advanced that **Wr^b** is a composed epitope containing the determinant glutamic acid residue at position 658 of band 3 and a peptide sequence of glycophorin A between amino acids 59 and 71 [6,26]. Recently, two glycophorin A variants with amino acid substitutions in this region have been detected, viz. Ala^{65} → Pro [53] and Glu^{63} → Lys [31]; the finding that these mutations markedly affect the expression of the **Wr^b** antigen corroborates the assumption that this part of the glycophorin A molecule is essentially involved in forming the **Wr^b** epitope.

The fact that **MNS** variants in no way affect the **Wr^a** antigen and the finding that anti-**Wr^a** does not co-precipitate any other membrane component [64] strongly suggest that the **Wr^a** epitope is formed solely by the band-3 protein.

20.2.3 Low-Incidence Blood Group Antigens Carried by the Band 3 Protein of the Erythrocyte Membrane

Thus far only one antigen has been located on the putative first ectoplasmic loop of band 3 protein: the antigen **ELO** (= **DI8**) [13] is characterised by a C → T mutation in exon 12 leading to an Arg → Trp substitution at position 432 [37,82].

Eight antigens are located in the putative third ectoplasmic loop of the band 3 protein:

The **Waldner** antigen (**Wd**^{a} = **DI5**), first described by Lewis & Kaita [44], was found only in members of the Hutterite community. Sequence analysis of the *AE1* gene from **Wd(a+)** individuals showed that a G → A mutation at nucleotide 1669 causes a Val → Met substitution at position 557 of the protein [7,33,34]. The close proximity of Met^{557} to the chymotrypsin cleavage sites on band 3 protein (Tyr^{553} and Leu^{558}) explains the inactivation of the **Wd**^{a} antigen by chymotrypsin.

The **Rb**^{a} (**Redelberger, DI6**) epitope [14] is characterised by a Pro → Leu exchange at position 548 [33,34].

The **WARR** (**Warrior, DI7**) antigen, which has been found only in two families [12]) is based on a C → T mutation in exon 14 which leads to a Thr → Ile substitution at position 552 [32].

The antigen **Tr**^{a} (**Traversu**) [57] is distinguished by a Lys → Asn at position 551 [34].

Further antigens on the third ectoplasmic loop are characterised by single amino acid exchanges: **Vg**^{a} (**Van Vugt, DI13**) [78] by Tyr^{555} → His [37], **BOW** (**Bowyer, DI15**) [10] by Pro^{561} → Ser [37], **Wu** (**Wulfsberg, DI9**) [41] by Gly^{565} → Ala [37,81], and **Bp**^{a} (**Bishop, DI10**) (Cleghorn, cit in [57]) by Asn^{569} → Lys [37].

Jn^a (**DI17**) and **KREP** (**DI18**) represent different amino acid substitutions at the same position of the band 3 protein, a Pro^{566} → Ser mutation giving rise to the antigen **Jn**^{a} and a Pro^{566} → Ala to the antigen **KREP** [54,55].

In addition to the **Wright** antigens three low-frequency characters have been located on the putative fourth ectoplasmic loop:

Hg^a (**Hughes, DI12**) [65] and **Mo**^{a} (**Moen, DI11**) [40] are characterised by an amino acid exchange at position 656, Arg → Cys substitution being characteristic for **Hg**^{a} and Arg → His substitution for **Mo**^{a} [37]. The **Sw**^{a} (**Swann, DI14**) character is determined by an Arg^{646} → Gln substitution [13a].

Two amino acid substitutions are associated with **NFLD** (**DI16**), viz. Glu^{429} → Asp and Pro^{561} → Ala [49].

The **Fr^a (Froese, DI20)** antigen is based on a G → A mutation in exon 13 which results in a Glu^{480} → Lys substitution [49a].

In all cases of low-incidence antigens the mutations do not have major effect on band 3 structure and function. Neither transcription nor RNA processing are affected. Further, the mutant proteins show normal glycosylation, electrophoretic mobility and anion transport.

Di^b and **Wr^b** are depressed in South East Asian ovalocytosis, which is fairly common in malaria-infested regions of Malaysia, Papua New Guinea, and the Philippines [3]. This disorder is caused by the heterozygous presence of a Memphis variant band 3 protein with a deletion of amino acids 400–408. The deletion is located at the boundary between the cytoplasmic and the first transmembrane domain of the protein [36]. The lack of part of a hydrophobic segment impairs the proper integration of the first membrane span into the membrane, results in misfolding of the membrane domain and probably affects the interaction of the band 3 protein with the cytoskeleton [9]. Homozygosity for this band 3 mutant is probably lethal [46].

Various reports suggested that ovalocytosis mediates resistance towards the malarial parasite *Plasmodium falciparum* [17,21,23].

References

1. ADAMS, J., BROVIAC, M., BROOKS, W., JOHNSON, N. R. & ISSITT, P. D. (1971): An antibody, in the serum of a Wr(a+) individual, reacting with an antigen of very high frequency. *Transfusion 11*, 290-291.
2. ANSTEE, D. J. & EDWARDS, P. A. W. (1982): Monoclonal antibodies to human erythrocytes. *Eur. J. Immunol. 12*, 228-232.
3. BOOTH, P. B., SERJEANTSON, S., WOODFIELD, D. G. & AMATO, D. (1977): Selective depression of blood group antigens associated with ovalocytosis among Melanesians. *Vox Sang 32*, 99-110.
4. BRUCE, L. J., ANSTEE, D. J., SPRING, F. A. & TANNER, M. J. A. (1994): Band 3 Memphis variant II. Altered stilbene disulfonate binding and the Diego (Di^a) blood group antigen are associated with the human erythrocyte band 3 mutation Pro^{854} → Leu. *J. Biol. Chem. 269*, 16155-16158.
5. BRUCE, L. J., GROVES, J. D., OKUBO, Y., THILAGANATHAN, B. & TANNER, M. J. A. (1994): Altered band 3 structure and function in glycophorin A- and B-deficient (M^{k}M^k) red blood cells. *Blood 84*, 916-922.
6. BRUCE, L. J., RING, S. M., ANSTEE, D. J., REID, M. E., WILKINSON, S. & TANNER, M. J. A. (1995): Changes in the blood group Wright antigens are associated with a mutation at amino acid 658 in human erythrocyte band 3: a site of interaction between band 3 and glycophorin A under certain conditions. *Blood 85*, 541-547.
7. BRUCE, L. J., ZELINSKI, T., RIDGWELL, K. & TANNER, M. J. A. (1996): The low-incidence blood group antigen, Wd^a, is associated with the substitution Val$_{557}$ → Met in human erythrocyte band 3 (AE1). *Vox Sang. 71*, 118-120.
8. CASEY, J. R. & REITHMEIER, R. A. F. (1991): Analysis of the oligomeric state of band 3, the anion transport protein of the human erythrocyte membrane, by size exclusion high performance liquid chromatography. *J. Biol. Chem. 266*, 15726-15737.

9. CHAMBERS, E. J., BLOOMBERG, G. B., RING, S. M. & TANNER, M. J. A. (1999): Structural studies on the effects of the deletion in the red cell anion exchanger (Band 3, AE1) associated with South East Asian ovalocytosis. *J. Mol. Biol. 285*, 1289-1307.

10. CHAVES, M. A., LEAK, M. R., POOLE, J. & GILES, C. M. (1988): A new low-frequency antigen BOW (Bowyer). *Vox Sang. 55*, 241-243.

11. CLEGHORN, T. E. (1959): A 'new' human blood group antigen, Sw^a. *Nature 184*, 1324.

12. COGHLAN, G., CROW, M., SPRUELL, P., MOULDS, M. & ZELINSKI, T. (1995): A 'new' low-incidence red cell antigen, WARR: unique to native Americans? *Vox Sang. 68*, 187-190.

13. COGHLAN, G., GREEN, C., LUBENKO, A., TIPPETT, P. & ZELINSKI, T. (1993): Low-incidence red cell antigen ELO (700.51): evidence for exclusion from 13 blood group systems. *Vox Sang. 64*, 240-243.

13a. COGHLAN, G., RUSNAK, A., MCMANUS, K.& ZELINSKI, T. (2000): Distinctive Swann blood group genotypes: molecular analyses (Abstract). *Vox Sang.* (in press).

14. CONTRERAS, M., STEBBING, B., MALLORY, D. M., BARE, J., POOLE, J. & HAMMOND, W. (1978): The Redelberger antigen Rb^a. *Vox Sang 35*, 397-400.

15. DAHR, W., WILKINSON, S., ISSITT, P. D., BEYREUTHER, K., HUMMEL, M. & MOIREL, P. (1986): High frequency antigens of human erythrocyte membrane sialoglycoproteins. III. Studies on the En^a FR, Wr^b, and Wr^a antigens. *Biol. Chem. Hoppe-Seyler 367*, 1033-1045.

16. DANIELS, G. (1992): Effect of enzymes on and chemical modifications of high-frequency red cells antigens. *Immunohaematology 8*, 53-57.

17. DLUZEWSKI, A. R., NASH, G. B., WILSON, R. J., REARDON, D. M. & GRATZER, W. B. (1992): Invasion of hereditary ovalocytes by Plasmodium falciparum in vitro and its relation to intracellular ATP concentration. *Mol. Biochem. Parasitol. 55*, 1-7.

18. FUKUDA, M., DELL, A., OATES, J. E. & FUKUDA, M. N. (1984): Structure of branched lactosaminoglycan, the carbohydrate moiety of band 3 isolated from adult human erythrocytes. *J. Biol. Chem. 259*, 8260-8273.

19. GAHMBERG, C. G., MYLLYLÄ, G., LEIKOLA, J., PIRKOLA, A. & NORDLING, S. (1976): Absence of the major sialoglycoprotein in the membrane of human En(a-) erythrocytes and increased glycosylation of band 3. *J. Biol. Chem. 251*, 6108-6116.

20. GARDNER, B., PARSONS, S. F., MERRY, A. H. & ANSTEE, D. J. (1989): Epitopes on sialyglycoprotein α: evidence for heterogeneity in the molecule. *Immunology 68*, 283-289.

21. GENTON, B., AL-YAMAN, F., MGONE, C. S., ALEXANDER, N., PANIU, M. M., ALPERS, M. P. & MOKELA, D. (1995): Ovalocytosis and cerebral malaria. *Nature 378*, 564-565.

22. GOLDFINGER, D., ZWICKER, H., BELKIN, G. A. & ISSITT, P. D. (1975): An autoantibody with anti-Wr^b specificity in a patient with warm autoimmune hemolytic anemia. *Transfusion 15*, 351-352.

23. HADLEY, T., SAUL, A., LAMONT, G., HUDSON, D. E., MILLER, L. H. & KIDSON, C. (1983): Resistance of Melanesian elliptocytes (ovalocytes) to invasion by Plasmodium knowlesi and Plasmodium falciparum malaria parasites in vitro. *J Clin Invest 71*, 780-782.

24. HOLMAN, C. A. (1953): A new rare human blood group antigen (Wr^a). *Lancet 2*, 119-120.

25. HSU, L. & MORRISON, M. (1985): A new variant of the anion transport proteins in human erythrocytes. *Biochemistry 24*, 3086-3090.

26. HUANG, C. H., REID, M. E., XIE, S. S. & BLUMENFELD, O. O. (1996): Human red blood cell Wright antigens: a genetic and evolutionary perspective on glycophorin A-band 3 interaction. *Blood 87*, 3942-3947.

27. IDEGUCHI, H., OKUBO, K., ISHIKAWA, A., FUTATA, Y. & HAMASAKI, N. (1992): Band 3-Memphis is associated with a lower transport rate of phosphoenolpyruvate. *Br. J. Haematol. 82*, 122-125.

28. ISSITT, P. D., PAVONE, B. G., GOLDFINGER, D. & ZWICKER, H. (1975): An En(a-) red cell sample that types as Wr(a-b-). *Transfusion 15*, 353-355.

29. ISSITT, P. D., PAVONE, B. G., GOLDFINGER, D., ZWICKER, H., ISSITT, C. H., TESSEL, J. A., KROOVAND, S. W. & BELL, C. A. (1976): Anti-Wr^b, and other autoantibodies responsible for positive direct antiglobulin tests in 150 individuals. *Br. J. Haematol. 34*, 5-18.

30. ISSITT, P. D., PAVONE, B. G., WAGSTAFF, W. & GOLDFINGER, D. (1976): The phenotypes En(a-),

Wr(a-b-) and En(a+), Wr(a+b-), and further studies on the Wright and En blood group systems. *Transfusion 16*, 396-407.

31. JAROLIM, P., MOULDS, J. M., MOULDS, J. J., RUBIN, H. L. & DAHR, W. (1997): Molecular basis of the MARS and AVIS blood group antigens. *Transfusion 37*, 90S.

32. JAROLIM, P., MURRAY, J. L., RUBIN, H. L., COGHLAN, G. & ZELINSKI, T. (1997): A Thr$_{552}$→Ile substitution in erythroid band 3 gives rise to the Warrior blood group antigen. *Transfusion 37*, 398-405.

33. JAROLIM, P., MURRAY, J. L., RUBIN, H. L., SMART, E. & MOULDS, J. M. (1995): Wd^a and Rb^a blood group antigens are located in the third ectoplasmic loop of the erythroid band 3 protein (abstract). *Blood 86 (Suppl.)*, 445a.

34. JAROLIM, P., MURRAY, J. L., RUBIN, H. L., SMART, E. & MOULDS, J. M. (1997): Blood group antigens Rb^a, Tr^a, and Wd^a are located in the third ectoplasmic loop of erythroid band 3. *Transfusion 37*, 607-615.

35. JAROLIM, P., MURRAY, J. L., RUBIN, H. L., TAYLOR, W. M., PRCHAL, J. T., BALLAS, S. K., SNYDER, L. M., CHROBAK, L., MELROSE, W. D., BRABEC, V. & PALEK, J. (1996): Characterization of 13 novel band 3 gene defects in hereditary spherocytosis with band 3 deficiency. *Blood 88*, 4366-4374.

36. JAROLIM, P., PALEK, J., AMATO, D., HASSAN, K., SAPAK, R., NURSE, G. T., RUBIN, H. L., ZHAI, S., SAHR, K. E. & LIU, S. C. (1991): Deletion in erythrocyte band 3 gene in malaria-resistant Southeast Asian ovalocytosis. *Proc. Natl. Acad. Sci. USA 88*, 11022-11026.

37. JAROLIM, P., RUBIN, H. L., ZAKOVA, D., STORRY, J. & REID, M. E. (1998): Characterization of seven low incidence blood group antigens carried by erythrocyte band 3 protein. *Blood 92*, 4836-4843.

38. JAROLIM, P., RUBIN, H. L., ZHAI, S., SAHR, K. E., LIU, S. C., MUELLER, T. J. & PALEK, J. (1992): Band 3 Memphis: a widespread polymorphism with abnormal electrophoretic mobility of erythrocyte band 3 protein caused by substitution AAG → GAG (Lys → Glu) in codon 56. *Blood 80*, 1592-1598.

39. KAITA, H., LUBENKO, A., MOULDS, M. & LEWIS, M. (1992): A serologic relationship among the NFLD, BOW, and Wu red cell antigens. *Transfusion 32*, 845-847.

40. KORNSTAD, L. & BROCTEUR, J. (1972): A new rare blood group antigen, Mo^a (MOEN). *XXV Annual Meeting and International Society of Blood Banks XIII International Congress*, American Association of Blood Banks, Washington DC, p. 58.

41. KORNSTAD, L., HOWELL, P., JORGENSEN, J., LARSEN, A. M. H. & WADSWORTH.L.D. (1976): The rare blood group antigen Wu. *Vox Sang. 31*, 337-343.

42. KORNSTAD, L., KOUT, M., LARSEN, A. M. H. & ØRJASAETER, H. (1967): A rare blood group antigen, Jn^a. *Vox Sang. 165-170*.

43. LAYRISSE, M., ARENDS, T. & DOMINGUEZ SISICO, R. (1955): Nuevo grupo sanguíneo encontrado en descendientes de Indios. *Acta Med. Venezol. 3*, 132-138.

44. LEWIS, M. & KAITA, H. (1981): A 'new' low incidence 'Hutterite' blood group antigen Waldner (Wd^a). *Am. J. Hum. Genet. 33*, 418-420.

45. LEWIS, M., KAITA, H., MCALPINE, P. J., FLETCHER, J. & MOULDS, J. J. (1978): A 'new' blood group antigen Fr^a: incidence, inheritance, and genetic linkage analysis. *Vox Sang. 35*, 251-254.

46. LIU, S. C., JAROLIM, P., RUBIN, H. L., PALEK, J., AMATO, D., HASSAN, K., ZAIK, M. & SAPAK, P. (1994): The homozygous state for the band 3 protein mutation in Southeast Asian Ovalocytosis may be lethal. *Blood 84*, 3590-3591.

47. LIU, S. C., PALEK, J., YI, S. J., NICHOLS, P. E., DERICK, L. H., CHIOU, S. S., AMATO, D., CORBETT, J. D., CHO, M. R. & GOLAN, D. E. (1995): Molecular basis of altered red blood cell membrane properties in Southeast Asian ovalocytosis: role of the mutant band 3 protein in band 3 oligomerization and retention by the membrane skeleton. *Blood 86*, 349-358.

48. LUX, S. E., JOHN, K. M., KOPITO, R. R. & LODISH, H. F. (1989): Cloning and characterization of band 3, the human erythrocyte anion-exchange protein (AE1). *Proc. Natl. Acad. Sci. USA 86*, 9089-9093.

49. MCMANUS, K., COGHLAN, G. & ZELINSKI, T. (1998): Substitution(s) in erythrocyte membrane protein, band 3, underlie(s) the BOW and NFLD blood group polymorphisms. *Transfusion 38 (Suppl.)*,

102S.

49a. McMANUS, K., LUPE, K., COGHLAN, G. & ZELINSKI, T. (2000): An amino acid substitution in the putative second extracellular loop of erythroid band 3 accounts for the Froese blood group polymorphism (Abstract). *Vox Sang.* (in press).

50. MOURANT, A. E. (1983): *Blood Relations - Blood Groups and Anthropology.* University Press, London, Oxford

51. MUELLER, T. J. & MORRISON, M. (1977): Detection of a variant of protein 3, the major transmembrane protein of the human erythrocyte. *J. Biol. Chem. 252*, 6573-6576.

52. PAVONE, B. G., PIRKOLA, A., NEVANLINNA, H. R. & ISSITT, P. D. (1978): Demonstration of anti-Wr^b in a second serum containing anti-En^a. *Transfusion 18*, 155-159.

53. POOLE, J., BANKS, J., BRUCE, L. J., RING, S. M., LEVENE, C., STERN, H., OVERBEEKE, M. A. M. & TANNER, M. J. A. (1999): Glycophorin A mutation Ala65 → Pro gives rise to a novel pair of MNS alleles ENEP (MNS39) and HAG (MNS41) and altered Wr^b expression: direct evidence for GPA/band 3 interaction necessary for normal Wr^b expression. *Transf. Med. 9*, 167-174.

54. POOLE, J., BRUCE, L. J., HALLEWELL, H., KUSNIERZ-ALEJSKA, G., ZUPANSKA, B., DANIELS, G. L. & TANNER, M. J. A. (1998): Erythrocyte band 3 mutation Pro^{561}→Ser gives rise to the BOW antigen and Pro566 → Ala to a novel antigen KREP. *Transf. Med. 8 (Suppl.1)*, 17.

55. POOLE, J., HALLEWELL, H., BRUCE, L., TANNER, M. J. A., ZUPANSKA, B. & KUSNIERZ-ALESKA, G. (1997): Indentification of two new Jn(a+) individuals and assigmant ofJn^a to erythrocyte band 3. *Transfusion 37 (Suppl.)*, 90S.

56. RACE, R. R. & SANGER, R. (1975): The Diego blood groups. In: *Blood Groups in Man.* Blackwell Scientific Publications, Oxford, pp. 372-378.

57. RACE, R. R. & SANGER, R. (1975): Some very infrequent antigens. In: *Blood Groups in Man,* Blackwell Scientific Publications, Oxford, pp. 431-446.

58. RANNEY, H. M., ROSENBERG, G. H., MORRISON, M. & MUELLER, T. J. (1990): Frequencies of band 3 variants of human red cell membranes in some different populations. *Br. J. Haematol. 76*, 262-267.

59. REARDEN, A. (1985): Phospholipid dependence of Wr^b antigen expression in human erythrocyte membranes. *Vox Sang. 49*, 346-353.

60. REITHMEIER, R. A. F. (1993): The erythrocyte anion transporter (band 3). *Curr. Opin. Strct. Biol. 3*, 515-523.

61. RIDGWELL, K., TANNER, M. J. A. & ANSTEE, D. J. (1983): The Wr^b antigen, a receptor for Plasmodium falciparum malaria, is located on a helical region of the major membrane sialoglycoprotein of human red blood cells. *Biochem. J. 209*, 273-276.

62. RIDGWELL, K., TANNER, M. J. A. & ANSTEE, D. J. (1984): The Wr^b antigen in St^a-positive and Dantu-positive human erythrocytes. *J. Immunogenet. 11*, 365-370.

63. RING, S. M., GREEN, C. A., SWALLOW, D. M. & TIPPETT, P. (1994): Production of a murine monoclonal antibody to the low-incidence red cell antigen Wr^a: characterisation and comparison with human anti-Wr^a. *Vox Sang. 67*, 222-225.

64. RING, S. M., TIPPETT, P. & SWALLOW, D. M. (1994): Comparative immunochemical analysis of Wr^a and Wr^b red cell antigens. *Vox Sang. 67*, 226-230.

65. ROWE, G. P. & HAMMOND, W. (1983): A new low-frequency antigen, Hg^a (Hughes). *Vox Sang. 45*, 316-319.

66. SCHOFIELD, A. E., MARTIN, P. G., SPILLETT, D. & TANNER, M. J. A. (1994): The structure of the human red blood cell anion exchanger (EPB3, AE1, band 3) gene. *Blood 84*, 2000-2012.

67. SOLOMON, E. & LEDBETTER, D. M. (1991): Report of the Committe on the Genetic Constitution of Chromosome 17. *Cytogenet. Cell Genet. 58*, 686-738.

68. SPRING, F. A., BRUCE, L. J., ANSTEE, D. J. & TANNER, M. J. A. (1992): A red cell band 3 variant with altered stilbene disulphonate binding is associated with the Diego (Di^a) blood group antigen. *Biochem. J. 288*, 713-716.

69. TANNER, M. J. A. (1993): Molecular and cellular biology of the erythrocyte anion exchanger (AE1). *Sem. Hematol. 30*, 34-57.

70. Tanner, M. J. A., Martin, P. G. & High, S. (1988): The complete amino acid sequence of the human erythrocyte membrane anion-transport protein deduced from the cDNA sequence. *Biochem. J. 256*, 703-712.

71. Telen, M. J. & Chasis, J. A. (1990): Relationship of the human erythrocyte Wr^b antigen to an interaction between glycophorin A and band 3. *Blood 76*, 842-848.

72. Thompson, P. R., Childers, D. M. & Hatcher, D. E. (1967): Anti-Di^b - first and second examples. *Vox Sang. 13*, 314-318.

73. Tokunaga, E., Sasakawa, S., Tamaka, K., Kawamata, H., Giles, C. M., Ikin, E. W., Poole, J., Anstee, D. J., Mawby, W. J. & Tanner, M. J. A. (1979): Two apparently healthy japanese individuals of type M^{k}M^k have erythrocytes which lack both the blood group MN and Ss-active sialoglycoproteins. *J. Immunogenet. 6*, 383-390.

74. Tsuji, T., Irimura, T. & Osawa, T. (1981): The carbohydrate moiety of band 3 glycoprotein of human erythrocyte membranes. Structure of lower molecular weight oligosaccharides. *J. Biol. Chem. 256*, 10497-10502.

75. Wainwright, S. D., Tanner, M. J., Martin, G. E., Yendle, J. E. & Holmes, C. (1989): Monoclonal antibodies to the membrane domain of the human erythrocyte anion transport protein. Localization of the C-terminus of the protein to the cytoplasmic side of the red cell membrane and distribution of the protein in some human tissues. *Biochem. J. 258*, 211-220.

76. Wang, D. N., Sarabia, V. E., Reithmeier, R. A. F. & Kühlbrandt, W. (1994): Three-dimensional map of the dimeric membrane domain of the human erythrocyte anion exchanger, band 3. *EMBO J. 13*, 3230-3235.

77. Yannoukakos, D., Vasseur, C., Driancourt, C., Blouquit, Y., Delaunay, J., Wajcman, H. & Bursaux, E. (1991): Human erythrocyte band 3 polymorphism (band 3 Memphis): characterization of the structural modification (Lys^{56} → Glu) by protein chemistry methods. *Blood 78*, 1117-1120.

78. Young, S. (1981): Vg^a: a new low incidence red cell antigen. *Vox Sang. 41*, 48-49.

79. Zelinski, T., Coghlan, G., White, L. & Philipps, S. (1993): The Diego blood group locus is located on chromosome 17q. *Genomics 17*, 665-666.

80. Zelinski, T., McKeown, I., McAlpine, P. J., Philipps, S. & Coghlan, G. (1996): Assignment of the gene(s) governing Froese and Swann blood group polymorphism to chromosome 17q. *Transfusion 36*, 419-420.

81. Zelinski, T., McManus, K., Punter, F., Moulds, M. & Coghlan, G. (1998): A Gly$_{565}$→Ala substitution in human erythroid band 3 accounts for the Wu blood group polymorphism. *Transfusion 38*, 745-748.

82. Zelinski, T., Punter, F., McManus, K. & Coghlan, G. (1998): The ELO blood group polymorphism is located in the putative first extracellular loop of human erythrocyte band 3. *Vox Sang. 75*, 63-65.

21 Cromer System

Seven high-frequency and three low-frequency antigens of the **Cromer** system have been described *(Table 21.1)*. Dimorphic variations are represented by **Tc^a** and **Tc^c** in Europids, **Tc^a** and **Tc^b** in Negrids, and **WES^a** and **WES^b** in Negrids and Finns.

The antigens of the **Cromer** system (see [7]) are absent from the red cells of the very rare **Inab** phenotype and are weakly expressed or lacking on **Dr(a–)** red cells. Further, erythrocytes of patients with paroxysmal nocturnal haemoglobinuria (see *Sect. 4.1.2)* also lack the **Cromer** antigens, or express them only very weakly.

The **Cromer** antigens are not confined to erythrocytes; they can also be found on peripheral blood leucocytes and platelets, as well as in several haematopoietic cell lines [36], and various tissue cells, such as kidney endothelium (Yendle et al. cited in [7]) or gastrointestinal mucosa [27]. Further, they have also been detected in serum and urine [7].

Table 21.1: Cromer antigens

Antigen	ISBT classification[1]	Frequency in Europids	References
Cr^a	CROM1	>99%	[37]
Tc^a	CROM2	>99%	[15]
Tc^b	CROM3	<1%	[14]
Tc^c	CROM4	<1%	[16]
Dr^a (Drori)	CROM5	>99%	[17,18]
Es^a	CROM6	>99%	[42]
IFC	CROM7	>99%	[8]
Wes^a	CROM8	<1%	[35]
Wes^b	CROM9	>99%	[9]
UMC	CROM10	>99%	[7]

[1] In numerical notation 021 001 to 021 010 (see *Chap. 1*)

From Lewis et al. [19], see *Chap. 1*, by permission of S. Karger AG, Basel.

Cromer antigens are destroyed by treatment of the red cells with α-chymotrypsin; but they are only partially degraded by pronase. In contrast to virtually all other protein-based blood group antigens, the serological activity of the **Cromer** antigens is not influenced by trypsin, papain, or ficin. The sulfhydryl-reagents 2-aminoethyl-iso-thiouronium bromide (AET) and dithiothreitol (DTT) cause only slight weakening of **Cromer** antigen expression on intact red cells [7].

Immunoblotting experiments with human alloantibodies have revealed that the **Cromer** antigens are located on the 'decay-accelerating factor' (DAF or CD55) [36,39]. DAF is widely distributed in peripheral blood cells [12,29,36], and in epithelial [27] and endothelial tissues [1]; soluble forms of DAF have further been detected in extracellular fluids, such as serum, saliva, urine, and tears [27].

DAF is one member of the complement regulatory proteins which are encoded by a group of linked genes on the long arm of human chromosome 1, band q32 [23,34]. The products of these genes down-regulate the activation of complement – they prevent the assembly and accelerate the decay of C3 and C5 convertases of both classical and alternative pathways, thus protecting the host tissue from complement-mediated damage [21].

The finding that all human erythrocytes can be agglutinated by fimbriae of gram-negative bacteria, such as 075X-positive *Escherichia coli*, suggested that **Cromer** antigens of kidney tissue function as receptors for the adhesion of uropathogenic *E. coli* strains [30]. The only exception are cells of the rare **Inab** phenotype (see below). It should also be mentioned that DAF acts as a receptor for several enteroviruses (e.g. [2,3,11].

DAF [21,28] is a 70 kDa glycophospholipid-anchored membrane protein which carries up to 30 highly sialylated O-linked carbohydrate chains (accounting for over 25 kDa in apparent molecular mass) and one N-linked carbohydrate unit of the complex type [22]. The molecular mass of the glycoprotein is lower than normal in **Tn** cells which are deficient in the O-glycosidically linked sialotetrasaccharides of red cell sialoglycoproteins (see *Sect. 10.6*), but is slightly above normal in **Cad** cells which are characterised by an N-acetylgalactosamine-containing pentasaccharide (see *Sect. 11.3*). These findings suggest that the O-linked carbohydrate chains are of the type found in the glycophorin molecules (see *Fig. 9.3.a*) [36].

The primary structure of the peptide moiety of the DAF molecule was determined by sequencing DAF cDNA clones isolated from HeLa and HL-60 cDNA libraries [5,26]. The deduced amino-acid sequence predicts a protein of 381 amino acids including a 34-amino-acid signal peptide (see *Fig. 21.1.a*). The N-terminus of the mature protein is composed of four adjacent short consensus repeat units (SCRs) of approximately

(a) GPI-anchored protein

```
-34                               mtva rpsvpaalpl lgelprllll vllclpavwg  -1

  1 DCGLPPDVPN AQPALEGRTS FPEDTVITYK CEESFVKIPG EKDSVTCLKG SQWSDIEEFC   60

 61 NRSCEVPTRL NSASLKQPYI TQNYFPVGTV VEYECRPGYR REPSLSPKLT CLQNLKWSTA  120

121 VEFCKKKSCP NPGEIRNGQI DVPGGILFGA TISFSCNTGY KLFGSTSSFC LISGSSVQWS  180

181 DPLPECREIY CPAPPQIDNG IIQGERDHYG YRQSVTYACN KGFTMIGEHS IYCTVNNDEG  240

241 EWSGPPPECR GKSLTSKVPP TVQKPTTVNV PTTEVSPTSQ KTTTKTTTPN AQATRSTPVS  300

301 RTTKHFHETT PNKGSGTTSG TTRLLSGHTC FTLTGLLGTL VTMGLLT              347
```

(b) 'Soluble protein':

```
301 RTTKHFHETT PNKGSGTTSG TTRLLSGSRP VTQAGMRWCD RSSLQSRTPG FKRSFHFSLP  360

361 SSWYYRAHVF HVDRFAWDAS NHGLADLAKE ELRRKYTQVY RLFLVS               406
```

Figure 21.1: The peptide sequence of human Decay Accelerating Factor.
N : N-glycosylation site. The serine / threonine-rich region (probable site for O-glycosylation) is underlined. The leader sequence (residues −34 to −1) is shown in small letters.
According to Caras et al. [5]. The sequence data of the DAF protein are deposited in the EMBL/Gen Bank data library (accession number M30142).

60 amino acids each (*Fig. 21.2*). These consensus repeat domains are followed by a 70-amino acid segment rich in serine and threonine carrying O-linked carbohydrate chains, and a hydrophobic segment which posttranslationally is replaced with the phosphatidylinositol-anchor. The N-glycan is linked to the asparagine residue at position 61, i.e. between the first and second consensus repeat unit (*Fig. 21.3*).

The SCR units (i.e. residues 1–63, 64–125, 126–187, 188–250) show significant homology to similar regions previously identified in C3b- and/or C4b-binding proteins and in some non-complement proteins, such as factor XIII, interleukin 2 receptor, and serum β_2-glycoprotein [26]. In analogy to the SCR domains of both β_2-glycoprotein and factor H, a structurally and functionally homologous serum protein, it has been suggested that each consensus repeat unit forms a compact domain with antiparallel β-sheets (see [21]); the four conserved cysteine residues in each SCR presumably being bonded Cys1 − Cys3 and Cys2 − Cys4 may provide a triple-loop structure (see *Sect. 22*).

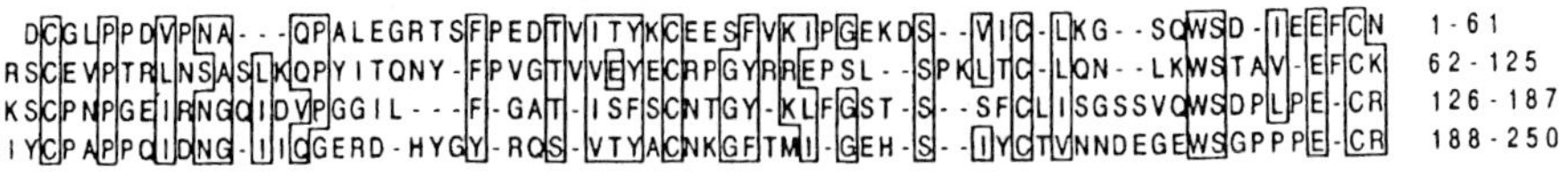

DCGLPPDVPNA---QPALEGRTSFPEDTVITYKCEESFVKIPGEKDS--VIC-LKG--SQWSD-IEEFCN 1-61
-RSCEVPTRLNSASLKQPYITQNY-FPVGTVVEYECRPGYRREPSL--SPKLTC-LQN--LKWSTAV-EFCK 62-125
KKSCPNPGEIRNGQIDVPGGIL---F-GAT-ISFSCNTGY-KLFGST-S--SFCLISGSSVQWSDPLPE-CR 126-187
EIYCPAPPQIDNG-IIQGERD-HYGY-RQS-VTYACNKGFTMI-GEH-S--IYCTVNNDEGEWSGPPPE-CR 188-250

Figure 21.2: Alignment of the four consensus repeat domains of DAF.
Conserved amino acids or conservative substitutions are indicated by boxes.
Reproduced from Lublin and Atkinson [21] by permission of the Annual Reviews Inc., Palo Alto, CA, USA.

NH_2
1
Tc^a
Es^a
WES^a
WES^b
BRIC 230
N-CHO
4 Consensus Repeat Domains
2
BRIC 110
3
Dr^a
BRIC 216
4
Cr^a
UMC
O-linked Carbohydrate Domain
CHO-O
O-CHO
CHO-O
O-CHO
CHO-O
O-CHO
Glycophospholipid Domain
Membrane

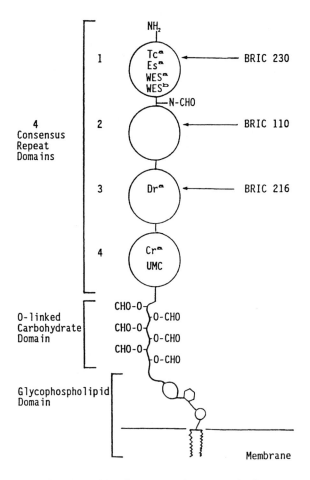

Figure 21.3: Localisation of the Cromer antigens on the four consensus repeat domains of the Decay Accelerating Factor.

According to Petty et al. [31] (modified) by permission of S. Karger AG, Basel.

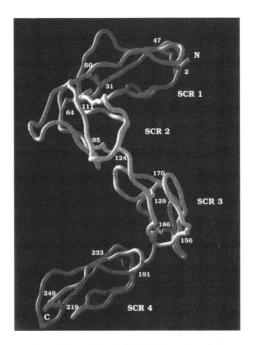

Figure 21.4: Three-dimensional model of the DAF molecule.
Backbone representation of the SCR domains.
Reproduced from Kuttner-Kondo et al. [13] by permission of Oxford University Press.

Further, based on the known coordinates of factor H, a three-dimensional model of the DAF molecule has been put forward by Kuttner-Kondo et al. [13]. According to this model the four SCRs are arranged in a helical mode. The 'pockets' formed by SCR2 and 3, as well as around the C-terminal disulphide bridges of SCR3 and 4 may function as ligands for the C3 convertases (*Fig. 21.4*). This assumption is corroborated by the finding that the regulatory function of the DAF molecule resides mainly within SCR-3 [4,6].

The O-glycosylated region serves as an important but non-specific spacer projecting the DAF functional domains above the plasma membrane. Although it represents only 20% of the peptide sequence, due to its linear rod-like structure it must be as long as all four highly folded SCR units together [6].

MAIEA assays and molecular biological studies have assigned the epitopes of the **Cromer** system to various domains of the DAF molecule – **Tc**[a], **WES**[b] (and **WES**[a]),

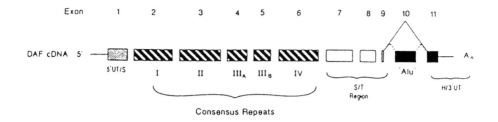

Figure 21.5: Structure of the DAF gene.
Reproduced from Post et al. [32] by permission of the authors and of The Journal of Immunology (© 1990).

and **Es**ᵃ to the first consensus repeat, **Dr**ᵃ to the third, and **Cr**ᵃ and **UMC** to the fourth consensus repeat or to the serine / threonine rich region of the DAF protein [25,31,40,41] (*Fig. 21.3*).

The human DAF gene has been thoroughly studied by Post and co-workers [32]. The gene spans ~40 kb and consists of 11 exons (*Fig. 21.5*): the first exon codes for the 5'-untranslated region and the signal peptide, the four consensus repeat units are encoded by exons 2, 3, 4 + 5, and 6 respectively, and the serine/threonine-rich domain is encoded by exons 7, 8, and 9. In the majority of mRNA's, exon 10 is removed by splicing, and exon 11 codes for the hydrophobic C-terminus, which is responsible for the attachment of the glycophospholipid anchor. The unspliced form of DAF mRNA codes for a 440-amino acid protein with a different and more hydrophilic C-terminus (*Fig. 21.1.b*); it has been proposed that this molecule represents the secreted form of DAF [5].

The rare **Cromer** phenotype **Dr(a–)** has been found as a recessive trait in individuals of four unrelated Israeli families (see [25]). Erythrocytes of this phenotype lack the **Dr**ᵃ antigen and have a reduced level of all the other **Cromer** antigens. Radioimmunoassay and flow cytometric analysis of **Dr(a–)** erythrocytes demonstrated a 60% reduction of DAF but normal levels of several other GPI-anchored proteins [25]. Despite the reduced DAF expression, **Dr(a–)** erythrocytes showed a near-normal sensitivity towards complement lysis, and no haematological abnormalities associated with the **Dr(a–)** phenotype have been reported.

Investigations on genomic DNA isolated from three unrelated **Dr(a–)** individuals showed that the **Dr(a–)** phenotype is characterised by a C to T change in nucleotide 661 [25,33]. This mutation, on the one hand, results in a serine to leucine substitution

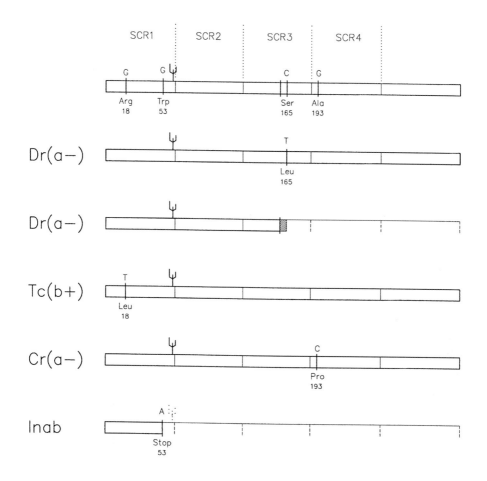

Figure 21.6: Schematic representation of the peptide sequence of DAF and the alleles responsible for Dr(a–), Tc(b+), Cr(a–), and Inab phenotypes.
⋎ : N-glycan;   ▨ : new sequence at the C-terminus.

at position 165[(1)] (*Fig. 21.6*) and, on the other hand, generates a cryptic branch point in the **Dr(a–)** allele that facilitates the use of a downstream cryptic acceptor splice site [24]. Alternative splicing using this cryptic splice site results in a 44-nucleotide deletion

---

[(1)] Though no respective antibody is known, a low-frequency allele **Dr**[b] antithetical to the high-frequency allele **Dr**[a] has been defined on the basis of this result (see *Chap. 1*).

starting at the 5'-end of exon 5; a concomitant alteration of the base sequence causes a shift in the reading frame and creates a premature termination codon six amino acids after the deletion (*Fig. 21.6*). **Dr(a–)** phenotype individuals are thus able to produce two species of mRNA from a single DAF gene – one encoding the full-length DAF molecule containing a single amino acid substitution and a second encoding a truncated version of DAF which lacks one and a half consensus repeat units as well as the glycosylphosphatidylinositol linkage site.

Semiquantitative estimates based on PCR experiments suggested that in **Dr(a–)** individuals only ~30–50% of the DAF mRNA encode the full-length molecule, whereas the majority of the mRNA results from alternative splicing [24]. These investigations thus explain on a molecular basis both the change in antigenicity and the reduced expression of DAF in the **Dr(a–)** phenotype.

Studies performed by Telen and co-workers revealed that an alanine → proline substitution at position 193 (i.e. caused by a G → C exchange in SCR-4) is responsible for **Cr(a–)** [41]. Further, **Tc^a** and **Tc^b** are characterised by arginine and leucine residues, respectively, at position 18 (i.e. caused by a G ↔ T exchange in SCR 1) [41] (*Fig. 21.6*).

Erythrocytes of **Inab** phenotype individuals lack all **Cromer** antigens [10]. This 'Cromer-null' phenotype is very rare – thus far only five **Inab** subjects have been reported [10,20,33,43]. The sera of **Inab** phenotype propositi contained an antibody (anti-**IFC** [8,33]) which represents a mixture of antibodies binding to different epitopes of the DAF protein.

In contrast to the erythrocytes of patients suffering from paroxysmal nocturnal haemoglobinuria (= PNH, see *Sect. 4.2.1*), **Inab** red cells and lymphocytes lack the Decay Accelerating Factor although they contain the other phosphatidylinositol-linked proteins in normal amounts [38]. **Inab** individuals show no clinical symptoms of increased haemolysis and have no documented haematological abnormalities. However, the finding that some **Inab** subjects were suffering from intestinal disorders (Crohn's disease [43] or protein-losing enteropathy [10], see also [33]) suggests that the DAF presence on the epithelial surface of gastrointestinal mucosa of healthy individuals [27] may play a critical role in human gastrointestinal physiology.

Molecular biological investigations on the DFA-encoding mRNA of an **Inab** propositus [24] revealed a G to A substitution which changes the codon for ^{53}Trp (= TGG) to a stop codon (= TGA). The protein is thus truncated after the SCR1 domain (see *Figs. 21.6* and *21.7*).

In another **Inab** proband a single nucleotide substitution has been found (C^{1579} → A) close to the 3'-end of exon 2 [44]. The substitution causes the activation

Normal

```
 -  G   S   Q   W   S   D   I   E   E   F   C   N   R   S   C   E   V  -
 - GGC AGT CAA TGG TCA GAT ATT GAA GAG TTC TGC AAT CGT AGC TGC GAG GTG -
```

Inab N.N. [1]

```
 -  G   S   Q   *
 - GGC AGT CAA TGA TCA GAT ATT GAA GAG TTC TGC AAT CGT AGC TGC GAG GTG -
```

Inab H.A. [2]

```
 -  G   S   Q   W                                        *
 - GGC AGT CAA TG. ... ... ... ... ... ... ... ... .G TAG CTG CGA GGT G -
```

```
 ——————————————————————— SCR1 (exon2) ——————————>| <— SCR2 (exon3) ——
```

Figure 21.7: Mutations in the DAF gene responsible for the Inab phenotype.
Fragment of the DAF peptide sequence between amino acid residues 50 and 66. *: stop code.
[1] Inab mutation according to Lublin et al. [24]; [2] Inab mutation according to Wang et al. [44].

of a cryptic splice site and results in the production of mRNA with a 26 bp deletion. This deletion introduces a reading frame shift and creates a stop codon immediately downstream from the deletion. Translation of mRNA would be terminated at the first amino acid residue of SCR2 (*Fig. 21.7*).

Thus, both defective genes encode a truncated version of the DAF protein in which the functional domains of complement regulatory activity (SCR2 – SCR4) are absent. Further, the lack of the C-terminal signal domain for glycosylphosphatidylinositol (GPI) anchoring explains the complete lack of membrane DAF in **Inab** phenotype erythrocytes.

References

1. ASCH, A. S., KINOSHITA, T., JAFFE, E. A. & NUSSENZWEIG, V. (1986): Decay-accelerating factor is present on cultured human umbilical vein endothelial cells. *J. Exp. Med. 163*, 221-226.
2. BERGELSON, J. M., CHAN, M., SOLOMON, K. R., JOHN, N. F. S., LIN, H. & FINBERG, R. W. (1994): Decay-accelerating factor (CD55), a glycosylphosphatidylinositol-anchored complement regulatory protein, is a receptor for several echoviruses. *Proc. Natl. Acad. Sci. USA 91*, 6245-6248.
3. BERGELSON, J. M., MOHANTY, J. G., CROWELL, R. L., JOHN, N. F. S., LUBLIN, D. M. & FINBERG, R. W. (1995): Coxsackievirus B3 adapted to growth in RD cells binds to decay-accelerating factor (CD55). *J. Virol. 69*, 1903-1906.

4. BRODBECK, W. G., LIU, D. C., SPERRY, J., MOLD, C. & MEDOF, M. E. (1996): Localization of classical and alternative pathway regulatory activity within the decay-accelerating factor. *J. Immunol. 156*, 2528-2533.

5. CARAS, I. W., DAVITZ, M. A., RHEE, G., WEDDELL, G. N., MARTIN, D. W. & NUSSENZWEIG, V. (1987): Cloning of decay-accelerating factor suggests novel use of splicing to generate two proteins. *Nature 325*, 545-549.

6. COYNE, K. E., HALL, S. E., THOMPSON, E. S., ARCE, M. A., KINOSHITA, T., FUJITA, T., ANSTEE, D. J., ROSSE, W. & LUBLIN, D. M. (1992): Mapping of epitopes, glycosylation sites, and complement regulatory domains in human decay accelerating factor. *J. Immunol. I49*, 2906-2913.

7. DANIELS, G. (1989): Cromer-related antigens – blood group determinants on decay-accelerating factor. *Vox Sang. 56*, 205-211.

8. DANIELS, G. & WALTHERS, L. (1986): Anti-IFC, an antibody made by Inab phenotype individuals. *Transfusion 26*, 117-118.

9. DANIELS, G. L., GREEN, C. A., DARR, F. W., ANDERSON, H. & SISTONEN, P. (1987): A 'new' Cromer-related high frequency antigen probably antithetical to WES. *Vox Sang. 53*, 235-238.

10. DANIELS, G. L., TOHYAMA, H. & UCHIKAWA, M. (1982): A possible null phenotype in the Cromer blood group complex. *Transfusion 22*, 362-363.

11. KARNAUCHOW, T. M., TOLSON, D. L., HARRISON, B. A., ALTMAN, E., LUBLIN, D. M. & DIMOCK, K. (1996): The HeLa cell receptor for enterovirus 70 is decay-accelerating factor (CD55). *J. Virol. 70*, 5143-5152.

12. KINOSHITA, T., MEDOF, M. E., SILBER, R. & NUSSENZWEIG, V. (1985): Distribution of decay-accelerating factor in the peripheral blood of normal individuals and patients with paroxysmal nocturnal hemoglobinuria. *J. Exp. Med. 162*, 75-92.

13. KUTTNER-KONDO, L., MEDOF, M. E., BRODBECK, W. & SHOHAM, M. (1996): Molecular modeling and mechanism of action of human decay-accelerating factor. *Protein Eng. 9*, 1143-1149.

14. LACEY, P. A., BLOCK, U. T., LAIRD-FRYER, B. J., MOULDS, J. J., BRYANT, L. R., GIANDELONE, J. A. & LINNEMEYER, D. R. (1985): Anti-Tc^b, an antibody that defines a red cell antigen antithetical to Tc^a. *Transfusion 25*, 373-376.

15. LAIRD-FRYER, B., DUKES, C. V., LAWSON, J., MOULDS, J. J., WALKER, E. M. & GLASSMAN, A. B. (1983): Tc^a: a high frequency blood group antigen. *Transfusion 23*, 124-127.

16. LAW, J., JUDGE, A., COVERT, P., LEWIS, N., SABO, B. & MCCREARY, J. (1982): A new low frequency factor proposed to be the product of an allele to Tc^a. (Abstract). *Transfusion 22*, 413.

17. LEVENE, C., HAREL, N., KENDE, G., PAPO, S., BRADFORD, M. F. & DANIELS, G. L. (1987): A second Dr(a-) proposita with anti-Dr^a and a family with Dr(a-) in 2 generations. *Transfusion 27*, 64-65.

18. LEVENE, C., HAREL, N., LAVIE, G., GREENBERG, S., LAIRD-FRYER, B. & DANIELS, G. L. (1984): A 'new' phenotype confirming a relationship between Cr^a and Tc^a. *Transfusion 24*, 13-15.

19. LEWIS, M., ANSTEE, D. J., BRID, G. W. G., BRODHEIM, E., CARTRON, J. P., CONTRERAS, M., CROOKSTON, M. C., DAHR, W., DANIELS, G. L., ENGELFRIET, C. P., GILES, C. M., ISSITT, P. D., JØRGENSEN, J., KORNSTAD, L., LUBENKO, A., MARSH, W. L., MCCREARY, J., MOORE, B. P. L., MOREL, P., MOULDS, J. J., NEVANLINNA, H., NORDHAGEN, R., OKUBO, Y., ROSENFIELD, R. E., ROUGER, P., RUBINSTEIN, P., SALMON, C., SEIDL, S., SISTONEN, P., TIPPETT, P., WALKER, R. H., WOODFIELD, G. & YOUNG, S. (1990): Blood group terminology 1990. *Vox Sang. 58*, 152-169.

20. LIN, R. C., HERMAN, J., HENRY, L. & DANIELS, G. L. (1988): A family showing inheritance of the Inab phenotype. *Transfusion 28*, 427-429.

21. LUBLIN, D. M. & ATKINSON, J. P. (1989): Decay-accelerating factor: biochemistry, molecular biology, and function. *Annu. Rev. Immunol. 7*, 35-58.

22. LUBLIN, D. M., KRSEK-STAPLES, J., PANGBURN, M. K. & ATKINSON, J. P. (1986): Biosynthesis and glycosylation of the human complement regulatory protein decay accelerating factor. *J. Immunol. 137*, 1629-1635.

23. LUBLIN, D. M., LEMONS, R. S., LE BEAU, M. M., HOLERS, V. M., TYKOCINSKI, M. L., MEDOF, M. E. & ATKINSON, J. P. (1987): The gene encoding decay-accelerating factor (DAF) is located in the complement-regulatory locus on the long arm of chromosome 1. *J. Exp. Med. 165*, 1731-1736.

24. LUBLIN, D. M., MALLINSON, G., POOLE, J., REID, M. E., THOMPSON, E. S., FERDMAN, B. R., TELEN, M. J., ANSTEE, D. J. & TANNER, M. J. A. (1994): Molecular basis of reduced or absent expression of decay-accelerating factor in Cromer blood group phenotypes. *Blood 84*, 1276-1282.
25. LUBLIN, D. M., THOMPSON, E. S., GREEN, A. M., LEVENE, C. & TELEN, M. J. (1991): Dr(a-) polymorphism of decay accelerating factor. Biochemical, functional, and molecular characterization and production of allele-specific transfectants. *J. Clin. Invest. 87*, 1945-1952.
26. MEDOF, M. E., LUBLIN, D. M., HOLERS, V. M., AYERS, D. J., GETTY, R. R., LEYKAM, J. F., ATKINSON, J. P. & TYKOCINSKI, M. L. (1987): Cloning and characterization of cDNAs encoding the complete sequence of decay-accelerating factor of human complement. *Proc. Natl. Acad. Sci. USA 84*, 2007-2011.
27. MEDOF, M. E., WALTER, E. I., RUTGERS, J. L., KNOWLES, D. M. & NUSSENZWEIG, V. (1987): Identification of the complement decay-accelerating factor (DAF) on epithelium and glandular cells and in body fluids. *J. Exp. Med. 165*, 848-864.
28. NICHOLSON-WELLER, A., BURGE, J., FEARON, D. T., WELLER, D. T. & AUSTEN, K. F. (1982): Isolation of human erythrocyte membrane glycoprotein with decay-accelerating activity for C3 convertases of the complement system. *J. Immunol. 129*, 184-189.
29. NICHOLSON-WELLER, A., MARCH, J. P., ROSEN, C. E., SPICER, D. B. & AUSTEN, K. F. (1985): Surface membrane expression by human blood leukocytes and platelets of decay-accelerating factor, a regulatory protein of the complement system. *Blood 69*, 1237-1244.
30. NOWICKI, B., MOULDS, J., HULL, R. & HULL, S. (1988): A hemagglutinin of uropathogenic Escherichia coli recognizes the Dr blood group antigen. *Infect. Immun. 56*, 1057-1060.
31. PETTY, A. C., DANIELS, G. L., ANSTEE, D. J. & TIPPETT, P. (1993): Use of the MAIEA technique to confirm the relationship between the Cromer antigens and decay-accelerating factor and to assign provisionally antigens to the short-consensus repeats. *Vox Sang. 65*, 309-315.
32. POST, T. W., ARCE, M. A., LISZEWSKI, M. K., THOMPSON, E. S., ATKINSON, J. P. & LUBLIN, D. M. (1990): Structure of the gene for human complement protein decay acceleration factor. *J. Immunol. 144*, 740-744.
33. REID, M. E., MALLINSON, G., SIM, R. B., POOLE, J., PAUSCH, V., MERRY, A. H., LIEW, Y. W. & TANNER, M. J. A. (1991): Biochemical studies on red blood cells from a patient with the Inab phenotype (decay-accelerating factor deficiency). *Blood 78*, 3291-3297.
34. REY-CAMPOS, J., RUBINSTEIN, P. & RODRIGUEZ DE CORDOBA, S. (1987): Decay accelerating factor. Genetic polymorphism and linkage to the RCA (regulator of complement activation) gene cluster in humans. *J. Exp. Med. 166*, 246-252.
35. SISTONEN, P., NEVANLINNA, H. R., VIRTARANTA-KNOWLES, K., TUOMINEN, I., PIRKOLA, A., GREEN, C. A. & TIPPETT, P. (1987): WES, a 'new' infrequent blood group antigen in Finns. *Vox Sang. 52*, 111-114.
36. SPRING, F. A., JUDSON, P. A., DANIELS, G. L., PARSONS, S. F., MALLINSON, G. & ANSTEE, D. J. (1987): A human cell-surface glycoprotein that carries Cromer-related blood group antigens on erythrocytes and is also expressed on leucocytes and platelets. *Immunology 62*, 307-313.
37. STROUP, M. & MCCREARY, J. (1975): Cr^a another high frequency blood group factor. (Abstract). *Transfusion 15*, 522.
38. TATE, G., UCHIKAWA , M., TANNER, M. J. A., JUDSON, P. A., PARSONS, S. F., MALLINSON, G. & ANSTEE, D. J. (1989): Studies on the defect which causes absence of decay accelerating factor (DAF) from the peripheral blood cells of an individual with the Inab phenotype. *Biochem. J. 261*, 489-493.
39. TELEN, M. J., HALL, S. E., GREEN, A. M., MOULDS, J. J. & ROSSE, W. F. (1988): Identification of human erythrocyte blood group antigens on decay-accelerating factor (DAF) and an erythrocyte phenotype negative for DAF. *J. Exp. Med. 167*, 1993-1998.
40. TELEN, M. J., RAO, N. & LUBLIN, D. M. (1995): Location of WES^b on decay-accelerating factor. *Transfusion 35*, 278.
41. TELEN, M. J., RAO, N., UDANI, M., THOMPSON, E. S., KAUFMAN, R. M. & LUBLIN, D. M. (1994): Molecular mapping of the Cromer blood group Cr^a and Tc^a epitopes of decay accelerating factor:

toward the use of recombinant antigens in immunohematology. *Blood 84*, 3205-3211.

42. TREGELLAS, W. M. (1984): Description of a new blood group antigen, Es^a. *18th Congress of the International Society of Blood Transfusion*, Karger, Basel, Abstracts p. 163.

43. WALTHERS, L., SALEM, M., TESSEL, J., LAIRD-FRYER, B. & MOULDS, J. J. (1983): The Inab phenotype: another example found. (Abstract). *Transfusion 23*, 423.

44. WANG, L., UCHIKAWA, M., TSUNEYAMA, H., TOKUNAGA, K., TADOKORO, K. & JUJI, T. (1998): Molecular cloning and characterization of decay-accelerating factor deficiency in Cromer blood group Inab phenotype. *Blood 91*, 680-684.

22 Dombrock System

The **Dombrock** blood group system originally comprised two antithetical antigens, **Do^a** [11] and **Do^b** [5]. More recent reports, however, showed that the **Dombrock** characters and the antigens **Jo^a (Joseph)** [4], **Hy (Holley)** [7], and **Gy^a (Gregory)** [10] are phenotypically closely related [6,13] for the following reasons:

- all these antigens are absent on **Gy(a–)** cells, suggesting that **Gy(a–)** is the null phenotype for these blood group characters. Lack of **Dombrock, Joseph,** or **Holley** antigens has a varying but nonetheless significant influence on the expression of the other antigens;
- the characters **Do^a, Hy,** and **Gy^a** are destroyed *in situ* by pronase and trypsin, and to a lesser extent by α-chymotrypsin [9]; they are resistant, however, to digestion by papain. Further, the antigens are sensitive to treatment with reducing agents;
- biochemical studies have provided evidence that all these characters reside on the same erythrocyte membrane glycoprotein [1,2,9].

Based on these observations, the 'ISBT Working Party on Terminology for Red Cell Surface Antigens' has assigned the antigens **Gy^a, Hy,** and **Jo^a** to the **Dombrock** blood group system[1].

The *DO* gene has been located on chromosome 12 in the region between 12p13.2 and 12p12.1 [3].

Immunoblotting experiments [8] show that the **Dombrock** antigens reside on an erythrocyte membrane glycoprotein which in SDS-polyacrylamidegel electrophoresis migrates as a broad band of an apparent molecular mass of 46,750–57,500 Da. Reduction in molecular mass to approximately 11 kDa after treatment with endoglycosidase F suggests that the molecule carries N-linked carbohydrate units. The fact that the **Dombrock** antigens are present in reduced quantity on red cell

---

[1] The ISBT numbers are **DO1**(= 014 001) for **Do^a**, **DO2** (= 014 002) for **Do^b**, **DO3** (= 014 003, formerly 206 001) for **Gy^a**, **DO4** (= 014 004, formerly 206 002) for **Hy**, and **DO5** (= 014 005, formerly 900 004) for **Jo^a** (see *Chap. 1*).

membranes from patients with paroxysmal nocturnal haemoglobinuria suggests that these characters reside on a phosphatidylinositol-linked membrane component (see *Sect. 4.1.2)* [8,12]. The protein, however, has not yet been further characterised.

References

1. BANKS, J. A., HEMMING, N. & POOLE, J. (1995): Evidence that the Gy^a, Hy and Jo^a antigens belong to the Dombrock blood group system. *Vox Sang. 68,* 177-182.
2. BANKS, J. A., PARKER, N. & POOLE, J. (1992): Evidence to show that Dombrock (Do) antigens reside on the Gy^a/Hy glycoprotein. (Abstract). *Transfus. Med. 2 (Supl. 1),* 68.
3. EIBERG, H. & MOHR, J. (1996): Dombrock blood group (DO): assignment to chromosome 12p. *Hum. Genet. 98,* 518-521.
4. JENSEN, L., SCOTT, E. P., MARSH, W. L., MACILROY, M., ROSENFIELD, R. E., BRANCATO, P. & FAY, A. F. (1972): Anti-Jo^a: an antibody defining a high-frequency erythrocyte antigen. *Transfusion 12,* 322-324.
5. MOLTHAN, L., CRAWFORD, M. N. & TIPPETT, P. (1973): Enlargement of the Dombrock blood group system: the finding of anti-Do^b. *Vox Sang. 24,* 382-384.
6. MOULDS, J. J., POLESKY, H. F., REID, M. E. & ELLISOR, S. S. (1975): Observations on the Gy^a and Hy antigens and the antibodies that define them. *Transfusion 15,* 270-274.
7. SCHMIDT, R. P., FRANK, S. & BAUGH, M. (1967): New antibodies to high incidence antigenic determinants (anti-So, anti-El, anti-Hy, and anti-Dp). (Abstract). *Transfusion 7,* 386.
8. SPRING, F. A. & REID, M. E. (1991): Evidence that the human blood group antigens Gy^a and Hy are carried on a novel glycosylphosphatidylinositol-linked erythrocyte membrane glycoprotein. *Vox Sang. 60,* 53-59.
9. SPRING, F. A., REID, M. E. & NICHOLSON, G. (1994): Evidence for expression of the Jo^a blood group antigen on the Gy^a/Hy-active glycoprotein. *Vox Sang. 66,* 72-77.
10. SWANSON, J., ZWEBER, M. & POLESKY, H. F. (1967): A new public antigenic determinant Gy^a (Gregory). *Transfusion 7,* 304-306.
11. SWANSON, J. L., POLESKY, H. F., TIPPETT, P. & SANGER, R. (1965): A 'new' blood group antigen, Do^a. *Nature 206,* 313.
12. TELEN, M. J., ROSSE, W. F., PARKER, C. J., MOULDS, M. K. & MOULDS, J. J. (1990): Evidence that several high-frequency human blood group antigens reside on phosphatidylinositol-linked erythrocyte membrane proteins. *Blood 75,* 1404-1407.
13. WEAVER, T., KAVITSKY, D., CARTY, L., DAH, L. K. E., MARCHESE, M., HARRIS, M., DRAPER, E. & BALLAS, S. K. (1984): An association between the Jo^a and Hy phenotypes. (Abstract). *Transfusion 24,* 426.

23 Yt System

The **Yt** blood group system[1] includes two antithetical antigens, **Yt^a** [4] and **Yt^b** [6][2] with the respective gene-frequencies of 0.96 and 0.04 [8].

The **Yt^a** character of human erythrocytes is destroyed by chymotrypsin but is resistant towards trypsin [3]. **Yt^a** and **Yt^b** are inactivated by sulfhydryl reagents [2,10].

When it was discovered that erythrocytes of patients with paroxysmal nocturnal haemoglobinuria lack the **Yt** antigens, this seemed to suggest that the carrier of the **Yt** antigens must be a phosphatidylinositol-anchored erythrocyte membrane protein (see *Sect. 4.1.2)* [12]. Indeed, more recent immunoprecipitation experiments using anti-**Yt^a** antibodies have confirmed this assumption: they show that the **Yt** antigens reside on the acetylcholine esterase of erythrocyte membranes [9,11,13].

Quantitative binding assays using Fab fragments of murine anti-acetylcholine esterase hybridoma antibodies showed 7000–10,000 sites per cell.

Acetylcholine esterase is a well-characterised enzyme which plays a pivotal role in cholinergic neurotransmission at neuromuscular junctions (terminating nerve impulse transmission by acetylcholine). A second isoform of the enzyme occurs on erythrocytes and leucocytes, where its function is still not known.

Acetylcholine esterase is a 72 kDa glycoprotein which in human erythrocytes is predominantly expressed as a disulphide-bonded dimer [7].

Sequence analysis of the cDNA encoding the haematopoietic form suggested a polypeptide of 607 amino acids, including a 31-amino-acid leader peptide and a 29-amino-acid hydrophobic domain which in the mature protein is replaced by the phosphatidylinositol anchor [1] *(Fig. 23.1.a)*. Acetylcholine esterase contains three N-linked carbohydrate chains.

The gene encoding acetylcholine esterase (= **ACHE**) has been located on chromosome 7, position q22.1–q22.3 [5,14]. It consists of six exons: exons 1–4

---

[1] The alternatively used term **Cartwright** has not been acknowledged by the ISBT working party.

[2] The ISBT numbers are **YT1** (= 011 001) for **Yt^a** and **YT2** (= 011 002) for **Yt^b**.

(a) Haematopoietic form:

```
 -31                                  m rppqcllhtp slasplllll lwllgggvga  -1

   1  EGREDAELLV TVRGGRLRGI RLKTPGGPVS AFLGIPFAEP PMGPRRFLPP EPKQPWSGVV  60

  61  DATTFQSVCY QYVDTLYPGF EGTEMWNPNR ELSEDCLYLN VWTPYPRPTS PTPVLVWIYG  120

 121  GGFYSGASSL DVYDGRFLVQ AERTVLVSMN YRVGAFGFLA LPGSREAPGN VGLLDQRLAL  180

 181  QWVQENVAAF GGDPTSVTLF GESAGAASVG MHLLSPPSRG LFHRAVLQSG APNGPWATVG  240

 241  MGEARRRATQ LAHLVGCPPG GTGGNDTELV ACLRTRPAQV LVNHEWHVLP QESVFRFSFV  300

 301  PVVDGDFLSD TPEALINAGD FHGLQVLVGV VKDEGSYFLV YGAPGFSKDN ESLISRAEFL  360

 361  AGVRVGVPQV SDLAAEAVVL HYTDWLHPED PARLREALSD VVGDHNVVCP VAQLAGRLAA  420

 421  QGARVYAYVF EHRASTLSWP LWMGVPHGYE IEFIFGIPLD PSRNYTAEEK IFAQRLMRYW  480

 481  ANFARTGDPN EPRDPKAPQW PPYTAGAQQY VSLDLRPLEV RRGLRAQACA FWNRFLPKLL  540

 541  SATASEAPST CPGFTHGeaa rrpglplpll llhqllllfl shlrrl
```

(b) Hydrophilic form:

```
 541  SATDTLDEAE RQWKAEFHRW SSYMVHWKNQ FDHYSKQDRC SDL
```

Fig. 23.1: Peptide sequence of human erythrocyte membrane acetylcholinesterase.
N : potential N-glycosylation sites. The signal peptide and the hydrophobic sequence removed during the attachment of the GPI anchor are written in small letters. The catalytic site of the enzyme is underlined.
According to Bartels et al. [1]. The sequence data of the hydrophilic form of the acetylcholinesterase molecule are deposited in the EMBL/Gen Bank data library (accession number M55040).

encode the signal peptide as well as the 558 amino acids shared by both isoforms of the enzyme; alternative splicing of exons 5 and 6 accounts for the structural divergence of the isoforms – exon 5 codes for the precursor form of glycolipid-anchored acetylcholine esterase expressed in haematopoietic cells, whereas exon 6 encodes the hydrophilic form of the enzyme found in mammalian brain and muscle [1] *(Fig. 23.1.b).*

Molecular genetic studies have revealed that **Yt**[a] and **Yt**[b] are characterised by a C ↔ A exchange in exon 2, which causes a histidine ↔ asparagine dimorphism at position 322 [1]. A silent mutation in codon 446 does not alter the amino acid sequence. Position 561 displays a proline ↔ arginine dimorphism independent of the **Yt** phenotype. It is situated within the 29 residue peptide domain which is cleaved from the precursor protein in the course of the attachment of the glycolipid anchor.

Thus far only one case of acquired **Yt(a–b–)** phenotype has been described [9]; this probably developed in connection with an antibody directed towards the acetylcholine esterase molecule.

References

1. BARTELS, C. F., ZELINSKI, T. & LOCKRIDGE, O. (1993): Mutation at codon 322 in the human acetylcholinesterase (ACHE) gene accounts for YT blood group polymorphism. *Amer. J. Hum. Genet. 52*, 928-936.
2. BRANCH, D. R., MUENSCH, H. A., SY SIOK HAN, A. L. & PETZ, L. D. (1983): Disulfide bonds are a requirement for Kell and Cartwright (Yt^a) blood group antigen integrity. *Brit. J. Haematol. 54*, 573-578.
3. DANIELS, G. (1992): Effect of enzymes on and chemical modifications of high-frequency red cells antigens. *Immunohaematology 8*, 53-57.
4. EATON, B. R., MORTON, J. A., PICKLES, M. M. & WHITE, K. E. (1956): A new antibody, anti-Yt^a, characterising a blood group antigen of high incidence. *Brit. J. Haematol. 2*, 333-335.
5. GETMAN, D. K., EUBANKS, J. H., CAMP, S., EVANS, G. A. & TAYLOR, P. (1992): The human gene encoding acetylcholinesterase is located on the long arm of chromosome 7. *Amer. J. Hum. Genet. 51*, 170-177.
6. GILES, C. M. & METAXAS, M. N. (1964): Identification of the predicted blood group antibody anti-Yt^b. *Nature 202*, 1122-1123.
7. OTT, P. (1985): Membrane acetylcholinesterases: purification, molecular properties, and interactions with amphiphilic environments. *Biochim. Biophys. Acta 822*, 375-392.
8. RACE, R. R. & SANGER, R. (1975): The Yt blood groups. In: *Blood Groups in Man*. Blackwell Scientific Publications, Oxford, pp. 379-382.
9. RAO, N., WHITSETT, C. F., OXENDINE, S. M. & TELEN, M. J. (1993): Human erythrocyte acetylcholinesterase bears the Yt^a blood group antigen and is reduced or absent in the Yt(a-b-) phenotype. *Blood 81*, 815-819.
10. SHULMAN, I. A., NELSON, J. M. & LAM, H. T. (1986): Loss of Yt^b antigen activity after treatment of red cells with either dithiothreitol or 2-mercaptoethanol. *Transfusion 26*, 214.
11. SPRING, F. A., GARDNER, B. & ANSTEE, D. J. (1992): Evidence that the antigens of the Yt blood group system are located on human erythrocyte acetylcholinesterase. *Blood 80*, 2136-2141.
12. TELEN, M. J., ROSSE, W. F., PARKER, C. J., MOULDS, M. K. & MOULDS, J. J. (1990): Evidence that several high-frequency human blood group antigens reside on phosphatidylinositol-linked erythrocyte membrane proteins. *Blood 75*, 1404-1407.
13. TELEN, M. J. & WHITSETT, C. F. (1992): Erythrocyte acetylcholinesterase bears the Cartwright blood group antigens. *Clin. Res. 40*, 170A.
14. ZELINSKI, T., WHITE, L., COGHLAN, G. & PHILIPPS, S. (1991): Assigment of the YT blood group locus to chromosome 7q. *Genomics 11*, 165-167.

24 Indian System

The **Indian** blood group system comprises two specificities, **In**^{a} (= **IN1**) and **In**^{b} (= **IN2**)[(1)]. The **In**^{a} antigen is extremely rare in Europids but is found in 2–4% of Indians [4] and in more than 10% of some Arab and Iranian populations [5]; it has been described by Badakere et al. [3,4] The antithetical antigen **In**^{b}, a common character in Europids, has been defined by Giles; it was found identical to the antigen originally specified by the serum 'Salis' [17].

Anti-In^a and **Anti-In**^{b} alloantibodies are commonly produced in **In**^{b} and **In**^{a} homozygotes respectively, as a response to blood transfusion or pregnancy [4,6,17].

A murine hybridoma antibody with anti-**In**^{b} specificity has been described [43]. Further, several hybridoma antibodies have been produced [11,40] towards **CD44**, which is the carrier of the **In**^{a}/**In**^{b} antigens.

The **In**^{a} and **In**^{b} antigens have been detected on erythrocytes, granulocytes, lymphocytes, and on various haematopoietic cell lines [40]. The expression of both antigens is down-regulated by the dominant inhibitor of **Lutheran** system antigens (= 'Lutheran inhibitor') which is inherited by the rare dominant suppressor gene *In(Lu)* (see *Chap. 18*) [44,46,47].

The **Indian** antigens are sensitive to digestion with pronase, trypsin, and chymotrypsin, and are readily destroyed by disulfide-reducing agents, such as 2-aminoethylisothiouronium bromide (= AET) and dithiothreitol [12,26,40].

Immunoblotting experiments [40,45] have located the **In**^{a} and **In**^{b} blood group antigens on an 80–kDa membrane glycoprotein which proved identical to the glycoprotein previously defined not only by hybridoma antibodies towards the **CD44** antigen, the 'lymphocyte homing receptor' [8], but also by the 'Hermes' antibodies towards a lymphocyte adhesion antigen [23,36]. The same molecule has also been described as 'human brain-granulocyte-T-lymphocyte antigen' [10], 'In(Lu)-related p80 glycoprotein' [44,46], 'Pgp-1 antigen' (= phagocytic glycoprotein-1') [22], 'p85

[(1)] The six-digit ISBT identification numbers (see *Chap. 1*) are 023 001 for **In**^{a} and 023 002 for **In**^{b}.

glycoprotein' [30,37], and the 'extracellular matrix receptor' (ECMRIII) or 'class III collagen receptor' (CRIII) [7,16].

The **CD44** glycoprotein is not only present on all circulating haematopoietic cells with the exception of platelets, but is also found in various tissue cells such as most mesenchymal cells (fibroblasts and smooth muscle), neuroectodermal cells, and endothelial and defined epithelial cells [15,34]. A highly homologous – if not identical – protein has also been detected in serum [31].

Moreover, the same glycoprotein is strongly expressed in different tumour tissues and in a spectrum of tumour cell lines of lymphoid, monocytic, epithelial, glial, and melanocytic origin [34,40].

CD44 is capable of binding to extracellular matrix components, such as hyaluronic acid [2,42], fibronectin [24], collagen types I and VI [7], heparan sulfate, [7], and the cytoskeletal protein ankyrin [27].

The broad tissue distribution of **CD44** and its sequence relationship to cartilage link proteins (see below) suggest that this membrane constituent plays a general role in the adhesion of cell surfaces to extracellular glycosaminoglycan or collagen matrices. In lymphocytes **CD44** is involved in mediating lymphocyte – endothelial cell interactions in the course of the recirculation of lymphocytes between the cardiovascular and lymphatic systems ('lymphocyte homing') [23,25]. Further studies have provided evidence that tumour cells in which **CD44** levels are highly elevated use this same mechanism for metastatic spreading [19,20]. It has also been suggested that **CD44** plays an important role in lymphocyte activation [13,21], control of lymphopoiesis [32], and cell – fibronectin interactions during erythropoiesis in haematopoietic progenitor cells [49].

CD44 is an acidic glycoprotein (pI ~4.2) in which the major part of the negative charge is due to the presence of neuraminic acid and, to a lesser degree to the presence of sulfate groups. When the protein was isolated from lymphocytes and treated with endoglycosidases, it was recognised that the molecule bears 3–4 N-linked and several O-linked carbohydrate chains, the structures of which have not yet been investigated [25]. Some forms of the **CD44** molecule contain covalently linked chondroitin-sulfate.

Using radioiodinated Fab fragments, Anstee et al. [1] estimated a number of 6,000 to 10,000 copies of **CD44** per red cell.

A large number of **CD44** isoforms have been detected [29]. These isoforms share the transmembrane region and the N-terminal 150 amino acids, but differ in the

```
-20                                           MDKFWWHAAW GLCLVPLSLA  -1

  1 QIDLNITCRF AGVFHVEKNG RYSISRTEAA DLCKAFNSTL PTMAQMEKAL SIGFETCRYG  60

 61 FIEGHVVIPR IHPNSICAAN NTGVYILTYN TSQYDTYCFN ASAPPEEDCT SVTDLPNAFD 120

121 GPITITIVNR DGTRYVQKGE YRTNPEDIYP SNPTDDDVSS GSSSERSSTS GGYIFYTFST 180

181 VHPIPDEDSP WITDSTDRIP ATRDQDTFHP SGGSHTTHES ESDGHSHGSQ EGGANTTSGP 240

241 IRTPQIPEWL IILASLLALA LILAVCIAVN SRRRCGQKKK LVINSGNGAV EDRKPSGLNG 300

301 EASKSQEMVH LVNKESSETP DQFMTADETR NLQNVDMKIG V 341
```

Figure 24.1: Amino acid sequence of the haematopoietic isoform of the CD44 protein.
N : sites of potential N-glycosylation, SG : sites of potential O-glycosylation, the hydrophobic segment
representing the transmembrane domain is underlined.
According to Stamenkovic et al. [41]. The sequence data are deposited in the EMBL/GenBank data library
(accession number M24915).

membrane-proximal extracellular domain. The main cell-specific variants of **CD44** are
the haematopoietic or lymphoid form (**CD44H** = CD44S), which is an 80–90 kDa
species expressed by cells of mesodermal origin and some carcinoma cell lines.
CD44E is the main epithelial form, a 160 kDa species which is weakly expressed by
normal epithelium but highly expressed in carcinomas [41]. Several minor isoforms
ranging from 52 kDa to 200 kDa have also been detected on different cell types.

The primary structure of the haematopoietic form of this glycoprotein has been
deduced from the nucleotide sequence of cDNA isolated from lymphoblastoid cell lines
[18,41]:

The amino acid sequence of the haematopoietic form as published by
Stamenkovich and co-workers [41] predicts a mature protein with a leader sequence
of 20 amino acids, an extracellular N-terminal domain of 248 residues, a hydrophobic
transmembrane domain of 21 amino acids, and a C-terminal cytoplasmic tail of 72
amino acids (*Fig. 24.1*). The sequence published by Goldstein and co-workers [18] is
identical to this structure except for two amino acid substitutions in the N-terminal
domain and the occurrence of a short cytoplasmic tail comprising only three residues.

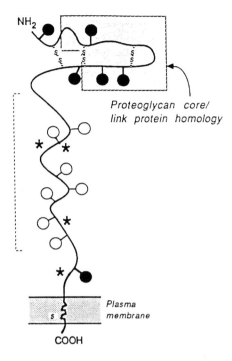

NH₂

Proteoglycan core/
link protein homology

Plasma
membrane

COOH

Figure 24.2: Schematic drawing of the predicted secondary structure of the CD44 protein.
The bracket indicates the probable site of attachment to endothelium, the region homologous to cartilage link protein and proteoglycan monomer is boxed.
● : site of potential N-glycosylation, ○ : site of potential O-glycosylation, ★: potential chondroitin sulfate linkage sites.
Reproduced from Goldstein et al. [18] by permission of the authors and of Cell Press, Cambridge, MA, USA.

The extracellular domain of the mature protein is composed of two distinct domains (*Fig. 24.2*):

– the section proximal to the plasma membrane (residues 162–16) which is characterised by a high content of hydroxy amino acids is probably the major site of O-glycosylation. Further, this segment contains four potential chondroitin sulphate linkage sites (i.e Ser–Gly dipeptides, see [14]);

– the distal section of the predicted mature protein (residues 20–149) contains six cysteine residues and five of the six potential N-glycosylation sites. It is likely to represent a globular domain, the secondary and tertiary structure of which is stabilised by disulfide linkages. A marked characteristic of this distal region is its significant sequence homology to domains in the cartilage link proteins [33] and to the proteoglycan core proteins [14].

In the extracellular portion of **CD44H**, two proposed hyaluronic acid binding domains containing positively charged amino acids (positions 29–46 and 150–162) have been located [35,50].

The gene encoding the **CD44** protein has been mapped to the short arm of chromosome 11 at position 11p13 [9]. It contains at least 19 exons spanning approximately 50 kb of DNA [38]. Ten exons encoding portions of the membrane-proximal extracellular domain of the molecule are variant and can be incorporated into the **CD44** mRNA by alternative splicing [29] (*Fig. 24.3*). Various combinations of these exons account for the observed heterogeneity of **CD44** molecules expressed in different tissues and tumour cells. As mentioned above, erythrocytes and non-activated hematopoietic cells express mainly the 80–90 kDa 'hematopoietic' isoform (**CD44H**) encoded by exons 1–5, 15–17, and 19 of the **CD44** gene (*Fig. 24.3.c*); spliceoforms containing exons 6–14 are expressed in various combinations under a variety of circumstances, including leukocyte activation and malignant transformation. The **CD44** isoform **CD44E** is encoded by exons 1–5, 12–17 and 19 (*Fig. 24.3.d*). One other isoform containing exon 18 shows a shortened cytoplasmic tail (see *Fig. 24.3.b*).

There is evidence that at least some of these variations cause functional changes in the molecule (cf. [29]). For example, splice variants containing exons v6 or v9 have been associated with increased metastatic potential of aggressive tumours (e.g. [19,20,28,39,51]).

The molecular basis of the **In^a/In^b** dimorphism has been established by Telen et al. [48], The authors demonstrate that a single point mutation at nucleotide 252 (G → C) causes an Arg → Pro dimorphism at position 46, arginine being responsible for **In^b** and proline for **In^a**. Two other point mutations A^{441} → C (Tyr^{109} → Ser) and A^{831} → G (Glu^{239} → Gly) do not influence the **In^a/In^b** antigenic phenotype as determined in site-directed mutagenesis studies.

Though Arg^{46} has previously been shown to be part of a binding motif for extracellular hyaluronate [35,50], the authors prove that the Arg^{46} → Pro substitution does not affect hyluronate binding to the intact **CD44** protein, and the **In(a+b-)** phenotype does not show any clinical abnormalities.

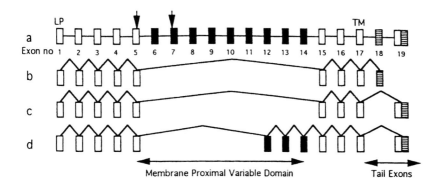

a
Exon no 1 2 3 4 5 6 7 8 9 10 11 12 13 14 15 16 17 18 19

b

c

d

Membrane Proximal Variable Domain   Tail Exons

Figure 24.3: Schematic map of the gene encoding the CD44 protein.
(a) is the complete CD44 gene with 19 exon,**(b)** and **(c)** are haemopoietic variants of CD44, **(d)** is the epithelial variant of the molecule. **LP** = leader peptide, **TM** = transmembrane domain.
Reproduced from Screaton et al. [38] by permission of the authors.

References

1. ANSTEE, D. J., GARDNER, B., SPRING, F. A., HOLMES, C. H., SIMPSON, K. L., PARSONS, S. F., MALLINSON, G., YOUSAF, S. M. & JUDSON, P. A. (1991): New monoclonal antibodies in CD44 and CD58: their use to quantify CD44 and CD58 on normal human erythrocytes and to compare the distribution of CD44 and CD58 in human tissues. *Immunology 74*, 197-205.
2. ARUFFO, A., STAMENKOVIC, I., MELNICK, M., UNDERHILL, C. B. & SEED, B. (1990): CD44 is the principal cell surface receptor for hyaluronate. *Cell 61*, 1303-1313.
3. BADAKERE, S. S., JOSHI, S. R., BHATIA, H. M., DESAI, P. K., GILES, C. M. & GOLDSMITH, K. L. G. (1973): Evidence for a new blood group antigen in the Indian population (a preliminary report). *Ind. J. Med. Res. 61*, 563.
4. BADAKERE, S. S., PARAB, B. B. & BHATIA, H. M. (1974): Further observations on the In (Indian) antigen in Indian populations. *Vox Sang. 26*, 400-403.
5. BADAKERE, S. S., VASANTHA, K., BHATIA, H. M., ALA, F., CLARKE, V. A., MOESRI, R., SOMMAI, S. & AMIN, A. B. (1980): High frequency of In antigen among Iranians and Arabs. *Hum. Hered. 30*, 262-263.
6. BHATIA, H. M., BADAKERE, S. S., MOKASHI, S. A. & PARAB, B. B. (1980): Studies on the blood group antigen In. *Immunol. Commun. 9*, 203-215.
7. CARTER, W. G. & WAYNER, E. A. (1988): Characterization of the class III collagen receptor, a phosphorylated, transmembrane glycoprotein expressed in nucleated human cells. *J. Biol. Chem. 263*, 4193-4201.
8. COBBOLD, S., HALE, G. & WALDMANN, H. (1987): Non-lineage, LFA-1 family, and leucocyte common antigens: new and previously defined clusters. In: *Leucocyte typing III: White Cell Differentiation Antigens* (A. J. McMichael et al., eds.), Oxford University Press, Oxford, pp. 788-798.
9. COUILLIN, P., AZOULAY, M., HENRY, I., RAVISE, N., GRISARD, M. C., JEANPIERRE, C., BARICHARD, F., METEZEAU, P., CANDELIER, J. J. & LEWIS, W. (1989): Characterization of a panel of somatic cell

hybrids for subregional mapping along 11p and within band 11p13. *Hum. Genet. 82*, 171-178.

10. DALCHAU, R., KIRKLEY, J. & FABRE, J. W. (1980): Monoclonal antibody to a human brain-granulocyte-T-lymphocyte antigen probably homologous to the W3/13 antigen of the rat. *Eur. J. Immunol. 10*, 745-749.

11. DANIELS, G. (1988): The Lutheran blood group system: monoclonal antibodies, biochemistry, and the effect of In(Lu). In: *Blood Group Systems: Duffy, Kidd, and Lutheran* (S. R. Pierce and C. R. MacPherson, eds.). American Association of Blood Banks, Arlington, Virginia, pp. 119-147.

12. DANIELS, G. (1992): Effect of enzymes on and chemical modifications of high-frequency red cells antigens. *Immunohaematology 8*, 53-57.

13. DENNING, S. M., LE, P. T., SINGER, K. H. & HAYNES, B. F. (1990): Antibodies against the CD44 p80 lymphocyte homing receptor molecule augment human peripheral blood T cell activation. *J. Immunol. 144*, 7-15.

14. DOEGE, K., SASAKI, M., HORIGAN, E., HASSELL, J. R. & YAMADA, Y. (1987): Complete primary structure of the rat cartilage proteoglycan core protein deduced from cDNA clones. *J. Biol. Chem. 262*, 17757-17767.

15. FLANAGAN, B. F., DALCHAU, R., ALLEN, A. K., DAAR, A. S. & FABRE, J. W. (1989): Chemical composition and tissue distribution of the human CDw44 glycoprotein. *Immunology 67*, 167-175.

16. GALLATIN, W. M., WAYNER, E., HOFFMAN, P. A., JOHN, T. S., BUTCHER, E. C. & CARTER, W. G. (1989): Structural homology between lymphocyte receptors for high endothelium and class III extracellular matrix receptor. *Proc. Natl. Acad. Sci. USA 86*, 4654-4658.

17. GILES, C. M. (1975): Antithetical relationship of anti-In[a] with the Salis antibody. *Vox Sang. 29*, 73-76.

18. GOLDSTEIN, L. A., ZHOU, D. F. H., PICKER, L. J., MINTY, C. N., BARGATZE, R. F., DING, J. F. & BUTCHER, E. C. (1989): A human lymphocyte homing receptor, the Hermes antigen, is related to cartilage proteoglycan core and link proteins. *Cell 56*, 1063-1072.

19. GÜNTHERT, U., HOFMANN, M., RUDY, W., REBER, S., ZÖLLER, M., HAUßMANN, I., MATZKU, S., WENZEL, A., PONTA, H. & HERRLICH, P. (1991): A new variant of glycoprotein CD44 confers metastatic potential to rat carcinoma cells. *Cell 65*, 13-24.

20. HERRLICH, P., ZÖLLER, M., PALS, S. T. & PONTA, H. (1993): CD44 splice variants: metastases meet lymphocytes. *Immunol. Today 14*, 395-399.

21. HUET, S., GROUX, H., CAILLOU, B., VALENTIN, H., PRIEUR, A. M. & BERNARD, A. (1989): CD44 contributes to T cell activation. *J. Immunol. 143*, 798-801.

22. ISACKE, C. M., SAUVAGE, C. A., HYMAN, R., LESLEY, J., SCHULTE, R. & TROWBRIDGE, I. S. (1986): Identification and characterization of the human Pgp-1 glycoprotein. *Immunogenetics 23*, 326-332.

23. JALKANEN, S., BARGATZE, R. F., HERRON, L. R. & BUTCHER, E. C. (1986): A lymphoid cell surface glycoprotein involved in endothelial cell recognition and lymphocyte homing in man. *Eur. J. Immunol. 16*, 1195-1202.

24. JALKANEN, S. & JALKANEN, M. (1992): Lymphocyte CD44 binds the COOH-terminal heparin-binding domain of fibronectin. *J. Cell Biol. 116*, 817-825.

25. JALKANEN, S., JALKANEN, M., BARGATZE, R., TAMMI, M. & BUTCHER, E. C. (1988): Biochemical properties of glycoproteins involved in lymphocyte recognition of high endothelial venules in man. *J. Immunol. 141*, 1615-1623.

26. JOSHI, S. R. & BHATIA, H. M. (1987): Effect of 2-aminoethylisothiouroniumbromide on In[a]/In[b] blood group antigens. *Ind. J. Med. Res. 85*, 420-421.

27. KALOMIRIS, E. L. & BOURGUIGNON, L. Y. W. (1988): Mouse T lymphoma cells contain a transmembrane glycoprotein (gp85) that binds ankyrin. *J. Cell Biol. 106*, 319-327.

28. LEGRAS, S., GÜNTHERT, U., STAUDER, R., CURT, F., OLIFERENKO, S., KLUIN-NELEMANS, H. C., MARIE, J. P., PROCTOR, S., JASMIN, C. & SMADJA-JOFFE, F. (1998): A strong expression of CD44-6v correlates with shorter survival of patients with acute myeloid leukemia. *Blood 91*, 3401-3413.

29. LESLEY, J., HYMAN, R. & KINCADE, P. W. (1993): CD44 and its interaction with extracellular matrix. *Adv. Immunol. 54*, 271-335.

30. LETARTE, M., ITURBE, S. & QUACKENBUSH, E. J. (1985): A glycoprotein of molecular weight 85,000 on human cells of B-lineage: detection with a family of monoclonal antibodies. *Mol. Immunol. 22*,

113-124.

31. LUCAS, M. G., GREEN, A. M. & TELEN, M. J. (1989): Characterization of the serum In(Lu)-related antigen: identification of a serum protein related to erythrocyte p80. *Blood 73*, 596-600.

32. MIYAKE, K., MEDINA, K. L., HAYASHI, S., ONO, S., HAMAOKA, T. & KINCADE, P. W. (1990): Monoclonal antibodies to Pgp-1/CD44 block lympho-hemopoiesis in long-term bone marrow cultures. *J. Exp. Med. 171*, 477-488.

33. NEAME, P. J., CHRISTNER, J. E. & BAKER, J. R. (1987): Cartilage proteoglycan aggregates. The link protein and proteoglycan amino-terminal globular domains have similar structures. *J. Biol. Chem. 262*, 17768-17778.

34. PALS, S. T., HOGERVORST, F., KEITZER, G. D., THEPEN, T., HORST, E. & FIGDOR, C. C. (1989): Identification of a widely distributed 90-kDa glycoprotein that is homologous to the Hermes-1 human lymphocyte homing receptor. *J. Immunol. 143*, 851-857.

35. PEACH, R. J., HOLLENBAUGH, D., STAMENKOVIC, I. & ARUFFO, A. (1993): Identification of hyaluronic acid binding sites in the extracellular domain of CD44. *J. Cell Biol. 122*, 257-264.

36. PICKER, L. J., DE LOS TOYOS, J., TELEN, M. J., HAYNES, B. F. & BUTCHER, E. C. (1989): Monoclonal antibodies against the CD44 [In(Lu)-related p80], and Pgp-I antigens in man recognize the Hermes class of lymphocyte homing receptors. *J. Immunol. 142*, 2046-2051.

37. QUACKENBUSH, E. J., VERA, S., GREAVES, A. & LETARTE, M. (1990): Confirmation by peptide sequence and co-expression on various cell types of the identity of CD44 and p85 glycoprotein. *Mol. Immunol. 27*, 947-955.

38. SCREATON, G. R., BELL, M. V., JACKSON, D. G., CORNELIS, F. B., GERTH, U. & BELL, J. I. (1992): Genomic structure of DNA encoding the lymphocyte homing receptor CD44 reveals at least 12 alternatively spliced exons. *Proc. Natl. Acad. Sci. USA 89*, 12160-12164.

39. SOUKKA, T., SALMI, M., JOENSUU, H., HÄKKINEN, L., SOINTU, P., KOULU, L., KALIMO, K., KLEMI, P., GRENMAN, R. & JALKANEN, S. (1997): Regulation of CD44v6-containing isoforms during proliferation of normal and malignant epithelial cells. *Cancer Res. 57*, 2281-2289.

40. SPRING, F. A., DALCHAU, R., DANIELS, G. L., MALLINSON, G., JUDSON, P. A., PARSONS, S. F., FABRE, J. W. & ANSTEE, D. J. (1988): The In^a and In^b blood group antigens are located on a glycoprotein of 80,000 MW (the CDw44 glycoprotein) whose expression is influenced by the In(Lu) gene. *Immunology 64*, 37-43.

41. STAMENKOVIC, I., AMIOT, M., PESANDO, J. & SEED, B. (1989): A lymphocyte molecule implicated in lymph node homing is a member of the cartilage link protein family. *Cell 56*, 1057-1062.

42. STAMENKOVIC, I., ARUFFO, A., AMIOT, M. & SEED, B. (1991): The hematopoietic and epithelial forms of CD44 are distinct polypeptides with different adhesion potentials for hyaluronate-bearing cells. *EMBO J. 10*, 343-348.

43. STOLL, M., DALCHAU, R. & SCHMIDT, R. (1989): Cluster report: CD44. In: *Leukocyte Typing IV.* (Knapp W, Dörken B, Gilks WR et al. eds.) Oxford University Press, Oxford, pp. 619-622.

44. TELEN, M. J., EISENBARTH, G. S. & HAYNES, B. F. (1983): Human erythrocyte antigens. Regulation of expression of a novel erythrocyte surface antigen by the inhibitor Lutheran In(Lu) gene. *J. Clin. Invest. 71*, 1878-1886.

45. TELEN, M. J. & FERGUSON, D. J. (1990): Relationship of In^b antigen to other antigens on In(Lu-)-related p80. *Vox Sang. 58*, 118-121.

46. TELEN, M. J., PALKER, T. J. & HAYNES, B. F. (1984): Human erythrocyte antigens. II. The In(Lu) gene regulates expression of an antigen on an 80-kilodalton protein of human erythrocytes. *Blood 64*, 599-606.

47. TELEN, M. J., ROGERS, I. & LETARTE, M. (1987): Further characterization of erythrocyte p80 and the membrane protein defect of In(Lu) Lu(a-b-) erythrocytes. *Blood 70*, 1475-1481.

48. TELEN, M. J., UDANI, M., WASHINGTON, M. K., LEVESQUE, M. C., LLOYD, E. & RAO, N. (1996): A blood group-related polymorphism of CD44 abolishes a hyaluronan-binding consensus sequence without preventing hyaluronan binding. *J. Biol. Chem. 271*, 7147-7153.

49. VERFAILLE, C. M., BENIS, A., IIDA, J., MCGLAVE, P. B. & MCCARTHY, J. B. (1994): Adhesion of committed human hematopoietic progenitors to synthetic peptides from the C-terminal

heparin-binding domain of fibronectin: cooperation between the integrin $\alpha4\beta1$ and the CD44 adhesion receptor. *Blood 84*, 1802-1811.

50. YANG, B., YANG, B. L., SAVANI, R. C. & TURLEY, E. A. (1994): Identification of a common hyaluronan binding motif in the hyaluronan binding proteins BHAMM, CD44, and link protein. *EMBO J. 13*, 286-296.

51. YASUI, W., KUDO, Y., NAKA, K., FUJIMOTO, J., UE, T., YOKOZAKI, H. & TAHARA, E. (1998): Expression of CD44 containing variant exon 9 (CD44v9) in gastric adenomas and adenocarcinomas: relation to the proliferation and progression. *Int. J. Oncol. 12*, 1253-1258.

25 Knops System

The blood group **Knops** system has only recently been established. It comprises thus far six antigen specificities, **Kn**[a], **Kn**[b], **McC**[a], **McC**[b], **McC**[c], and **Yk**[a] (*Table 25.1*) which, based on biochemical data, were located on one protein; the characters **Kn**[a] and **Kn**[b], and **McC**[a] and **McC**[b], are probably antithetical.

The formation of anti-**Knops** antibodies is stimulated by transfusion or pregnancy. These mainly IgG antibodies are characterised by high titre but low avidity (= HTLA antibodies), thus showing optimum reaction only in the indirect antiglobulin test [24]. Erythrocytes from different individuals show great variations in agglutinability when tested with anti-**Knops** antibodies; the extremely weak agglutination observed in ~10% of the cases may lead to difficulties in differentiating weakly positive from negative reactions ([19], see below).

The **Knops** antigens *in situ* are resistant towards ficin and papain but are readily destroyed by trypsin and chymotrypsin [4]. They are sensitive towards disulfide-reducing agents, such as 2-aminoethylthiouronium bromide (AET) or dithiothreitol (DTT) [4,26].

Table 22.1: Antigens of the Knops blood group system.

ISBT nr.[1]	Antigen	Frequency (%)[2] in Europids	Negrids	References
KN1	Kn[a] (Knops-Helgeson)	99.8	n.t	[7]
KB2	Kn[b]	4.5	0	[14]
KN3	McC[a] (McCoy)	98.5	96.7	[17]
	McC[b]	0	42	[15]
KN4	McC[c] (Sl[a])	98	52	[13]
KN5	Yk[a] (York)	92	92	[16]

[1] The six-digit identification numbers (see *Chap. 1*) are 022 001 to 022 005.
[2] According to Swanson [24].

Recently the **Knops** antigens have been located on the complement regulator molecule CR1 (also termed 'complement receptor type 1', C3b/C4b receptor, or CD35) [20,21]. CR1 is expressed mainly on mature haematopoietic cells (erythrocytes, monocytes/macrophages, granulocytes, B cells, and a subset of T cells), but can also be found on other cell types, such as glomerular podocytes and follicular dendritic cells [1]. As a receptor for C3b- and C4b-coated ligands, the major biological function of CR1 in erythrocytes is the transport of C3b/C4b-coated immune complexes and particles to the liver and spleen for clearance ('immune adherence' [23]).In nucleated cells, in particular in neutrophils and monocytes, CR1 plays a pivotal role in phagocytosis and endocytosis of 3b-coated particles [9]. The second function of CR1 is the down-regulating ofthe complement activation by accelerating the dissociation of the C3 convertases which produce the C3b fragment, as well as mediating the proteolysis of C3b and C4b [1,9].

As concerns the function of the CR1 molecule it is of interest to mention that certain pathogens (e.g. *Trypanosoma cruzi*, herpes simplex virus, and Epstein-Barr virus) protect themselves from complement-induced destruction by producing proteins, which are homologous to human CR1 and able to destroy the C3 convertase [2]. Further, bacteria such as *Babesia rodhaini*, *Legionella pneumophilia* and *Mycobacterium leprae* are coated with C3b in the host; they utilise the C3b binding to CR1 to initiate infection of host cells (see [18]).

The CR1 molecule itself is a glycoprotein containing six to eight N-linked oligosaccharide chains of the complex type per molecule [12]; the receptor lacks O-linked chains [9]. CR1 shows a characteristic size polymorphism [6,32]: four codominantly inherited allelic size variants have been identified, viz. A (M_r 220 kDa) and B (M_r 250 kDa) representing 82% and 18% of the total CR1 molecules, and two rare forms, C (M_r 190 kDa) and D (M_r 280 kDa)[1]. No functional difference between these molecular weight variants has thus far been found. A soluble form of CR1 has been detected in plasma [34].

The gene encoding the CR1 molecule has been mapped to chromosome 1, position q32 [28]; it is part of the **RCA** (= 'Regulators of Complement Activation') cluster (*Fig. 25.1.a*) [9,32]. The most common form, the A allele, comprises 39 exons spanning approximately 160 kb. It encodes a single polypeptide chain of 2039 amino acid residues. The protein is composed of a 41-amino acid signal peptide, a 1930-amino acid extracellular domain, a 25-amino acid transmembrane region, and a 43-amino acid cytoplasmic tail ([1], *Fig. 25.1.b*). The extracellular

---

[1] Nomenclature used by Dykman et al. [5]; the corresponding nomenclature proposed by Wong et al. [33] is F for A, S for B, and F' for C.

(a)

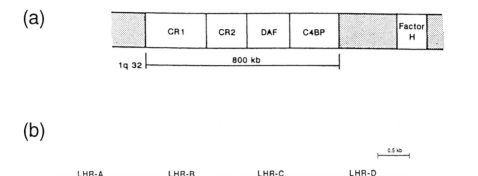

(b)

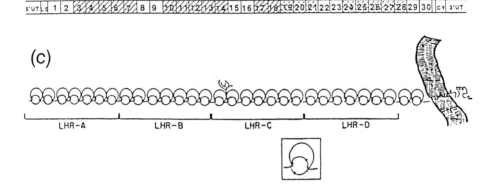

(c)

Figure 25.1: Human CR1 gene and proposed secondary structure of the CR1 protein.
(a) The gene cluster of the 'regulators of complement action' on chromosome 1q32 (according to Moulds [18].
(b) Schematic representation of the cDNA encoding the A allotype of CR1. **UT** : untranslated sequence, **LS** : leader sequence, 1–30 : short consensus repeat units, **TM** : transmembrane region, **CY** : cytoplasmic domain. The four long homologous repeats (LHR-A to LHR-D) are indicated by brackets, regions of >90% sequence homology are shaded by the same patterns. (From Wong et al. [32]).
(b) Model of the CR1 protein. The inset shows the triple-loop structure of the short consensus repeat units. The branched structure within LHR-C represents one of the 17 potential N-linked carbohydrate units. (From Klickstein et al. [12]).

567

domain of the CR1 protein contains 30 short consensus repeat units (SCRs), each comprising 60 to 70 amino acids. It has been proposed that disulfide linkages between the four conserved internal cysteine residues (Cys^2–Cys^{45} and Cys^{31}–Cys^{58}) provide a triple-loop structure for each single consensus repeat unit ([12], *Fig. 25.1.c*). A fairly regular homology pattern between the consensus repeat units further shows that groups of seven units constitute 'long homologous repeats' (= LHR) which represent distinct segments in the *CR1* gene (comprising 8 exons each). The size polymorphism of the CR1 molecule may be accounted for by the presence of variable numbers of these long homologous repeat units [8,32]. The three amino-terminal LHRs (A, B, and C) each have independent binding sites for C3b and C4b [11].

In addition to the size polymorphism, an expression polymorphism of CR1 has been described. Although the number of CR1 molecules on most cells is fairly constant, CR1 expression on erythrocytes may range from <50 to >1000 CR1 molecules per erythrocyte among different individuals of a population [19]. Cells expressing 20–100 CR1 molecules/cell were negative with all antibodies and have been included in the previously designated null-phenotype of this system (**Helgeson** phenotype), cells with 100–150 CR1 molecules/cell reacted weakly or negatively depending on the antibody used, and cells with >200 CR1 molecules/cell were usually positive [19,20].

The number of CR1 molecules per erythrocyte is a genetically regulated trait independent of size polymorphism: it obviously represents a variability within the promoter region of the CR1 gene [9,32]. Restriction fragment length polymorphism analysis on CR1 cDNA showed that a low level of CR1 expression on red cells was correlated with a 6.9-kb genomic HindIII restriction fragment, whereas a high level of expression was correlated with a 7.4-kb fragment [30,32].

It should be noted that reduction or loss of red cell CR1 has been observed in various diseases involving autoantibodies or complement activation [22,27]. Acquired CR1 deficiencies have also been described in some malignant tumours [3], AIDS [25], and systemic lupus erythematosus [10,22,29,31].

The location of the **Knops** determinants on the CR1 protein has not yet been investigated.

References

1. AHEARN, J. M. & FEARON, D. T. (1989): Stucture and function of the complement receptors, CR1 (CD35) and CR2 (CD21). *Adv. Immunol. 46*, 183-219.

2. COOPER, N. R. (1991): Complement evasion strategies of microorganisms. *Immunol. Today 12,* 327-331.
3. CURRIE, M. S., VALA, M., PISETSKY, D. S., GREENBERG, C. S., CRAWFORD, J. & COHEN, H. J. (1990): Correlation between erythrocyte CR1 reduction and other blood proteinase markers in patients with malignant and inflammatory disorders. *Blood 75,* 1699-1704.
4. DANIELS, G. (1992): Effect of enzymes on and chemical modifications of high-frequency red cells antigens. *Immunohaematology 8,* 53-57.
5. DYKMAN, T. R., COLE, J. L., IIDA, K. & ATKINSON, J. P. (1983): Polymorphism of human erythrocyte C3b/C4b receptor. *Proc. Natl. Acad. Sci. USA 80,* 1698-1702.
6. DYKMAN, T. R., HATCH, J. A., AQUA, M. F. & ATKINSON, J. P. (1985): Polymorphism of the C3b/C4b receptor (CR1): characterization of a fourth allele. *J. Immunol. 134,* 1787-1789.
7. HELGESON, M., SWANSON, J. G. & POLESKY, H. F. (1970): Knops-Helgeson (Kn^a), a high-frequency erythrocyte antigen. *Transfusion 10,* 137-138.
8. HOLERS, V. M., CHAPLIN, D. D., LEYKAM, J. F., GRUNER, B. A., KUMAR, V. & ATKINSON, J. P. (1987): Human complement C3b/C4b receptor (CR1) mRNA polymorphism that correlates with the CR1 allelic molecular weight polymorphism. *Proc. Natl. Acad. Sci. USA 84,* 2459-2463.
9. HOURCADE, D., HOLERS, V. M. & ATKINSON, J. P. (1989): The regulators of complement activation (RCA) gene cluster. *Adv. Immunol. 45,* 381-416.
10. IIDA, K., MORNAGHI, R. & NUSSENZWEIG, V. (1982): Complement receptor (CR1) deficiency in erythrocytes from patients with systemic lupus erythematosus. *J. Exp. Med. 155,* 1427-1438.
11. KLICKSTEIN, L. B., BARTOW, T. J., MILETIC, V., RABSON, L. D., SMITH, J. A. & FEARON, D. T. (1988): Identification of distinct C3b and C4b recognition sites in the human C3b/C4b receptor (CR1, CD35) by deletion mutagenesis. *J. Exp. Med. 168,* 1699-1717.
12. KLICKSTEIN, L. B., WONG, W. W., SMITH, J. A., WEIS, J. H., WILSON, J. G. & FEARON, D. T. (1987): Human C3b/C4b receptor (CR1). Demonstration of long homologous repeating domains that are composed of the short consensus repeats characteristic of C3/C4 binding proteins. *J. Exp. Med. 165,* 1095-1112.
13. LACEY, P., LAIRD-FRYER, B., BLOCK, U., LAIR, J., GUILBEAU, L. & MOULDS, J. J. (1980): A new high-incidence blood group factor, SI^a, and its hypothetical allele. (Abstract). *Transfusion 20,* 632.
14. MALLAN, M. T., GRIMM, W., HINDLEY, L., KNIGHTON, G., MOULDS, M. K. & MOULDS, J. J. (1980): The Hall serum: detecting Kn^b, the antithetical allele to Kn^a. *Transfusion 20,* 630.
15. MOLTHAN, L. (1983): The status of the McCoy/Knops antigens. *Med. Lab. Sci. 40,* 59-63.
16. MOLTHAN, L. & GILES, C. M. (1975): A new antigen, Yk^a (York), and its relationship to Cs^a (Cost). *Vox Sang. 29,* 145-153.
17. MOLTHAN, L. & MOULDS, J. (1978): A new antigen, McC^a (McCoy), and its relationship to Kn^a (Knops). *Transfusion 18,* 566-568.
18. MOULDS, J. M. (1992): Structure and function of the Chido/Rodgers and Knops/McCoy/York blood groups. In: *Blood Groups: Chido/Rodgers, Knops/McCoy/York and Cromer* (J. M. Moulds and B. Laird-Fryer, eds.). American Association of Blood Banks, Bethesda, Maryland, pp. 13-30.
19. MOULDS, J. M., MOULDS, J. J., BROWN, M. & ATKINSON, J. P. (1992): Antiglobulin testing for CR1-related (Knops/McCoy/Swain-Langley/ York) blood group antigens: negative and weak reactions are caused by variable expression of CR1. *Vox Sang. 62,* 230-235.
20. MOULDS, J. M., NICKELLS, M. W., MOULDS, J. J., BROWN, M. C. & ATKINSON, J. P. (1991): The C3b/C4b receptor is recognized by the Knops, McCoy, Swain-Langley, and York blood group antisera. *J. Exp. Med. 173,* 1159-1163.
21. RAO, N., FERGUSON, D. J., LEE, S. F. & TELEN, M. J. (1991): Identification of human erythrocyte blood group antigens on the C3b/C4b receptor. *J. Immunol. 146,* 3502-3507.
22. ROSS, G. D., YOUNT, W. J., WALPORT, M. J., WINFIELD, J. B., PARKER, C. J., FULLER, C. R., TAYLOR, R. P., MYONES, B. L. & LACHMANN, P. J. (1985): Disease-associated loss of

erythrocyte complement receptors (CR1, C3b receptors) in patients with systemic lupus erythematosus and other diseases involving autoantibodies and/or complement activation. *J. Immunol. 135*, 2005-2014.

23. ROTHMAN, I. K., GELFAND, A. S., FAUCI, A. S. & FRANK, M. M. (1975): The immune adherence receptor: dissociation between the expression of erythrocyte and mononuclear cell C3b receptors. *J. Immunol. 115*, 1312-1315.

24. SWANSON, J. L. (1992): Serology and genetics of Chido/Rodgers and Knops/McCoy/York blood groups. In: *Blood Groups: Chido/Rodgers, Knops/McCoy/York and Cromer* (J. M. Moulds and B. Laird-Fryer, eds.). American Association of Blood Banks, Bethesda, Maryland, pp. 1-11.

25. TAUSK, F. A., McCUTCHAN, A., SPECHKO, P., SCHREIBER, R. D. & GIGLI, I. (1986): Altered erythrocyte C3b receptor expression, immune complexes, and complement activation in homosexual men in varying groups for acquired deficiency syndrome. *J. Clin. Invest. 78*, 977-982.

26. TOY, E. M. (1986): Inactivation of high-incidence antigens on red blood cells by dithiothreitol. *Immunohematology 2*, 57-59.

27. WALPORT, M. J. & LACHMANN, P. J. (1988): Erythrocyte complement receptor 1, immune complexes, and the rheumatic diseases. *Arthritis Rheum. 31*, 153-158.

28. WEIS, J. H., MORTON, C. C., BRUNS, G. A., WEIS, J. J., KLICKSTEIN, L. B., WONG, W. W. & FEARON, D. T. (1987): A complement receptor locus: genes encoding C3b/C4b receptor and C3d/Epstein-Barr virus receptor map to 1q32. *J. Immunol. 138*, 312-315.

29. WILSON, J. G., JACK, R. M., WONG, W. W., SCHUR, P. H. & FEARON, D. T. (1985): Autoantibody to the C3b/C4b receptor and absence of this receptor from erythrocytes of a patient with systemic lupus erythematosus. *J. Clin. Invest. 76*, 182-190.

30. WILSON, J. G., MURPHY, E. E., WONG, W. W., KLICKSTEIN, L. B., WEIS, J. H. & FEARON, D. T. (1986): Identification of a restriction fragment length polymorphism by a CR1 cDNA that correlates with the number of CR1 on erythrocytes. *J. Exp. Med. 164*, 50-59.

31. WILSON, J. G., WONG, W. W., MURPHY, E. E., SCHUR, P. H. & FEARON, D. T. (1987): Deficiency of the C3b/C4b receptor (CR1) of erythrocytes in systemic lupus erythematosus: analysis of the stability of the defect and of a restriction fragment length polymorphism of the CR1 gene. *J. Immunol. 138*, 2708-2710.

32. WONG, W. W., CAHILL, J. M., ROSEN, M. D., KENNEDY, C. A., BONACCIO, E. T., MORRIS, M. J., WILSON, J. G., KLICKSTEIN, L. B. & FEARON, D. T. (1989): Structure of the human CR1 gene. Molecular basis of the structural and quantitative polymorphisms and identification of a new CR1-like allele. *J. Exp. Med. 169*, 847-863.

33. WONG, W. W., WILSON, J. G. & FEARON, D. T. (1983): Genetic regulation of a structural polymorphism of human C3b receptor. *J. Clin. Invest. 72*, 685-693.

34. YOON, S. H. & FEARON, D. T. (1985): Characterization of a soluble form of the C3b/C4b receptor (CR1) in human plasma. *J. Immunol. 134*, 3332-3338.

26 Kidd System

The **Kidd** system comprises three antigen characters, **Jk**ᵃ [1], **Jk**ᵇ [19], and **Jk3** (or **Jk**ᵃᵇ) (see [13])[1].

The expression of the **Kidd** antigens is controlled by a gene locus with three alleles, *Jk*ᵃ, *Jk*ᵇ, and the silent allele *Jk*. The gene frequencies in Europids are 0.5142 for *Jk*ᵃ and 0.4838 for *Jk*ᵇ [20]; *Jk* is extremely rare in Europids but occurs relatively frequently in Mongolids (e.g. the **Kidd**-null or **Jk**$_{null}$ phenotype is found in 0.9% of Polynesians [7]). Furthermore, discussions of the possible existence of a dominant suppressor gene, *In(Jk)*, and the occurrence of intermediate genes which give weak expression of **Jk**ᵃ and **Jk**ᵇ still continue to date [13].

The **Kidd** gene has been assigned to the long arm of chromosome 18 at position q12–q21.1 [4,16].

Anti-**Kidd** are mostly IgG1 and IgG3 antibodies [5] stimulated by **Kidd**-incompatible blood transfusions or pregnancies. Though only weakly reactive *in vitro*, anti-**Kidd** antibodies often cause severe haemolytic transfusion reactions. **Jk(a–b–)** (= **Kidd**-null) individuals produce an antibody, anti-**Jk3**, which reacts with a determinant present on both **Jk**ᵃ and **Jk**ᵇ erythrocytes [13]. The logical conclusion is drawn that **Jk3** represents an epitope common to **Jk**ᵃ and **Jk**ᵇ substances.

More recently anti-**Jk**ᵃ [23] and anti-**Jk**ᵇ [8] specific hybridoma antibodies have been prepared.

The **Kidd** characters are restricted to human erythrocytes and are well developed on the red cells of very young fetuses [24]. They are not detected on other blood cells or in any body fluid or secretion [2]. Molecular biological investigations also revealed the expression of **Kidd** substance in the medulla of the human kidney, where it is confined to the endothelial cells of the vasa recta [18,25].

The number of **Jk**ᵃ antigen sites on erythrocytes obtained from **Jk**ᵃ**Jk**ᵃ individuals has been estimated at 14,000 [12,14].

Sinor et al. [22] localised the **Kidd** antigens on a 45 kDa protein which more recently has been identified as the urea transporter of the erythrocyte membrane [14,16]. This protein catalyses the rapid transport of urea across the membrane and

---

[1] The ISBT numbers are 009 001 for **Jk**ᵃ, 009 002 for **Jk**ᵇ, and 009 003 for **Jk3**; in abbreviated form **JK1**, **JK2**, and **JK3** (see *Chap. 1*).

```
  1 MEDSPTMVRV DSPTMVRGEN QVSPCQGRRC FPKALGYVTG DMKKLANQLK DKPVVLQFID 60
 61 WILRGISQVV FVNNPVSGIL IVVGLLVQNP WWALTGWLGT VVSTLMALLL SQDRSLIASG 120
121 LYGYNATLVG VLMAVFSDKG DYFWWLLLPV CAMSMTCPIF SSALNSMLSK WDLPVFTLPF 180
181 NMALSMYLSA TGHYNPFFPA KLVIPITTAP NISWSDLSAL ELLKSIPVGV GVGQIYGCDN 240
241 PWTGGIFLGA ILLSSPLMCL HAAIGSLLGI AAGLSLSAPF EDIYFGLWGF NSSLACIAMG 300
301 GMFMALTWQT HLLALGCALF TAYLGVGMAN FMAEVGLPAC TWPFCLATLL FLIMTTKNSN 360
361 IYKMPLSKVT YPEENRIFYL QAKKRMVESP L
```

Figure 26.1: Peptide sequence of the human urea transporter.

N: potential N-glycosylation site; underlined: proposed membrane-spanning domains.
From Olivès et al. [18]. The sequence data are deposited in the EMBL/GenBank library (accession number Q13336).

thus obviously helps to preserve the osmotic stability of the red cells as they traverse the part of the kidney where the urea concentration varies considerably between cortex and medulla [10]. The protein However, the role of the urea transporter is obviously not essential, since **Jk(a–b–)** individuals whose cells lack the urea transport protein [16] show no haematological abnormalities [2].

The human blood group **Kidd** (= urea transporter) gene is organized into 11 exons distributed over 30 kilobase pairs, the mature protein being encoded by exons 4–11 [9]. Two equally abundant transcripts (4.4 and 2.0 kb) detected in erythroid cells were shown to arise from the usage of two different polyadenylation signals.

The primary structure of the urea transporter protein has been determined by sequencing a cDNA from a human bone marrow library [18] (*Fig. 26.1*). The predicted 391-amino acid protein is identical in sequence to 62.4% with the 397-amino acid vasopressin-regulated urea transporter of rabbit kidneys [26].

The protein presumably forms 10 membrane-traversing segments, the N- and C-termini being localised in the cytoplasm. The proposed membrane topology of the human urea transporter suggests two large hydrophobic domains with virtually no hydrophilic regions. Thus, a large portion of the protein is entirely embedded in the membrane. Assuming this topology, only one potential N-glycosylation site (Asn^{211}) is situated on the extracellular side. The occurrence of 10 cysteines further suggests a secundary and tertiary structure of the protein stabilised by disulfide bonds.

More recently a sequence variant of the urea transport protein has been detected: an A → G exchange resulting in a Lys^{44} → Glu substitution, and a hexanucleotide deletion which does not change the reading frame and results in the absence of the dipeptide Val–Gly at positions 228–229 [21]. The authors provide evidence that this variant is the proper urea transporter of human erythrocytes.

The **Jk^a/Jk^b** polymorphism is caused by a G → A exchange at nucleotide 838, resulting in an Asp → Asn dimorphism at residue 280 in the fourth extracellular loop of the erythrocyte urea transporter molecule [17].

The erythrocytes of **Jk(a–b–)** individuals are characterised by their resistance towards lysis in 2 M urea [6]. Subsequent investigations revealed a urea transport defect in **Jk(a–b–)** erythrocytes [3]. In kidneys the deficiency of this protein is compensated by the presence of a second urea transporter, which shares 61.1% sequence identity with the **Kidd** substance but is restricted to renal tissue [11,15].

Different mutations have thus far been found responsible for the **Jk$_{null}$** variant:

(a) Two splice site mutations – in one case the invariant G residue of the 3'-acceptor splice site of intron 5 was affected (donor *B.S.*), whereas in another case the invariant G residue of the 5'-donor splice site of intron 7 was changed (donor *L.P.*) [9]. These mutations caused the skipping of exon 6 and 7, respectively. Expression studies in *Xenopus* oocytes demonstrated that the truncated proteins encoded by the spliced transcripts did not mediate urea transport and were not expressed on the oocyte's plasma membrane.

(b) Molecular biological analysis of Finnish **Jk$_{null}$** individuals revealed a T^{871} → C transition at the *JK*B* allele resulting in a S → P substitution at position 291 (i.e. in the 8th predicted transmembrane domain) [9a]. Transfection experiments in erythroleukemic K562 cells have shown that this mutant was less efficiently expressed than the wild-type **Jk** protein. It has been suggested that the presence of the destabilising S → P mutation impairs the transit of the mutant protein to the plasma membrane. In human red cells the mutant protein was absent.

(c) In a Tunisian proband a genomic deletion of ~1.4 kb has been detected in the *JK*A* allele. The deletion encompasses parts of exons 4 and 5 and results in the absence of these exons [1a]. Due to cryptic donor/acceptor splice sites which are not used in the normal *JK* allele, the two exons are replaced by 136 bp of intron 3 sequence. No protein was produced *in vitro* by the variant cDNA.

(d) In a Swiss family a nonsense mutation has been detected in exon 7 which changes the tyrosine residue at position 194 into a premature stop codon [6a].

References

1. ALLEN, F. H., DIAMOND, L. K. & NIEDZIELA, B. (1951): A new blood group antigen. *Nature 167*, 482.

1a. BAILLY, P., LUCIEN, N., CHIARONI, J. & CARTRON, J.P. (2000): Partial deletion of the Jk locus causing a Jk_{null} phenotype (Abstract). *Vox Sang.* (in press).

2. EDWARDS-MOULDS, J. (1988): The Kidd blood group system: drug-related antibodies and biochemistry. In: *Blood Group Systems: Duffy, Kidd, and Lutheran* (S. R. Pierce & C. R. Macpherson, eds.). American Association of Blood Banks, Arlington, Virginia, pp. 73-92.

3. FRÖHLICH, O., MACEY, R. I., EDWARDS-MOULDS, J., GARGUS, J. J. & GUNN, R. B. (1991): Urea transport deficiency in Jk(a-b-) erythrocytes. *Am. J. Physiol. 260*, C778-C783.

4. GEITVIK, G. A., HOYHEIM, B., GEDDE-DAHL, T., GRZESCHIK, K. H., LOTHE, R., TOMTER, H. & OLAISEN, B. (1987): The Kidd (JK) blood group locus assigned to chromosome 18 by close linkage to a DNA-RFLP. *Hum. Genet. 77*, 205-209.

5. HARDMAN, J. T. & BECK, M. L. (1981): Hemagglutination in capillaries: correlation with blood group specificity and IgG subclass. *Transfusion 21*, 343-346.

6. HEATON, D. C. & McLOUGHLIN, K. (1982): Jk(a-b-) red blood cells resist urea lysis. *Transfusion 22*, 70-71.

6a. IRSHAID, N.M., HUSTINX, H. & OLSSON, M.L. (2000): A novel molecuar basis of the Jk(a-b-) phenotype in a Swiss family (Abstract). *Vox Sang.* (in press).

7. ISSITT, P. D. (1985): The Kidd blood group system. In: *Applied Blood Group Serology*. Montgomery Scientific Publications, Miami, Florida, USA, pp. 308-315.

8. LECOINTRE-COATMELEC, M., BOUREL, D., FERRETTE, J. & GENETET, B. (1991): A human anti-Jk[b] monoclonal antibody. *Vox Sang. 61*, 255-257.

9. LUCIEN, N., SIDOUX-WALTER, F., OLIVÈS, B., MOULDS, J., LE PENNEC, P. Y., CARTRON, J. P. & BAILLY, P. (1998): Characterization of the gene encoding the human Kidd blood group urea transporter protein. Evidence for splice site mutations in Jk_{null} individuals. *J. Biol. Chem. 273*, 12973-12980.

9a. LUCIEN, N., SIDOUX-WALTER, F., SISTONEN, P. CARTRON, J.P. & BAILLY, P. (2000): Functional analysis of the Jk(S291P) mutation found in Finnish Jk_{null} donors (Abstract). *Vox Sang.* (in press).

10. MACEY, R. I. & YOUSEF, L. W. (1988): Osmotic stability of red cells in renal circulation requires rapid urea transport. *Am. J. Physiol. 254*, C669-C674.

11. MARTIAL, S., OLIVES, B., ABRAMI, L., COURIAUD, C., BAILLY, P., YOU, G., HEDIGER, M. A., CARTRON, J. P., RIPOCHE, P. & ROUSSELET, G. (1996): Functional differentiation of the human red blood cell and kidney urea transporters. *Am. J. Physiol. 271*, F1264-F1268.

12. MASOUREDIS, S. P., SUDORA, E., MAHAN, L. & VICTORIA, E. L. (1980): Quantitative immunoferritin microscopy of Fy[a], Fy[b], Jk[a], U, and Di[b] antigen site numbers on human red cells. *Blood 56*, 969-977.

13. MOUGEY, R. (1988): The Kidd blood group system: serology and genetics. In: *Blood Group Systems: Duffy, Kidd, and Lutheran* (S. R. Pierce & C. R. Macpherson, eds.). American Association of Blood Banks, Arlington, Virginia, pp. 53-71.

14. NEAU, P., DEGEILH, F., LAMOTTE, H., ROUSSEAU, B. & RIPOCHE, P. (1993): Photoaffinity labeling of the human red-blood-cell urea-transporter polypeptide components. Possible homology with the Kidd blood group antigen. *Eur. J. Biochem. 218*, 447-455.

15. OLIVÈS, B., MARTIAL, S., MATTEI, M. G., MATASSI, G., ROUSSELET, G., RIPOCHE, P., CARTRON, J. P. & BAILLY, P. (1996): Molecular characterization of a new urea transporter in the human kidney. *FEBS Lett. 386*, 156-160.

16. OLIVÈS, B., MATTEI, M. G., HUET, M., NEAU, P., MARTIAL, S., CARTRON, J. P. & BAILLY, P. (1995): Kidd blood group and urea transport function of human erythrocytes are carried by the same protein. *J. Biol. Chem. 270*, 15607-15610.

17. OLIVÈS, B., MERRIMAN, M., BAILLY, P., BAIN, S., BARNETT, A., TODD, J., CARTRON, J. P. & MERRIMAN, T. (1997): The molecular basis of the Kidd blood group polymorphism and its lack of association with type 1 diabetes susceptibility. *Hum. Mol. Genet. 6*, 1017-1020.

18. OLIVÈS, B., NEAU, P., BAILLY, P., HEDIGER, M. A., ROUSSELET, G., CARTRON, J. P. & RIPOCHE, P.

(1994): Cloning and functional expression of a urea transporter from human bone marrow cells. *J. Biol. Chem. 269*, 31649-31652.

19. PLAUT, G., IKIN, E. W., MOURANT, A. E., SANGER, R. & RACE, R. R. (1953): A new blood group antibody, anti-Jk^b. *Nature 171*, 431.

20. RACE, R. R. & SANGER, R. (1975): The Kidd blood groups. In: *Blood Groups in Man.* Blackwell Scientific Publications, Oxford, pp. 364-371.

21. SIDOUX-WALTER, F., LUCIEN, N., OLIVES, B., GOBIN, R., ROUSSELET, G., KAMSTEEG, E. J., RIPOCHE, P., DEEN, P. M. T., CARTRON, J. P. & BAILLY, P. (1999): At physiological expression levels the Kidd blood group/ urea transporter protein is not a water channel. *J. Biol. Chem. 274*, 30228-30235.

22. SINOR, L. T., EASTWOOD, K. L. & PLAPP, F. V. (1987): Dot-blot purification of the Kidd blood group antigen. *Med. Lab. Sci. 44*, 294-296.

23. THOMPSON, K., BARDEN, G., SUTHERLAND, J., BELDON, I. & MELAMED, M. (1991): Human monoclonal antibodies to human blood group antigens Kidd Jk^a and Jk^b. *Transfus. Med. 1*, 91-96.

24. TOIVANEN, P. & HIRVONEN, T. (1973): Antigens Duffy, Kell, Kidd, Lutheran, and Xg^a on fetal red cells. *Vox Sang. 24*, 372-376.

25. XU, Y., OLIVÈS, B., BAILLY, P., FISCHER, E., RIPOCHE, P., RONCO, P., CARTRON, J. P. & RONDEAU, E. (1997): Endothelial cells of the kidney vasa recta express the urea transporter HUT11. *Kidney Internat. 51*, 138-146.

26. YOU, G., SMITH, C. P., KANAI, Y., LEE, W. S., STELZNER, M. & HEDIGER, M. A. (1993): Cloning and characterization of the vasopressin-regulated urea transporter. *Nature 365*, 844-847.

27 Xg System

Discovered in 1962 [26], **XG** is the only known blood group system to be submitted by an X-borne gene. It comprises only one antigen specificity, **Xgᵃ**, approximately 89% of all females and 66% of all males being **Xg(a+)** [28]. Two alleles of the **XG** gene have been postulated – the structural form **Xg ᵃ**, which controls the formation of **Xgᵃ** (= **XG1**) character (ISBT No. 012 001), and the silent form **Xg**, which is assumed to be responsible for the absence of **Xgᵃ** in the **Xg(a–)** phenotype [28].

The **Xgᵃ** antigen is characteristic for haematopoietic tissue, but has also been found on cultured skin fibroblasts [6,10].

Most examples of anti-**Xgᵃ** are IgG antibodies, which can be detected only by the indirect haemagglutination technique [28]. Though some examples bind complement, anti-**Xgᵃ** is not considered clinically significant.

In females only one of the two X chromosomes is active, the other being inactivated early in embryonic development. In any one cell it is a matter of chance whether the X chromosome of maternal or that of paternal origin is in-operative. Once inactivated, an X chromosome usually remains permanently down-regulated in all descendants of that cell. This so-called X-chromosome inactivation [25], however, does not influence the expression of all X-borne genes. The human **X** chromosome is divided into two functionally distinct regions, i.e. sex chromosome-specific and pseudoautosomal sequences [14,30]. The latter are located at the distal ends of the chromosome and are not affected by chromosome inactivation.

It was soon recognised that the **XG** locus escaped inactivation [12,13]. Subsequent investigations have mapped the **XG** blood group gene (originally described as **PBDX**[1] [8]) to the distal part of the short arm of the X-chromosome (position pter–p22.32) [5,11]. The gene consists of ten exons – the three exons at the 5' end of the gene are located in the pseudoautosomal region immediately centromeric to **MIC2**, with the remaining exons situated in the X-specific region [9] (*Fig 27.1*).

---

[1] PBDX = 'pseudoautosomal boundary divided on the X chromosome'

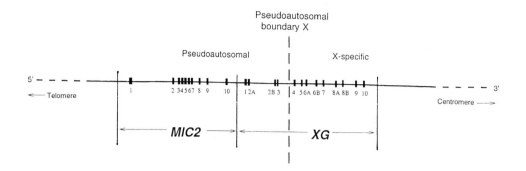

Figure 27.1: Arrangement of the *MIC2* and *XG* genes on the X-chromosome.
According to Ellis et al. [9].

The neighboring **MIC2** gene which encodes the **CD99** antigen is located completely within the pseudoautosomal region of the X chromosome and is therefore not subject to inactivation [3,17,18,31]. The fact that the gene is homologous to **XG** with regard to base sequence and overall gene structure suggests that the genes **XG** and **MIC2** evolved from a common source. One gene which may be identical to **MIC2** has been found in the euchromatic region of the Y chromosome at position pter–p11.2 [3,5,16].

A *cis*-acting mechanism correlates the expression in erythrocytes of **XG** and **MIC2** [19,20]: all **Xg(a+)** individuals express **MIC2** at a high level and all **Xg(a–)** females individuals express **MIC2** at a low level, but **Xg(a–)** males may have high or low **MIC2** expression. The postulated **XGR** locus which should control this regulation is pseudoautosomally inherited [19,32]. The molecular mechanism of this phenomenon is not yet understood.

The human erythrocyte **Xg^a** antigen is susceptible to protease treatment: it is destroyed by papain, ficin, trypsin, α-chymotrypsin, pronase, and bromelin [21], but is not affected by neuraminidase.

Immunoblot studies with anti-**Xg^a** showed that the **Xg^a** character is associated with a 24.5 kDa red cell membrane component [22,27]. Treatment of the erythrocytes with neuraminidase reduces the apparent molecular weight by about 1500.

Molecular biological investigations revealed that the **XG** gene encodes a polypeptide of 180 amino acids (*Fig. 27.2*). It is a transmembrane protein with an extracellular N-terminal domain containing 16 potential O-glycosylation sites and no N-glycosylation site.

```
  1 MESWWGLPCL AFLCFLMHAR GQRDFDLADA LDDPEPTKKP NSDIYPKPKP PYYPQPENPD  60

 61 SGGNIYPRPK PRPQPQPGNS GNSGGYFNDV DRDDGRYPPR PRPRPPAGGG GGGYSSYGNS 120

121 DNTHGGDHHS TYGNPEGNMV AKIVSPIVSV VVVTLLGAAA SYFKLNNRRN CFRTHEPENV 180
```

Figure 27.2: Peptide sequence of the Xgª protein (PBDX) as deduced from cDNA analysis.
Underlined: putative membrane-spanning domain.
According to Ellis et al. [9].The sequence data are deposited in the EMBL/GenBank library (accession number P55808).

In accordance with the observed tissue distribution of the **Xgª** antigen (see above), RNA expression of **XG** has been found only in haematopoietic tissues (adult bone marrow, as well as fetal liver, spleen, thymus and adrenal) and cultured skin fibroblasts [8,9]. **XG**-derived RNA molecules could not be detected in other tissues and cell lines.

CD99, the product of the **MIC2** gene [29], was originally described as *12E7*[(2)] antigen [24] but proved identical to **E2**, a T-cell surface adhesion molecule [2,4,15]. **CD99** is highly expressed on precursor cells of the haematopoietic system [7]. Further, an elevated level of **CD99** is found in Ewing's sarcomas and some neuroectodermal tumours [1,23].

Immunochemical analysis of **CD99** has revealed a sialoglycoprotein of an approximate molecular weight of 32.5 kDa [3]. This protein is sensitive to protease but resistant to sialidase treatment. **CD99** has been detected on the surface of virtually all human cells, including erythrocytes [31].

The primary structure of **CD99** as deduced from the nucleotide sequence of the **MIC2** gene predicts a polypeptide of 185 amino acid residues ([2,15,31], *Fig. 27.3*). 48% of its peptide sequence is homologous to that of the protein encoded at the **XG** locus. The protein consists of a putative 22-amino-acid signal sequence, an extracellular 100-residue region, a 25-residue hydrophobic transmembrane region, and a C-terminal cytoplasmic domain of 38 residues. The protein is highly glycosylated, which accounts for ~14 kDa of its molecular weight. It lacks N-glycosylation sites but contains O-glycosylation sites [2]

[(2)] Defined by its reaction with the hybridoma antibody, *12E7*, which has been produced by immunising mice with a human T-cell acute lymphocytic leukemia cell line.

```
  1 MARGAALALL LFGLLGVLVA APDGGFDLSD ALPDNENKKP TAIPKKPSAG DDFDLGDAVV  60

 61 DGENDDPRPP NPPKPMPNPN PNHPSSSGSF SDADLADGVS GGEGKGGSDG GGSHRKEGEE 120

121 ADAPGVIPGI VGAVVVAVAG AISSFIAYQK KKLCFKENAE QGEVDMESHR NANAEPAVQR 180

181 TLLEK
```

Figure 27.3: Peptide sequence of the MIC2 protein as deduced from cDNA analysis.
Underlined: putative membrane-spanning domain.
According to Gelin et al. [15]. The sequence data are deposited in the EMBL/GenBank library (accession number P14209).

The presence of a functional homologue of *MIC2* on the Y chromosome shows that expression of **CD99** on the cell surface is controlled by both the X-linked (*MIC2X*) and the Y-linked genes (= *MIC2Y*), which act independently of each other [3,18]. The presence of the Y-linked gene offers a clear explanation for the presence of either high or low **CD99** expression in **Xg(a–)** males [20].

References

1. AMBROS, I. M., AMBROS, P. F., STREHL, S., KOVAR, H., GADNER, H. & SALZER-KUNTSCHIK, M. (1991): MIC2 is a specific marker for Ewing's sarcoma and periferal primitive neuroectodermal tumors. Evidence for a common histogenesis of Ewing's sarcoma and peripheral primitive neuroectodermal tumors from MIC2 expression and specific chromosome aberration. *Cancer 67*, 1886-1893.
2. AUBRIT, F., GELIN, C., PHAM, D., RAYNAL, B. & BERNARD, A. (1989): The biochemical characterization of E2, a T cell surface molecule involved in rosettes. *Eur. J. Immunol. 19*, 1431-1436.
3. BANTING, G. S., PYM, B. & GOODFELLOW, P. N. (1985): Biochemical analysis of an antigen produced by both human sex chromosomes. *EMBO J. 4*, 1967-1972.
4. BERNARD, A., AUBRIT, F., RAYNAL, B., PHAM, D. & BOUMSELL, L. (1988): A T cell surface molecule different from CD2 is involved in spontaneous rosette formation with erythrocytes. *J. Immunol. 140*, 1802-1807.
5. BUCKLE, V., MONDELLO, C., DARLING, S., CRAIG, I. W. & GOODFELLOW P.N (1985): Homologous expressed genes in the human sex chromosome pairing region. *Nature 317*, 739-741.
6. CAMPANA, T., SZABO. P, PIOMELLI, S. & SINISCALCO, M. (1978): The Xg^a antigen on cells and fibroblasts. *Cytogenet. Cell Genet. 22*, 524-526.
7. DWORZAK, M. N., FRITSCH, G., BUCHINGER, P., FLEISCHER, C., PRINTZ, D., ZELLNER, A., SCHOLLHAMMER, A., STEINER, G., AMBROS, P. F. & GADNER, H. (1994): Flow cytometric assessment of human MIC2 expression in bone marrow, thymus, and peripheral blood. *Blood 83*, 415-425.
8. ELLIS, N. A., TIPPETT, P., PETTY, A., REID, M., WELLER, P. A., YE, T. Z., GERMAN, J., GOODFELLOW, P. N., THOMAS, S. & BANTING, G. (1994): PBDX is the XG blood group gene. *Nat. Genet. 8*, 285-290.

9. ELLIS, N. A., YE, T. Z., PATTON, S., GERMAN, J., GOODFELLOW, P. N. & WELLER, P. (1994): Cloning of PBDX, an MIC2-related gene that spans the pseudoautosomal boundary on chromosome Xp. *Nat. Genet. 6*, 394-399.

10. FELLOUS, M., BENGTSSON, B., FINNEGAN, D. & BODMER, W. F. (1974): Expression of the Xg[a] antigen on cells in culture and its segregation in somatic cell hybrids. *Ann. Hum. Genet. 37*, 421-430.

11. FERGUSON-SMITH, M. A., SANGER, R., TIPPETT, P., AITKEN, D. A. & BOYD, E. (1982): A familial t(X;Y) translocation which assigns the Xg blood group locus to the region Xp22.3→pter. *Cytogenet. Cell Genet. 32*, 273-274.

12. FIALKOW, P. J. (1970): X-chromosome inactivation and the Xg locus. *Am. J. Hum. Genet. 22*, 460-463.

13. FIALKOW, P. J., USKER, R., GIBLETT, E. R. & ZAVALA, C. (1970): Xg locus: failure to detect inactivation in females with chronic myelocytic leukaemia. *Nature 226*, 367-368.

14. FREIJE, D., HELMS, C., WATSON, M. S. & DONIS-KELLER, H. (1992): Identification of a second pseudoautosomal region near the Xq and Yq telomeres. *Science 258*, 1784-1787.

15. GELIN, C., AUBRIT, F., PHALIPON, A., RAYNAL, B., COLE, S., KACZOREK, M. & BERNARD, A. (1989): The E2 antigen, a 32 kd glycoprotein involved in T-cell adhesion processes, is the MIC2 gene product. *EMBO J. 8*, 3253-3259.

16. GOODFELLOW, P., BANTING, G., SHEER, D., ROPERS, H. H., CAINE, A., FERGUSON SMITH, M. A., POVEY, S. & VOSS, R. (1983): Genetic evidence that a Y-linked gene in man is homologous to a gene on the X chromosome. *Nature 302*, 346-349.

17. GOODFELLOW, P., PYM, B., MOHANDAS, T. & SHAPIRO, L. J. (1984): The cell surface antigen locus, MIC2X, escapes X-inactivation. *Am. J. Hum. Genet. 36*, 777-782.

18. GOODFELLOW, P. J., DARLING, S. M., THOMAS, N. S. & GOODFELLOW, P. N. (1986): A pseudoautosomal gene in man. *Science 234*, 740-743.

19. GOODFELLOW, P. J., PRITCHARD, C., TIPPETT, P. & GOODFELLOW, P. N. (1987): Recombination between the X and Y chromosomes: implications for the relationship between MIC2, XG, and YG. *Ann. Hum. Genet. 51*, 161-167.

20. GOODFELLOW, P. N. & TIPPETT, P. (1981): A human quantitative polymorphism related to Xg blood groups. *Nature 289*, 404-405.

21. HABIBI, B., TIPPETT, P., LEBESNERAIS, M. & SALMON, C. (1979): Protease inactivation of the red cell antigen Xg[a]. *Vox Sang. 36*, 367-368.

22. HERRON, R. & SMITH, G. A. (1989): Identification and immunochemical characterization of the human erythrocyte membrane glycoproteins that carry the Xg[a] antigen. *Biochem. J. 262*, 369-371.

23. KOVAR, H., DWORZAK, M., STREHL, S., SCHNELL, E., AMBROS, I. M., AMBROS, P. F. & GADNER, H. (1990): Overexpression of the pseudoautosomal gene MIC2 in Ewing's sarcoma and peripheral primitive neuroectodemmal tumor. *Oncogene 5*, 1067-1070.

24. LEVY, R., DILLEY, J., FOX, R. I. & WARNKE, R. (1979): A human thymus-leukemia antigen defined by hybridoma monoclonal antibodies. *Proc. Natl. Acad. Sci. USA 76*, 6552-6556.

25. LYON, M. F. (1972): X-chromosome inactivation and developmental patterns in mammals. *Biol. Rev. 47*, 1-35.

26. MANN, J. D., CAHAN, A., GELB, A.G., FISHER, N., HAMPER, J., TIPPETT, P., SANGER, R. & RACE, R.R. (1962): A sex-linked blood group. *Lancet 1*, 8-10.

27. PETTY, A. C. & TIPPETT, P. (1995): Investigation of the biochemical relationship between the blood group antigens Xg[a] and CD99 (12E7 antigen) on red cells. *Vox Sang. 69*, 231-235.

28. RACE, R. R. & SANGER, R. (1975): The Xg blood groups. In: *Blood Groups in Man*. Blackwell Scientific Publications, Oxford, pp.578-593.

29. SCHLOSSMAN, S. F., BOUMSELL, L., GILKS, W., HARLAN, J. M., KISHIMOTO, T., MORIMOTO, C., RITZ, J., SHAW, S., SILVERSTEIN, R. L., SPRINGER, T. A., TEDDER, T. F. & TODD, R. F. (1994): CD antigens 1993. *Blood 83*, 879-880.

30. SIMMLER, M. C., ROUYER, F., VERGNAUD, G., NYSTRÖM-LAHTI, M., NGO, K. Y., DE LA CHAPELLE, A. & WEISSENBACH, J. (1985): Pseudoautosomal DNA sequences in the pairing region of the human sex chromosomes. *Nature 317*, 692-697.

31. SMITH, M. J., GOODFELLOW, P. J. & GOODFELLOW, P. N. (1993): The genomic organisation of the human pseodoautosomal gene MIC2 and the detection of a related locus. *Hum. Mol. Genet. 2*, 417-422.
32. TIPPETT, P., SHAW, M. A., GREEN, C. A. & DANIELS, G. L. (1986): The 12E7 red cell quantitative polymorphism: control by the Y-borne locus, Yg. *Ann. Hum. Genet. 50*, 339-347.

28 Colton System

In the **Colton** blood group system three specificities are defined: **Co^a** (= **CO1**) [6], its antithetical antigen **Co^b** (= **C02**) [5], and **Co^{ab}** (= **Co3**), a specificity common to **Co^a** and **Co^b** [23][1].

The expression of the **Colton** antigens is controlled by a gene locus with three alleles, viz. *Co^a*, *Co^b*, and the inactive allele *Co*. In Europids the gene frequencies are 0.957 for *Co^a* and 0.0427 for *Co^b* [21]. The *Co* allele is extremely rare – thus far only five **Co(a–b–)** kindreds have been published [20].

The *CO* locus (= *AQP1* or aquaporin-1 locus, see below) has been localised on the short arm of chromosome 7, position p14 [14,30].

Both anti-**Co^a**, which is frequently found, and the relatively rare anti-**Co^b** antibody are usually IgG in nature and react best in antiglobulin test or with protease-treated cells [7]. Individuals of the **Co(a–b–)** phenotype produce anti-**Co3** antibody, which reacts equally well with **Co(a+b–)** and **Co(a–b+)** red cells [9].

Smith et al. [24] found that the **Colton** antigens reside on aquaporin-1 (= AQP1), the 'CHannel-forming Integral Protein' (= CHIP or CHIP28) of the erythrocyte membrane. This 28-kDa protein is a member of the large family of homologous intrinsic membrane constituents responsible for the high water-permeability of plasma membranes of a variety of cell types[2]. The 'water-selective pore' aquaporin-1 establishes a highly specific water-channel; this allows the unhindered passage of water but is not permeable for uncharged solutes, ions, and other small hydrophilic molecules. The passage of water is reversibly inhibited by $HgCl_2$ and p-chloromercuribenzoate [10,18].

Aquaporin-1 occurs in relatively high concentration on erythrocytes – 120,000 to 160,000 copies per cell have been determined [4].

---

[1] The six-digit ISBT identification numbers (see *Chap. 1*) are 015 001 for **Co^a**, 015 002 for **Co^b**, and 015 003 for **Co^{ab}**.

[2] Water transporters highly homologous to aquaporin-1 are widely distributed in animals, plants and microorganisms [22].

1 MASEFKKKLF WRAVVAEFLA TTLFVFISIG SALGFKYPVG NNQTAVQDNV KVSLAFGLSI 60

61 ATLAQSVGHI SGAHLNPAVT LGLLLSCQIS IFRALMYIIA QCVGAIVATA ILSGITSSLT 120

121 GNSLGRNDLA DGVNSGQGLG IEIIGTLQLV LCVLATTDRR RRDLGGSAPL AIGLSVALGH 180

181 LLAIDYTGCG INPARSFGSA VITH*N*FSNHW IFWVGPFIGG ALAVLIYDFI LAPRSSDLTD 240

241 RVKVWTSGQV EEYDLDADDI NSRVEMKPK

Figure 28.1: Amino acid sequence of human red cell aquaporin-1.
N : potential N-glyosylation site.
After Preston & Agre [16]. The peptide sequence is deposited in the EMBL/GenBank data library (accession number M77829).

In humans (or, more generally speaking, in mammals) aquaporin-1 is not restricted to erythrocyte membranes but is found in the water-permeable epithelia of many tissues. This indicates its major role in transcellular water transport within multiple organs. In the kidney, aquaporin-1 is localised in the basolateral plasma membranes of both renal proximal tubules and descending thin limbs of Henle's loop, where it is obviously essential in the process of re-absorption of glomerular filtrate. The protein is not expressed in other nephron segments or in the collecting ducts [11,15]. Considerable amounts of aquaporin-1 are also found in the continuous endothelia of capillaries and post-capillary venules of cardiac muscle, skeletal muscle, and lung, as well as in the plasma membranes of many secretory and resorptive epithelia. In these cases aquaporin-1 is found in the epithelium of salivary glands and hepatobiliary ductules in particular. It is also expressed in the choroid plexus epithelium of the brain, in ciliary epithelium, lens epithelium, and corneal endothelium of the eye. However, aquaporin-1 is absent from glandular epithelia of mammary, salivary, and lacrimal glands, as well as from the epithelium of gastrointestinal mucosa, and this fact seems to suggest the existence of other water transport systems [16].

Aquaporin-1 is encoded by a single 17 kb gene consisting of four exons [14]. Its primary structure as deduced by cDNA analysis (*Fig. 28.1*) predicts a protein with six membrane-spanning domains [17]. Both the amino and the carboxy terminus are at the cytoplasmic side of the membrane [19] (*Fig. 28.2.a*). The asparagine residue at position 42 is glycosylated [19], and the polylactosaminoglycan chain carries **ABH**

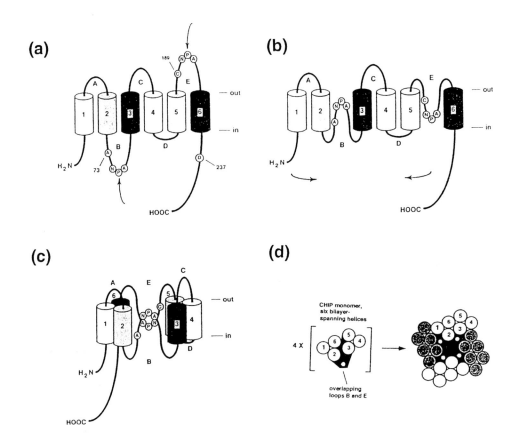

Figure 28.2: Membrane topology and proposed structure of human red cell aquaporin-1 (hourglass model).

(a) Orientation of the aquaporin-1 molecule within the erythrocyte membrane – the six membrane-spanning domains are arranged in opposite symmetry.

(b) The predicted folding of loops B and E (which contain the NPA motives) into the bilayer.

(c) Loops B and E form a single aqueous pore (the 'hourglass').

(d) AQP1 subunits form a tetramer with the four sets of B and E loops constituting four central water pores.

Reproduced from Jung et al. [8] by copyright permission of the authors and of the American Society for Biochemistry and Molecular Biology Inc.

determinants [24]. The mercury-sensitive site critical for the function of the water pore is a cysteine residue at position 189, close to the exofacial leaflet of the lipid bilayer [18].

Further analysis of the amino acid sequence of aquaporin-1 disclosed a significant internal homology between the amino- and carboxy-terminal halves of the protein [17]. The polypeptide is composed of two internal tandem repeats, which comprise three putative transmembrane helices each and span the bilayer in opposite orientations [19]. The most highly conserved domains are the connecting loops B and E, both of which contain the sequence Asn–Pro–Ala (NPA). This tripeptide sequence is preserved in all members of the aquaporin gene family from diverse animal and plant species, and suggests that the the sequence is a pivotal structural motif in forming a water-channel [22].

The three-dimensional structure of aquaporin-1 has been investigated by biochemical and molecular biological methods [8,19] and subsequently by electron crystallography [2,13,27,28], freeze fracturing [25], and cryo-electronmicroscopy [26].
These investigations have revealed that the water-channel of the erythrocyte membrane is a complex of four non-covalently associated aquaporin-I units. According to the current hypothesis advanced by Jung et al. [8], the individual subunits of the water-channel acquire a highly asymmetric, hourglass-like structure, within which a single water-conducting pore is formed by the fairly hydrophobic loops B and E, which fold from opposite sides into the membrane bilayer (*Fig. 28.2.b-d*). The experiments also suggested that, although each molecule has a functioning water pore, oligomerisation is essential for the formation of an active water channel [18].

Sequencing of aquaporin-specific cDNA from **Colton**-typed individuals revealed an amino acid dimorphism at position 45 caused by a C ↔ T exchange at nucleotide 134 – here, the alanine residue characteristic for the **Co(a+b–)** phenotype is replaced by a valine residue in the case of **Co(a–b+)** phenotype [24]. The dimorphic site is located on the first extracellular loop of the molecule, close to the N-glycosylation site at position 42.

Genomic DNA analyses on **Co(a–b–)** individuals have revealed four different mutations thus far:

– deletion of exon 1 [20];
– a single base insertion at nucleotide 307 provoking a shift in the reading frame starting after Gly[104]; the mutant gene, if at all transcribed, encodes a mutilated protein [20];

- a C → T mutation at nucleotide 113 caused a Pro → Leu substitution at position 38 [20]. Transfection studies revealed that the latter gene encoded an aquaporin molecule with extremely low water transfer activity. Although this mutation does not alter the initial level of the CHIP protein synthesis, the Pro → Leu substitution affects the glycosylation of the mutant polypeptide and the amount of protein within the cell decreases over time (see [29]);
- a C → A transition at nucleotide 614 resulting in an Asn → Lys substitution at position 192 [3]; this mutation destroys the NPA motif essential for channel function.

Though the red cell membrane and the renal tubules of **Co(a–b–)** individuals totally lack aquaporin-1 or contain an unsatisfactorily working water-channel, none of the three **Co(a–b–)** individuals investigated has suffered a **Colton**-related clinical disorder [20]. Clinical laboratory studies revealed only a slightly reduced red cell life span *in vivo*; further, the red cell morphology appeared normal with only an insignificant reduction in membrane surface area. Studies reported by Mathai et al. [12] revealed that **Co(a–b–)** erythrocytes still exhibit ~36% of the diffusional water permeability of control cells, about one third being inhibited by p-chloromercuri-benzenesulfonate. These experiments suggested that diffusion through the lipid bilayer and water transfer through transport proteins other than water-pores has to compensate for the lack of aquaporin-1 in **Colton**-null cells.

References

1. AGRE, P., BROWN, D. & NIELSEN, S. (1995): Aquaporin water channels: Unanswered questions and unresolved controversies. *Curr. Opin. Cell Biol. 7*, 472-483.
2. CHENG, A. C., VAN HOEK, A. N., YEAGER, M., VERKMAN, A. S. & MITRA, A. K. (1997): Three-dimensional organization of a human water channel. *Nature 387*, 627-630.
3. CHRÉTIEN, S., CARTRON, J. P. & DE FIGUEIREDO, M. (1999): A single mutation inside the NPA motif of aquaporin-1 found in a Colton-null phenotype. *Blood 93*, 4021-4023.
4. DENKER, B. M., SMITH, B. L., KUHAJDA, F. P. & AGRE, P. (1988): Identification, purification, and partial characterization of a novel M_r 28,000 integral membrane protein from erythrocytes and renal tubules. *J. Biol. Chem. 263*, 15634-15642.
5. GILES, C. M., DARNBOROUGH, J., ASPINALL, P. & FLETTON, M. W. (1970): Identification of the first example of anti-Co[b]. *Br. J. Haematol. 19*, 267-269.
6. HEISTÖ, H., VAN DER HART, M., MADSEN, G. F., MOES, M., NOADES, J., PICKLES, M. M., RACE, R. R., SANGER, R. & SWANSON, J. (1967): Three examples of a new red cell antibody, anti-Co[a]. *Vox Sang. 12*, 18-24.
7. ISSITT, P. D. (1985): The Colton blood group system. In: *Applied Blood Group Serology.* Montgomery Scientific Publications, Miami, Florida, USA, pp. 391-392.
8. JUNG, J. S., PRESTON, G. M., SMITH, B. L., GUGGINO, W. B. & AGRE, P. (1994): Molecular structure of the water channel through aquaporin CHIP: the hourglass model. *J. Biol. Chem. 269*, 14648-14654.
9. LACEY, P. A., ROBINSON, J., COLLINS, M. L., BAILEY, D. G., EVANS, C. C., MOULDS, J. J. & DANIELS,

G. L. (1987): Studies on the blood of a Co(a-b-) proposita and her family. *Transfusion 27*, 268-271.

10. MACEY, R. I. & FARMER, R. E. I. (1970): Inhibition of water and solute permeability in human red cells. *Biochim. Biophys. Acta 211*, 104-106.

11. MAEDA, Y., SMITH, B. L., AGRE, P. & KNEPPER, M. A. (1995): Quantification of aquaporin-CHIP water channel protein in microdissected renal tubules by fluorescence-based ELISA. *J. Clin. Invest. 95*, 422-428.

12. MATHAI, J. C., MORI, S., SMITH, B. L., PRESTON, G. M., MOHANDAS, N., COLLINS, M., VAN ZIJL, P. C. M., ZEIDEL, M. L. & AGRE, P. (1996): Functional analysis of aquaporin-1 deficient red cells. The Colton-null phenotype. *J. Biol. Chem. 271*, 1309-1313.

13. MITRA, A. K., VAN HOEK, A. N., WIENER, M. C., VERKMAN, A. S. & YEAGER, M. (1995): The CHIP28 water channel visualized in ice by electron crystallography. *Nat. Struct. Biol. 2*, 726-729.

14. MOON, C., PRESTON, G. M., GRIFFIN, C. A., JABS, E. W. & AGRE, P. (1993): The human Aquaporin-CHIP gene. Structure, organization, and chromosomal location. *J. Biol. Chem. 268*, 15772-15778.

15. NIELSEN, B. M., SMITH, B. L., CHRISTENSEN, E. I., KNEPPER, M. A. & AGRE, P. (1993): CHIP28 water channels are localized in constitutively water-permeable segments of the nephron. *J. Cell Biol. 120*, 371-383.

16. NIELSEN, S., SMITH, B. L., CHRISTENSEN, E. I. & AGRE, P. (1993): Distribution of the aquaporin CHIP in secretory and resorptive epithelia and capillary endothelia. *Proc. Natl. Acad. Sci. USA 90*, 7275-7279.

17. PRESTON, G. M. & AGRE, P. (1991): Isolation of the cDNA for erythrocyte integral membrane protein of 28 kilodaltons: member of an ancient channel family. *Proc. Natl. Acad. Sci. USA 88*, 11110-11114.

18. PRESTON, G. M., JUNG, J. S., GUGGINO, W. B. & AGRE, P. (1993): The mercury-sensitive residue at cysteine 189 in the CHIP28 water channel. *J. Biol. Chem. 268*, 17-20.

19. PRESTON, G. M., JUNG, J. S., GUGGINO, W. B. & AGRE, P. (1994): Membrane topology of Aquaporin CHIP. Analysis of functional epitope-scanning mutants by vectorial proteolysis. *J. Biol. Chem. 269*, 1668-1673.

20. PRESTON, G. M., SMITH, B. L., ZEIDEL, M. L., MOULDS, J. J. & AGRE, P. (1994): Mutations in aquaporin-1 in phenotypically normal humans without functional CHIP water channels. *Science 265*, 1585-1587.

21. RACE, R. & SANGER, R. (1975): The Colton blood groups. In: *Blood Groups in Man*. Blackwell Scientific Publications, Oxford, pp. 391-394.

22. REIZER, J., REIZER, A. & SAIER, M. H. (1993): The MIP family of integral membrane channel proteins: sequence comparisons, evolutionary relationships, reconstructed pathway of evolution, and proposed functional differentiation of the two repeated halves of the proteins. *Crit. Rev. Biochem. Mol. Biol. 28*, 235-257.

23. ROGERS, M. J., STILES, P. A. & WRIGHT, J. (1974): A new minus-minus phenotype: three Co(a-b-) individuals in one family (Abstract). *Transfusion 14*, 508.

24. SMITH, B. L., PRESTON, G. M., SPRING, F. A., ANSTEE, D. J. & AGRE, P. (1994): Human red cell aquaporin CHIP.1. Molecular characterization of ABH and Colton blood group antigens. *J. Clin. Invest. 94*, 1043-1049.

25. VERBAVATZ, J. M., BROWN, D., SABOLIC, I., VALENTI, G., AUSIELLO, D. A., VAN HOEK, A. N., MA, T. & VERKMAN, A. S. (1993): Tetrameric assembly of CHIP28 water channels in liposomes and cell membranes: a freeze-fracture study. *J. Cell Biol. 123*, 605-618.

26. WALZ, T., HIRAI, T., MURATA, K., HEYMANN, J. B., MITSUOKA, K., FUJIYOSHI, Y., SMITH, B. L., AGRE, P. & ENGEL, A. (1997): The three-dimensional structure of aquaporin-1. *Nature 387*, 624-627.

27. WALZ, T., SMITH, B. L., AGRE, P. & ENGEL, A. (1994): The three-dimensional structure of human erythrocyte aquaporin CHIP. *EMBO J. 139*, 2985-2993.

28. WALZ, T., TYPKE, D., SMITH, B. L., AGRE, P. & ENGEL, A. (1995): Projection map of aquaporin-1 determined by electron crystallography. *Nat. Struct. Biol. 2*, 730-732.

29. YAZDANBAKHSH, K., LEE, S., YU, Q. & REID, M. E. (1999): Identification of a defect in the intracellular trafficking of a Kell blood group variant. *Blood 94*, 310-318.
30. ZELINSKI, T., KAITA, H., GILSON, T., COGHLAN, G., PHILIPPS, S. & LEWIS, M. (1990): Linkage between the Colton blood group locus and ASSP11 on chromosome 7. *Genomics 6*, 623-625.

29 OK system

The only antigen of the **OK** blood group system **Ok^a** [1] is a very high incidence blood group character [11]. The **Ok(a–)** phenotype has thus far been found only in eight Japanese families [12].

The **Ok^a** antigen is not restricted to erythrocytes but is expressed on all haematopoietic cells. It is also found on many tissue cells (epithelial cells in particular) and various malignant cell lines [9,12,14].

The gene encoding the **Ok^a** protein maps to chromosome 19 at position p13.2–pter [4,14].

First investigations revealed that the **Ok^a** antigen is carried by a glycoprotein of 35–69 kDa [11,14]. The discovery soon followed [12] that the **Ok^a** protein is identical to the M6 leukocyte activation antigen [6] – also described as human basigin [10] and EMMPRIN ('extracellular matrix metalloproteinase inducer') [3][2].

The M6 protein is a member of the immunoglobulin superfamily. The predicted translation product of the M6 cDNA is a polypeptide consisting of 269 amino acids, the first 21 residues of which represent a putative leader sequence [3,6,10]. The mature protein of 251 amino acids contains a single transmembrane region, a cytoplasmic domain of 40 amino acids and an extracellular domain of 187 amino acids *(Fig. 29.1)*. The latter is arranged in two immunoglobulin-related domains, an N-terminal C$_2$-SET and a V-SET (other Ig-related proteins see *Chap. 19*). The M6 protein has three potential N-glycosylation sites where the carbohydrate moiety makes up to 50% of the molecular mass of the molecule. The presence of a glutamic acid residue within the putative hydrophobic sequence of the transmembrane sequence indicates an association of M6 with other polypeptides (cf. [13]).

The **Ok(a–)** phenotype is caused by a G → A mutation at nucleotide 331. This gives rise to a Glu → Lys exchange at position 92, which is in the first immuno-

---

[1] ISBT number 0245 001 or **OK1**, previously 901 006.

[2] M6 was designated CD147 at the 6th International Workshop and Conference on Human Leukocyte Differentiation antigens (Kobe, Japan, November 1996).

```
  1 MAAAALFVLLG FALLGTHGAS GAAGTVFTTV EDLGSKILLT CSLNDSATEV TGHRWLKGGV  60

 61 VLKEDALPGQ KTEFKVDSDD QWGEYSCVFL PEPMGTANIQ LHGPPRVKAV KSSEHINEGE 120

121 TAMLVCKSES VPPVTDWAWY KITDSEDKAL MNGSESRFFV SSSQGRSELH IENLNMEADP 180

181 GQYRCNGTSS KGSDQAIITL RVRSHLAALW PFLGIVAEVL VLVTIIFIYE KRRKPEDVLD 240

241 DDDAGSAPLK SSGQHQNDKG KNVRQRNSS                                    269
```

Figure 29.1: Amino acid sequence of the human leukocyte activation antigen M6.
N : potential N-glycosylation sites, underlined: putative signal peptide (residues 1–19) and predicted membrane-spanning domain(residues 206–229).
According to Kasinrerk et al. [6] and Biswas et al. [3]. The sequence data of the polypeptide are deposited in the EMBL/GenBank data library (accession numbers X64364, S40605,and R36804).

globulin-like domain of the protein [12]. This result has been confirmed by transfection of mouse NS-0 cells with normal or **Ok(a–)** cDNA.

Earlier investigations have shown that the M6 molecule interacts with fibroblasts to stimulate the expression of matrix metalloproteases, such as collagenase, gelatinase, and stromelysin-1, all of which degrade the extracellular matrix components of the basement membrane and interstitial matrix [3,7]. On this basis the M6 molecule may be an important mediator of matrix remodelling in wound healing and during embryonic development in normal tissues.

The expression of M6 is substantially increased in certain pathological conditions. Elevated levels of M6 on the granulocytes of rheumatoid and reactive arthritis patients [5] may induce matrix degradation in the arthritic joint. A further, well-documented function of the M6 molecule in humans is its involvement in tumour cell metastasis [2,7]. The stimulation of MMPs is obviously a pivotal step in tumour cell invasion through basement membranes and interstitial matrices [1,8].

References

1. Biswas, C. (1982): Tumor cell stimulation of collagenase production by fibroblasts. *Biochem. Biophys. Res. Commun. 109*, 1026-1034.
2. Biswas, C. (1984): Collagenase stimulation in cocultures of human fibroblasts and human tumor

cells. *Cancer Lett. 24*, 201-207.

3. BISWAS, C., ZHANG, Y., DeCASTRO, R., GUO, H., NAKAMURA, T., KATAOKA, H. & NABESHIMA, K. (1995): The human tumor cell-derived collagenase stimulatory factor (renamed EMMPRIN) is a member of the immunoglobulin superfamily. *Cancer Res. 55*, 434-439.

4. BROOK, J. D., BERESFORD, H. R., SHAW, D. J., OLD, L. J. & RETTIG, W. J. (1987): Localisation on human chromosome 19 of three genes for cell surface antigens defined by monoclonal antibodies. *Cytogenet. Cell Genet. 45*, 156-162.

5. FELZMANN, T., GADD, S., MAJDIC, O., MAURER, D., PETERA, P., SMOLEN, J. & KNAPP, W. (1991): Analysis of function-associated receptor molecules on peripheral blood and synovial fluid granulocytes from patients with rheumatoid and reactive arthritis. *J. Clin. Immunol. 11*, 205-212.

6. KASINRERK, W., FIEBIGER, E., STEFANOVA, I., BAUMRUKER, T., KNAPP, W. & STOCKINGER, H. (1992): Human leukocyte activation antigen M6, a member of the Ig superfamily, is the species homologue of rat OX-47, mouse basiginin, and chicken HT7 molecule. *J. Immunol. 149*, 847-854.

7. KATAOKA, H., DeCASTRO, R., ZUCKER, S. & BISWAS, C. (1993): Tumor cell-derived collagenase-stimulatory factor increases expression of interstitial collagenase, stromelysin and 72-kDa gelatinase. *Cancer Res. 53*, 3154-3158.

8. LIOTTA, L. A., STEEG, P. S. & STETLER-STEVENSON, W. G. (1991): Cancer metastasis and angiogenesis: an imbalance of positive and negative regulation. *Cell 64*, 327-336.

9. MATTES, M. J., CAIRNCROSS, J. G., OLD, L. J. & LLOYD, K. O. (1983): Monoclonal antibodies to three widely distributed human cell surface antigens. *Hybridoma 2*, 253-264.

10. MIYAUCHI, T., MASUZAWA, Y. & MURAMATSU, T. (1991): The basigin group of the immunoglobulin superfamily: complete conservation of a segment in and around transmembrane domains of human and mouse basigin and chicken HT7 antigen. *J. Biochem. 110*, 770-774.

11. MOREL, P. A. & HAMILTON, H. B. (1979): Ok^a: an erythrocyte antigen of high frequency. *Vox Sang. 36*, 182-185.

12. SPRING, F. A., HOLMES, C. H., SIMPSON, K. L., MAWBY, W. J., MATTES, M. J., OKUBO, Y. & PARSONS, S. F. (1997): The Ok^a blood group antigen is a marker for the M6 leukocyte activation antigen, the human homolog of OX-47 antigen, basigin and neurothelin, an immunoglobulin superfamily molecule that is widely expressed in human cells and tissues. *Eur. J. Immunol. 27*, 891-897.

13. WILLIAMS, A. F. & BARCLAY, A. N. (1988): The immunoglobulin superfamily - domains for cell surface recognition. *Annu. Rev. Immunol. 6*, 381-405.

14. WILLIAMS, B. P., DANIELS, G. L., PYM, B., SHEER, D., POVEY, S., OKUBO, Y., ANDREWS, P. W. & GOODFELLOW, P. N. (1988): Biochemical and genetic analysis of the Ok^a blood group antigen. *Immunogenetics 27*, 322-329.

30 JMH Antigen

The **John Milton Hagen** antigen **JMH** is a high-incidence serological character [7][(1)]. It is expressed on erythrocytes and weakly on peripheral blood lymphocytes [2,3] as well as on a series of non-haematopoietic human tissues, such as neurons of the central nervous system and in respiratory epithelium [5].

Anti-**JMH** autoantibodies occur fairly frequently in elderly persons with reduced expression of **JMH** [4]. Allo-immune **JMH** antibodies have also been produced by **JMH**-positive persons whose cells expressed variant **JMH** antigens [4,6]. A **JMH**-specific hybridoma antibody (*H8*) has also been produced [2,3].

Bobolis et al. [1] performed a study which showed that the erythrocyte **JMH** antigen is carried on a 76-kDa membrane protein. The protein is absent from complement-sensitive erythrocytes of patients with paroxysmal nocturnal haemo-globinuria [9]. This and its partial release from the intact erythrocyte membrane by treatment with phosphatidylinositol-specific phospholipase C clearly indicate that a phosphatidylinositol anchor attaches the protein to the cell membrane (see *Chap. 20*).

Mudad et al. [5] have identified the protein carrying the **JMH** determinant as **CDw108**, a 'cluster-of-differentiation' antigen defined by its reactivity with hybridoma antibodies MEM-121 and MEM-150 [8]. The finding that **CDw108** is associated with the *src*-family of tyrosine kinases suggests that **CDw108** plays an important role in lymphocyte development and activation.

The peptide sequence of the **CDw108** protein has only recently been determined [10]. The cDNA clone containing the entire coding sequence of the *CDw108* gene encodes a protein of 666 amino acids, including a signal peptide (46 residues) and the GPI-anchor motif (19 residues). The protein contains five putative N-glycosylation sites *(Fig. 30.1)*.

CDw108 mRNA was found highly expressed in activated lymphocytes, spleen, testis, and placenta, and only weakly expressed in thymus and brain, but not expressed in any other tissues tested [10].

The gene encoding the **CDw108** protein was located in the middle of the long arm of chromosome 15, position q23–24.

[(1)] ISBT nr. 901 007 or, alternatively, JMH1.

```
  1 MTPPPPGRAA PSAPRARVPG PPARLGLPLR LRLLLLLWAA AASAQGHLRS GPRIFAVWKG  60

 61 HVGQDRVDFG QTEPHTVLFH EPGSSSVWVG GRGKVYLFDF PEGKNASVRT VNIGSTKGSC 120

121 LDKRDCENYI TLLERRSEGL LACGTNARHP SCWNLVNGTV VPLGEMRGYA PFSPDENSLV 180

181 LFEGDEVYST IRKQEYNGKI PRFRRIRGES ELYTSDTVMQ NPQFIKATIV HQDQAYDDKI 240

241 YYFFREDNPD KNPEAPLNVS RVAQLCRGDQ GGESSLSVSK WNTFLKAMLV CSDAATNKNF 300

301 NRLQDVFLLP DPSGQWRDTR VYGVFSNPWN YSAVCVYSLG DIDKVFRTSS LKGYHSSLPN 360

361 PRPGKCLPDQ QPIPTETFQV ADRHPEVAQR VEPMGPLKTP LFHSKYHYQK VAVHRMQASH 420

421 GETFHVLYLT TDRGTIHKVV EPGEQEHSFA FNIMEIQPFR RAAAIQTMSL DAERRKLYVS 480

481 SQWEVSQVPL DLCEVYGGGC HGCLMSRDPY CGWDQGRCIS IYSSERSVLQ SINPAEPHKE 540

541 CPNPKPDKAP LQKVSLAPNS RYYLSCPMES RHATYSWRHK ENVEQSCEPG HQSPNCILFI 600

601 ENLTAQQYGH YFCEAQEGSY FREAQHWQLL PEDGIMAEHL LGHACALAAS LWLGVLPTLT 660

661 LGLLVH
```

Figure 30.1: Peptide sequence of the CDw108 protein as deduced from cDNA analysis.
The GPI-anchor motif is underlined, **H**: N-terminus of the native molecule, **N** : potential N-glycosylation sites.
According to Yamada et al. [10]. The sequence data are deposited in the EMBL / GenBank data library (accession number AF069493).

References

1. BOBOLIS, K. A., MOULDS, J. J. & TELEN, M. J. (1992): Isolation of the JMH antigen on a novel phosphatidylinositol-linked human membrane protein. *Blood 79*, 1574-1581.
2. DANIELS, G. L. & KNOWLES, R. W. (1982): A monoclonal antibody to the high frequency red cell antigen JMH. *J. Immunogenet. 9*, 57-62.
3. DANIELS, G. L. & KNOWLES, R. W. (1983): Further analysis of the monoclonal antibody H8 demonstrating a JMH-related specificity. *J. Immunogenet. 10*, 257-265.
4. MOULDS, J. J., LEVENE, C. & ZIMMERNMAN, S. (1982): Serological evidence for heterogeneity among antibodies compatible with JMH-negative red cells. *17th Congress of the International Society of Blood Transfusion, Budapest, Hungary 1982*, Abstract 287.
5. MUDAD, R., RAO, N., ANGELISOVA, P., HOREJI, V. & TELEN, M. J. (1995): Evidence that CDwl08 membrane protein bears the JMH blood group antigen. *Transfusion 35*, 566-570.
6. MUDAD, R., RAO, N., ISSITT, P. D., ROY, R. B., COMBS, M. R. & TELEN, M. J. (1995): JMH variants: serologic, clinical, and biochemical analyses in two cases. *Transfusion 35*, 925-930.
7. SABO, B., MOULDS, J. J. & MCCREARY, J. (1978): Anti-JMH: another high titer low avidity antibody against a high frequency antigen. *Transfusion 18*, 387-389.
8. SCHLOSSMAN, S. F., BOUMSELL, L., GILKS, W., HARLAN, J. M., KISHIMOTO, T., MORIMOTO, C., RITZ, J., SHAW, S., SILVERSTEIN, R. L., SPRINGER, T. A., TEDDER, T. F. & TODD, R. F. (1994): CD antigens 1993. *Blood 83*, 879-880.
9. TELEN, M. J., ROSSE, W. F., PARKER, C. J., MOULDS, M. K. & MOULDS, J. J. (1990): Evidence that

several high-frequency human blood group antigens reside on phosphatidylinositol-linked erythrocyte membrane proteins. *Blood 75*, 1404-1407.

10. YAMADA, A., KUBO, K., TAKESHITA, T., HARASHIMA, N., KAWANO, K., MINE, T., SAGAWA, K., SUGAMURA, K. & ITOH, K. (1999): Molecular cloning of a glycosylphosphatidylinositol-anchored molecule CDw108. *J. Immunol. 162*, 4094-4100.

31 Antigens of the Major Histocompatibility Complex (HLA Antigens)

The **HLA** (= 'Human Leucocyte antigen locus **A**') antigens are transmitted by a group of genes representing the **Major Histocompatibility Complex** (= MHC) of humans.

The first recognised feature of the **HLA** antigens was their function as targets for antibodies and cytotoxic T lymphocytes during rejection of foreign transplants. Due to their extremely high polymorphism they provide a sort of serological identity pattern for a human individual and thus act as an essential prerequisite for discriminating between 'self' and 'non-self'. In this context the **HLA** antigens probably also play a pivotal role in cell-cell interaction.

More recently the essential function of the **HLA** gene products in regulating the immune response has been recognised. It has become apparent that the T cell receptors which are responsible for stimulating the lymphocyte recognise intracellularly processed foreign antigens only when combined with a particular 'self'-**HLA** molecule (so-called 'MHC-restricted' T cell recognition).

Comprehensive surveys on genetics, serology, and biology of the **HLA** antigens are found in review articles by Mayr [14,15,16] and in 'Immunobiology of HLA' edited by Dupont [7].

31.1 Genetics

The **HLA** antigens are encoded by a series of closely linked genes clustered in two separate regions (see *Fig. 31.1*[(1)]):

- the **class-I complex** comprising the genes *HLA-A*, *HLA-B*, and *HLA-C* and the recently detected loci *HLA-E*, *HLA-F*, and *HLA-G*,
- the **class-II complex** comprising the genes of the *HLA-D* group, i.e. *HLA-DR*, *HLA-DQ* (formerly *DS* or *MB*), and *HLA-DP* (formerly *SB*).

[(1)] A more exact gene map of the Human Major Histocompatibility Complex was published by Trowsdale et al. [21].

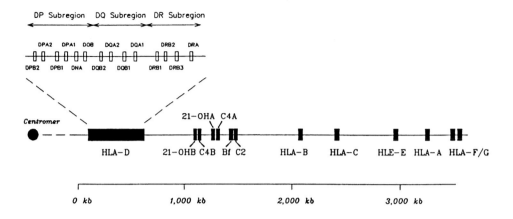

Figure 31.1: Gene map of the human Major Histocompatibility Complex.
According to Mayr [14,16]. Redrawn by permission of Springer-Verlag, Heidelberg.

The *HLA* genes show a uniquely high degree of polymorphism when compared to other mammalian genes. A survey on the specificities officially recognised by the **HLA** nomenclature commission is found in *Table 31.1* (class-I) and *Table 31.2* (class-II). The *HLA* alleles identified by nucleotide sequencing are listed in the reports of the 'WHO Nomenclature Committee for Factors of the **HLA** System' (see [4]).

The *HLA* loci are located on the short arm of chromosome 6, band 6p21.1 (**HLA-D** region) and band 6p21.3 (**HLA-ABC**) [14]. Situated between the gene complexes *HLA-ABC* and *HLA-D* are the loci for the complement factors C2 and C4 (see *Chap. 14*, **Chido/Rodgers** blood groups), for the properdin factor B (Bf), and for the structural gene encoding the steroid 21-hydroxylase; these substances are sometimes designated as class-III antigens, though there is no functional connection to the **HLA** antigens.

31.2 HLA Antisera

Almost all antisera towards **HLA** antigens thus far described are immune sera induced by pregnancies, blood transfusions, or organ transplantations. They normally contain mixtures of antibodies with different spectra of specificity and varying degrees of cross-reactivity. Serum absorption only in rare cases yielded antisera with adequate specificities.

Table 31.1: Complete listing of HLA Class-I specificities encoded by the genes *HLA-A*, *HLA-B*, and *HLA-C*.

HLA-A		HLA-B		HLA-C	
A1		B5 {	B51	Cw1	
A2			Bw52	Cw2	
A3		B7			Cw9
				Cw3 {	Cw10
A9 {	A23	B8			
	A24	B12 {	B44	Cw4	
	A25		B45	Cw5	
A10 {	A26	B13		Cw6	
	Aw34	B14 {	Bw64	Cw7	
	Aw66		Bw65	Cw8	
A11			Bw62	Cw11	
A28 {	Aw68		Bw63		
	Aw69	B15 {	Bw75		
A29			Bw76		
A30			Bw77		
A31 }	Aw19	B16 {	B38		
A32			B39		
A33		B17 {	Bw57		
Aw74			Bw58		
Aw36		B18			
Aw43		B21 {	B49		
			Bw50		
		Bw22 {	Bw54		
			Bw55		
			Bw56		
		B27			
		B35			
		B37			
		B40 {	Bw60		
			Bw61		
		Bw41			
		Bw42			
		Bw46			
		Bw47			
		Bw48			
		Bw53			
		Bw59			
		Bw67			
		Bw70 {	Bw71		
			Bw72		
		Bw73			
		Bw4			
		Bw6			

Specificities recognised by the HLA Nomenclature Committee. From Mayr [14], by permission of Springer Verlag, Heidelberg.

Table 31.2: Complete listing of HLA-D Class-II specificities.

HLA-D	HLA-DR	HLA-DQ	HLA-DP
Dw1	DR1	DQw1 { DQw5	DPw1
Dw2	DR2 { DRw15	DQw1 { DQw6	DPw2
Dw3	DR2 { DRw16	DQw2	
			DPw3
Dw4	DR3 { DRw17	DQw3 { DQw7	DPw4
Dw5	DR3 { DRw18	DQw3 { DQw8	DPw5
		{ DQw9	DPw6
Dw6 { Dw18	DR4	DQw4	
Dw6 { Dw19	DR5 { DRw11		
Dw7 { Dw11	DR5 { DRw12		
Dw7 { Dw17	DRw6 { DRw13		
Dw8	DRw6 { DRw14		
Dw9			
Dw10	DR7		
	DRw8		
Dw12	DR9		
Dw13	DRw10		
Dw14			
Dw15	DRw52		
Dw16	DRw53		
Dw20			
Dw21			
Dw22			
Dw23			
Dw24			
Dw25			
Dw26			

Specificities recognised by the 'HLA Nomenclature Committee'. From Mayr [14], by permission of Springer Verlag, Heidelberg.

More recently a variety of hybridoma antibodies towards **HLA-A, HLA-B,** and **HLA-D** antigens have been produced; antibodies towards **HLA-C** antigens, however, are not yet known. Further information on monoclonal antibodies is found in review articles by Kennedy et al. [13] and Nelson et al. [19].

31.3 HLA Gene Products

As discussed above, the *HLA* genes control the formation of two types of substances, the class-I molecules encoded by the genes *HLA-A*, *HLA-B*, and *HLA-C*, and the class-II molecules encoded by the *HLA-D* gene complex. Class-I substances occur on almost all nucleated cells of the human organism, whereas the class-II substances are confined to only a few cell types, such as B lymphocytes, activated T lymphocytes, monocytes and macrophages, thymus epithelium, and most endothelial cells etc. They are also found on certain melanoma cells.

The two types of **HLA** substances show distinct differences in function:

- the class-I molecules are involved in the recognition process for membrane-bound antigens (i.e. endogenously synthesised proteins, such as histocompatibility antigens and viral antigens); they stimulate cytotoxic (killer) T cells which express the **CD8** cell surface protein,
- the class-II molecules bind exogenous antigens processed by macrophages and are involved in activating **CD4+** helper T cells.

Class-I HLA substances [3] are composed of two polypeptide units, the α-chain with a molecular mass of 44 kDa and β_2-microglobulin (β_2m) with 12 kDa [18] (*Fig. 31.2*). The α-chain is encoded by the *HLA-A*, *HLA-B*, or *HLA-C* genes; β_2-microglobulin, however, is not polymorphic and is encoded by a gene located on chromosome 15 (band 15q21–22) [9]. The α-chain is inserted into the cell membrane, and its quarternary structure maintained by the non-covalently attached β_2-microglobulin molecule.

The α-chain contains 338 amino acids. It is an intrinsic membrane protein with an extracellular portion consisting of approximately 278 amino acids, a transmembrane region of some 28 amino acids, and an intracellular segment of 32 amino acids. The extracellular portion is organised in three domains, α_1 (position 1–90), α_2 (position 91–182), and α_3 (position 183–274); the configuration of these domains is partly stabilised by disulphide bonds. The asparagine residue at position 86 which is located in the loop connecting α_1 and α_2 carries an N-linked carbohydrate unit.

The peptide sequences of class-I molecules thus far determined reveal a high degree of amino acid polymorphism at about 17 positions in α_1 and α_2 domains [2]. The α_3 domain is situated close to the cell membrane and is essentially invariable. As also found for β_2-microglobulin, it shows substantial sequence homologies to the constant domains of immunoglobulin molecules (especially to the c_H3 domain of IgG_1).

X-ray crystallographic data obtained from investigations on, thus far, two different **HLA** class-I molecules (**HLA-A2** [1,20] and **HLA-Aw68** [8]) revealed that the

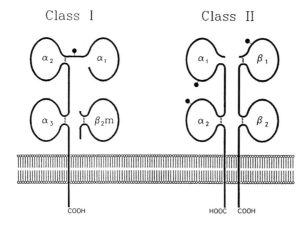

Figure 31.2: Schematic model of Class-I and Class-II HLA substances.
🍴 : Oligosaccharide unit
According to Kaufman [12].

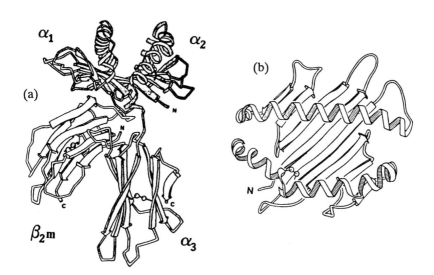

Figure 31.3: HLA-A2 molecule.
Schematic representation of the structure of the extracellular portion of the molecule.
(a) Lateral view. **(b)** Molecule viewed from above.
Reproduced from Bjorkman et al. [1] by permission of the authors and of MacMillan Press Ltd, Basingstoke, UK
(©1987).

600

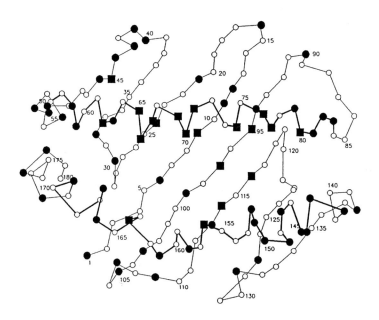

Figure 31.4: Location of polymorphic amino acids in the HLA-A2 molecule.
O : Conserved residues.
● : Low variability residues (variability <5).
■ : High variability residues (variability =5).
Schematic representation according to Colombani [6]. Redrawn by permission of the author and of Munksgaard International Publications, Copenhagen, Denmark.

secondary and tertiary structures of the α_1 and α_2 domains are practically identical (*Fig. 31.3*): Each domain consists of a long α-helix and four β-strain segments. In the native molecule the α_1 and α_2 domains pair to form a 'platform' of a single eight-stranded β-pleated sheet structure which is topped by the two α-helices. Between the α-helices and the central β-strands of the platform a characteristic cleft is formed − about 25 Å long by 10 Å wide by 11 Å deep. Most polymorphic amino acid side chains face into the cleft (*Fig. 31.4*, [2,20]) thus determining the detailed shape of this putative antigen binding site and, further, the distribution of electrostatic charge within the cleft (*Fig. 31.5*). Some variable amino acids located at the surface of the molecule probably interact with T-cell receptors or represent binding sites for **HLA** subtype-specific hybridoma antibodies (see also [6]).

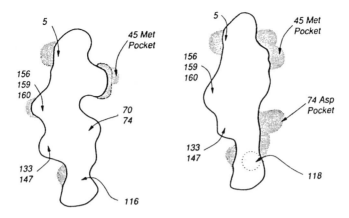

Figure 31.5: Schematic outline of the putative antigen recognition site of HLA Class-I molecules.

Reproduced from Garrett et al. [8] by permission of the authors and of MacMillan Press Ltd, Basingstoke, UK (©1989).

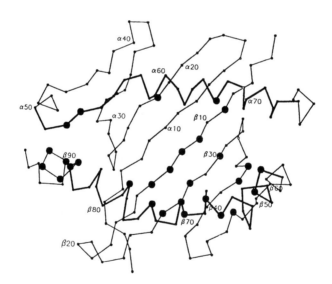

Figure 31.6: Antigen recognition site of the Class-II histocompatibility antigen HLA-DR1.

● : Polymorphic positions.

Redrawn from Brown et al. [5] by permission of the authors and of MacMillan Press Ltd, Basingstoke, UK (©1993).

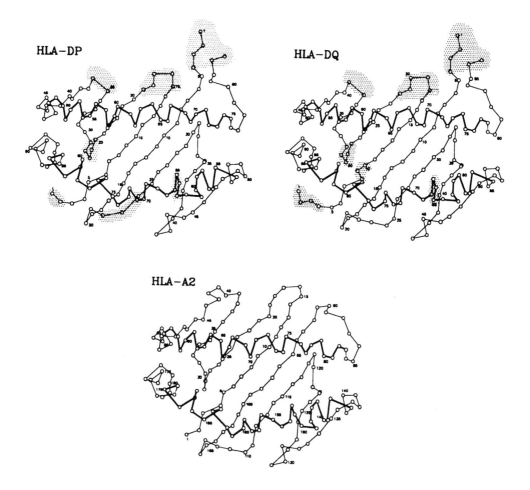

Figure 31.7: Proposed structures of the peptide-binding domains of HLA-DP and HLA-DQ molecules.

The models are based on the crystal structure of HLA-A2. Areas expected to differ from the class-I molecules are shaded.

Redrawn according to Gorga et al. [10] by permission of CRC Press Inc., Boca Rota, FL, USA.

The secondary and tertiary structures of α_3 domain and β_2-microglobulin, as was expected on the basis of significant homologies in amino acid sequence, closely resemble the conformation of the constant domains in immunoglobulins [22].

603

Class-II HLA substances [11] are also composed of two non-covalently associated polypeptide subunits, the α-chain (34 kDa) and the β-chain (29 kDa), which, in this case, both span the lipid bilayer (*Fig. 31.2*). The α-chain contains two N-linked carbohydrate units and the β-chain only one.

In contrast to class-I antigens, both subunits are encoded by genes of the *HLA-D* complex on chromosome 6. **HLA-DP** and **HLA-DR** substances have more or less invariant α-chains and highly polymorphic β-chains, whereas in **HLA-DQ** substances both α- and β-chains are polymorphic.

Sequence analyses of the class-II proteins revealed an organisation similar to that of the class-I α-chains: the α- and β-chains of the class-II molecules contain two extracellular domains composed of 90 – 100 amino acids each, a transmembrane portion of 20 – 25 and an intracellular portion of 8 – 15 residues. The allelic variations are confined to very limited regions in the outer domains; the α_2 and β_2 domains, which are located close to the cell membrane, are invariant and show significant homology to the α_3-domain of the class-I antigens.

The three-dimensional structure of the class-II histocompatibility antigen **HLA-DR1** has recently been determined by X-ray crystallography [5]. The conformation of the molecule proved closely similar to that of the class-I antigens. In contrast to class-I molecules, however, the antigen binding cleft built of a platform of eight β-strands and two α-helices is formed by the N-terminal domains of the α and β chain. Also in the **HLA-DR1** molecule most of the polymorphic amino acids are located at the antigen binding site (*Fig. 31.6*).

The conformations of **HLA-DP** and **HLA-DQ** molecules have not yet been determined. However, models of the antigen recognition sites based on characteristic features in the peptide sequences have been proposed by Gorga [10] (see also [17]). The pattern of conserved and polymorphic residues and the position of amino acids essential for tertiary structure (e.g. location of disulphide bonds and salt bridges) suggested also in these cases a configuration of the distal part of the molecule fairly similar to that of **HLA** class-I substances (*Fig. 31.7*).

References

1. Bjorkman, P.J., Saper, M.A., Samraoui, B., Bennett, W.S., Strominger, J.L. & Wiley, D.C. (1987): Structure of the human class I histocompatibility antigen, HLA-A2. *Nature 329*, 506-512.
2. Bjorkman, P.J., Saper, M.A., Samraoui, B., Bennett, W.S., Strominger, J.L. & Wiley, D.C. (1987): The foreign antigen binding site and T cell recognition regions of class I histocompatibility antigens. *Nature 329*, 512-518.
3. Bjorkman, P.J. & Parham, P. (1990): Structure, function, and diversity of class I major histocompatibility complex molecules. *Annu. Rev. Biochem. 59*, 253-288.
4. Bodmer, J.G., Marsh, S.G.E., Albert, E.D., Bodmer, W.F., Bontrop, R.E., Charron, D., Dupont, B., Erlich, H.A., Fauchet, R., Mach, B., Mayr, W.R., Parham, P., Sasazuki, T., Schreuder, G.M.T., Strominger, J.L., Svejgaard, A., Terasaki, P.I. (1997): Nomenclature for factors of the

HLA system, 1996. *Vox Sang. 73*, 105-130.

5. BROWN, J.H., JARDETZKY, T.S., GORGA, J.C., STERN, L.J., URBAN, R.G., STROMINGER, J.L. & WILEY, D.C. (1993): Three-dimensional structure of the human class II histocompatibility antigen HLA-DR1. *Nature 364*, 33-39.

6. COLOMBANI, J. (1990): Conserved and variable structures in HLA class I molecules: a review. *Tissue Antigens 35*, 103-113.

7. DUPONT, B. (1989): *Immunobiology of HLA* (Vols. I and II). Springer, New York - Berlin - Heidelberg.

8. GARRETT, T.P.J., SAPER, M.A., BJORKMAN, P.J., STROMINGER, J.L. & WILEY, D.C. (1989): Specificity pockets for the side chains of peptide antigens in HLA-Aw68. *Nature 342*, 692-696.

9. GOODFELLOW, P.N., JONES, E.A., VAN HAYNINGEN, V., SOLOMON, E., BOBROW, M., MIGGIANO, V. & BODMER, W.F. (1975): The β_2 microglobulin gene is on chromosome 15 and not in the HLA region. *Nature 254*, 267-269.

10. GORGA, J.C. (1992): Structural analysis of class II major histocompatibility complex proteins. *Crit. Rev. Immunol. 11*, 305-335.

11. KAPPES, D. & STROMINGER, J.L. (1988): Human class II major histocompatibility complex genes and proteins. *Annu. Rev. Biochem. 57*, 991-1128.

12. KAUFMAN, J.F., AUFFRAY, C., KORMAN, A.J., SHACKELFORD, D.A. & STROMINGER, J. (1984): The class II molecules of the human and murine major histocompatibility complex. *Cell 36*, 1-13.

13. KENNEDY, L.J., MARSH, S.G.E. & BODMER, J. (1987): Cytotoxic monoclonal antibodies. In: *Immunobiology of HLA. Vol. I. Histocompatibility testing 1987*. (B. Dupont, ed.), Springer Verlag, Berlin, pp. 301-305.

14. MAYR, W.R. (1988): Das HLA-System. In: *Transfusionsmedizin* (C. Mueller-Eckhardt, ed.), Springer Verlag, Berlin-Heidelberg-New York, pp. 173-193.

15. MAYR, W.R. (1989): HLA polymorphisms. In: *Two-dimensional electrophoresis*. Proceedings of the International Two-Dimensional Electrophoresis Conference, Vienna, Nov. 1988. (A.T. Endler & S. Hanash, eds.), pp. 112-119.

16. MAYR, W.R. (1990): HLA 1990. *Blut 61*, 207-212.

17. MARSH, S.G.E. & BODMER, J.G. (1989): HLA-DR and -DQ epitopes and monoclonal antibody specificity. *Immunol. Today 10*, 305-312.

18. NAKAMURO, K., TANIGAKI, N. & PRESSMAN, D. (1973): Multiple common properties of human β_2-microglobulin and the common portion fragment derived from HL-A antigen molecules. *Proc. Natl. Acad. Sci. USA 70*, 2863-2865.

19. NELSON, K., BODMER, J., MARTIN, A., NAVARETTE, C.G. & STRONG, D.M. (1987): Micro EIA and monoclonal antibodies. In: *Immunobiology of HLA. Vol. I. Histocompatibility Testing 1987* (B. Dupont, ed.), Springer Verlag, Berlin, pp. 292-301.

20. SAPER, M.A., BJORKMAN, P.J. & WILEY, D.C. (1991): Refined structure of the human histocompatibility antigen HLA-A2 at 2.6 Å resolution. *J. Mol. Biol. 219*, 277-319.

21. TROWSDALE, J., RAGOUSSIS, J. & CAMPBELL, R.D. (1991): Map of the human MHC. *Immunol. Today 12*, 443-446.

22. WILLIAMS, A.F. & BARCLAY, A.N. (1988): The immunoglobulin superfamily - domains for cell surface recognition. *Annu. Rev. Immunol. 6*, 381-405.

32 HEMPAS (CDA-II)

Congenital dyserythropoietic anaemias (CDAs) are a group of inherited genetic disorders characterised by mild to moderate anaemia, ineffective erythropoiesis, and morphologic abnormalities of the erythrocytes and their precursors in the bone marrow. Three main types of CDA (CDA-I, CDA-II, and CDA-III) and a number of variants are known to date.

The most common and best investigated CDA type II will be discussed here. On the basis of its key characteristics CDA-II is also termed HEMPAS (= 'Hereditary Erythroblastic Multinuclearity associated with a Positive Acidified-Serum test' [7]).

Histological investigations of the bone marrow obtained from HEMPAS patients revealed an increase of the number of erythroblasts 5–10 times above normal [7,20]. Sometimes almost 40% of these cells are multinuclear – in most cases they contain two nuclei, some cells, however, have up to seven. Many phagocytes are found filled with PAS-positive material.

Electronmicroscopical investigations on the erythroblasts of the bone marrow showed significant changes in the cell membranes: up to 95% of the cells show 'double membranes'. These linear structures lie parallel to the membranes of the endoplasmic reticulum in a distance of 40–60 nm. Double membranes are also found in about 2% of the circulating erythrocytes [9,19,35]. The presence of reticulo-endothelial proteins (such as calreticulin, glucose regulated protein 78, protein disulphide isomerase) in erythrocyte ghosts from HEMPAS patients suggested that the additional membrane derives from the smooth endoplasmic reticulum [1].

The essential serological characteristics of HEMPAS erythrocytes is lysis by human serum at acid pH values ('Ham test') [7,8]. The lysis is induced by an IgM alloantibody present in most normal human sera. This antibody is selectively adsorbed on HEMPAS erythrocytes and is absent from the serum of HEMPAS patients; it reacts obviously with an antigen characteristic for HEMPAS erythrocytes ('anti-HEMPAS').

A further characteristic of HEMPAS cells is the elevated **i** (and to a lesser extent **I**) activity [7], and a decreased **H** and **P₁** activity [29]. Moreover, significantly altered reactions with various lectins (esp. concanavalin A) have also been reported [2].

Chemical investigations of the erythrocyte membrane constituents from HEMPAS patients revealed abnormalities in membrane proteins and in lipid composition:

Analysis of the **membrane proteins** showed a slightly elevated electrophoretic mobility of band 3 and 4.5 proteins in SDS polyacrylamidegel electrophoresis due to a lowered molecular mass [1,17,27]. The reduced PAS stain of glycophorins, the relatively weak labelling of bands 3 and 4.5 by galactose oxidase / NaB[^{3}H$_4$], as well as various peptide-chemical investigations suggest a defective glycosylation of these membrane proteins [17,30]. The elevated reaction of native HEMPAS cells and their membrane proteins with concanavalin A also seems to indicate a change in the alkali-stable chains.

Further structural studies on HEMPAS band 3 and band 4.5 carbohydrates revealed that these membrane proteins are not glycosylated by poly-lactosamino-glycans but contain tri-mannosyl and penta-mannosyl hybrid-type oligosaccharides [12,15].

The altered glycosylation did not change the anion and glucose transport activities of band 3 and band 4.5 proteins, respectively [25]. However, it causes structural changes in band 3 which result in increased cytoskeletal interaction and abnormal clustering of this membrane constituent [14]. The band 3 protein plays a critical role in the organisation of the membrane skeleton for determining red cell flexibility and shape. Thus, anomalous clustering of this protein may cause disturbance in the membrane skeleton network and improper distribution of band 3-associated proteins and lipids through aberrant localisation of band 3-associated proteins.

Analysis of the **membrane lipids** of HEMPAS erythrocytes revealed a normal amount of phospholipid and cholesterol, the phospholipids, however, showing a significant increase in phosphatidyl choline [24].

In contrast to membrane glycoproteins, however, membrane glycolipids appear overglycosylated [3,36]. In HEMPAS patients the average glycosphingolipid content of the red cells is 2 – 3 times above normal [3,23,24]. Exact analysis revealed a drastic increase in the glycolipids of the lacto-series (*Table 32.1*). For example: lacto-N-triaosyl-ceramide (**GlcNAcβ1→3Galβ1→4Glc→Cer**) cannot be detected in normal cells, but is equimolar to globotriaosyl-ceramide (**Galα1→4Galβ1→4Glc→Cer**) in HEMPAS cells. Furthermore, the greatly increased level of poly-N-acetyllactosaminyl-ceramides in HEMPAS cells (up to 10-times above normal [11,36]) explains the aberrant reactivity of HEMPAS erythrocytes with anti-**i** sera.

Table 32.1: Glycosphingolipid composition of the erythrocyte membrane of a HEMPAS patient [1].

Glycosphingolipid	HEMPAS patient	control
Glycosyl-ceramide	1.5	0.3
Lactosyl-ceramide	8.6	1.7
Globotriaosyl-ceramide	6.3	1.7
Lactotriaosyl-ceramide	6.4	–
Globoside	18.4	9.6
Paragloboside	4.6	0.4
Sialoyl-paragloboside	1.6	0.5
Total glycosphingolipid	47.4	14.2

[1] µmol glycosphingolipid in the membranes of 100 ml packed erythrocytes.
After Bouhours et al. [4].

Moreover, the glycosphingolipids of HEMPAS erythrocyte showed an altered fatty acid composition – they contained higher amounts of long chain fatty acids (C_{22} - C_{24}) as compared to controls. Finally, the content of free ceramide is about twice that of normal [3].

Enzymatic Investigations revealed that HEMPAS is a heterogeneous group of diseases caused by defects of different enzymes involved in the synthesis of N-linked oligosaccharides. Three HEMPAS types have been described thus far:

(a) Deficiency in α-mannosidase II (variant *G.C.*) [15],
(b) deficiency in β1,2-N-acetylglucosaminyltransferase II (variant *T.O.*) [12], and
(c) deficiency in the membrane-bound form of galactosyltransferase (variant *G.K.*) [16].

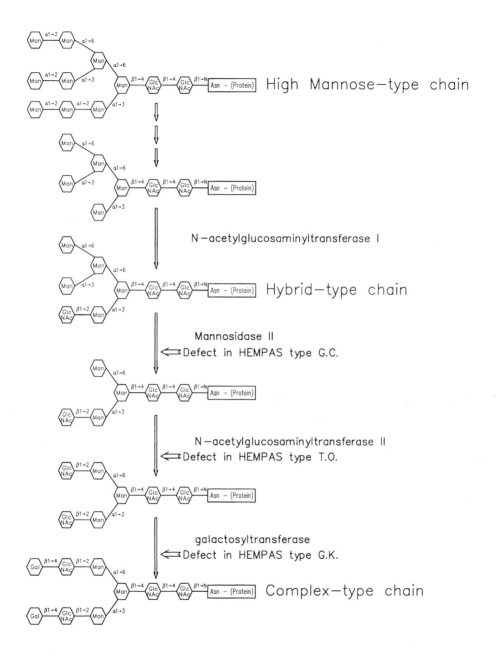

High Mannose—type chain

N—acetylglucosaminyltransferase I

Hybrid—type chain

Mannosidase II
⇐ Defect in HEMPAS type G.C.

N—acetylglucosaminyltransferase II
⇐ Defect in HEMPAS type T.O.

galactosyltransferase
⇐ Defect in HEMPAS type G.K.

Complex—type chain

Figure 32.1: Biosynthetic pathway for the biosynthesis of N-glycans.
Location of the genetic blocks in different HEMPAS types.

N-Acetylglucosaminyltransferase II and α-mannosidase II are essential enzymes involved in the formation of N-linked oligosaccharide chains catalysing pivotal steps in the conversion of oligomannose type to complex type N-glycans (*Fig. 32.1*). The lowered activity of these enzymes results in lack of poly-N-acetyllactosamine chains and formation of hybrid-type oligosaccharides on the proteins; it also leads to an accumulation of polylactosaminyl-glycolipids.

The galactosyl-transferase is involved at various stages of oligosaccharide synthesis, its absence severely affects the formation of complex-type chains and results in accumulation of high mannose-type oligosaccharide structures on erythrocyte glycoproteins [16]. The HEMPAS variant G.K. also lacks poly-N-acetyllactosamine units on lipids; further, the concentration of keratansulfate (= sulfated polylactosamine) in the patient's serum was abnormally low.

Since the variant G.K. demonstrates high levels of soluble β1,4-galactosyl-transferase but a reduced amount of membrane-bound enzyme, the mutation may have occurred close to the membrane-binding domain, thus increasing the susceptibility of the proteolytic cleavage site at Arg[77] [26].

These three enzymes are encoded by so-called house keeping genes. Lack of these genes influences the N-glycan synthesis in various tissues and organs of HEMPAS patients. Accordingly, aberrant glycosylation has also been detected in serum glycoproteins synthesised by hepatocytes, such as transferrin and α_1-acid glycoproteins [13,16]. Experiments on rats showed that aberrant transferrin is rapidly cleared from the plasma by the liver. Accumulation of enormous amounts of aberrantly glycosylated serum glycoproteins may lead to liver cirrhosis and secondary tissue siderosis observed in HEMPAS patients.

Other cell lineages, however, are not affected to the same extent in the HEMPAS phenotype, mainly because the glycoproteins have different oligosaccharide structures on their surfaces: erythrocyte glycoproteins exclusively contain biantennary units [10,32-34], whereas other cell types, such as granulocytes, macrophages, platelets, and megacaryocytes contain tri- and tetraantennary core N-glycans. For example: since the N-acetylglucosaminyltransferase II defect affects only the formation of poly-N-acetyllactosamine chains of biantennary core N-glycans [12], these cells should be able to form N-glycan units more closely resembling their normal counterparts.

Additional evidence for cell type-specific variations in N-linked oligosaccharide biosynthesis was obtained by Kameh et al. [25]. In this investigation on three HEMPAS patients, nonerythroid cell types (cells of the lymphoid lineage) were shown to contain complex-type glycans [25]. This finding suggests that some cell types either are not deficient in the processing enzyme(s) or use an alternative oligosaccharide processing pathway [6].

Genetics

The HEMPAS phenotype is inherited as an autosomal recessive trait. The biochemical data support the hypothesis that the primary defect of HEMPAS lies in the genes encoding the enzymes responsible for the biosynthesis of asparagine-linked oligosaccharides.

Molecular biological studies on the α-mannosidase II gene have shown a substantially reduced expression of the gene (less than 10% of normal); however, no abnormal mRNA species has been detected [15]. This result suggested that the α-mannosidase II gene in the HEMPAS variant *G.C.* contains a mutation in the promoter region which results in reduced transcription of the gene or in message instability.

In contrast to the original assumption, more recent genetic analyses have revealed that the molecular lesions causing HEMPAS are not primarily located in N-acetylglucosaminyltransferase II and α-mannosidase II genes [21][1].

In another case HEMPAS has been associated with microsatellite markers on the long arm of chromosome 20 at position q11.2 [18]. No linkage to chromosome 20, however, could be detected in other probands [22].

In consideration of the biochemical data on reduction of activity of these enzymes, it has been suggested that the disease is due to a defect in an unknown transcriptional factor regulating both N-acetylglucosaminyltransferase II and α-mannosidase II genes [18,21].

It should be noted here that a deficiency of N-acetylglucosaminyltransferase II has also been detected in two cases of carbohydrate-deficient glycoprotein syndrome (CDGS) type-II [5]. As found in HEMPAS patients, here too the erythrocyte membrane glycoproteins show increased reactivities with concanavalin A, thus demonstrating the presence of hybrid- or oligomannose-type carbohydrate structures. However, CDGS type-II patients have a totally different clinical presentation and their erythrocytes do not show the serology typical of HEMPAS, suggesting that the genetic lesions responsible for these two diseases are different.

These investigations clearly confirm the genetic heterogeneity of the HEMPAS phenotype, and the actual genetic lesions remain elusive.

---

[1] The gene encoding the α-mannosidase II gene has been located at 5q21–22 [28] and another closely related gene at 15q25; the gene encoding the N-acetylglucosaminyltransferase II has been assigned to chromosome band 14q21 [31].

References

1. ALLOISIO, N., TEXIER, P., DENOROY, L., BERGER, C., MIRAGLIA DEL GIUDICE, E., PERROTTA, S., IOLASCON, A., GILSANZ, F., BERGER, G., GUICHARD, J., MASSÉ, J. M., DEBILI, N., BRETON-GORIUS, J. & DELAUNAY, J. (1996): The cisternae decorating the red blood cell membrane in congenital dyserythropoietic anemia (type II) originate from the endoplasmic reticulum. *Blood 87*, 4433-4439.

2. BLANCHARD, D., PILLER, F., DOINEL, C., SEVERAC, P., TESTA, U., ROCHANT, H. & CARTRON, J. P. (1983): Surface modifications of erythrocytes in congenital dyserythropoietic anaemia, type II (Hempas). *Blood Transfus. Immunohaematol. 26*, 412-413.

3. BOUHOURS, J. F., BOUHOURS, D. & DELAUNAY, J. (1985): Abnormal fatty acid composition of erythrocyte glycosphingolipids in congenital dyserythropoietic anemia type II. *J. Lipid Res. 26*, 435-441.

4. BOUHOURS, J. F., BOUHOURS, D., DELAUNAY, J. & BRYON, P. A. (1983): Les glycosphingolipides des érythrocytes de trois enfants atteints d'anémie dyserythropoiétique de type II (CDA-II ou Hempas). *Blood Transfus. Immunohaematol. 26*, 410-411.

5. CHARUK, J. H. M., TAN, J., BERNARDINI, M., HADDAD, S., REITHMEIER, R. A. F., JAEKEN, J. & SCHACHTER, H. (1995): Carbohydrate-deficient glycoprotein syndrome type II. An autosomal recessive N-acetylglucosaminyltransferase II deficiency different from typical hereditary erythroblastic multinuclearity, with a positive acidified-serum lysis test (HEMPAS). *Eur. J. Biochem. 230*, 797-805.

6. CHUI, D., OH-EDA, M., LIAO, Y. F., PANNEERSELVAM, K., LAL, A., MAREK, K. W., FREEZE, H. H., MOREMEN, K. W., FUKUDA, M. N. & MARTH, J. D. (1997): α-Mannosidase-II deficiency results in dyserythropoiesis and unveils an alternate pathway in oligosaccharide biosynthesis. *Cell 90*, 157-167.

7. CROOKSTON, J. H., CROOKSTON, M. C., BURNIE, K. L., FRANCOMBE, W. H., DACIE, J. V., DAVIS, J. A. & LEWIS, S. M. (1969): Hereditary erythroblastic multinuclearity associated with a positive acidified-serum test: a type of congenital dyserythropoietic anaemia. *Brit. J. Haematol. 17*, 11-26.

8. CROOKSTON, J. H., CROOKSTON, M. C. & ROSSE, W. F. (1972): Red-cell abnormalities in HEMPAS (hereditary erythroblastic multinuclearity with a positive acidified-serum test). *Br. J. Haematol. 23 (Suppl.)*, 83-91.

9. FRESCO, R. (1981): Electron microscopy in the diagnosis of the bone marrow disorders of the erythroid series. *Semin. Hematol. 18*, 279-292.

10. FUKUDA, M., DELL, A., OATES, J. E. & FUKUDA, M. N. (1984): Structure of branched lactosaminoglycan; the carbohydrate moiety of band 3 isolated from adult human erythrocytes. *J. Biol. Chem. 259*, 8260-8273.

11. FUKUDA, M. N., BOTHNER, B., SCARTEZZINI, P. & DELL, A. (1986): Isolation and characterization of poly-N-acetyllactosaminylceramides accumulated in the erythrocytes of congenital dyserythropoietic anemia type-II patients. *Chem. Phys. Lipids 42*, 185-197.

12. FUKUDA, M. N., DELL, A. & SCARTEZZINI, P. (1987): Primary defect of congenital dyserythropoietic anemia type II. Failure in glycosylation of erythrocyte lactosaminoglycan proteins caused by lowered N-acetylglucosaminyltransferase II. *J. Biol. Chem. 262*, 7195-7206.

13. FUKUDA, M. N., GAETANI, G. F., IZZO, P., SCARTEZZINI, P. & DELL, A. (1992): Incompletely processed N-glycans of serum glycoproteins in congenital dyserythropoietic anaemia type-II (HEMPAS). *Br. J. Haematol. 82*, 745-752.

14. FUKUDA, M. N., KLIER, G., YU, J. & SCARTEZZINI, P. (1986): Anomalous clustering of underglycosylated band 3 in erythrocytes and their precursor cells in congenital dyserythropoietic anemia type II. *Blood 68*, 521-529.

15. FUKUDA, M. N., MASRI, K. A., DELL, A., LUZZATTO, L. & MOREMEN, K. W. (1990): Incomplete synthesis of N-glycans in congenital dyserythropoietic anemia type II caused by a defect in the gene encoding α-mannosidase II. *Proc. Natl. Acad. Sci. USA 87*, 7443-7447.

16. FUKUDA, M. N., MASRI, K. A., DELL, A., THONAR, E. J. M., KLIER, G. & LOWENTHAL, R. M. (1989): Defective glycosylation of erythrocyte membrane glycoconjugates in a variant of congenital

dyserythropoietic anemia type-II. Association of low level of membrane-bound form of galactosyltransferase. *Blood 73*, 1331-1339.

17. FUKUDA, M. N., PAPAYANNOPOULOU, T., GORDON-SMITH, E. C., ROCHANT, H. & TESTA, U. (1984): Defect in glycosylation of erythrocyte membrane proteins in congenital dyserythropoietic anaemia type II (HEMPAS). *Brit. J. Haematol. 56*, 55-68.

18. GASPARINI, P., MIRAGLIA DEL GIUDICE, E., DELAUNAY, J., TOTARO, A., GRANATIERO, M., MELCHIONDA, S., ZELANTE, L. & IOLASCON, A. (1997): Localization of the congenital dyserythropoietic anemia II locus to chromosome 20q11.2 by genomewide search. *Amer. J. Hum. Genet. 61*, 1112-1116.

19. HUG, G., WONG, K. Y. & LAMPKIN, B. C. (1972): Congenital dyserythropoietic anemia type II. Ultrastructure of erythroid cells and hepatocytes. *Lab. Invest. 26*, 11-21.

20. IOLASCON, A., D'AGOSTARO, G., PERROTTA, S., IZZO, P., TAVANO, R. & MIRAGLIA DEL GIUDICE, B. (1996): Congenital dyserythropoietic anemia type II: molecular basis and clinical aspects. *Haematologica 81*, 543-559.

21. IOLASCON, A., MIRAGLIA DEL GIUDICE, E., PERROTTA, S., GRANATIERO, M., ZELANTE, L. & GASPARINI, P. (1997): Exclusion of three candidate genes as determinants of congenital dyserythropoietic anemia type II (CDA-II). *Blood 90*, 4197-4200.

22. IOLASCON, A., DE MATTIA, D., PERROTTA, S., CARELLA, M., GASPARINI, P. & DELILIERS, G. L. (1998): Genetic heterogeneity of congenital dyserythropoietic anemia type II. *Blood 92*, 2593-2594.

23. JOSEPH, K. C. & GOCKERMAN, J. P. (1975): Accumulation of glycolipids containing N-acetylglucosamine in erythrocyte stroma of patients with congenital dyserythropoietic anaemia type II (HEMPAS). *Biochem. Biophys. Res. Commun. 65*, 146-152.

24. JOSEPH, K. C., GOCKERMAN, J. P. & ALVING, C. R. (1975): Abnormal lipid composition of the red cell membrane in congenital dyserythropoietic anaemia type II (HEMPAS). *J. Lab. Clin. Med. 85*, 34-40.

25. KAMEH, H., LANDOLT-MARTICORENA, C., CHARUK, J. H. M., SCHACHTER, H. & REITHMEIER, R. A. F. (1998): Structural and functional consequences of an N-glycosylation mutation (HEMPAS) affecting human erythrocyte membrane glycoproteins. *Biochem. Cell Biol. 76*, 823-835.

26. MASRI, K. A., APPERT, H. E. & FUKUDA, M. N. (1988): Identification of the full-length coding sequence for human galactosyltransferase (β-N-acetylglucosaminide : β1,4-galactosyltransferase. *Biochem. Biophys. Res. Commun. 157*, 657-663.

27. MAWBY, W. J., TANNER, M. J. A., ANSTEE, D. J. & CLAMP, J. R. (1983): Incomplete glycosylation of erythrocyte membrane proteins in congenital dyserythropoietic anaemia type II (CDA II). *Brit. J. Haematol. 55*, 357-368.

28. MISAGO, M., LIAO, Y. F., KUDO, S., ETO, S., MATTEI, M. G., MOREMEN, K. W. & FUKUDA, M. N. (1995): Molecular cloning and expression of cDNAs encoding human α-mannosidase II and a previously unrecognized α-mannosidase IIx isozyme. *Proc. Natl. Acad. Sci. USA 92*, 11766-11770.

29. ROCHANT, H., TONTHAT, H., MAN NGO, M., LEFAOU, J., HENRI, A. & DREYFUS, B. (1973): Etude quantitative des antigènes érythrocytaires I et i en pathologie. *Nouv. Rev. Fr. Hematol. 13*, 307-318.

30. SCARTEZZINI, P., FORNI, G. L., BALDI, M., IZZO, C. & SANSONE, G. (1982): Decreased glycosylation of band 3 and band 4.5 glycoproteins of erythrocyte membrane in congenital dyserythropoietic anaemia type II. *Brit. J. Haematol. 51*, 569-576.

31. TAN, J., D'AGOSTARO, A. F., BENDIAK, B., RECK, F., SARKAR, M., SQUIRE, J. A., LEONG, P. & SCHACHTER, H. (1995): The human UDP-N-acetylglucosamine: α-6-D-mannoside-β-1,2-N-acetylglucosaminyltransferase II gene (MGAT2). Cloning of genomic DNA, localization to chromosome 14q21, expression in insect cells and purification of the recombinant protein. *Eur. J. Biochem. 231*, 317-328.

32. TSUJI, T., IRIMURA, T. & OSAWA, T. (1980): The carbohydrate moiety of band-3 glycoprotein of human erythrocyte membranes. *Biochem. J. 187*, 677-686.

33. TSUJI, T., IRIMURA, T. & OSAWA, T. (1981): The carbohydrate moiety of band 3 glycoprotein of human erythrocyte membranes. Structure of lower molecular weight oligosaccharides. *J. Biol. Chem. 256*, 10497-10502.

34. Tsuji, T., Irimura, T. & Osawa, T. (1981): Heterogeneity in the carbohydrate moiety of band 3 glycoprotein of human erythrocyte membranes. *Carbohydr. Res. 92*, 328-332.

35. Vainchenker, W., Guichard, J. & Breton-Gorius, J. (1979): Morphological abnormalities in cultured erythroid colonies (BFU-E) from the blood of two patients with HEMPAS. *Brit. J. Haematol. 42*, 363-369.

36. Zdebska, E., Anselstetter, V., Pacuska, T., Krauze, R., Chelstowska, A., Heimpel, H. & Koscielak, J. (1987): Glycolipids and glycopeptides of red cell membranes in congenital dyserythropoietic anaemia type II (CDA II). *Br. J. Haematol. 66*, 385-391.

Note: A review article on HEMPAS has recently been published by **M.N. Fukuda** (Biochim. Biophys. Acta 1455 (1999) 231-239.

33 Carcinoembryonic Antigen (CEA) Family

The carcinoembryonic antigen (**CEA**) was first described as a tumor-associated antigen found in extracts from human colon cancer tissue [9]. Originally **CEA** was assumed to be an oncofetal antigen, expressed during fetal development, but deficient in healthy individuals, and re-expressed only in cancer tissues. Later investigations have shown, however, that **CEA** is also expressed in normal adult tissue. It remains an important tumour marker, particularly because sera of patients with colorectal carcinomas or tumours of epithelial origin show increased levels of **CEA** [10,31].

To its amazement, the scientific establishment later discovered an entire family of proteins closely related to the **CEA** molecule originally described.

Table 33.1: Nomenclature of the human CEA family.

CEA subgroup		PSG subgroup	
Original name	New name	Original name	New name
BGP1, CD66a	CEACAM1	PSG1	PSG1
CGM1, CD66d	CEACAM3	PSG2	PSG2
CGM7	CEACAM4	PSG3	PSG3
CEA, CD66e	CEACAM5	PSG4,PSG9	PSG4
NCA-50/90, CD66c	CEACAM6	PSG5	PSG5
CGM2	CEACAM7	PSG6, PSG10	PSG6
CGM6, CD66b	CEACAM8	PSG7	PSG7
		PSG8	PSG8
		PSG11	PSG9
		PSG12	PSG10
		PSG13	PSG11

According to Beauchemin et al. [2].

Molecular biological investigations revealed that the human **CEA** family locus comprises 18 structural genes [26] – seven belonging to the **CEA** subgroup and eleven to the **PSG** ('pregnancy-specific glycoprotein') subgroup (*Table 33.1*). The genes are clustered on chromosome 19q13.2 within a region of 1.8 Mb [6,26,32,41].

The primary structures showed that the proteins of the **CEA** family belong to the immunoglobulin superfamily [1,23,27]. The molecules contain two types of immunoglobulin-like domains: a single N-terminal 108 amino acid domain homologous and thus related to the Ig variable domain (IgV-like) and up to six domains homologous and thus related to the Ig constant domain of the C2 set (IgC2-like) ([36], see *Fig. 33.1*). The IgV-like domain lacks an intra-chain disulfide linkage, which in this case is replaced by a salt bridge. The IgC2 domains consist either of type A and contain 93 amino acids, or of type B with 85 amino acids.

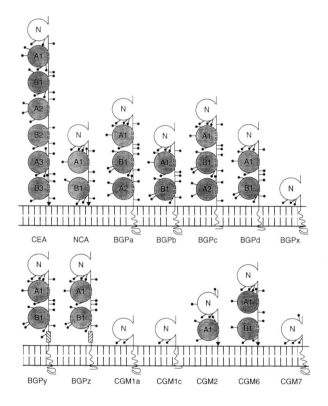

Figure 33.1: Models of the molecules of the CEA subgroup.
▨ : non-Ig domains, ↓: GPI linkage, ↑ : potential glycosylation sites.
Reproduced from Hammarström [12] by permission of Academic Press Ltd.

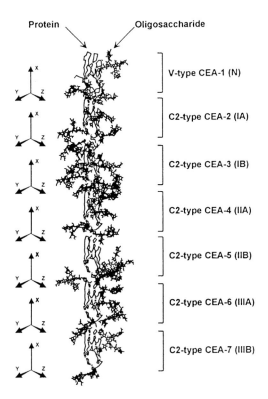

Protein   Oligosaccharide

V-type CEA-1 (N)

C2-type CEA-2 (IA)

C2-type CEA-3 (IB)

C2-type CEA-4 (IIA)

C2-type CEA-5 (IIB)

C2-type CEA-6 (IIIA)

C2-type CEA-7 (IIIB)

Figure 33.2: A linear model of the CEA molecule.
Reproduced from Boehm et al.[4] by permission of Academic Press Ltd.

The molecules are heavily N-glycosylated, containing up to 50% carbohydrate [7,13,16,35]. The oligosaccharide chains are almost exclusively multiantennary of the complex type [37,38]

The solution arrangement of the seven domains of the **CEA** molecule has been determined by X-ray and neutron scattering ([4], *Fig. 33.2).*

The proteins of the **CEA** subgroup are bound to the cell membrane by two types of membrane connections – **BGP**, **CGM1**, and **CGM7** (= CEACAM1, CEACAM3, and CEACAM4), which are inserted into the lipid bilayer of the cell membrane by a hydrophobic transmembrane domain, and **CEA**, **NCA**, **CGM2**, and **CGM6** (CEACAM5–CEACAM8), which are connected to the cell membrane via a glycosylphosphatidyl inositol moiety [14]. The members of the **PSG** subgroup are secreted molecules.

617

Tissue distribution

In normal adult tissue the molecules of the **CEA** family show a characteristic distribution pattern:

CEA and **CGM2** have a fairly limited tissue expression. They are present only in selected epithelial and glandular cells of the gastrointestinal and urogenital tract and in duct cells of sweat glands [19,20,28].
BGP is broadly distributed in normal tissues. It is expressed in different epithelia of the gastrointestinal and urogenital tract, in sweat and sebaceous glands, as well as in granulocytes and lymphocytes [28].
NCA, too, has a fairly broad tissue distribution, being present in epithelial cells of different organs and in granulocytes and monocytes [15].
CGM1 and **CGM6** are expressed only in granulocytes (cf. [12]).
The members of the **PSG** branch are expressed at a high rate in the placenta during embryonic development [39]. **PSG** synthesis is limited to the syncytio-trophoblast. However, **PSG**-specific cDNA clones have also been isolated from other tissues like fetal liver, salivary gland, pancreas, testis, uterus, and myeloid cells [33].

CEA as tumour marker

CEA is produced in high quantity in adult colon [17]. Normal colon epithelial cells are polarised and express **CEA** only on the apical surface. In healthy adults most of the **CEA**) is released into the gut lumen and only minuscule amounts are found in the blood. In colonic adenocarcinoma cells, however, **CEA** is distributed over the entire cell surface. Since the malignant cells have no basal lamina and are multiplying within the tissue, **CEA** accumulates in the blood [12]. It is for this reason that **CEA** is widely used as a tumour marker, particularly in post-surgical surveillance of colon cancer.

Biological functions of the CEA family molecules

In vitro experiments have shown that several members of the **CEA** subfamily (**CEA** [3], **BGP** [22,29], **NCA** [25], and **CGM6** [24]) may act as homophilic and heterophilic cell adhesion molecules. It has therefore been suggested that **CEA** family molecules have some function as contact mediating devices.
[21,30].[8,40].
BGP, **CGM1** and the **PSGs** occur in alternatively spliced forms differing in their cytoplasmic domains ([18], see *Fig. 33.1*). The longer cytoplasmic forms contain two tyrosine residues which may be phosphorylated. These tyrosines are part of modified

'immunoreceptor tyrosine-based activation/inhibition motives' (ITAM/ITIM) which may participate in signalling events [21].

The N-domains of **CEA** and **NCA** are recognised by the virulence associated *Opa* ('opacity') proteins of *Neisseria gonorrhoeae* and *N. meningitidis* [5,34]. It has therefore been suggested that members of the **CEA** family constitute the cellular receptors for *N. gonorrhoeae* in urinary bladder [11]. In the alimentary tract, however, where the **CEA** is generally released into the lumen, the molecule may play an essential role in innate immune defense against microbial attack.

References

1. BEAUCHEMIN, N., BENCHIMOL, S., COURNOYER, D., FUKS, A. & STANNERS, C. P. (1987): Isolation and characterization of full-length functional cDNA clones for human carcinoembryonic antigen. *Mol. Cell Biol. 7*, 3221-3230.
2. BEAUCHEMIN, N., DRABER, P., DVEKSLER, G., GOLD, P., GRAYOWEN, S., GRUNERT, F., HAMMARSTROM, S., HOLMES, K. V., KARLSSON, A., KUROKI, M., LIN, S. H., LUCKA, L., NAJJAR, S. M., NEUMAIER, M., OBRINK, B., SHIVELY, J. E., SKUBITZ, K. M., STANNERS, C. P., THOMAS, P., THOMPSON, J. A., VIRJI, M., VONKLEIST, S., WAGENER, C., WATT, S. & ZIMMERMANN, W. (1999): Redefined nomenclature for members of the carcinoembryonic antigen family. *Exp. Cell Res. 252*, 243-249.
3. BENCHIMOL, S., FUKS, A., JOTHY, S., BEAUCHEMIN, N., SHIRORTA, K. & STANNERS, C. P. (1989): Carcinoembryonic antigen, a human tumor marker, functions as an intercellular adhesion molecule. *Cell 57*, 327-334.
4. BOEHM, M. K., MAYANS, M. O., THORNTON, J. D., BEGENT, R. H., KEEP, P. A. & PERKINS, S. J. (1996): Extended glycoprotein structure of the seven domains in human carcinoembryonc antigen by X-ray and neutron solution scattering and an automated curve fitting procedure: implications for cellular adhesion. *J. Mol. Biol. 259*, 718-736.
5. BOS, M. P., GRUNERT, F. & BELLAND, R. J. (1997): Differential recognition of members of the carcinoembryonic antigen family by Opa variants of Neisseria gonorrhoeae. *Infect. Immun. 65*, 2353-2361.
6. BRANDRIFF, B. F., GORDON, L. A., TYNAN, K. T., OLSEN, A. S., MOHRENWEISER, H. W., FERTITTA, A., CARRANO, A. V. & TRASK, B. J. (1992): Order and genomic distances among members of the carcinoembryonic antigen (CEA) gene family determined by fluorescence in situ hybridization. *Genomics 12*, 773-779.
7. COLIGAN, J. E., LAUTENSCHLEGER, J. T., EGAN, M. L. & TODD, C. W. (1972): Isolation and characterization of carcinoembryonic antigen. *Immunochemistry 9*, 377-386.
8. GAIDA, F. J., PIEPER, D., RODER, U. W., SHIVELY, J. E., WAGENER, C. & NEUMAIER, M. (1993): Molecular characterization of a cloned idiotypic cascade containing a network antigenic determinant specific for the human carcinoembryonic antigen. *J. Biol. Chem. 268*, 14138-14145.
9. GOLD, P. & FREEMAN, S. O. (1965): Specific carcinoembryonic antigens of the human digestive system. *J. Exp. Med. 122*, 467-481.
10. GRAHAM, R. A., WANG, S., CATALANO, P. J. & HALLER, D. G. (1998): Postsurgical surveillance of colon cancer. *Ann. Surg. 228*, 59-63.
11. GRAY-OWEN, S. D., DEHIO, C., HAUDE, A., GRUNERT, F. & MEYER, T. F. (1997): CD66 carcinoembryonic antigens mediate interactions between Opa- expressing Neisseria gonorrhoeae and human polymorphonuclear phagocytes. *Embo J. 16*, 3435-3445.
12. HAMMARSTRÖM, S. (1999): The carcinoembryonic antigen (CEA) family: structures, suggested functions and expression in normal and malignant tissues. *Sem. Cancer Biol. 9*, 67-81.
13. HAMMARSTRÖM, S., ENGVALL, E., JOHANSSON, B. G., SVENSSON, S., SUNDBLAD, G. & GOLDSTEIN, I. J. (1975): Nature of the tumor-associated determinant(s) of carcinoembryonic antigen. *Proc. Natl.*

Acad. Sci. USA 72, 1528-1532.

14. HEFTA, S. A., HEFTA, L. J. F., LEE, T. D., PAXTON, R. J. & SHIVELY, J. E. (1988): Carcinoembryonic antigen is anchored to membranes by covalent attachment to a glycosylphosphatidylinositol moiety: Identification of the ethanolamine linkage site. *Proc. Natl. Acad. Sci. USA 85*, 4648-4652.

15. KODERA, Y., ISOBE, K., YAMAUCHI, M., SATTA, T., HASEGAWA, T., OIKAWA, S., KONDOH, K., AKIYAMA, S., ITOH, K., NAKASHIMA, I. & TAAKAGI, H. (1993): Expression of carcinoembryonic antigen (CEA) and nonspecific crossreacting antigen (NCA) in gastrointestinal cancer; the correlation with degree of differentiation. *Brit. J. Cancer 68*, 130-136.

16. KRUPEY, J., GOLD, P. & FREEDMAN, S. O. (1967): Purification and characterization of carcinoembryonic antigens of the human digestive system. *Nature 215*, 67-68.

17. MATSUOKA, Y., MATSUO, Y., OKAMOTO, N., KUROKI, M. & IKEHARA, Y. (1991): Highly effective extraction of carcinoembryonic antigen with phosphatidylinositol-specific phospholipase C. *Tumour Biol. 12*, 91-98.

18. NAGEL, G., GRUNERT, F., KUIJPERS, T. W., WATT, S. M., THOMPSON, J. & ZIMMERMANN, W. (1993): Genomic organization, splice variants and expression of CGM1, a CD66- related member of the carcinoembryonic antigen gene family. *Eur. J. Biochem. 214*, 27-35.

19. NAP, M., HAMMARSTROM, M. L., BORMER, O., HAMMARSTROM, S., WAGENER, C., HANDT, S., SCHREYER, M., MACH, J. P., BUCHEGGER, F., VON KLEIST, S. & ET AL. (1992): Specificity and affinity of monoclonal antibodies against carcinoembryonic antigen. *Cancer Res. 52*, 2329-2339.

20. NAP, M., MOLLGARD, K., BURTIN, P. & FLEUREN, G. J. (1988): Immunohistochemistry of carcino-embryonic antigen in the embryo, fetus and adult. *Tumour Biol. 9*, 145-153.

21. ÖBRINK, B. (1997): CEA adhesion molecules: multifunctional proteins with signal-regulatory properties. *Curr. Opin. Cell Biol. 9*, 616-626.

22. OCKLIND, C. & OBRINK, B. (1982): Intercellular adhesion of rat hepatocytes. Identification of a cell surface glycoprotein involved in the initial adhesion process. *J. Biol. Chem. 257*, 6788-6795.

23. OIKAWA, S., IMAJO, S., NOGUCHI, T., KOSAKI, G. & NAKAZATO, H. (1987): The carcinoembryonic antigen (CEA) contains multiple immumoglobulin-like domains. *Biochem. Biophys. Res. Commun. 144*, 634-642.

24. OIKAWA, S., INUZUKA, C., KUROKI, M., ARAKAWA, F., MATSUOKA, Y., KOSAKI, G. & NAKAZATO, H. (1991): A specific heterotypic cell adhesion activity between members of carcinoembryonic antigen family, W272 and NCA, is mediated by N-domains. *J. Biol. Chem. 266*, 7995-8001.

25. OIKAWA, S., INUZUKA, C., KUROKI, M., MATSUOKA, Y., KOSAKI, G. & NAKAZATO, H. (1989): Cell adhesion activity of non-specific cross-reacting antigen (NCA) and carcinoembryonic antigen (CEA) expressed on CHO cell surface: homophilic and heterophilic adhesion. *Biochem. Biophys. Res. Commun. 164*, 39-45.

26. OLSEN, A., TEGLUND, S., NELSON, D., GORDON, L., COPELAND, A., GEORGESCU, A., CARRANO, A. & HAMMARSTRÖM, S. (1994): Gene organization of the pregnancy-specific glycoprotein region on human chromosome 19: assembly and analysis of a 700 kb cosmid contig spanning the region. *Genomics 23*, 659-668.

27. PAXTON, R. J., MOOSER, G., PANDE, H., LEE, T. D. & SHIVELY, J. E. (1987): Sequence analysis of carcinoembryonic antigen: identification of glycosylation sites and homology with the immunoglobulin supergene family. *Proc. Natl. Acad. Sci. USA 84*, 920-924.

28. PRALL, F., NOLLAU, P., NEUMAIER, M., HAUBECK, H. D., DRZENIEK, Z., HELMCHEN, U., LONING, T. & WAGENER, C. (1996): CD66a (BGP), an adhesion molecule of the carcinoembryonic antigen family, is expressed in epithelium, endothelium, and myeloid cells in a wide range of normal human tissues. *J. Histochem. Cytochem. 44*, 35-41.

29. ROJAS, M., FUKS, A. & STANNERS, C. P. (1990): Biliary glycoprotein, a member of the immunoglobulin supergene family, functions in vitro as a Ca2(+)-dependent intercellular adhesion molecule. *Cell Growth Differ. 1*, 527-533.

30. SIPPEL, C. J., SHEN, T. & PERLMUTTER, D. H. (1996): Site-directed mutagenesis within an ectoplasmic ATPase consensus sequence abrogates the cell aggregating properties of the rat liver canalicular bile acid transporter/ecto-ATPase/cell CAM 105 and carcinoembryonic antigen. *J. Biol. Chem. 271*, 33095-33104.

31. THOMPSON, D. M. P., KRUPEY, J., FREEDMAN, S. O. & GOLD, P. (1966): The radioimmunoassay of circulating carcinoembryonic antigen of the human digestive tract.*Proc. Natl. Acad. Sci. USA*, 161-167.

32. THOMPSON, J., ZIMMERMANN, W., OSTHUSBUGAT, P., SCHLEUSSNER, C., EADESPERNER, A. M., BARNERT, S., VONKLEIST, S., WILLCOCKS, T., CRAIG, I., TYNAN, K., OLSEN, A. & MOHRENWEISER, H. (1992): Long-range chromosomal mapping of the arcinoembryonic antigen (CEA) gene family cluster. *Genomics 12*, 761-772.

33. THOMPSON, J. A., GRUNERT, F. & ZIMMERMANN, W. (1991): Carcinoembryonic antigen gene family: molecular biology and clinical perspectives. *J. Clin. Lab. Anal. 5*, 344-366.

34. VIRJI, M., WATT, S. M., BARKER, S., MAKEPEACE, K. & DOYONNAS, R. (1996): The N-domain of the human CD66a adhesion molecule is a target for Opa proteins of Neisseria meningitidis and Neisseria gonorrhoeae. *Mol. Microbiol. 22*, 929-939.

35. WESTWOOD, J. H., BESSEL, E. M., BUKHARI, M. A., THOMAS, P. & WALKER, J. M. (1974): Studies on the structure of the carcinoembryonic antigen. I. Some deductions on thebasis of chemical degradations. *Immunochemistry 11*, 811-820.

36. WILLIAMS, A. F. & BARCLAY, A. N. (1988): The immunoglobulin superfamily - domains for cell surface recognition. *Annu. Rev. of Immunol. 6*, 381-405.

37. YAMASHITA, K., TOTANI, K., IWAKI, Y., KUROKI, M., MATSUOKA, Y., ENDO, T. & KOBATA, A. (1989): Carbohydrate structures of nonspecific cross-reacting antigen-2; a glycoprotein purified from meconium as an antigen cross-reacting with anticarcinoembryonic antigen antibody. Occurrence of complex-type sugar chains with the Galβ1→3GlcNAcβ1→3Galβ1→4GlcNAcβ1→ outer chains. *J. Biol. Chem. 264*, 17873-17881.

38. YAMASHITA, K., TOTANI, K., KUROKI, M., MATSUOKA, Y., UEDA, I. & KOBATA, A. (1987): Structural studies of the carbohydrate moieties of carcinoembryonic antigens. *Cancer Res. 47*, 3451-3459.

39. ZHOU, G. Q., BARANOV, V., ZIMMERMANN, W., GRUNERT, F., ERHARD, B., MINCHEVA-NILSSON, L., HAMMARSTRÖM, S. & THOMPSON, J. (1997): Highly specific monoclonal antibody demonstrates that pregnancy- specific glycoprotein (PSG) is limited to syncytiotrophoblast in human early and term placenta. *Placenta 18*, 491-501.

40. ZHOU, H., FUKS, A., ALCARAZ, G., BOLLING, T. J. & STANNERS, C. P. (1993): Homophilic adhesion between Ig superfamily carcinoembryonic antigen molecules involves double reciprocal bonds. *J. Cell. Biol. 122*, 951-960.

41. ZIMMERMANN, W., WEBER, B., ORTLIEB, B., RUDERT, F., SCHEMPP, W., FIEBIG, H. H., SHIVELY, J. E., VON KLEIST, S. & THOMPSON, J. (1988): Chromosomal localization of the carcinoembryonic antigen gene family and differential expression in various tumours. *Cancer Res. 48*.

Subject Index

SpringerMedicine

Mario Campanacci

Bone and Soft Tissue Tumors

Clinical Features, Imaging,
Pathology and Treatment

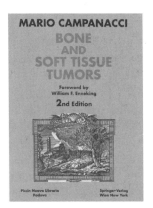

Second, completely revised edition
Foreword by William F. Enneking.
1999. XX, 1319 pages. 1120 figures.
Hardcover DM 598,–, öS 4186,–
(recommended retail price)
Jointly published with
Piccin Nuova Libraria, Padova
ISBN 3-211-83235-1

This second english edition is an entirely new book. It has been thoroughly rewritten, from the first to the last word. About 30% of the pictures are new. The new book incorporates the accumulated personal experience of the author, covering over 20.000 inpatients and many more outpatients, the perusal of the literature of the last 10 years, the recent developments in imaging (particularly MRI), microscopic diagnosis (especially immunohistochemistry and electron microscopy) and the ultimate progress in surgical and non-surgical treatment modalities.

Mario Campanacci (1932–1999) was an orthopaedic surgeon and a pathologist with 40 years of experience (started in 1958 in the Laboratory of Pathology and Tumor Center of the Rizzoli Orthopaedic Institute) focused on musculoskeletal oncology. He was Professor of Orthopaedic Surgery and Pathology, University of Bologna, Director of the 1st Orthopaedic Clinic and of the Tumor Centre, Rizzoli Orthopaedic Institute, Bologna and Director of the Graduate School of Orthopaedics, University of Bologna.

 SpringerWienNewYork

A-1201 Wien, Sachsenplatz 4–6, P.O.Box 89, Fax +43.1.330 24 26, e-mail: books@springer.at, Internet: **www.springer.at**
D-69126 Heidelberg, Haberstraße 7, Fax +49.6221.345-229, e-mail: orders@springer.de
USA, Secaucus, NJ 07096-2485, P.O. Box 2485, Fax +1.201.348-4505, e-mail: orders@springer-ny.com
Eastern Book Service, Japan, Tokyo 113, 3–13, Hongo 3-chome, Bunkyo-ku, Fax +81.3.38 18 08 64, e-mail: orders@svt-ebs.co.jp

Springer-Verlag
and the Environment

WE AT SPRINGER-VERLAG FIRMLY BELIEVE THAT AN international science publisher has a special obligation to the environment, and our corporate policies consistently reflect this conviction.

WE ALSO EXPECT OUR BUSINESS PARTNERS – PRINTERS, paper mills, packaging manufacturers, etc. – to commit themselves to using environmentally friendly materials and production processes.

THE PAPER IN THIS BOOK IS MADE FROM NO-CHLORINE pulp and is acid free, in conformance with international standards for paper permanency.